Power Systems

Electrical power has been the technological foundation of industrial societies for many years. Although the systems designed to provide and apply electrical energy have reached a high degree of maturity, unforeseen problems are constantly encountered, necessitating the design of more efficient and reliable systems based on novel technologies. The book series Power Systems is aimed at providing detailed, accurate and sound technical information about these new developments in electrical power engineering. It includes topics on power generation, storage and transmission as well as electrical machines. The monographs and advanced textbooks in this series address researchers, lecturers, industrial engineers and senior students in electrical engineering.

Power Systems is indexed in Scopus

More information about this series at http://www.springer.com/series/4622

Hassan Haes Alhelou · Almoataz Y. Abdelaziz ·
Pierluigi Siano
Editors

Wide Area Power Systems Stability, Protection, and Security

Springer

Editors
Hassan Haes Alhelou
Department of Electrical Power Engineering
Tishreen University
Lattakia, Syrian Arab Republic

Almoataz Y. Abdelaziz
Faculty of Engineering, Electrical Power and Machines Engineering Department
Ain Shams University
Cairo, Egypt

Pierluigi Siano
Department of Management and Innovation Systems
University of Salerno
Fisciano, Italy

ISSN 1612-1287 ISSN 1860-4676 (electronic)
Power Systems
ISBN 978-3-030-54274-0 ISBN 978-3-030-54275-7 (eBook)
https://doi.org/10.1007/978-3-030-54275-7

This Springer imprint is published by the registered company Springer Nature Switzerland AG
The registered company address is: Gewerbestrasse 11, 6330 Cham, Switzerland

Preface

Power systems are moving toward the smart grid concept, which enables the high penetration levels of renewable energies, due to environmental concerns and energy security risks. Renewable energy resources can be mainly divided into two categories, i.e., synchronous machine-based and inverter-based integration. The inverter-based renewable energy sources, such as wind turbines and photovoltaic panels, introduce new problems and technical issues to modern power systems stability, control, protection, and security. Such energy sources highly decrease the system total inertia which puts the system stability and security in danger. Likewise, the high stochastic nature of renewable resources due to their generation variation leads to increasing the power imbalances and consequently frequency fluctuations that might threaten the overall stability and security of the system. On the other hand, increasing the renewable-based distributed generating units in the distribution systems introduces new challenges to the voltage stability and the phase-balancing concerns. The recent advances in the measurements and communications systems, i.e., phasor measurement units (PMUs) and wide-area monitoring systems (WAMSs) open a promising environment for developing new control and protection approaches that can help solving the renewable energy and modern power system's technical issues. Wide-area monitoring systems provide the modern power systems with the real-time electric quantities. Based on these quantities, the power system stability and security can be monitored and assessed in real-time framework. Therefore, the stability and security of the modern energy system can be guaranteed in a suitable level using new wide-area control and protection schemes. This book introduces chapters for developing such schemes that can bring several advantages to both the system operators and consumers. Furthermore, the book suggests smart wide-area sensor network (WSN) utilization beside WAMS that can enable the deployment of smart grid features, such as demand response, demand-side management, smart electric transportation and electric vehicle deployments, and active distribution networks. This book proposed new control and protection schemes that can improve the overall stability and security of wide-area future power systems.

Lattakia, Syrian Arab Republic — Hassan Haes Alhelou
Cairo, Egypt — Almoataz Y. Abdelaziz
Fisciano, Italy — Pierluigi Siano

Introduction

The book is principally focused on wide-area power system stability, protection, and security. The book is sorted out and organized in twenty-two chapters. Each chapter begins with the fundamental structure of the problem required for a rudimentary understanding of the methods described.

Chapter "A Comprehensive Review on Wide-Area Protection, Control and Monitoring Systems": The damping enhancement of power system oscillations remains one of the challenging current interests for secure and reliable operation. This paper presents a comprehensive overview of a novel control scheme that considers synchrophasors and Power System Stabilizers in coordination with an optimized Load Frequency Control loop in order to resolve the undamped local and wide-area oscillatory troubles. Accordingly, a Robust Fuzzy PSS using local signals is first examined. Additionally, an Inter-Area PSS based on high-sampling rate phasor measurement unit is investigated. In fact, using time-synchronized measurements as control input signals will participate effectively in monitoring the energy management process. Thus, another configuration mixing local and remote control inputs of a Mixed-PSS is proposed. Performances of these PSSs are evaluated in coordination with a tuned PI-based load frequency control design under different operating conditions. Results on a modified 9-Bus IEEE test system including DFIG wind turbines are reported in order to justify the proposal's applicability.

Chapter "Introduction to WAMS and Its Applications for Future Power System": Nowadays, the increase in electrical energy consumption and power system restructuring have posed new challenges to the operation, control, and monitoring of power systems. In this situation, the supervisory control and data acquisition (SCADA) system is not enough to ensure power system security and stability. The SCADA is often unable to measure data of all buses simultaneously. In addition, the sampling rate in this system is not enough for some power system applications. Therefore, the information obtained from SCADA does not show power system dynamics properly. In order to improve the power system monitoring, wide-area measurement system (WAMS) has been developed to overcome the problems of SCADA system. Phasor measurements units (PMUs) are the main part of WAMS system and it basically consists of three essential processes including collecting, transmitting, and analyzing data. WAMS receives obtained data via a high-speed communication links. After data processing and extracting appropriate information, decisions are made to improve the power system performance. Efficient use of power system data to achieve a secure operation strategy is targeted using the WAMS system. Due to the effective role of WAMS and PMUs in the reliable operation of power system, it is necessary to study their concept and applications in this chapter. The history of PMU and its structure is presented in this chapter. In addition, the necessity of WAMS for future power system and its difference from SCADA system have been investigated in this chapter. Different algorithms and application of the WAMS are also introduced in this chapter which can be implemented to improve the performance of the future power system.

Chapter "Information and Communication Infrastructures in Modern Wide-Area Systems": Information and communication infrastructures (ICIs) in modern wide-area systems handle the transmitting, receiving, and storing of high-speed, large-volume synchrophasor data. Such infrastructures are important components in modern wide-area systems. Although power systems are becoming rather complex, with the introduction of synchrophasor technology, the highly accurate, high-speed, widely deployed, and time-synchronized phasor measurement units (PMUs) are providing operators and auditors an unprecedented way to understand the complex power systems. Consequently, these advanced PMUs challenge the current information and communication infrastructures. This chapter reviews the basics, challenges, and visions of the ICIs in wide-area monitoring systems (WAMS). Authors will first overview some important wide-area ICI topics and share our experience in building an efficient, reliable, and secure distribution-level WAMS, FNET/GridEye. Authors will introduce some key technologies that ensure the efficiency and reliability of such a WAMS. Finally, some outstanding challenges and future directions of the contemporary ICIs are discussed and envisioned.

Chapter "Wide-Area Measurement Systems and Phasor Measurement Units": Wide-Area Measurement Systems (WAMS) is a collective technology to monitor power system dynamics in real-time, identify system stability related weakness, and helps to design and implement counter measures. It uses a global positioning system (GPS) satellite signal to time synchronize from phasor measurement units (PMUs) at important nodes in the power system, sends real-time phasor (angle and magnitude) data to a Control Center. The acquired phasor data provide dynamic information on power systems, which help operators to initiate corrective actions to enhance the power system reliability. The goals of WAMS are real-time monitoring, post disturbance analysis, adaptive protection, and power system restoration. The major components of WAMS are Phasor Measurement Unit (PMU), Phasor Data Concentrator (PDC), Global Positioning System (GPS for Time Synchronization of the phasors), Communication channel (Preferably optical fiber cable), Visualization and analysis tools, Wide-area situational awareness system, and Wide-area protection and control.

Chapter "Optimal Selection of Phasor Measurement Units": Phasor Measurement Unit (PMU) is an important device for the power system operation as it provides the synchronized data required for the monitoring, protection, and control of the power system. So, to deploy the PMUs for the power system, their optimal locations are needed to be identified. This paper presents the optimal selection of PMU set from the available sets of PMUs. Firstly, it obtains all possible sets of PMUs required for the complete observability of the power system. Then, it defines four criteria such as System Observability Index (SOI), Restorable Islands Observability Index (RIOI), Critical Bus Observability Index (CBOI), and Critical Line Observability Index (CLOI) for the selection of best PMU set. Later, the Multi-Attribute Decision Making (MADM) techniques such as Analytic Hierarchy Process (AHP), Technique for Order Preference by Similarity to Ideal Solution (TOPSIS), Preference Ranking Organization Method for Enrichment Evaluations (PROMETHEE), and Compromise Ranking Method (VIKOR) have been used for the optimal selection of PMU set. This selection has been tested on some of the IEEE test systems. The results are then compared to analyze the performance of these four methods.

Chapter "Coordinated Designs of Fuzzy PSSs and Load Frequency Control for Damping Power System Oscillations Considering Wind Power Penetration": The damping enhancement of power system oscillations remains one of the challenging current interests for secure and reliable operation. This paper presents a comprehensive overview of a novel control scheme that considers synchrophasors and Power System Stabilizers in coordination with an optimized Load Frequency Control loop in order to resolve the undamped local and wide-area oscillatory troubles. Accordingly, a Robust Fuzzy PSS using local signals is first examined. Additionally, an Inter-Area PSS based on high-sampling rate phasor measurement unit is investigated. In fact, using time-synchronized measurements as control input signals will participate effectively in monitoring the energy management process. Thus, another configuration mixing local and remote control inputs of a Mixed-PSS is proposed. Performances of these PSSs are evaluated in coordination with a tuned PI-based load frequency control design under different operating conditions. Results on a modified 9-Bus IEEE test system including DFIG wind turbines are reported in order to justify the proposal's applicability.

Chapter "Wide-Area Monitoring of Large Power Systems Based on Simultaneous Processing of Spatio-Temporal Data": Accurate identification of electromechanical oscillations on power systems and determination of its stability condition is a fundamental process in order to carry out an appropriate control action to prevent the partial loss or complete blackout of the system. However, the nonlinear characteristics of measured variables often lead to incorrect information about the development of the electromechanical oscillations, making wide-area monitoring a challenging task. In addition, significant amount of information in extra large power systems is produced, which has to be stored on local servers requiring large amounts of central processing unit (CPU) storage. For these reasons, algorithms for Big Data problems in power systems are required and the methods presented on this chapter introduce some potential solutions. In this context, different data-driving methods based on spectral analysis of linear operator are presented for the analysis of electromechanical oscillations from a spatio-temporal perspective. These algorithms have the ability to process spatio-temporal data simultaneously, making possible to characterize inter-area and global oscillations (from 0.1 Hz to 1.0 Hz). To validate the effectiveness of the proposed approaches, two test systems with different structural and generation capacities are analyzed: the Mexican Interconnected (MI) system and the initial dynamic model of Continental Europe from ENTSO-E. First, data collected from a transient stability study on the MI system are used to illustrate the ability of data-driving methods to characterize modal oscillations on longitudinal systems; where several inter-area modes produce interactions of different electrical areas. Then, simulation results from the initial dynamic model of ENTSO-E are analyzed to characterize the propagation of its global electromechanical modes across Europe, which have been denominated as the North-South and East-West modes with frequencies of approximately 0.15 Hz and 0.25 Hz, respectively. The second analysis include the interconnection of Turkey (TR) to Continental Europe in December 2010, which is derived on the grow of size and complexity of the original system having as result a decrease in the frequency value for the East-West mode and the introduction of a third inter-area mode on the system. The chapter concludes comparing the results of the proposed approaches against conventional methods available in the literature.

Chapter "Electromechanical Mode Estimation in Power System Using a Novel Nonstationary Approach": The modern power grid protection system should have considerable operational flexibility and resiliency to hedge the variability and uncertainty of high dimensional dependencies. The use of wide-area monitoring systems (WAMS) in the smart grid enables the real-time supervision of power system oscillations. With the help of advanced signal processing methods and big data analytics, time-synchronized phasor measurements can be used to extract valuable information concerning the electromechanical modal properties of power system oscillations. This chapter introduces a novel method for identifying electromechanical inter-area oscillation modes with the help of wide-area measurement data. Variational mode decomposition (VMD) can be considered as a flexible signal processing technique on the wide-area phasor measurements in power oscillation analysis. For the real-time operation, it is challenging to preset the value of the mode number in the VMD process. This issue has been addressed by improving the strategy for VMD, which is presented in this chapter. The first stage involves the use of Complete ensemble empirical mode decomposition with adaptive noise (CEEMDAN) technique to generate intrinsic mode functions and gives indexing based on the correlation factor. Depending on the indexing, the mode number is selected for the second stage VMD process. Techniques such as spectral analysis and Hilbert transform are quite suitable for the estimation of modal parameters. The study is based on significant features of power oscillations, such as determination of damping ratio, amplitude, and frequency. The identification and estimation of low frequency modes have been performed using this improvised VMD technique, and the results have been compared with those obtained using empirical mode decomposition approaches. The proposed approach is also validated using real-time data obtained from load dispatch centers. The results indicate the effectiveness of nonlinear, nonstationary analysis methods for analyzing the low frequency modes and provide reliable validation of these algorithms in analyzing real-time data patterns.

Chapter "Small Signal Stability Improvement of Pumped Storage Hydropower Using Wide Area Signal Considering Wind Farm": The electromechanical oscillations in the power system, known as local and inter-area modes, as well as power system oscillations, in presence of wind turbines due to its inherent stochastic are two important cases in small signal stability study. Fixed speed (FS) pumped storage power plants (PSHP) similar to other power plants based on synchronous machine experience low frequency power oscillations. Therefore, a power system stabilizer (PSS) is developed for damping these oscillations. However, insufficient damping of these oscillations limited the capacity of energy transfer. On the other hand, state-of-the-art PSHP based on doubly fed induction machine (DFIM) known as variable speed (VS) have different effect on both small signal and transient stabilities of power system. Moreover, PSSs can be more important in multi-machine power grid to be tuned in a precise method. Nowadays, in smart power grids, PSS with wide-area signal (WAS) instead of local signal is attended to decrease low frequency power oscillations, and therefore, improve the small signal stability of the power system. This chapter intends to consider effect of DFIM and SM-based PSHP with different PSS tuning methods. Aiming at this purpose, a case of 343 MW hydro pump-turbine (HPT) coupled to DFIM with 381 MVA in comparison to the SM with same capacity, i.e., 381 MVA, as well as an aggregated wind farm are applied as the study case. Calculation and simulations are conducted in Digsilent 15.1 under diverse conditions. Also, modified New England test system, including10-machine and 39-bus system, is adopted as a large power network in presence of a wind farm. The results show using PSS with WAS can be a good option for FS-PSHP to improve damping low frequency oscillations.

Chapter "Impact Analysis and Robust Coordinated Control of Low Frequency Oscillations in Wind Integrated Power System": With rapid proliferation of wind generation in current generation mix, the issue of low frequency oscillations (LFOs) may get escalated in the modern power grids. The eigenvalue and dynamic sensitivity analysis have been employed to examine the effect of wind integration on system damping. Further, a wide-area based robust damping improvement control is suggested. It involves the coordinated control of power system stabilizers (PSSs) of synchronous generators (SGs) and power oscillation dampers (PODs) of doubly fed induction generators (DFIGs). The robust control is attained by employing a new fitness function based on eigenvalue and damping ratio and optimized by Whale Optimization Algorithm (WOA). The wide-area POD inputs are selected using modal observability criterion, obtained using phasor measurement units (PMUs) located optimally in the system. The results are verified on IEEE benchmark 68 bus NY–NE (New York–New England) test system. The simulation results highlight the robustness of proposed control to changing system conditions and shows its effectiveness in augmenting system damping, and thus small signal stability with high level of wind penetration.

Chapter "Frequency Stability of Two-Area Interconnected Power System with Doubly Fed Induction Generator Based Wind Turbine": This chapter presents a comparison of the performance of integral (I) and proportional-integral-derivative (PID) controllers in frequency stabilization or load frequency control (LFC) of two-area interconnected power system considering generation rate constraints (GRCs) with Doubly fed induction generator (DFIG)-based wind energy. Two mathematically models are identified for investigations. Power system model 1 is two-area interconnected power system which contains two identical non-reheat thermal plants without DFIG participation. Whereas, power system model 2 contains two identical non-reheat thermal plants with dynamic participation of DFIG at both areas. Moreover, Harris Hawks Optimizer (HHO), Salp Swarm Algorithm (SSA), and Sine Cosine Algorithm (SCA) are applied to find the optimal values of the controller settings mentioned above. The effectiveness of the proposed controllers, which are optimally designed by several optimization techniques (i.e., HHO, SSA, and SCA) is tested and verified through an interconnected power system comprising of two identical non-reheat thermal power plants with/without DFIG participation. Time-domain simulation results of the studied power system with all mentioned optimization techniques are carried out using Matlab/Simulink® software to validate the robustness of the proposed controllers.

Chapter "Wide-Area Measurement-Based Voltage Stability Assessment by Coupled Single-Port Models": As the power system becomes more stressed and the penetration of intermittent renewable energies increase, voltage stability assessment (VSA) becomes a key concern for maintaining and enhancing the security of bulk power systems. Physically, the phenomenon of voltage instability is indeed caused by an uncontrollable drop in system voltage after being subjected to a disturbance. This deterioration may ultimately result in voltage collapse that has been responsible for several blackout incidents. So far, a vast number of methods ranging from simple static techniques to complex dynamic methods have been proposed for performing VSA. More recently, with wide deployment of synchronized phasor measurement unit (PMUs), PMU-based wide-area measurement system (WAMS) has attracted lots of interests from both academia and industry. In this chapter, recent developments of measurement-based coupled single-port models will be presented for VSA. Generally speaking, the concept of the coupled single-port model is to decouple a mesh power grid into several single-port local equivalent models with considering extra coupling impedances. By collecting real-time PMU measurements in each individual load bus, the reactive power response derived from the extended Ward-type equivalent model can be applied to eliminate the reactive power mismatch of the existing single-port model. Meanwhile, these parameters of the Thevenin equivalent circuit in the existing single-port model will be modified by a mitigation factor to improve the model accuracy of VSA. Since the proposed method is simple, several voltage stability indicators can be easily extended with slight modifications. Simulations are conducted on two test systems, including IEEE 57-bus and IEEE 118-bus test systems, to validate the accuracy of the proposed method.

Chapter "Adaptive WAMS-Based Secondary Voltage Control": Voltage instability is a growing threat to the security and the reliability of power grids, especially as the penetration level of intermittent renewable energies increase significantly in recent years. Voltage instability and even voltage collapse will take place as the loss of control of the voltage profiles in a power system. To achieve more efficient voltage regulation in power systems, the hierarchical three-level coordinated voltage control mechanism has been developed recently to prevent voltage collapse through the appropriate management of reactive power sources. This chapter presents recent developments in adaptive secondary voltage control (SVC) by utilizing real-time measurements of power systems obtained from the wide-area measurement system (WAMS). These methods are adaptive in the sense that load disturbances are estimated from synchronized phasors of WAMS in nearly real-time. Thus, these control inputs of SVC can be synthesized to minimize deviations in load voltage profiles under the worst-case scenario. Uncertainties in measurement are also taken into considerations by exploring the maximum likelihood (ML) method to further improve SVC performance. Comprehensive simulations on a variety of IEEE benchmark systems have been performed to verify the feasibility and the effectiveness of these schemes.

Chapter "Applications of Decision Tree and Random Forest Methods for Real-Time Voltage Stability Assessment Using Wide Area Measurements": Traditionally, voltage stability assessment (VSA) are widely investigated by model-based appraches. Several achievements have been developed along this direction, including continuation power flow methods (CPFLOW), direct methods, and optimal power flow methods. Since precise model information are required and their computations are very demanding, their applications to real-time VSA are challenging, especially when network operating conditions and/or network topology may be always changed. In recent years, with wide deployment of synchronized phasor measurement unit (PMUs), PMU-based wide-area measurement system (WAMS) has already attracted lot of interests in investigating VSA in advanced artificial intelligence approaches. By collecting real-time big data from power grids and studying these historical data through statics analytics, some prediction models can be constructed for VSA of the current operation conditions. This chapter presents some recent advances in data mining framework for power system VSA under real-time environments. The proposed framework adapts a new enhanced online random forest (EORF) algorithm to update decision trees (DTs), such as tree growth and replacement. By means of weighted majority voting, one of the ensemble learning skills, DTs in the random forest are able to reach consesus to deal with power system changes. The proposed EORF framework is first tested on IEEE 57-bus power systems, and then is applied to Taiwan 1821-bus power system. Through comprehensive computer simulations, the robustness, the computation speed, as well as the assessment accuracy, of the proposed EORF framework are justified for assessing the power system voltage stability in real-time.

Chapter “Superseding Mal-Operation of Distance Relay Under Stressed System Conditions”: Mal-operation of distance relay imposes serious threats to system stability and a big reason for large-scale blackouts. These relays operate in its third zone due to the inability of detecting fault during stressed system conditions. These stressed conditions are load encroachment, power swing, voltage instability conditions, extreme contingencies, etc. Conventional distance relay operates on the basis of local measurements. It calculates the impedance from the relay to the fault point for its operation. Load encroachment and power swing are very similar to the symmetrical fault condition and it is difficult for these conventional relays to distinguish these stressed conditions from symmetrical faults. It is, therefore, important to make the distance relay intelligent enough so that it will be able to discriminate between a fault and stressed system condition. With the advancement in synchrophasor technology, the drawbacks of conventional relays have been overcome. The wide-area monitoring system (WAMS) is capable of development of online intelligent techniques that can segregate the stressed system condition from any fault. With these advanced techniques, the mal-operation of distance relays can be avoided and thus wide-area blackouts can be stopped. In this chapter, a new scheme for detecting the zone-III operation of distance relay is proposed to discriminate the stressed system conditions such as voltage instability, power swing, or load encroachment from fault. The proposed scheme is based on the monitoring of active and reactive power of the load buses using WAMS. Various cases are created on WSCC-9, IEEE-14, and IEEE-30 bus system to test the performance of the proposed algorithm. The simulations have been done on the MATLAB Simulink platform. Results shows that the proposed method is helpful to avoid the unwanted distance relay operation under stressed system conditions.

Chapter “Real-Time Voltage Stability Monitoring Using Machine Learning-Based PMU Measurements”: This chapter proposes a new MLP-based MSA algorithm for power system voltage stability monitoring using the information provided by PMU devices. In the proposed model, MSA algorithm is integrated to MLP network to optimize the connection weights and biases of the network. In the second approach, a novel hybrid model combining the adaptive neuro-fuzzy inference system (ANFIS) and MSA is proposed to monitor the voltage stability. In the proposed ANFIS–MSA model, authors adopt MSA algorithm to obtain proper parameter settings for the ANFIS-based subtractive clustering (SC) technique. The results of the proposed models have been validated and compared with existing methods for IEEE 30-bus and IEEE 118-bus standard test systems considering different operating conditions.

Chapter “Wide-area Transmission System Fault Analysis Based on Three-Phase State Estimation with Considering Measurement Errors”: In this chapter, a wide-area integrated method including a set of algorithms for transmission lines fault analysis is introduced. The proposed method is based on extension and modification of state estimation formulation. Thus, the method is applicable to both symmetrical and asymmetrical networks, as well as all fault types, including symmetrical and asymmetrical ones. The method exploits the capacities of state estimation formulation and the solution algorithm of weighted least squares (WLS) to reduce the effect of inherent errors on the fault location accuracy and detection and elimination of bad data in the measurement vectors. For this purpose, an error model of the measurement chain including instrument transformers and PMUs is proposed. This model is used to design measuring errors covariance matrix in the state estimation formulation. The performance of the proposed method has been investigated through numerous fault events simulated on different locations of all transmission lines of the IEEE 118-bus test system.

Chapter "Data-Driven Wide-Area Situation Analyzer for Power System Event Detection and Severity Assessment": Real-time power system monitoring and assessment leads to two major concern, prediction and evaluation of security and stability of power system. This assists in determination of in-time probable anomaly of the system. However, at the same time it requires real-time technological applications to measure network data at all strategic geographical locations. Synchrophasor technology based wide-area situational awareness ensures real-time monitoring and assessment of power system. This chapter proposes real-time data-driven Wide-area Situation Analyzer (WASA). WASA first detects an event in the system using synchrophasor measurements and then assesses its vulnerability posed to power network. The vulnerability is measured as severity in terms of first swing transient instability. Level of severity index is developed in terms of generator going out of step. The bus voltage trajectories going away with rest of the system due to generator(s) transient instability are considered. The proposed new approach is based on Center of Frequency (COF) formulated from limited (Phasor Measurement Unit) PMU measurements. To check for an event existence in the system, a new decision-based COF concept is defined. In order to determine the severity of the identified event, a new Predictor Indices (PI) is proposed using COF and PMU measurements. These predictor indices are used in assessment methodology, based on Adaptive Boosting (AdaBoost) of decision estimators. Furthermore, comparative results of proposed wide-area situational analyzer with other machine learning algorithms are also shown. The proposed WASA is instigated on IEEE New England 39 Bus system, successfully validating the performance of the proposed analyzer. The different type of events considered are generation outage, bus outage, load outage, and line events. Additionally, if any bus outage occurs due to line faults then it is considered as single severity. The results reflect the efficacy of the proposed analyzer in the power system event detection and its assessment efficiently and effectively with very less computational burden. The ability of the proposed analyzer to identify events quickly and correctly makes it suitable for real-time applications.

Chapter "Techno-Economic Analysis of WAMS Based Islanding Detection Algorithm for Microgrids with Minimal PMU in Smart Grid Environment": With large-scale deployment of non dispatchable renewable energy sources, distributed generators (DGs) have paved way for multiple microgrids. Existing standards stipulate disconnection of DGs in the event of any fault in either utility or microgrid side. However, to ensure reliable power supply from these microgrids, the proposed scheme operates an islanding detection algorithm (IDA) employed at the microgrid control center (MGCC) that acts on the switch at the utility point of common coupling (PCC). The whole microgrid is switched over to an islanded mode on detection of an islanding event. Phasor measurement unit (PMU) data from the utility PCC enables accurate islanding detection using islanding detection monitoring factor (IDMF) and rate of change of inverse hyperbolic cosecant function of voltage (ROCIHCF) along with voltage at the PCC. Mathematical morphological filters are employed to detect any persistent short circuit fault in the microgrid side which may island the DG. For such faults, decision for disconnecting the DG or islanded operation is based on probability of power balance (PoB) and probability of islanding duration in the sub-microgrids, computed at the MGCC. Further, performance of the IDA in microgrids is assessed using a proposed, microgrid performance index (MGPI) considering the uncertainties in NDRES. Suitability of the proposed indices to predict events leading to islanded operation in real-time is also validated using decision tree (DT) method. The discrimination capability between islanding and other transient events of DTs, yielded an accuracy of approximately 99.9% for minimum detection time, which proves the prowess of the method in real-time scenario. Compared to existing methods, the proposed method promises reduced islanding detection time. Another distinct feature is that time for islanding detection remains same irrespective of power mismatch ratios. The proposed method prevents false alarms for critical non-islanding events with zero non detection zone. Utilizing the proposed method, any redundant DG outage can be avoided, minimizing their down time. An economic analysis of the proposed islanding detection method using wide-area monitoring system (WAMS) with minimal PMU deployment has also been studied to reduce the cost of PMU installation without sacrificing the reliability benefit for the customers.

Chapter “Independent Estimation of Generator Clustering and Islanding Conditions in Power System with Microgrid and Inverter-Based Generation”: The use of phasor measurement units (PMU) allows us to obtain synchronized measurements of various points in the network and whit them analyze the stability of power systems. This chapter presents an algorithm based on participation factors to estimate generator clustering and to evaluate its application on controlled islanding on a power system, with distributed generation, using the data from PMUs after a severe disturbance. The proposed islanding detection method uses the data obtained from PMUs to represent the dynamics of the entire power system and form a measurement matrix, updated using a sliding window, containing the angles of the voltage phasors. Then, a covariance matrix is computed, and the eigenvalues and eigenvectors of this matrix are obtained. Subsequently, the most energetic eigenvalue is identified, and its participation factors are calculated. The participation factors are used as a contribution measurement of each generator into the most energetic eigenvalue, i.e., they will show the contribution made by each one after a disturbance. The clusters will be formed by generators sharing the same participation level. Controlled islanding condition of the system will be evaluated by using the clustering schemes proposed in the literature.

Chapter "Resilience in Wide Area Monitoring Systems for Smart Grids": WAMS infrastructures consist of various elements such as digital metering devices, communication, and processing systems, in order to facilitate the operation, monitoring, and control of power grids. For smart grids, resilience is a high-priority design requirement, since they must be able to resist in failures at any layer, caused by intentional attacks or unintentional events. In this chapter, authors review the existing approaches in the literature for WAMS resilience. Based on our recent work on dependency analysis for WAMS resilience, authors describe methodologies that take into consideration both optimization and resilience metrics during WAMS design. Authors explain how WAMS resilience can be increased by reducing the dependencies of WAMS components and by selectively adding controlled redundancy of measurement units and communication links. Finally, authors describe how this resilience model can be extended to also take into account the dynamic structure of the smart grid caused by the integration of renewable energy sources.

Chapter "Cyber Kill Chain-Based Hybrid Intrusion Detection System for Smart Grid": Today's electric power grid is a complex, automated, and interconnected cyber-physical system (CPS) that relies on supervisory control and data acquisition (SCADA)-based communication infrastructure for operating wide-area monitoring, protection, and control (WAMPAC) applications. With a push towards making the grid smarter, the critical SCADA infrastructure like power system is getting exposed to countless cyberattacks that necessitate the development of state-of-the-art intrusion detection systems (IDS) to provide comprehensive security solutions at different layers in the smart grid network. While considering the continuously evolving attack surfaces at physical, communication, and application layers, existing conventional IDS solutions are insufficient and incapable to resolve multi-dimensional cybersecurity threats because of their specific nature of the operation, either a data-centric or protocol-centric, to detect specific types of attacks. This chapter presents a hybrid intrusion detection system framework by integrating a network-based IDS, model-based IDS, and state-of-the-art machine learning-based IDS to detect unknown and stealthy cyberattacks targeting the SCADA networks. Authors have applied the cyber kill model to develop and demonstrate attack vectors and their associated mechanisms. The hybrid IDS utilizes attack signatures in grid measurements and network packets, as well as leverages secure phasor measurements to detect different stages of cyberattacks, while following the kill-chain process. As a proof of concept, authors present the experimental case study in the context of centralized wide-area protection (CWAP) cybersecurity by utilizing resources of the Power Cyber testbed at Iowa State University (ISU). Authors also describe different classes of implemented cyberattacks and generated heterogeneous datasets using the IEEE 39 bus system. Finally, the performance of the hybrid IDS is evaluated based in terms of detection rate in real-time cyber-physical environment.

Lattakia, Syrian Arab Republic — Hassan Haes Alhelou
Cairo, Egypt — Almoataz Y. Abdelaziz
Fisciano, Italy — Pierluigi Siano

Contents

About the Editors

Dr. Hassan Haes Alhelou is a faculty member at Tisheen University, Lattakia, Syria. He received the B.Sc. degree (ranked first) from the Tishreen University, Lattakia, Syria, in 2011; the M.Sc. degree from the Isfahan University of Technology (IUT), Isfahan, in 2016, all in Electrical Power Engineering, power systems (with honors). Since 2016, He started his Ph.D. at the Isfahan University of Technology, Isfahan, Iran. He is included in the 2018 & 2019 Publons's list of the top 1% best reviewer and researchers in the field of engineering. He was the recipient of the best young researcher in the Arab Student Forum Creative among 61 researchers from 16 countries at Alexandria University, Egypt, 2011. He has published more than 75 research papers in high-quality peer-reviewed journals and international conferences. His major research interests are Power systems, Power system dynamics, Power system operation and control, Dynamic state estimation, Frequency control, Smart grids, Micro-grids, Demand response, Load shedding, and Power system protection.

Prof. Almoataz Y. Abdelaziz completed his M.Sc. in 1990 and his Ph.D. in 1996, both at Ain Shams University. He then held numerous academic positions, before his current role as a Professor at the Department of Electric Power and Machines at Ain Shams University. He has numerous professional affiliations, serving as Chairman of the IEEE Education Society Chapter in Egypt since 2014, as an Editor of Electric Power Components and Systems Journal, and as a Senior Editor of Ain Shams Engineering Journal. He has published over 200 journal articles, over 150 conference papers, 16 books, and 25 book chapters.

Prof. Pierluigi Siano received the M.Sc. degree in electronic engineering and the Ph.D. degree in information and electrical engineering from the University of Salerno, Salerno, Italy, in 2001 and 2006, respectively. He is a Professor and Scientific Director of the Smart Grids and Smart Cities Laboratory with the Department of Management & Innovation Systems, University of Salerno. His research activities are centered on demand response, on the integration of distributed energy resources in smart grids and on planning and management of power systems. He has co-authored more than 450 papers including more than 250 international journal papers that received more than 8400 citations with an H-index equal to 46. He received the award as 2019 Highly cited Researcher by ISI WoS Group.

A Comprehensive Review on Wide-Area Protection, Control and Monitoring Systems

Valabhoju Ashok, Anamika Yadav, and Almoataz Y. Abdelaziz

Abstract In recent days electrical power system experiencing a rapid change thereby inclusion of advanced equipment and expansion of transmission/distribution network. Besides the modernisation of existing power system network, a number of renewable energy sources such as wind parks and solar parks etc., have been integrated to balance growing power demand due to the industrialization and digitalization. However, a secure and dependable operation of power system network is not so easy because of its complex nature in terms of control, operation and maintenance of various components in wide-area network. Inspite of gigantic developments in power system operations, components and protection technologies, today's power systems are more susceptible to blackouts than ever before. In this context some of the recorded major blackout incidents have been classified according to the time and location as of research reference purpose. As per the survey the frequency of occurrence of blackouts is increased over time. Practically, it is not possible to avoid blackouts completely; though, various research studies and number of research articles were acknowledged that by taking some rationally gainful measures, incidence of the blackouts could be abated and/or their effects could be mitigated. The forthright approach is to minimize the peril of unplanned disturbances thereby extenuating opportune paradigms to the extent that possible, the root causes of the disturbances through analyses and audits followed by initiating various preventive and corrective actions. In view of the preventive and corrective actions to avoid wide-area disturbances, a new paradigm has been embraced with Wide-Area Protection, Control and Monitoring (WAPCAM) system. With the rapidly growing capabilities in advanced computer and communication technologies such as Intelligent Electronic Devices

V. Ashok · A. Yadav (✉)
Department of Electrical Engineering, National Institute of Technology, Raipur, Chhattisgarh, India
e-mail: ayadav.ele@nitrr.ac.in

V. Ashok
e-mail: ashokjntuk@gmail.com

A. Y. Abdelaziz
Faculty of Engineering and Technology, Future University in Egypt, Cairo, Egypt
e-mail: almoatazabdelaziz@hotmail.com

H. Haes Alhelou et al. (eds.), *Wide Area Power Systems Stability, Protection, and Security*, Power Systems, https://doi.org/10.1007/978-3-030-54275-7_1

(IEDs) enabled Remote Terminal Units (RTUs), Global Positioning System (GPS) enabled Phasor Measurement Units (PMUs); opportunities are now being available to adopt the Wide-Area Protection, Control and Monitoring (WAPCAM) system. Such systems receive wide-range of data or information e.g. system-wide bus voltages, angles, active and reactive power flows, etc., and by analysing them, can estimate whether the system is at stressed condition or not. By taking coordinated actions, the power system network can be saved from proceeding to total collapse, or even, mitigate the wide-area disturbance effects upon the system. WAPCAM system has different level of hierarchies in realization of preventive and corrective actions such as local feeder level, sub-station level and central/regional level. One of the recommended preventive plans against the wide-area disturbances and the blackouts is Wide-Area Protection and Control (WAPC) system that includes Special Protection Schemes (SPS) (or) System Integrated Protection Schemes (SIPS) (or) Remedial Action Schemes (RAS) based on an advanced communication infrastructure etc. To mitigate the impact of wide-area disturbances, the remedial/corrective actions have been initiated by implementing Wide-Area Stability and Control (WASC) system that embraces power system stabilizers (PSS) and ON-Load Tap Changers (OLTC) and Wide-Area Monitoring and Control (WAMC) system. It also includes out-of-step (OSS) bus splitting and optimal islanding schemes etc. Although; the power system exhibits unstable dynamic phenomena at stressed conditions such as Transient Angle Instability, Voltage Instability and Frequency Instability, the WAPCAM has to bring back the power system to normal restorative condition as soon as possible. This chapter enlightens a comprehensive research review and explicates different type of WAPCAM systems that can address the major blackouts to improve stability, reliability and security of power system networks. A comparative assessment has been explicated by summarizing various recently reported conventional and intelligent schemes. It also enlightens the research insights to power system researchers and protection engineers while planning and designing of stable, reliable and secured power system networks.

Keywords Wide-area power systems · Blackouts · Monitoring systems · Power system protection · WAMC · PMUs

1 Introduction

Topologically speaking, power system networks are may be the widest interconnected networks in service today [1] and frequently exposing to thought-provoking engineering challenges. In addition, unfortunately, growth of generation and transmission networks is limited due to right-of-way restrictions and the public resist to building neighbouring plants, substations, lines and any other bulk facilities. Optimization of the available resources is then a must, leading to the operation of the power system close to its stability and security limits. Furthermore, the power system deregulation categorized different operating regions, that confounds the enactment of

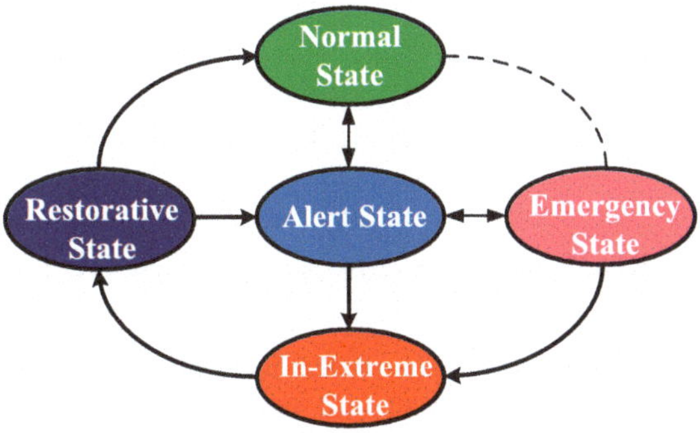

Fig. 1 Typical operating statuses of power system [2]

network-wide policies. An appropriate power system preserves operating frequency and voltage within tolerable range at both normal and abnormal scenarios, while also not beyond the thermal limits of the different power system equipment.

Typical Operating States of Wide-area Power System Network: For better understanding of how blackouts evolve from different disturbances and contingencies, it is very helpful to theoretically categorize the system-operating scenarios into five statuses [2, 3]. Figure 1 shows different operating statuses of wide-area power system.

- **Normal State**: The entire system parameters are within the ordinary ranges and no electrical component is burdened. The system is operating in a safe way and is able to survive an exigency without encroaching any of the constrictions. Normally, the power system is planned for (N-1) exigency that means it is able to withstand the disconnection of one main element.
- **Alert State**: The system comes in the alert state if the safety/security threshold crosses above or below to range of adequacy, or if the likelihood of a disturbance upsurges because of hostile weather circumstances such as wind storms etc. In the alert state, entire system parameters lies within the tolerable limits and all constrictions are fulfilled. Nevertheless, the system has been subjected to disturbance; the exigency may lead to overloading of components that pushes the system in an emergency state. If the disturbance is upsurges further, the alert state moves into extreme emergency (in-extremis) state.
- **Emergency State**: In this state, electrical components loadings and system variables surpass short-range emergency ratings. However, the system in this state is still undamaged, and may return to the alert state by the instigation of emergency control actions, e.g. fault clearance, excitation control, generation-rejection, HVDC modulation, load shedding, etc.
- **In-Extremis State**: If the above-mentioned emergency actions were not pragmatic or unsuccessful, the system proceeds to the in-extremis state, where the results are cascading outages and probably shut-down of a most important part of the system. Last resort remedial actions like bulk load rejection and optimal system separation are intended to except as much of the system as possible from a prevalent blackout. The system may develop into partial or total blackouts if corrective and saving actions are not taken or are in-effective.

- **Restorative State**: This state represents the conditions in wherein remedial actions (or) special protection schemes [3] are instigated to recouple all the services and bring back system load. If the fast restorations are successful, the system proceeds to either the alert state or the normal state.

2 Blackouts in Wide-Area Power System Networks

The continuous supply of electrical power is very essential for modern human life that is taken for granted until, unexpectedly and without any preceding cautioning, a blackout happens, ensuing in a complete interruption of power supply in an area, or whole regions. Most of this blackout incidents led by cascaded trippings owing to the short circuit faults, over loading, equipment malfunction and other external disturbances such as thunderstorms, bad weather condition and fire accidents.

Progression of Blackout Incidents: Foremost power grid blackout is originated by an incident or else manifold incident for example a fault and a relay maloperation that are frontrunners to cascading outages and subsequent failure of the whole system. Figure 2 illustrates progression of cascading events with different stages. This cascading events causes large disturbances to spread and exaggerates as blackouts. When a fault (or) disturbance happens or in case of the system is under stress and equipment is disconnected without following proper procedures, the series of incidents may arise. For illustration, few generators and/or lines are disconnected for maintenance, and another line may trip owing to a fault. The existing lines due to sag, touches a tree, and trips. There may be unnoticed failures e.g. obsolete relay sets or malfunction of hardware equipment that may lead to additional line or alternator to trip. Thereby the power system is confronted with overloaded components and multiple disturbances growing in due course of time. If speedy actions are not taken, the system experiences uneven distribution of load and generation, cascades into unintended islands and may lead to blackouts if load/generation rejection actions are

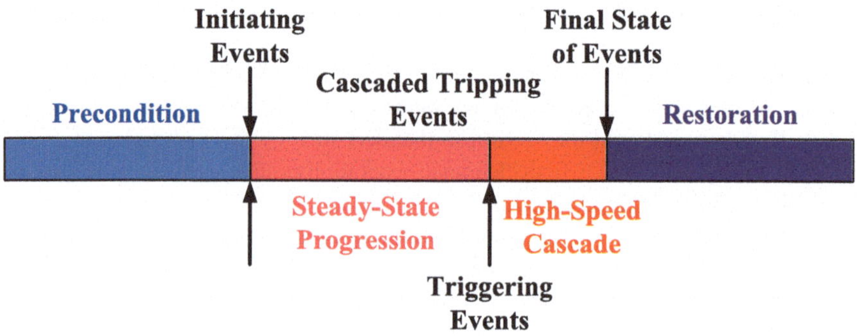

Fig. 2 Progression stages of Blackout Incidents [4]

not commenced. Results from previous studies [4] suggest that the progressions of cascaded blackouts may be categorized into different stages, detailed as follows:

- Preconditions
- Initiating events
- Cascade events
- Final state of events
- Restoration

However, the fact is that not all blackouts have all the phases listed above, some may be skipped or quickly developed. Conferring to these different stages, the phenomena of major incidents could be analysed and assessed, to identify mutual characteristics and features, summarizing their mechanisms and thus greatly helping to prevent future blackouts.

- **Preconditions**: Different preconditions that may exists before the initiating events of blackout, includes weather condition, aging equipment, inadequate reactive power reserve, significant equipment out of service. Also some natural condition such as abnormal wind speed, high temperatures, rainstorms/thunderstorms, fog, fires, drought, flocks of migrating birds that encounter vital overhead transmission lines, etc., are often conditions triggering the cascading disturbances and blackouts.
- **Initiating events**: Different initiating events may unswervingly lead to blackouts or deteriorate the system conditions. Any of the following initiating events may start the chain of events and introduce the disturbance(s) viz. short-circuits, overloading, relay maloperation and loss of generation etc.
- **Cascade events**: The cascading is a vigorous phenomenon. It is prompted by the originating events causing power oscillations and voltage fluctuations, and overloads, (propagating disturbances) as a consequence the current increases and voltage reduces. The high current and low voltage experienced by the lines and generators could lead them to trip according to their pre-decided settings. This would lead to more and more oscillations, fluctuations and overloads and thus more tripping of generators and lines making the system out of order. According to analysis [4], we can categorize the cascade events as follows: -

  Steady-state progression: In this period, the progression of cascading events or trips is slow and the system can still maintain the equilibrium between the generation and the consumption. The situation worsens slowly in this period, and thus there is still a good opportunity for the system operator and emergency control systems to take corrective actions to halt the spread of the cascade overload, and thus preventing the blackout from occurring.
  Triggering events: The separating line between slow steady-state progression of cascading and the high-speed cascade. There is no point of return.
  High-speed cascade: In this period, it is mostly too-late for the operating engineer to initiate any remedial actions to stop this fast progression of blackout, and the system could collapse in very short time. As the equilibrium between the

generation and the loads may be violated and thus series of equipment, numerous line and generators, may be tripped rapidly.

- **Final State of Events:** The final state of the power system reached after the previously mentioned chain of stages. The system is mostly separated into unplanned islands, some with shortage of generation and some others with excessive generation. Both will introduce tripping of loads and/or generation. This will leave plenty of islands in partial effective blackouts, if not total.
- **Restoration**: This is the post-process of attempting to reconnect all system facilities together and bring back the system to its original operating state.

2.1 Analysis of Major Blackouts in Wide-Area Power System Network

In this section, analysis of several major blackout incidents is reported with root cause of the blackout followed by progression of blackout incident such as preconditions, initiating events, cascading events and final state of events are discussed along with summary of recommendations.

New York, USA and Canada, November 9th, 1965 [5, 6]*:* The foremost major power blackout properly reported in the US and Canada that affected 30 million people. Restoration of service was hindered for several hours and in parts of NYC for periods up to 13 h.

- **Preconditions**: Substantial loading situations all over the transmission system. There was no enough generation. Spinning reserve set aside at that time. Backup relay settings were too low relative to the overload that was carried by the line.
- **Initiating events**: The operation of a backup relay on one of the 5 primary transmission lines transferring power to Toronto from Ontario Hydro's Sir Adam Beck no. 2 hydroelectric plant on the Niagara River.
- **Cascaded events**: After disconnection of that line, power flow reshuffled to the remaining four lines, causing overloading and progressively tripping. After opening of the remaining four lines, approximately 1500 MW of power generated at Ontario's Beck plant and the Niagara plant of the power authority of the state of NY, that was serving the Canadian loads in the Toronto area, reversed the power flow and endeavored to get to the loads through the only remaining US-Canadian tie at Massena. This caused Massena intertie to overload and opened, totally isolating the Canadian system. As a result, the American system was left with a surplus of about 1700 MW of generation, that surged into the US. Transmission flow capabilities, across the system, were exceeded and thus the breakup of the systems in the northeastern US was triggered. A widespread separation of systems through New York and New England monitored in a matter of seconds. Islands in New York and New England were left without sufficient generation to

meet their loads, resulting in ceasing all of the power generation within three to twelve minutes (except the Maine and eastern New Hampshire areas).

- **Root Cause**: The major reason was the weakest transmission corridor between the northeast and southwest.
- **Summary of recommendations**: New EHV (Extra high voltage) transmission lines were planned to be constructed. Not as much of critical load shedding was presented for emergency cases. More dispersed spinning reserve was kept into practice.

Con Edison System, July 13th, 1977 [5, 7]: Approximately 8 million people were left in dimness, with New York City. Blackout remained for stages of 5–25 h.

- **Preconditions**: The Con Edison system was in normal stable operation, with all elements operating within allowable continuous ratings. Several major elements were out of service for maintenance or repair. The load on the system was 6091 MW, below the day's 4 pm peak of 7264 MW; of which 3891 were supplied by Con Edison local generation, and about 2200 MW was being received over interconnections to adjoining systems. Con Edison's stated operating reserve generation was 2021 MW, with 738 MW in 10-min reserve and 1283 MW in 30-min reserve.
- **Initiating events**: At 8:37:17 pm, a lighting stroke caused a fault on both circuits of a double-circuit 345 kV TL between the Buchanan South and Millwood West Substations. CBs at Buchanan south tripped, but due to a design error in the protective scheme, a transfer trip signal was sent to Laden town substation. The 345 kV line carrying 427 MW from Buchanan South to Laden town opened. Indian Point no. 3 generating unit that was generating 883 MW, without a transmission path for its output, tripped off the line and shut down. Relay operation prevented (by protective system design) the normal reclosure of the Buchanan South breakers. The result was a total loss of 1310 MW through this connection. Power flow into Con Edison's system immediately increased by 1265 MW. The remaining deficiency of 45 MW was made up by a slight increase in output of various generators and by a slight drop in load related to a very small drop in system frequency. One transmission line, from Pleasant Valley to Millwood West was carrying power over its normal limit, but within its long-term emergency limit, and system conditions were still stable. At 8:55:53 pm, another lightning stroke caused simultaneous faults and breaker openings on two more 345 kV transmission circuits; from Buchanan North to Sprain Brook and from Millwood West to Sprain Brook, both via Eastview. In 2 s, tie between Millwood West and Sprain Brook was successfully auto-reclosed. On the Buchanan North-Sprain Brook tie, the breaker at Buchanan North end did not reclose due to a phase angle difference that developed across the open breaker contacts, that did not permit reclosing. Transmission tie to Ramapo substation was isolated from the Con Edison system. Flows on other interconnections increased to make up this loss (1044 MW).
- **Cascaded events**: The Pleasant Valley-Millwood West tie, tripped due to an improper relay operation (due to bent contact), due to overload. Con Edison

system reached a very serious emergency state, with the two remaining vital interconnections seriously overloaded. At 9:19:11 pm, the 345 kV line from Niagara Mohawk's Leeds substation to Con Edison's Pleasant Valley substation opened due to sagging of the thermally expanded (due to overload) conductors and their contact with a tree. This resulted in an increase of its 1202 MW load on the rest of the remaining lines. At 9:19:53 pm, a transformer at Pleasant Valley tripped out on an overload, resulting in the further loss of 415 MW that was flowing into Con Edison. About 2 min later, the Long Island Lighting Co. (LILCO) operator manually opened the tie between LILCO and Con Edison, as an effort to avert a major emergency on the LILCO system, thus, reducing another 520 MW from flowing to Con Edison. At 9:22:47 pm, the attempt to manually restore one 345 kV transmission line from Pleasant Valley to Millwood West failed. There was no net change in system condition resulting from this incident. At 9:29:41 pm, the 230 kV interconnection from Goethals to Linden opened as a result of a failure in the heavily overloaded phase-angle-regulating transformer that was an integral part of this connection. This resulted in the loss of another 1150 MW. The last remaining tie from Con Edison to external sources, two 138 kV feeders from Pleasant Valley to Millwood West, immediately tripped from the accompanying power swing. Now, the Con Edison system was isolated, with net load of 5981 MW and net generation of 4282 MW. System frequency began dropping, and UFLS relays operated, however voltage transients occurred during this period, causing the tripping of Ravenswood no. 3 line carrying 884 MW. Over the next 4 min, outputs of other generators deteriorated as low frequency affected their auxiliary equipment. By 9:36 pm, the Con Edison system was completely shut down.
- **Root Causes**: Natural thunderstorms conditions, Maloperation of equipment, Inappropriate system design topographies and Operational errors as lack of groundwork for foremost emergencies.
- **Summary of recommendations**: Proper attention to matters affecting reliability and emergency preparedness (design of the reliability criteria to identify the extreme sensitivity of the city network, i.e. operating reserve policy). Develop and implement aiding tools for operational staff, providing quick and complete information about services accessible, operating conditions, dynamic line loadings and computer-generated alternative actions to facilitate accurate and timely decision making. Con Edison should install load-shedding controls to allow single-action activation of large-block load shedding. Enhancement is needed for communications associated with system operating stations. Critical review of all the extra high voltage relay schemes and settings.

Tokyo, Japan, July 23rd, 1987 [5, 8, 9]: Affected 2.8 million customers. Approximately 8 GW of power supply was interrupted for near about 3.35 h.

- **Preconditions**: Unusual hot weather, leading to the increase in power demand, i.e. air-conditioning loads. Maximum power demand forecast (38500 MW) was updated, and 950 MW were added and 570 MW were exchanged from other utilities to guarantee the transfer capability of 41520 MW of the new power demand prediction. Subsequently, the operational standby became 1520 MW, ratio of 3.8%, that was adequate to supply the growing demand. Tokyo Electric Power Company's (TEPCO) major power sources are located in the eastern regions, and thus large amounts of power flow from east to west are common on the 500 kV trunk transmission lines.

- **Initiating events**: Demand increased at a very unprecedented speed (400 MW/minute) owing to the oddly hot climate. To sustain the voltage level of the 500 kV transmission TEPCO switched on shunt capacitors and increased the generators' reactive power supply. By 13:07, all shunt capacitors were switched on, while the power demand stretched to 39000 MW.
- **Cascaded events**: The voltage of the trunk network progressively go down. Finally, at 13:19, protective relays tripped (zone-4 of distance relays) due to voltage decrease and current increase, leading to shutdown of two 500 kV substations (Shin-Fuji and Shin-Hatano, and one 275 kV substation, Kita-Tokyo).
- **Root Causes**: Inadequate knowledge of load characteristics. Insufficient monitoring tools. Lack of policies in contradiction of voltage instability. Uneven distribution of power plants.
- **Summary of recommendations**: The operation of system voltage at high level, within 5%. To decrease power flow on the East–West lines by introducing a new generation source. To reduce heavy loads across the 500 kV transmission network by altering the sub-transmission network connections at 275 kV substations. To request to customers in emergency, with interruptible load contracts. Setting up of Static VAR compensators and shunt capacitors. Improvement of demand forecast techniques.

Egypt, April 24, 1990 [10]: This blackout was occurred all over Egypt. Supply was interrupted from all customers for several hours.

- **Preconditions**: The heavy loading conditions all over the transmission system, on the 500 kV system in particular, during the peak load making the system operating at critical voltages.
- **Initiating events**: A short circuit at Kommombo (near Aswan), lead to traditional transient instability, voltage instability and eventually voltage collapse.
- **Cascaded events**: The officially announced fault is a 3-phase short circuit on the Aswan zone bus, cleared by opening one circuit of the two 500 kV circuits between Aswan and Nag-Hammady.
- **Root Causes**: The loss of synchronism between the northern pool and the southern pool, of the system. The voltage collapses due to sudden opening of double circuit lines. The short-circuit recoveries at loads of major induction motors content. Concurrent starting of some induction motor loads. The operation of network nodes near critical voltages. Machine operation near their steady-state limit. Uncontrolled flow of reactive power. The failure of proper results from control and protection actions.
- **Summary of recommendations**: The avoidance of upcoming blackouts may be mostly realized by continuous stabilization of voltages at load-buses. The use

of fast controllable VAR compensators by SVC at Nag-Hammady can mitigate voltage instability and reserve machine synchronism, i.e. avoid transient inertial instability. The power factor correction at individual loads components must be provided.

Western North America, July 2, 1996 [5, 11]*:* The most disruptive breakup, far ever experienced by the western system. Few GW of power were interrupted.

- **Preconditions**: Heavy loading conditions in southern Idaho and Utah since the adverse weather conditions. The power flow on the Pacific ac and dc interties were high (4300 MW and 2800 MW) that exports power from the Pacific Northwest to California.
- **Initiating events**: At 14:24, a flashover on a tree on the Jim Bridger-Kinport 345 kV line. The maloperation of a ground protection relay tripped the parallel Jim Brodger-Goshen 345 kV line. The special stability controls tripped two Jim Bridger units (1000 MW) for loss of two of the three Jim Bridger 345 kV outlet lines. This tripping was correct, and should have confirmed stability and avoided supplementary outages. A 230 kV line relayed in Eastern Oregon, about 500 km away from Jim Bridger.
- **Cascaded events**: The main root cause of this event was voltage depressed in Southern Idaho. Within 24 s, various small hydro generators near Boise tripped due to high field current. In central Oregon (500 km away) voltage gradually reduced along the Pacific intertie, due to reduced generation at the Dalles power plant. After the 24 s, a key 230 kV intertie line (Amps line) linking western Montana and southern Idaho tripped by zone-3 relay operating on insignificant overload and insignificant voltage depression. Interruptions of around 300 MW on this line initiated power swing in eastern Washington and eastern Oregon. Lines, between Hells Canyon generation and the Boise load area in southern Idaho, causing further overloading. The loading on the summer Lake-Midpoint line also augmented. Anaconda-Amps-Antelope line (Amps line) tripping caused quick overload, voltage collapse and angular instability across several tie lines. About 3 s later four 230 kV lines tripped linking Snake River Hells Canyon generation to Boise. After 2 s, the Pacific ac intertie also tripped. Further, cascading led to five separate electrical islands.
- **Root Causes**: The root cause of the disturbance was voltage collapse and Improper action of Zone-3 of some distance relays.
- **Summary of recommendations**: To prevent this kind of blackouts in the forthcoming the defense in depth method was employed. In this context, a depth analysis of zone-3 protection, where outage detection-based stability control was altered with response-based control for quicker response and further operating ranges were upgraded that helped to curtail generator tripping for voltage and frequency expeditions.

North America, August 14, 2003 [12, 13]*:* This blackout was affected approximately 50 million people whereas fully affected 8 US states and 2 Canadian provinces.

Approximately 70 GW of power were interrupted. It caused around $7–10 billion of economic loss. It took over a day to re-establish power supply to NYC and other pretentious regions; while some areas in the US lost power for two days and some regions of Ontario faced revolving blackouts for up to 15 days.

- **Preconditions**: Prior to 15:05: System was being functioning in comply with NERC operating guidelines. Prior to noon: Substantial reactive power supply complications in the states of Indiana and Ohio. From 12:15 to 16:04: Midwest ISO (MISO) state estimator and real-time contingency analysis software was not working properly owing to problems in software that prevented the MISO from carrying out early warnings. From 14:14 to 15:45: Energy management system software experienced failures at the FirstEnergy (FE) control center, contributing to the insufficient situational alertness at FE.
- **Initiating events**: At 13:31: the first foremost incident that occurred, was the outage of the FE's Eastlake unit 5 generators. Owing to large reactive power, the unit's AVR (Automatic Voltage Regulator) tripped to manual since the response of over-excitation, and as the operator tried to reinstate the AVR, the generator finally tripped.
- **Cascaded events**: At 16:10 owing to the concurrent loss of foremost tie lines between Ohio and Michigan, the power supply between the US and Canada on the Michigan border removed, i.e. the power flow reversed, i.e. anticlockwise from Pennsylvania, through NY and then Ontario and finally into Michigan and Ohio (3700 MW). At this instant voltage collapsed, owing to enormously loaded transmission lines and cascading outage of fewer hundred lines and generators resulted in a power failure of the whole region.
- **Root Causes**: The root cause of this event was identified as an insufficient system considerate, insufficient situational alertness, the lack of tree trimming and the lack of problem-solving sustenance from reliability establishments.
- **Summary of recommendations**: Make reliability standards compulsory and enforceable, with penalties for non-compliance. Enhancing institutional attention to reliability. Start enforceable standards for preservation of electrical consents in right-of-way areas. Improve training of operators and associated personnel and wider use of system protection measures. Reinforce reactive power and voltage control practices. Advance the quality of system state estimation, modeling data and data exchange practices. Require use of time-synchronized data recorders.

Southern Sweden and Eastern Denmark, September 23, 2003 [12, 14]*:* The most severe disturbance experienced by the Nordic power system in 20 years. Approximately 4700 MW load were interrupted in Sweden and 1.6 million people affected and 1850 MW load were lost in Denmark and 2.4 million people affected.

- **Preconditions**: The operating circumstances were steady and within the limitations mentioned in the operational scheduling and grid security calculation. Power requirement in Sweden was 15000 MW, that was, comparatively, quite moderate owing to usually warm climate conditions. Nuclear generation was inadequate owing to on-going yearly over-haul plans, and lately pick-up for some generators due to nuclear safety necessities. Approximately 400 MW of generation export was scheduled from Zealand to Sweden. Two 400 kV lines in the region were interrupted owing to planned maintenance work. The HVDC links with Poland and Germany were removed because of yearly inspection and some minor repairs.

- **Initiating Events**: At 12:30- interruption of a single 1250 MW nuclear generation unit (due to failure in containing inside valve complications in the feed-water routes). This was regarded as a standard contingency, and conditions were still well at hand at this moment. Spinning reserves at Norway, northern Sweden and Finland were activated, enabling the system to resume stable operation, within a minute. Voltages in the southern region had decreased to around 5 kV, but persisted within the acceptable 405–409 kV level. Frequency was somewhat below the typical operating limit of 49.99 Hz, thus initiating further actions to advance the frequency. Power flow was reallocated in the grid, as more power was transferred to the western side to adjust the demand in the south; however, transmission levels were still within the pre-programmed security limitations. At 12:35- a double busbar fault (due to damage to one disconnector) happened in a 400 kV substation on the western coast of Sweden, causing the operation of the busbar differential protection, thus, isolating the supply from two 900 MW units. At this moment, total of 1750 MW generation was tripped and the grid lost its transmission corridor through the west coast. Mainly, this triggered high power fluctuations in the network, very low voltages and an additional decrease in frequency fallen to a level somewhat around 49.00 Hz, where UFLS relays started to function. The grid was then severely overloaded on another part of south-east and south-central regions in view of ability to withstand the voltages. This region/area of the grid had no foremost generation associated and therefore the reactive power backup was poor. One and half minute after the busbar fault, the fluctuations vanished and the system bring-back to steady-state. The following actions of numerous transformers tap-changers dropped the voltage more on the 400 kV grid miserable to a critical level. The condition established into a voltage collapse in a segment of the grid south-west in region of the capital of Stockholm.
- **Cascaded events**: In case of voltage collapsed to very lower limits, CBs in the sensitive grid segments were tripped by reaching the low impedance measures. Then grid fragmented into two separate sections, covering south of Sweden and eastern Denmark, primarily continued interconnected but suffered from an immense shortage of generation. The residual generation in Denmark was incapable to supply the power demand. Eventually within seconds, the dropped voltages and frequencies caused generator and other grid defenses to operate and the whole subsystem collapsed. The northern part of the split grid stayed intact, with the interlinks to Norway and Finland.
- **Root Causes**: The local accident of major faults causes a problem to the system far beyond the exigencies observed in typical system strategy and operational security standards.
- **Summary of determined recommendations**: The assessment of the scheduling and functioning reliability policies employed within the support between the Nordic transmission system operators. Compulsory technical necessities on power

stations will be imposed, in specific to accomplish the shift to domestic-load operation on outside grid instabilities. Strengthening of the transmission capability to the south of Sweden by constructing a new 400 kV line in the present right-of-way of 220/135 kV existing line. Courtesy taken to the need of establishing new power plant in the region. Additionally, the adaption of innovative system protection/remedial schemes. Restricting of the switchgear in the fault-struck substation with reverence to the menace of flashovers among the focal busbars. Revision of other substations of like status. Compulsory checkup of disconnectors and scheduled auxiliary equipment at sensitive sections. Study of the activities to secure distant control schemes at the abnormal and transient situations.

Italy, September 28, 2003 [15, 16]*:* This black out was affected 57 million people in Italy. It was affected the whole of Italy, and parts of Switzerland. The power supply was not provided for a time period of 1.5–19 h.

- **Preconditions**: Italian consumers, especially industrial, import the maximum amount of cheaper power from foreign states. The interlink between the Italian system and the UCTE grids is through six 380 kV lines and nine 220 kV lines. During 2002 an extra 500 MW DC undersea cable is put in service, between Italy and Greece.
- **Initiating events**: At 3:01:22, a tree flashover tripped a major tie-line between Italy and Switzerland (380 kV). Auto-reclosing of the line, and manual reclosing both failed owing to the great phase angles difference across the line breaker. Total Italian importation unaffected, decreased from Switzerland, however increased from France, Austria and Slovenia. At 3:21, the import of Italy was decreased by 300 MW; but was not adequate to overcome the overload of the Sils-Soazza line that tripped at 3:25.22, due to unclear reason, either overload protection or another tree flashover.
- **Cascaded events**: Few seconds later Mettlen-Airolo line tripped (220 kV). Another 3 s later, two other 220 kV lines tripped, making the southern part of Switzerland now separate from the remaining part of the Swiss grid and supplied only by the Italian system. The overload caused significant and fast voltage decreases and instabilities at French buses. The mixture of low voltage and high currents triggered distance relay to trip a main 380 kV line, at 3:25.32. This activated the loss of synchronism of the Italian system. Loss of synchronism instantly caused tripping of another double circuit 380 kV line with France, in addition to other 220 kV lines. The last lines to disconnect were with Slovenia. Disconnection was complete at 3:26.24.
- **Root Causes**: There was not enough load shedding to equalize the generation and load in the isolated (Islanding Italian) system.
- **Summary of recommendations**: should be worked to guarantee the N-1 operation and planning criterion, meaning that each/every grid has to be secured with reverence to every single exigency. Putting an updated special programmed load shedding strategy that monitors the overloading of such critical sections, and operates before the protection devices operate, to prevent parting of the interlinked

grid. Revising of the traditional load shedding schemes, and resetting the stages in a different perspective.

India 30 and 31 July, 2012 [17]: This blackout in India affected approximately 620 million people in India. It was affected the whole northern, north-eastern, eastern parts of India, and few parts of packets (areas) survived from blackout. The first blackout was occurred at 02:33 am on 30 July 2012 and power supply was restored at 16:00 pm on the same day and the total 36,000 Mw was interrupted. The second blackout was occurred at 13:00 pm in the afternoon on 31 July 2012 and total load of 48,000 MW was interrupted however, the supply was restored after 2–8 h approximately.

- **Preconditions**: Due to multiple outages experienced by the grid, the transmission system becomes weak. In addition, Northern region grid drew more power from western grid. Therefore, the corridor linking between north and western region grid was overloaded. Inspite of absence of any zone-3 fault, distance relay on the Bina-Gwalior line was tripped and western region separated from the northern region.
- **Initiating events**: At 02:33:11, on 30 July 2012, maloperation of distance relay on zone-3 for load encroachment condition in the line linking between Bina-Gwalior that led to parting of Northern grid to remaining part of the Western grid and in due course Northern grid was collapsed and lead to blackouts in the western region, eastern region and north-eastern region. At 13:00:13, on 31 July 2012, maloperation of distance relay on zone-3 for load encroachment on the same line and led to collapse of the northern, north-eastern, eastern parts of India.
- **Cascaded events:** First event was occurred at 02:33:11 on 400 kV Bina-Gwalior line and Few seconds later at 02:33:13, 220 kV Bhinmal-Sanchor line was tripped on Zone-1 due to power swing. After few seconds at 02:34:00 220 kV, Gwalior-Malanpur was tripped. In the second of blackout, first event was occurred at 13:00:13, on 400 kV Nina-Gwalior line and within few milliseconds 220 kV Bina-Gwalior line-1 and line-2 were tripped. And further 220 kV Shivpuri-Sabalgarh line-1 and 132 kV Pichhore-Shivpuri and 132 kV Picchore-Chanderi was simultaneously tripped. Simultaneously some of the interconnected lines were tripped from 13:00:13 to 13:00:20 that are steady progressed and a sequence of high-speed tripping was occurred from 13:00:25 to 13:03:18 on different lines at different places in the most part of Indian grid except southern grid.
- **Root Causes**: There was weak inter-state transmission corridors owing to several outages. Heavy loading on 400 kV Bina-Gwalior-Agra link whereas operator utilized unscheduled interchange. Maloperation of Zone-3 at load encroachment condition.
- **Summary of recommendations**: There is a need to review of protection schemes. Frequency band needs to be tightened further and brought close to 50 Hz. Ensuring prior function of defense mechanism by enhancing under-frequency and df/dt-based load shedding strategies. Coordinated outage planning of transmission lines with proper reactive power planning. Optimum utilization of available assets and deployment of WAMS. There is a need of dynamic security assessment and

analysis of state estimation. Execution of Islanding schemes and development of intra-state transmission system. Reduction of start-up time of generators and review of transmission planning criteria. Strengthening of system study groups and formation of task force to study the grid security issues including cyber security challenges in future.

2.2 Assessment of Major Blackouts

In this section, several major blackout incidents are assessed in Table 1 wherein 1 million people have experienced blackout for minimum 1 h in the last decade, and a brief summary of their root causes and impacts are presented. The major blackouts

Table 1 Assessment of major blackouts occurred in wide-area power system networks in the last decade [22]

Sl. No	Country/Region	Date	Affected people/Outage hours	Root cause
1	Mexico and USA [18, 19]	September 8 2011	2.7 million, 12 h	Transmission line trip
2	Brazil [20]	February 4 2011	53 million, 16 h	Line fault and fluctuated power flow
3	India [17]	July 30 2012	620 million, 14 h	Zone-3 malfunction at overload condition
4	India [17]	July 31 2012	700 million, 2–8 h	Zone-3 malfunction at overload condition
5	Vietnam [21]	May 25 2013	8 million, 1–8 h	Tree Fault, Crane operator
6	Philippines [21, 22]	August 6 2013	8 million, 12 h	Voltage collapse
7	Thailand [21, 23]	May 21 2013	8 million, 10 h	Lightning strike
8	Bangladesh [24, 25]	November 1 2014	150 million, 24 h	HVDC station outage
9	Pakistan [16]	January 26 2015	140 million, 2 h	Plant technical fault
10	Holland [26, 27]	March 27 2015	1 million, 1.5 h	Bad weather conditions, short circuit trip and overload

(continued)

Table 1 (continued)

Sl. No	Country/Region	Date	Affected people/Outage hours	Root cause
11	Turkey [28, 29]	March 31 2015	70 million, 4–7 h	Heavy loading, Lines and capacitor bank out of service
12	Ukraine [13, 22]	November 21 2015	1.2million, 6 h	Power system failure
13	Ukraine [22, 30],	December 23 2015	230 million, 6 h	Cyber-attack
14	Kenya [31, 32]	June 7 2016	10 million, 4 h	Animal shorted the transformer
15	Sri Lanka [33, 34]	March 3 2016	10 million, 16 h	Severe thunderstorm
16	South Australia [35, 36]	September 28 2016	1.7 million, 6.1 h	Transmission infrastructure damage due to heavy storm, cascading events
17	The USA(NY) [37, 38]	March 1 2017	21 million, 11 h	Cascading failure in transmission system
18	Uruguay [39]	August 26 2017	3.4 million, 4 h	Bad weather condition causing to cascading failure
19	US (southeast) [37, 40]	September 10 2017	7.6 million, 5 h	Cascading events and transmission tripping
20	Sudan [41–44]	January 10 2018	41.5 million, 24 h	Cascading failure
21	Azerbaijan [45–47]	July 3 2018	8 million, 8 h	Unexpectedly high temperature
22	Brazil [48]	March 21 2018	10 million, 1 h	Transmission line failure

during the year 2011–2020 have been assessed in Table 1. The blackout duration, number of affected people in millions and its root cause are specified in Table 1. It can be understood that the blackout which occurred in Bangladesh on dated 1 November 2014 has the maximum power interruption duration of 24 h [17, 25]. The Indian blackout on 31 July 2012 has affected the greatest number of people (700 million) [17]. The main root causes of the blackout are associated to severe overloading of lines and bad weather conditions, inappropriate control and protection schemes. Figure 3 illustrates the affected people during the blackout in millions and Fig. 4 shows the outage duration of blackouts in hours.

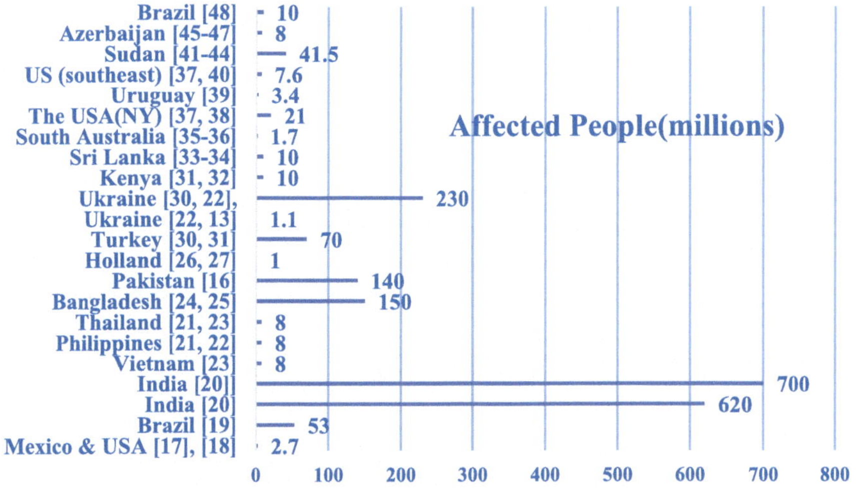

Fig. 3 Blackout affected country (region) versus Blackout affected people in millions

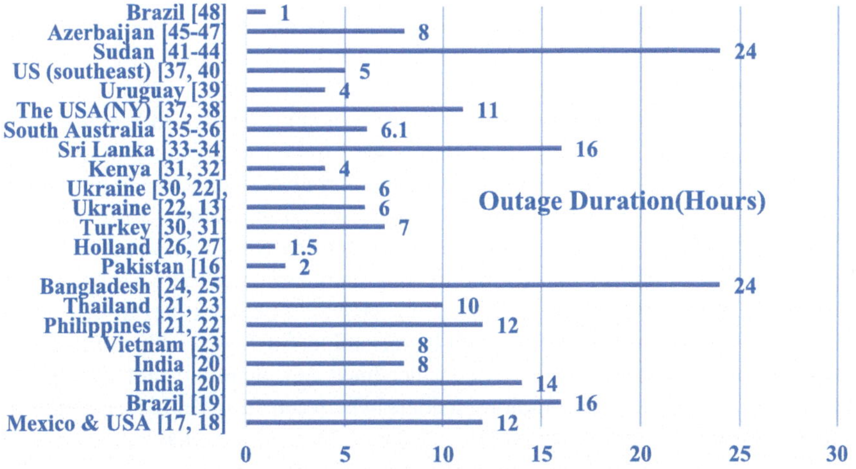

Fig. 4 Blackout affected country (region) versus Outage duration in hours

Prevention Mechanism for Blackouts: It is not persuasively imaginable to fully eradicate blackouts; however, it can be revealed that by enforcing some rationally cost-effective procedures, incidence of the blackouts can be curtailed and/or their effects could be mitigated. Obviously, the most forthright means is to minimalize the menace of unintentional disturbances by vindicating, the root causes of system disturbances by following three steps [49–51], the first one is analyses and assessments of blackouts, second step is taking preventive and corrective actions and third step is revision of public policies and forthcoming investments.

3 Challenges in Wide-Area Power System Operation Against Blackouts

Any of the unstable dynamic phenomena may causes to the separation of the intact power system into smaller islands of source and load combinations. The resulting partial or complete blackouts are thus developed and always associated by the different disturbance phenomena, that the power system experiences. The most significant disturbance phenomena associated with power system instability generally occur in wide-area power system networks are discussed as follows [2, 52];

- Transient Angle Instability (loss of synchronism).
- Voltage Instability (voltage collapse).
- Frequency Instability (frequency collapse).

3.1 Transient Angle Instability

This is also called as loss of synchronism and this disturbance phenomena can occur when group of generators at different parts/regions in wide-area power system network accelerate at dissimilar speeds, creates a scenario where the system is likely to split. The initial large disturbance (e.g. transmission line fault) will lead to eccentricities in generator rotor angles. If this is then trailed by insufficient coupling between cluster of generators (owing to removal of transmission lines), it may cause to an absence of synchronizing power and therefore an incapability for generators at various parts/regions in wide-area power system network to keep in synchronism. It results an uninterruptedly progressing angular change between groups of generators. As the two provinces/regions in wide-area network part in angle, the voltage in between the two regions will be miserable. Such miserable voltage may cause to protective relays tripping of further lines and, therefore probable subsequent splitting of all AC transmission corridors in between the two provinces/regions. This phenomenon happens within a few seconds. Typical remedial actions can be employed against the transient angle instability are braking resistors (or) FACTS devices and Fast-valving to rejection of generation (or) to reduce generation.

Small-signal Angle Instability: This kind of disturbance occurs repetitively on the system owing to the small discrepancies in loads and generation. This kind of instability may happen in two forms; steady upsurge in rotor angle owing to absence of adequate synchronizing torque and rotor oscillations of growing amplitude owing to absence of adequate damping torque. The weakening of the transmission system, coupled with high power transmission levels, lead to an unrestrained increasing electromechanical fluctuation between cluster of generators. The pragmatic response are unrestrained increasing power oscillations on transmission passages till protecting relay upshot in further separating of the system. The most outdated way of damping those oscillations is by means of Power System Stabilizers (PSS). FACTS devices are also used for the damping of small-signal angle oscillations.

3.2 Voltage Instability

A system come in a state of voltage instability when the system is subjected to a disturbance or upsurge in load demand leading to incontrollable drop in voltage. Thus, the system is not able to keep stable voltage profile. The main factor contributing to this instability is upsurge of reactive power demand in the system. However, in most cases, voltage collapse (wide-area voltage instability) may occur within several minute range. The emergency corrective actions usually taken are resetting of generator voltage set point (or) fast re-dispatch of generation, automatic control of shunt switching (or) series compensation and blocking of tap changer of transformers.

3.3 Frequency Instability

When the power system fails to keep its operating frequency stable within operational limits, it is known as frequency instability. Usually the operating frequency can be maintained within a narrow range by considering the system safety and stability. In case of large disturbance, for illustration, loss of generation, loss of load or loss of interconnection between AC or DC, an emergency control and corrective actions should be taken to maintain frequency stability such as tripping of generators (or) Load shedding, fast-valving to control generation, HVDC power transfer control and controlled islanding of network (or) load.

3.4 The Need of Wide-Area Protection, Control and Monitoring (WAPCAM) System

In view of the major blackouts and wide-area power system disturbances, the present conservative method of protection, control and monitoring system is unsuitable due to lack of system dynamic view with absence of real-time measurements. The major blackouts, wide-area disturbances, etc., may occur in a very short time span, faster than the time the operator could take to analyse and determine the action to initiate. Henceforth the need of smart, wide-area-perspective system with advanced communication and computer architectures is justified. Particularly application of cutting-edge communication technology along with global positioning system (GPS) enhances the protection and control system of wide-area networks [53, 54]. The advanced real-time measurement elements such as Intelligent Electronic Devices

(IEDs) and Phasor measurement units (PMUs) now permit precise assessment of measuring quantities over extensively parted locations as well as possible real-time measurement-based control schemes. Henceforth wide-area protection control and monitoring (WAPCAM) system should be adopted to fulfil the main purposes of not only growing transmission capacity but also improving system stability, reliability and security.

4 Wide-Area Protection, Control and Monitoring Philosophy

The revelation of the power system to wide area collapse and partial or total black-outs has amplified in the recent ages, as the system has been pushed to operate near to its stability and security limits. The power system networks are exposed to major disturbances, that causes the disruption of the power service to the consumers. Even for the well-designed system, erratic incidents can stress the system outside the pre-determined limits. The operation and maintenance of the power system is substantially diverse from the expectancy of the system planners/engineers, predominantly in an emergency condition [55]. With the rapidly growing competences in computer and communication technologies, opportunities are now being available for the introduction of advanced WAPCAM that shows a great potential. Such systems would receive wide-span data, e.g. system-wide voltages, angles, active and reactive power flows, etc., and analyse them, indicating whether the system is on the urge of a transformation into an unstable state or not. Furthermore, issuing wide-span, coordinated actions that will save the system from proceeding to total collapse, or even, mitigate the wide-area disturbance effects.

The major blackouts and other disturbances worldwide have highlighted the requirement of implementation of WAPCAM systems as a gainful solution to improve power system network planning, operation, maintenance in view of reliability, stability, and security of wide-area power system networks. The WAPCAM systems has adopted state-of-the-art technology developments in sensing, high-speed communication, advanced computing, dynamic visualization and optimized algorithms. The synchronised measurement technology (SMT) and System Integrated Protection (SIP) Schemes are main core elements of WAPCAM that enables optimized coordinated functionalities at wide-area perspective. Figure 5 demonstrates basic view of WAPCAM and its core elements.

The synchronised measurement technology (SMT) that provides real-time synchronised measurements using a synchronized clock enabled with Global Positioning System (GPS) and can transmit wide-area measurements for real-time monitoring of power system network [56]. This synchronized measurement technology can be realized by Intelligent Electronic devices (IEDs) at local bay level and Phasor Measurement Units (PMUs) at regional level (or) wider level. These IEDs/PMUs processes a real-time measured signal and estimates its magnitude and phase angle

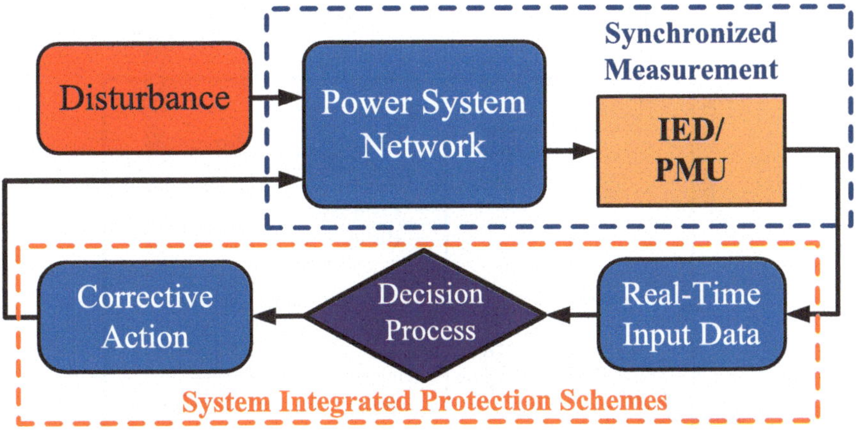

Fig. 5 WAPCAM: System Integrated Protection Scheme (SIPS) with Synchronized measurement

in time-synchronized form using GPS clock signal. Figure 6 elucidates conversion of a sine wave and its phasor with in a time-frame.

Intelligent Electronic Device (IED): The IED is most significant intelligent device at local level in the basis of WAPCAM, as shown in Fig. 5. The 'IED" incorporates the basic functionalities such as merging unit, intelligent unit and protection relay unit in a sole inlet/bay [57]. The basic structure of the "IED" is revealed in Fig. 7. The monitoring and control of circuit breakers and switches are employed by digital input/output edges. This IED can provisions Sample Value (SV) communication based on IEC61850-9-2 and the Generic Object-Oriented Substation Event (GOOSE). Some operations of protection and control, that fits to the bay digital/logic level, are realized in IED devices in case of communication system fails. The IED devices can be mounted close to the main apparatus. According to the principle of WAPCAM, the existing functionalities of the local protection are optimized. Only main protections based on local fault information are reserved in local IEDs.

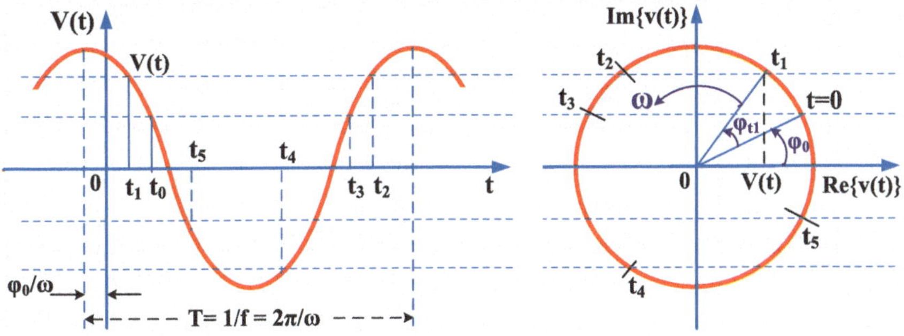

Fig. 6 Conversion of sine wave to its phasor form

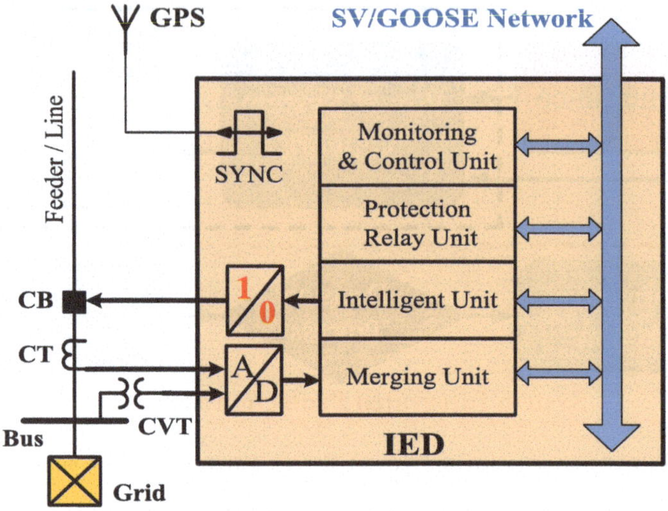

Fig. 7 Architecture of Intelligent Electronic Device (IED)

The standby backup protection and controls are available only in the condition of a substation protection failure.

Phasor Measurement Unit (PMU): The main concept of PMU is to measure the voltage and current phasors at the identical time on the designated locations, transmit them into a central location, where they can be compared, assessed and further processed. The device performed measurements in real-time is called Phasor measurement unit (PMU). Figure 8 shows basic architecture of PMU. This PMU is embedded with the receiver of GPS signal synchronizing the measurements and

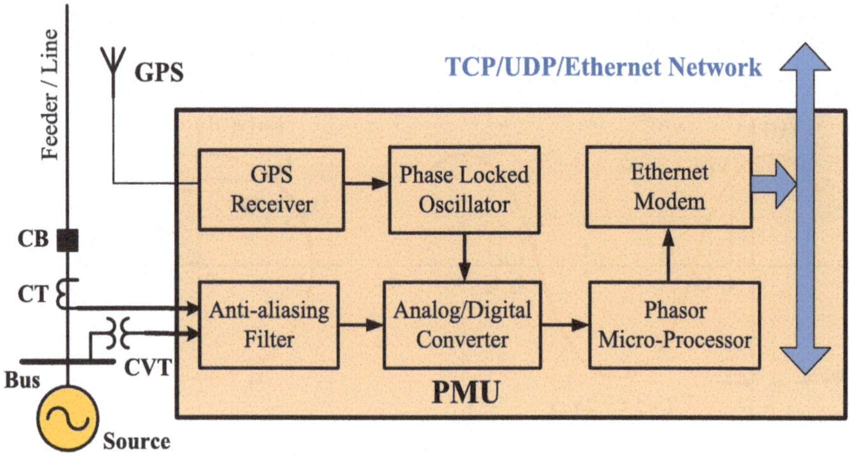

Fig. 8 Architecture of Phasor Measurement Unit (PMU)

labelling the time stamp on them. The PMU is able to pre-processing of data. The basic functionalities of PMUs are as follows:

- System monitoring/state estimation
- Event recording
- Analysis system/load characteristics
- System controls

PMU can estimate/process synchronized voltage, current signals, frequency and rate of change of frequency (ROCOF) according to united coordinated universal time (UTC) [58]. These PMUs generally installs at optimal locations in substations and also electrical network with a standard instrument transformer such as CT and CVT. The PMU processes a real-time measured signal and estimates its magnitude and phase angle in time-synchronized form. A fixed time-frame is used with synchronized GPS-clock signal. This GPS signal having 2.046 MHz band width and a 1575.42 MHz centre frequency. However, it has some advantages to wide-area power system networks such as wide-area monitoring with global coverage, free accessibility and microseconds level of timing accuracy. A GPS receiver along with phase-locked oscillator are used to time tag the measured signals after passing through an anti-aliasing filter and an analog-to-digital converter [59]. Further this digital signal is sent to a micro-processor to compute the phasors and frequency, rate of change of frequency and binary data. This whole data is transformed from PMU to phasor data concentrators (PDCs) that are setup in local substation to archiving data for online monitoring of system condition and offline assessment of wide-area power system network. This PMUs will process the data as per the IEEE1344, IREG_B, and IEEEC37.118 standards and these PMUs are having two classes in application point of view such as measuring PMU(M-class) and protection PMU(P-class).The M-class PMU has low response time with high accuracy but P-class PMU has high response time with less accuracy. So, the P-class PMUs are appropriate for real-time protection and control applications where fast response and low latency is needed.

WAPCAM and its Functionalities embedded with SPS: In this WAPCAM an existing protection, control and monitoring schemes are reviewed to achieve an optimal coordination between the three-level system integrated protection schemes (SIPS) (or) special protection schemes (SPS) (or) remedial action schemes (RAS). The latest research and developments are reviewed and reassessed in this section, including wide-area protection and control, wide-area stability and control and wide-area monitoring and control that is based on wide-area communication infrastructure etc. to improve stability, reliability and security of power grid. These three-level hierarchy is a concept of system integrated protection schemes (SIPS) for wide area power system networks [57]. The hierarchy of three-level functionalities of wide-area protection, control and monitoring (WAPCAM) systems is illustrated in Fig. 9. From the Fig. 9 the most significant is the first level of protection and control system that assures the power system transient stability by rapidly removing the faulty component (or) system using a protection relaying equipment. The second level of stability and control system stabilizes the power system after post-fault condition (or) other disturbance condition by using stability control actions such as generator tripping

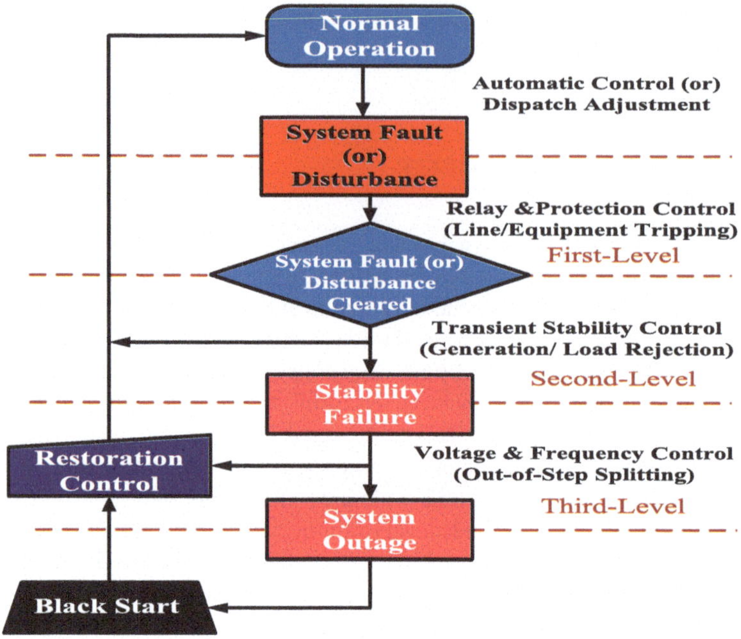

Fig. 9 Three-level hierarchy of WAPCAM

and load shedding. The third level of control and monitoring system secures the power system thereby balancing the separated power system after splitting, that can prevent the accidents from causing a system collapse by using frequency and voltage emergency control actions when serious damage is occurring.

The conventional three-level system integrated protection for wide-area power system is based on local information and static protection and control strategy, that is not suitable any more to wide-area power system networks. Its limitations include: information islanding caused by separated stability control systems, in-coordination between protection and control for different devices, static setting and configuration of protection and control that cannot comply with the new necessities of power system security and stability. The advancement of PMU is significantly enhanced the observability of the power system dynamics. The planned automatic control action that could be engaged to secure the power system can be divided into either preventive or corrective. The ultimate objective would be keeping as much of the system intact and generators coupled to the power grid. The customized WAPCAM systems are intended to enhance power system reliability and/or upgrade the transmission capability in the future [56, 60]. The architecture for such systems should be developed conferring to what kind of technologies the operators/utilities holds at the particular time period, in addition to the choose of fitting the technology adaptation method that the operator/utility in request will take. Figure 10 illustrates operational

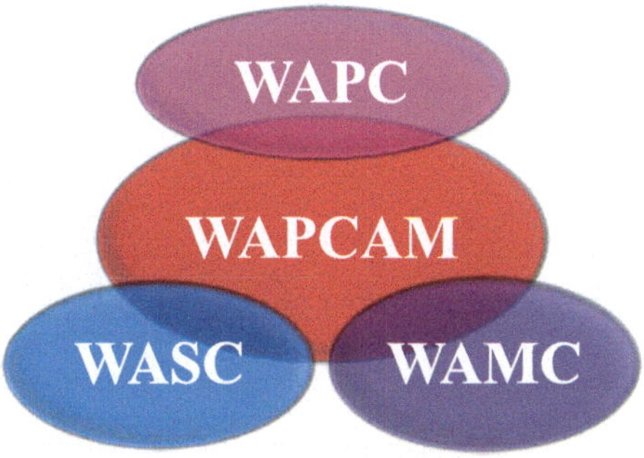

Fig. 10 Operational coordination of WAPCAM

coordination of WAPCAM where WAPC, WASC and WAMC are optimally coordinated each other to operate the wide-area power system network safely and reliably. The WAPC employs locally measured data to instigate special protection schemes and WASC and WAMC operates based on wide-area measurement data to initiate remedial action schemes at regional level.

4.1 Wide-Area Protection and Control (WAPC) System

This WAPCAM system basically consists of three-level SPS with protection and control equipment such as IEDs at the local bay level, PMUs at the sub-station feeder level and reginal level. The basic architecture of wide-area protection, control system (WAPC) is shown in Fig. 11.

Local Bay Level Protection and Control: As per the WAPC philosophy the existing functionalities of local bay level protection are optimised to main protection based on local measurements such as fault data measured by IEDs. The standby backup protection and controls are activated when the substation protection fails.

Substation Level Protection and Control: As per the WAPC philosophy, the substation level protection and control does not only includes the backup protection of transmission lines, bus and transformer, but also aids functionalities of the substation control system, that includes automatic reclosing, automatic bus transfer, circuit breaker failure protection, under frequency and under voltage load shedding, overload tripping and so. The substation backup protection is employed by using the whole substation data. The conventional protection system like over current protection and breaker failure protection are replaced with extended current differential protection. In contrast to enhance the reliability of main protection, the functionalities of local IEDs are replaced by standby functionalities of the substation level protection if in case of local IEDs fails or undergo any maintenance.

Regional Level (or Wide-Area Level) Protection and Control: In the context of WAPC philosophy, the region level protection and control incorporated advanced functionalities such as backup protection and control. Some functionalities can be employed at both substation level and region level protection and control centre such

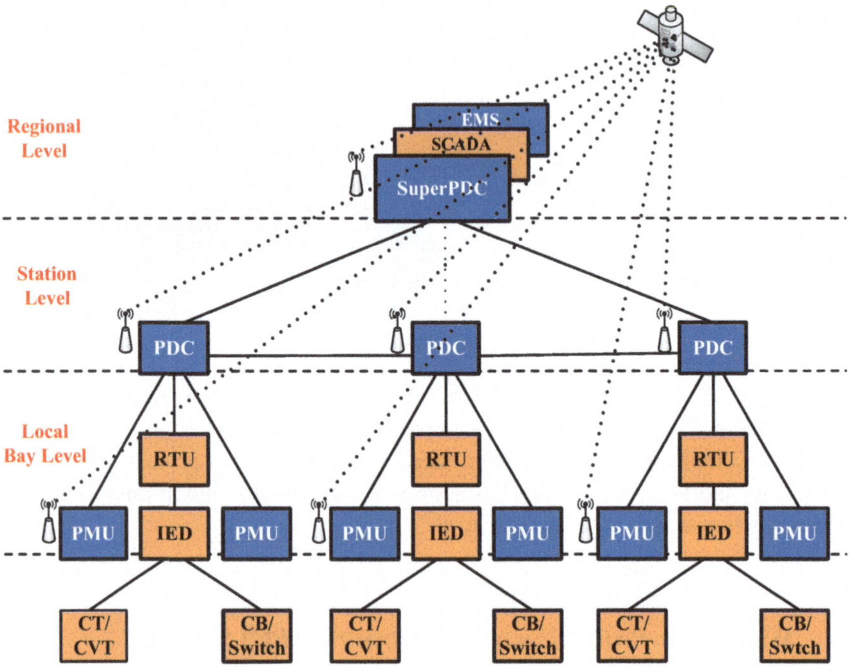

Fig. 11 Architecture of wide-area protection and control systems

as under frequency and under voltage load shedding and automatic bus transfer. The transmission line section safety, protection and control across wide-area network can be done by sharing data of PMUs at regional level that includes oscillation detection and out-of-step separation. Further, it can coordinate the substation level protection and control with regional level to enhance wide- area protection and control.

The physical architecture of wide-area protection and control system has been discussed [60] and the three types of architectures such as flat architecture, central (or) hub-based architecture and multi-layered architecture are explained for power system networks. The WAPC with system protection terminal methodology has been proposed along with few case studies of practical wide-area power system networks in [55]. The design and implementation of wide-area special protection schemes were proposed thereby introducing a physical structure and logical structure separately. Further testing and validation of proposed schemes have been illustrated [61]. New wide-area relaying protection has been proposed based on two main approaches such as online-adaptive relay settings and fault element identification that aims to enhance the performance of wide-area backup protection [62]. The physical architecture of wide-area protection systems such as flat architecture with system protection terminuses, central architecture and multi-layered architecture were discussed. The adaptive controllers for wide-area measurement systems such as de-centralized controllers, centralized controllers and Multi-agent controllers have been emphasized. The existing technology and their implications in wide-area protection and control was reported [63] as follows:

- A complete defense plan to deal with all kinds of power system instabilities to be provided by topological and numerical observability of power system.
- A suitable physical architecture for complete defense plan to tackle with all kinds of power system security problems whether local or global in nature.
- Dynamics of loads in power system has to be considered while dealing with frequency instability in WAPC.
- WAMS based adaptive relaying to deal malfunction of relays at stressed condition.
- WAMS can be used to deal uncertainties in distance protection schemes by getting control parameters of FACTS in a precise time.
- The real-time adaptive protection scheme should be designed by enhancing time delay in information exchange from PDC to control center.

The administrative control of backup protection and enhancement of system integrity protection schemes thereby making more adaptive and intelligent because of 70% of major blackouts were occurred due to malfunction of relays. The some of the important issues related to hidden failure and non-hidden failure such as load encroachment, under frequency load shedding, power swing and out-of-step tripping.[64, 65]. The design, development and implementation of a centralized Remedial Action Scheme (CRAS) in a practical power system network (Southern California Edison) was emphasized in [66]. This implementation of CRAS has enlightened two major contributions; (a) How the wide-area monitoring and protection system can be designed by using advanced information and communication technology (ICT) with the future expanding functionalities for RAS. (b) The experience of conceptualization and implementation of proposed wide-area protection and control system was given valuable insights to adopt for existing power system network. An adoptive out-of-step protection scheme has been proposed based on extended equal area criterion (EEAC) to regulate critical clearing angle (CCA) and critical clearing time (CCT). The parameters of dynamic model and coherency between the group of generators/system are determined in real-time investigation on 16- machine, 68-bus network model. A logical scheme based new wide-area protection and control system has been designed to predict transient voltage instability and the proposed algorithm response time was 10 s after the post-disturbance that may lead to system collapse if not predicted properly [67]. To define parameters for alarm triggering and other protection and control modules, a key performance Indices (KPIs) can be integrated as transient angle stability function in WAMPAC [68]. The proposed scheme was employed on two standard network models such as IEEE 9 bus system and Chilean's CRO 6 bus system. A combination of current differential scheme and phase comparison scheme are used to design a novel centralized wide-area protection system that enhances the performance of pilot wire relays and it has many advantages such as invulnerable to power swing, mutual coupling, sequence impedance imbalances, independent of load currents, good selectivity for external faults and zone coordination is not required [69]. The coordinated substation protection system has been elaborated using inter-substation information and IEDs embedded with three basic modules such as function module (FM), local coordinated module (LCM) and remote coordinated module (RCM). The FM explores protection functionalities like sharing data with other local IED and performs coordinated tripping actions, LCM is used to monitor need of local IED and sends appropriate commands [70]. The RCM module, receives data from other remote IEDs and analyse to local use. The coordinated substation protection system doesn't require any third-party for control and monitoring and it is redundant in case of IED damages. The range of functionalities of IEDs can be extended through standard module interface.

A multi-function line protection (MFLP) model has been designed with a combination of system integrated protection and backup protection functionalities using wide-area measurements that tested for out-of-step condition in MATLAB/Simulink environment [71]. Development of hardware-in-loop (HIL) test bed for traditional protection relays and latest protection relays that operates on IEC61850-9-2LE in laboratory environment. A precise model of IEEE 123 bus test system has been used to simulate a large number of test sequences for confirmation of relay response times at different substation topologies [72]. The DC power flow model with synchronized measurements based an optimal bus-splitting scheme has been proposed in [73] and the same tested on IEEE 14 bus test system. A centralized busbar differential protection scheme has been developed by considering data desynchronization scenarios. The first-order Lagrange interpolation polynomial and voltage angle difference between IEDs are used to recompense data desynchronization errors. And the same protection philosophy was tested on real-time digital simulator by using a IEC 61850-based IED that are employed on Smart Grid infrastructure evaluation Module of MMS-EASE lite library [74]. Travelling wave theory-based fault location estimation using wide-area measurements and the same proposed scheme has been validated on a practical Korean power transmission network [75]. The zone division based wide-area protection system for fault identification using positive sequence fault component in [76] and a case study has been performed on IEEE 10 machine 39 bus network to confirm the efficacy of the proposed scheme.

A wide-area backup protection scheme has been developed based on fault component voltage distribution. The voltage and currents at sending end of the line are used to estimate fault component voltage at the receiving end. The proposed scheme was validated on IEEE 39-bus system and the fault component can be detected by the ratio between measured values and estimated value [77]. Detection of transmission line outage using wide-area measurements were proposed based on bus voltage phasor measurements and aiding to change in system topology. Additionally, optimal placement of PMUs has been done to use minimum use of wide-area measurement data [78]. A centralized multi-functional WAMPAC system has been developed for out-of- step (OOS) tripping using wide-area measurements and the Croatian 400 kV transmission system has modelled in MATLAB/Simulink to validate proposed scheme [79]. A multi-objective model for optimal placement of PMUs by considering line outages has been investigated where ε-constraint method and fuzzy satisfying approach were used to solve the optimization problem. The developed scheme has been validated successfully on IEEE-57 bus test system [80]. Power differential protection scheme based on wide-area measurements has been designed and in addition CT saturation, data mismatch, latency in data transfer has been performed. The proposed algorithm was validated on a practical Iranian transmission system in simulation environment [81]. A novel wide-area protection scheme based on 3D-phase surface method using second order differential equations have been explained. The solutions curves will move either close to the equilibrium point or move away from the equilibrium point indicates whether transmission line is being faulted or not [82]. This 3D-Phase surface method doesn't need any zone settings and can be adaptable to single and double circuit transmission lines without any major changes. The realization of real-time Hardware-in -the-loop (HIL) test bed for wide-area protection scheme was done by modelling 1100 bus TNB network on Opal-RT real-time simulator. This comprehensive real-time study helps to reduce risk of design errors and conforms the adoptability of proposed wide-area protection in practical power system network [83]. The real-time analysis of wide-area protection scheme has been developed by deriving three types of relaying indices such as distance relay index, over current relay index and directional relay index [84]. The proposed scheme has been tested on New England 39 bus system. A system Integrity Protection Scheme (SIPS) was proposed to improve wide-area protection security during stressed condition by detecting vulnerable points to prevent maloperation of relay [85].

The proposed scheme includes two algorithms; the first one, it updates relay operating characteristics with respect to the system conditions and the second one, it identifies the node at which load can be reduced to avoid relay maloperation. A wide-area supervisory protection scheme has been designed for islanded networks by calculating two indices; one is power flow between the line and angle difference between the bus voltages [86]. This proposed scheme is robust to the high frequency oscillations and changes in system configurations. The wide-area protection system for transmission line has been developed by considering time delay in data transfer and change in different communication links and devices as well. A comprehensive testing method have been followed to enhance WAMPAC functionalities [87]. The wide-area transmission line protection has been proposed by considering unattainability of data of critical buses. Herein this proposed scheme three indices are derived; the index1 identifies disturbance area by detecting changes in current injections and then after a subsystem has been chosen. The index2 detects fault in the network by calculating input and output power difference in the subsystem. Further the index3 helps to identify fault in the line based on the voltage feature. This proposed scheme is thoroughly tested on NEW England 39 bus test system in OPAL RT real-time platform and effect of stressed conditions of power system also evaluated [88].

4.2 Wide-Area Stability and Control (WASC) System

The power system stability includes electromechanical (rotor angle) stability amongst group of synchronous machines and voltage stability concerning load response to disturbances. Most of the stability controls are unremitting feedback controls at power stations such as automatic voltage regulators and power system stabilizers for generation excitation control and prime mover control [89]. The installed wide-area stability controls are mostly based on straight detection of designated outages. This emergency controls are called as Special protection schemes (SPS) or remedial action schemes (RAS). The wide-area stability controls deal with feedforward/feedback approach with the help of unremitting wide-area observability and controllability. In present scenario wide-area stability controls measures power system response to disturbances. Some of the wide-area stability control practices are reviewed as fol1ows;

An adaptive wide-area damping control method has been designed by considering signal time delay for smart grid environment. The subspace system identification (SIT) is adopted to low-order nonlinear model using wide-area measurements. Further a new wide-area damping control method has been designed by online tuning of wide-area damping controller (WADC) parameters with the residue method. A simple time delay compensation algorithm was incorporated to validate effectiveness of the method [90]. Based on wide-area measurements a load shedding method is offered to prevent frequency instabilities by keeping line flows within acceptable limits and rejecting the most vulnerable loads in a single instant/step [91]. Though conventional under frequency load shedding is most significant to deal with frequency instabilities some of the draw backs are identified as follows;

- It suffers with the delay in response because the conventional scheme needs the frequency drop to some threshold value to initiate corrective actions.
- This scheme designed based on offline assumptions of load variations followed by frequency deviations and corresponding system inertia. The conventional scheme doesn't contemplate the actual system state, topology, magnitude and location.

In the proposed wide-area load shedding approach, a single machine equivalent model was designed based on wide-area measurement. The optimal load shedding is proposed based on the estimated magnitude of disturbance. The progression of frequency stability after a disturbance is estimated and stabilizing load shedding has been proposed in a coordinated approach using wide-area measurements [92]. This proposed scheme overcomes delayed response of conventional load shedding mechanism. A wide-area out-of-step (OOS) protection scheme has been proposed by detecting stable and unstable power swings for several coherent group of generators [93]. This proposed scheme detects transient angle instability after occurrence of first swing of oscillation. The fault-masking emergency control to maintain stability by considering system reconfiguration using wide-area measurements. The concept of a virtual actuator for lure-type system has been presented based on the solution of linear matrix inequality [94]. A new scheme for load shedding for self-healing of power system under emergencies to deal with frequency and voltage instabilities. An appropriate load shedding has been calculated based on the disturbance power and voltage stability condition using wide-area measurements. In this proposed scheme voltage stability risk index (VSRI) has been assessed to find appropriate load shedding at each/every load bus [95]. The proposed scheme has been successfully verified on

New England 39 bus system and a real-world 246 bus Indian power system network. The proposed adaptive under frequency load shedding scheme was compared with traditional under frequency load shedding to conform the superiority. The parallel optimization scheme to enhance voltage stability margin has been proposed based on voltage stability (QV) index that helps to detect week buses and further proposed scheme has been validated on IEEE 118 bus system and IEEE 300 bus system to confirm the accuracy [96]. A new methodology Vector Analysis (VA) is proposed that derives new instability index to detect voltage stability for wide-area power system networks. This voltage stability index calculated based on active and reactive power measured at generator bus. The network graph, and zoning is carried out if proposed scheme detects the voltage collapse and disconnects weak lines in the power system. After separating zones such as damaged and undamaged zones, load shedding has been employed using ANFIS-TIK (AN-T). This proposed scheme is verified on IEEE-39 bus system to conform the speed and accuracy [97].

4.3 Wide-Area Monitoring and Control (WAMC) System

When the conventional control strategy fails to predict or prevent critical exigencies that may causes power system collapse, wide-area monitoring and control (WAMC) system that offers an unfailing security estimation and optimized coordinated remedial actions is able to moderate or avoid wide-area disturbances. The foremost challenge is early detection of power system instabilities, enhances power system reliability, system operates closer to the stability limit, improves power transfer capability without any compromise in security, and optimal load shedding. The main drawback of traditional system comprising of inappropriate steady-state system view such as SCADA (or) EMS, or uncoordinated local control actions. A combination of stability assessment and stabilization algorithms based on synchronized measurements has been introduced as wide-area monitoring and control (WAMC) system. The power system instability assessment and various control schemes are accomplished in WAMC as follows [98, 99]:

- Based on full observation of power system network: frequency instability assessment, voltage instability assessment of the interconnected networks
- Based on customized observation of power system network: Oscillation detection assessment, voltage instability assessment of the transmission corridor, line temperature monitoring.

The controlled islanding is an alternative solution for wide-area power system networks when the system experiences major cascading incidents. The main objective of controlled islanding is to deal with power system stability of each/every island and to maintain the minimum loss of load from the whole power system network. A WAMS-Integrated adaptive controlled islanding scheme [100] has been proposed based on the dynamic post-disturbance trajectories considering various failure modes. An Independent component analysis (ICA) based coherency identification in wide-area power system network has been proposed in [101]. The proposed scheme is employed to generator speed and bus angle data to find the coherent regions of the power system. This proposed ICA applied/validated based on simulated data of 16- machine, 68-bus power system model and also real-time data collected from U.K University using wide-area measurement system. This proposed scheme successfully identified the group of generators and buses followed by a disturbance in the system. A dynamic mode decomposition (DMD) algorithm to extract the global attributes of transients that are measured by wide-area sensors has been proposed. This global multi-scale based DMD infers global dynamic behaviour in the form of both spatial patterns and temporal patterns that were associated with dynamic modes consisting single frequency components [102]. The proposed scheme is adoptable for fast wide-area monitoring and assessment of global instability in view of recent data fusion-based estimation method. The integration of wind parks, solar parks and the HVDC transmission causes inter-harmonics in the wide-area power system network that may increase power system security problem. These inter-harmonics will lead to the sub-synchronous oscillation in the phasor measurements. The impact of inter-harmonics on wide-area measurements has been investigated there by employing frequency transmutation principle. Further a practical case study was carried on china's field PMUs data and the correctness of fundamental derivations with an appropriate monitoring scheme is verified [103]. The estimation of actual inertia of power system based on wide-area measurements such as ambient frequency and active power. This proposed method helps to monitoring of inertia in continuous manner in the real-time domain like minute and tens of minutes. In this method whole system separated as number of different regions and operative inertia of each/every area was calculated individually by perceiving the dynamic changes between active power and corresponding frequency deviations. This method is validated on practical Icelandic power system network [104]. Most of the oscillations were damps out in the power system instead few of undamped oscillations causes whole system collapse. A practical case study has been carried out by employing seven advanced signal processing techniques such as Fast Fourier Transform (FFT), Wigner Ville Distribution (WVD), Prony Analysis (PA), Hilbert-Hung Transform (HHT), S-Transform (ST), Matrix-pencil Method (MMT) and Estimation of Signal Parameters by Rational Invariance Technique (ESPRIT) that helps to identify low frequency mode oscillations. The wide-area measurement data of practical Indian power system network has been used to demonstrate the methodology. Among all seven Signal processing techniques FFT and ESPRIT exploring good performance by extracting low frequency modes without losing its attributes [105]. A real-time wide-area monitoring and analysis on a hardware setup has been demonstrated to define power system security and stability indices [106]. The online-power flow followed by a contingency analysis helps to determine security and stability indices that can initiate corrective actions in power system at emergency conditions. The online dynamic security assessment requires real-time computation of dynamic equivalents. The wide-area measurements can give exact resolutions to compute real-time dynamic equivalent. It can be realized in two different approaches [107].

The proposed methodology has been tested on EPRI's 8-machine, 36-bus system and outcomes demonstrates that dynamic equivalent model of external system can replicate the dynamic behaviours of the practical power system accurately. Some of the existing methods for detection of inter-area oscillation modes are assessed based on calculation speed, accuracy, criterion and adoptability of Prony analysis method and ESPRIT method have been compared. Based on the investigations and comparative assessment of existing methods a novel scheme for monitoring low-frequency oscillations using wide-area measurements are elucidated [108]. An extensive study on dominant inter-area oscillation path signals and its adaptability in wide-area damping control (WADC) has been carried out [109]. Since the performance of WADC depends on selection of appropriate feedback signal such as dominant inter-area oscillation path signals. The case studies have been carried out for open PSS, fixed PSS and returned PSS. The proposed methodology has been employed on large power system network, i.e., KTH-NORDIC32 system that confirms the applicability for real-time power system networks. The imperialist competitive algorithm (ICA) was used to design a multi-stage optimal under frequency load shedding (UFLS) scheme based on wide-area measurements thereby deriving a system frequency response model (SFR) [110]. The applicability and effectiveness of proposed scheme is evaluated on a practical SESCO grid transmission system. A novel UFLS scheme has been proposed by considering time-domain simulations-based system frequency response model using wide-area measurements [111]. A review on wide-area monitoring systems (WAMS) has been reported wherein centralized architecture, decentralized architecture and distributed architectures are explained for realization of WAMPAC in future power system networks [112]. The decision tree algorithm aided islanding disturbance detection and warning scheme [113] has been proposed by considering islanding database of simulated records, islanding records of practical system and system geographical and topological data. The proposed islanding scheme has been employed in Dominion Virginia Power System for practical utilization.

5 Implementation of WAPCAM Systems in Practical Power System Network

In this section, several cases that were encountered by the author have been mentioned. Brief summaries of WAPCAM systems, or any SPS/RAS that classifies to a wide-area nature (from the point of view of the author), are specified below as a part of exploration. In 1985 [59] the Florida Power and Light utility installed a system-wide fast acting load shedding system. After the addition of two new 500 kV lines with neighbouring utilities, system studies showed that concurrent outage of two or more large generators of FPL will affect the excessive power import into Florida, that will result in a stable but overloaded state. However severe thermal and reactive overloads might lead to voltage collapse and uncontrolled system partitions after approximately 30 s. The installed FALS package runs in the system control centre computers in Miami, Florida. Using state-wide communications, of SCADA, to identify a "stable system overload" that results in lines functioning above their nominal ratings, low system voltages at pre-planned buses and heavy reactive power demands upon generators, FALS initiates a trip signal to shed pre-planned quantity of load to prevent uncontrolled outages of lines and generators that might cause blackouts.

In France [59] in the early 1980s, several contingencies highlighted the cons and weaknesses of the traditional out-of-step schemes already implemented, for example: local processing of measurements while the fact that out-of-step affects large areas. A wide-area system, called "Syclopes", was introduced, coming closer to the root of the loss of synchronism problem, that are angular and frequency differences between two points. Syclopes realizes detection of loss of synchronism, controlled islanding and separation of different homogeneous areas and orders of load shedding, if required to balance load and generation in the areas. The Tokyo Electric Power Co. Inc. (TEPCO) [114] employed an out-of-step (OOS) wide-area relaying scheme, based on power-angle estimation method. A big nuclear generator was selected as a locus generator and its instantaneous power and voltages data were transferred to each pumped-storage plant via a microwave linkage. At each plant, the received data were processed to predict out-of-step conditions and accordingly estimate the optimum shedding capacity, ordering selective tripping of some local pumped-storage generators.

Again in 1989 [114], TEPCO applied another similar scheme based on Voltage-angle estimation method. It measures voltage waveforms at four locations on the bulk 500 kV transmission system and communicates using microwave communication system. Phasor variances between the western region (load concentration) and each of the other regions (containing most generation) of the TEPCO system is estimated from the voltage waveforms. Phase angles for the next 10 cycles are estimated based on extrapolation. If phasor variances surpass a pre-defined threshold phase angle, out-of-step is detected, and system partitioning and load shedding are instigated. In [115] a wide area protection system using Artificial Neural Networks (ANN) for transient stability detection is proposed. Rate of change of bus voltages and angles for six cycles after fault tripping and/or clearing is used to edify a two-layer ANN. Coherent groups of generators, that swing together, are recognized through an algorithm using PMUs data. Islanding and under-frequency load shedding actions are taken by the system to counteract the system instability.

WAP systems schemes were designed in the Pacific NW and northern California to detect 500 kV line outages and instigate suitable remedial actions to avoid overloads, low-voltages and out-of-step situations in the WECC system [1]. In [116], an on-line demonstration of a Wide-Area stability and voltage Control System (WACS) is described, at the Bonneville Power Administration (BPA). Phasor measurements are acquired from eight BPA stations. Existing transfer trip circuits are accessible for generator tripping and 500 kV capacitor/reactor bank switching. WACS first processes the phasors to attain the required voltage magnitudes and active and reactive powers. The voltage magnitude-based algorithm computes a weighted average voltage from the measurements at seven 500 kV stations. Non-linear accumulators calculate volt-seconds under voltage threshold settings. Control actions are instructed when volt-second accumulation reaches a set point. For generator tripping the weighted average voltage must also be below a set point, taking into consideration blocking of accumulation during voltage recovery. The voltage magnitude and generator reactive power-based algorithm combines both fuzzy logic and weighted average 500 kV voltage magnitude from twelve phasor measurements at seven locations and weighted average reactive power from fifteen transmission lines emanating from six large power plants. With accumulator set point reached, capacitor/reactor bank switching is commanded and then generator tripping only in severe conditions. Also under voltage load shedding is utilized for voltage stability problems. Simulations [117] showed that famous disturbances/blackouts of August 10, 1996 and June 14, 2004, could have been circumvented using these algorithms of the WACS.

In [118], a scheme of wide-area protection in contrast to chain over-load trip based on multi-agent technology is proposed. Following a line outage, the possibility of a chain over-load trip is analyzed by applying the elements of node impedance matrix, that is updated from the dispatching (control) center through WAN. Data from calculation agent is analyzed to judge whether a fault or an overload can appear, and thus chain over-load trip can be blocked and maintain the power supply reliability and continuity. In [119], an innovative method for the detection of voltage instability and the corresponding control in the existence of voltage-dependent loads has been proposed. Local voltage and current phasor measurements provide basis for calculating voltage-stability load bus index (VSLBI), in addition to the system-wide information on reactive power reserves, will provide suggestions of control actions for deployment when the stability margin is small and the reactive power reserves are nearly exhausted.

Palo Verde is the largest nuclear power station in the western hemisphere [120]. The power station has 3 reactors that initially produced 1287 MW. The first unit was further upgraded by 120 MW, and thus, restudies showed that the system will no longer be stable after the loss of two of the generating units. It was found that the loss of 2 units plus the added margin will cause overload problems on flows of the California Oregon Intertie (COI-Path 66), and slight decay in frequency to less than 59.75 Hz. Moderation studies showed that shedding load, equivalent to the surplus portion of the generation, would mitigate the problem. A scheme was developed based on SRP's digital communication infrastructure to permit load shedding in small blocks at several locations all over the Phoenix metro area.

In [121] a WAP system design is proposed based on phasor measurement units reassuring applications against different kinds of instabilities of frequency, small signal angle and voltage. The proposed algorithm uses sliding windows of voltages V at each bus and feeder loads P & Q. Mean values of gradients between timely neighboring measurement points compose a set of load equations for different time steps within a predictive window. These equations are then solved with a non-linear solver algorithm that is calculated for all the system feeders to define the behavior of all loads. Defined load parameters are then fed into the basic system model to be solved for determining the equilibrium point. If no equilibrium is found, then collapse will be the predicted end. Therefore, a more accurate and early prediction of instabilities is probable and thus corrective actions can be taken earlier and more optimized than in traditional systems. A new prophetic method for emergency frequency control is described in [122]. By creating control on PMU measurements, a single machine equivalent for each power system island is computed online, modeling the evolution of frequency stability after an exigency is detected and stabilizing load shedding is defined in a coordinated way for the administered region. The exact quantity of load or generation to be shed is determined in order to reinstate the frequency to a desired value. An advanced technology has been designed by ABB [123] for the identification of voltage instability. "PsGuard" system comprises a number of phasor measurement units (PMUs) that are placed at premeditated points all over the power system and a central evaluation unit. They comprise an algorithm that computes the stability of the several power line transfer circumstances based on locally measured data of voltage/current phasors. Using the voltage instability prediction (VIP), based on the Thevenin principle. In case of incipient voltage instability detection, the system alerts the operator by indication of the outstanding security margin and provides online supervision to counter this condition. Furthermore, corresponding data is processed as input to the EMS. Also, automatic regulator actions are started if the safety/security margin crosses a pre-defined critical level to prevent occurrence of voltage instability; FACTS, blocking of OLTCs, maximize reactive power output from generators and shed load as last defense measure. Another wide area protection system [124] based on "three defensive lines" is proposed utilizing IEDs that are in service throughout the system. The IEDs realize four functionalities. First, IED record analog and digital information and calculate fault location and direction if any is detected. Secondly, the fault affected IED communicates with other IEDs inside and outside the substation transmitting and receiving fault calculation results and information. Third, IED implements the WAP criterion to determine whether it shall trip its local CB or send inter trips to other remote CBs. Lastly IED Transmit electrical information to remote master station that will implement the WASAC controls.

The indicated WASAC (Wide Area safety and automation control) provides function of load shedding, switching off generators, and controlling steam valve of generator etc. If the power system still suffers from major disturbances, the WAP system will disperse the interconnected power system network into several islanded networks to preserve stable and reliable operation.

6 Conclusion

This chapter presents a comprehensive overview of wide-area protection, control and monitoring systems to prevent major blackouts in complex interconnected power system. The typical operating states of wide-area power system network have been categorized such as normal state, alert state, emergency state, in-extreme state and restoration sate. However, this wide-area power system often experiences major blackouts owing to inappropriate system planning in terms of protection, stability, reliability and security. These major blackouts events being evolved in progress of incidents are emphasized as preconditions, initiating events, cascading events, final state of events and restoration. Further the analysis and assessment of major blackouts in wide-area power system network have been elucidated followed by root causes of blackouts and summary of recommendation to prevent black outs. The prevention mechanisms of blackouts are highlighted along with the need of wide-area protection, control and monitoring (WAPCAM) systems. The most important challenges in wide-area power system operation against blackouts (or) disturbances are explicated such as transient angle instability, voltage instability and frequency instability. The WAPCAM is designed with a combination of special protection schemes (SPS) and synchronized measurement technology (SMT) wherein advanced technologies such as IEDs, PMUs and high-speed communication infrastructure are most significant. The three-level hierarchy of WAPCAM is illustrated thereby describing optimized coordinated operational functionalities. The WAPCAM functionalities have been emphasized as WAPC, WASC and WAMC that are integrated with remedial action schemes to maintain stable, reliable and secure power system. By adapting advanced innovative technologies implementation of WAPCAM systems in practical power system have been discussed. This comprehensive review of literature will be very useful for the power engineers, researcher, scientist working in the area of the wide-area protection, control and monitoring system to prevent blackouts.

References

1. A. Guzman, D. Tziouvaras, E.O. Schweitzer, Ken Martin, Local and wide-area network protection systems improve power system reliability, in *Power Systems Conference: Advanced Metering, Protection, Control, Communication, and Distributed Resources* (2006)
2. P. Kundur, *Power System Stability and Control* (1994)
3. V. Madani, M. Admiak, M. Thakur, Design and implementation of wide area special protection schemes, in *57th Annual Conference for Protective Relay Engineers*, 2004, College Station, TX, USA, pp. 392-402
4. W. Lu, Y. Besanger, E. Zamai, D. Radu, Blackouts: description, analysis and classification, in *Proceedings of the 6th WSEAS International Conference on Power Systems*, Lisbon, Portugal, 22–24 September 2006
5. A. Atputharajah, T. Kumar Saha, Power system blackouts-literature review, in *2009 International Conference on Industrial and Information Systems (ICIIS)*, Sri Lanka, 2009, pp. 460–465
6. L.C. White, J.A. Carver, L.1 Connor, C.R. Ross., C.E. Bagge, Prevention of power failures, Volume 1- Report of the commission submitted to the President by Federal Power Commission. July 1967
7. US Department of Energy federal energy regulatory commission, The Con Edison power failure of July 13 and 14, 1977. Final staff report. June 1978
8. A.Kurita, T. Sakurai, The power system failure on July 23, 1987 in Tokyo, in *Proceedings of the 27th IEEE Conference on Decision and Control*, Austin, TX, USA, 1988, vol. 3, pp. 2093–2097
9. T. Ohno, S. Imai, The 1987 Tokyo blackout, in *Power Systems Conference and Exposition* (2006)

10. M.Z. El-Sadek, Prevention of repetitive blackouts in the Egyptian power system, in *First International conference on Engineering Research, Development and Application ERDA*, Port-Said, Egypt, 26–28 November 1992
11. C.W. Taylor, D.C. Erickson, Recording and analyzing the July 2 cascading outage, in *IEEE Computer Applications in Power*, United States: N. p. (1997). https://doi.org/10.1109/67.560830
12. G. Andersson, P. Donalek, R. Farmer, N. Hatziargriou, I. Kamwa, P. Kundur, N. Martins, J. Paserba, P. Pourbeik, J. Sanchez-Gasca, R. Shulz, A. Stankovic, C. Taylor, V. Vittal, Causes of the 2003 major grid blackouts in North America and Europe and recommended means to improve system dynamic performance. IEEE Trans. Power Syst. **20**(4), 1922–1928 (2005). (Nov)
13. U.S.-Canada Power System Outage Task Force, Final Report on the August 14, 2003 Blackout in the United States and Canada: Causes and Recommendations (2004). https://www.nerc.com
14. S. Larsson, A. Danell, The blackout in southern Sweden and eastern Denmark, September 23, 2003, in *IEEE Power Engineering Society General Meeting*, Denver, CO, 2004
15. A. Berizzi, The Italian 2003 blackout, in *IEEE Power Engineering Society General Meeting*, 2004., Denver, CO, 2004, Vol. 2, pp. 1673–1679
16. O.P. Veloza, F. Santamaria, Analysis of major blackouts from 2003 to 2015: Classification of incidents and review of main causes. Electr. J. **29**, 42–49 (2016)
17. Report of the Enquiry Committee on Grid Disturbance in Northern Region on 30 July 2012 and in Northern, Eastern & North-Eastern Region on 31 July 2012. https://powermin.nic.in/sites/default/files/uploads/GRID_ENQ_REP_16_8_12.pdf
18. E.C. Portante, S.F. Folga, J.A. Kavicky, L.T. Malone, Simulation of the september 8, 2011, san diego blackout, in *Proceedings of the 2014 Winter Simulation Conference (WSC)*, Savannah, GA USA, 7–10 December 2014, pp. 1527–1538
19. M. Papic, Pacific southwest blackout on September 8, 2011 at 15: 27, in *Proceedings of the IEEE PES General Meeting*, Vancouver, BC, Canada, 21–25 July 2013
20. W. Lin, Y. Tang, H. Sun, Q. Guo, H. Zhao, B. Zeng, Blackout in Brazil power grid on February 4, 2011 and inspirations for stable operation of power grid. Autom. Electr. Power Syst. **35**, 1–5 (2011)
21. S.P. Trung, N. Voropai, The major outage in South Vietnam in 2013: the nature of blackout, security measures and strategy of national power system modernization, in *Proceedings of the International Conference on Problems of Critical Infrastructures (ICCIP 2015)*, Arlington, VA, USA, 16–18 March 2015
22. H. Haes Alhelou, M.E. Hamedani-Golshan, T. Cuthbert Njenda, P. Siano, A survey on power system blackout and cascading events: research motivations and challenges. Energies **12**, 682, 1–28 (2019)
23. P. Gomes, New strategies to improve bulk power system security: lessons learned from large blackouts, in *Proceedings of the 2004 IEEE Power Engineering Society General Meeting*, Denver, CO, USA, 6–10 June 2004, pp. 1703–1708
24. N. Phuangpornpitak, S. Tia, Opportunities and challenges of integrating renewable energy in smart grid system. Energy Procedia **34**, 282–290 (2013)
25. M.A. Kabir, M.M.H. Sajeeb, M.N. Islam, A.H. Chowdhury, Frequency transient analysis of countrywide blackout of Bangladesh power system on 1st November, 2014, in *Proceedings of the 2015 International Conference on Advances in Electrical Engineering (ICAEE)*, Dhaka, Bangladesh, 17–19 December 2015, pp. 267–270
26. M. Xiang, Yu. Dongzhen, W. Wang, J. Zuo, Y. Shen, X. Xie, The Reason analysis of holland blackout on March 27, 2015 and lessons for hunan power grid. J. Clean Energy Technol. **6**(3), 263–267 (2018)
27. B. Liu, B. Zhou, D. Jiang, Z. Yu, X. Yang, X. Ma, Distributed accommodation for distributed generation—from the view of power system blackouts, in *ICSEE2018/IMIOT2018, CCIS925* (2018), pp. 236–246

28. L. Baojie, L. Jinbo, L. Hongjie, Analysis of Turkey blackout on March 31 2015 and lessons on China power grid. Proc. CSEE **36**, 5788–5795 (2016)
29. Y. Shao, Y. Tang, J. Yi, A. Wang, Analysis and lessons of blackout in Turkey power grid on March 31, 2015, Autom. Electr. Power Syst. **40**, 9–14 (2016)
30. G. Liang, S.R. Weller, J. Zhao, F. Luo, Z.Y. Dong, The 2015 Ukraine blackout: Implications for false data injection attacks. IEEE Trans. Power Syst. **32**, 3317–3318 (2017)
31. E. Miguel, C. Wolfram, K. Lee, Experimental evidence on the demand for and costs of rural electrification (National Bureau of Economic Research, Cambridge, MA, USA, 2016)
32. J.O. Okumu, A. Hood, *Strategic Management Factors Affecting Performance of Thermal Power Generation Companies in Kenya* (Elixir Publisher, Tamilnadu State, India, 2017)
33. T. Bambaravanage, S. Kumarawadu, A. Rodrigo, Comparison of three under-frequency load shedding schemes referring to the power system of Sri Lanka. Eng. J. Inst. Eng. **49**, 41–52 (2016)
34. N. Marikkar, T. Jayath, K. Egodawatta, M. Vierling, M. Aboujaib, D. Sokolov, D. Meskers, R. Russell, M. Moliere, A review of the experience achieved at the Yugadanavi 300 MW CCGT in Sri Lanka: increasing the firing temperature of gas turbines using a novel vanadium inhibitor, in *Proceedings of the ASME Turbo Expo 2017: Turbomachinery Technical Conference and Exposition*, Charlotte, NC, USA, 26–30 June 2017; American Society of Mechanical Engineers: New York, NY, USA, 2017; p. V003T03A005
35. R. Yan, N. Al Masood, T.K. Saha, F. Bai, H. Gu, The anatomy of the 2016 South Australia blackout: a catastrophic event in a high renewable network. IEEE Trans. Power Syst. (2018)
36. A. Lucas, Confected conflict in the wake of the South Australian blackout: diversionary strategies and policy failure in Australia's energy sector. Energy Res. Soc. Sci. **29**, 149–159 (2017)
37. B.T. Eaton, United States Annual Report 2013; Technical Report, 2013. https://powerquality.eaton.com/blackouttracker/default.asp. Accessed 19 Jan 2018
38. I. Dobson, D.E. Newman, Cascading blackout overall structure and some implications for sampling and mitigation. Int. J. Electr. Power Energy Syst. **86**, 29–32 (2017)
39. T. Apagó, N. Restituyen EnergíA, OSE Dice Que Se Solucionaron Fallas; Technical Report, 26 August 2017. https://www.elpais.com.uy/informacion/apagon-sur-ute-lluvias.html. Accessed 10 Dec 2017
40. Major Power Outage Events; Technical Report, 2018. https://poweroutage.us/about/majorevent. Accessed 5 Dec 2018
41. Sudan, Complete power blackout hits Sudan; Technical Report, 26 August 2018. https://xinhuanet.com/english/2018/02/28/c137004277.html. Accessed 10 Jan 2019
42. M.A.A. Osman, Contingency analysis in presence of SVC controller of Sudan national grid for electricity. Ph.D. Thesis, Sudan University of Science and Technology, Khartoum, Sudan (2018)
43. M.A. Yagoub, Z. Tao, Modelling and mitigation of geomagnetically induced currents (GICs) for single-phase power transformer, in *Proceedings of the 2018 International Conference on Computer, Control, Electrical, and Electronics Engineering (ICCCEEE)*, Khartoum, Sudan, 12–14 August 2018, pp. 1–6
44. K.E. Mohammed, An intelligent load shedding system application in sudan national grid. Ph.D. Thesis, Sudan University of Science and Technology, Khartoum, Sudan (2018)
45. Power Reported Restored after Baku Hit with Partial Second Blackout; Technical Report, 4 July 2018. https://www.rferl.org/amp/29334268.html. Accessed 3 Jan 2019
46. J. Hyndle-Hussein, S. Kardas´, K. Kłysin´ski, Troublesome INVESTMent. The Belarusian nuclear power plant in astravyets. OSW Studies 74, July 2018; University of Pittsburgh: Pittsburgh, PA, USA, 2018
47. S. Carlson, Powerful ties: Eu-turkey energy relations. Turk. Policy Q. **17**, 105–115 (2018)
48. J.D. Hunt, D. Stilpen, M.A.V. de Freitas, A review of the causes, impacts and solutions for electricity supply crises in Brazil. Renew. Sustain. Energy Rev. **88**, 208–222 (2018)
49. D. Novosel, M.M. Begovic, V. Madani, Shedding light on blackouts. IEEE Power Energy Mag. **2** (2004)

50. P. Pourbeik, P.S. Kudnur, C.W. Taylor, The anatomy of a power grid blackout. IEEE Power Energy Mag. **4** (2006)
51. D. Zhong Meng, Recommendations to prevent power system cascading blackout, in *IEEE PES Power Systems Conference and Exposition*, New York, NY, 2004, vol. 1, pp. 533–542
52. P. Kundur, J. Paserba, V. Ajjarapu, G. Anderson, A. Bose, C. Canizares, N. Hatziargyriou, D. Hill, A. Stankovic, C. Taylor, T. Van Cutsem, V. Vittal, Definition and classification of power system stability IEEE/CIGRE joint task force on stability terms and definitions. IEE Trans. Power Syst. **19**(3), 1387–1401 (2004)
53. Z.Q. Bo, J.A.S.A.B. Jayasinge, R.K. Aggrawal, Anew Scheme for monitoring and protection of power transmission system based on global positioning system, in *Proceeding of the 31st University Power Engineering Conference*, Crete, Greece, 1996, pp. 21–24
54. Z.Q. Bo, G. Weller, T. Lomas, M.A. Redfem, Positional protection of transmission systems using global positioning system. IEEE Trans. Power Deliv. **15**(04), 1163–1168 (2000). (Oct)
55. M. Begovic, D. Novosel, D. Karlsson, C. Henville, G. Michel, Wide area protection and emergency control. Proc. IEEE **93**(5), 867–891 (2005)
56. D. Yan, Wide-area protection and control system with WAMS based, in *Proceeding of International Conference on Power System Technology-2006*, 2006, pp. 1–5
57. Q. Wang, Z. Bo, Y. Zhao, X. Ma, M. Zhang, H. Zheng, L. Wang, Integrated wide-area-protection and control for power grid security. CSEE J. Power Energy Syst. **5**(2) (June 2019)
58. IEEE Standard for Synchrophasor Measurements for Power Systems; IEEE: Piscataway, NJ, USA (2011)
59. S.T. Mini, J.D. McDonald, *Power System SCADA and Smart Grids*, 1st edn. (CRC Press, Boca Raton, FL, USA, 2015)
60. J. Bertsch, C. Carnal, D. Karlsson, J. Mcdaniel, K. Vu, Wide-area protection and power system utilization. Proc. IEEE **93**(5), 997–1003 (May 2005)
61. V. Madani, M. Adamiak, M. Thakur, Design and implementation of wide-area special protection schemes, in *57th Annual Conference for Protective Relay Engineers* (1 April 2004), pp. 35–44
62. X. Yin, Z. Zhang, Z. Li, X. Qi, W. Cao, Q. Guo, The research and the development of the wide-area relaying protection based on fault element identification, in *Protection and Control of Modern Power Systems* (2016), pp. 1–12
63. K. Seethalekshmi, S.N. Singh, S.C. Srivastava, Wide-area protection and control: present status and key challenges, in *Fifteenth National Power System Conference (NPSC)*, IIT Bombay, December 2008, pp. 169–175
64. A.G. Phadke, P. Wall, L. Ding, V. Terzija, Improving the performance of power system protection using wide area monitoring. J. Mod. Power Syst. Clean Energy **4**(3), 319–331 (2016)
65. J. Wen, W.-H. Edwin Liu, P.L. Arons, S.K. Pandey, Evolution pathway towards wide-area monitoring and protection- a real-world implementation of centralized RAS system. IEEE Trans. Smart Grid **5**, 1506-1513 (May 2014)
66. M.A.M. Ariff, B.C. Pal, Adaptive protection and controlling power system for wide-area blackout prevention. IEEE Trans. Power Deliv. **31**(4), 1815–18251-10 (Aug 2016)
67. C. Rehtanz, J. Bertsch, A new wide-area protection system, in *2001 IEEE Porto Power Tech Proceedings*, 10–13 Sept 2001, pp. 1–5
68. I. Ivankovic, I. Kuzle, N. Holjevac, Key performance indices for angle stability protection function in WAMPAC system, in *2018 IEEE Power & Energy Society General Meeting (PESGM)*, 5–10 Aug 2018, pp. 1–6
69. M.M. Eissa, A novel centralized wide-area protection "CWAP" in phase portrait based on pilot wire including phase comparison. IEEE Trans. Smart Grid **10**(3), 2671–2682 (May 2019)
70. J. He, L. Liu, W. Li, M. Zhang, Development and research on integrated protection system based on redundant Information analysis. J. Protect. Control Modern Power Syst. 1–13 (2016)
71. I. Ivankovic, I. Kuzle, N. Holjevac, Wide-area information-based transmission system centralized out-of-step protection scheme. Energies **10**, 633 (2017)

72. G. Jurisic, J. Havelka, T. Capuder, S. Sucic, Laboratory test bed for analysing fault detection reaction times of protection relays in different substation topologies. Energies **11**(9), 2482, 1–14 (2018)
73. Z. Zbunjak, I. Kuzle, System Integrity Protection Scheme (SIPS) development and optimal bus splitting scheme supported by Phasor Measurement Units (PMUs). Energies, 1–21 (Sept 2019)
74. M.-H. Song, S.-H. Kang, N.-H. Lee, S.-R. Nam, IEC 61850_ based centralized busbar differential protection with data desynchronization compensation. Energies, 1–15 (Feb 2020)
75. J. Won Lee, W.-K. Kim, J. Han, W.-H. Jang, Fault area estimation using travelling wave for wide-area protection. J. Mod. Power Syst. Clean Energy **4**(3), 478–486 (2016)
76. X. Liu, Y. Wang, Hu. Hub, Fault identification algorithm based on zone-division wide-area protection system. J. Eng. Sci. Technol. Rev. **7**(2), 80–86 (2014)
77. Z. He, Z. Zhang, W. Chen, O.P. Malik, X. Yin, Wide-area backup protection algorithm based on fault componenet voltage distribution. IEEE Trans. Power Deliv. **26**(4) (Oct 2011)
78. M.S. Srikumar, T. Ananthapadmanbha, F. Zafar Khan, V. Girish, Line outage detection using phasor measurement units. Procedia Technol. **21**, 88–95 (2015)
79. I. Ivankovic, I. Kuzle, N. Holjevac, Multi-functional WAMPAC system concept for out-of-step protection based on syschrophasor. Int. J. Electr. Power Energy **87**, 77–88 (2017)
80. Y. Huang, S. Li, X. Liu, Y. Zhang, L. Sun, K. Yang, A probabilistic multi-objective model for phasor measurement units placement in the presence of line outage. Sustainability, 1–12 (Dec 2019)
81. F. Namdari, S. Jamali, P.A. Crossley, Power differentiation. Iran. J. Electr. Electron. Eng. **1**(3), 53–65 (2005)
82. M.M. Eissa, A new wide-area protection scheme for single and double circuit lines using 3D-phase surface. IEEE Trans. Power Deliv. **33**(6), 2613–2623 (Dec 2018)
83. M. Tamizi Azmi, N. Sofizan Nik Yusuf, Ir., S. Kamar, S. Abdullah, M. Khairun Nizam Mohd Sarmin, N. Saadun, N. Nadia Nor Khairul Azha, Real-time haedware-in-the-loop testing platform for wide-area protection system in large-scale power systems, in *IEEE International Conference on Automatic Control and Intelligent Systems (I2CACIS 2019)*, 29 June 2019, Selangor, Malaysia, pp. 210–215
84. H.H. Alhelou, M.E. Golshan, N.D. Hatziargyriou, A decentralized functional observer based optimal LFC considering unknown inputs, uncertainties, and cyber-attacks. IEEE Trans. Power Syst. **34**(6), 4408–4417 (2019). (May 13)
85. P. Kundu, A.K. Pradhan, Enhanced protection security using system integrity protection scheme (SIPS), in *2016 IEEE Power and Energy Society General Meeting (PESGM)*, Boston, MA, 17–21 July 2016, pp. 1–1
86. P. Kundu, A.K. Pradhan, Supervisory protection of islanded network using synchrophasor data. IEEE Trans. Smart Grid **10**(2), 1772–1780 (March 2019)
87. I. Ivankovic, D. Brnobic, S. Skok, I. Sturlic, R. Rubesa, Time delay aspect for basic line protection function with synchrophasor in WAMPAC system, in *2018 IEEE International Energy Conference (ENERGYCON)*, Limassol, 2018, pp. 1–6
88. P. Kundu, A.K. Pradhan, Power network protection using wide-area measurements considering uncertainty in data availability. IEEE Syst. J. **12**(4), 3358–3368 (Dec 2018)
89. C.W. Taylor, D.C. Erickson, K.E. Martin, R.E. Wilson, V. Venkatasubramanian, WACS-wide-area stability and voltage control system: R & D and online demonstration. Proc. IEEE **93**(5), 892–906 (2005)
90. G. Cai, D. Yang, C. Liu, adaptive wide-area damping control scheme for smart grids with considering of signal time delay. Energies **6**, 4841–4858 (2013). (Sept)
91. S. Mirzazad-Barijough, M. Mashhuri, A. M. Ranjbar, A predictive approach to control frequency instabilities in a wide-area system, in *2009 IEEE/PES Power Systems Conference and Exposition*, Seattle, WA, 2009, pp. 1–6
92. M. Larsson, C. Rehtanz, Predictive frequency stability control based on wide-area phasor measurement, in *IEEE Power Engineering Society Summer Meeting*, July 2002, pp. 233–238

93. F.G. Velez, V.A. Centeno, A.G. Phadke, Multiple swing transient stability assessment with phasor measurements, in *2017 IEEE Manchester PowerTech* (Manchester, 2017), pp. 1–6
94. H.H. Alhelou, M.H. Golshan, J. Askari-Marnani, Robust sensor fault detection and isolation scheme for interconnected smart power systems in presence of RER and EVs using unknown input observer. Int. J. Electr. Power Energy Syst. **1**(99), 682–694 (2018). (Jul)
95. K. Seethalekshmi, S.N. Singh, S.C. Srivastava, A Synchrophasor assisted frequency and voltage stability based load shedding scheme for self-healing of power system. IEEE Trans. Smart Grid **2**, 221–230 (June 2011)
96. H. Li, A. Bose, V. Venkatasubramanian, Wide-area voltage monitoring and optimization. IEEE Trans. Smart Grid **7**(2), 785–793 (2016). (March)
97. M. Mohammadniael, F. Namdari, M.R. Shakarami, A fast voltage collapse detection and prevention based on wide-area monitoring and control. J. Oper. Autom. Power Eng. 1–11 (Dec 2019)
98. M. Zima, M. Larsson, P. Korba, C. Rehtanz, G. Andersson, Design aspects for wide-area monitoring and control systems. Proc. IEEE **93**(5), 980–996 (May 2005)
99. H.H. Alhelou, M.E. Golshan, N.D. Hatziargyriou. Deterministic dynamic state estimation-based optimal lfc for interconnected power systems using unknown input observer. IEEE Trans. Smart Grid (9 Sep 2019)
100. H. Song, J. Wu, K. Wu, A wide-area measurement system- based adaptive strategy for controlled islanding in bulk power systems. Energies **7**, 2631–2657 (2014)
101. M.A.M. Ariff, B.C. Pal, Coherency identification in interconnected power system-an independent component analysis approach. IEEE Trans. Power Syst. **28**(2), 1747–1755 (2013). (May)
102. E. Baroco, B.C. Pal, N.F. Thornhill, A. Roman Messina, A dynamic mode decomposition framework for global power system oscillations analysis. IEEE Trans. Power Syst. **30**(6), 2902–2912 (Nov 2015)
103. T.B. Liu, X. Chang, X. Guo, L. Wang, C. Cao, Q. Yan, J. Li, Impact of subsynchronous and Supersynchronous frequency components on synchrophasor measurements. J. Mod. Power Syst. Clean Energy **4**(3), 362–369 (2016)
104. K. Tuttelberg, J. Kilter, D. Wilson, Uhlen, Estimation of power system inertia from ambient wide-area measurements. IEEE Trans. Power Syst. **33**(6), 7249–7257 (Nov 2018)
105. P. Ray, Power system low frequency oscillations mode estimation using wide area measurement systems. Int. J. Eng. Sci. Technol. **20**(2), 598–615 (2017)
106. V. Salehi, A. Mazloomzadeh, J.F. Fernandez, O.A. Mohammed, Real-time power system analysis and security monitoring by WAMPAC systems, in *IEEE PES Innovative Smart Grid Technologies (ISGT)*, Washington, DC, 2012, pp. 1-8
107. Z. Yang, W. Kui, X. Yumin, Z. Buhan, Study of power system online dynamic equivalent based on wide area measurement system. Energy Proceedia **16**, 1768–1775 (2012)
108. D. Xinwei, H. Qian, C. Yong, The monitor of inter-area oscillation based on wide area measurement system. Energy Proceedia **16**, 2033–2043 (2012)
109. Y. Chompoobutrgoo, L. Vanfretti, Using PMU signals from dominant paths in power system wide-area damping control. Sustain. Energy Grids Netw. **4**, 16–28 (2015)
110. H. Haes Alhelou, M. Esmail Hamedani-Golshan, T. Cuthbert Njenda, P. Siano, Wide-area measurement system-based optimal multi-stage under-frequency load-shedding in interconnected smart power systems using evolutionary computing techniques. Appl. Sci. 1–23 (1 Feb 2019)
111. T. Cuthbert Njenda, M.E.H Golshan, H. Haes Alhelou, WAMS based intelligent under frequency load shedding considering online disturbance estimation, in *Smart Grid Conference (SGC-18)*, Sanandaj, Iran, 2018, pp. 1–5
112. A. Nageswara Rao, P. Vijaya Priya, M. Kowsalya, R. Gnanadass, Wide area monitoring for energy system: a review. Int. J. Ambient Energy **40**(5), 537–553 (2019)
113. R. Sun, V.A. Centeno, Wide area system islanding contingency detecting and warning scheme. IEEE Trans. Power Syst. **29**(6), 2581–2589 (2014). (Nov)
114. Working Group C6, Wide area protection and emergency control. Final report (2002)

115. F. Hashiesh, H.E. Mostafa, M.M. Mansour, A.-R. Khatib, I. Helal, Wide area transient stability prediction using on-line artificial neural networks, in *2008 IEEE Canada Electric Power Conference*, Vancouver, BC, 2008, pp. 1–7
116. H.H. Alhelou, M.E. Golshan, T.C. Njenda, N.D. Hatziargyriou, An Overview of UFLS in conventional, modern, and future smart power systems: challenges and opportunities. Electr. Power Syst. Res. **1**(179), 106054 (2020). (Feb)
117. C. Taylor, WACS-Wide Area stability and voltage control system: R&D and online demonstration. Proc. IEEE **93**(5), 892–906 (2005). (May)
118. Y. Wang, Y.-Xia Zhang, S.-Xiao Xu, Wide-area protection against chain over-load trip based on multi-agent technology, in *2008 International Conference on Machine Learning and Cybernetics*, Kunming, 2008, pp. 1548–1552
119. B. Milosevic, M. Begovic, Voltage-stability protection and control using a wide-area network of phasor measurements. IEEE Trans. Power Syst. **18**(1), 121–127 (2003). (Feb)
120. J. Sykes, M. Adamiak, G. Brunello, Implementation and operational experience of a wide area special protection scheme on the SRP system, in *2006 Power Systems Conference: Advanced Metering, Protection, Control, Communication, and Distributed Resources*, Clemson, SC, 2006, pp. 145-158
121. C. Rehtanz, Wide area measurement and protection system for emergency voltage stability control, in *2002 IEEE Power Engineering Society Winter Meeting. Conference Proceedings* (Cat. No.02CH37309), NY, USA, 2002, vol. 2, pp. 842–847
122. M. Larsson, C. Rehtanz, Predictive frequency stability control based on wide-area phasor measurements, in *2002 IEEE Power Engineering Society Summer Meeting*, Chicago, IL, USA, 2002, vol. 1, pp. 233-238
123. V. Lohmann, Wide area protection: a strategy to counteract large area disturbances (2003)
124. W. Cong, A wide area protection system based on "three defensive lines" and its application in power system, in *2004 International Conference on Power System Technology, Power Con 2004*, Singapore, 2004, vol. 1, pp. 718–722

Introduction to WAMS and Its Applications for Future Power System

Reza Zamani, Habib Panahi, Arash Abyaz, and Hassan Haes Alhelou

Abstract Nowadays, the increase in electrical energy consumption and power system restructuring have posed new challenges to the operation, control, and monitoring of power systems. In this situation, the supervisory control and data acquisition (SCADA) system is not enough to ensure power system security and stability. The SCADA is often unable to measure data of all buses simultaneously. In addition, the sampling rate in this system is not enough for some power system applications. Therefore, the information obtained from SCADA does not show power system dynamics properly. In order to improve the power system monitoring, wide area measurement system (WAMS) has been developed to overcome the problems of SCADA system. Phasor measurements units (PMUs) are the main part of WAMS system and it basically consists of three essential processes including collecting, transmitting, and analyzing data. WAMS receives obtained data via a high speed communication links. After data processing and extracting appropriate information, decisions are made to improve the power system performance. Efficient use of power system data to achieve a secure operation strategy is targeted using the WAMS system. Due to the effective role of WAMS and PMUs in the reliable operation of power system, it is necessary to study their concept and applications in this chapter. The history of PMU and its structure is presented in this chapter. In addition, the necessity of WAMS for future power system and its difference from SCADA system have been investigated

R. Zamani (✉)
Faculty of Electrical and Computer Engineering, Tarbiat Modares University, Tehran, Iran
e-mail: r_zamani@modares.ac.ir

H. Panahi · A. Abyaz
Department of Electrical and Computer Engineering, University of Tehran, Tehran, Iran
e-mail: habibpanahi@ut.ac.ir

A. Abyaz
e-mail: arash.abyaz@ut.ac.ir

H. Haes Alhelou
Department of Electrical Power Engineering, Tishreen University, Lattakia, Syria
e-mail: h.haesalhelou@gmail.com

H. Haes Alhelou et al. (eds.), *Wide Area Power Systems Stability, Protection, and Security*,
Power Systems, https://doi.org/10.1007/978-3-030-54275-7_2

in this chapter. Different algorithms and application of the WAMS are also introduced in this chapter which can be implemented to improve the performance of the future power system.

Keywords WAMS · SCADA · PMU · Power system monitoring

1 Introduction

The need to visibility, controllability, and send remote command power system has been felt, from the beginning of the creation it. In order to operate the network securely and with high reliability, it is necessary to have the information of different places of the network in a center and according to that information, the necessary decisions should be made. Telecommunication system with following specifications was necessary for this aim [1]:

- Fast: A fast telecommunication system is required to be able to send signals to the control center with minimum delay. Suppose that load increase in a substation in the south of the country and control center will find out five seconds later. In this case, there may not be enough time for preventive control.
- Bandwidth: It has a high bandwidth to be able to send a lot of information. This information includes buses voltage, current and power of lines, and keys status.

With the passage of time, bandwidth and speed of telecommunication systems increased. With this progress, system operators thought of creation of a control center for network observability. At first, supervisory control and data acquisition (SCADA) was created. This system received the information of different point of the network in control center. This development allows the operator to control and better operate the network in the control center.

Disadvantages of SCADA were found over the time. The lack of synchronization of data obtained from different substations is the most important disadvantage of SCADA. To extract the signal phase, a time reference is required, and for data reference time, data synchronization is required. For this reason signal phase was not calculated in SCADA. In first, state estimation (SE) was used to extract phase signal in SCADA. SE has a lot of errors due to bad data and error of measurement device. Sometimes, error of SE has so much such that many researchers believed that the blackout of US occurs because of it. For this reason, the need for a system that can measure synchronized data became very tangible.

With the help of a global positioning system (GPS), the synchronized data of different places can be measured. With deploying this technology, researchers focused on build a device that it could measure synchronized data. In the early 1980s, a device that could sample synchronized data from different point of the network was built in Virginia Tech. This device called phase measurement unit (PMU). An example of PMU is shown in Fig. 1.

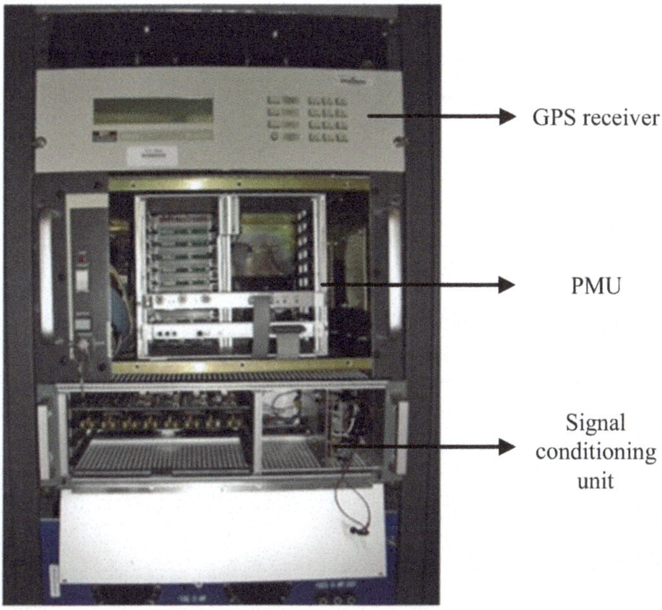

Fig. 1 An example of first PMU that was built at 1980s

The first built PMU was a huge improvement in power system because it could measure synchronized data from different Point. Therefore, IEEE published a standard for PMU [2–4]. PMU could extract signal phase because of synchronized measurements. Nevertheless, the first PMU had some disadvantages such as large scale and less input. Primary PMU was difficulty installed in panel because of large dimensions [5–9].

After this introduction, The SCADA has been described to make it clearer why PMU was built. Then structure of PMU and how to extract phase and frequency of signal are described. Finally, various WAMs and PMU applications have been discussed.

2 SCADA

As aforementioned in the introduction, the SCADA system has been used in the last decades for the visibility of the network purpose. The SCADA system consists of the following three parts [10–14]:

(1) Remote terminal unit (RTU)
(2) Concept and object modeling notation (COMN)
(3) Master terminal units (MTU)

2.1 RTU

RTUs are installed in substations and measure network parameters. The variables that RTUs can measure usually are listed as follows:

- Voltage magnitude
- Active and reactive power
- Tap changer position
- Circuit breaker status

The first three cases have continuous values, but the circuit breaker statuses are sent to the control center by binary variables. To increase the accuracy and security, status is sent with two bits instead of one bit, for example, 01 means that the circuit breaker is closed and 10 means that it is open. Also, these first three cases are transformed into DC currents using transducers in the range of 4–20 mA.

2.2 COMN

The telecommunication system is stablished between the substations and the control center. Power transmission lines have been used to send signals in some cases. In this method, the signal is added to the transmission line in the location of capacitor voltage transformer and the signal moves along the line to reach the upstream bus. Signals were sent on two different phases to ensure that transmission faults did not cause problems. The operation of this system would be disrupted if there was a fault on the transmission line (for example, a three-phase short circuit fault). For this reason, other methods of sending, for example, radio, etc., gradually replaced this method.

2.3 MTU

It is the control center to which signals are sent by a special protocol from RTUs. The most common protocol is IEC 60870-5-101 (or 104). There are several applications in MTU, the most common of which is SE, which is designed for two purposes:

- Signal phase extraction
- Reducing measuring signal error and sending data

In addition to performing these applications, the data is displayed on the diagram mimic and the operator sees information about different locations of the network on a large screen.

It should be noted that if only the time error for receiving signals assumed to be 1 ms, in the power network with a frequency of 50 Hz, this 1 ms error will caused to 18 degree error in measuring the signals phasor.

2.4 SCADA Problems

After explaining the different concepts of SCADA, its problems are described so that by knowing these problems, the requirements for constructing an efficient PMU system can be better understood. In general, the most shortages of SCADA system can be summarized as bellow:

(1) Measured data are not real-time and polling rate is about 2–10 s.
(2) Measurements do not exactly belong to a given time, i.e. time deviation exists.
(3) There is not any information on phase angle while this is a necessary tool for power system stability assessment.
(4) SE runs every few minutes, if any
(5) Dynamic of the system is not observable.

These problems were lead to several blackouts in the world such as:

- August 14, 2003 Northeast blackout due to lack of wide-area visibility of the system.
- A report prepared by U.S.-Canada investigation stated that if a phasor system was available at that time, this blackout can be prevented. Particularly, the voltage problem in Ohio could be identified and fixed earlier.
- In a few minutes before the cascade event, a divergence was occurred in phase angle between Cleveland and Michigan.

That's why researchers have developed a device that can simultaneously measure signals.

3 What Are PMU and WAMS?

Active power is closely related to the angle of the bus voltage. Many of the controllers of power plants depend on the active power of the system. For this reason, the need to measure the voltage and current angle of the buses has been considered to attract attentions. The main issue for this aim is existence of a reference for the angle and synchronization in the measurements. In 1980s, first PMU has been manufactured which was able to synchronize the measured angles. By developing the GPS, a huge change was made in synchronizing measurements. With the help of GPS, data can be synchronized with an accuracy of 1 μs that this accuracy is sufficient and appropriate for most of the power system applications. In the past, satellites were low enough to send signals to GPS, and the synchronization was keeping using a crystal for up

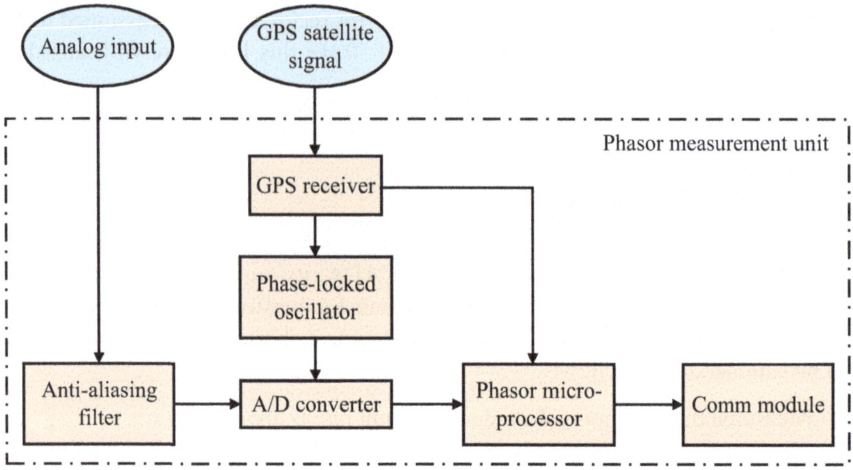

Fig. 2 Structure of the PMU

to 4 h. But today, the number of satellites has increased so that between 5 and 8 of these satellites are visible for GPS system.

The prototype PMU has been employed at some of substations of Bonneville power administration, American Electric Power Service Corporation, and New York power authority. The commercial version of the PMU was manufactured by Macrodyne in 1991. Nowadays, more than 50 vendors are offering PMU devices in the world. In addition, the related standards have been published by IEEE in 1991 and revised in 2005 and 2011.

The structure of the PMU is as shown in Fig. 2. As can be seen from the Fig. 2, A PMU consists of following essential elements:

(1) GPS receiver
(2) Phase-locked oscillator
(3) Anti-aliasing filter
(4) A/D converter
(5) Phasor microprocessor
(6) Communication module

The analog signals are imported into the PMU. First, the high frequency component of the signal will be removed by an anti-aliasing filter. The reason for this is that according to the sampling theorem, if the frequency of a signal is F and it is sampled with F_S frequency, this signal is similar to another signal with the frequency of

$$F \pm N \times F_S \tag{1}$$

where N can be an arbitrary natural number. Figure 3 shows interface between two signals.

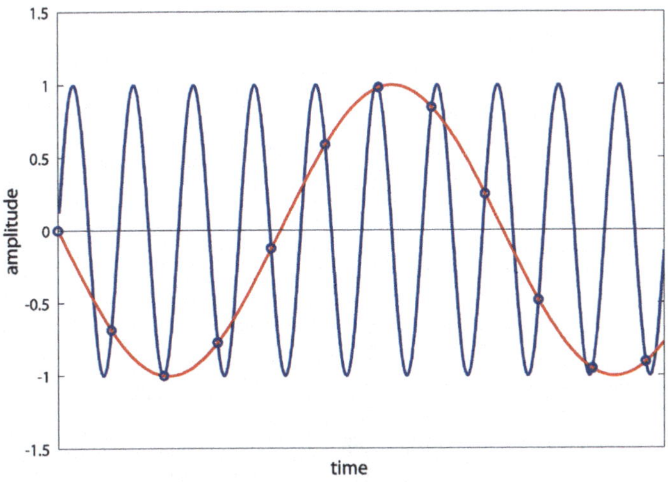

Fig. 3 Interfacing between two signals with low sampling rate

To prevent this phenomenon, a low-pass filter is considered at the beginning, which removes high frequencies components. The signals are then imported to the analog-to-digital (A/D) converters and converted to digital signals. In some PMUs, the anti-aliasing filter is also digitally located inside the A/D unit.

The task of the GPS is to be able to create synchronization with high accuracy, and the phase-locked unit is also designed to keep the clock in good accuracy until the next GPS signal is received.

The main idea of wide area monitoring systems (WAMS) is to create centralized data analysis capabilities that data are collected from different area of the power system simultaneously and in synchronized with each other. The purpose of analyzing this data is to evaluate the actual operating conditions of the system at any time and to compare the network parameters (voltage and current values and angles, temperature, active and reactive power) with the standard or predetermined limits. Also, with the help of certain algorithms, the security margin of stability or the distance of the system from the stability boundary is determined. Figure 4 shows the simple structure of the WAMS. In general, each WAMS is consist of the following processes:

- Data measuring and collecting
- Data delivering and communication
- Data analysis

The WAMS structure comprises following infrastructures:

(1) Phase measurement units (PMUs), which must be installed at key points in the network.
(2) Synchronizing system, which is the primary element to have a simultaneous image of the system variables, the effect of events, and the general nature of

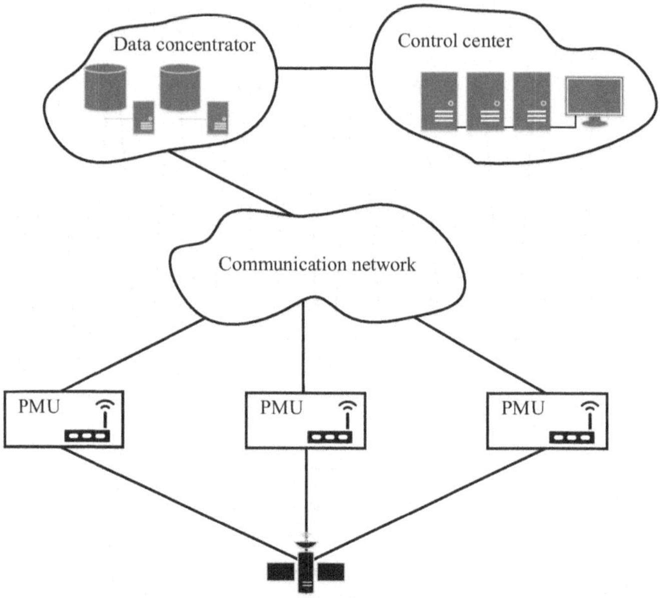

Fig. 4 Simple WAMS structure of power system

the system. This system synchronized the sampling frequency/time of PMUs to collect and send data.

(3) Data communication system, which must have the appropriate speed, reliability, and security for data transferring.
(4) Data collection and analysis center, which should be equipped with appropriate software for data analysis.

PMUs serve as the mainstay of the WAMS structure. The microprocessor is the brain of the PMU, which performs all the necessary actions on the signal, such as amplitude and phase extraction, as well as calculating the frequency and so on. After performing the calculations, the data is sent to the WAMS center by the communication module. At the WAMS center, data from all PMUs of the network is available, and because simultaneous data is available on all buses, dynamic and observable studies of the system can be performed at the same time, unlike SCADA. The WAMS center can be connected to different PMUs due to conduct following analysis by receiving data:

(1) Events analysis of power system using recorded event signals
(2) The behavior of the different bus phases at different times to predict future conditions
(3) Frequency behavior of different areas to predict generation shortages
(4) Different bus voltage behavior to improve operating conditions.

Also, due to the simultaneous data of different network buses, dynamic studies can be performed in the WAMS center and the network stability conditions can be evaluated and preventive actions can be executed if the network is going to be unstable. In addition, the protection system configuration can be considered and performed by using wide area data and remote buses that current transformers are not saturated and do not suffer from voltage fluctuations instead of local measurements.

4 Amplitude and Phase Angle Calculation Using PMU Data

The amplitude and phase angle of the signal can be calculated using different algorithms based on the PMUs measured data. In the following, three algorithms have been described for the purpose of amplitude and phase angle computation [7–9].

4.1 *Man and Morison Algorithm*

This algorithm is one of the short window algorithms that uses three samples to find the amplitude and phase of the signal. Suppose the input signal is represented as follows:

$$V = V_P \sin(\omega t + \varphi) \longrightarrow t = 0 V = V_P \sin\varphi \tag{2}$$

The derivative of the signal with respect to φ can be obtained as follows:

$$V^{'} = V_P \cos\varphi \tag{3}$$

Therefore, the amplitude and phase angle of the signal can be calculated as follows:

$$V_P = \sqrt{V^2 + V'^2} \tag{4}$$

$$\varphi = \tan^{-1}\left(\frac{V}{V'}\right) \tag{5}$$

The derivative of a signal in the adjacency of t = 0 can be achieved regarding to Fig. 5 and using (6).

$$V'(t = 0) = \frac{V^+ - V^-}{2\omega_0 \Delta t} \tag{6}$$

where

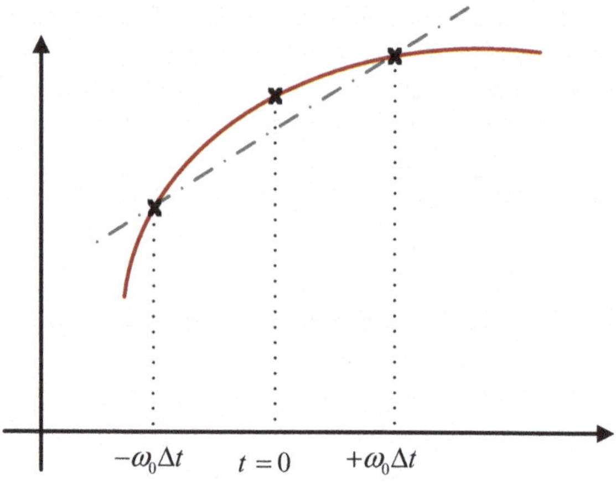

Fig. 5 Three samples for calculating derivative at t = 0

$$\omega_0 = 2\pi f_s \tag{7}$$

$$\Delta t = \frac{1}{f_s} \tag{8}$$

The f_s is the sampling frequency.

Therefore, using three samples of the signal, the V and V' can be obtained. If the V^-, V^0, and V^+ assumed to be three consecutive samples of the signal, the Man and Morison estimation algorithm can be implemented as shown in Fig. 6. Hence, the

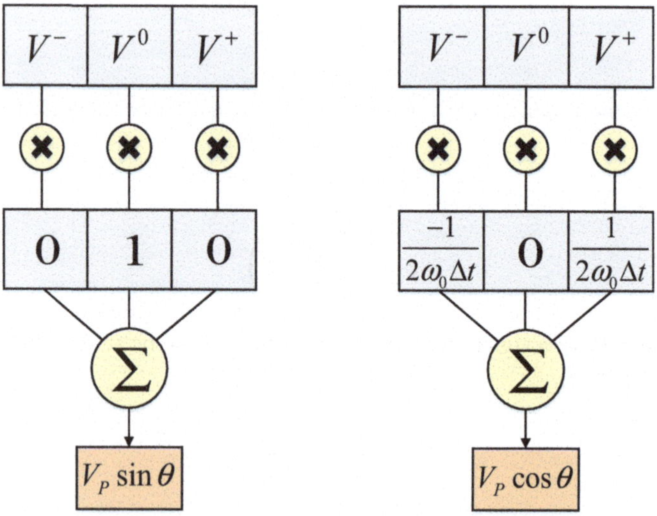

Fig. 6 Man and Morison estimation algorithm implementation

amplitude and phase angle of the input signal can be calculated as follows:

$$V_P = \sqrt{(V_P \sin\theta)^2 + (V_P \cos\theta)^2} \tag{9}$$

$$\theta = \tan^{-1}\left(\frac{V_P \sin\theta}{V_P \cos\theta}\right) \tag{10}$$

Despite the simplicity and ease implementation of this algorithm, following are the main drawbacks of this method:

(1) Has weak performance against noise.
(2) Due to the frequency response of this algorithm, in the presence of harmonics, the obtained result of this algorithm confronts with high error.
(3) The presence of a DC component in the input signal can cause error in this algorithm.
(4) The obtained amplitude for the main harmonic signal in this algorithm is 95% of the actual value of the signal and indicates error in the nature of this algorithm.

The Prodar 70 algorithm was developed to cope with the problem of having a smaller calculated amplitude by the Man and Morrison algorithm with a real signal amplitude.

4.2 *Prodar 70 Algorithm*

In this algorithm, the first and second derivatives of the input signal are used to extract the amplitude and phase angle of the signal. If the input signal and its derivative are assumed to be as shown in (2) and (3). Also, the second order derivative of the input signal can be obtained as follows:

$$V'' = -V_P \sin\varphi \tag{11}$$

Therefore, the amplitude and phase angle of the signal can be calculated as follows:

$$V_P = \sqrt{V'^2 + V''^2} \tag{12}$$

$$\varphi = \tan^{-1}\left(\frac{V''}{V'}\right) \tag{13}$$

In addition, the first and second order derivative of the signal can obtained regarding to Fig. 7. Accordingly, the first order derivative of the signal can be calculated as follows:

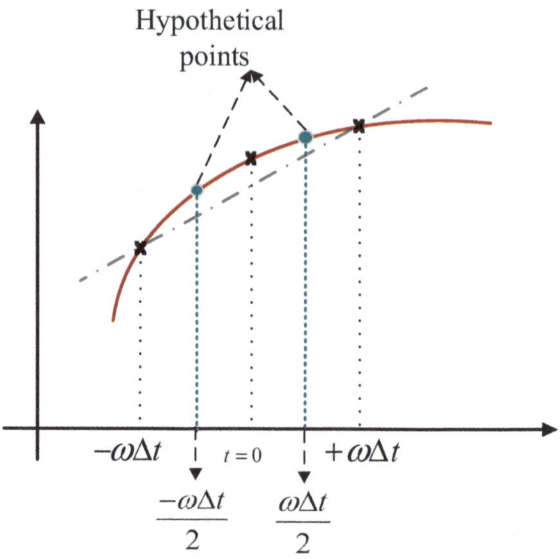

Fig. 7 First and second order derivative of the signal

$$V'(t=0) = \frac{V^{+1} - V^{-1}}{2\omega\Delta t} \tag{14}$$

The derivative of the signal in two hypothetical points at times $\frac{\omega\Delta t}{2}$ and $\frac{\omega\Delta t}{2}$ can be obtained as follows:

$$V'\left(t = \frac{\omega\Delta t}{2}\right) = \frac{V^{+1} - V^{0}}{\omega\Delta t} \tag{15}$$

$$V'\left(t = \frac{-\omega\Delta t}{2}\right) = \frac{V^{0} - V^{-1}}{\omega\Delta t} \tag{16}$$

Therefore, the second order derivative of the input signal can be obtained at $t = 0$ as follows:

$$V''(t=0) = \frac{V''\left(t = \frac{\omega\Delta t}{2}\right) - V'\left(t = \frac{-\omega\Delta t}{2}\right)}{\omega\Delta t} = \frac{V^{+1} + V^{-1} - 2V^{0}}{(\omega\Delta t)^2} \tag{17}$$

Likewise, the implementation of the Prodar 70 algorithm to calculate the amplitude and phase angle of the signal can be seen from the Fig. 8 similar to the Man and Morison algorithm. Although the problem of estimating the main harmonic amplitude error has been modified in this method, this algorithm is still susceptible to noise, harmonics, and DC component.

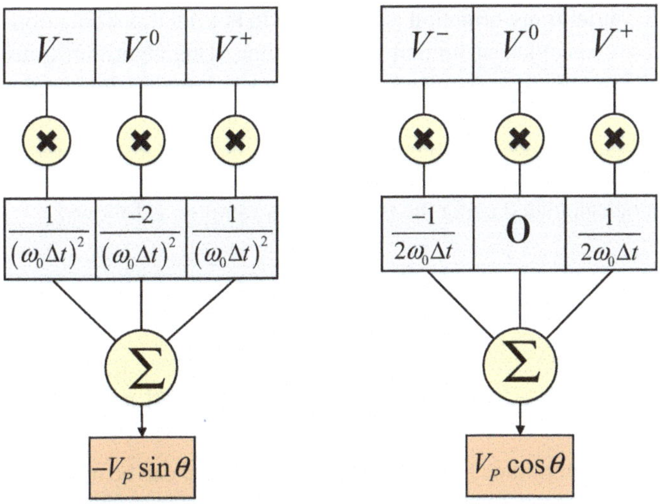

Fig. 8 Prodar 70 algorithm implementation

4.3 *Fourier Algorithm*

To improve the estimation of the phase angle and reduce the error caused by the presence of harmonics, DC component, and noise, the Fourier long window algorithm has been considered. If a periodic signal $V(t)$ with period T is assumed, then for all the values of $K = 0, 1, 2, \ldots$, $V(t) = V(t + kT)$. According to the Fourier algorithm, any periodic function can be written in the form of a Fourier series according to the following relationships [15–18]:

$$V(t) = \frac{a_0}{2} + \sum_{k=1}^{\infty} a_k \cos\left(\frac{2\pi kt}{T}\right) + \sum_{k=1}^{\infty} b_k \sin\left(\frac{2\pi kt}{T}\right) \tag{18}$$

where

$$a_0 = \frac{1}{T}\int_{t_0}^{t_0+T} V(t)dt \tag{19}$$

$$a_k = \frac{2}{T}\int_{t_0}^{t_0+T} V(t)\cos\left(\frac{2\pi kt}{T}\right)dt,\ k = 1,\ 2,\ 3, \ldots \tag{20}$$

$$b_k = \frac{2}{T}\int_{t_0}^{t_0+T} V(t)\sin\left(\frac{2\pi kt}{T}\right)dt,\ k = 1,\ 2,\ 3,\ \ldots \tag{21}$$

Discrete Fourier transformation is equivalent to Fourier transformation for digital signals that are specified at limited points in time. This algorithm transforms the input signal from the time domain to the frequency domain while retaining all the information in it. Therefore, by considering a signal in one of the domains, it is possible to obtain the signal in another domain. If the number of samples in a period of time is equal to M, the real and imaginary value of the signal in the frequency domain can be calculated using the time domain samples as follows:

$$\Re\{V\} = V_p \sin\varphi = \frac{2}{M}\sum_{n=0}^{M-1} V\left(\frac{n}{M}\right)\cos\left(\frac{2\pi n}{M}\right) \tag{22}$$

$$\Im\{V\} = V_P \cos\varphi = \frac{2}{M}\sum_{n=0}^{M-1} V\left(\frac{n}{M}\right)\sin\left(\frac{2\pi n}{M}\right) \tag{23}$$

Therefore, the amplitude and phase angle of the signal can be obtained using (24) and (25).

$$V_p = \sqrt{\left(\frac{2}{M}\sum_{n=0}^{M-1} V\left(\frac{n}{M}\right)\cos\left(\frac{2\pi n}{M}\right)\right)^2 + \left(\frac{2}{M}\sum_{n=0}^{M-1} V\left(\frac{n}{M}\right)\sin\left(\frac{2\pi n}{M}\right)\right)^2} \tag{24}$$

$$\varphi = \tan^{-1}\left(\frac{\frac{2}{M}\sum_{n=0}^{M-1} V\left(\frac{n}{M}\right)\cos\left(\frac{2\pi n}{M}\right)}{\frac{2}{M}\sum_{n=0}^{M-1} V\left(\frac{n}{M}\right)\sin\left(\frac{2\pi n}{M}\right)}\right) \tag{25}$$

For instance, if the system frequency is equal to 60 Hz and the sampling frequency is 720 Hz, the frequency response of sinusoidal and cosinusoidal filters (real and imaginary part of the Fourier algorithm) can be seen in Fig. 9.

It can be seen that, this algorithm unlike the previous two presented algorithms, can reduce the impact of the DC and other harmonics.

5 Frequency Calculation Using PMU

One of the important parameters of the system is frequency. PMUs measure the network frequency in a variety of ways. Some of these methods are listed as follows:

- Zero-crossing
- Least square
- Phase deviation

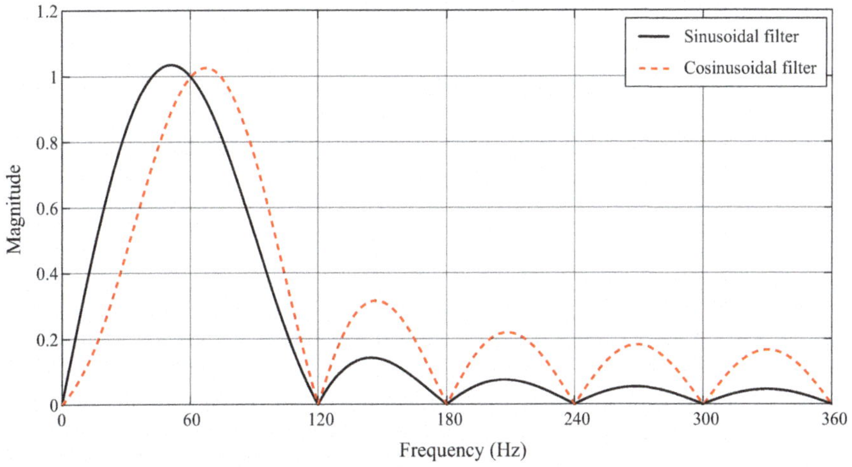

Fig. 9 Frequency response of the sine and cosine filters

5.1 *Zero-crossing*

This method takes the times that the signal passes to zero. For example, if the first pass occurs at time t_1 and the second time become t_2, half the period of the signal can be obtained as follows

$$\frac{T}{2} = t_2 - t_1 \tag{26}$$

Therefore the frequency of the system can be easily achieved as follows:

$$f = \frac{1}{2 \times (t_2 - t_1)} \tag{27}$$

The aforementioned frequency estimation method is considerably weak against harmonics and noise hence its accuracy is low.

5.2 *Least Square*

This approach is a statistical method that is used to solve the set of equations which the number of equations is greater than its unknown variables. In fact, this method is based on curve fitting of the data set. In this method, the best fit model where the sum of the squares of the difference between the data and the values obtained from the fitted curve is minimal.

Defining of the curve fitting problem:

$$e = ax - y \tag{28}$$

This definition of the error leads to the fact that if the error is positive and negative, the true value of the error is not shown correctly the accuracy of the fitted curve. Since, the positive and negative values may be removed together. Therefore, one way is to use absolute magnitude of the error as follows:

$$|e| = |ax - y| \tag{29}$$

Since it is difficult to apply absolute magnitude in mathematics, the other proposed solution is using the second power of the error, which is essentially the same as the second norm concept of the error or, more fully, the energy of error.

Now by minimizing the squares of error, we are actually looking for the ax vector to be as similar as possible to the y vector. If the input signal is assumed to be as follows:

$$V(t) = V_m \sin(\omega t + \theta) = V_m \cos\theta \sin\omega t + V_m \sin\theta \cos\omega t \tag{30}$$

If the three-terms of Taylor series are used instead of the $\sin\omega t$ and $\cos\omega t$, the above relation will be rewritten as follows:

$$\begin{aligned} V(t) &= \text{Sin}(\omega t)V_p\text{Cos}\theta + (\Delta\omega)t\text{Cos}(\omega t)V_p\text{Cos}\theta + \text{Cos}(\omega t)V_p\text{Sin}\theta - \\ &(\Delta\omega)t\text{Sin}(\omega t)V_p\text{Sin}\theta - \frac{t^2}{2}\text{Sin}(\omega t)(\Delta\omega)^2\text{Cos}\theta - \frac{t^2}{2}\text{Cos}(\omega t)(\Delta\omega)^2\text{Sin}\theta \end{aligned} \tag{31}$$

where $\Delta\omega$ is the change of angular speed. Six unknown variables can be found in (Eq. 31). Hence, seven equations are needed to obtain these variables using the least square error method. The measured voltage signal in (Eq. 31) can be rewritten as follows:

$$V(t) = S_{11}X_1 + S_{12}X_2 + S_{13}X_3 + S_{14}X_4 + S_{15}X_5 + S_{16}X_6t \tag{32}$$

where

$$\begin{aligned} &S_{11} = V_p\text{Cos}\theta \; S_{12} = tV_p\text{Cos}\theta \; S_{13} = V_p\text{Sin}\theta \\ &S_{14} = -tV_p\text{Sin}\theta \; S_{15} = -\frac{t^2}{2}\text{Cos}\theta \; S_{16} = -\frac{t^2}{2}\text{Sin}\theta \end{aligned} \tag{33}$$

$$\begin{aligned} &X_{11} = \text{Sin}(\omega t) \; X_{12} = (\Delta\omega)\text{Cos}(\omega t) \; X_{13} = \text{Cos}(\omega t) \\ &X_{14} = (\Delta\omega)\text{Sin}(\omega t) \; X_{15} = \text{Sin}(\omega t)(\Delta\omega)^2 \; X_{16} = \text{Cos}(\omega t)(\Delta\omega)^2 \end{aligned} \tag{34}$$

Therefore, based on the least square theorem following relation can be developed:

$$
\begin{aligned}
&[S]_{N\times 6}[X]_{6\times 1} = [V]_{N\times 1}\\
&[X] = [S]^{*}[V]\\
&[S]^{*} = \left[[A]^{T}[A]\right]^{-1}[A]^{T}
\end{aligned} \tag{35}
$$

After calculating the unknown variables, using the following relationships, it is possible to calculate changes of frequency and frequency estimation:

$$
\Delta\omega = \frac{X_2 + X_4}{X_1 + X_3} \tag{36}
$$

$$
f = f_{zero} + \frac{\Delta\omega}{2\pi f_0} \tag{37}
$$

5.3 Phase Deviation

If the angular speed deviation from the nominal value, and the rate of change of angular speed at $t = 0$ are $\Delta\omega$ and ω', respectively, the angular speed at any time (t) is given by:

$$
\omega(t) = \left(\omega_0 + \Delta\omega + t\omega'\right) \tag{38}
$$

The phase of signal can be obtained from the integral of angular speed as follows:

$$
\phi(t) = \int \omega dt = \int \left(\omega_0 + t\Delta\omega + t\omega'\right)dt = \phi_0 + t\omega_0 + t\Delta\omega + \frac{1}{2}t^2\omega' \tag{39}
$$

It can be rewritten in the matrix form as follows:

$$
\phi(t) = a_0 + a_1 t + a_2 t^2 \tag{40}
$$

The phase of the signal can be calculated from the algorithms in the previous section preferably by Fourier algorithm. Afterward, three unknown variables a_0, a_1, and a_2 can be achieved using the least square approach. If the phase of the signal at $t = 0$, $t = \Delta T$, $t = 2\Delta T$, $t = 3\Delta T$ are available, following relation can be derived based on the least square error algorithm:

$$
\underbrace{\begin{bmatrix}\phi_0\\ \phi_1\\ \phi_2\\ \phi_3\end{bmatrix}}_{\varphi} = \underbrace{\begin{bmatrix}100\\ 1\ \Delta\mathrm{t}\ \Delta t^2\\ 12\Delta\mathrm{t}\ 4\Delta t^2\\ 13\Delta\mathrm{t}\ 9\Delta t^2\end{bmatrix}}_{B}\underbrace{\begin{bmatrix}a_0\\ a_1\\ a_2\end{bmatrix}}_{A} \tag{41}
$$

Therefore, the unknown vector A can be calculated using four samples. This matrix contains the frequency and rate of change of frequency according to (39). The unknown vector A can be obtained as follows:

$$[A] = \left[B^T B\right]^{-1} B^T [\phi] \tag{42}$$

6 Communication Infrastructure of WAMS

As mentioned, in order to implement WAMS, it is necessary to transfer the measured data provided by different PMUs to the control center using a high-speed communication infrastructure. This infrastructure must have the required reliability and capability. The various high-speed communication systems have been developed with an increasing need for data transferring. Therefore, there are different communication systems such as power line carrier (PLC), fiber optic, radio, microwave, and satellite to data transmission between PMUs and the control center [3].

The communication methods of WAMS that have different advantages and disadvantages are classified in two categories of wired and wireless communication. So, according to different capability of each method, the most suitable method for transferring data securely, reliably and economically is found according to network conditions. Transmission lines are used to data transferring in PLC method. The implement cost of this method is low because it does not require a separate structure for signal transmission. However the attenuation and distortion of signal while it propagate through the transmission line, noise caused by high voltage equipment around the transmission lines and limited band width are some of the problems of this method. In addition, data transferring is disrupted when a faults happens on the transmission line [19–23].

Fiber optic is another method used to data transmission in the WAMS. The high capability of data transmission along with the wide bandwidth attracts attentions to this method. A small portion of its bandwidth is used to data transmission and the rest of it can be used for other proposed. The transmission delay and attenuation are low in this method and it is not affected by noise and environmental factors. Thus, the security of this method is high. However, the high implementation time and initial cost are disadvantages of this method. Compare to fiber optic, the microwave and radio methods are low cost. The radio method has limited band width for data transmission. So, the data of PMUs may interfere with other data which may be sent in this frequency band. Due to the lack of need for wiring, the implementation time and cost of this method are low [24–28].

Due to the easy installation, relatively high security and lack of need for wiring, using microwave to data transferring between two points with a relatively large distance seems to be a good option. The microwave is a relatively low cost method

Table 1 Comparison of different communication methods in WAMS

Features	PLC	Fiber optic	Radio	Microwave	Satellite
Connection type	Wired	Wired	Wireless	Wireless	Wireless
Security	Low (bit error rate $<10^{-2}$)	high (bit error rate 10^{-15})	Low (questionable)	Medium (bit error rate 10^{-7})	Medium (bit error rate 10^{-7})
Bandwidth	Limited	Not limited	Limited	Limited	Limited
Transmission delay	Low	Low	Low	Low	High
Attenuation	High	Almost zero	Medium	Medium	Medium (sometimes High)
Implementation time	Low	High	Low	Medium	High
Implementation cost	Low	High	Low	Low	Very High
Environmental condition	Low affected	Not affected	Affected	Affected	Affected

for data transmission. However, signal fading and multipath propagation are the challenges of using this technology. Also, weather condition may affect the performance of this method.

Satellite communication can be used to data transmission between PMUs and control center. The data transmission is not depend on the distance between two points. The high initial cost of a satellite communication structure is one of the limitation of this method. The transmission delay in this method is higher than others and weather conditions and electromagnetic interference may effect on this. Important data transferring method used in WAMS were introduced. The summery of these is presented in Table 1.

7 WAMS Applications Based on the PMU Measured Data

The PMUs were manufactured to measure the phase angle, but this ability, which could simultaneously measure the data of different locations and send it to a center, can create many capabilities and applications [29–32]. Nowadays, these capabilities are more and more developed so that there are several applications based on the WAMS which are briefly described in this section.

7.1 State Estimation

State estimation is assigning a value to an unknown variable of the system based on a special criterion that is obtained by measured data of the system. Measurements can be incomplete or over samples, and the state estimation of the system performed based on statistical methods, which the actual values of the state variables are estimated with maximum or minimum of specific criteria.

Monitoring of power systems is achieved by having information about different parts of the network. Therefore, estimating the state and knowing the information in all parts of the network is very important for the safe operation of the network. State estimation methods in the decade have been developed in 1970s. In these methods, active power and reactive power flows of transmission lines and bus voltage were obtained using measurements and the telecommunication system which is sent to a central unit for computational process. Many countries around the world still use the same method to estimate the state of the network. Due to the slowness of the telecommunications network, the frequency limitations, and the asynchronous data collection, the measured data from different parts of the network had time delay for several seconds to some minutes. Therefore, the estimated states had good accuracy only in the steady state condition. If a change is occurred in the system and leads to activation of dynamic modes, the result of the state estimation using these methods were only approximate to the actual state of the system, which in the most optimistic view was average vales of the actual state of the system and is therefore known and called as static state estimation [33].

Nowadays, with the expansion of telecommunication system and the implementation of PMUs, it is possible to have dynamic state estimation of the system. The important point of the system in using phase measurement for state estimate of the system is that it is not necessary to implement PMUs for measurement in all the points of the system and only its observability is sufficient. Having a limited number of PMUs at key points in the network, with the help of existing software, makes the entire system visible, so there is no chronic concern for installing new PMUs due to the implementation of network development projects.

7.2 Frequency Stability

As aforementioned before, PMUs are able to measure the frequency and rate of change of frequency and send them to the control center. The frequency is very important parameter in the network and its accurate measurement in different parts of the network is a very important and vital achievement. By measuring the frequency, we can get the loss of generation and the area where the generation is lost. It can be seen that there is a shortage of generation or surplus generation and by controlling it, the balance between generation and load is established. It is also possible to inform that if a part of the network becomes islanded. For example, if the frequency of a part of the network is 49 Hz and the other part is 51 Hz, and this command remains stable for a while, this indicates the separation of the two parts in the system [34–36].

7.3 Situational Awareness

Observability means having network information in different points of the system. This information can be the amplitude and phase angle of the bus voltage and line current. In addition, the frequency of different parts of the network is one of the parameters that help the observability of the network. By simultaneously measuring network data and sending it to the control center, PMUs have created the ability to access information from different locations. One of the special features of PMU and WAMS center is that it can draw information in illustrated form (different diagrams) to convey better vision and more accurate information to the reader (operator).

7.4 Power System Oscillation Detection

In the power grid, a group of generators may start to oscillate against another group. If these fluctuations do not stop, they will cause network instability and even global blackouts. For example, if 500 MW flows through two parallel transmission lines and one of these lines goes out of circuit due to a short circuit fault and its load responsibility falls on the other line, this will cause the power oscillation which will cause to cascade line outage and blackout if it does not damped. That's why it's very important to detect power oscillation. In SCADA system, it was impossible to detect power oscillation because the information was sent every few seconds, and this could be done in a suitable manner using WAMS structure. There are two ways to detect power oscillation in the WMAS system:

- Oscillation detection: When power oscillations are created and its amplitude increases, fluctuations are detected by processing the measured signals from the PMUs and this problem is solved by appropriate control measures or opening the lines.
- Mode meters: The various frequencies and harmonics of the signal are extracted before the occurrence of high amplitudes of the power oscillations is detected. And it can be prevented by taking preventive measures. In other hand, these methods detect the power oscillations before the fluctuations become severe when the system become closer to the critical situation.

7.5 Voltage Monitoring

Voltage instability can occur very quickly. After growing penetration of distributed generations, the issue of voltage stability has become much more important. Measurement of voltage and measurement of reactive power reserve value are among the vital issues in voltage stability that are measured and calculated by PMUs and sent to the WAMS center. To prevent voltage instability, there are under voltage load shedding (UVLS) relays that were operated locally in the past decades. In this approach, UVLS relays measure the voltage of a bus and according to it values; the necessary decisions have been made to disconnect the load. Nowadays, PMU data are deployed to improve the performance of the UVLS relays using wide area measured data.

7.6 Alarming

By having angles of different buses, several alarms can be implemented on the system. For example, the difference between the angles of the two ends of a line can be criterion of an alarm that if exceeds from a certain value, the alarm will be sent. It is also possible to record different buses voltage angles in different regions and then use them during the post event analysis.

7.7 Dynamic Line Ratings

Using phasor data of the system, the lines flow can be measured instantly and their loading status can be assessed. By implementing PMU and achieving instantaneous data, it is possible to determine the loading of the lines using the flow information of the lines and weather conditions to allow more power pass through the lines under certain conditions. There are two ways for dynamic line rating:

(1) The CAT method, which uses environmental information such as temperature and wind velocity and conductor information such as voltage and current measured data by a PMU, determines the dynamic loading limit of the lines.
(2) At both ends of the line PMUs are installed and simultaneously measure the voltage and current data to calculate the loading limit of the line.

7.8 Security and Stability Analysis

The security of the power system means that the system will be able to successfully pass through the disturbances without disruption in servicing to a customer group of customers. This is largely due to the system robustness during abrupt disturbances, and is dependent on operation conditions as well as the likelihood of disturbances. In other words, the security of the system means that the system is resistant to disturbances after the power system has been forced to change its status during the event of a disturbance.

Correct understanding of the disturbance, the place of its occurrence and the type of happened fault help to carry out these studies. Using WAMS features, the online snapshot is provided from the network structure, operation of protection elements, status of control devices, load and generation characteristics, condition of reactive power sources and etc. Then, it is examined at control center using the security analysis software. Also, the state estimation of power system is available at the pre-fault moment, which can be used as initial states for starting network analysis software. Finally, the results of power system dynamic security analysis are expressed as a set of necessary control proceedings. These proceedings are either to prevent faults and increase the security margin of the power system or to prevent instability and blackouts.

7.9 Event Analysis

Extensive events occur on power system every year. To find the source of these events and solve them, it is necessary to check the network status before and during the event. In the SCADA system, due to the lack of synchronization of data and the slowness of the telecommunication link, the network dynamics are not available. Therefore, it is not possible to check the network behavior during the disturbance. Due to the synchronization of data and large implementation of PMUs in the WAMS system, network dynamics can be observed and it is possible to investigate network events.

8 Conclusion

Over years, with the expansion of power system, its complexity has also increased. In this situation, monitoring of the dynamic behavior of power system is vital for secure operation. So, the focus of this chapter is on the concept and importance of WAMS in power systems. Before development of WAMS, the SCADA was used to monitor the power system condition. This chapter describes different features of SCADA. It shows that the SCADA system is not able to follow power system dynamics due to its features. Therefore, the advent of PMU along with increasing the data processing speed led to the development of WAMS. By collecting and processing the data, WAMS provides a set of results to the operators that facilitate system monitoring. Different application of WAMS which increase system performance, have been illustrated in this chapter. A part of this chapter is dedicated to a brief survey of the telecommunication systems used in WAMS because of the data transmission importance.

In this chapter, especially attention has been given to PMUs as the main components of WAMS. Thus, the developments, phasor calculation and frequency estimation methods used in PMUs are covered. Finally, the main purpose of this chapter is to show the importance of WAMS in modern power systems.

References

1. R. Cresap, J. Hauer, Emergence of a new swing mode in the western power system. IEEE Trans. Power Appar. Syst. **4**, 2037–2045 (1981)
2. A.G. Phadke, J.S. Thorp, *Synchronized Phasor Measurements and their Applications* (Springer, 2008)
3. M. Patel, S. Aivaliotis, E. Ellen, Real-time application of synchrophasors for improving reliability, *NERC Report, Oct,* vol. 1 (2010)
4. H.H. Alhelou, M.E. Hamedani-Golshan, R. Zamani, E. Heydarian-Forushani, P. Siano, Challenges and opportunities of load frequency control in conventional, modern and future smart power systems: a comprehensive review. Energies **11**(10), 2497 (2018)
5. R. Zamani, M.E. Hamedani Golshan, H.H. Alhelou, N. Hatziargyriou, A novel hybrid islanding detection method using dynamic characteristics of synchronous generator and signal processing technique. Elect. Power Syst. Res.**175**, 105911 (2019)
6. R. Zamani, M. Parsa Moghaddam, M. Imani, H.H. Alhelou, M.E. Hamedani-Golshan, P. Siano, A novel improved hilbert-huang transform technique for implementation of power system local oscillation monitoring, in *2019 IEEE Milan PowerTech* (IEEE, 2019), pp. 1–6
7. H.H. Alhelou, M.E.H. Golshan, R. Zamani, M.P. Moghaddam, C.N. Takawira, P. Siano, M. Marzband, An improved UFLS scheme based on estimated minimum frequency and power deficit, in *2019 IEEE Milan PowerTech* (IEEE, 2019), pp. 1–6
8. G. Zhang, P. Hirsch, S. Lee, Wide area frequency visualization using smart client technology, in *2007 IEEE Power Engineering Society General Meeting* (IEEE, 2007), pp. 1–8
9. I. Power, IEEE Standard for Synchrophasor Measurements for Power Systems–Amendment 1: Modification of Selected Performance Requirements, in *IEEE Std C37. 118.1 a-2014 (Amendment to IEEE Std C37. 118.1–2011),* vol. 2014, 2014 (pp. 1–25)
10. S. Bhide, *Digital Power System Protection* (PHI Learning Pvt. Ltd., 2014
11. A. Abyaz, H. Panahi, R. Zamani, H.H. Alhelou, P. Siano, M. Shafie-khah, M. Parente, An effective passive islanding detection algorithm for distributed generations. Energies **12**(16), 3160 (2019)
12. D.W. Roop, Power system SCADA and smart grids [Book Reviews]. IEEE Power Energ. Mag. **14**(1), 115–116 (2015)
13. J. De La Ree, V. Centeno, J.S. Thorp, A.G. Phadke, Synchronized phasor measurement applications in power systems. IEEE Trans. Smart Grid **1**(1), 20–27 (2010)
14. A. Phadke, J. Thorp, History and applications of phasor measurements, in *2006 IEEE PES Power Systems Conference and Exposition* (IEEE, 2006), pp. 331–335
15. S. Kumar, M.K. Soni, D.K. Jain, Requirements and challenges of PMUs communication in WAMS environment. Far East J. Electron. Commun. **13**(2), 121 (2014)
16. D.B. Ünsal, T. Yalçınöz, Applications of new power line communication model for smart grids. (2015)
17. M. Hojabri, U. Dersch, A. Papaemmanouil, P. Bosshart, A comprehensive survey on phasor measurement unit applications in distribution systems. Energies **12**(23), 4552 (2019)
18. P. Bonanomi, Phase angle measurements with synchronized clocks-principle and applications. IEEE Trans. Power Appar. Syst. **12**, 5036–5043 (1981)

19. M. Parashar et al., Wide-Area monitoring and situational awareness. Power Syst. Stab. Contr., 1–46 (2012)
20. A.G. Phadke, Synchronized phasor measurements-a historical overview, in *IEEE/PES Transmission and Distribution Conference and Exhibition*, vol. 1: IEEE, 2002, pp. 476–479
21. A. Monti, C. Muscas, F. Ponci, *Phasor Measurement Units and Wide Area Monitoring Systems* (Academic Press, 2016)
22. J.Y. Cai, Z. Huang, J. Hauer, K. Martin, Current status and experience of WAMS implementation in North America, in *2005 IEEE/PES Transmission & Distribution Conference & Exposition: Asia and Pacific*, IEEE, 2005, pp. 1–7
23. H. Li, A. Bose, V.M. Venkatasubramanian, Wide-area voltage monitoring and optimization. IEEE Trans. Smart Grid **7**(2), 785–793 (2015)
24. H.H. Alhelou, M.E. Hamedani-Golshan, R. Zamani, E. Heydarian-Forushani, P. Siano, Challenges and opportunities of load frequency control in conventional, modern and future smart power systems: a comprehensive review. Energies **11**(10), 2497 (2018 Oct)
25. H.H. Alhelou, M.E. Golshan, J. Askari-Marnani, Robust sensor fault detection and isolation scheme for interconnected smart power systems in presence of RER and EVs using unknown input observer. Int. J. Electr. Power Energy Syst. **1**(99), 682–694 (2018). (Jul)
26. H. Haes Alhelou, M.E. Hamedani Golshan, M. Hajiakbari Fini, Wind driven optimization algorithm application to load frequency control in interconnected power systems considering GRC and GDB nonlinearities. Electric Power Compon. Syst. **46**(11–12), 1223–38 (21 Jul 2018)
27. H.H. Alhelou, M.E. Golshan, N.D. Hatziargyriou, Deterministic dynamic state estimation-based optimal lfc for interconnected power systems using unknown input observer. IEEE Trans. Smart Grid (9 Sep 2019)
28. H.H. Alhelou, M.E. Golshan, N.D. Hatziargyriou, A decentralized functional observer based optimal LFC considering unknown inputs, uncertainties, and cyber-attacks. IEEE Trans. Power Syst. **34**(6), 4408–4417 (2019). (May 13)
29. H.H. Alhelou, M.E. Golshan, T.C. Njenda, N.D. Hatziargyriou, An overview of UFLS in conventional, modern, and future smart power systems: challenges and opportunities. Electr. Power Syst. Res. **1**(179), 106054 (2020). (Feb)
30. H.S.H. Alhelou, M.E.H. Golshan, M. Hajiakbari Fini, Multi agent electric vehicle control based primary frequency support for future smart micro-grid. in *2015 Smart Grid Conference (SGC)* (IEEE, 2015), pp. 22–27
31. T.C. Njenda, M.E. Golshan, H.H. Alhelou, WAMS based intelligent under frequency load shedding considering online disturbance estimation, in *2018 Smart Grid Conference (SGC)* (IEEE, 28 Nov 2018), pp. 1–5
32. T.C. Njenda, M.E. Golshan, H.H. Alhelou, WAMS based under frequency load shedding considering minimum frequency predicted and extrapolated disturbance magnitude, in *2018 Smart Grid Conference (SGC)* (IEEE, 2018 Nov 28), pp. 1–5
33. H. Haes Alhelou, M.E. Hamedani-Golshan, T.C. Njenda, P. Siano, A survey on power system blackout and cascading events: research motivations and challenges. Energies **12**(4), 682 (Jan 2019)
34. H. Haes Alhelou, M.E. Hamedani Golshan, T.C. Njenda, P. Siano, Wams-based online disturbance estimation in interconnected power systems using disturbance observer. Appl. Sci. **9**(5), 990 (Jan 2019)
35. H. Haes Alhelou, M.E. Hamedani-Golshan, T.C. Njenda, P. Siano, Wide-area measurement system-based optimal multi-stage under-frequency load-shedding in interconnected smart power systems using evolutionary computing techniques. Appl. Sci. **9**(3), 508 (Jan 2019)
36. H.H. Alhelou, S.J. Mirjalili, R. Zamani, P. Siano, Assessing the optimal generation technology mix determination considering demand response and EVs. Int. J. Electr. Power Energy Syst. **1**(119), 105871 (2020). (Jul)

Information and Communication Infrastructures in Modern Wide-Area Systems

Weikang Wang, Kaiqi Sun, Chujie Zeng, Chang Chen, Wei Qiu, Shutang You, and Yilu Liu

Abstract Information and communication infrastructures (ICIs) in modern wide-area systems handle the transmitting, receiving, and storing of high-speed, large-volume synchrophasor data. Such infrastructures are important components in modern wide-area systems. Although power systems are becoming rather complex, with the introduction of synchrophasor technology, the highly accurate, high-speed, widely deployed, and time-synchronized phasor measurement units (PMUs) are providing operators and auditors an unprecedented way to understand the complex power systems. Consequently, these advanced PMUs challenge the current information and communication infrastructures. This chapter reviews the basics, challenges, and visions of the ICIs in wide-area monitoring systems (WAMS). We will first overview some important wide-area ICI topics and share our experience in building an efficient, reliable, and secure distribution-level WAMS, FNET/GridEye. We will introduce some key technologies that ensure the efficiency and reliability of such a WAMS. Finally, some outstanding challenges and future directions of the contemporary ICIs are discussed and envisioned.

Keywords ICI · Phasor measurement units · WAMs · Wide-area systems · infrastructures · Wide-area power systems · Measurement systems

W. Wang · K. Sun · C. Zeng · C. Chen (✉) · W. Qiu · S. You · Y. Liu
The University of Tennessee, Knoxville, TN, USA
e-mail: cchen75@utk.edu

Y. Liu
Oak Ridge National Laboratory, Oak Ridge, TN, USA

H. Haes Alhelou et al. (eds.), *Wide Area Power Systems Stability, Protection, and Security*, Power Systems, https://doi.org/10.1007/978-3-030-54275-7_3

1 Communication Infrastructures in Modern Wide-Area System

1.1 Introduction

Wide-area communication infrastructures (WACIs) are a series of cyber-physical components which act as the backbone of data transmission. In the WACIs, there are three fundamental elements, which are protocol, method, and scheme. A communication method defines the physical medium on which the data is transmitted. A protocol defines the way in which the data is formatted. Finally, a scheme specifies the overall architecture of a communication system and the directions of data streams in between. These components are integral to a WACI and they work together to ensure efficient and reliable transmission of wide-area measuring data. In the following sections, these components will be illustrated in detail.

1.2 Communication Protocols

Communications protocols in wide-area systems define the formats in which the data are transmitted between a PMU and a phasor data concentrator (PDC). In the modern wide-area systems, several protocols including IEEE C37.118, IEC-61850-90-5, and Streaming Telemetry Transport Protocol (STTP) have been proposed and adopted in real operations. Communication protocols vary on their reliability, efficiency, security, etc. and are subject to specific use cases. This section will focus on three widely accepted communication protocols and give a comparison to demonstrate their pros and cons.

1.2.1 IEEE C37.118

The synchrophasor data exchange protocol IEEE C37.118 was published in 2005 [1]. Later in 2011, the standard was divided into two parts where IEEE C37.118.1 [2] defines the requirement for synchrophasor measurement and IEEE C37.118.2 [3] defines the synchrophasor data transmission format. In 2014, the IEEE C37.118.1a amendment [4] was published, which updated some performance requirements.

A PMU communicates with a PDC via binary frames. The IEEE C37.118.2 standard defines four frame types: configuration frame, header frame, command frame, and data frame. A configuration frame defines the format of the synchrophasor data stream. Hence, it must be received before data parsing. There are three types of configuration frames: CFG-1, CFG-2 and CFG-3. The first two are identical in structure but are used in different contexts. CFG-1 provides information about the device's reporting capability, indicating all the data that the device reports. CFG-2 indicates synchrophasor measurements that are currently being transmitted. CFG-3 is optional

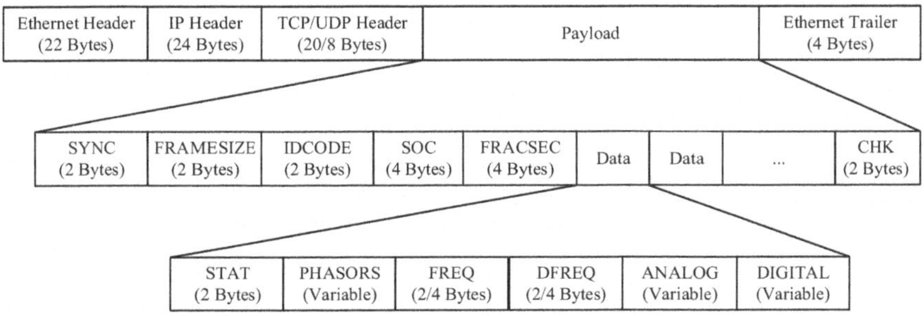

Fig. 1 Data frame structure of IEEE C37.118.2 protocol

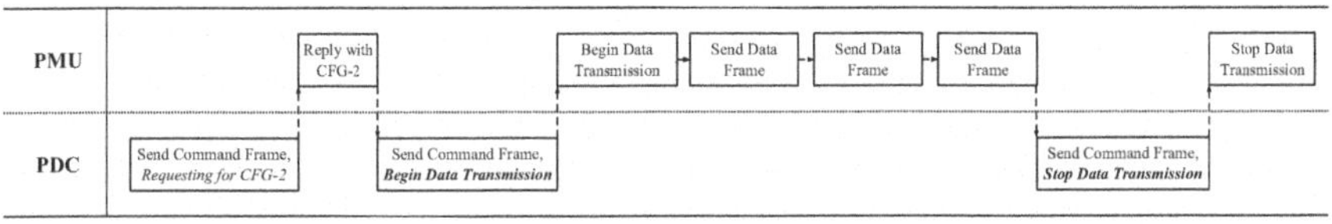

Fig. 2 Message exchange flow of IEEE C37.118.2 protocol

and indicates PMU characteristics and quantities being sent. Then, a header frame is supposed to transmit human-readable ancillary information, such as data sources, scaling, algorithms used and other related information. A command frame is used to control the behavior of an established connection. Possible commands include turn off transmission of data frames, turn on the transmission of data frames, send the header frame, send CFG-1 frame, etc. Finally, a data frame is used to transmit measurement data and a set of status bits. A data frame can be properly parsed only when an active configuration frame is present. A sample structure of the message in the IEEE C37.118.2 protocol is demonstrated in Fig. 1.

A typical IEEE C37.118.2 message exchange flow between a PMU and a PDC is shown in Fig. 2. First, the PDC sends a command message requesting the configuration frame. Then, the PMU replies with the requested configuration frame. Afterward, the PDC sends out the command frame asking for data transmission. The PMU responds by flushing data frames to the PDC continuously until it receives the turn-off command from the PDC.

1.2.2 IEC 61850–90-5

IEC 61850 standard [5] is the de-facto standard for substation automation. In 2012, IEC 61850-90-5 standard [6] was introduced as the synchrophasor data transmission protocol within IEC 61850 stack. IEC 61850-90-5 standard is usually preferred when IEC 61850 is already adopted. It employs existing elements of IEC 61850 and adopts IEEE C37.118.1, which defines the measurement requirements.

In IEC 61850, sample values (SV) and generic object-oriented substation events (GOOSE) are two types of real-time communication services. SV is used to exchange streaming data like phasor measurements while GOOSE is used to transmit status and control commands. SV and the GOOSE can only be used within a local area network (LAN) as they are Ethernet layer messages [7]. However, a WAMS requires communication over wide-area network (WAN). The routable SV and GOOSE then are introduced and termed as R-SV and R-GOOSE respectively. User Datagram Protocol (UDP) is usually used due to the need for multicasting. Control blocks are used to control the message flow. The control blocks for R-SV and R-GOOSE are termed as Routed Multicast Sampled Value Control Block (R-MSVCB) and Routed GOOSE Control Block (R-GoCB) respectively.

The structure of R-SV message is shown in Fig. 3, where the Sample Value Application Protocol Data Unit (SV APDU) and Application Service Data Units (ASDU) are expanded. A typical communication procedure is demonstrated in Fig. 4. First, the PDC sends a Manufacturing Message Specification (MMS) request message asking for PMU information. After the PDC receives the response, it sends out an R-MSVCB to start the data transmission process. After receiving this message, the PMU continuously sends R-SV messages to the PDC until it receives another R-MSVCB requesting the end of data transmission.

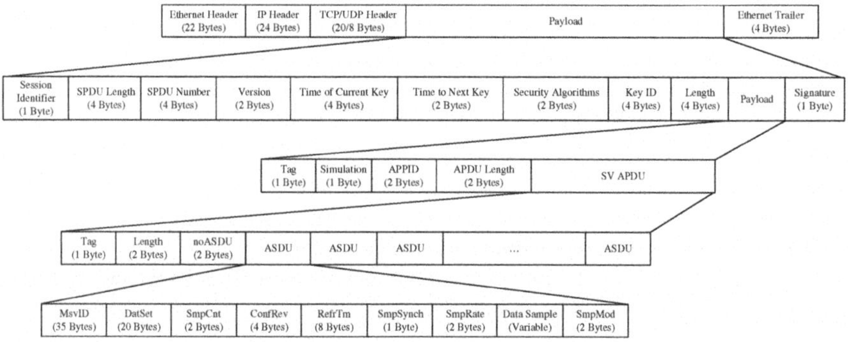

Fig. 3 R-SV frame structure of IEC 61850-90-5 protocol

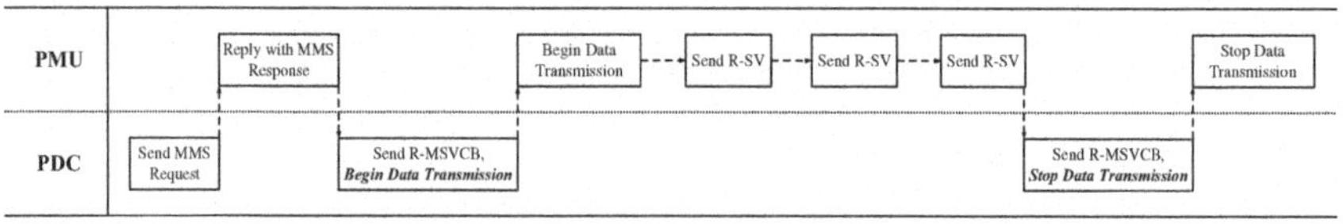

Fig. 4 Message exchange flow of IEC 61850-90-5 protocol

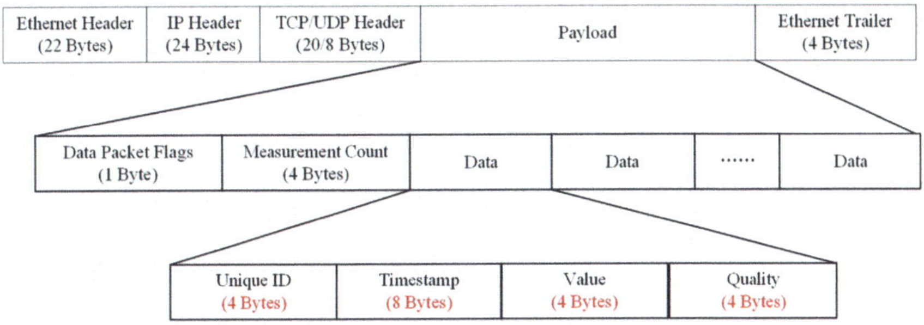

Fig. 5 STTP data frame structure

1.2.3 Streaming Telemetry Transport Protocol

The data losses and delivery latencies of IEEE C37.118 and IEC TR 61850 will greatly increase when the frame size approaches 32 K byte [8]. To address this issue, a signal-based protocol—the Streaming Telemetry Transport Protocol (STTP) is designed, which can send the compressed data instead of the raw data in binary. Figure 5 shows the latest version of data frame of STTP, where the length of each element before compression is colored in red. In this version, more than 65 K measurement values can be transmitted in one frame. In the payload of the frame, the data packet flag is the indicator representing the data payload format, which is omitted when UDP encryption is enabled with the UPDATE CIPHER KEYS command. The number of measurements is recorded in the measurement count field. The data block consists of four parts- a Unique ID that indicates a measurement, a timestamp, a measurement value, and a quality flag.

STTP uses lossless algorithms to compress the raw data. When it is used over UDP, the Gzip algorithm is used to compress the payload. For STTP TCP compression, a time-series special compression (TSSC) is implemented to compress the data [9]. The TSSC can compress streaming time-series data quickly. It works by first find the different bits between two values with XOR calculation. Then, Only the bits that have changed since the last measurement and the code word that represent the length of the calculated bits are stored. This method performs well when the data has a certain trend, for example, when the sampling rate is very high so the difference between every two values is small.

1.2.4 Comparison of Communication Protocols

In terms of structure, both IEEE C37.118.2 and IEC 61850-90-5 are frame-based and must transmit the configuration prior to the data transmission. However, STTP is measurement-based and its configuration differ from one data packet to another. Since variable configuration introduces extra overhead, STTP consumes higher bandwidth than the other two. However, STTP supports compression. The experiment shows

Table 1 Comparison of wide-area communication protocols

Feature	IEEE C37.118.2	IEC 61850-90-5	STTP
Structure	Frame-based	Frame-based	Dynamic
Efficiency	Good	Fair	Good with TCP Fair with UDP
Scalability	Limited	Limited	Good
Compression	Yes/No	No	Yes/No
Encryption	No	Yes	Yes
Multicast supported	Yes	Yes	Limited

that when using TCP and stateful compression, STTP consumes bandwidth at least 30% less than IEEE C37.118 [9]. Both IEEE C37.118.2 and IEC 61850-90-5 have a 65 K bytes frame size limitation, as a result, no more than 6700 uniquely identifiable measurements can be transmitted in one connection [9]. On the contrary, STTP does not have such a limitation, making it more scalable. In terms of security, IEEE C37.118.2 only utilizes a cyclic redundancy check (CRC) codes to ensure data integrity, which can be easily modified by the intruder [10]. IEC 61850-90-5 ensures integrity with asymmetric cryptography and ensures confidentiality with symmetric encryption [6, 10]. STTP uses Transport Layer Security (TLS) to provide security. Table 1 summarizes the differences of three protocols.

1.3 Communication Methods

For now, the communication methods that could be used in synchrophasor data transfer could be basically classified into two categories: wired communication and wireless communication.

The wired communication is the most common communication technology used in the world. The main media in use for wired communication are power supply cable and optical fiber [11]. The wired communication relies on the physical circuit to exchange data, thus it could offer high reliability, huge bandwidth and high protection capacity against interference [12]. However, the wired communication also has some disadvantages due to its physical constraint. With the development of communication technology, wireless communication witnesses a continuous increasing share in the communication mix. The most popular media for wireless communication are cellular, microwave, and satellite. Table 2 shows the feature comparison between the wired communication and wireless communication on mobility, cost, expansion, and remote-access capability.

According to the different media adopting, the communication methods for synchrophasor data transfer could be summarized in Table 3.

Table 2 The feature comparison between the wired communication and wireless communication

	Mobility	Cost	Expansion	Remote access capacity
Wired communication	Limited by cable	High cost (physical media depending & installation cost)	Hard	weak
Wireless communication	Not limited to the physical location	Low cost (easy deployment & maintenance-free)	Easy	Strong (could be deployed in forest or hilly terrains)

Table 3 The communication method for synchrophasor applications

Synchrophasor communication method	Wired communication	Power line communication	 or 
		Optical fiber communication	
	Wireless communication	Cellular communication	
		Microwave-based Communication	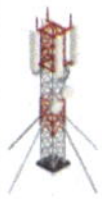
		Satellite communication	

1.3.1 Power Line Communication (PLC) for Synchrophasor Data Transfer

The PLC technology uses standard power supply cables to realize Synchrophasor data transfer between two PMUs. The structure of the WAMS using PLC technology to achieve synchrophasor data transfer could be depicted as Fig. 6.

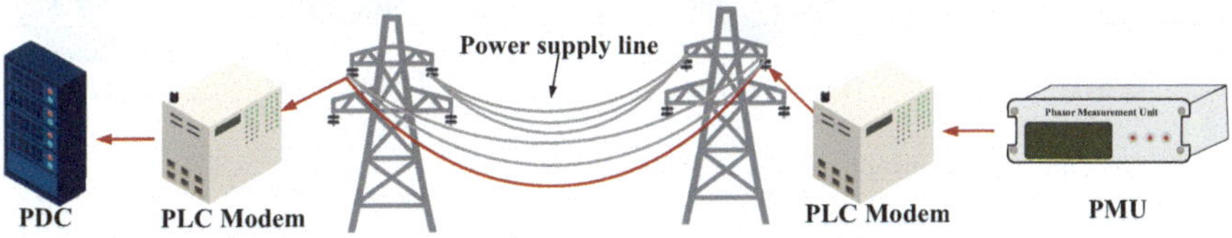

Fig. 6 The PLC for synchrophasor application

PLC technology avoids additional network cables installation, it provides the easiest and economical means for the installation and deployment of the construction system. There are two types of PLC technologies: narrowband PLC (NB-PLC) and broadband PLC (BB-PLC). The NB-PLC, which is also named low-speed PLC (LS-PLC), could offer a nominal speed of few kilobytes per second and normally could be connected to cost-effective electronic equipment in a simple way. The BB-PLC, which is also named high-speed PLC (HS-PLC), could offer a nominal speed from Mb to hundreds of Mb per second. In most of the synchrophasor applications, the BB-PLC is adopted as its communication method for data rates requirements. However, PLC technology also has some disadvantages. The noisy background is serious in the power supply cables. Thus, the communication channel of the PLC is difficult to be modeled. In addition, the fading and interference of the PLC are severe in practice so that it is not suitable for higher bandwidth synchrophasor applications. For now, the PLC methods are usually working with other wireless communication methods such as cellular communication or microwave communication to provide a hybrid communication solution for synchrophasor applications.

1.3.2 Optical Fiber-Based Communication (OFC) for Synchrophasor Data Transfer

The OFC has been widely used in telephone signals transmitting, cable television signals transmitting, and internet communication. The OFC uses the optical fiber to send the pulses of infrared light to realize the data transfer from PMUs to PDC. The typical structure of the WAMS using OFC technology to achieve synchrophasor data transfer could be depicted as Fig. 7.

Compared to the PLC method, the OFC has higher data rates, lower attenuation, higher reliability, and negligible interference. In spite of having several advantages, due to its physical constraints, the OFC technology still suffers from many disadvantages, such as high installation and maintenance costs, potential risk of stolen or damaged, and expansion issues.

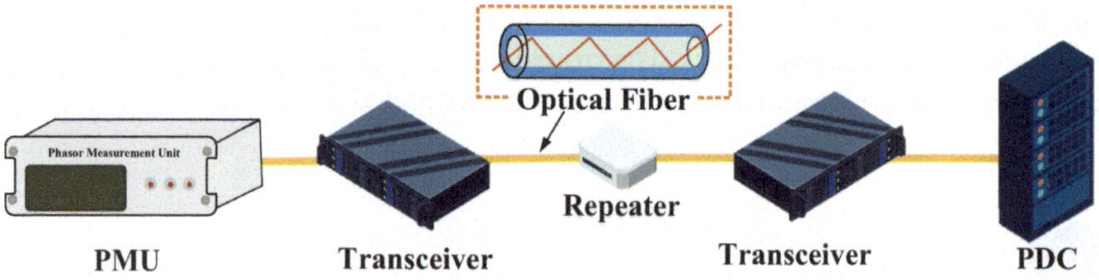

Fig. 7 The OFC for synchrophasor application

Table 4 Data rates of different cellular technologies for synchrophasor applications

Cellular technology	Data rates
General Packet Radio Service (GPRS)	≤114 Kbps
Enhanced Data rates for GSM Evolution (EDGE)	≤384 Kbps
Universal Mobile Telecommunications (UMTS)	≤2 Mbps
High-Speed Packet Access (HSPA)	Up to 384 Kbps
Long Term Evolution-Advanced (LTE-A)	Up to 384 Kbps

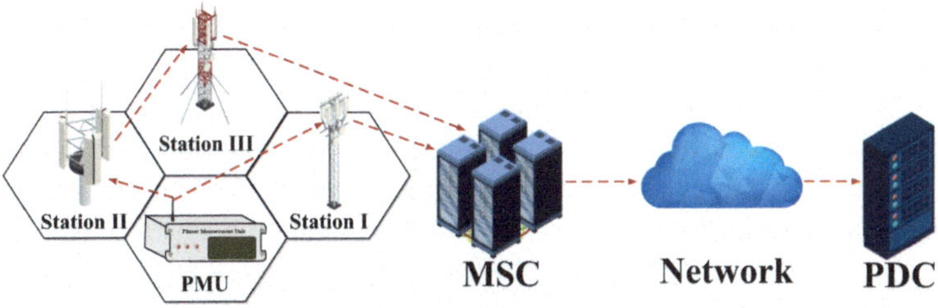

Fig. 8 The cellular communication for synchrophasor application

1.3.3 Cellular Communication for Synchrophasor Data Transfer

The cellular communication method is the most common wireless communication method in the world. The data cellular communication network has been deployed over most of the inhabited land area of Earth. Due to the high proliferation of the cellular communication infrastructure, it has been regarded as an economic alternative for synchrophasor applications. The data rates of different cellular technologies for synchrophasor applications are shown in Table 4 [13].

The typical structure of the WAMS using cellular communication technology to achieve synchrophasor data transfer could be depicted as Fig. 8.

As shown in Fig. 8, the cellular communication system is comprised of cellular stations, mobile switching centers (MSC) and cellular networks. The shared nature of the cellular communication system is an advantage. However, it is unacceptable for the synchrophasor applications due to security considerations. In addition, the uninterrupted communication of cellular communication is difficult to be guaranteed, but it is a forced requirement for mission-critical applications.

1.3.4 Microwave-Based Communication for Synchrophasor Data Transfer

The microwave-based communication system could realize the several Gbps data rates, which could cater to the demands of the synchrophasor data transfer. The

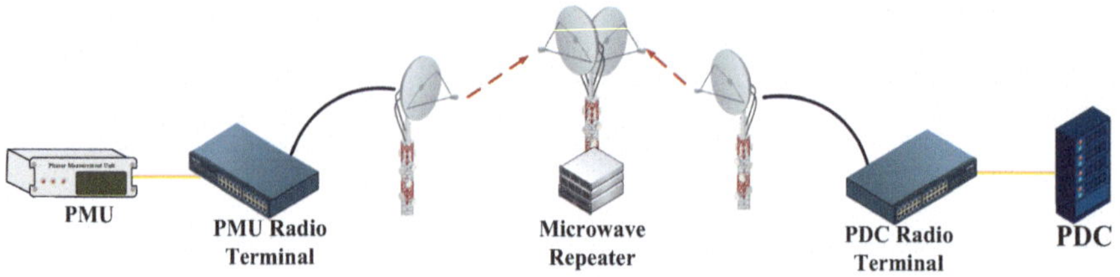

Fig. 9 The microwave-based communication for synchrophasor application

typical structure of WAMS using microwave-based communication technology to achieve synchrophasor data transfer could be depicted as Fig. 9.

The process of microwave-based communication is similar to the OFC except it is wireless communication. Microwave-based communication has a very large information-carrying capacity due to its high-frequency characteristic. In addition, the interference of microwave-based communication is also negligible. However, the main disadvantage of microwave-based communication is that the signal propagates in space is susceptible to cyber-physical attacks, which significantly affects its reliability and security.

1.3.5 Satellite Communication for Synchrophasor Data Transfer

Satellite communication is a prospective solution for synchrophasor data transfer. Compared to other communication methods, satellite communication could provide uninterrupted communication for unaffecting by natural disasters due to its communication equipment is in space. The typical structure of the WAMS using satellite technology to achieve synchrophasor data transfer could be depicted as Fig. 10.

The communicating process of the satellite communication system is similar to microwave-based communication. The PMUs collect the phasor measurements data from the power system and compress the data as packages. The main disadvantage of the satellite communication system is its communication delay, besides, the antenna for satellite signal receiving is quite expensive, which limits the large-scale promotion of the satellite communication methods in the synchrophasor applications.

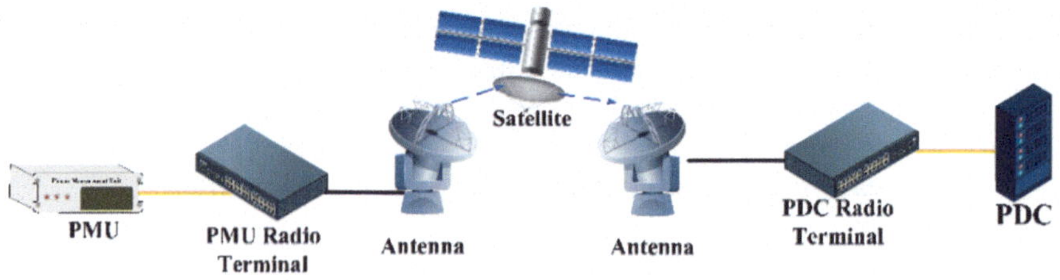

Fig. 10 The satellite communication for synchrophasor application

1.4 Communication Requirement

The synchrophasor applications require high reliability of the communication channel, to provide uninterrupted synchrophasor data measurement for the system operators to help with monitoring the status of the modern power grid in real-time. The primary requirements of communication methods choosing for synchrophasor data transfer in WAMS are throughput, bandwidth, time delay, and reliability.

- Throughput

The throughput is a bottleneck for the synchrophasor data transfer in the WAMS. The throughput represents the average data delivered per second, which is determined by measuring the data transfer speed at a specific time. The throughput must be considered as primary factors in the WAMS communication system design to build a reliable communication channel while controlling the cost in a reasonable scope.

- Bandwidth

The higher bandwidth of communication methods adopted for synchrophasor applications is required to guarantee the large volume synchrophasor data transfer in the WAMS. The bandwidth is defined as how much data can be sent over a specific network connection per give unit time [14]. The bandwidth design in the WAMS should keep enough redundancy for future synchrophasor applications development.

- Time delay

Time delay, i.e. latency, is one of the most stringent requirements for the communication system in the WAMS. Time delays are caused by communication disturbances or data alignment. The time delay could be classified as, according to the root cause, transducer delays, the propagation delay, processing delay, communication link transmission delay and data alignment delay. In general, the permissible time from PMUs to PDC is 20 ms, If the PDCs also need to transfer data to Central PDCs, an extra 40 ms is permitted from PMUs to the central PDCs. The WAMS requires a short time delay to support the local area and wide area real-time response control and protection [15]. However, the time delay requirements are subject to the actual application types. Table 5 lists the communication time delay requirement of some typical synchrophasor applications [17].

Table 5 Communication time delay requirement for synchrophasor applications

Synchrophasor application	Communication delay (ms)
State estimation	100
Generator synchronization	50
Intelligent scheduling	50
Oscillation control	200
Islanding	50

Table 6 Reliability comparison between different communication methods

Communication method	Throughput	Time delay (ms)	Bit error rate
Power line communication	256 kbps ~ 2.7 Mbps	150–350	$<10^{-2}$
Optical Fiber communication	10 Gbps	100–150	10^{-15}
Wireless communication	75 mbps	100–150	10^{-7}–10^{-12}
Satellite communication	256 kbps–1 Mbps	1000–1400	10^{-7}

- Reliability

The reliability of communication is the backbone of the WAMS for providing uninterrupted, real-time monitoring, protection, and control of the power system. The throughout, time delay and bandwidth rate support the WAMS operating at a prescribed level of reliability. The bit error rate is usually to measure WAMS reliability. Table 6 indicates the throughput, time delay, and bit error rate of various communication methods in the WAMS to show their reliability comparison.

1.5 Advanced Topic for Communication Infrastructure: Cybersecurity

1.5.1 Challenges of Cyber Attacks

Nowadays, with an increasing number of PMUs, PDC and other types of power electronic devices in the power system, the communication infrastructure of WAMS is confronted more challenges including safe operation, cybersecurity, and data quality. For example, it is reported that more than 360 cyber-physical attacks happened during 2011 and 2014 according to the Energy Department in the USA [18]. Additionally, the safe operation of the power system is threatened by multiple aspects. At the physical level, the GPS signal can be spoofed thereby increasing the error of synchronous measurement. At the communication level, the denial of service and man-in-the-middle attacks cause erroneous failures by leveraging the knowledge of grid structure.

The synchrophasors support a two-way communication channel. Synchrophasor data measurement values flow from the power devices to the control center and server. Then the control signals flow in the other direction. The protocol used by the device helps the transmission and integration of data. However, the vulnerability of the protocol puts the data security in the public eye. In the IEEE C37.118, the packet and the CRC code can be modified and transmitted to the receiver. Thereafter, the availability and confidentiality of the data will be changed. In the communication system, some primary attacks include

- Denial of service (DoS)

The power system network resources, such as the IP address and bandwidth, are controlled by the DoS attackers. Legitimate users will be denied access to the server. The DoS attacks control multiple devices and machines, making the network channel blocked or dropped.

- Man-in-the-middle (MITM)

The attackers impersonate the other end of a legitimate protocol session between the server and a legitimate client. In the WAMS, the MITM would happen between PDC and PMUs. The destinations and the packet of the PMU can be modified by the MITM attacker. Meanwhile, the PDC can be deceived using fake certificates. In this case, the entire PMU data system will be chaotic.

- Delay

When there is a lot of information redundancy in the network, the useful bandwidth and throughput of the routers will be limited. In this case, the measurement synchrophasor data need to wait longer to transfer to the target device, resulting in the data loss and real-time control of the grid.

1.5.2 Remedies to Cyber Attacks

To address the challenge of cybersecurity, different strategies are proposed to decline the potential damage. Specifically, the solutions can be summarized as follows:

- DoS attack Countermeasures

To detect the DoS attack, the "air-gapped" network may provide a solution because it is completely isolated from the local machine. However, the construction of such a separate network requires separate infrastructure, thereby increasing costs. Some other methods, such as anomaly detection method using wavelet analysis and cumulative sum [16], is used to detect anomalous traffic.

- Man-in-the-middle (MITM)

The primary means to prevent this MITM attack is to check and authenticate the client and the server. One of the commonly used certificates is X.509, the devices and system can communication only after passing authentication. The public key cryptography is used to prevent MITM attack [19].

- Delay

To address the delay attack, it is reasonable to use the new IP Multicast protocols. This IP multicast can minimize packet replication, which can provide higher bandwidth.

Here, a tree construction model is used to minimize the invalid synchrophasor data, this tree structure is sensitive to the delay [20].

1.6 Advanced Topic for Communication Infrastructure: Data Compression

1.6.1 Challenges of Communication Efficiency from Advanced PMUs

The performance of communication systems has significant effects on the WAMs. To build an efficient and secure PMUs communication, some factors that challenge PMU communication should be addressed. These factors include

- The high-density deployment of PMUs

The number of PMUs has been growing rapidly with the development of WAMS [21]. The traffic of data stream increases with the increasing deployment of the PMUs, which challenges the network bandwidth limit of WAMSs.

- High reporting rate

The reporting rate of PMUs can reach up to 1440 Hz, depending on the phasor calculating algorithms. In an extreme case, several terabytes of data may be generated per day, causing an unprecedented burden for servers to digest it [22].

- Data communication delay

Since most of the devices accommodated in the WAMS are executed in real-time, communication delay occurs in PMU measurement. The large volume of data puts pressure on bandwidth, which greatly increases the delay [23]. Large communication delays will deteriorate the reliability and accuracy of data [24].

- Data loss

Data losses are becoming serious due to the old PMUs and the incapable of the network to accommodate high sampling rate PMUs [25]. This can lead to harmful consequences such as affecting the performance of PMU-based control and closed-loop control [25].

1.6.2 Data Compression for Efficient Communication

Data compression is the process of efficiently encoding data to reduce the number of bits required to transmit or store data. An intelligent data compression algorithm requires a prior understanding of the characteristics of the original data, in which some patterns, e.g. repeated data, can be reduced [26]. The compression ratio (CR) is the primary indicator to represent the efficiency of data compression, which can be calculated by (1).

$$CR = \frac{Original\ Data}{Compressed\ Data} \tag{1}$$

Data compression methods can generally be divided into lossy and lossless data compression. The lossy data compression may incur an irreversible loss of data information, but it may achieve a much higher CR. Therefore, lossy compression methods are most used to compress multimedia data, where some loss of data quality is tolerable. Lossless data compression methods allow data to be reconstructed without any loss, although the CR may be sacrificed. These methods are adopted when a high data accuracy is required. Since the WAMS ICIs have stringent requirements for data accuracy, here, we only discuss some lossless methods.

- Entropy coding

Entropy coding, such as Huffman coding and arithmetic coding, is independent of the specific characteristics of the data. It replaces each symbol with a sequence whose length depends on how frequently the symbol appears in the original data set [27]. Huffman coding is the mostly used entropy coding method. However, entropy coding does not work well with streaming data since it needs to get the whole data and construct the coding tree [28]. Additionally, errors in the coded sequence of bits will tend to propagate when decoding.

- Bit-wise difference coding

As STTP provides a bit-wise coding method, which compares each value with previous values by bit., it performs well with streaming data. However, since the values are dependent on each other, the error propagation will be even serious compared with entropy coding as all the values followed by the wrong bits will be damaged.

To gain higher CR, a preprocessing is usually necessary for raw data before compression [29]. This step is known as "data prediction", which can reduce the variance of data by the prior knowledge of the system [30, 31]. By doing this, a to-compress value is compared with a predicted value instead of the original value that is supposed to be compared. If the predictor is good enough, the errors will fall within a tight range near zero, yielding a highly repetitive pattern so that is can be reduced. The prediction step can be roughly divided into linear prediction and non-linear prediction. The linear prediction is used when a linear relationship exists between consecutive data points, e.g. phase angle. As opposed to it, non-linear predictions may fit other signals, e.g. point-of-wave, considering time and other parameters [30, 32].

1.6.3 Open Topics for Data Compression

Although many efforts have been done for PMU data compression, there are still some open issues such as real-time compression and the trade-off between CR and data accuracy. First, offline compression could result in severe congestion in the communication system due to the huge data volume. However, the real-time data compression methods generally have large sampling window and inaccurate measurement during disturbances. This can lead to long delays or packet loss [33]. How to implement and optimize real-time compression should be deeply investigated. Lossless data compression algorithms are preferred in the PMU level to maintain data reliability [30]. However, the Lossy algorithms can have much higher CRs. How to maintain data accuracy without scarifying CR for high-resolution data still needs to be studied.

2 Information Infrastructures in Modern Wide-Area Systems

2.1 Introduction

Wide-area information infrastructures (WAII) provide critical functionalities to collect, process, store, and distribute information. In this verse, the data center architecture, data management, and data center security are discussed. A data center architecture specifies a structure of the information infrastructures from a high-level and defines the data streams from the PMUs to control centers. Data management includes the storage and distribution of measurement data and analytical results. Finally, the data center security covers basic topics including firewalls, access control, etc.

2.2 Information Infrastructure Architecture

Information architectures mainly fall into two categories, centralized and decentralized [13]. In a centralized architecture, all PMUs reports to one central control station. The advantages of a centralized communication architecture are simple topology and low cost. However, the reliability of centralized communication architecture is worse than the decentralized architecture since the failure of the single control station can be unaffordable. In decentralized communication architecture, there is more than one control stations. Compared to the centralized communication architecture, decentralized communication architecture is more robust, but the installation and maintenance cost may be much higher due to the complexity of the communication architecture. Generally, the choice of communication architecture should consider requirements such as efficiency, reliability, security, etc. Figures 11 and

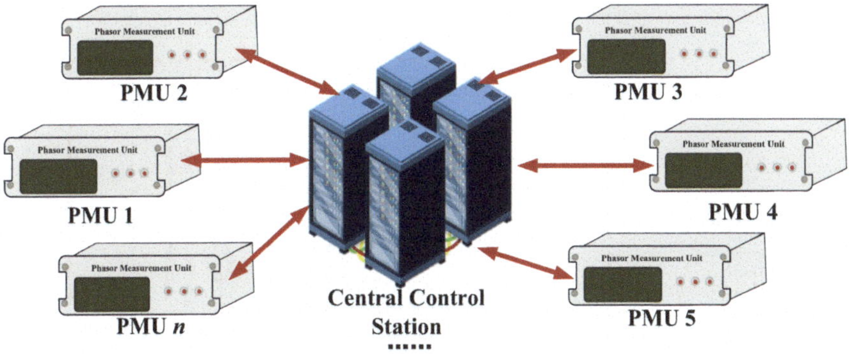

Fig. 11 Centralized information infrastructure

12 demonstrate the centralized and the decentralized communication architectures, respectively.

A generic information infrastructure architecture is demonstrated in Fig. 13. In Fig. 13 a decentralized architecture is employed, where each local control station operates independently and streams data to others as a client. Heterogeneous methods are adopted for the PMU to control center communication. Within the control center, multiple PDCs that communicate directly with the data server, and a load balancer may increase the capability of digesting large amounts of data in a distributed manner. For security, PDCs only push data into a short-term data storage server, then such a server periodically flushes data into another server for long-term data archive. Moreover, a caching server acts as a high-performance interface for data access. For performance, PDCs also push the same data into the data cache server while they write into the short-term data server.

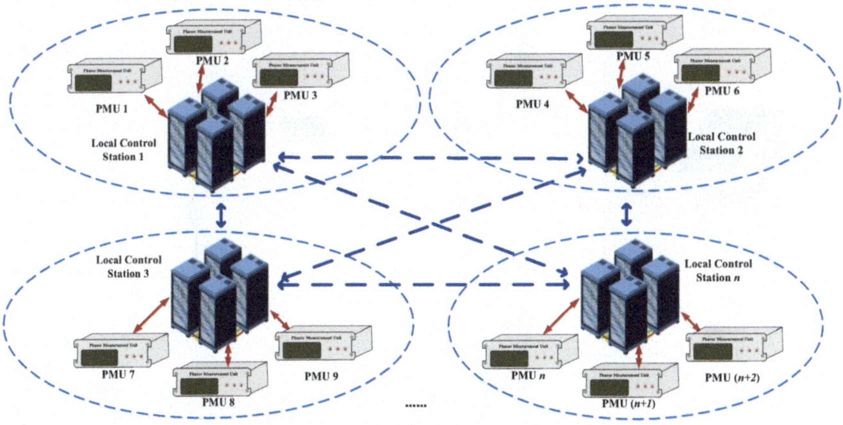

Fig. 12 Decentralized information infrastructure

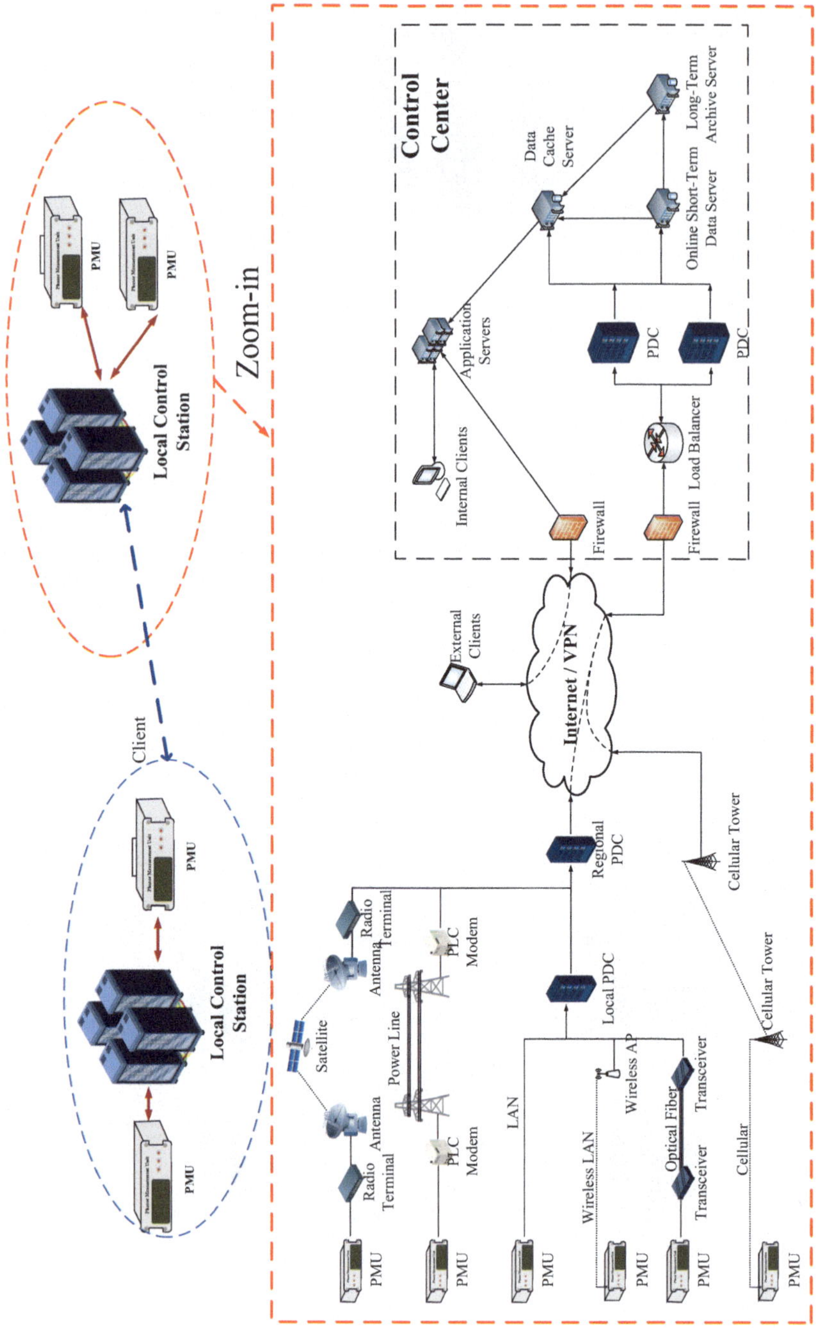

Fig. 13 A generic architecture of a wide-area information infrastructure

2.3 *Data Center Management*

The main purpose of the contemporary WAMSs is to provide accurate and large-volume measurements of electrical quantities to demonstrate the power system dynamics. With the fast development of synchrophasor technology, to properly manage the synchrophasor data becomes a rather important issue for control centers nowadays. From the functionality standpoint, data center management consists of data storage, data warehousing, and data center security.

2.3.1 Data Storage

Storage Method

Storage method defines the binary representation of the data in a computer system. In general, data storage can be implemented in three methods, relational database, non-relational database, and formatted files. The relational database is a mature technology, which is primarily used to permanently store structured data. Relational databases have advantages including well-defined structure, efficient data manipulation (small volume), etc. Relational databases can provide reliable storage, but its data is not directly readable by human-being. A database management system (DBMS) may be required to manage such a database and its data. A known issue of relational databases is the deterioration of insertion efficiency under large-volume data due to the reading of a whole large page [34], which makes it difficult to storage large-volume PMU data in WAMSs. However, formatted files can provide straightforward views of the collected data and are directly readable by human-being. Another advantage of this method is it provides better extensibility. For example, once defined, a relational database may be difficult to change due to its defined relational structure, which, in return, makes adding new measurement types difficult. However, new measurements can be easily added into the formatted files by creating new columns without altering the relational structure. The disadvantages of the formatted files are also obvious. Since they usually store data in plain, formatted files, the sacrifice of storage efficiency is usually inevitable. Moreover, the plain text format usually takes up larger space due to a lack of encoding. It is worth to note, in recent years, non-relation databases have been proposed and developed to combine the advantages of the relational databases and formatted files. Non-relational database exploits advanced data structures to efficiently store the time-series data but also keep acceptable extensibility. The insertion efficiency in non-relational databases is ultra-high because it does not require the reading of a whole large page, but a small immutable batch [34]. To sum up, relational databases provide good features to handle the storage of structured and small-in-size data, which is an ideal storage method for the analytical results [35]. As opposed to it, non-relational databases provide ultra-high

Table 7 Performance comparison of storage methods

	Relational database	Non-relational database	Formatted files
Reliability	High	High	High
Efficiency	Medium	High	Medium
Extensibility	Low	Medium	High
Space	Low	Low	Medium
Scenario	Analytical results	Timeseries measurements	Timeseries measurements

data manipulation efficiency and they are widely adopted to store real-time, large-in-size PMU data [46]. Table 7 compares formatted files and databases on various matrices.

Storage Scheme

The storage scheme defines how the data is stored across the information system. Typically, there are mainly 3 storage schemes in contemporary WAMS. First, a single machine scheme can provide the easiest and most direct way to store the data. This scheme is usually adopted in a simple client–server paradigm. Its advantage is simplicity since the administrator maintains only one machine. However, its disadvantage it cannot reliably store the data and the failure of the machine can be immediately unaffordable to power companies. Second, a multi-machine scheme provides better reliability by storing the data in a simultaneous, fully redundant manner. However, the scheme brings some other issues. First, the cost to set up a multiple-machine data storage system can be big. Since each machine stores a complete copy of data, it is expected to have advanced configurations to properly handle the data. When the number of machines increases, the cost could be unaffordable for users. Secondly, the fully redundant strategy may be unnecessary and could be a waste of storage resources. Finally, a cluster scheme can more efficiently utilize the storage resources. It utilizes a series of conventional computers, employing a partial overlapping strategy to store massive-volume data. A single machine is referred to as a node in the cluster and it only stores a small chunk of data. Furthermore, a chunk of data is replicated by n times and sent to n node when it is received from the PMU. Figure 14 demonstrates the effectiveness of the cluster scheme. In Fig. 14, the input stream is segmented as 4 data chunks (A, B, C, D) and each data chunk is replicated by 3 times. As is seen, although Node #1 goes offline due to a hardware failure, other nodes can still ensure the integrity of the input stream. The cluster scheme resolves the issues brought by the naïve multi-machine scheme and it is widely adopted in many industries nowadays [36].

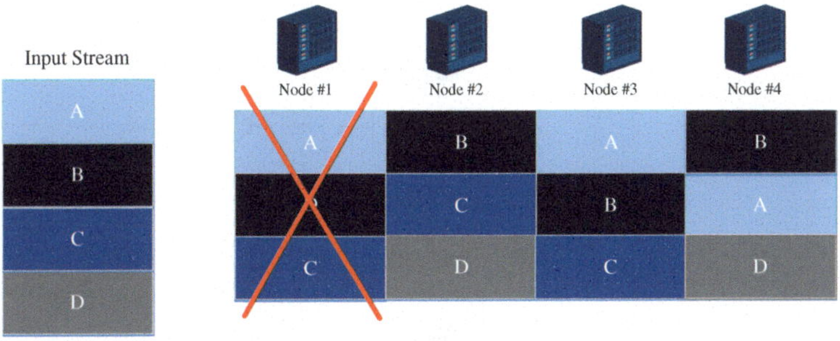

Fig. 14 The cluster data storage scheme

2.3.2 Data Warehousing

A data warehouse is a system that provides functionalities including data reporting, data analysis, and business intelligence. Data warehousing is an integral part of the modern WAMS, and it serves as a fundamental component for monitoring, operation, analysis, and compliance for various entities.

Data Reporting

A data reporting system receives a query request from a user and prepares the requested data. A data reporting system is an important component in a WAII since many operations in the control room are data-driven. In general, data reporting in modern WAMSs can be categorized as online reporting and offline reporting. Online reporting can support various real-time WAMS applications including monitoring, control, analytics, alarms, etc. To support such real-time WAMS applications, an online reporting system is required to have low latency, high availability, and high resiliency, although its throughput can be small. Offline reporting supports other analytical applications including post-event analysis, compliance determination, transmission planning, etc. As opposed to an online reporting system, an offline reporting system is required to have a large throughput to efficiently query large-volume data. However, there are usually low requirements for latency, availably or resiliency.

In modern WAMS, data reporting systems are implemented via various schemes to meet business requirements. An online data query usually requests a small amount of data but the tolerance for latency is low. To meet this requirement, an online data reporting system may utilize a random-access memory (RAM) to enable low-latency data reporting. In this scenario, when the PDC receives the data, it writes the data into a RAM-based cache instead of a read-only member (ROM). When an application files a data query request for real-time measurement data, the data reporting system fetches the requested data from the RAM and sends it out. On the other hand, an offline data query usually involves large-volume, historical measuring data and there

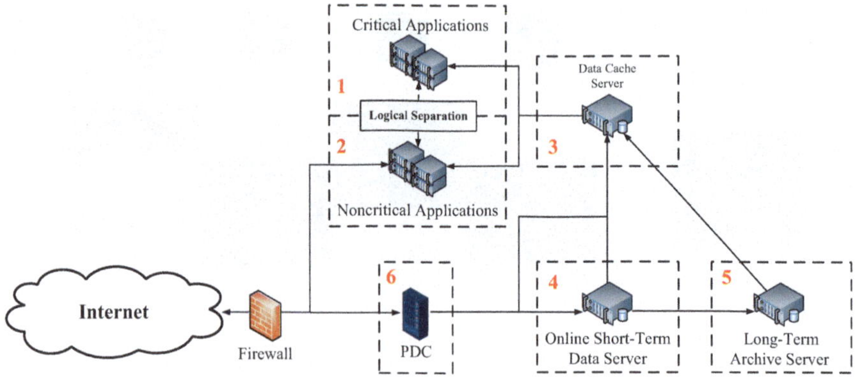

Fig. 15 Control station subnets separation

is usually a loose requirement for latency. In this scenario, the data reporting system directs a data query request to the ROM, where the historical data archives reside. Moreover, solid-state drives (SSD) can be used to replace the conventional hard disk drive (HDD) to improve the query speed, although their cost for large-volume, high-accuracy PMU data can be big [37].

2.3.3 Wide-Area Data Center Security

Network Access Control

Network access control plays an important role in securing information systems, including WAMS. In WAMS, PMUs and some clients must communicate with the control center through the public Internet, which raises the concern regarding security and therefore shall be restricted. Least functionality, the separation of duties and role-based access control (RBAC) [38, 39] shall be considered. For example, PMUs shall have only access to PDC. And the IEEE 1686 standard [40] suggests that the intelligent electronic devices shall be protected with ID/password pairs and have the ability to provide RBAC. The control center LAN can be divided into 6 subnets according to different duties they serve.

- Critical applications subnet
- Noncritical applications subnet
- Data cache subnet
- Real-time data subnet
- Archive data subnet
- Data concentration subnet

Figure 15 shows the connection between subnets. Because subnet 1 usually consists of real-time control and protection applications, it shall be isolated from the Internet and should be logically isolated from subnet 2: communication between subnet 1 and 2 should be blocked with few necessary exceptions. Subnet 1 and 2 can only access data via subnet 3, which improves both security and performance. PMUs on the Internet only push data to subnet 6, then subnet 6 will forward data into subnet 3 and 4 for short-term storage. For long-term storage, subnet 4 will forward data to subnet 5.

For individual components in WAMS, the IEEE 1686 standard [40] suggests that the intelligent electronic devices shall be protected with ID/password pairs and have the ability to provide RBAC.

2.4 Advanced Topic for Information Infrastructure: False Data Injection Attack

In the smart grid, the False Data Injection Attack (FDIA) is applicable to various layers and structures. Generally, these FDIA occur at the physical layer, network layer, and data center [41]. Particularly, the FDIA can manipulate the measurement value without the need to modify the code program. By exploiting communication protocol vulnerabilities or attacking server permissions, false data can directly replace and disturb measurement value stored in the data center.

The FDIA has a wide range of impacts in the following two parts. The first is economic impacts, such as energy theft, resulting in the electrical bill drop. The FDIA attack could impact the topology of the smart grid. Under normal operating conditions, the FDIA can cause erroneous control, leading to a significant loss. The second is stability impacts. By injecting fake measurements, the power grid will produce false responses, causing unstable conditions.

FDIA mainly implements attacks by tampering with data, so it is difficult to rule out the attack using the device operating conditions or data delay. In addition, there are many FDIA attack methods, including ramp attack, scale attack, noise attack, and replacement attack, etc. To remedy this, two types of methods are used including model-based and data-driven detection. In the model-based method, the real-time measurements of the data center are used to model the static and dynamic system parameters and configuration. For example, the Weighted Least Squares (WLS) is utilized to find the system estimated states. However, the WLS is built based on the assumption of stable power system state modeling. Thereafter, some dynamic estimation methods such as distributed and extended Kalman filter, are used to simulate a non-linear system model, which can eventually estimate and detect FDIA at a more accurate level. Some other estimation-free model-based methods have also developed to detect the FDIA. The cooperative vulnerability factor and matrix separation are introduced according to the normal and anomalies power grid under FDIA. Different from the model-based method, the data-driven method does not depend on the model of the power system. Using the characteristics of measurement data, these methods are mainly divided into machine learning and data mining algorithms. Machine learning methods learn the characteristics of the data to determine whether the data has been attacked. For example, the Support Vector Machine (SVM) is one of the most common FDIA detection methods. The normal and attacked data belong to different hyperplanes in SVM so that FDIA can be distinguished. Not only that, the artificial neural networks, K-nearest neighbor, decision tree and random forests can also be used to detect abnormal behavior. However, these machine learning methods have insufficient learning ability on the one hand, and limited ability to process huge data on the other. To eliminate this defect, the deep learning methods such as the Recurrent Neural Networks (RNN), Deep Belief Network (DBN) and Convolutional Neural Network (CNN) provide new workarounds for FDIA detection [42]. Typically, they have better recognition ability and accuracy when facing different FDIA attacks.

The prerequisite for using machine learning is that the difference in attack data is known, namely, which type of FDIA attack method is already clear. The data mining method provides another perspective to deal with the hidden patterns or attributes of false data. Since it's a kind of unsupervised method, it is not necessary to know the label of the data in advance. For example, the Principal Component Analysis (PCA) is used to detect the FDIA using the covariance of the different measurement data. The rest methods include Hoeffding adaptive trees, non-nested generalized exemplars, and common path mining. The data mining method has low computational complexity, so they are especially suitable for big data detection in the data center of WAMS.

3 Development of a Distribution-Level Wide-Area Monitoring System-FNET/GridEye: Infrastructures and Applications

3.1 Introduction

As an advanced technology, a WAMS measures critical electrical quantities, providing the system operators an unprecedented way to monitor and control the electric power systems to meet the challenge brought by low inertia power systems. The PMU is the most important component in WAMS. The PMUs provide high-resolution, high-accuracy, and time-synchronized phasor measurements, which are generally known as synchrophasors. Based on the extraordinary ability of synchrophasor technology, FNET/GridEye, the first distribution-level wide-area phasor measurement system, is developed in 2003. FNET/GridEye is a frequency monitoring network. The FNET/GridEye mainly adopts two types of low-cost and high-accuracy PMU variants, frequency disturbance recorders (FDRs) and universal grid analyzers (UGAs), which are referred as synchronized measurement devices (SMDs), to collect power grid quantities including but not limited to frequency, voltage magnitude, voltage phase angle, harmonics. During the past 17 years,

FNET/GridEye has been helping utilities, balancing authorities (BAs), regional coordinators (RCs), electric reliability organizations (EROs), and the U.S. federal government on many critical aspects including situation awareness, operations, post-event analysis, compliance, etc., and is widely acknowledged by the power industry.

3.2 *FNET/GridEye Communication and Information Infrastructures*

FNET/GridEye is developed as a pilot wide-area phasor measurement system that can cover the national or continental level power grid at a much lower cost before the universal PMU installations can be achieved [43]. The SMDs transmit the collected phasor measurements to two data centers located at the University of Tennessee, Knoxville (UTK), and the Oak Ridge National Lab (ORNL). The FNET/GridEye data center employs a multi-layer architecture, it is designed to receive, process, utilize, and archive a large volume of phasor measurements in real-time [44]. The structure of the FNET/GridEye data center is shown in Fig. 16. As shown in Fig. 16, the FNET/GridEye data center consists of four fundamental layers: data collection layer, real-time analysis layer, data storage layer, and non-real-time analysis layer. In the data collection layer, the FDRs and UGAs collect the phasor measurements data from the power system and compress the data as packages. Then, the FDRs and UGAs send connecting signals to two PDCs through Ethernet via TCP/IP protocol. Once connected, the compressed data package will be as data frames and send to PDCs via standard PMU communication protocols (such as IEEE C37.118.2-2011). For protecting the information security, the Firewall is also configured on the PDCs' server. When the PDCs receive the data frames, the main PDCs will de-compress the data and send it to the real-time analysis layer and data storage layer, where a data cluster is deployed. The real-time analysis layer hosts various FNET/GridEye applications that utilize the field-collected synchrophasor measurement data to monitor the

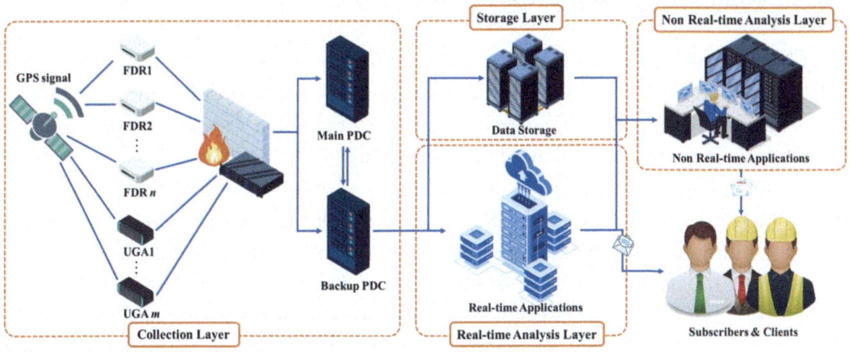

Fig. 16 The structure of the FNET/GridEye data center

operational status of worldwide power grids. Frequency disturbance events including inter-area oscillation, generator trip, and load disconnection, etc. will be detected by the developed real-time application modules. Then the modules send disturbance alerts to the subscribers and clients of FNET/GridEye for warning. On the other hand, the data storage layer archives phasor measurement data streams from the main PDC for offline applications in the non-real-time analysis layer. The non-real-time analysis layer runs an offline application to further investigate the archived data from the real-time analysis layer and the data storage layer. With the increasing deployment of SMDs, FNET/GridEye has evolved its data center towards reliability, availability, and security. The multi-layer structure of the FNET/GridEye data center facilitates the concentrating, processing, and archiving of a large volume of phasor measurements to successfully meet the timeliness requirements of various applications.

3.3 FNET/GridEye Advanced Applications

3.3.1 Real-Time Visualizations for Situation Awareness

The purpose of the FNET/GridEye real-time visualizations is to provide control-room situation awareness tools for industry consortium and cooperative partners. The real-time visualizations query the real-time frequency and phase angle, which are collected from high-resolution SMDs located across the North American continent and the world [45]. Afterward, both data are processed by dedicated algorithms to generate insightful visualizations. Figure 17a, b demonstrate some examples of the real-time visualizations. In general, the FNET/GridEye real-time visualizations use the FNETVision [46], world-wide frequency table display, world-wide frequency map, and U.S. relative angle contour map. In addition, some sample events are also provided to all interested researchers for advancing studies.

3.3.2 Frequency Disturbance Detection

The FNET/GridEye system exploits the real-time frequency and phase angle measurements to detect frequency disturbances [47, 48] and determine their location [49] and magnitude [50]. To locate the source of a frequency disturbance, FNET/GridEye exploits the time delay of arrival (TDOA) characteristics of phase angle data, using a triangulation algorithm to estimate the source of a generation event from the first several PMUs. Furthermore, the estimated event source is further aligned with the location of power plants and pump storage units to help improve the accuracy of disturbance source estimation. Apart from the source location, FNET/GridEye also employs an disturbance magnitude estimation algorithm to determine the size of a generation event. When an event happens, the primary frequency response will be activated to stabilize the frequency and, where the frequency change is proportional to the disturbance magnitude. Accordingly,

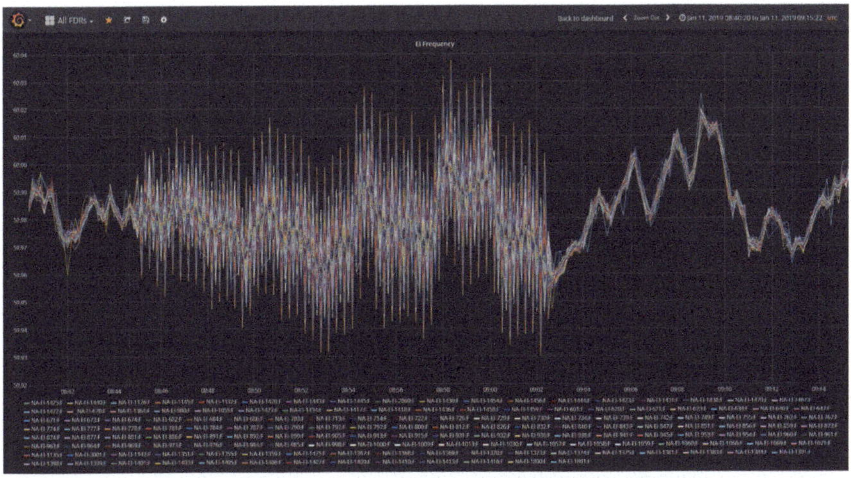

(a) Frequency visualization of a forced oscillation event

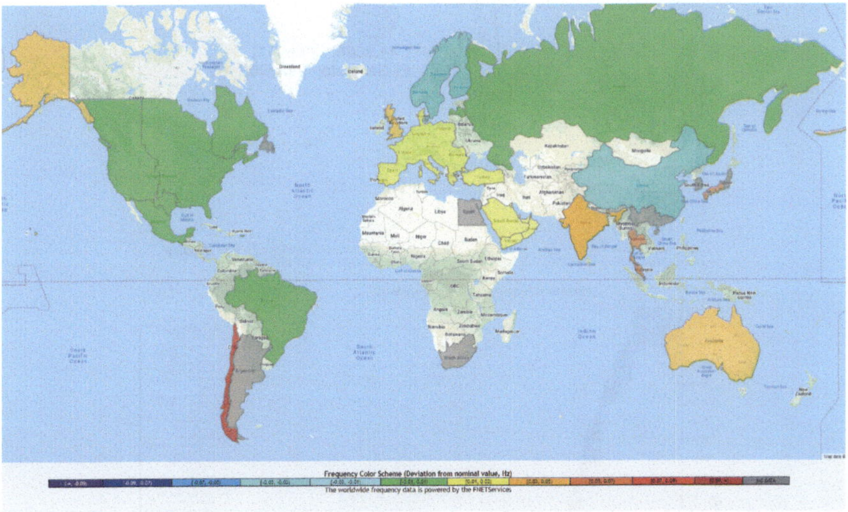

(b) Geological frequency map for world major power grids

Fig. 17 FNET/GridEye real-time visualizations

the magnitude can be calculated with FNET/GridEye frequency measurements. Figure 18 shows an example of a generator trip event captured by the FNET/GridEye system.

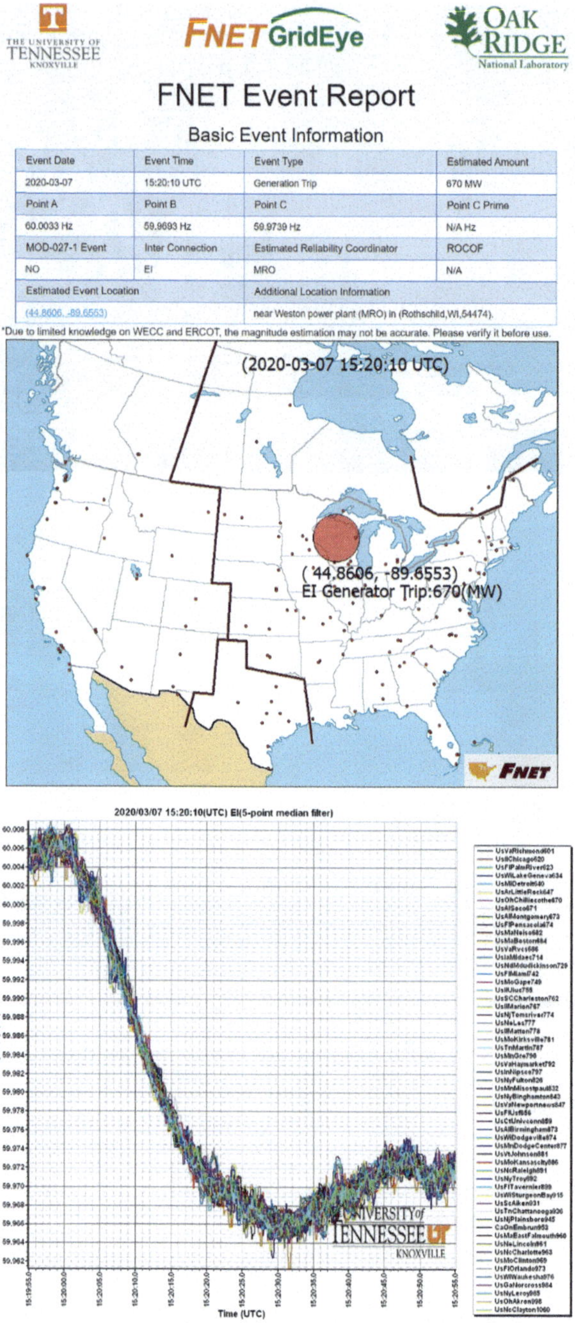

FNET Event Report

Basic Event Information

Event Date	Event Time	Event Type	Estimated Amount
2020-03-07	15:20:10 UTC	Generation Trip	670 MW
Point A	Point B	Point C	Point C Prime
60.0033 Hz	59.9693 Hz	59.9739 Hz	N/A Hz
MOD-027-1 Event	Inter Connection	Estimated Reliability Coordinator	ROCOF
NO	EI	MRO	N/A
Estimated Event Location		Additional Location Information	
(44.8606, -89.6553)		near Weston power plant (MRO) in (Rothschild,WI,54474).	

*Due to limited knowledge on WECC and ERCOT, the magnitude estimation may not be accurate. Please verify it before use.

Fig. 18 FNET/GridEye generation event reports

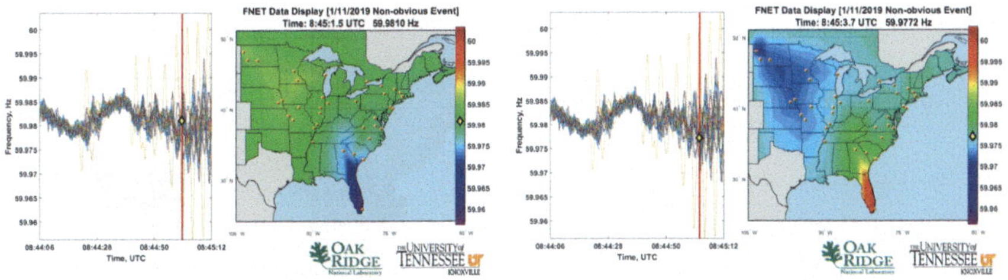

(a) Replay snapshot (event start) (b) Replay snapshot #2 (event propagation)

Fig. 19 Event replay for forced oscillation 01/11/2019

3.3.3 Dynamic Event Replay

An important offline application of FNET/GridEye is to replay power system disturbance events. Taking the forced oscillation event on January 11, 2019 as an example. An oscillation with 0.25 Hz dominant frequency happened in Tampa, Florida area due to a steam turbine control failure, which lasted for 18 min and caused a unit being removed from service by the plant operator. After such an event, FNET/GridEye utilized high-resolution frequency data to generate a video replay to help operators and regulators to investigate the causes and effects of the event, then take actions to avoid similar events thereafter. Figure 19 demonstrates the propagation of the electromechanical wave at the start of the event. As seen, the oscillatory energy evolves from FRCC, then travels to the test of the grid. The event replay function of the FNET/GridEye was also used for the post-event analysis of many other similar events with their replay videos posted on the project website [51] and the YouTube channel [52].

3.3.4 Model-Less Forced Oscillation Source Location

Low frequency forced oscillation is one of the major threats to the security and stability of power systems. Major state-of-the-art forced oscillation source location methods require a known model to locate the source of the forced oscillation [53, 54]. With highly accurate synchrophasor collected from the field, FNET/GridEye employs a data-driven approach to achieve model-less source location for forced oscillation events [55]. Figure 20 demonstrates the observation-time maps of two forced oscillation events. In Fig. 20, regions that are colored with shorter-wavelength colors (purple, blue, etc.) are closer to the oscillatory source, while other regions that are colored with longer-wavelength colors (red, orange, etc.) are farther from the oscillatory source. Finally, an FFT-based algorithm utilizes two-cycle measurements to locate the oscillatory source, which makes it robust to the change of dominant frequency. As Fig. 20 indicates, the FFT-based algorithm successfully locates the oscillatory sources for two events.

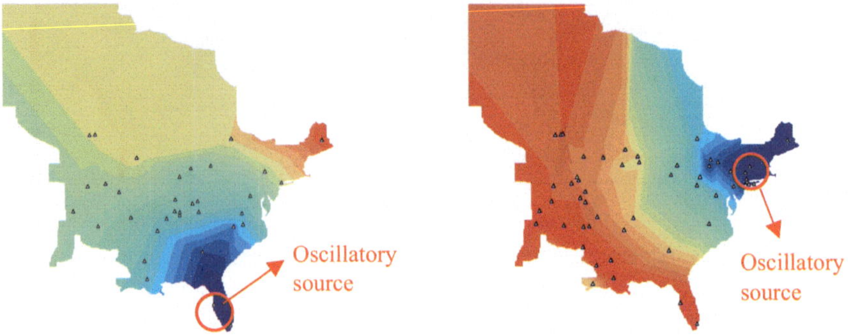

(a) Real forced oscillation (0.25Hz) on 01/11/2019, at Tampa, FL

(b) Simulated forced oscillation (0.5Hz) in ISO-NE, Boston, MA

Fig. 20 Model-less ultra-wide-area forced oscillation source location

3.3.5 Load Control

Since PMUs can provide accurate frequency measurement in real-time and communicate with the control center, they can be used to help use distribution-level resources to improve system reliability. Figure 21a shows the framework of using local frequency measurement to control loads at the distribution level for frequency regulation. A mobile-device-PMU (MDPMU) is connected to the Energy Management Circuit Breaker (EMCB) to selectively trip load when the system frequency is low [56]. The location and amount of load to be tripped is determined in a centralized manner based on the offered price of load response and current generation-load imbalance calculated using the ROCOF value. Figure 21b shows the simulation result of the frequency control performance using FNET/GridEye sensors for distribution load response in a fictitious ERCOT system. It is seen the frequency crosses the under-frequency load shedding (UFLS) at 59.3 Hz when no load control is involved. Other frequency curves are the system frequencies with distribution-level load response using MDPMUs with different data reporting rates.

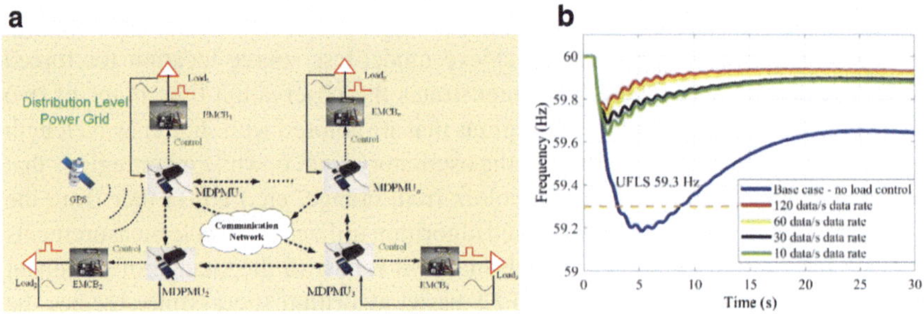

Fig. 21 (**a**) Distribution load response using (**b**) Frequency responses of ERCOT

3.4 Summary

This chapter first goes through a section of wide-area communication topics including communication protocol, method, schemes/architecture. For the wide-area communication, major wide-area communication protocols including IEEE C37.118, IEC61850-90-5, and STTP are explained and compared. Then, communication methods including power line, optical fiber, cellular, microwave, etc. are introduced and compared. Last but not the least, two types of communication schemes are demonstrated. Advanced topics including data compression and cybersecurity are also introduced and some potential applications are envisioned. Then, this chapter introduces some basics on the information infrastructure topics including data center paradigm, data center management, and data center security. This section first illustrates different data center paradigms and how various paradigms work together. Then, it introduces important data center management topics including data storage and warehousing. Different data storage schemes and data reporting schemes are explained and compared. Finally, security issues including the implementation of firewall and subdomains, and how they ensure the security of a WAMS data center are introduced. An advanced topic on false data injection attack (FDIA) is introduced and some effective methods to address FDIA are discussed. Lastly, the authors share their experience on the communication and information infrastructures using the distribution-level wide-area monitoring system, FNET/GridEye as an example. It introduces some key technologies that ensure the efficiency and reliability of such a WAMS. Various applications are demonstrated to illustrate the effectiveness of the FNET/GridEye system.

In modern WAMS, the actual communication and information infrastructures can be heterogeneous and rather complex due to the increasing integration of distributed energy resources [57, 58], the adoption of microgrid [59], etc. In fact, blended wide-area systems bring many new challenges in terms of compatibility, transparency, efficiency, security, etc. Some efforts have been made to establish ultra-wide-area communication and information infrastructures to promote interconnection-wise efficient, secure data communications [60, 61]. WAMS has several vital applications in smart grids to improve its operation, control, stability, and security [62–67]. However, the ever-growing WAMS demand continuous innovation in the communication and information infrastructures to support the need for the next five or ten years.

References

1. IEEE, IEEE Std C37.118-2005. IEEE standard for synchrophasors for power systems (2005)
2. IEEE, IEEE Std C37.118.1-2011. IEEE standard for synchrophasor measurements for power systems (2011)
3. IEEE, IEEE Std C37.118.2-2011. IEEE standard for synchrophasor data transfer for power systems (2011)
4. IEEE, IEEE Std C37.118.1a-2014. IEEE standard for synchrophasor measurements for power systems—Amendment 1: modification of selected performance requirements (2014)

5. IEC, IEC 61850 Communication networks and systems for power utility automation (2013)
6. IEC, IEC 61850-90-5: 2012. Communication networks and systems for power utility automation—Part 90–5: use of IEC 61850 to transmit synchrophasor information according to IEEE C37.118 (2012)
7. IEEE, IEEE Std 802.3-2018. IEEE Standard for Ethernet, pp. 1–5600 (2018)
8. R. Robertson, *Advanced Synchrophasor Protocol (ASP) Development and Demonstration Project* (Department of Energy, United States, 2020), pp. 1–105
9. R. Carroll, F. Robertson, *A Comparison of Phasor Communications Protocols* (No. PNNL-28499). (Pacific Northwest National Lab, Richland, WA, United States, 2019), pp. 1–86
10. I. Ali, M.A. Aftab, S. S. Hussain, Performance comparison of IEC 61850-90-5 and IEEE C37. 118.2 based wide area PMU communication networks. J. Modern Power Syst. Clean Energy **4**(3), 487–495 (2016)
11. I. Edward, Chapter 4—Network Physical Layer Technologies, TCP/IP embedded internet applications (Newnes, Boston, MA, 2003), pp. 69–107
12. R. Zurawski, From wireline to wireless networks and technologies. IEEE Trans. Ind. Informat. **3**(2), 93–94 (2007)
13. B. Appasani, D.K. Mohanta, A review on synchrophasor communication system: communication technologies, standards and applications. Protect. Contr. Modern Power Syst. **3**(2), 37 (2018)
14. S. Kumar, M.K. Soni, D.K. Jain, Requirements and challenges of PMUs communication in WAMS environment. Far East J. Electron. Commun. **13**(2), 121–135 (2014)
15. K. Zhu, J. Song, M. Chenine, L. Nordström, Analysis of phasor data latency in wide area monitoring and control systems, in *Proceedings of 2010 IEEE International Conference on Communications Workshops*, Capetown (2010)
16. C. Callegari, S. Giordano, M. Pagano, T. Pepe, WAVE-CUSUM: Improving CUSUM performance in network anomaly detection by means of wavelet analysis. Comput. Secur. **31**(5), 727–735 (2012)
17. A.G. Phadke, J.S. Thorp, Communication needs for wide area measurement applications, in *Proceedings of 2010 5th International Conference on Critical Infrastructure*, Beijing (2010)
18. A.S. Musleh, G. Chen, Z.Y. Dong, A survey on the detection algorithms for false data injection attacks in smart grids, IEEE Trans. Smart Grid (2019)
19. C.T. Beasley, Electric power synchrophasor network cyber security vulnerabilities. Dissertation. Clemson University, (2014)
20. R. Kateb, P. Akaber, M.H.K. Tushar, A. Albarakati, M. Debbabi, C. Assi, Enhancing WAMS communication network against delay attacks. IEEE Trans. Smart Grid **10**(3), 2738–2751 (2019)
21. R. Klump, P. Agarwal, J. E. Tate, H. Khurana, Lossless compression of synchronized phasor measurements, in *Proceedings of 2010 IEEE PESGM* (2010), pp. 1–7
22. P.H. Gadde, M. Biswal, S. Brahma, H. Cao, Efficient compression of PMU Data in WAMS. IEEE Trans. Smart Grid **7**(5), 2406–2413 (2016)
23. Z. Hu, Y. Li, J. Wu, J. Guo, H. Gu, Research of PMU data transmission mechanism in smart grid based on NDN, in *Proceedings of 2017 IEEE Conference on Energy Internet and Energy System Integration (EI2)* (Beijing, 2017), pp. 1–6
24. M. Asprou, E. Kyriakides, The effect of time-delayed measurements on a PMU-based state estimator, in *Proceedings of 2015 IEEE Eindhoven* (PowerTech, Eindhoven, 2015), pp. 1–6
25. M. Wang, Data quality management of synchrophasor data in power systems by exploiting low-dimensional models, in *Proceedings of 2017 51st Annual Conference on Information Sciences and Systems (CISS)* (Baltimore, MD, 2017), pp. 1–2
26. B. Calabrese, *Data Reduction, Encyclopedia of Bioinformatics and Computational Biology*, (Academic Press, 2019), pp. 480-485
27. Z.M. Lu, S.Z. Guo, Chapter 1—*Introduction, Lossless Information Hiding in Images* (Syngress, 2017), pp. 1–68
28. S.B. Choi, M.H. Lee, High speed pattern matching for a fast Huffman decoder. IEEE Trans. Consum. Electron. **41**(1), 97–103 (1995)

29. A. Balaji, G. Sharma, M. Shaw, R. Guay, Preprocessing methods for improved lossless compression of color look-up tables. J. Imag. Sci. Technol. **52**(4) (2008)
30. J.E. Tate, Preprocessing and Golomb-Rice encoding for lossless compression of phasor angle data. IEEE Trans. Smart Grid **7**(2), 718–729 (2016)
31. T.J. Richardson, R.L. Urbanke, Efficient encoding of low-density parity-check codes. IEEE Trans. Inf. Theory **47**(2), 638–656 (2001)
32. T. Izumi, Y. Iiguni, Data compression of nonlinear time series using a hybrid linear/nonlinear predictor. Signal Process. **86**(9), 2439–2446 (2006)
33. W. Ren, T. Yardley, K. Nahrstedt, ISAAC: intelligent synchrophasor data real-time compression framework for WAMS, in *Proceedings of 2017 IEEE International Conference on Smart Grid Communications* (Dresden, 2017), pp. 430–436
34. Timescale, Time-series data: Why (and how) to use a relational database instead of NoSQL (2020). https://blog.timescale.com/blog/time-series-data-why-and-how-to-use-a-relational-database-instead-of-nosql-d0cd6975e87c/. Accessed 13 Mar 2020
35. A. Armenia, J.H. Chow, A flexible phasor data concentrator design leveraging existing software technologies. IEEE Trans. Smart Grid **1**(1), 73–81 (2010)
36. S. Ghemawat, H. Gobioff, S. Leung, The Google file system, in *Proceedings of the 19th ACM Symposium on Operating Systems Principles* (ACM, Bolto cn Landing, NY, 2003), pp. 20–43
37. I. Kosen, C. Huang, Z. Chen, X. Zhang, L. Min, D. Zhou, L. Zhu, Y. Liu, UPS: Unified PMU-Data storage system to enhance T+D PMU Data usability. IEEE Trans. Smart Grid **11**(1), 739–748 (2020)
38. J.T. Force, T. Initiative, Security and privacy controls for federal information systems and organizations. NIST Spec. Publ. **800**(53), 8–13 (2013)
39. D.F. Ferraiolo, R. Sandhu, S. Gavrila, D.R. Kuhn, R. Chandramouli, Proposed NIST standard for role-based access control. ACM Trans. Inf. Syst. Secur. **4**(3), 224–274 (2001)
40. IEEE, IEEE Std 1686-2013 IEEE Standard for Intelligent Electronic Devices Cyber Security Capabilities (2014), pp. 1–29
41. G. Liang, J. Zhao, F. Luo, S.R. Weller, Z.Y. Dong, A review of false data injection attacks against modern power systems. IEEE Trans. Smart Grid **8**(4), 1630–1638 (2017)
42. W. Qiu, Q. Tang, Y. Wang, L. Zhan, Y. Liu, W. Yao, Multi-view convolutional neural network for data spoofing cyber-attack detection in distribution synchrophasors, IEEE Trans. Smart Grid (2020)
43. Y. Liu, S. You, W. Yao, Y. Cui, L. Wu, D. Zhou, J. Zhao, H. Liu, Y. Liu, A distribution level wide area monitoring system for the electric power grid–FNET/GridEye. IEEE Access **5**, 2329–2338 (2017)
44. D. Zhou, J. Guo, Y. Zhang, J. Chai, H. Liu, Y. Liu, C. Huang, X. Gui, Y. Liu, Distributed data analytics platform for wide-area Synchrophasor measurement systems. IEEE Trans. Smart Grid **7**(5), 2397–2405 (2016)
45. X. Deng, H. Li, W. Yu, W. Wang, Y. Liu, Frequency observations and statistic analysis of worldwide main power grids using FNET/GridEye, in *Proceedings of 2019 IEEE PESGM* (IEEE, Atlanta, GA, 2019), pp. 1–5
46. W. Wang, J. Zhao, W. Yu, Y. Liu, FNETVision: A WAMS big data knowledge discovery system, in *Proceedings of 2018 IEEE PESGM* (IEEE, Portland, OR, 2018), pp. 1–5
47. W. Wang, H. Yin, C. Chen, A. Till, W. Yao, X. Deng, Y. Liu, Frequency disturbance event detection based on synchrophasors and deep learning. IEEE Trans. Smart Grid (2020)
48. W. Wang, W. Yao, C. Chen, X. Deng, Y. Liu, Fast and accurate frequency response estimation for large power system disturbances using second derivative of frequency data. IEEE Trans. Power Syst. (2020)
49. R. Gardner, J. Bank, J. Wang, A. Arana, Y. Liu, Power system event location analysis using wide-area measurements, in *Proceedings of 2006 IEEE PESGM* (IEEE, Montreal, QB, 2006), pp. 1–7
50. Y. Zhang, X. Deng, S. You, J. Dong, W. Yu, Y. Liu, Measurement-driven disturbance magnitude estimations for bulk power systems, in *Proceedings of 2019 IEEE PES GTD Grand International Conference and Exposition Asia* (2019)

51. PowerIT Group, FNET/GridEye Web Display (Sample Events) (2020). https://fnetpublic.utk.edu/sample_events.html. Accessed 11 Mar 2020
52. YouTube, FNET/GridEye UTK ORNL (2020). https://www.youtube.com/channel/UC40n2KTjwRhC9_CvtIasaWA, Accessed 11 Mar 2020
53. R. Jha, N. Senroy, Forced oscillation source location in power systems using system dissipating energy. IET Smart Grid **2**(4), 514–521 (2019)
54. S. Feng, B. Zheng, P. Jiang, J. Lei, A Two-level forced oscillations source location method based on phasor and energy analysis. IEEE Access **6**, 44318–44327
55. W. Wang, C. Chen, L. Zhu, W. Qiu, K. Sun, Y. Liu, Model-less source location for forced oscillation based on synchrophasor and moving fast fourier transformation, in *Proceedings of 2020 IEEE PES ISGT-Europe* (IEEE, Hague, The Netherlands, 2020)
56. W. Yao, S. You, W. Wang, X. Deng, Y. Li, L. Zhan, Y. Liu, A fast load control system based on mobile distribution-level phasor measurement unit. IEEE Trans. Smart Grid **11**(1), 895–904 (2020)
57. T. Li, Y. Li, M. Liao, W. Wang, C. Zeng, A new wind power forecasting approach based on conjugated gradient neural network. Math. Probl. Eng. **2016**, 1–8 (2016)
58. W. Wang, K. Sun, C. Chen, W. Qiu, H. Yin, S. You, W. Yao, J. Dong, C. Zeng, X. Deng, Y. Liu, Advanced synchrophasor-based application for potential distributed energy resources management: key technology, challenge and vision, in *Proceedings of 2020 IAS I&CPS Asia* (IEEE, Weihai, China, 2020)
59. S. Parhizi, H. Lotfi, A. Khodaei, S. Bahramirad, State of the art in research on microgrids: a review. IEEE Access **3**, 890–925 (2015)
60. EIDSN, Our Mission (2020). https://eidsn.org. Accessed 11 Mar 2020
61. WECC Data Exchange Work Group, Guideline for WECC operations network node responsibilities and procedures, WECC (2012)
62. H.H. Alhelou, M.H. Golshan, J. Askari-Marnani, Robust sensor fault detection and isolation scheme for interconnected smart power systems in presence of RER and EVs using unknown input observer. Int. J. Electr. Power Energy Syst. **99**, 682–694 (2018)
63. H.H. Alhelou, M.E. Hamedani-Golshan, R. Zamani, E. Heydarian-Forushani, P. Siano, Challenges and opportunities of load frequency control in conventional, modern and future smart power systems: a comprehensive review. Energies **11**(10), 2497 (2018)
64. H. Haes Alhelou, M.E. Hamedani Golshan, M. Hajiakbari Fini, Wind driven optimization algorithm application to load frequency control in interconnected power systems considering GRC and GDB nonlinearities. Electric Power Compon. Syst. **46**(11–12), 1223–1238 (2018)
65. H. Haes Alhelou, M.E. Hamedani-Golshan, T.C. Njenda, P. Siano, A survey on power system blackout and cascading events: research motivations and challenges. Energies **12**(4), 682 (2019)
66. H.H. Alhelou, M.E.H. Golshan, N.D. Hatziargyriou, A decentralized functional observer based optimal LFC considering unknown inputs, uncertainties, and cyber-attacks. IEEE Trans. Power Syst. **34**(6), 4408–4417 (2019)
67. H.H. Alhelou, M.E.H. Golshan, N.D. Hatziargyriou, Deterministic dynamic state estimation-based optimal lfc for interconnected power systems using unknown input observer. IEEE Trans. Smart Grid (2019)
68. H.H. Alhelou, S.J. Mirjalili, R. Zamani, P. Siano, Assessing the optimal generation technology mix determination considering demand response and EVs. Int. J. Electr. Power Energy Syst. **119**, 105871 (2020)

Wide-Area Measurement Systems and Phasor Measurement Units

M. Maheswari, N. Suthanthira Vanitha, and N. Loganathan

Abstract Wide Area Measurement Systems (WAMS) is a collective technology to monitor power system dynamics in real time, identify system stability related weakness and helps to design and implement counter measures. It uses a global positioning system(GPS) satellite signal to time synchronize from phasor measurement units (PMUs) at important nodes in the power system, sends real-time phasor (angle and magnitude) data to a Control Centre. The acquired phasor data provide dynamic information on power systems, which help operators to initiate corrective actions to enhance the power system reliability. The goals of WAMS are real time monitoring, post disturbance analysis, adaptive protection and power system restoration. The major components of WAMS are Phasor Measurement Unit (PMU), Phasor Data Concentrator (PDC), Global Positioning System (GPS for Time Synchronization of the phasors), Communication channel (Preferably optical fiber cable), Visualization and analysis tools, Wide area situational awareness system and Wide area protection and control. This chapter is going to discuss about the goals and benefits of using PMUs, comparison between PMUs and SCADA system, Detailed description of WAMS components, synchronized PMUS, different kind of applications of WAMS in power sector, Components and operation of PMUS, real time examples of WAMS in power system operation and control.

Keywords PMUs · Wide-area monitoring systems · GPS · Phasor Data Concentrator · SCADA · Smart grid · Monitoring systems

---

M. Maheswari (✉) · N. Loganathan
Engineering Department, Nizwa College of Technology, Nizwa, Sultanate of Oman
e-mail: maheswarikgd@gmail.com

N. Loganathan
e-mail: loganathan.nachimuthu@nct.edu.om

N. Suthanthira Vanitha
Department of EEE, Muthayammal Engineering College (Autonomous), Rasipuram, Tamilnadu, India
e-mail: varmans03@gmail.com

H. Haes Alhelou et al. (eds.), *Wide Area Power Systems Stability, Protection, and Security*, Power Systems, https://doi.org/10.1007/978-3-030-54275-7_4

1 Introduction

The objectives of the power sector are good quality, high stability and cheap. In conventional power system, there are lot of possibilities to switch to instability mode. It's mainly due to the occurrence faults and the outages [1]. The stability of the conventional power system is improved by means of interconnection. Hence the modern power system interconnects the regional electric networks such as interconnection of Europe networks (UCTE) [2, 3], the japan power grid to improve the stability of the power system [4, 5]. Interconnected power system effectively utilize the distributed resources from different locations and accomplish the optimal scheduling of energy sources. In addition to this, it provides support during fault conditions which improves the reliability of the whole power system [6]. But low frequency oscillations (LFOs) are excited quickly when the faults or disturbance exists in the interconnected systems. There is no predefined characteristics for these oscillations in each system.

2 Stability Problems of Interconnected Systems

The conventional power systems are interconnected to optimize the generation of distributed energy sources from different areas, to augment the global optimal scheduling for power utilization, also to elevate the power economic dispatch. Apart from this, it provides the mutual power support from one area to another area which improves reliable and security of large scale power systems. But the developments happened in the grid structure and operating mode of interconnected systems creates stability problems due to its complexity. The stability of power system can be classified into three classifications as mentioned by IEEE guide [7], namely rotor angle stability, frequency stability and voltage stability as shown in Fig. 1. The ability of the power system to sustain steady frequency followed by a severe system upset due to the imbalance between generation and load is known as frequency stability. It

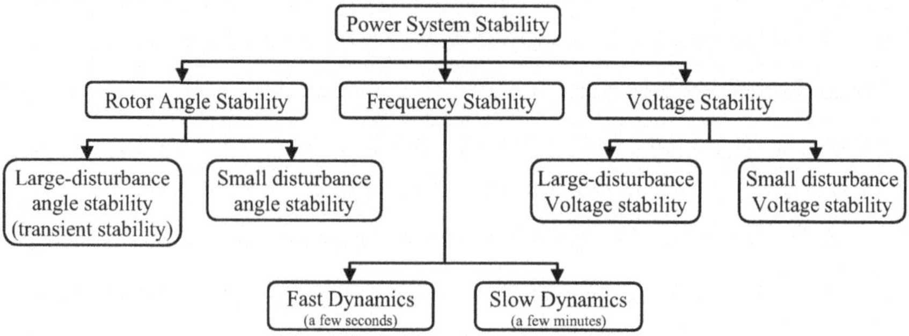

Fig. 1 Classification of power system stability

is determined by the ability to restore the equilibrium between the generation and load without much affecting the load. The constant frequency swings leading to the instability of the system by tripping of generating units and loads. Voltage stability refers the ability of the power system to maintain the constant voltage at all buses even after it is subjected to a disturbance. It is influenced by the ability to maintain the equilibrium between the demand and the supply.

The continuous voltage fall or raise in buses will lead to instability of the system. The ability of the synchronous machines of an interconnected power system to maintain in synchronism after it is subjected to a disturbance is called as rotor angle stability. It is governed by the capability of maintaining equilibrium between electromagnetic torque and mechanical torque of synchronous machine in the system. When there is an increase in the angular swings of few generators will lead to the loss of synchronism with other generators, it leads to the instability of the system [8]. The rotor angle stability is reduced by means of proper damping arrangements. In case of insufficient damping, these oscillations will start to spread to the transmission facilities and to the tie lines also. The frequency of power oscillations in the interconnected systems lies between 0.2 and 3 Hz and it is called as low frequency oscillations (LFO) [9]. It may be represented as fast dynamics with the duration of a few seconds in the absence of effective oscillation damping control.

In interconnected systems, LFO is an existing rotor angle stability problem. In case of IAO, the oscillation frequency lies between 0.2 and 0.8 Hz and it leads to the threatening to the reliability operation, reduction in the transmission power capacity and limiting the ability of interconnection between large electric networks. The voltage swings may also be created by the undamped oscillations in some buses and also leads to the frequency stability.

3 WAMS Technology

The concept of wide area measurement has been introduced by Bonneville Power Administration in the late 1980s. WAMS has made up of advanced measurement technology, information tools and operational infrastructure to manage the complexity in the behavior of large power system. It can be used as standalone infrastructure or as a complementary system. The first one will replace the function of conventional SCADA system and the later one designed to improve the operator's real time situational awareness to improve the performance of the grid operation. WAMS is the combination of synchronized phasor measurement unit, communication engineering and information technology in electric power systems. The objective of WAMS is to realize dynamic monitoring, analysis and control for the stability and the efficient operation of the global power system. WAMS is mainly applied in the two fields namely, smart grid construction and stability analysis and control based on wide area measurements. The major components of WAMS are

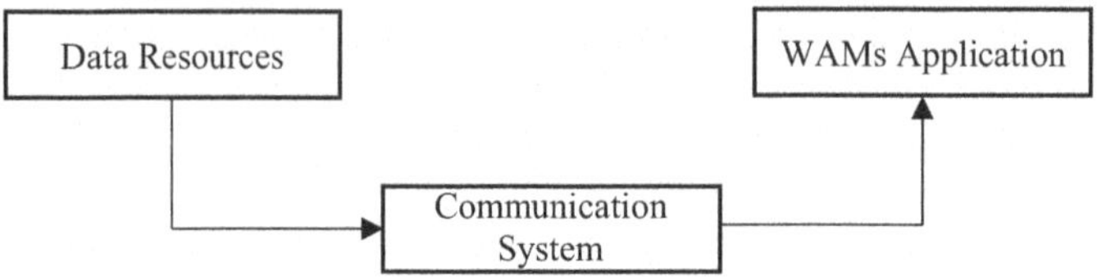

Fig. 2 WAMs elementary process

- Phasor Measurement Units (PMUs) placed at different areas for measuring local variables like voltage and current
- A monitoring and control system located in the power system dispatch center
- A digital communication network for information exchange

The application of WAMS in the power system, used to implement online monitoring of remote operating variables. These variables are used to disclose the dynamic behavior of the power systems for energy management and decision making of transmission system operators (TSOs). The concept of wide area measurement has been introduced by Bonneville Power Administration in the late 1980s. WAMS has made up of advanced measurement technology, information tools and operational infrastructure to manage the complexity in the behavior of large power system. It can be used as standalone infrastructure or as a complementary system. The first one will replace the function of conventional SCADA system and the later one designed to improve the operator's real time situational awareness to improve the performance of the grid operation. There are three sub processes which are interconnected in the WAMS, namely, data acquisition, data transmission and data processing. Each process is performed by measurement systems, communication systems and energy management systems respectively [10, 11]. It accumulates the data from the conventional or advanced measurement devices and transmit it through proper communication systems to the control centers. The WAMS elementary process is shown in Fig. 2.

3.1 Data Resources

The data resources are classified into two categories based on its nature such as operational data and non-operational data. The data which is transmitted by the installed device continuously to the control centers such as voltage, current and breaker status. If the data is transmitted once in a while during the events then it is called as nonoperational data like faults, lightning strokes and power fluctuations etc., The data resource in the power system can be classified into operational and non-operational similar to WAMs. The operational data includes supervisory control and data acquisition, synchronized phasor measurement system. Circuit breaker monitor, digital fault recorder and digital protective relays are coming under non-operational data resources [12].

3.1.1 Supervisory Control and Data Acquisition (SCADA)

SCADA is used to collect and process the data over a long distance through the application of operational controls and it is completely assisted by the computer. There are three major functions are executed by SCADA viz, data acquisition, supervisory control and alarm display and control. These functions are carried out by the hardware devices namely, Master terminal unit (MTU), remote terminal unit (RTU) and communication channel. MTU is located in the control centers, whereas RTU is located in the remote field site and it may be programmable logic controllers (PLC) or intelligent electronic devices (IED) and the communication channel connecting both remote site and the control center.

MTU acts as the heart of the SCADA system and installed at the control centers. The major roles played by MTU are,

- Manage all communications
- Gathers data of RTU
- Stores obtained data and information
- Send information to other systems
- Commands the actuators connected with the RTU and
- Interfaces with operators.

RTU is used for monitors and controls the equipment through the data acquisition and control at the remote sites and send the collected data to MTU. It may be microprocessor based devices and can be used as a relay. The size of RTU is named as small, medium and large, based on the number of analog/digital inputs. PLC can be used to realize the operations of electrical equipment such as relays, drum switches and mechanical timer/counters at the remote sites. PLC is a small computer which is suitable for industries because it is more economical, versatile, flexible and compact than RTU. The communication system provides the channel between the master station and the remote sites through fiber optic or wireless or satellite. SCADA employ the physical communication architectures such as point to point, multi point and relay station architecture. The structure of SCADA architecture is shown in Fig. 3.

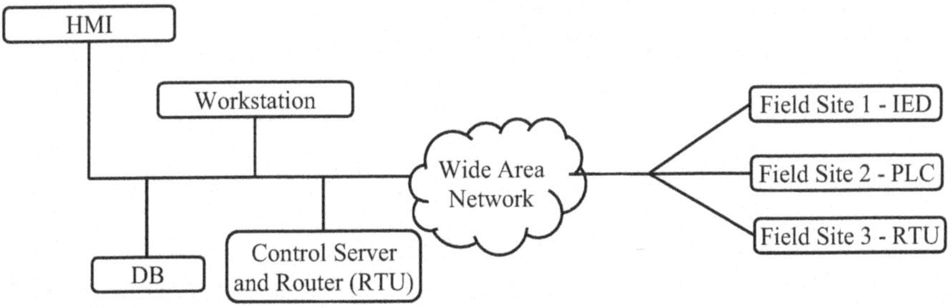

Fig. 3 Structure of SCADA architecture

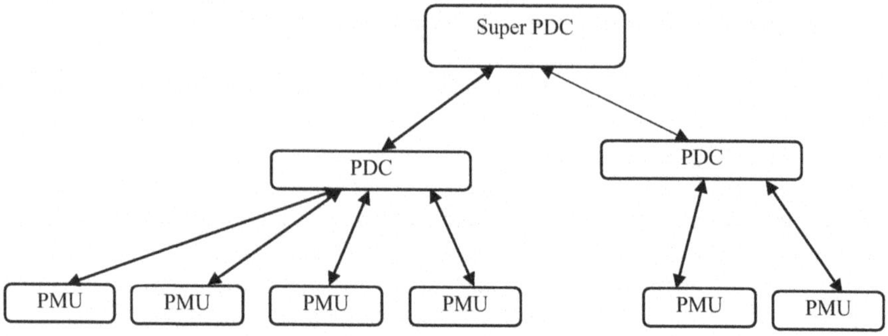

Fig. 4 Structure of synchronized phasor measurement system

3.1.2 Synchronized Phasor Measurement System (SPMS)

SPMS is an advanced device employs sample clock of common time source like GPS and is used to measure currents, voltage and angle between them. Apart from these, it can be customized to measure the parameters like local frequency, rates of frequency changes, harmonics measurements, negative and zero sequence quantities [13]. Phasor measurement unit (PMU), phasor data concentrator (PDC) and communication system are the three main parts of SPMS. PMU is used to collect the electrical signals like voltage and current at the rate of 48 samples per cycle. PMU is a kind of microprocessor based device. The output of PMU is a time-tagged phasors also called as synchro phasor. Synchro phasor is transmitted to the local or remote receiver at the rate of 60 samples per second. PMU plays the role of RTU in a SCADA system.

PDC plays a very significant role in SPMS. It collects the synchro phasors from several PMU or PDC and fed out them as a single stream of data after several checks on data consistency. Rejection of bad data, aligning the time stamps and creating the coherent record of continuous recorded data. There are two major distinguish between the SCADA and WAMs in case of streamed data. Synchro phasor data is continuous and streaming in nature whenever it transmits the data to the master station either in specified time intervals or required by master station and synchro phasor data has more sensible than the data provided by RTU which requires high bandwidth, low latency communication systems. The structure of SPMS is shown in Fig. 4.

3.1.3 Digital Fault Recorder

Digital fault recorder is used to record the waveforms accurately during fault conditions. It collects huge amount of data such as analog or status data in the different stages of faults like pre fault, fault and post fault conditions [14]. Maximum current, fault sequence, type of faults and the circuit breaker operation sequence are some

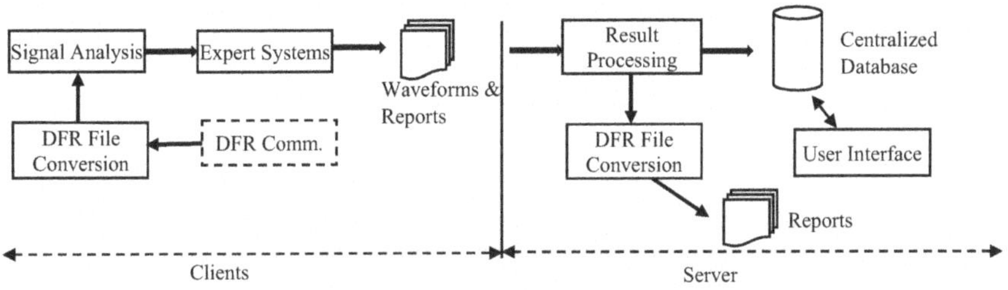

Fig. 5 Digital fault recorder software modules

examples of data collected. These collected data are sampled at the very high rate of 64 to 356 samples per cycle and it cannot be used in real time applications, hence it is stored as samples for offline processing. The above said task is achieved by using SPMS directly. The software module for digital fault recorder is shown in Fig. 5 [15].

3.1.4 Digital Protective Relay

The most important thing in power system is to isolate the faulty area from the healthy one to minimize the impact. This kind of functions are taken care by protective relays, these are controlled by digital signal recently and are known as digital protective relays (DPRs). It is a microprocessor based device and is capable of detecting faults by means of current and voltage waveforms. It is also enabled to measure and record analog and status data which is communicated to the centralized location. The sampling rates of DPR is quite low and varies from 64 to 128 samples per cycle. Obviously this sampling rate reports lower accuracy of DPR data compared with the other data resources. The digital fault relay software module is shown in Fig. 6 [15].

3.1.5 Circuit Breaker Monitor (CBM)

CBM is an electronic device used to monitor the status of circuit breakers. It is designed in such a way that it works in real time and capturing information about each breaker for its proper operation either by manually or automatically. The Circuit Breaker Monitor Analysis Software Module is shown in Fig. 7 [15].

3.2 WAMS Communication Systems

Communication system perform a vital role in power system operation and control because it enhances the security of the system. Ensuring the data delivery both from data resources to the control centers and from the control center to the system actuators are the two major task performed by the communication system in WAMS.

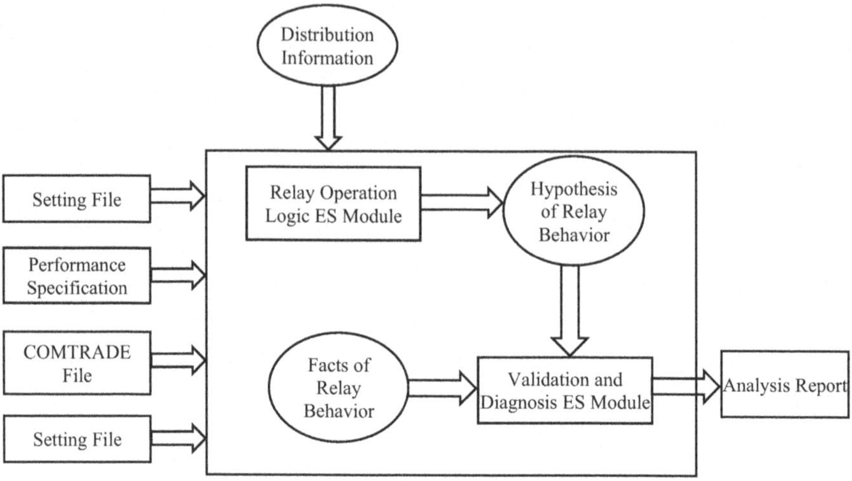

Fig. 6 Software module of digital fault relay

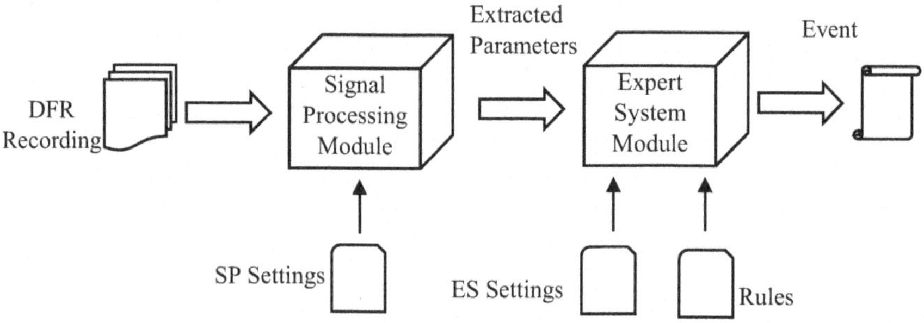

Fig. 7 Circuit breaker monitor analysis software modules

A new developing model in communication system is open system interconnection (OSI) layer. It deals about the comprehensive architecture for explanation, designing, implementation, standardization and use of communication networks. There are seven layers are present in the OSI reference model namely physical, data link, network, transport, session, presentation and application layer as shown in Fig. 8. It also emphasis the link between the OSI layers and data resources, applications and communication systems.

The performance of WAMS communication systems are based on the characteristics of the transmission media such as the cost, bandwidth, propagation delay, security and reliability. Transmission data can be classified into two kinds such as guided and unguided [16]. If the information is transferred through a solid medium such as optical fiber, twisted cable pair, coaxial cable and power transmission lines then it is known as guided transmission media. The transmission of information is done by electromagnetic waves like atmosphere and outer space then it is called as

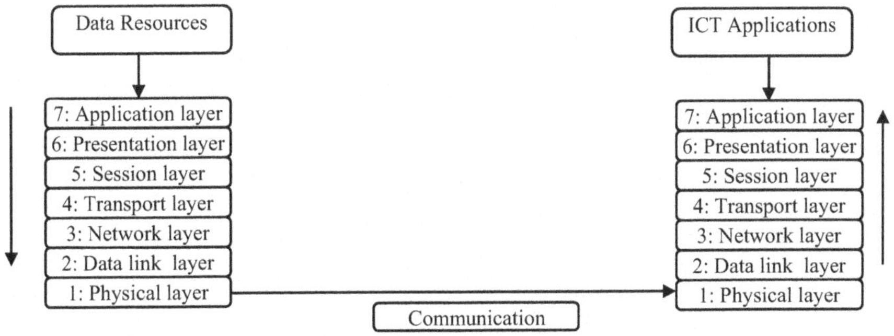

Fig. 8 Layers in WAMS based OSI reference model

unguided transmission media. The comparison between the characteristics of most common media used in WAMS both in guided and unguided media is listed in Tables 1 and 2 [17].

Table 1 Parameters comparison between different types of guided media

Type of Media	Description	Bandwidth	Latency	Security
Optical fiber	It is a flexible, transparent fiber made by glass or plastic of thickness smaller than human hair. It transmits the light between two ends with wider bandwidth for long distance. It employs visible light or infra-red rays for transmission. Optical power ground wire and all dielectric self-supporting types are mostly used in industries	High	Low	High
Power line carrier	Power lines are used for transmission medium and may cause issues while the line outages. It is classified into two types as narrow band and broad band as per data rate values	Medium	Low	High
Leased line	LL means using some other communication lines for transmitting wide area signals	Medium	Low - Medium	High

Table 2 Parameters comparison between different types of unguided media

Type of Media	Description	Bandwidth	Latency	Security
WPAN	Wireless Personal Area Network connects the device around a person over a very short distance i.e., 10 m through wireless mode. One such frequently used technology is Bluetooth	Low-Medium	Low-Medium	Low
WLAN	Wireless Local Area Network connects the devices around through wireless distribution method. The most frequently adopted WLAN is Wi-Fi	Low-Medium	Medium	Low
WMAN	WiMAX, GPRS, GSM, CDMA and 3G mobile carrier services are the frequently used techniques in Wireless Metropolitan area Network	Medium	Medium	Low
WWAN	It is Wireless Wide Area Network using satellite communications. It is employed under two situations like, when a guided medium is not possible between the remote site and the control center and there is no line of sight between the remote site and pre-installed communication network	Low-Medium	High	Low

3.3 WAMS Applications

The software based tools which are used to process the raw data from data resources to extract the useful information for the power system operators are called as WAMS Applications. In general, the WAMS applications are specific as generation, transmission and distribution. In case of generation applications, it is used to control the generator operation in real time and also monitoring the stability condition of the generators. The group of computer aided tools which is used to control and monitor transmission and sub transmission system is called as EMS. State estimation, load flow, optimal power flow, load forecast and economical dispatch are the few conventional EMS applications. Apart from this, the application of WAMS is extended to

- Integrated phasor data platform
- Wide area dynamic monitoring and analysis
- Power stability prediction and alarming
- Fault analysis support
- Comprehensive system load monitoring
- Power system restoration support tool.

In distribution side, systems that enable a distribution company to monitor, coordinate and operate distribution components from the remote area in real time. It is termed as automation applications as per IEEE community. The major objective of this idea is to reduce the cost, enhance the service availability and better services to the customers. It is again sub divided into substation automation, feeder automation and consumer side automation.

WAMS applications also categorized further into two kinds in general. The application of WAMS in measurement of quantities on only their own, area, region or specific TSO becomes the first category. WAMS is used to measure the quantities from neighboring national WAMS becomes the second kind. All these applications may use online or off line to collect the information about the power system.

4 Phasor Measurement Units (PMUs)

4.1 Historical Development of PMUs

Power system engineers are interested to find the phase angles of voltage phasors of power network buses because the real power flow in the power line is proportional to the sine of the angle between voltages at the two terminals of the line. The real power and the phasor angle between the voltages are the two parameters which takes major role in power system planning and operation. The use of modern application in the measurement of phase angle difference [18, 19]. They have used LORAN-C, GOES satellite transmissions and the HBG radio transmissions to obtain synchronization of reference time at different locations in power system. In general, the immediate next positive going zero crossing of a phase voltage was used to determine the local phase angle with respect to the time reference. By finding the difference between the measured phase angles of voltages on a common reference at two locations was determined. The accuracy of the measured values were achieved in the order of 40 μs. These measurement techniques are not suitable for the implementation of wide area measurement system in power system.

The origin of present era of phasor measurement technology has started when the research on computer relaying of transmission lines has started. Before these works, the above said research was conducted by microprocessor and its capacity was not sufficient to carry out the calculations. In early days, the fault in the transmission lines were identified by solving six fault loop equations at each instant. The research was carried out to eliminate the solving of six loop equations for finding the fault and derived a new technique based on symmetrical component analysis of line voltages and currents. In the above said method, it is possible to identify the types of fault and all types of fault calculation from the single equation. A new symmetrical component based algorithm for finding symmetrical components of three phase voltage and currents [20]. This paper played a vital role in defining the algorithm and methods to calculate symmetrical components and positive sequence voltages

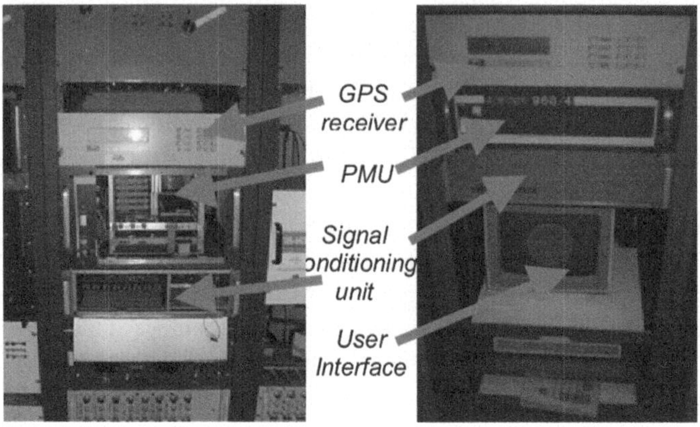

Fig. 9 The first phasor measurement units (PMUs) built at the Power Systems Research Laboratory at Virginia Tech [13]

and currents. Later the research was recognized that the measurement of positive sequence components in power system is of great value and it is the fundamental importance. The importance of positive sequence voltages and currents and its uses and it acts as starting point of synchronized phasor measurement technology [21].

The Global Positioning system (GPS) has been used to synchronize power system measurements over a long distance. The first prototype model of PMU using GPS was built by Virginia Tech in the year 1980 and the prototype has shown in Fig. 9. The first commercial product of PMU was started by Macrodyne in association with Virginia Tech [22]. Nowadays there are lot of manufacturer producing PMUs worldwide. IEEE recommended the standards for format of data files created and transmitted by PMUs in the year 1991 and also the same was reviewed in the year 2005 [23, 24]. There are lot of research works are carried in the worldwide regarding the applications of PMUs.

4.2 Introduction

The PMU historical development has been discussed in Sect. 4.1 in detail. This section aims to cover the aspects of the PMUs regarding its architecture and management system in practical implementation. The measurements made by PMUs are timestamped at the source and there is no issues of transmission speed. The PMU measurements are used to identify the state of the power system at each instant because of its time stamp. The data from PMUs will reach the central location based on the delay in the communication channel used. But the time tags associated with each measurement will help the operators to get the clear picture of the power system operation. This section address about the GPS method suitable for PMU measurements, PDCs, Communication systems and the file structures for PMUS to ensure the interoperability of the power system.

4.3 Block Diagram of PMU

The basic block diagram of PMU based on the first module developed by Virginia Tech is shown in Fig. 10. The basic difference between the first model and the new ones are the symmetrical component relays are replaced by computer relays. The analog inputs like voltages and currents are obtained from the secondary of the potential and current transformers respectively. In general, three phase voltages and currents are measured so that it is easy to determine the positive sequence measurements. PMU collects the currents from the several feeders and voltages of the various buses in the substation. The measured values of currents and voltages are converted into voltages usually in the range of ± 10 V so as to match with the requirement of the analog to digital converters. The frequency response of the anti-aliasing filters are decided by the sampling rate chosen. The cut-off frequency of analog filters always chosen less than the sampling frequency to satisfy the Nyquist criterion. In relay design process, the sampling rate is chosen high than its high cut off frequency of the analog anti-aliasing filters. This stage is followed by a digital decimation filter. It converts the sampled data to a lower sampling rate and this combination provides the concatenation of digital anti aliasing filter with the analog anti-aliasing filters. This will enhance the stable operation of the filters under aging and temperature variation conditions. This guarantees that all analog signals have the same phase shift and attenuation, hence the phase angle differences and relative magnitudes of the different signals are unchanged. If the raw data from samples of analog signal are stored, then it will be very much useful as high bandwidth digital fault recorders. It is an added advantage of using oversampling technique.

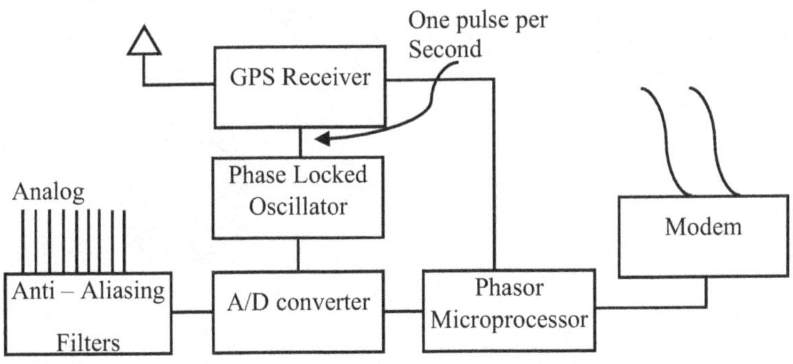

Fig. 10 Block diagram of PMU

The sampling clock is synchronized with the GPS clock pulse through phase locked loop. The sampling rates are started with 12 samples per cycle of the nominal frequency in the early days and it has improved to 96 or 128 samples per cycle due to faster analog to digital converters and processors in the modern PMUs. The sampling rate will improve in the future so as to improve the estimation accuracy further [13]. The positive sequence estimates are calculated by microprocessors for all currents and voltages using the frequency detection techniques. The other techniques are determination of frequency and rate of change of frequency measured locally and included in the output of PMU. Then the time stamp signal is formed from the two of the signals derived from the GPS receiver and it will help to identify the universal time coordinated (UTC) second and the instant defining the boundary of the power frequency as prescribed by IEEE standard. The output of PMU will be time stamped and it is transferred to the suitable modems in proper file structure through proper communication channels.

4.4 Global Positioning System (GPS)

In the year 1978, US department of defense originated the GPS by using the launch of first Block I satellites. There were 24 modern satellites was put in place in the year 1994. These are arranged in six orbital planes displaced from each other by 60° and inclined about 55° with respect to equatorial plane as shown in Fig. 11. The orbital radius of the satellites are 16,500 miles and they go around the earth twice per day. It is arranged in such a way that at least six satellites has to be visible from the earth at a time but there are 10 satellites are visible as of now. The major function of the GPS is to find out the coordinates of the receiver and finding the time of pulse per second in PMUs. The pulse received from the receiver will coincide with all other

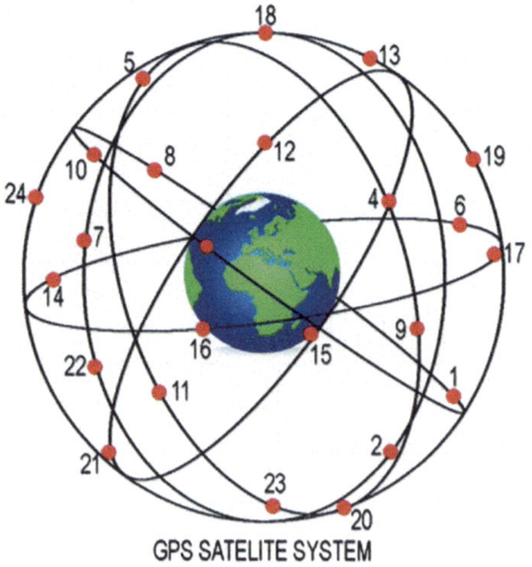

Fig. 11 GPS satellite disposition

received pulses in the earth within 1 microsecond. Nowadays the synchronization is achieved within few hundred nano seconds.

The accurate clocks available in the GPS will provide one pulse per second signal. The time they use is known as GPS time and it will not account the earth's rotation. The corrections in the received signal is done at the receivers to provide UTC clock time. PMU standard uses UNIX time base with a second of century (SOC) counter which began to count at midnight on January 1, 1970. Nowadays there are lot of GPS systems are deployed by other nations also for the same objective. GPS system will play a major role in the synchronization for PMUs in the future also.

4.5 *Hierarchy for Phasor Measurement Systems*

In power system, the PMUs are installed in the substations and it depends on the provisions made for measurements in that location. The measured phasor data is used in the remote locations from the PMUs. Hence there should be some predefined architecture has to be followed to realize the full benefit of the PMU measurement system. The most commonly used architecture of the PMU measurement system is shown in Fig. 12. PMUs are used to measure time stamped positive sequence voltages, currents, frequency and rate of change of frequency of all buses and feeders in the substation. The measured data are stored in the local storage devices and can be retrieved from remote locations for diagnostic purpose. The storage capacity of the local storage is always limited and the data belongs to important events needs to be stored in the permanent storage to avoid overwritten. The measured phasor data will be immediately available for real time applications either in locally or higher level operations.

In the hierarchy of PMUs measurement, the next device is Phasor Data Concentrator (PDCs). The major function of the PDCs is to collect data from many PMUs, discard bad data, line up the time stamps and make a record of data from wider part of

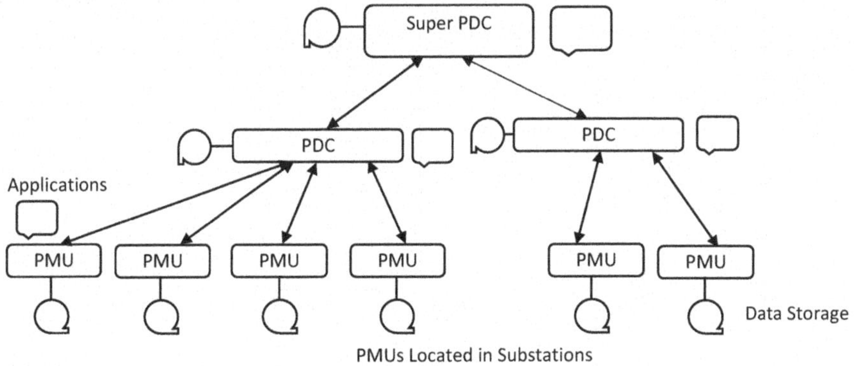

Fig. 12 Hierarchy of Phasor Measurement System

power system. The local storage facilities are also available in the PDCs along with the application functions which made PDCs exists for the real time applications. In the next level of hierarchy, the device available is called as super data concentrator as shown in Fig. 12. Its function is also similar to PDCs but in large scale for the entire power system. From the Fig. 12, it is clear that the communication between each level should be bidirectional. In most of the cases, the data flow required is only upward except few instants. There is a standard format available for upward and downstream communications among the hierarchy.

4.6 *Functional Requirements of PMUs and PDCs*

4.6.1 Synchro Phasor Evolution

PMUs are manufactured by different manufacturers in the various parts of the World. But it should be ensured that the interoperability among PMUs of different manufacturers should perform a common standard. The standard for synchro phasor is first developed by IEEE in the year 1995 [23]. The early models of PMUs were tested for interoperability based on this standard and it was found that the performance is not identical at off—nominal frequencies [25]. Hence it was realized that the existing standard was not enough to define the performance of PMUs at off-nominal frequencies. It was resolved by IEEE team in the year 2005.

The phasor of a single is independent of the signal frequency. If any pure sinusoid with any frequency is applied at the input of PMU then it is represented as a phasor of magnitude equal to its rms value of the signal and its phase angle (θ), which is the angle between the reporting instant and the peak of the sinusoid as shown in Fig. 13. There are many filters are employed at the input of the PMUs and the phase delays produced by these should be compensated before it estimates the phasor. The positive sequence produced by the PMU must be correct at all frequencies irrespective of the balanced or unbalanced input signal. PMU standard provides the specification for frequency

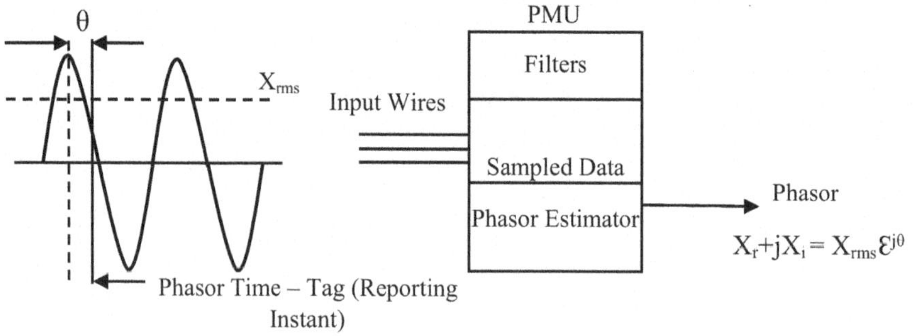

Fig. 13 PMU input signal and output Phasor estimate

deviation of ±5 Hz from the nominal frequency. The other accuracy specifications provided by PMU standards are standardized reporting time for phasors which is phase locked to the GPs is 1 pps and is at intervals are multiples of nominal power frequency.

4.7 File Structure of 'Synchro Phasor' Standard

Synchro phasor file structure is similar to that of COMTRADE, which describes about the files for transient data collection and dissemination. Initially COMTRADE standard has been modified by International Electro Technical Commission (IEC) and now it has been followed by computer relays, digital fault recorders, other producers and users of power system transient data. There are four types of files are defined by synchro phasor standard for data transmission to and from PMUs. In which three files are created by PMUs and the other one is produced by PDC. Header files, Configuration files and Data files are the files created by PMUs and Command file is created by PDC. The common structure of all files are shown in Fig. 14. The two bytes of first word is for synchronization of the data transfer. The second word describes the size of the total record, data originator is identified by third word, Second of Century (SOC) and fraction of a second (FRACSEC) is delivered by the next two words. The FRACSEC data word length is specified by the configuration file. The check sum is the last word used to find any errors in data transmission.

Header file contains the information need to be conveyed to the user from the producer of the data and it is in human readable format. The configuration and data files are in machine readable format. The Configuration file delivers the information about the explanation of the data contained in the data file. PMU will send the Header and configuration files once the nature of data to be transmitted is defined. The data file comprises the phasor data which is the output of the PMU and it may be in rectangular or polar form. The performance of the PMUs are controlled by the Command files from the higher level hierarchy. There are several commands have been defined and are available for current usage.

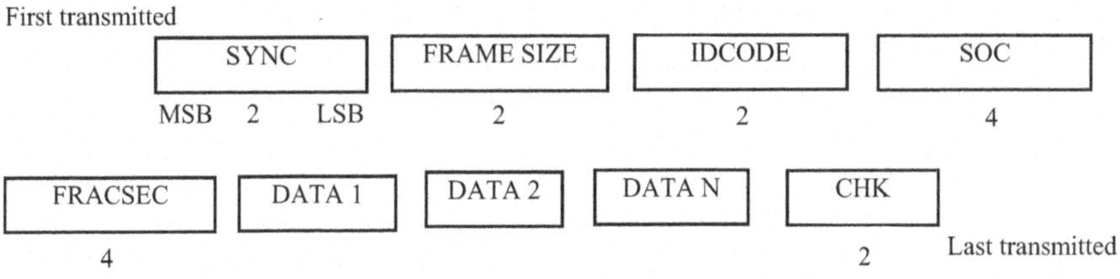

Fig. 14 Common structure of files transmitted from and to PMUs

Table 3 Comparison between SCADA and PMUs

S.No	Parameters	SCADA	PMUs
	Resolution	1 sample at every 2–4 s and used for only steady state observability	It can be done 10–60 samples per second and can be used for transient and steady state observability
	Measurement	It measures only magnitudes of the quantities	It measures both magnitude and phase angles of the quantities
	Time synchronization	There is no possibility for synchronization	Synchronization is done with the help of GPS
	Number of Input/Output channels	There are more than 100 analog and digital inputs and output channels are available	There are 10 channels for phasors, more than 16 channels for each digital and analog signals are available
	Focus	It is designed to do local monitoring and control	It is designed to carry out wide area monitoring and control

4.8 Comparison between SCADA and PMUs

PMUs stands at the top while comparing with its counterparts like SCADA. There are few added advantages of PMUs compared to SCADA as described in the Table 3.

4.9 PMU scenario in Indian Power Grid

Indian power grid is one among the largest power grids in the World. The operation and monitoring of Grids in Indian power grid is carried out by National Load Dispatch Centre (NLDC), five Regional Load Dispatch Centers (RLDC) and State Load Dispatch Centers (SLDC). The entire grid has subdivided into five regions such as Northern region, Southern region, Eastern region, North East region and Western region. To enhance the advanced technologies in the power network to improve the service and reliability of the supply to the customers, Synchro Phasor measurements are deployed in the Indian Power Grid. The first pilot project on PMU was installed in Northern region in the year 2010. The said project consists of PMUs along with GPS connected at 9 selected substations in the grid. A PDC and other accessories were located at NRLDC at New Delhi as published in an article of electrical India.

The present scenario of PMUs in India as there is one PDC is installed at the NLDC and it is integrated with other five PDCs connected at the RLDC [15]. There are 60 number of PMUs are installed as of now in various regions respectively 12 PMUs in Eastern region, 8 in North –east region, 14 in Norther region, 10 in Southern region and 16 in Western region. In most of the locations, the PMUs are used to measure the voltages in 220 kV and 400 kV buses. There are other pilot projects are also carried out throughout the grid to enhance the power system reliability and security.

4.10 Applications of PMU in Power System

The implementation of PMUs in the power system and collecting the magnitude and phase angle of voltages and currents of various buses in the substation will help us to undergo the following applications,

- Post disturbance analysis
- Stability monitoring
- Thermal overload monitoring
- Power system restoration
- State estimation
- Real time control
- Adaptive protection

4.11 Challenges in PMU implementation

Wide area measurement and controlling leads for robust control of power system from remote area and also it enhances the power system stability and reliability. There are lot of benefits behind the invention of new technologies as well as the challenges are also existing everywhere. The challenges in the implementation of PMUs in the power system are as follows,

- Selecting suitable location for PMU placement
- Integration of synchro phasor technology with SCADA
- Communication delays
- Low frequency oscillation monitoring
- Distorted power system waveforms make prediction difficult
- High computational requirement
- Developing tools for in depth post facto analysis.

5 Challenges of Wide Area Dynamic Monitoring and Control

There are many theoretical research works are still going on regarding WAMS and it should be carried forward to the practical application of wide area stability control in the large interconnected system. The challenges available in front of the researchers are summarized as below,

- *Problem of interconnection between local controllers and wide area controllers*: Large interconnected power systems in practice employs lot of conventional stability control devices such as PSS devices. PSS devices are using lot of local controllers and these are designed to damp the local oscillation modes. If the wide area damping control is employed in power system then it may reduce the damping capability of local controllers. Hence, it is important to maintain the overall stability of the power system by simultaneously optimizing the local and wide are controllers. The new techniques are proposed to tune large number of local PSSs as but only few works are carried out for the tuning on both local and wide area controllers [26–28].
- *Problem of control coordination among multiple wide-area damping controllers*: The wide area stability control employs devices such as HVDC and FACTS for power system control and control signals with multi-channel. It is the recent trends used in the smart transmission grids. There are lot of LFO modes are inbuilt in the control devices to provide multiple damping on dominant modes and improve the overall stability of the interconnected systems.

But the interaction between the different control loops may exists if the coordination designs are not taken care in multiple wide area controllers. These interactions may reduce the damping performance of multiple controllers or leads to instability of the power system. Hence it is noted that the coordination of wide area control is must for the practical application of wide area stability control.

- *Problem of how to choose optimal control-input for multiple wide-area damping controllers*: The performance of the controllers depends on the selection on the feedback signal. Some methods are proposed for the selection of feedback signals for local controllers [29–34]. But in WAMS application, it is possible to select the control signals in both local and wide area range. On comparing with the local controllers there are more local and wide area signals are existing to select as a feedback control input. But the research work related to these selections are only few.
- *Problem of design and implementation for practical application of online wide area monitoring and control*: Many of the research works related to wide area stability control stays only at the theoretical stage. To implement in the practical case, there are lot of problems encountered still such as design of advanced control algorithm to reduce the delay effect of wide area control, embedding the wide area control in the power system [35–37].

6 Conclusion

WAMS is a collective technology to monitor power system dynamics in real time, identify system stability related weakness and helps to design and implement counter measures. This chapter has described about the need for WAMS in the present scenario power grid to manage LFO and enhance the stability of the power system.

It also illustrates the major components of WAMS and its operation in brief. It also emphasis on the application of WAMS in the power system generation, transmission and distribution. The most important part of WAMS is PMU devices which are located at the remote locations to measure the physical quantities. This chapter covered the importance of PMU in smart grid, its components, application and limitations in detail.

References

1. N. Anandan, S. Sivanesan, S. Rama, T. Bhuvaneswari, Wide area monitoring system for an electrical grid. Energy Procedia **160**, 381–388 (2019)
2. C.D. Vournas, A. Metsiou, B.M. Nomikos, Analysis of intra-area and interarea oscillations in South-Eastern UCTE interconnection, in *IEEE Power Energy Society General Meeting*, 2 (2009)
3. J. Lehner, M. Kaufhold, M. Treuer, T. Weissbach, Monitoring of inter-area oscillations within the European interconnected network based on a wide area measuring system. in *IEEE PES Transmission and Distribution Conference and Exposition* (2010)
4. T. Hashiguchi, Y. Mitani, O. Saeki, K. Tsuji, M. Hojo, H. Ukai, Monitoring power system dynamics based on phasor measurements from demand side outlets developed in Japan Western 60 Hz system, in *IEEE Power Systems Conference and Exposition*, 4 (2004)
5. Y. Ota, T. Hashiguchi, H. Ukai, M. Sonoda, Y. Miwa, A. Takeuchi, Monitoring of interconnected power system parameters using PMU based WAMS, in IEEE Lausanne Power Tech (2007)
6. Y. Li, D. Yang, F. Liu, Y. Cao, C. Rehtanz, *Interconnected Power Systems* (Springer Press, 2016)
7. IEEE Power Engineering Society, *IEEE Guide for Synchronous Generator Modeling Practices and Applications in Power System Stability Analyses* (The Institute of Electrical and Electronics Engineers, Inc., 2003)
8. P. Kundur, J. Paserba, V. Ajjarapu, G. Andersson, A. Bose, C. Canizares, T. Van Cutsem, Definition and classification of power system stability IEEE/CIGRE joint task force on stability terms and definitions. IEEE Trans. Power Syst. **19**(3), 1387–1401 (2004)
9. G. Rogers, *Power System Oscillations* (Kluwer, Norwell, 2000)
10. D. Yan, Wide-area protection and control system with WAMS based. In: *International Conference on Power System Technology (PowerCon2006)*, China, 22–26 October, Chongqing, China, 2006, pp. 1–5
11. M. Shahraeini, M.H. Javidi, Wide area measurement systems, in *Advanced Topics in Measurements* (InTech, 2012). https://www.intechopen.com/books/advanced-topics-in-measurements/wide-area-measurement-systems
12. A. Vaccaro, A.F. Zobaa, Wide area monitoring, protection and control systems: the enabler for smarter grids (The Institution of Engineering and Technology, 2016)
13. A.G. Phadke, J.S. Thorp, *Synchronized Phasor Measurements and Their Applications* (Springer, New York, USA, 2008)
14. M. Kezunovic, Integration of substation IED information into EMS functionality, final project report (Power Systems Engineering Research Center (PSERC), 2008)
15. P. Jena, Introduction to Samrt Grid, NPTEL online certification courses (IIT Roorkee, 2017) [Lecture 11 & 12]
16. M. Shahraeini, M.H. Javidi, M.S. Ghazizadeh, A new approachfor classification ofdata transmission media in power systems, in *International Conference on Power System Technology*, 24–28 October, Hangzhou, China, 2010, pp. 1–7
17. C. Pisani, Real time tracking of electromechanical oscillations in ENTSO-e Continental European Synchronous Area, PhD Dissertation, University of Naples, May 2014

18. G. Missout, P. Girard, Measurement of bus voltage angle between Montreal and Sept-Iles. IEEE Trans. PAS. **99**(2), 536–539. 2 (March/April 1980)
19. P. Bonanomi, Phase angle measurements with synchronized clocks—Principles and applications. IEEE Trans. PAS **100**(11), 5036–5043 (1981)
20. A.G. Phadke, T. Hlibka, M. Ibrahim, Fundamental basis for distance relaying with symmetrical components. IEEE Trans. PAS. **96**(2), 635–646 (March/April 1977)
21. A.G. Phadke, J.S. Thorp, M.G. Adamiak, A new measurement technique for tracking voltage phasors, local system frequency, and rate of change of frequency. IEEE Trans. PAS **102**(5), 1025–1038 (1983)
22. Macrodyne Model 1690 PMU Disturbance Recorder", Macrodyne Inc. 4 Chelsea Place, Clifton Park, NY, 12065
23. IEEE Standard for Synchrophasors for Power Systems, C37.118–2005, pp 56–57, IEEE 1344–1995. Sponsored by the Power System Relaying Committee of the Power Engineering Society, pp 56–57
24. IEEE Standard Common Format for Transient Data Exchange (COMTRADE) for Power Systems, IEEE C37.111–1991, Sponsored by the Power System Relaying Committee of the Power Engineering Society
25. J. Depablos, V. Centeno, A.G. Phadke, M. Ingram, Comparative testing of synchronized phasor measurement units, in *Power Engineering Society General Meeting*, 2004. IEEE, vol. 1, 6–10 June 2004, pp. 948–954
26. A.L.B. Do Bomfim, G.N. Taranto, D.M. Falcao, Simultaneous tuning of power system damping controllers using genetic algorithms. IEEE Trans. Power Syst. **15**(1), 163–169 45 (2000)
27. G. Missout, J. Beland, G. Bedard, Dynamic measurement of the absolute voltage angle on long transmission Lines, IEEE Trans. PAS. **100**(11), 4428–4434. 3 (November 1981)
28. L.J. Cai, I. Erlich, Simultaneous coordinated tuning of PSS and FACTS damping controllers in large power systems. IEEE Trans. Power Syst. **20**(1), 294–300 (2005)
29. M.M. Farsangi, Y.H. Song, K.Y. Lee, Choice of FACTS device control inputs for damping inter-area oscillations. IEEE Trans. Power Syst. **19**(2), 1135–1143 50 (2004)
30. M.M. Farsangi, H. Nezamabadi-pour, Y.H. Song, K.Y. Lee, Placement of SVCs and selection of stabilizing signals in power systems. IEEE Trans. Power Syst. **22**(3), 1061–1071 (2007)
31. H.H. Alhelou, M.H. Golshan, J. Askari-Marnani, Robust sensor fault detection and isolation scheme for interconnected smart power systems in presence of RER and EVs using unknown input observer. Int. J. Electr. Power Energy Syst. **1**(99), 682–694 (2018). (Jul)
32. H.H. Alhelou, M.E. Hamedani-Golshan, R. Zamani, E. Heydarian-Forushani, P. Siano, Challenges and opportunities of load frequency control in conventional, modern and future smart power systems: A comprehensive review. Energies **11**(10), 2497 (2018). (Oct)
33. H. Haes Alhelou, M.E. Hamedani Golshan, F.M. Hajiakbari, Wind driven optimization algorithm application to load frequency control in interconnected power systems considering GRC and GDB nonlinearities. Electric Power Compon. Syst. **46**(11–12), 1223–1238 (2018). (Jul 21)
34. H. Haes Alhelou, M.E. Hamedani-Golshan, T.C. Njenda, P. Siano, A survey on power system blackout and cascading events: Research motivations and challenges. Energies **12**(4), 682 (2019). (Jan)
35. H.H. Alhelou, M.E. Golshan, N.D. Hatziargyriou, A decentralized functional observer based optimal LFC considering unknown inputs, uncertainties, and cyber-attacks. IEEE Trans. Power Syst. **34**(6), 4408–4417 (2019). (May 13)
36. H.H. Alhelou, M.E. Golshan, N.D. Hatziargyriou, Deterministic dynamic state estimation-based optimal lfc for interconnected power systems using unknown input observer. IEEE Trans. Smart Grid. 2019 Sep 9
37. H.H. Alhelou, S.J. Mirjalili, R. Zamani, P. Siano, Assessing the optimal generation technology mix determination considering demand response and EVs. Int. J. Electr. Power Energy Syst. **1**(119), 105871 (2020). (Jul)

Optimal Selection of Phasor Measurement Units

N. V. Phanendrababu

Abstract Phasor Measurement Unit (PMU) is an important device for the power system operation as it provides the synchronized data required for the monitoring, protection, and control of the power system. So, to deploy the PMUs for the power system, their optimal locations are needed to be identified. This paper presents the optimal selection of PMU set from the available sets of PMUs. Firstly, it obtains all possible sets of PMUs required for the complete observability of the power system. Then, it defines four criteria such as System Observability Index (*SOI*), Restorable Islands Observability Index (*RIOI*), Critical Bus Observability Index (*CBOI*) and Critical Line Observability Index (*CLOI*) for the selection of best PMU set. Later, the Multi-Attribute Decision Making (MADM) techniques such as Analytic Hierarchy Process (AHP), Technique for Order Preference by Similarity to Ideal Solution (TOPSIS), Preference Ranking Organization Method for Enrichment Evaluations (PROMETHEE) and Compromise Ranking Method (VIKOR) have been used for the optimal selection of PMU set. This selection has been tested on some of the IEEE test systems. The results are then compared to analyze the performance of these four methods.

Keywords PMU · Observability · MADM · SOI · RIOI · CBOI · CLOI · AHP · TOPSIS · PROMITHEE · VIKOR

1 Introduction

The Phasor measurement unit (PMU) is becoming an important tool for monitoring, controlling, and protecting the electric networks [1]. Hence its deployment for the present and future power system networks has become a great challenge for the planning engineers. The optimal PMU placement (OPP) is quite an important strategy for deploying the PMUs, optimally. With the invention of the Phasor Measuring Unit

N. V. Phanendrababu (✉)
Department of EEE, Chaitanya Bharathi Institute of Technology, Hyderabad 500075, Telangana, India
e-mail: phanendrababu_eee@cbit.ac.in

H. Haes Alhelou et al. (eds.), *Wide Area Power Systems Stability, Protection, and Security*, Power Systems, https://doi.org/10.1007/978-3-030-54275-7_5

(PMU), the power system protection schemes such as backup protection, adaptive protection, and model analysis based network redesigning and highly-secured power networks have been made adapted [2]. Starting from the introduction of optimal PMU placement [3], many authors [4–11] have suggested many algorithms to identify optimal locations of PMUs. Most of these placements were purely based on observability constraints. And, giving the same number of PMUs, the PMU locations obtained were different from author to author. But, the power utilities are not clear about how to select the best set that serves most of their purpose (criteria), out of them.

Later, the additional constraints were started to be modeled as the objective functions, to model a multi-constrained optimization into a multi-objective optimization algorithm. But, this method found to be yielding more number of PMUs. Moreover, this is highly expensive by means of convergence since it involves the selection of Pareto fronts. And, the issues such as complexity in modeling multi-constrained objective function, and increased computational time have made this method to find limited application.

Towards answering these issues, in [12], authors have suggested a new technique of placing PMUs based on three criteria such as bus voltage observability index, voltage control area observability index and tie-line oscillation observability indices. But, it was found that the number of PMUs obtained is almost double the number of PMUs required for complete observability. Later, authors [13] have proposed a method of PMU placement using Fuzzy Weighted Average (FWA) algorithm using 4 criteria. But, they have ranked only PMU installed buses using FWA technique. Recently, paper [14] has used Revised Analytical Hierarchy Process (RAHP) to rank the PMU installed buses. Firstly, it identifies the PMU locations required for system observability under N-1 contingencies. Then they have applied RAHP to rank them. But, they have not considered normal operating condition without ZIB effect. And, none of them have considered the criteria that could help the power utilities in improving the system security, and to reduce the chances of blackouts. Unfortunately, both of these aspects are very important for the power companies.

As a response to all these issues, this chapter suggests a methodology that applies the Multi-Attribute Decision-Making methods for selection criteria such as System Observability Index (*SOI*), Restorable Islands Observability Index (*RIOI*), Critical Bus Observability Index (*CBOI*) and Critical Line Observability Index (*CLOI*). MADM helps the decision-makers in taking decisions in the presence of multiple criteria. The PMU-sets obtained for observability problem [15, 16] considering normal operating conditions with no Zero Injection Bus effect are considered as alternatives for this selection.

This chapter is sectionalized as follows: After introducing the importance of MADM techniques in Sect. 1, Sect. 2 formulates the criteria based on their importance. Section 3 gives an introduction to MADM techniques and then explains AHP, TOPSIS, PROMITHEE and VIKOR methods. After obtaining the results from Sect. 3, the comparison amongst AHP, TOPSIS, PROMITHEE and VIKOR will be discussed in Sect. 4. The Sect. 5 will conclude the work.

2 Formulating the Criterion

Here, as selection criteria, four indices have been proposed. They are System Observability Index (*SOI*), Critical Bus Observability Index (*CBOI*), Critical Line Observability Index (*CLOI*), and Restorable Islands Observability Index (*RIOI*). These can be defined as below.

*2.1 Criterion 1: System Observability Index (*SOI*)*

It is a parameter that tells the level of system observability. It can be defined as the ratio of the sum of a number of buses observed for at least one time and the number of buses observed for more than one time to the total number of buses in the system. It is given by,

$$SOI = 1 + \frac{\text{No. of buses observed for more than one time } (\mathrm{n_p})}{\mathrm{N}} \tag{1}$$

Since the PMU placement is done only on the basis of system complete observability, there we get '1' in the above formula. Its value is typically greater than or equal to (≥) unity. High is the *SOI*, maximum will be the system redundancy level. So, the solution that corresponds to high *SOI* can be considered as the best solution to the OPP problem.

*2.2 Criterion 2: Restorable Islands Observability Index (*RIOI*)*

This index gives the observability of the islands resulting from the system islanding studies [17]. This section presents a network partition algorithm before determining the observability of islands.

2.2.1 Proposed Network Partition

This partition will be processed in two phases [18]. Initial partition gives stable islands which may or may not be observable. The final partition assures observable islands obtained by moving the unobservable bus to suitable islands. After the initial partition, it is to be noted that all the transformers must be allocated to respective regions to which the transformer buses are incident. This enables each region to run independently.

2.2.2 Initial Partition

For partitioning the network, the first step starts with identifying the number of islands equal to the minimum of the number of generator buses, load buses, and PMU installed buses. Then, formulate a regional matrix R in which each row represents one region. Initially, all the elements of a row are set to zero (0) except for one generator bus and a load bus that could be afforded by the generator. After, the next order regional bus can be calculated using the system connectivity matrix as (2):

$$R_k = R'_{k-1} \times A,\ k > 1 \tag{2}$$

Here, R'_{k-1} is R_{k-1} after corrections. This equation has to be executed iteratively until all the buses are considered. The following instructions are needed to be followed while constructing the higher-order regional:

1. No node should be considered for the next iteration once it is assigned.
2. No entry should be greater than unity. If they exist, correct them to unity.
3. A node will be assigned to a particular region if and only if the region satisfies the generation-load balance constraint. Otherwise, reset the entry and consider it for the next iteration.
4. In any iteration, if a node appears in two or more regions then do the assignment based on the nature of the bus. The process of assigning a bus based on its type is explained below:

Load Bus:

- Assign it to a region that could afford it sparing less generation-load balance.
- If no single region could afford the load alone then join the two regions that are incident it so that they could supply the load.

Zero-Injection Bus (ZIB):

- Assign to any region incident to it.

Generator bus:

- Assign it to a region with less generation-load balance. This improves the stability of an island.

2.2.3 Final Partition of Network

This section ensures the observability of the islands. For this, first, check whether all the buses are observable or not. If any bus in a particular island is not observed, make it observable by taking suitable actions. If a Load/Generator bus is unobserved, move it to an island which could make it observable without disturbing the generation-load balance. Else if, a ZIB bus is unobserved, move to an island that would make it observed.

2.2.4 Calculation of RIOI

After partitioning the network into islands, calculate *SOI* for the only islands which are completely observable. Let us consider the vector *S* with each element as SOI of the respective island which is completely observable.

$$S = [s_1,\ s_2,\ s_3,\ \ldots s_r],\ r \in c$$

Here, c is the number of completely observable islands. Then, calculate *RIOI* by multiplying it with the set of number of buses of completely observable islands (3).

$$RIOI = \sum_{i=1}^{r} s_r \cdot N_r \tag{3}$$

Here, N_r is the number of buses in r^{th} island. The regional information for the test systems considered of case study is given in the Tables 1, 2 and 3, below.

Table 1 Regional Information for IEEE-30

IEEE-30	
Region	Buses
r1	1, 3
r2	2, 4, 5, 6, 7, 8, 9, 10, 11, 12, 13, 16, 17, 19, 20, 21, 22, 24, 25, 26, 27, 28, 29, 30

Table 2 Regional Information for IEEE-57

IEEE-57	
Region	Buses
r1	1, 2, 16
r2	3, 4, 5, 14, 15, 18, 19, 20, 21, 22, 23, 24, 25, 30, 31, 38, 44, 45, 46
r3	12, 17
r4	6, 7, 8, 9, 10, 11, 13, 26, 27, 28, 29, 32, 33, 34, 35, 36, 37, 39, 40, 41, 42, 43, 47, 48, 49, 50, 51, 52, 53, 54, 55, 56, 57

Table 3 Regional Information for IEEE-118

IEEE-118	
Region	Buses
r1	1, 2, 3, 4, 5, 6, 7, 8, 9, 10, 11, 12, 14, 16, 17, 18, 19, 20, 21, 22, 34, 36, 43, 113, 117
r2	28, 29, 31
r3	23, 24, 25, 26, 27, 32, 114, 115
r4	44, 45, 46
r5	13, 15, 33, 35, 37, 38, 39, 40, 41, 42, 47, 48, 49, 50, 51, 55, 56, 57, 58, 59, 60, 61, 62, 63, 64, 65, 66, 67, 68, 52
r6	68, 69, 70, 71, 72, 73, 74, 75, 76, 77, 78, 116, 118, 82
r7	53, 54
r8	79, 80, 81, 96, 97, 98, 99
r9	92, 93, 94, 95, 100, 101, 102, 103, 104, 105, 106, 107, 108
r10	83, 84, 85, 88, 89, 90, 91
r11	86, 87
r12	109, 110, 111, 112

2.3 *Criterion 3: Critical Bus Observability Index* (CBOI)

The critical buses are the system buses whose outage causes network instable, and may lead to power system blackout. So, it is desirable to monitor them dedicatedly. This index is modelled here as a non-beneficial attribute. There are two cases that may arise while calculating *CBOI*. They are:

1. If the bus is directly observable, its *CBOI* is zero (0).
2. If the bus is indirectly observable, then it's CBOI becomes 0.5 * n. Where, n is the number of PMU installed buses incident to it. And, 0.5 is the proposed *CBOI* index of an indirectly observed bus from a PMU installed bus.

Here, the identification of critical buses and critical lines is done by using Newton–Raphson load flow method. As per the definition of critical element, the Newton–Raphson would get diverged for during the outage of critical elements. The details of critical elements identified are listed in Table 4.

Table 4 Critical elements identified for different test systems

System	Critical lines	Critical buses
IEEE-30	1–2	2, 6
IEEE-57	1–15, 3–15, 41–43, 49–50, 50–51	8, 12
IEEE-118	68–65, 38–65	10, 26, 65, 66

2.4 *Criterion 4: Critical Line Observability Index* (CLOI)

As similar to critical buses, the critical lines are also defined as system branches whose failure leads to unstable oscillations in large systems, and then to a system power-halt. So, it is most important to monitor the critical lines exclusively. The *CLOI* can be determined as the sum of the *CBOI* of the critical line buses ignoring their physical connection. This helps in restoring the critical line after its outage. Means, if *i* and *j* are the two buses of critical lines, then *CLOI* will be (4),

$$CLOI = CBOI_i + CBOI_j \tag{4}$$

There are three possible situations that may arise while calculating *CLOI*.

Case1: *Directly observed critical line*: As shown in Fig. 1, if the two ends of critical lines have PMU installed buses it's *CLOI* becomes, $CLOI = 0$.
Case2: *Partially observed critical line:* Here, only one end of the critical line will have a PMU installed bus as shown in Fig. 2. The *CLOI* becomes, $CLOI = 0 + 0.5 * n$, where *n* is the number of PMU installed buses connected to *j*th bus.
Case3: *Indirectly observed critical line:* In this case, none of the ends of the critical line will have PMU. But, as shown in Fig. 3, they may have PMU installed buses incident to them. So, $CLOI = 0.5*n_i + 0.5 * n_j$. Here, n_i and n_j are the number of PMU installed buses incident to the buses *i* and *j* respectively.

From this, it is clearly understood that these indices would be valued in ascending order from the directly observed critical line case to indirectly observed critical line case. In order to keep them ascending order from case1 to case3, the *CLOI* of case2 and case3 are multiplied with the multiplication factors $(0.5)^N$ and $(0.5)^{-N}$,

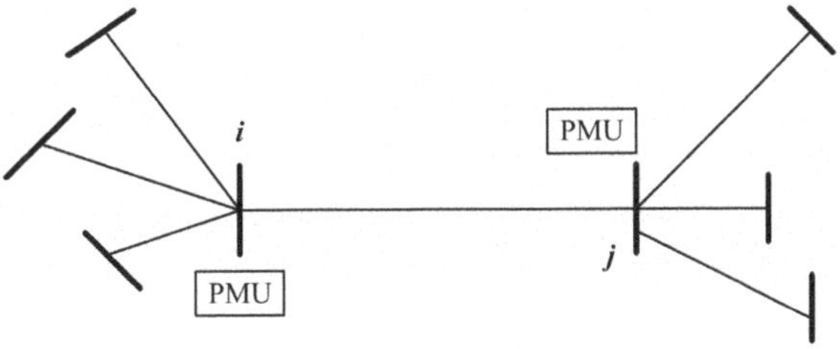

Fig. 1 Directly observed critical line

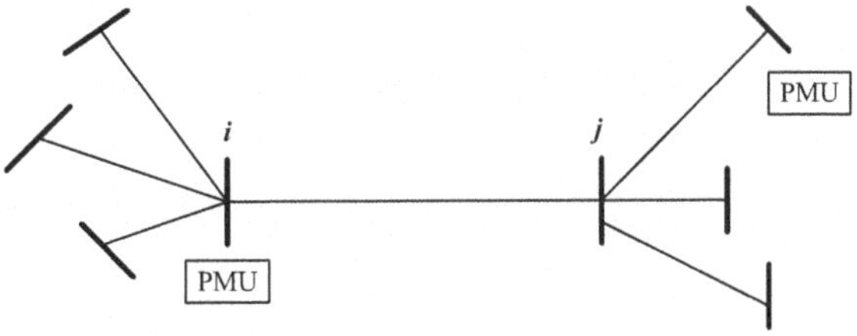

Fig. 2 Partially observed critical line

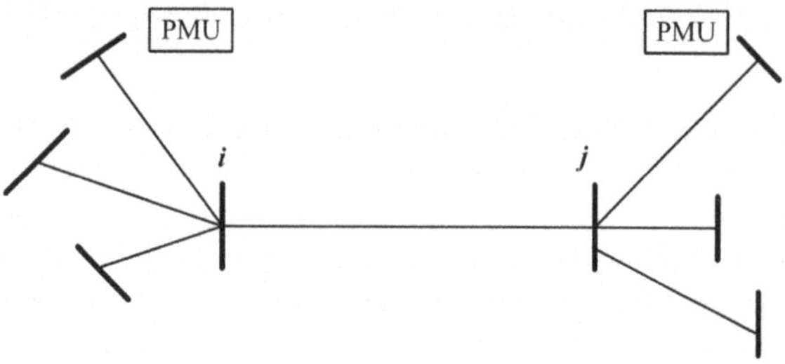

Fig. 3 Indirectly observed critical line

respectively. Here, N is the maximum connectivity of a bus with the highest degree. The values of N are 7, 6 and 9 for IEEE-30, 57 and 118 test systems, respectively.

The possible maximum and minimum values of *CLOI* for different IEEE test systems are given in Table 5. Now, the new *CLOI* factors of case2 and case3 are listed in Table 6. The criteria and the respective sets of weights we have considered for decision making are tabulated in Table 7. The list of alternatives and their attribute values, for different test systems, are given Tables 8, 9 and 10.

In this selection, except System observability index (*SOI*) all the remaining attributes are considered to be non-beneficial. Since the PMU-set has to ensure system observability completely, *SOI* is considered as a beneficial attribute.

Table 5 Maximum and Minimum values of *CLOI*

Test system	Case1		Case2		Case3	
	Maximum value	Minimum value	Maximum value	Minimum value	Maximum value	Minimum value
IEEE-30	0	0	1	0.0078125	1	0.015625
IEEE-57	0	0	1	0.015625	1	0.03125
IEEE-118	0	0	1	0.001953125	1	0.00390625

Table 6 New *CLOI* factors of case2 and case3

Test system	Case1		Case2		Case3	
	Maximum value	Minimum value	Maximum value	Minimum value	Maximum value	Minimum value
IEEE-30	0	0	0.0078125	0.000061035156	128	1
IEEE-57	0	0	0.015625	0.000244140625	64	1
IEEE-118	0	0	0.001953125	0.000003814697	512	1

Table 7 The criteria and the respective sets of weights

Weight_set (WS)	*SOI*	*CBOI*	*CLOI*	*RIOI*
Weight_set1 (WS1)	0.2	0.2	0.2	0.2
Weight_set2 (WS2)	0.5	0.2778	0.1667	0.0556
Weight_set3 (WS3)	0.5	0.1667	0.2778	0.0556
Weight_set4 (WS4)	0.0556	0.1667	0.2778	0.5
Weight_set5 (WS4)	0.0556	0.2778	0.1667	0.5

3 Multiple Attribute Decision Making (MADM) Methods

MADM helps the decision-maker in the process of evaluation. Even though many conventional methods were introduced earlier, they have confronted with quite different situations while solving MADM problems. So, all the fields have started developing new methods to serve their purposes particularly. And, their intention was to understand and estimate decision behavior but not to guide for accurate decision-making. But, if we define MADM, narrowly, as a decision that could identify the best alternative to maximize more than one attribute, many of the above methods would fail in identifying the best alternative [19]. The most common aspects in MCDM are described below:

Alternatives: These are the choices available for the decision-maker in making a decision. These are supposed to be prioritized and ranked ultimately.

Attributes: Attributes can be referred as goals or decision criteria. Even though alternatives can be viewed from the attributes, they may conflict with each other. The decision-maker may also determine the weights of the relative importance of criteria. In order to prioritize the criteria, MADM methods assign criteria with the weights of importance. Based on the ideal values of the attributes, they will be classified as; beneficial that is desired to be maximized, and non-beneficial that is desired to be maximized.

Decision matrix: It is a m × n matrix whose element d_{ij} represents the performance of ith alternative with respect to jth criteria.

3.1 *Analytic Hierarchy Process (AHP)*

The Analytic Hierarchy Process (AHP) [20, 21] is one of the best techniques for solving MADM problems. AHP enables us to define as many levels as possible to formulate a decision criterion. Its multiple functional characteristics have made AHP a more powerful tool in MADM. This includes handling subjective judgments, multiple decision makers, etc. It handles both subjective and objective criteria effectively. The steps involved in the AHP method are been explained below [22].

Table 8 Alternatives and their attribute values for IEEE-30 bus system

S. No	Alternatives	System Observability Index (SOI)	Critical Line (CL)-1	Critical Bus (CB)-1	Critical Bus-2	Island observability (RIOI)
1	3, 5, 6, 10, 11, 12, 19, 24, 26, 29	1.3667	64	0.25	0	36.0008
2	1, 2, 6, 9, 10, 12, 15, 19, 25, 27	1.6667	0	0	0	43.9992
3	2, 4, 6, 10, 11, 12, 19, 24, 26, 29	1.5000	0.015625	0	0	39
4	1, 2, 6, 9, 10, 12, 19, 23, 26, 27	1.5333	0	0	0	2
5	2, 4, 6, 9, 10, 12, 15, 18, 25, 27	1.7333	0.015625	0	0	6
6	2, 4, 6, 9, 10, 12, 15, 19, 25, 27	1.7333	0.015625	0	0	46
7	2, 4, 6, 9, 10, 12, 15, 20, 25, 27	1.7333	0.015625	0	0	47.0008
8	2, 4, 6, 9, 10, 12, 18, 24, 25, 27	1.7000	0.015625	0	0	0
9	2, 4, 6, 9, 10, 12, 19, 24, 25, 27	1.7000	0.015625	0	0	45
10	2, 4, 6, 10, 11, 12, 15, 20, 25, 27	1.6667	0.015625	0	0	44.9992
11	2, 4, 6, 10, 11, 12, 19, 24, 25, 27	1.6333	0.015625	0	0	43.0008
12	2, 3, 6, 9, 10, 12, 15, 20, 25, 27	1.6667	0.007813	0	0	45
13	1, 5, 6, 9, 10, 12, 15, 20, 25, 27	1.6000	0.003906	0.25	0	44
14	1, 2, 6, 10, 11, 12, 15, 18, 25, 27	1.6000	0	0	0	8
15	1, 2, 6, 10, 11, 12, 15, 20, 25, 27	1.6000	0	0	0	43
16	1, 5, 6, 9, 10, 12, 18, 24, 25, 27	1.5667	0.003906	0.25	0	2
17	1, 2, 6, 10, 11, 12, 19, 24, 25, 27	1.5667	0	0	0	41
18	2, 4, 6, 9, 10, 12, 19, 23, 26, 29	1.5333	0.015625	0	0	0
19	1, 5, 6, 9, 10, 12, 18, 23, 26, 27	1.4667	0.003906	0.25	0	2
20	3, 6, 7, 10, 11, 12, 15, 19, 25, 29	1.4667	64	0.25	0	39

Table 9 Alternatives and their attribute values for IEEE-57 bus system

S. No	Alternatives	SOI	CL-1	CL-2	CL-3	CL-4	CL-5	CB-1	CB-2	RIOI
1	2, 6, 12, 14, 19, 22, 25, 27, 32, 36, 39, 41, 44, 48, 50, 52, 54	1.1053	32.00	32.00	0.313	0.0156	0.313	0.5	0	2
2	1, 4, 9, 13, 20, 23, 27, 29, 30, 32, 36, 39, 41, 44, 47, 51, 54	1.1930	0.0156	32.00	0.313	32.00	0.313	0.5	0.25	3
3	1, 4, 7, 9, 14, 20, 22, 25, 27, 32, 36, 41, 45, 47, 50, 53, 57	1.1579	0.0078	24	0.313	0.0156	0.313	0.25	0.5	3
4	2, 6, 12, 13, 19, 22, 25, 26, 29, 32, 36, 39, 41, 44, 47, 50, 54	1.1579	32.00	32.00	0.313	0.0078	0.313	0.5	0	2
5	1, 4, 9, 13, 19, 22, 26, 29, 30, 32, 36, 41, 44, 47, 50, 54, 57	1.2105	0.0156	32.00	0.313	0.0078	0.313	0.5	0.25	3
6	1, 4, 7, 9, 19, 22, 25, 27, 32, 36, 38, 39, 41, 45, 46, 51, 53	1.1930	0.0156	0.0156	0.313	32.00	0.313	0.25	0.5	3
7	1, 4, 9, 14, 20, 23, 25, 27, 29, 32, 36, 39, 41, 44, 47, 50, 53	1.1404	0.0156	32.00	0.313	0.0156	0.313	0.5	0.5	3
8	1, 4, 9, 20, 22, 25, 27, 29, 32, 36, 38, 39, 41, 44, 46, 51, 53	1.1930	0.313	48	0.313	32.00	0.313	0.5	0.5	3
9	1, 4, 9, 14, 19, 22, 26, 29, 30, 32, 36, 39, 41, 44, 48, 50, 53	1.1754	0.0156	32.00	0.313	0.0156	0.313	0.5	0.5	3
10	1, 4, 9, 14, 19, 22, 25, 27, 29, 32, 36, 41, 45, 48, 51, 54, 57	1.1754	0.0078	24	0.313	32.00	0.313	0.5	0.5	3
11	1, 6, 12, 15, 19, 22, 26, 29, 30, 32, 36, 38, 39, 41, 46, 50, 54	1.2281	0.00	0.313	0.313	0.0156	0.313	0.5	0	5
12	1, 6, 9, 15, 19, 22, 25, 27, 28, 32, 36, 38, 39, 41, 47, 50, 53	1.2281	0.00	0.313	0.313	0.0078	0.313	0.25	0.5	41
13	1, 4, 6, 10, 20, 23, 27, 30, 32, 36, 39, 41, 45, 46, 49, 52, 54	1.1404	0.0156	32.00	0.313	0.313	32.00	0.5	0.5	38
14	1, 4, 7, 9, 13, 19, 22, 25, 27, 32, 36, 39, 41, 45, 47, 50, 53	1.2105	0.0078	24	0.0078	0.313	0.313	0.25	0.5	43
15	1, 4, 9, 15, 20, 24, 26, 29, 31, 32, 36, 38, 39, 41, 47, 50, 54	1.2456	0.00	0.0156	0.313	0.0156	0.313	0.5	0.5	3

Table 10 Alternatives and their attribute values for IEEE-118 bus system

S. No	Alternatives	SOI	CL-1	CL-2	CB-1	CB-2	CB-3	CB-4	RIOI
1	2, 5, 9, 12, 13, 17, 21, 23, 26, 29, 34, 37, 42, 45, 49, 53, 56, 62, 64, 71, 75, 77, 80, 85, 86, 90, 94, 101, 105, 110, 115, 116	1.313	256	256	0.5	0	0.5	0.25	83
2	2, 5, 9, 12, 15, 17, 21, 23, 28, 30, 35, 40, 43, 46, 51, 54, 57, 62, 64, 71, 75, 77, 80, 85, 87, 90, 94, 101, 105, 110, 115, 116	1.254	256	256	0.5	0.5	0.5	0.5	37
3	3, 6, 9, 11, 12, 17, 21, 25, 29, 34, 37, 40, 45, 49, 52, 56, 62, 64, 68, 70, 71, 75, 77, 80, 85, 87, 90, 94, 102, 105, 110, 114	1.339	0.002	256	0.5	0.5	0.25	0.25	104
4	1, 5, 9, 12, 15, 17, 20, 23, 25, 29, 34, 37, 42, 45, 49, 52, 56, 62, 63, 68, 71, 75, 77, 80, 85, 87, 90, 94, 101, 105, 110, 114	1.356	0.004	384	0.5	0.5	0.5	0.25	53
5	2, 5, 9, 11, 12, 17, 21, 24, 26, 28, 34, 37, 41, 45, 49, 53, 56, 62, 63, 68, 71, 75, 77, 80, 85, 86, 91, 94, 101, 105, 110, 114	1.322	0.004	384	0.5	0	0.5	0.25	64
6	2, 5, 10, 11, 12, 17, 21, 25, 29, 34, 37, 41, 45, 49, 52, 56, 62, 64, 68, 70, 71, 76, 78, 85, 86, 90, 92, 96, 100, 105, 110, 114	1.331	0.004	256	0	0.5	0.25	0.25	55
7	2, 6, 9, 11, 12, 17, 21, 25, 28, 34, 37, 40, 45, 49, 52, 56, 62, 64, 68, 71, 72, 75, 77, 80, 85, 87, 91, 94, 102, 105, 110, 114	1.297	0.002	256	0.5	0.5	0.25	0.25	84
8	1, 5, 9, 12, 15, 17, 21, 23, 26, 28, 34, 37, 42, 45, 49, 52, 56, 62, 64, 71, 75, 77, 80, 85, 87, 90, 94, 102, 105, 110, 115, 116	1.313	256	256	0.5	0	0.5	0.25	75
9	2, 6, 9, 11, 12, 17, 21, 25, 28, 34, 37, 42, 45, 49, 53, 56, 62, 63, 68, 70, 71, 76, 77, 80, 85, 86, 91, 94, 102, 105, 110, 114	1.288	0.004	384	0.5	0.5	0.5	0.25	57
10	3, 5, 9, 11, 12, 17, 20, 23, 28, 30, 34, 37, 40, 45, 49, 52, 56, 62, 63, 68, 71, 75, 77, 80, 85, 86, 90, 94, 102, 105, 110, 115	1.364	0.004	384	0.5	0.5	0.5	0.25	38
11	1, 5, 10, 12, 15, 17, 21, 25, 29, 34, 37, 42, 45, 49, 52, 56, 62, 64, 68, 70, 71, 76, 78, 85, 87, 90, 92, 96, 100, 105, 110, 114	1.339	0.002	256	0	0.5	0.25	0.25	56
12	3, 7, 9, 11, 12, 17, 21, 23, 26, 28, 34, 37, 41, 45, 49, 52, 56, 62, 64, 68, 71, 75, 77, 80, 85, 86, 91, 94, 101, 105, 110, 115	1.322	0.002	256	0.5	0.5	0.25	0.25	74
13	2, 5, 9, 12, 15, 17, 20, 23, 28, 30, 35, 40, 43, 47, 49, 52, 56, 62, 63, 68, 71, 75, 77, 80, 85, 86, 91, 94, 102, 105, 110, 114	1.305	0.004	0.004	0.5	0.5	0.5	0.25	21
14	1, 5, 9, 11, 12, 17, 20, 23, 25, 28, 34, 37, 40, 45, 49, 53, 56, 62, 64, 71, 75, 77, 80, 85, 87, 90, 94, 102, 105, 110, 115, 116	1.322	384	256	0.5	0.5	0.5	0.25	74
15	1, 5, 10, 12, 15, 17, 20, 23, 28, 30, 35, 40, 43, 47, 49, 52, 56, 62, 64, 68, 71, 75, 77, 80, 85, 87, 90, 94, 102, 105, 110, 114	1.305	0.002	0.002	0	0.5	0.25	0.25	35

Step 1: Identify the attributes and the alternatives. Formulate the hierarchy among objectives, attributes, and alternatives.

Step 2: Formulate a pair-wise comparison matrix ($C_{n\times n}$) (5), where n is the number of available alternatives. The element (c_{ij}) of C represents the importance of attribute i over attribute j. The integers 1, 2 …9 will be used for representing comparative importance between two attributes.

$$c_{ij} = \begin{cases} 1, \ if\ i = j \\ {}^{1}/{C_{ji}}, \ if\ i \neq j \end{cases} \tag{5}$$

Step 3: Determine the relative normalized weights (W_j) (6). It uses geometric mean method for identifying these weights.

$$W_j = M_j \Big/ \sum_{j=1}^{n} M_j \tag{6}$$

where

$$M_j = \left[\prod_{j=1}^{n} c_{ij} \right]^{1/n}$$

Step 4: Calculate C_1 and C_2, using $C_1 = C * W^T$ and $C_2 = C_1 / W^T$.

Step 5: Calculate λ_{max} which is the average of C_2.

Step 6: Determine the consistency index using, $CI = (\lambda_{\max} - n)/(n-1)$. Lower the *CI* better is the consistency.

Step 7: Obtain random index (*RI)* for all attributes available. The Table 11 gives the *RI* values.

Step 8: Calculate the consistency ratio *CR* = *CI/RI*. A typical value of *CR* should be less than equal to 0.1.

Step 9: Do the pair-wise comparison of alternatives based on their dominance.

Table 11 *RI* values

Attributes	3	4	5	6	7	8	9	10
RI	0.52	0.89	1.11	1.25	1.35	1.4	1.45	1.49

Step 10: Calculate the scores of all alternatives by multiplying W_j of each attribute with the corresponding normalized weight, and then sum it over attributes for each alternative.

3.2 Technique for Order Preference by Similarity to Ideal Solution (TOPSIS)

It is developed by Hwang and Yoon in 1981. It is developed based on the idea that the best alternative should always be close to the ideal value and far away from the non-ideal value [23]. For this, they have used the concept of Euclidean distance. The best solution alternative should always be close to the ideal (best) solution and far away from the negative ideal (worst) solution. The ideal solution is the solution for which all the attributes mean maximum values. But, practically, it is not feasible to have that kind of solution from the set of alternatives. Similarly, a negative ideal solution is also an imagined solution for which all attributes mean minimum values. So, the solution that will be obtained from this method is not only closest to an ideally best solution but also farthest from ideally worst solution. The steps involved in the TOPSIS method are given below:

Step 1: Determine the attributes.

Step 2: Construct the decision table A_T. And, it is to be noted that the subjective attributes must be represented with a ranked value on a scale.

Step 3: Obtain the normalized decision matrix (7), N_{Tij}:

$$N_{Tij} = E_{Tij} \bigg/ \left[\sum_{j=1}^{m} E_{Tij}^2\right]^{1/2} \tag{7}$$

where, $i = (1, 2, \ldots, n)$, $j = (1, 2, \ldots, m)$, E_{Tij} is the performance of ith alternative for jth attribute, m is the number of attributes,

Step 4: Estimate the weights (W_{Tj}) of the different attributes according to their objective. It is very important to remember that their cumulative sum should be unity.

Step 5: Calculate the weighted normalized matrix, M_{Tij}, using (8),

$$M_{Tij} = W_{Tj} \cdot N_{Tij} \tag{8}$$

Step 6: Identify the best and worst solutions from (9) and (10), given below:

$$S^{+} = \left\{ \left(\sum_{i}^{\max} M_{Tij} \Big/ j \in J \right), \left(\sum_{i}^{\min} M_{Tij} \Big/ j \in J' \right) / i = 1, 2, \ldots, n \right\} \\ = \left\{ S_1^{+}, S_2^{+}, \ldots, \ldots, S_m^{+} \right\} \tag{9}$$

$$S^{-} = \left\{ \left(\sum_{i}^{\min} M_{Tij} \Big/ j \in J \right), \left(\sum_{i}^{\max} M_{Tij} \Big/ j \in J' \right) / i = 1, 2, \ldots, n \right\} \\ = \left\{ S_1^{-}, S_2^{-}, \ldots, \ldots, S_m^{-} \right\} \tag{10}$$

where, $J = (j = 1, 2 \ldots m)$; j is associated with beneficial attributes.

$J' = (j = 1, 2 \ldots m)$; j is associated with no-beneficial attributes.

Let S_j^{+} and S_j^{-} are the best and worst values of the jth attribute in table A_T. Means, S_j^{+} becomes the higher value of the attribute j if it is a beneficial attribute and lower value otherwise. Similarly, S_j^{-} becomes the lower value of the attribute j if it is a beneficial attribute and higher value otherwise.

Step 7: Using the following Euclidean distance formula (11) and (12), measure the separation of each alternative from the ideal one.

$$D_i^{+} = \left\{ \sum_{j=1}^{m} \left(M_{ij} - S_j^{+} \right)^2 \right\}^{1/2} \tag{11}$$

$$D_i^{-} = \left\{ \sum_{j=1}^{m} \left(M_{ij} - S_j^{-} \right)^2 \right\}^{1/2} \tag{12}$$

Step 8: Calculate the relative closeness (13) of alternatives to the best solution using:

$$R_i = D_i^{-} / \left(D_i^{+} + D_i^{-} \right) \tag{13}$$

Step 9: Arrange the alternatives in the descending order of their R_i values to identify the most and the least preferred alternatives.

3.3 *PROMETHEE (Preference Ranking Organization Method for Enrichment Evaluations)*

It is an outranking-based method introduced by Brans in [24]. The strength of preference of each alternative for a particular attribute will be identified by comparing it over other alternatives. The steps involved in decision making using PROMETHEE are given below.

Step 1: Identify the alternatives and attributes from the data. Let x is the set of alternatives and y is the set of attributes.

Step 2: Construct the decision making matrix, A_E.

Step 3: Estimate the weights, W_{Ej}.

Step 4: Define preference function (14) as,

$$F_j(d_j) = \begin{cases} 0, \; d_j \le s_j \\ (|d_j| - s_j)/(p_j - s_j), \; s_j < d_j \le p_j \\ 1, \; d_j > p_j \end{cases} \tag{14}$$

where, p_j and s_j are the preference and indifference cut-offs.

Step 5: After getting the preference degree for each pair of alternatives for each criterion, calculate the overall preference (15) as,

$$p(A_x, A_y) = \sum_{j=1}^{n} \omega_j \, F_j(A_x, A_y) \tag{15}$$

where, ω_j is the weight of *j*th criterion.

Step 6: Calculate the incoming (16) and outgoing flow (17) for each alternative.

The inflow of alternative A_x is,

$$\Phi^-(A_x) = \sum p(A_y, A_x) \tag{16}$$

The outflow of the alternative A_x is,

$$\Phi^+(A_x) = \sum p(A_x, A_y) \tag{17}$$

Step 7: Determine the net flow (18), $\Phi(A_x)$, from

$$\Phi(A_x) = \Phi^+(A_x) - \Phi^-(A_y) \tag{18}$$

Step 8: Later, the ranking will be done from best to worst using this net flow values.

Alternative A_i is indifferent from A_j, if $\Phi(A_i) = \Phi(A_j)$, and Alternative A_i is preferential to A_j if $\Phi(A_i) > \Phi(A_j)$.

3.4 Compromise Ranking Method (VIKOR)

This method was developed by Yu and Zeleny in [25, 26]. This method is helpful in the situations when the decision-maker is unable and does not have an idea to give the preference for the attributes. This method works based on L_p-metric suggested in [22]. This method gives the most feasible solution as it yields a maximum utility of majority and minimum of the individual regret of the opponent. The procedure of the VIKOR method is explained below:

Step 1: Identify the attributes.

Step 2: Calculate the best $((m_{ij})_{max})$ and the worst $((mij)_{min})$ values of all attributes.

Step 3: Determine E_i and F_i values using (19) and (20), given below.

$$E_i = \sum_{j=1}^{M} \omega_j \left[(m_{ij})_{\max} - (m_{ij}) \right] / \left[(m_{ij})_{\max} - (m_{ij})_{\min} \right] \tag{19}$$

$$F_i = Max\ of\ \left\{ \omega_j \left[(m_{ij})_{\max} - (m_{ij}) \right] / \left[(m_{ij})_{\max} - (m_{ij})_{\min} \right],\ j = 1, 2, 3, \ldots M \right\} \tag{20}$$

Step 4: Calculate P_i using (21) as,

$$\begin{aligned} P_i = {} & v\left((E_i - E_{i-\min}) / (E_{i-\max} - E_{i-\min}) \right) \\ & + (1 - v)\left((F_i - F_{i-\min}) / (F_{i-\max} - F_{i-\min}) \right) \end{aligned} \tag{21}$$

where, $E_{i-\min}$ and $E_{i-\max}$ are the minimum and maximum values of E_i. Similarly, $F_{i-\min}$ and $F_{i-\min}$ are the minimum and maximum values of F_i. v is the weight of the strategy, and it can take values from 0 to 1. Generally, it can be taken as 0.5.

Step 5: Obtain the three ranking list by arranging the alternatives in the descending order of their P_i, E_i and F_i values respectively. For a given value of v, compromise the ranking list based on P value.

Step 6: Compromise the solution.

For the Alternative A_k to become the best by the measure P it should follow two conditions [27]:

1. *Acceptable advantage*: $P(A_1) - P(A_k) \geq 1/(m-1)(22)$

   where, A_l is the second best in the rank list by P.

2. *Acceptable stability in decision*: alternative A_k should also be ranked best by E and F. This compromise solution is stable I the DM process. The compromise solution contains:

   - Alternatives A_k and A_l if the condition 2 is not satisfied.

- Alternatives $A_{k,}A_{l...}A_p$ if the condition 1 is not satisfied. where A_p can be calculated from (23),

$$P(A_p) - P(A_l) \approx 1/(m-1) \tag{23}$$

4 Results and Discussions

The rankings obtained using AHP for the available alternatives, for the three test systems, are given in the Tables 12, 13 and 14 respectively.

The rankings obtained, using TOPSIS, for the available alternatives are given in the Tables 15, 16 and 17 respectively.

Using PROMITHEE, the rankings for the available alternatives are given in the Tables 18, 19 and 20 respectively. The Table 21 gives the rankings for the available alternatives obtained using VIKOR.

After the alternatives are ranked using AHP, TOPSIS, PROMITHEE and VIKOR, the performance of those methods is compared. For weight_set1, for IEE-30 bus system, AHP has ranked the alternative-20 followed by the alternatives 4 and 18, whereas TOPSIS identifies the alternatives 20, 6 and 10, PROMITHEE identifies the alternatives 20, 4 and 16, and VIKOR finds the alternatives 20, 6 and 10 as the first three best alternatives respectively. For IEE-57 bus system, AHP has ranked the alternative-14 followed by the alternatives 9 and 8, whereas TOPSIS identifies the alternatives 14, 11 and 5, PROMITHEE identifies the alternatives 15, 10 and 5, and VIKOR finds the alternatives 14, 13 and 9 as first three best alternatives respectively. Similarly, for IEEE-118 bus system, AHP has ranked the alternative-9 followed by the alternatives 15 and 3, whereas TOPSIS identifies the alternatives 12, 14 and 6, PROMITHEE identifies the alternatives 12, 15 and 11, and VIKOR finds the alternatives 10, 15 and 1 as first three best alternatives respectively. This comparison is tabulated in the Table 22.

For effective comparison, the 2-D charts are drawn for all the alternatives with their ranks obtained from these four methods, in the same plot for a particular test system, as shown in Figs. 4, 5 and 6. These are with respect to weight-set1.

Tables 23, 24, 25 and 26 list the rankings identified these four MADM techniques for the weight-sets 2, 3, 4 and 5 respectively. For the weight_set2, the 2-D charts are shown in the Figs. 7, 8 and 9.

The Figs. 10, 11 and 12 give the rank analysis for weight_set3, for all the test systems. The comparisons of performance for the remaining weight_sets are given in the Figs. 13, 14, 15, 16, 17 and 18.

As a future work and research gaps, the applications of PMU in modern power systems operation, control, stability, protection and security should be well-investigated [28–34]. PMUs are the backbone of wide are monitoring system that can

Table 12 AHP results for IEEE-30 bus system

Alternatives	Weight_set1		Weight_set2		Weight_set3		Weight_set4		Weight_set5	
	Score	Rank	Score	Rank	Score	Rank	Score	Rank	Score	Rank
1	0.0898	6	0.0966	6	0.1008	6	0.0921	2	0.1009	2
2	0.0898	7	0.0966	7	0.1008	7	0.0816	12	0.0819	15
3	0.087	5	0.0938	5	0.0867	5	0.0738	15	0.0788	12
4	0.0802	2	0.0735	9	0.0761	2	0.0679	13	0.0781	13
5	0.0666	9	0.072	2	0.0737	9	0.065	6	0.0677	17
6	0.0645	8	0.0713	8	0.0717	12	0.065	7	0.0594	14
7	0.0599	15	0.0622	12	0.0628	8	0.0615	10	0.0584	6
8	0.0581	12	0.0583	10	0.0626	10	0.0602	17	0.0584	7
9	0.058	14	0.0517	15	0.0496	15	0.0523	14	0.0551	10
10	0.0533	17	0.0498	14	0.0479	13	0.052	9	0.0451	4
11	0.0515	10	0.0464	11	0.0466	11	0.0496	11	0.0451	9
12	0.0478	4	0.0451	17	0.0397	14	0.0479	3	0.0451	20
13	0.0402	13	0.0397	4	0.0381	17	0.0433	20	0.0428	11
14	0.0396	11	0.0321	13	0.0275	3	0.0386	4	0.0411	3
15	0.0302	16	0.0274	18	0.0248	4	0.0342	5	0.0312	1
16	0.0228	19	0.0273	3	0.0246	20	0.03	1	0.0262	5
17	0.0206	18	0.0221	16	0.019	16	0.0284	8	0.0243	16
18	0.0205	3	0.0147	19	0.0189	18	0.0244	18	0.0236	19
19	0.0115	20	0.0113	20	0.0165	1	0.0165	16	0.0203	8
20	0.0082	1	0.0081	1	0.0116	19	0.0158	19	0.0165	18

Table 13 AHP results for IEEE-57 bus system

Alternatives	Weight_set1		Weight_set2		Weight_set3		Weight_set4		Weight_set5	
	Score	Rank	Score	Rank	Score	Rank	Score	Rank	Score	Rank
1	0.1319	11	0.135	11	0.1276	12	0.1196	14	0.1266	12
2	0.1276	15	0.1145	12	0.1246	11	0.1155	12	0.1186	14
3	0.1242	12	0.1145	15	0.1181	14	0.111	11	0.1065	11
4	0.1054	14	0.1076	14	0.1122	15	0.0894	13	0.0903	13
5	0.089	5	0.0898	5	0.0886	5	0.0711	4	0.0708	15
6	0.0641	4	0.0672	4	0.0546	6	0.0583	3	0.0674	3
7	0.055	6	0.0558	6	0.0538	2	0.0583	15	0.0653	4
8	0.0548	3	0.0538	2	0.0502	13	0.0573	1	0.0531	5
9	0.0506	2	0.0465	1	0.0499	8	0.0542	5	0.0502	6
10	0.0425	8	0.0452	3	0.0441	4	0.0511	6	0.0463	9
11	0.0403	9	0.0451	8	0.0399	9	0.049	2	0.0457	2
12	0.0333	10	0.0376	9	0.0392	3	0.0433	9	0.0453	7
13	0.0287	7	0.0345	10	0.0386	10	0.0422	7	0.0393	10
14	0.0281	1	0.0267	13	0.0303	1	0.0399	10	0.0376	8
15	0.0246	13	0.026	7	0.0283	7	0.0397	8	0.037	1

Table 14 AHP results for IEEE-118 bus system

Alternatives	Weight_set1		Weight_set2		Weight_set3		Weight_set4		Weight_set5	
	Score	Rank	Score	Rank	Score	Rank	Score	Rank	Score	Rank
1	0.1532	10	0.1511	10	0.1462	10	0.1591	3	0.1677	3
2	0.1242	4	0.1221	4	0.1297	3	0.1419	1	0.1282	1
3	0.0987	3	0.0947	11	0.1189	4	0.0877	14	0.0891	14
4	0.0928	11	0.0944	3	0.0839	11	0.0804	8	0.0847	7
5	0.074	6	0.0749	6	0.0817	1	0.0797	7	0.0706	12
6	0.0715	15	0.0711	5	0.0652	14	0.0739	5	0.065	8
7	0.0644	12	0.0663	1	0.0651	6	0.0663	12	0.0644	15
8	0.0579	5	0.0605	12	0.0627	12	0.0605	11	0.0592	5
9	0.0506	1	0.0592	8	0.0592	5	0.0571	6	0.0572	11
10	0.0489	13	0.0569	15	0.0535	8	0.0523	15	0.0549	6
11	0.0444	8	0.0432	14	0.0446	7	0.0343	4	0.0409	13
12	0.0418	14	0.0366	7	0.0344	15	0.0323	10	0.0357	4
13	0.0405	7	0.0353	13	0.0244	13	0.029	9	0.0334	10
14	0.0236	9	0.0214	9	0.0196	9	0.0285	13	0.031	9
15	0.0135	2	0.0124	2	0.0109	2	0.017	2	0.018	2

Table15 TOPSIS results for IEEE-30 bus system

Alternatives	Weight_set1		Weight_set2		Weight_set3		Weight_set4		Weight_set5	
	Score	Rank	Score	Rank	Score	Rank	Score	Rank	Score	Rank
1	0.9997	7	0.9997	12	0.9997	7	0.9997	7	0.9997	7
2	0.9992	6	0.9997	7	0.9996	6	0.998	6	0.9987	6
3	0.9986	12	0.9996	6	0.9995	12	0.9961	12	0.9975	12
4	0.9986	9	0.9996	2	0.9995	9	0.996	9	0.9975	9
5	0.9986	10	0.9996	9	0.9995	10	0.996	10	0.9975	10
6	0.9979	2	0.9996	10	0.9993	2	0.9941	2	0.9963	2
7	0.9973	15	0.9994	15	0.9991	15	0.9921	15	0.995	15
8	0.9972	11	0.9994	11	0.9991	11	0.9921	11	0.995	11
9	0.9959	17	0.9992	17	0.9987	17	0.9883	17	0.9926	17
10	0.9945	3	0.9988	3	0.9982	3	0.9844	3	0.9901	3
11	0.9739	14	0.9946	14	0.9915	14	0.9281	14	0.9534	14
12	0.9726	5	0.9943	5	0.991	5	0.9247	5	0.9512	5
13	0.97	4	0.9938	4	0.9902	4	0.918	4	0.9467	4
14	0.9687	8	0.9935	8	0.9897	8	0.9146	8	0.9444	8
15	0.9687	18	0.9935	18	0.9897	18	0.9146	18	0.9444	18
16	0.827	13	0.8885	13	0.7415	13	0.7422	13	0.8886	13
17	0.827	20	0.8884	20	0.7414	20	0.7418	20	0.8882	20
18	0.8254	16	0.8883	16	0.7414	16	0.7346	16	0.8789	16
19	0.8254	19	0.8883	19	0.7414	19	0.7346	19	0.8789	19
20	0.0241	1	0.005	1	0.0079	1	0.0667	1	0.0431	1

Table 16 TOPSIS results for IEEE-57 bus system

Alternatives	Weight_set1		Weight_set2		Weight_set3		Weight_set4		Weight_set5	
	Score	Rank	Score	Rank	Score	Rank	Score	Rank	Score	Rank
1	0.9747	12	0.985	12	0.9597	12	0.9584	12	0.9811	12
2	0.9664	14	0.9738	11	0.9571	11	0.9573	14	0.9729	14
3	0.8817	11	0.972	14	0.9537	14	0.7132	11	0.8055	11
4	0.8745	15	0.9689	15	0.9423	5	0.7013	15	0.797	15
5	0.8728	3	0.9612	3	0.9406	15	0.7013	5	0.7961	3
6	0.8728	5	0.9579	5	0.9366	3	0.7008	3	0.7957	5
7	0.8716	9	0.9561	9	0.9335	9	0.7005	9	0.7955	9
8	0.8716	7	0.9561	7	0.9335	7	0.7005	7	0.7955	7
9	0.7901	6	0.8172	6	0.8129	2	0.6716	6	0.7451	6
10	0.7893	10	0.8162	10	0.8127	6	0.6715	2	0.7444	10
11	0.7892	2	0.8157	2	0.8117	10	0.6711	10	0.7442	2
12	0.787	8	0.8133	8	0.8089	8	0.6701	8	0.7428	8
13	0.6442	4	0.6542	4	0.6536	4	0.5813	4	0.6233	4
14	0.6442	1	0.6542	1	0.6536	1	0.5813	1	0.6233	1
15	0.3811	13	0.3745	13	0.3748	13	0.4263	13	0.3955	13

Table 17 TOPSIS results for IEEE-118 bus system

Alternatives	Weight_set1		Weight_set2		Weight_set3		Weight_set4		Weight_set5	
	Score	Rank	Score	Rank	Score	Rank	Score	Rank	Score	Rank
1	0.9294	15	0.9577	15	0.8917	15	0.8803	11	0.9455	15
2	0.9205	11	0.9416	11	0.8857	11	0.8802	15	0.9371	11
3	0.9204	6	0.9416	6	0.8857	6	0.8801	6	0.9369	6
4	0.9097	5	0.941	13	0.8782	5	0.8748	5	0.9294	3
5	0.9033	13	0.9293	3	0.852	13	0.8488	3	0.9287	7
6	0.8988	3	0.9293	7	0.8485	3	0.848	7	0.9285	13
7	0.8986	7	0.9293	12	0.8485	7	0.8472	12	0.9278	12
8	0.8984	12	0.9268	5	0.8485	12	0.8411	13	0.9243	5
9	0.8897	9	0.9169	9	0.8433	9	0.84	9	0.914	9
10	0.8896	4	0.9169	4	0.8433	4	0.8394	4	0.9135	4
11	0.8891	10	0.9169	10	0.8433	10	0.8368	10	0.9113	10
12	0.3376	1	0.3348	1	0.3438	1	0.3465	1	0.3359	1
13	0.3375	8	0.3348	8	0.3438	8	0.3457	8	0.3356	8
14	0.3306	2	0.3324	2	0.3261	2	0.3254	2	0.3321	2
15	0.0268	14	0.0218	14	0.0236	14	0.0501	14	0.0349	14

Table 18 PROMITHEE results for IEEE-30 system

Alternatives	Weight_set1		Weight_set2		Weight_set3		Weight_set4		Weight_set5	
	Score	Rank	Score	Rank	Score	Rank	Score	Rank	Score	Rank
1	38	12	156	12	170	7	188	12	204	12
2	32	7	142	7	168	6	186	7	158	7
3	30	6	140	6	142	5	168	6	140	6
4	27	2	121	2	140	12	127	9	129	2
5	21	9	114	5	119	9	125	2	111	13
6	16	10	91	9	117	2	122	10	99	9
7	15	15	60	8	88	8	81	15	94	10
8	9	13	46	10	74	10	49	17	85	15
9	7	17	45	15	41	15	47	13	53	17
10	4	5	39	13	30	14	37	11	9	11
11	4	14	34	14	29	11	−18	14	−14	14
12	3	11	1	11	−7	17	−33	3	−61	3
13	−10	8	−3	17	−25	13	−66	5	−74	20
14	−11	4	−27	16	−57	4	−81	4	−77	4
15	−17	16	−53	4	−91	16	−106	20	−83	16
16	−19	3	−117	19	−110	18	−147	16	−93	19
17	−27	19	−138	18	−121	3	−152	8	−94	5
18	−32	18	−149	3	−181	19	−157	19	−180	8
19	−32	20	−186	20	−218	20	−174	18	−202	18
20	−58	1	−316	1	−308	1	−196	1	−204	1

Table 19 PROMITHEE results for IEEE-57 bus system

Alternatives	Weight_set1		Weight_set2		Weight_set3		Weight_set4		Weight_set5	
	Score	Rank	Score	Rank	Score	Rank	Score	Rank	Score	Rank
1	78	12	392	12	357	15	261	14	296	12
2	75	15	389	15	348	12	252	12	293	14
3	69	11	347	11	321	11	237	15	269	15
4	57	14	237	14	205	14	225	11	251	11
5	7	3	80	5	51	6	49	3	39	3
6	7	6	31	6	44	5	27	6	16	5
7	6	5	−9	3	7	10	23	10	7	6
8	1	10	−23	10	1	3	−20	5	−7	10
9	−23	9	−105	2	−85	2	−87	9	−89	9
10	−27	2	−105	9	−103	9	−109	2	−129	2
11	−39	7	−196	4	−199	7	−119	7	−137	7
12	−42	4	−217	7	−204	4	−143	13	−148	4
13	−47	13	−225	8	−205	8	−156	4	−177	13
14	−57	8	−257	13	−223	13	−211	1	−235	1
15	−65	1	−339	1	−315	1	−229	8	−249	8

Table 20 PROMITHEE results for IEEE-118 bus system

Alternatives	Weight_set1		Weight_set2		Weight_set3		Weight_set4		Weight_set5	
	Score	Rank	Score	Rank	Score	Rank	Score	Rank	Score	Rank
1	51	11	245	11	241	11	233	15	293	15
2	43	15	170	3	196	3	185	11	189	11
3	36	3	125	15	126	10	140	3	114	3
4	23	6	121	6	111	6	98	12	102	13
5	22	12	86	12	98	4	79	6	89	6
6	2	13	64	10	98	12	76	13	86	12
7	−2	7	40	4	65	15	26	7	38	7
8	−2	10	−5	5	3	5	−47	1	−19	1
9	−6	4	−50	13	−70	7	−47	8	−19	8
10	−11	5	−58	7	−76	13	−74	10	−106	14
11	−13	1	−59	1	−87	1	−86	4	−109	5
12	−13	8	−59	8	−87	8	−94	14	−136	10
13	−26	14	−106	14	−94	14	−101	5	−138	2
14	−50	2	−248	9	−238	9	−158	2	−144	4
15	−54	9	−266	2	−286	2	−230	9	−240	9

Table 21 VIKOR results for different test systems

Alternative	Rank														
	IEEE-30					IEEE-57					IEEE-118				
	WS1	WS2	WS3	WS4	WS5	WS1	WS2	WS3	WS4	WS5	WS1	WS2	WS3	WS4	WS5
1	7	7	7	7	7	12	15	15	14	14	3	10	11	3	3
2	6	6	6	6	6	14	12	12	12	12	11	11	10	7	7
3	9	9	9	9	9	15	11	11	13	13	5	4	3	5	12
4	12	5	5	12	12	11	14	14	11	11	6	3	6	1	5
5	10	12	12	10	10	5	5	5	15	15	12	6	4	12	11
6	2	10	10	2	2	6	6	6	5	5	7	5	5	11	6
7	11	2	2	11	11	9	9	9	6	6	4	12	12	8	1
8	15	8	8	15	15	13	10	10	9	9	10	15	15	6	8
9	17	11	11	17	17	3	3	3	3	3	15	7	1	15	4
10	3	15	15	3	3	10	2	2	7	7	1	13	8	4	9
11	5	14	14	13	13	7	7	7	10	10	8	1	7	9	15
12	8	17	17	20	20	4	4	4	4	4	9	8	13	14	10
13	14	4	4	14	14	2	13	13	2	2	13	9	9	10	14
14	13	3	3	5	5	1	8	8	1	1	14	14	14	13	13
15	4	18	18	8	8	8	1	1	8	8	2	2	2	2	2
16	18	13	13	4	4										
17	20	16	16	18	18										
18	16	20	20	1	1										
19	19	19	19	16	16										
20	1	1	1	19	19										

Table 22 Performance for the four methods with Weight_Set1, for different test systems

ALTERNATIVES	IEEE-30				ALTERNATIVES	IEEE-57				ALTERNATIVES	IEEE-118			
	AHP	TOPSIS	PROMITHEE	VIKOR		AHP	TOPSIS	PROMITHEE	VIKOR		AHP	TOPSIS	PROMITHEE	VIKOR
1	6	7	12	7	1	11	12	12	12	1	10	15	11	3
2	7	6	7	6	2	15	14	15	14	2	4	11	15	11
3	5	12	6	9	3	12	11	11	15	3	3	6	3	5
4	2	9	2	12	4	14	15	14	11	4	11	5	6	6
5	9	10	9	10	5	5	3	3	5	5	6	13	12	12
6	8	2	10	2	6	4	5	6	6	6	15	3	13	7
7	15	15	15	11	7	6	9	5	9	7	12	7	7	4
8	12	11	13	15	8	3	7	10	13	8	5	12	10	10
9	14	17	17	17	9	2	6	9	3	9	1	9	4	15
10	17	3	5	3	10	8	10	2	10	10	13	4	5	1
11	10	14	14	5	11	9	2	7	7	11	8	10	1	8
12	4	5	11	8	12	10	8	4	4	12	14	1	8	9
13	13	4	8	14	13	7	4	13	2	13	7	8	14	13
14	11	8	4	13	14	1	1	8	1	14	9	2	2	14
15	16	18	16	4	15	13	13	1	8	15	2	14	9	2
16	19	13	3	18										
17	18	20	19	20										
18	3	16	18	16										
19	20	19	20	19										
20	1	1	1	1										

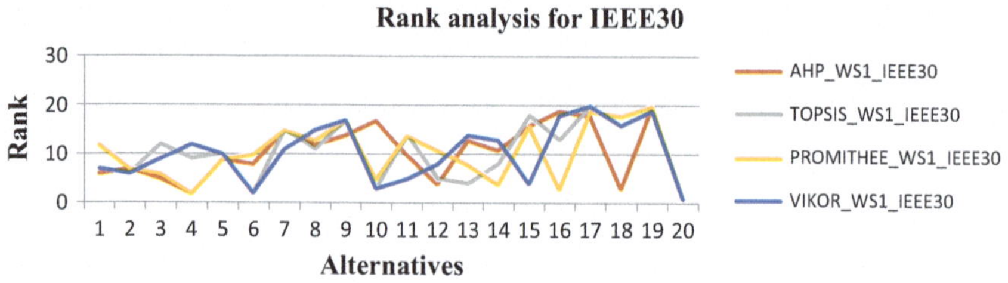

Fig. 4 Rank analysis for the four methods with Weight_Set1, for IEEE-30 test system

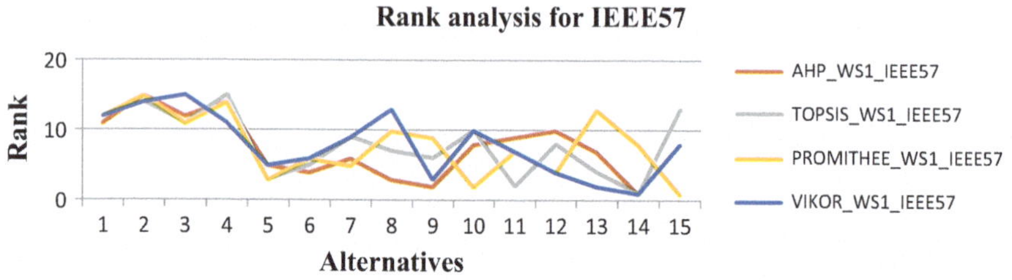

Fig. 5 Rank analysis for the four methods with Weight_Set1, for IEEE-57 test system

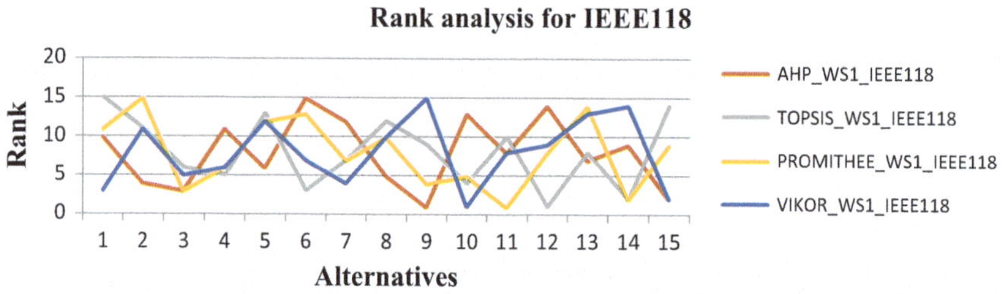

Fig. 6 Rank analysis for the four methods with Weight_Set1, for IEEE-118 test system

improve the overall security and stability of power systems. Therefore, it is suggested to make in-depth evaluation of the power system security under the new environment control based one PMUs data.

5 Conclusions

The selection of better alternatives from the set of best results is always the best policy. In the decision making, the alternative that outperforms the remaining alternatives will become the optimum alternative (or solution to our decision-making problem).

Table 23. Performance of the four methods with Weight_Set2, for different test systems

ALTERNATIVES	IEEE-30				ALTERNATIVES	IEEE-57				ALTERNATIVES	IEEE-118			
	AHP	TOPSIS	PROMITHEE	VIKOR		AHP	TOPSIS	PROMITHEE	VIKOR		AHP	TOPSIS	PROMITHEE	VIKOR
1	6	12	12	7	1	11	12	12	15	1	10	15	11	10
2	7	7	7	6	2	12	11	15	12	2	4	11	3	11
3	5	6	6	9	3	15	14	11	11	3	11	6	15	4
4	9	2	2	5	4	14	15	14	14	4	3	13	6	3
5	2	9	5	12	5	5	3	5	5	5	6	3	12	6
6	8	10	9	10	6	4	5	6	6	6	5	7	10	5
7	12	15	8	2	7	6	9	3	9	7	1	12	4	12
8	10	11	10	8	8	2	7	10	10	8	12	5	5	15
9	15	17	15	11	9	1	6	2	3	9	8	9	13	7
10	14	3	13	15	10	3	10	9	2	10	15	4	7	13
11	11	14	14	14	11	8	2	4	7	11	14	10	1	1
12	17	5	11	17	12	9	8	7	4	12	7	1	8	8
13	4	4	17	4	13	10	4	8	13	13	13	8	14	9
14	13	8	16	3	14	13	1	13	8	14	9	2	9	14
15	18	18	4	18	15	7	13	1	1	15	2	14	2	2
16	3	13	19	13										
17	16	20	18	16										
18	19	16	3	20										
19	20	19	20	19										
20	1	1	1	1										

Table 24 Performance of the four methods with Weight_Set3, for different test systems

ALTERNATIVES	IEEE-30				ALTERNATIVES	IEEE-57				ALTERNATIVES	IEEE-118			
	AHP	TOPSIS	PROMITHEE	VIKOR		AHP	TOPSIS	PROMITHEE	VIKOR		AHP	TOPSIS	PROMITHEE	VIKOR
1	6	7	7	7	1	12	12	15	15	1	10	15	11	11
2	7	6	6	6	2	11	11	12	12	2	3	11	3	10
3	5	12	5	9	3	14	14	11	11	3	4	6	10	3
4	2	9	12	5	4	15	5	14	14	4	11	5	6	6
5	9	10	9	12	5	5	15	6	5	5	1	13	4	4
6	12	2	2	10	6	6	3	5	6	6	14	3	12	5
7	8	15	8	2	7	2	9	10	9	7	6	7	15	12
8	10	11	10	8	8	13	7	3	10	8	12	12	5	15
9	15	17	15	11	9	8	2	2	3	9	5	9	7	1
10	13	3	14	15	10	4	6	9	2	10	8	4	13	8
11	11	14	11	14	11	9	10	7	7	11	7	10	1	7
12	14	5	17	17	12	3	8	4	4	12	15	1	8	13
13	17	4	13	4	13	10	4	8	13	13	13	8	14	9
14	3	8	4	3	14	1	1	13	8	14	9	2	9	14
15	4	18	16	18	15	7	13	1	1	15	2	14	2	2
16	20	13	18	13										
17	16	20	3	16										
18	18	16	19	20										
19	1	19	20	19										
20	19	1	1	1										

Table 25 Performance of the four methods with Weight_Set4, for different test systems

ALTERNATIVES	IEEE-30				ALTERNATIVES	IEEE-57				ALTERNATIVES	IEEE-118			
	AHP	TOPSIS	PROMITHEE	VIKOR		AHP	TOPSIS	PROMITHEE	VIKOR		AHP	TOPSIS	PROMITHEE	VIKOR
1	2	7	12	7	1	14	12	14	14	1	3	11	15	3
2	12	6	7	6	2	12	14	12	12	2	1	15	11	7
3	15	12	6	9	3	11	11	15	13	3	14	6	3	5
4	13	9	9	12	4	13	15	11	11	4	8	5	12	1
5	6	10	2	10	5	4	5	3	15	5	7	3	6	12
6	7	2	10	2	6	3	3	6	5	6	5	7	13	11
7	10	15	15	11	7	15	9	10	6	7	12	12	7	8
8	17	11	17	15	8	1	7	5	9	8	11	13	1	6
9	14	17	13	17	9	5	6	9	3	9	6	9	8	15
10	9	3	11	3	10	6	2	2	7	10	15	4	10	4
11	11	14	14	13	11	2	10	7	10	11	4	10	4	9
12	3	5	3	20	12	9	8	13	4	12	10	1	14	14
13	20	4	5	14	13	7	4	4	2	13	9	8	5	10
14	4	8	4	5	14	10	1	1	1	14	13	2	2	13
15	5	18	20	8	15	8	13	8	8	15	2	14	9	2
16	1	13	16	4										
17	8	20	8	18										
18	18	16	19	1										
19	16	19	18	16										
20	19	1	1	19										

Table 26 Performance of the four methods with Weight_Set5, for different test systems

ALTERNATIVES	IEEE-30				ALTERNATIVES	IEEE-57				ALTERNATIVES	IEEE-118			
	AHP	TOPSIS	PROMITHEE	VIKOR		AHP	TOPSIS	PROMITHEE	VIKOR		AHP	TOPSIS	PROMITHEE	VIKOR
1	2	7	12	7	1	12	12	12	14	1	3	15	15	3
2	15	6	7	6	2	14	14	14	12	2	1	11	11	7
3	12	12	6	9	3	11	11	15	13	3	14	6	3	12
4	13	9	2	12	4	13	15	11	11	4	7	3	13	5
5	17	10	13	10	5	15	3	3	15	5	12	7	6	11
6	14	2	9	2	6	3	5	5	5	6	8	13	12	6
7	6	15	10	11	7	4	9	6	6	7	15	12	7	1
8	7	11	15	15	8	5	7	10	9	8	5	5	1	8
9	10	17	17	17	9	6	6	9	3	9	11	9	8	4
10	4	3	11	3	10	9	10	2	7	10	6	4	14	9
11	9	14	14	13	11	2	2	7	10	11	13	10	5	15
12	20	5	3	20	12	7	8	4	4	12	4	1	10	10
13	11	4	20	14	13	10	4	13	2	13	10	8	2	14
14	3	8	4	5	14	8	1	1	1	14	9	2	4	13
15	1	18	16	8	15	1	13	8	8	15	2	14	9	2
16	5	13	19	4										
17	16	20	5	18										
18	19	16	8	1										
19	8	19	18	16										
20	18	1	1	19										

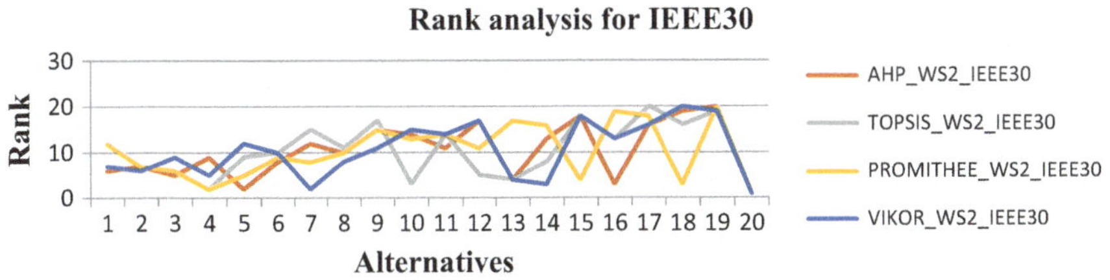

Fig. 7 Rank analysis for the four methods with Weight_Set2, for IEEE-30 test system

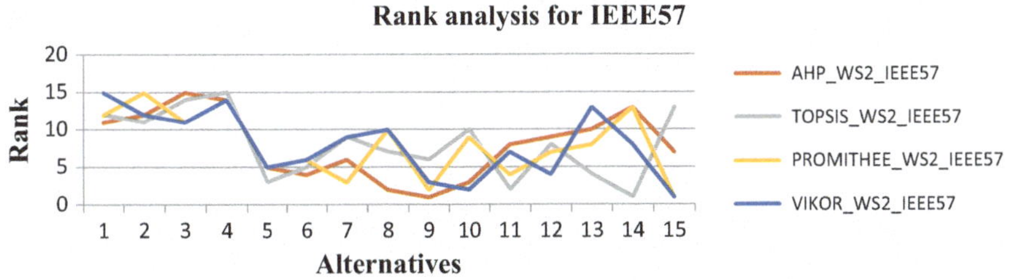

Fig. 8 Rank analysis for the four methods with Weight_Set2, for IEEE-57 test system

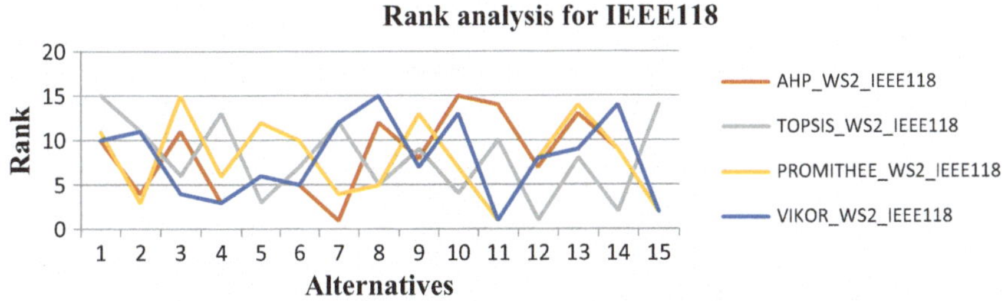

Fig. 9 Rank analysis for the four methods with Weight_Set2, for IEEE-118 test system

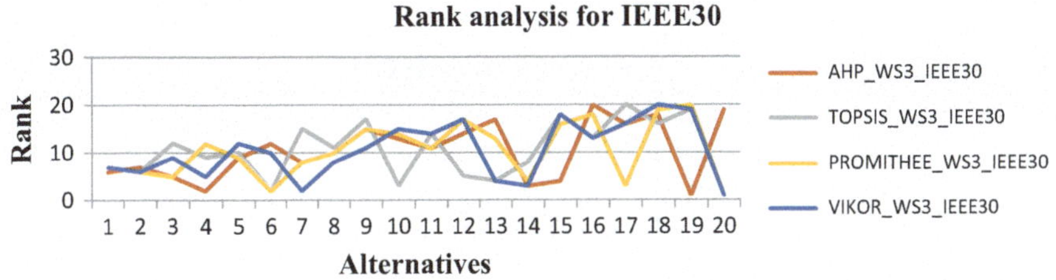

Fig. 10 Rank analysis for the four methods with Weight_Set3, for IEEE-30 test system

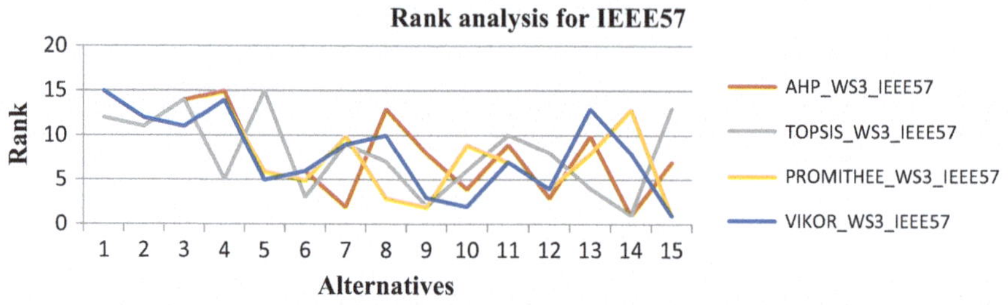

Fig. 11 Rank analysis for the four methods with Weight_Set3, for IEEE-57 test system

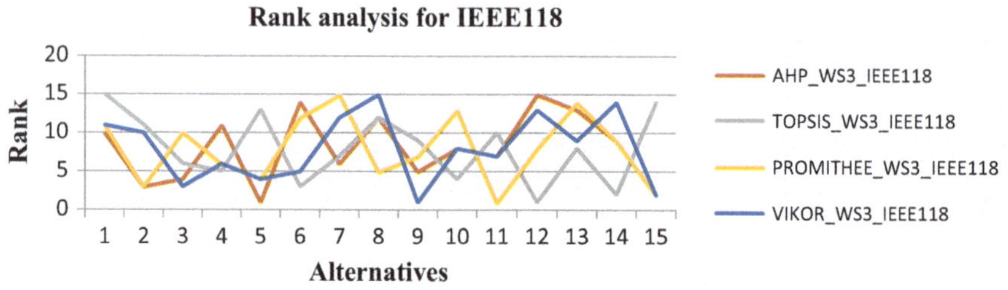

Fig. 12 Rank analysis for the four methods with Weight_Set3, for IEEE-118 test system

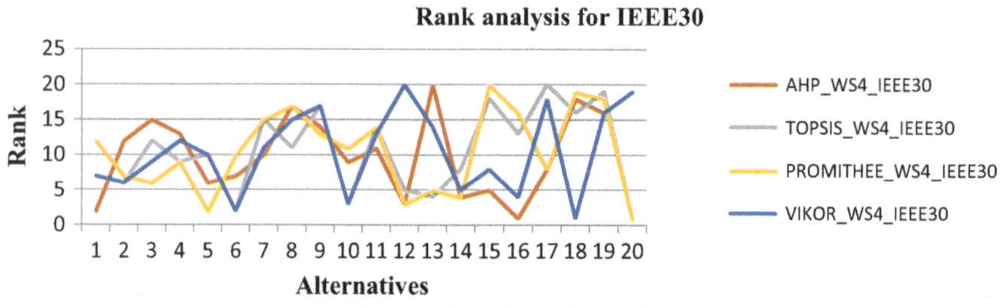

Fig. 13 Rank analysis for the four methods with Weight_Set4, for IEEE-30 test system

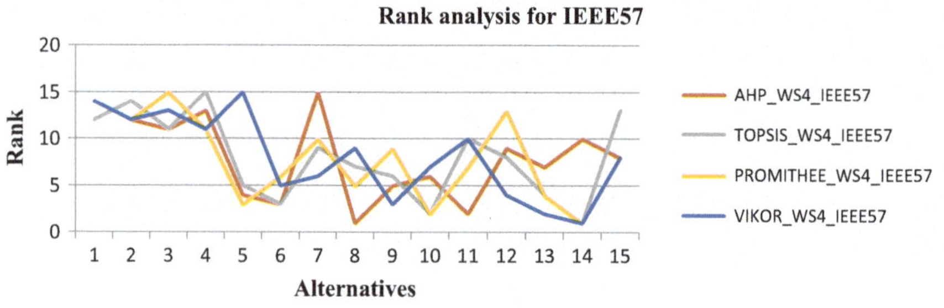

Fig. 14 Rank analysis for the four methods with Weight_Set4, for IEEE-57 test system

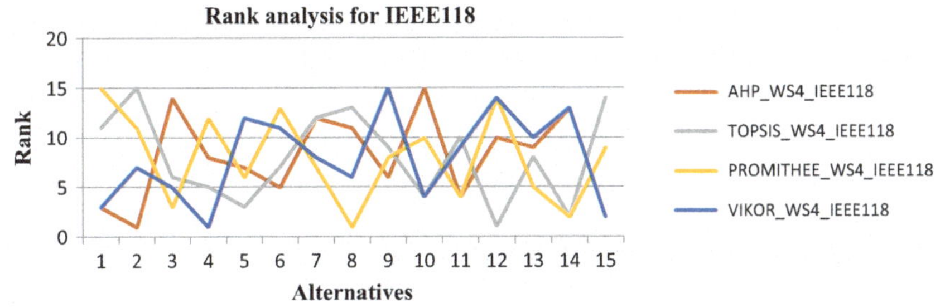

Fig. 15 Rank analysis for the four methods with Weight_Set4, for IEEE-118 test system

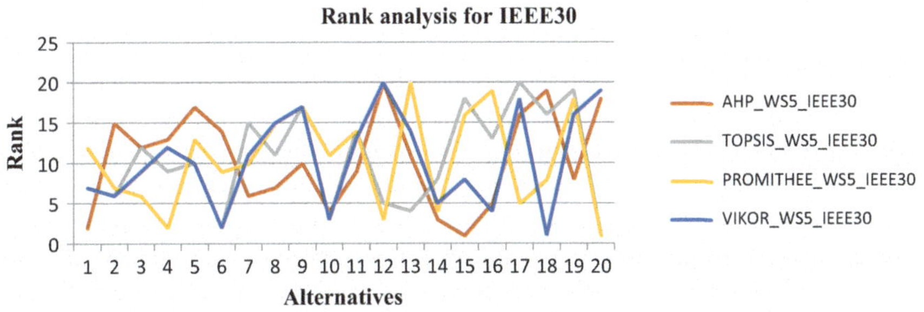

Fig. 16 Rank analysis for the four methods with Weight_Set5, for IEEE-30 test system

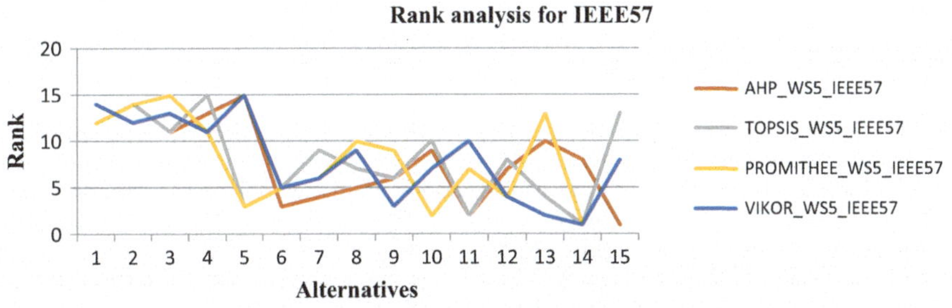

Fig. 17 Rank analysis for the four methods with Weight_Set5, for IEEE-57 test system

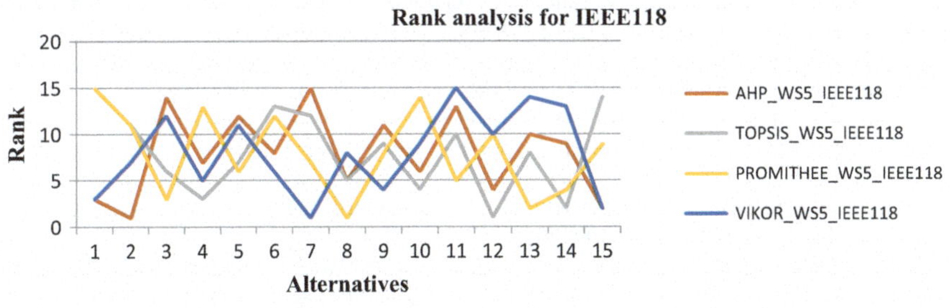

Fig. 18 Rank analysis for the four methods with Weight_Set5, for IEEE-118 test system

This paper first defines four indices that are essential for the power system operation and security. Then, using AHP, TOPSIS, PROMITHEE, and VIKOR, it finds the best alternative from the set of available alternatives. The miscellaneous features of the proposed algorithm will be as follows:

- Simplest method to identify best PMU-set that could serve the system many functions
- Since this approach ignores ZIB effect, the reliability of the solution is high.
- Apart from the collections of alternatives, it is fast and easiest method to identify best PMU-set
- Since the criteria consider critical element selection, the solution set assures prevention of power blackouts
- The islanding criterion considered for attributes formulation ensures a faster restoration scheme.

Based on the performance of MADM techniques, it is the utility's own choice to choose a suitable MADM technique. Interestingly, this approach gives a wider scope for the utilities who want their PMUs to serve system operation, control, and protection schemes. The only effort that they have to put for this is to formulate their criteria with relative priority. And, for more service-specific, the suitable PMU-set can be even identified by the selection of weights using any of the Artificial Neural Network, Fuzzy Logic, and the heuristic or meta-heuristic optimization techniques.

References

1. A.G. Phadke, J.S. Thorp, *Synchronized Phasor Measurements and Their Applications*, vol. 1 (Springer, New York, 2008), p. 81
2. N.P. Babu, P.S. Babu, D.V.S.S. SivaSarma, A wide-area prospective on power system protection: a state-of-art, in *2015 International Conference on Energy, Power and Environment: Towards Sustainable Growth (ICEPE)*. (IEEE, 2015), pp. 1–6
3. T.L. Baldwin, L. Mili, M.B. Boisen, R. Adapa, Power system observability with minimal phasor measurement placement. IEEE Trans. Power Syst. **8**(2), 707–715 (1993)
4. B. Xu, A. Abur, Observability analysis and measurement placement for systems with PMUs, in *IEEE PES Power Systems Conference and Exposition, 2004*. (IEEE, 2004), pp. 943–946
5. M. Khodadadi, H.H. Alhelou, P. Siano, A novel multi-objective OPP for power system small signal stability assessment considering WAMS uncertainties. IEEE Trans. Ind. Inf. (2019)
6. S. Chakrabarti, E. Kyriakides, D.G. Eliades, Placement of synchronized measurements for power system observability. IEEE Trans. Power Delivery **24**(1), 12–19 (2008)
7. F. Aminifar, M. Fotuhi-Firuzabad, M. Shahidehpour, A. Khodaei, Observability enhancement by optimal PMU placement considering random power system outages. Energy Syst. **2**(1), 45–65 (2011)
8. V.K. Tran, H.S. Zhang, Optimal PMU placement using modified greedy algorithm. J. Control Autom. Electr. Syst. **29**(1), 99–109 (2018)
9. A.A. Abdelsalam, K.M. Hassanin, A.Y. Abdelaziz, H.H. Alhelou, Optimal PMUs placement considering ZIBs and single line and PMUs outages. AIMS Energy **8**(1), 122 (2020)
10. S. Ghosh, J.K. Das, C.K. Chanda, Placement of phasor measurement unit for complete observability of an isolated microgrid system. Microsyst. Technol. **25**(12), 4671–4674 (2019)

11. N.P. Babu, P.S. Babu, D.S. Sarma, A reliable wide-area measurement system using hybrid genetic particle swarm optimization (HGPSO). Int. Rev. Electr. Eng. **10**(6), 747–763 (2015)
12. R. Sodhi, S.C. Srivastava, S.N. Singh, Multi-criteria decision-making approach for multi-stage optimal placement of phasor measurement units. IET Gener. Transm. Distrib. **5**(2), 181–190 (2011)
13. O. Linda, M. Manic, A. Giani, M. McQueen (2013). Multi-criteria based staging of optimal PMU placement using fuzzy weighted average, in *2013 IEEE International Symposium on Industrial Electronics*. IEEE, pp. 1–8
14. K. Sadanandan Sajan, A. Kumar Mishra, V. Kumar, B. Tyagi, Phased optimal PMU placement based on revised analytical hierarchy process. Electr. Power Compon. Syst. **44**(9), 1005–1017 (2016)
15. T.K. Maji, P. Acharjee, Multiple solutions of optimal PMU placement using exponential binary PSO algorithm for smart grid applications. IEEE Trans. Ind. Appl. **53**(3), 2550–2559 (2017)
16. V. Rajasekaran, K. Chandrasekaran, A multi-objective placement of phasor measurement units considering observability and measurement redundancy using firefly algorithm. J. Electr. Eng. Technol. **10**(2), 474–486 (2015)
17. Enquiry Committee (2012) Report of the enquiry committee on grid disturbance in northern region on 30th July 2012 and in northern, eastern & north-eastern region on 31st July, 2012. Enquiry Committee, New Delhi, Tech. Rep.
18. N.P. Babu, P.S. Babu, D.S. Sarma, A new power system restoration technique based on WAMS partitioning. Eng. Technol. Appl. Sci. Res. **7**(4), 1811–1819 (2017)
19. C.L. Hwang, A.S.M. Masud, *Multiple Objective Decision Making—Methods and Applications: A State-of-the-Art Survey*, vol. 164. (Springer Science & Business Media, 2012)
20. E. Triantaphyllou, Multi-criteria decision making methods, in *Multi-Criteria Decision Making Methods: A Comparative Study* (Springer, Boston, MA, 2000, pp. 5–21)
21. T.L. Saaty, *The Analytic Hierarchy Process* (McGraw Hill, New York, 1980)
22. T.L. Saaty, *Fundamentals of Decision Making and Priority Theory with the Analytic Hierarchy Process*, vol. 6. (RWS publications, 2000)
23. R.V. Rao, *Decision Making in the Manufacturing Environment: Using Graph Theory and Fuzzy Multiple Attribute Decision Making Methods*. (Springer Science & Business Media, 2007)
24. J.P. Brans, P. Vincke, B. Mareschal, How to select and how to rank projects: the PROMETHEE method. Eur. J. Oper. Res. **24**(2), 228–238 (1986)
25. P.L. Yu, A class of solutions for group decision problems. Manage. Sci. **19**(8), 936–946 (1973)
26. M. Zeleny, Multiple criteria decision making: eight concepts of optimality. Human Syst. Manag. **17**(2), 97–107 (1998)
27. G.H. Tzeng, C.W. Lin, S. Opricovic, Multi-criteria analysis of alternative-fuel buses for public transportation. Energy Policy **33**(11), 1373–1383 (2005)
28. H.H. Alhelou, M.H. Golshan, J. Askari-Marnani, Robust sensor fault detection and isolation scheme for interconnected smart power systems in presence of RER and EVs using unknown input observer. Int. J. Electr. Power Energy Syst. **99**, 682–694 (2018)
29. H.H. Alhelou, M.E. Hamedani-Golshan, R. Zamani, E. Heydarian-Forushani, P. Siano, Challenges and opportunities of load frequency control in conventional, modern and future smart power systems: a comprehensive review. Energies **11**(10), 2497 (2018)
30. H. Haes Alhelou, M.E. Hamedani Golshan, M. Hajiakbari Fini, Wind driven optimization algorithm application to load frequency control in interconnected power systems considering GRC and GDB nonlinearities. Electr. Power Compon. Syst. **46**(11–12), 1223–1238 (2018)
31. H.H. Alhelou, M.E. Hamedani-Golshan, E. Heydarian-Forushani, A.S. Al-Sumaiti, P. Siano, Decentralized fractional order control scheme for LFC of deregulated nonlinear power systems in presence of EVs and RER, in *2018 International Conference on Smart Energy Systems and Technologies (SEST)*. (IEEE, 2018, September), pp. 1–6
32. H.H. Alhelou, M.E.H. Golshan, N.D. Hatziargyriou, Deterministic dynamic state estimation-based optimal lfc for interconnected power systems using unknown input observer. IEEE Trans. Smart Grid (2019)

33. H. Haes Alhelou, M.E. Hamedani-Golshan, T.C. Njenda, P. Siano, A survey on power system blackout and cascading events: research motivations and challenges. Energies **12**(4), 682 (2019)
34. H.H. Alhelou, S.J. Mirjalili, R. Zamani, P. Siano, Assessing the optimal generation technology mix determination considering demand response and EVs. Int. J. Electr. Power Energy Syst. **119**, 105871 (2020)

Coordinated Designs of Fuzzy PSSs and Load Frequency Control for Damping Power System Oscillations Considering Wind Power Penetration

Nesrine Mekki and Lotfi Krichen

Abstract The damping enhancement of power system oscillations remains one of the challenging current interests for secure and reliable operation. This paper presents a comprehensive overview of a novel control scheme that considers synchrophasors and Power System Stabilizers in coordination with an optimized Load Frequency Control loop in order to resolve the undamped local and wide-area oscillatory troubles. Accordingly, a Robust Fuzzy PSS using local signals is first examined. Additionally, an Inter-Area PSS based on high-sampling rate phasor measurement unit is investigated. In fact, using time synchronized measurements as control input signals will participate effectively in monitoring the energy management process. Thus, another configuration mixing local and remote control inputs of a Mixed-PSS is proposed. Performances of these PSSs are evaluated in coordination with a tuned PI-based load frequency control design under different operating conditions. Results on a modified 9-Bus IEEE test system including DFIG wind turbines are reported in order to justify the proposal's applicability.

Keywords Phasor measurement unit · Load frequency control · Inter-area oscillations · Power system stabilizer · Fuzzy logic controller · Damping enhancement

Nomenclature

LFC	Load Frequency Control
IAPSS	Inter-Area Power System Stabilizer
RFPSS	Robust Fuzzy Power System Stabilizer
MPSS	Mixed-Power System Stabilizer

N. Mekki (✉) · L. Krichen
Department of Electrical Engineering, National School of Engineers of Sfax, Sfax, Tunisia
e-mail: mekki_nesrine@hotmail.fr

L. Krichen
e-mail: lotfi.krichen@enis.rnu.tn

H. Haes Alhelou et al. (eds.), *Wide Area Power Systems Stability, Protection, and Security*, Power Systems, https://doi.org/10.1007/978-3-030-54275-7_6

DFIG WTs	Doubly Fed Induction Generator Wind Turbines
ACE	Area Control Error
TG	Turbine Governor
PI	Proportional-Integral
PID	Proportional-Integral-Derivative
PSS	Power System Stabilizer
MB-PSS	Multi Band-Power System Stabilizer
FLC	Fuzzy Logic Controller
WACS	Wide Area Control Systems
PMU	Phasor Measurement Unit

1 Introduction

Recently, with the continuous increasing of electricity demand and high complexity of power grid interconnections, power systems will be eventually vulnerable to several problems during the system operation mainly caused not only by low frequency oscillations but also by inter-area oscillations. In fact, poorly-damped oscillations may lead considerably to system wide breakups especially under severe operating conditions. As local measurements based controls have restricted modal observability [1], using wide-area signals control determinations in such circumstance could be considerably beneficial in damping inter-area oscillations. Hence, in order to enhance the stability of power system swings prompted by these oscillatory modes, Power System Stabilizer (PSS) installation is economically and effectively adopted. Particularly, conventional PSS is not able enough to damp sufficiently inter-area oscillations. Getting in real time the system state variables would certainly be effective. Thus, the use of fast communication devices like PMUs will facilitate mostly the Wide Area Control Systems functionalities [2] and allow consumers to participate rationally in the electricity markets especially during peak demands when the system is more susceptible to experience the instability problem.

Traditionally, candidate input signals which are all locally available at the power plant ensure supplementary control action over the generator excitation system. Later, several advents use remote signals that are derived from PMUs as PSS inputs including local signals of the remaining plants [3]. Primarily, this technology affords globally-synchronized measurements of current and voltage phasors, phase angles, line flows and frequency resulting in generating accurate system-wide data sets mainly appropriate for damping targets. Over the past decades, several approaches have been suggested to damp local and inter-area oscillations. Some of them consider different control devices like PSS and FACTS [4] while others are suitable for operational conditions. Literately, various researchers have been presented a global PSS [5] based on multiple methods such as mixed integer non-linear programming, genetic algorithm-based dynamic search spaces and deviation-error vector and dynamic feedback control signal scaling [1, 5]. Likewise, another review has been specified several basis in PSS design including frequency response method, modal decomposition method [6], adaptive wide area PSS, and Eigen-structure-based performance index [7]. Other techniques like adaptive time delay compensator approach and fuzzy logic wide-area damping controller [8] have been stated for signal delays compensation purposes. In the same regard, damping inter-area oscillations stills as a challenging issue oriented primarily for system dynamic security enhancement as reported by Murali and Rajaram [9]. Currently, power system utility uses practically conventional PSS in their control process even it cannot provide satisfactory results over wider ranges of operating conditions [9, 10]. Equally, other types of PSS have been proposed such as proportional-integral PSS and proportional-integral derivative PSS. Subsequently, several studies regarding the PSS design have been preferred Fuzzy logic based technique. In fact, experienced human operators afford qualitative rules for effective control purposes especially in case of the unavailability of a mathematical model for the power plant [9]. Similarly, in order to improve the performance of Fuzzy logic based PSSs, Hybrid PSSs using fuzzy logic and/or neural networks or Genetic Algorithms have been illustrated in some literature surveys [10].

One of the most important aspects for damping unstable inter-area oscillations is to select the appropriate control signals. Over the last years, many procedures have been industrialized and verified to damp these oscillations by using both of local and global signals. Some of them focused on the signal type while the others taken into account the signal selection methods. Commonly, generator speed, active power and terminal-bus frequency are used widely as PSS input signals. For local control, generator speed deviations is mostly selected as an input signal [11]. Further studies in [12] have been chosen angle differences between buses as input signal. In fact, several methods based on modal observability have been developed for input signal selection in the literature. In [13], the interchanged active power of lines have been proposed as a stabilizer global signal for damping improvement aims. This input signal ensures high inter-area modes observability under different operating conditions. Using wide-area signals, the power transfer could be maximized. Another review that is based on sensitivity analysis of eigenvalues has been investigated in [8, 9, 14] to identify the transmission lines that are involved in each swing mode while analyzing the modal observability related to the network variables like voltage and current phasors which are delivered by PMUs. In [15], the clustering algorithm is recommended. Correspondingly, a comparison between geometric and residue approaches for selecting the global signal has been mentioned in [16]. As well, other methodologies based on virtual generators concept and trajectory-based supplementary damping control have been literately reported [15, 17]. Additionally, thorough attempts have been studied the PSS control signals using the center of inertia concept and calculating the frequency damping ratio for each inter-area mode [18].

In addition to PSSs control, modern energy management systems associate another multi-level supervisory schemes such as importantly the load–frequency control (LFC). Actually, it is considered as an additional secondary control which increases damping the system oscillations. The first step in the LFC mechanism is to add an Area Control Error (ACE) that acts on the Turbine Governor's (TG) load reference settings [19]. It is a function of the exchanged real power deviations that the LFC desires to make zero. For a multi area system, different generation control approaches have been proposed literately since the 1970s [19, 20]. In [21], an overview regarding the LFC issue have been stated, it describes its schemes, system models, control procedures, load characteristics, and the mutual interaction with the renewable resources. Similarly, various approaches in relation with Proportional–Integral–Derivative (PID), adaptive structure, intelligent, and networked control schemes have been revealed. Pandey et al. suggested an LFC's survey for distribution and conventional power systems [22]. Rerkpreedapong et al. proposed two decentralized control designs [23] using the linear matrix inequalities method and Proportional-Integral (PI) controller which is tuned by the genetic algorithm. Likewise, the fuzzy controller is considered in [24]. Particularly, PI controllers are commonly used in industrial applications which aim to reduce the steady-state error to zero [25].

Almost, previous studies have been proven that optimally-tuned PSSs with local signals are not able enough to damp inter-area swings. Thus, the main contribution of this paper is to investigate a novel control scheme that combines a Robust Fuzzy Power System Stabilizer (RFPSS) with an optimized PI-based LFC in order to increase damping of power system oscillations and improve the dynamic performances. The availability of PMUs signals affords new accurate measurement sets that are evaluated alongside traditional local signals. Hence, a new Inter-Area PSS (IAPSS) design is proposed taking into account both of local and wide-area signals in order to regulate the generator excitation system and then provide better system response. In addition, another configuration mixing local and remote control inputs of a Mixed-PSS (MPSS) is suggested. The rest of paper carries out the simulation studies of a modified IEEE 9-bus test system including DFIG wind turbines to judge the control scheme applicability under different operating conditions.

2 Description of the Power System Under Study

The undamped oscillations problem is analyzed by considering a 3-machine, 9-bus IEEE test system, whose single line diagram is represented in Fig. 1. The system frequency is 50 Hz. The generator 'Gen 1' is connected to the reference bus. The standard system state is reconfigured by replacing the generators 'Gen 2' and 'Gen 3' with new power plants which involve new control schemes. Each area is equipped with the performed PSSs and an optimized LFC process. Effectiveness of these controllers will be tested whenever incorporating DFIG wind turbines into the power system with different rates under different operating conditions in an advanced step.

- F_i: Frequency of a PMU-Equipped Generator, in Hz.

With I = 1, 2 (when Considering a Two-Area System Where Each Area Includes One Generator).

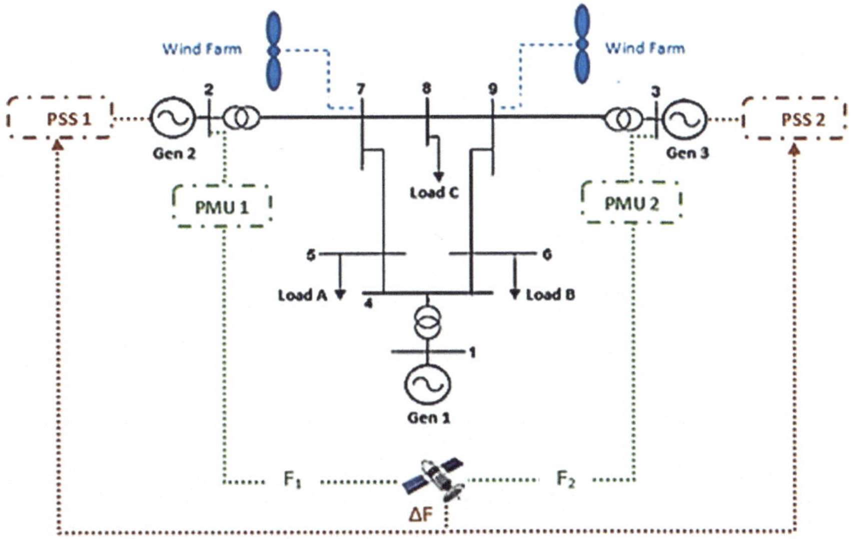

Fig. 1 Single line diagram of a 3-machine, 9-bus IEEE test system

3 Proposed Control Scheme

3.1 PSS-Based Local Control

Literately, compared to conventional PSSs, fuzzy logic based PSSs show better operational performances. In fact, by utilizing the output values of the power plant, it could apply the appropriate control actions in conformity with the rule-base. The key feature of a fuzzy controller is that it handles approximate data in a systematic way. It is ideal for modeling complex systems where an inexact model or ambiguous knowledge exist and for controlling non-linear systems [26]. Typically, the main configuration of a FLC consists of four basic components: selection and fuzzification of the input variables, fuzzy rule identification, rule inference and defuzzification. The dynamic performance of the system is primarily defined through its state variables which are taken as input signals to the proposed FLC. The fuzzified input and output variables are expressed into linguistic variables which are distinguished by different labels and membership functions. In fact, several studies have been reported that seven linguistic variables are practically enough, but, although the control accuracy will be improved as much as the number of variables increase, the computational time will rise too. In addition, choosing the appropriate membership function shape is crucial for a particular problem in relation with the fuzzy inference system. Comparisons between the different shapes such as triangular, trapezoidal, Gaussian and sigmoidal have been proved that both of the Gaussian and the triangular membership functions ensure similarly the most effective performance of the fuzzy logic based PSS design for all test conditions [27].

Table 1 Decision table for FLC

		Accelerating power						
		NB	NM	NS	Z	PS	PM	PB
Speed deviation	NB	NB	NB	NB	NB	NM	NS	Z
	NM	NB	NB	NM	NM	NS	Z	PS
	NS	NB	NM	NM	NS	Z	PS	PM
	Z	NM	NM	NS	Z	PS	PM	PM
	PS	NM	NS	Z	PS	PM	PM	PB
	PM	NS	Z	PS	PM	PM	PB	PB
	PB	Z	PS	PM	PB	PB	PB	PB

In the present work, the performed control scheme based on a RFPSS design uses the *rotor speed deviation* ($\Delta\omega$) and *the accelerating power (Pa)* as local input linguistic variables. The output linguistic variables elected for this controller is the *stabilizing voltage* (V_s). Each variable is assigned seven linguistic fuzzy subsets. Each subset is associated with a triangular membership function, which is chosen for reasons of simplicity in terms of implementation and fast computation [28]. Particularly, the trapezoidal shape is designated only for the upper and the lower boundaries. The fuzzy sets that define the relation between the inputs and the output is done through a rule-base by utilizing seven linguistic terms which are Negative Large (NL), Negative Medium (NM), Negative Small (NS), Zero (Z), Positive Small (PS), Positive Medium (PM) and Positive Large (PL) such as given in Table 1. The rules are framed taking into account the nature of the system performance and the communal sense.

Generally, the *i*th rule can be presented as:

If *speed deviation* ($\Delta\omega$) is NB and *active power deviation* $(Pa)_i$ is NB then $(V_s)_i$ is NB.

Commonly, these control rules are consequent results from past experiences, intuitions, off-line simulations and expert operator decisions. The Mamdani inference mechanism and the Centroid method of defuzzification are selected for the proposed controller. Accordingly, the input–output control surface is obtained as displayed in Fig. 2.

Afterward, in order to justify the effectiveness of the proposed control scheme, the damping performance of the RFPSS is compared to a tuned Multiband-PSS (MB-PSS) that has already shown great improvements of the system response when compared to other types of PSSs. Its conceptual representation is depicted in Fig. 3. Actually, the tuning strategy of the MB-PSS is based on the symmetrical approach. Therefore, only six parameters are required. The center frequency and gain of each band of the lead-lag compensation block are suitably varied so as to get a nearly flat phase response at the frequency of interest. In fact, the differential filters are supposed to be symmetrical bandpass filters tuned at the center frequencies $\boldsymbol{F_L}$,$\boldsymbol{F_I}$ and $\boldsymbol{F_H}$.

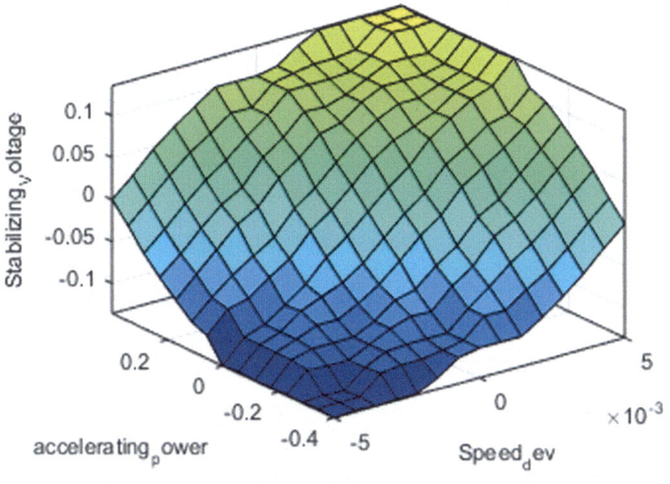

Fig. 2 Input–output control surface of RFPSS

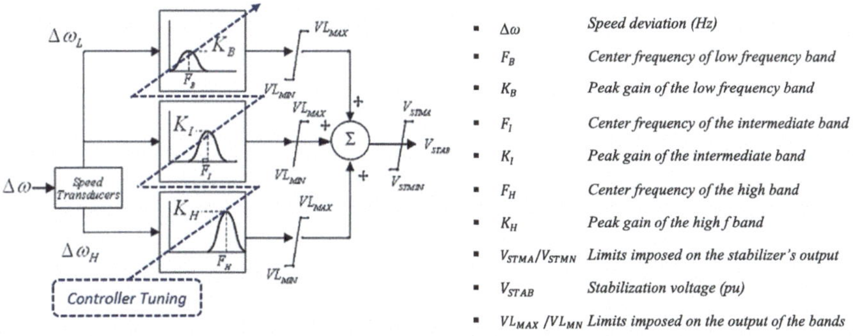

Fig. 3 Conceptual representation of the tuned MB-PSS

The peak magnitude of the frequency responses is settled individually through the three gains $\boldsymbol{K}_L$, $\boldsymbol{K}_I$ and $\boldsymbol{K}_H$.

Based on the literature review [29], the MB-PSS parameters are optimally tuned and summarized in Table 2.

The PSS structure is based on three separate working bands. The first band presents the low band which is associated with global oscillation modes. The intermediate one is related to inter-area modes while the high band deals typically with the local modes. Each band is composed by a differential bandpass filter, a gain, and a limiter. In order to guarantee robust damping of the existing electromechanical oscillations, this PSS may include at all frequencies of interest reasonable phase advance that will compensate the inherent lag between the resultant electrical torque and the field excitation.

Table 2 Optimal MB-PSS settings

R	V_{smax}	V_{smin}	V_{Imax}	V_{Hmax}	V_{Imin}	V_{Hmin}	V_{Lmin}	V_{Lmax}	K_L	F_L	K_I	F_I	K_H	F_H
1.2	0.1	−0.1	0.6	0.6	−0.6	−0.6	−0.02	0.02	6	0.19	30	1.1	150	12

3.2 PMU-Based Wide-Area Control

Depending on the control design purposes, some signals are better competitive candidates than others. Lately, global signals measured directly by PMUs are considered favorably as new alternatives to local signals. The $\Delta\omega$ signals must be synchronized whenever used [30]. Till date, the aspect of PSSs using PMU control input signals has not been addressed in the literature but currently being investigated by numerous researchers. PMU is a high-accuracy platform that could contribute valuably to the dynamic monitoring of transient processes in modern electric power systems. Actually, it samples real-time values of currents and voltages. The time synchronization through GPS allows comparing the measured synchrophasors from different locations far apart and providing situational awareness which diminishes uncertainty in the decision-making.

The PMU block used in this paper is inspired by the IEEE Std C37.118.1-2011. It is based mainly on a Phase-Locked Loop which calculates the positive-sequence component of the input signal. In fact, the three-phase PLL tracks evidently the phase and frequency of the sinusoidal three-phase signal via an internal frequency oscillator which is regulated through the control system in order to keep the phase difference at zero. Afterwards, the positive-sequence components including the magnitude and phase of the input signal are computed over a running window of one cycle of the fundamental frequency that is tracked previously throughout the PLL closed-loop control system. Importantly, the reference frame needed for the computation is set via the angle given also by the PLL and varying between 0 and 2*pi, synchronized on zero crossings of the positive-sequence of the fundamental. The PMU outputs define the magnitude, the phase, the frequency and the rate of change of frequency of the positive-sequence component of the input signal at the fundamental frequency. The sample time T_S is expressed as follows:

$$T_S = \frac{1}{f_n \times N_{sr}} \tag{1}$$

f_n Nominal frequency, in Hz.

N_{sr} Sampling rate, in point/cycle.

3.2.1 PSS with Two Remote Inputs

In the present study, a new inter-area fuzzy based PSS design (IAPSS) is suggested considering two input signals which are the *frequency difference (Δf)* and its *differentiation ($\dot{\Delta f}$)*, and a single output which represents *the stabilizing signal (V_s)*. Beforehand, each generator terminal is equipped with one PMU to measure accurately the desired input signals which will vary as well as the consumed power changes. The exciter input of each generator is controlled using the generators frequencies given

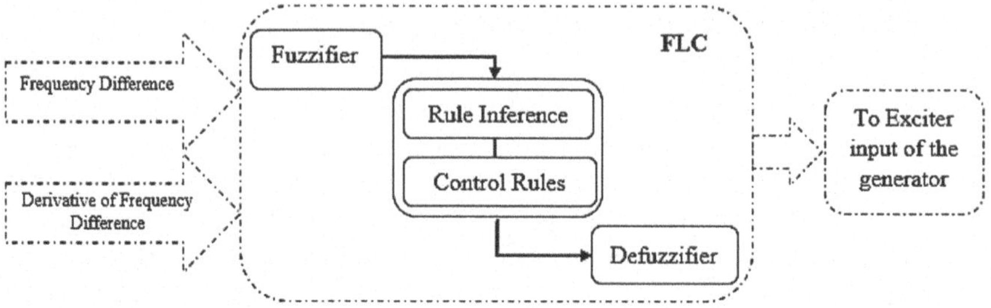

Fig. 4 Block diagram of FLC for one area

by the existing PMUs. These PMUs outputs are evidently considered as wide area measurements that communicate each generator to the remaining areas constituting the power system. Based on this scheme, inter-area controllers are performed referring to the fuzzy inference mechanism mentioned in the previous section. The basic configuration of a Mamdani FLC is presented in Fig. 4.

3.2.2 PSS with Remote and Local Inputs

As global signal's involvement shows satisfactorily results regarding the damping of system oscillations, further examinations combining one local PSS input and one remote input obtained by PMU are carried out. In fact, in the present work, another fuzzy based PSS design is performed considering a mixture between *the frequency difference* (Δf) as a global input signal and the *accelerating power (Pa)* as a local input signal.

The main objective is to investigate superior local mode damping with higher inter-area mode observability by using simultaneously local and global control input signals. Accordingly, performances of the suggested Mixed-PSS (MPSS) configuration are compared later against the previously-mentioned designs of PSSs Viz. RFPSS and IAPSS controllers.

4 Involvement of LFC Design

The aim of this section is to discuss the efficiency of combining an enhanced LFC design with the performed PSSs in order to increase the damping of system oscillations. In fact, the LFC scheme will involve an appropriate control loop that is able enough to adjust the system frequency to the scheduled set point values effectively after any load change or fault occurrence. Particularly, PI controller is chosen to be applied for the current LFC system due to its simplicity in execution and its ability to afford fast response and zero steady state error. Actually, choosing randomly the controller parameters will not guarantee the best dynamic performances regarding the system stability. Hence, special attentions are given for establishing an optimal tuning approach of the PI parameters in order meet the desired requirements whenever exposed to load demand variations or sudden disturbances.

The Nonsmooth H_∞ minimization technique is adopted in the present paper for optimal tuning purposes. This approach is adopted in multiple control designs for damping of large power systems [31]. It solves in general the following constrained problem:

$$Minimize \max_i f_i(x) \tag{2}$$

$$Subject\ to \max_j g_j(x) < 1,\ for\ x_{\min} < x < x_{\max} \tag{3}$$

The optimization is reached by resolving a sequence of unconstrained subproblems given by:

$$\min_x max(\alpha f(x), g(x)) \tag{4}$$

$f_i(x)$*and* $g_j(x)$: *Normalized values of soft and hard tuning requirements.*

x: vector of PI parameters to tune.

x_{min} *and* x_{max}: *Minimum and maximum values of the free parameters of the controller.*

α: *A multiplier which is adjusted until convergence to the solution of the original constrained problem.*

The auxiliary action of the LFC based-tuned PI controller is incorporated into the studied power plants in coordination with the performed designs of PSSs Viz. RFPSS, IAPSS and MPSS as depicted in Figs. 5, 6 and 7, respectively, in order to restrain the system deviations efficiently while maintaining fast response and robust stability. In an advanced step, the simulation results will address mainly the effectiveness of

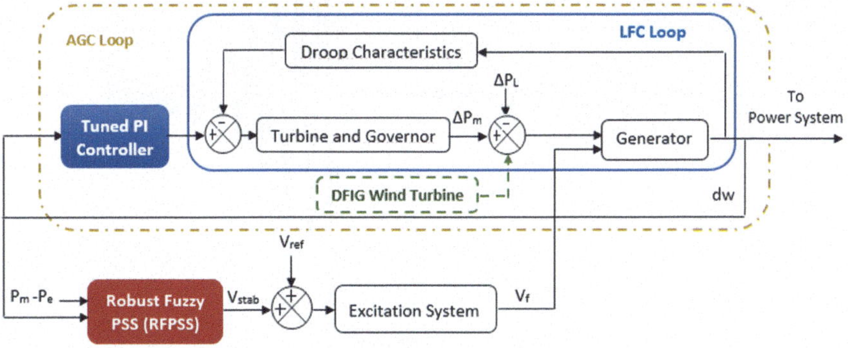

Fig. 5 Block diagram of the LFC-RFPSS control scheme in a single area power plant

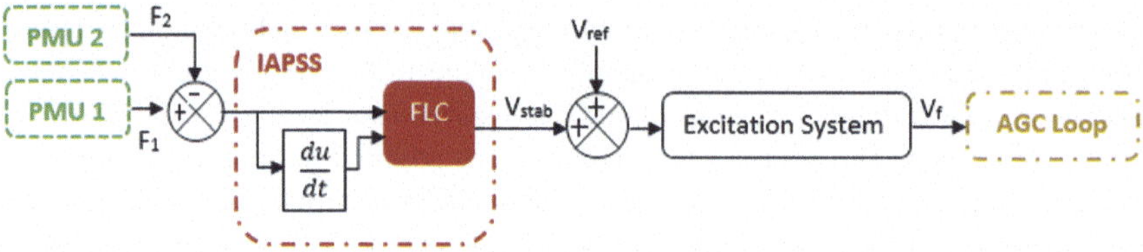

Fig. 6 Block diagram of the LFC-IAPSS control scheme in a single area power plant

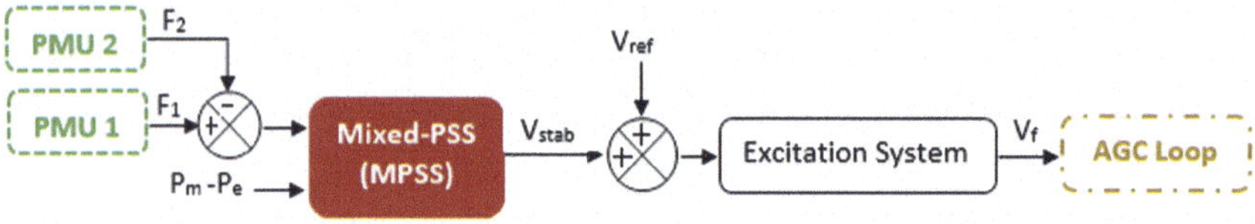

Fig. 7 Block diagram of the LFC-MPSS control scheme in a single area power plant

the suggested control scheme whenever integrating the wind energy sources in the power system.

5 Simulation and Discussion

For reasons of simplicity, this study elects as example of concern the generator number 2, 'Gen 2', and a variable load located at the nearest bus, 'Bus 8'. The system under study is subjected to an increase in load demand at 3 s and a three-phase fault

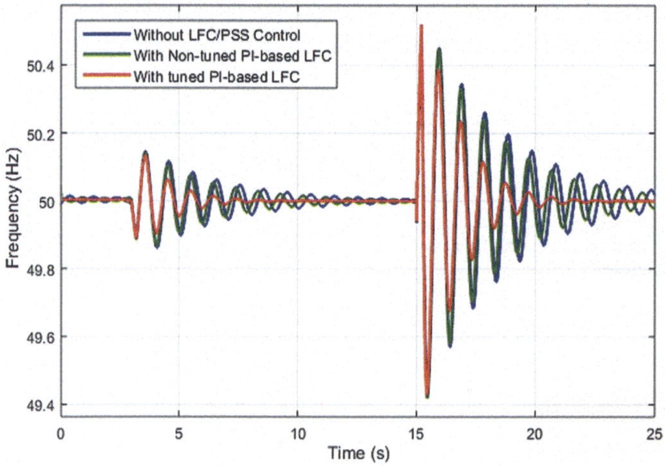

Fig. 8 Frequency response measured at generator 'Gen 2'

of 0.2 s duration on that bus at 15 s. The main interest first is to evaluate the efficiency of applying an optimized LFC loop in reaction to the changing loading conditions. Hence, Fig. 8 which presents different frequency responses of 'Gen 2', from which, it can be noted that the tuned PI-Based LFC shows better damping for the frequency oscillations than the non-tuned controller. Actually, during overloading and fault conditions, the frequency fluctuations persist for a long period, but, the application of the tuning process helps the governor system to absorb effectively the arising swings which are attenuated gradually in number and amplitude and maintain finally a frequency recovered close to the scheduled values.

5.1 Local Versus Wide-Area Damping Control

Generally, local PSSs are not able enough to provide global and real-time vision of power systems. Hence, the inclusion of accurate remote inputs in the PSS's control loop becomes a key factor in considering better dynamic performances for wide-area oscillations. The major goal of the present work is to investigate the efficiency of the designed controllers in combination with the suggested LFC loop whenever exposed to sudden disturbances. For instance, several comparisons between the performed PSSs among different operating conditions are made to achieve further damping enhancement. The simulation results obtained for the frequency response of the machine of concern are shown in Fig. 9.

It is noticeable that the system operating without any control scheme is highly oscillatory according to Fig. 8. Yet, the MPSS with LFC integration shows the most effective output response for reducing the overshoot and settling time. Actually, all

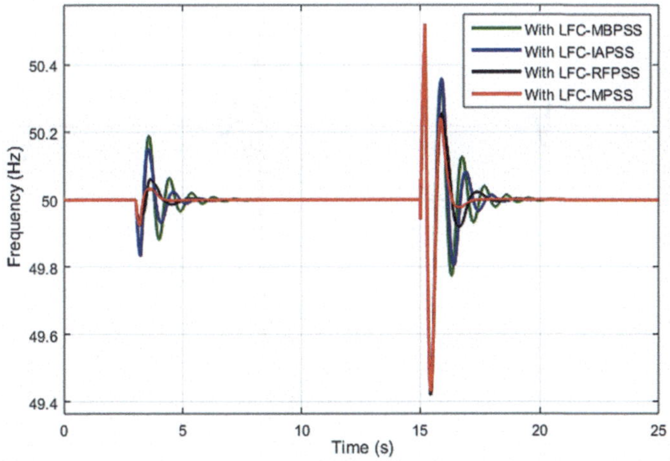

Fig. 9 Frequency response measured at generator 'Gen 2'

the oscillations decay much faster by utilizing the mixed configuration of local and remote inputs than using the tuned MBPSS or RFPSS or IAPSS designs.

The numerical values of the frequency response for different cases are mentioned in Table 3. Under normal operating conditions, when the machine is equipped with the proposed MPSS scheme in coordination with the LFC design, the peak overshoot and settling time are reduced reaching the value of 0.04 Hz and time of 1 s. Compared to the other responses, this combination shows more stable performances too when the system is forced by a disturbance.

As well, compared to the remaining types of PSSs, the designed MPSS demonstrates powerful capabilities in regaining the nominal values and keeping the system equilibrium as shown in Figs. 10 and 11 which display the electromagnetic torque and the rotor angle deviation of the generator of concern 'Gen 2', respectively.

These results confirm the critical role that the MPSS plays in damping the maximum of undesirable fluctuations at a short period of time in reaction to load

Table 3 Frequency responses for different scenarios

	Normal operating conditions		Fault operating condition	
	Peak overshoot (Hz)	Settling time (s)	Peak overshoot (Hz)	Settling time (s)
With LFC-MBPSS	0.18	3.8	0.34	4.7
With LFC-IAPSS	0.14	2.35	0.32	3.23
With LFC-RFPSS	0.08	1.76	0.24	2.9
With LFC-MPSS	0.04	1.17	0.22	2.05

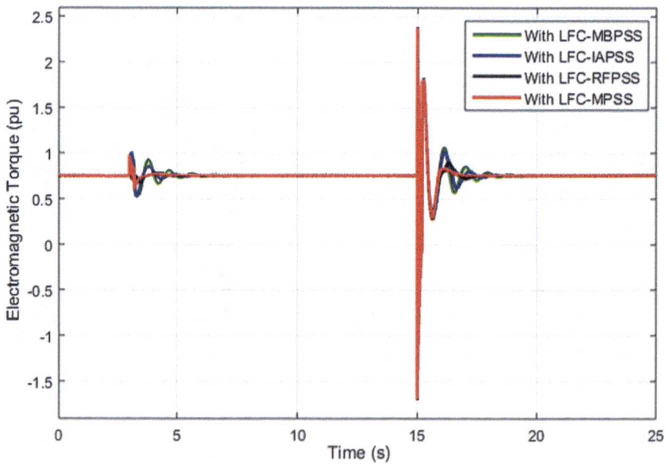

Fig. 10 Electromagnetic torque measured at generator 'Gen 2'

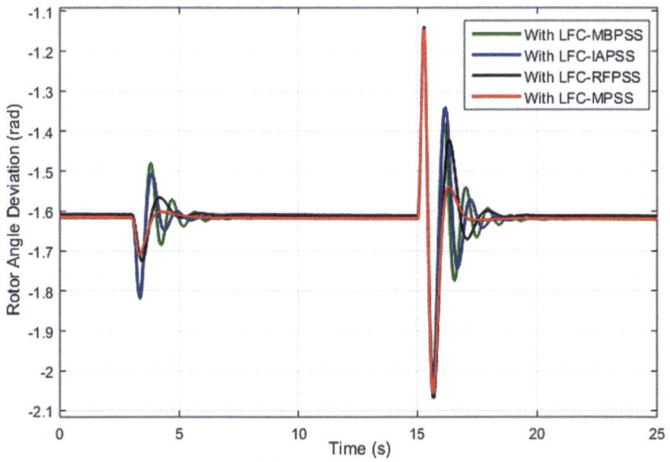

Fig. 11 Rotor angle deviation measured at generator 'Gen 2'

Table 4 Electromagnetic torque responses for different scenarios

	Normal operating conditions		Fault operating condition	
	Peak overshoot (pu)	Settling time (s)	Peak overshoot (pu)	Settling time (s)
With LFC-MBPSS	0.25	2.9	0.4	3.8
With LFC-IAPSS	0.2	1.47	0.33	2.9
With LFC-RFPSS	0.18	0.58	0.16	2.35
With LFC-MPSS	0.16	0.29	0.08	1.7

demand variations and under fault condition. Correspondingly, the different electromagnetic torque responses are improved significantly as summarized in Table 4.

Referring to these numerical results, the performed control design of the combined LFC-MPSS produces effectively not only the least peak overshoot values but also the least settling time of all responses. Therefore, an important increase in damping system oscillations is obviously checked under different operating conditions. Furthermore, the stability enhancement is achieved by accelerating the settling time and reducing the peak overshoot of rotor angle deviation as shown in Table 5.

Compared to the responses of the remaining cases, it is proven that coordinated LFC-MPSS control scheme is able enough to provide the best performance while maintaining reduced values of the peak overshoot and settling time under normal and fault operating conditions.

Table 5 Rotor angle deviation responses for different scenarios

	Normal operating conditions		Fault operating condition	
	Peak overshoot (rad)	Settling time (s)	Peak overshoot (rad)	Settling time (s)
With LFC-MBPSS	0.12	3.75	0.27	4.75
With LFC-IAPSS	0.1	2.75	0.22	4
With LFC-RFPSS	0.04	2.5	0.2	3.5
With LFC-MPSS	0.01	2	0.07	2.5

5.2 Wind Power Integration

In order to test the efficiency of the proposed control scheme in improving the system stability in presence of DFIG wind turbines, diverse penetration levels ranging from 10 to 30% are examined. For that, two wind farms are connected to bus 9 and bus 7, near to the performed power plants ('Gen 3' and 'Gen 2'), respectively. Each wind farm is consisted of DFIG wind turbines which are connected to a local 25 kV distribution system and rated 9 MW each. Initially, the simulation results are carried out considering only the LFC loop. The dynamical frequency responses of the generator of concern, 'Gen 2', for three levels of wind power penetration are presented in Fig. 12 under normal and fault conditions.

Clearly, injecting an intermittent wind energy with rates of 10 and 20% reveal relatively similar results. However, compared to these cases, when wind penetration increases up to 30%, the frequency stability is significantly degraded. Hence, the rest of analysis will be conducted chiefly for a penetration level of 20% of wind power

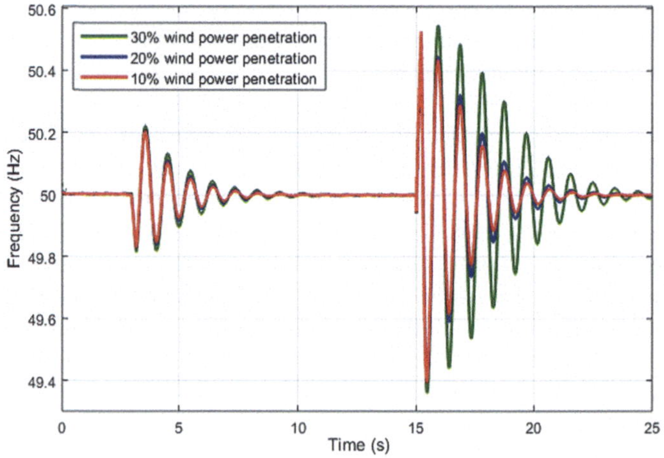

Fig. 12 Frequency response of generator 'Gen 2' with three levels of wind penetration

to keep stable operation of generators. The installed DFIG wind turbines export to the test system a total power of 72 MW.

Almost, wind energy incorporation affects the power system functionality. As much as the electrical power is produced, small fluctuations will definitely occur. Additionally, fault operation condition may force the system to lose its dynamic stability. Accordingly, DFIG wind turbines will operate till the fault is cleared. During the disturbance, the wind farms generate sufficient reactive power that supports the system outputs. The performed controllers try then to increase damping the appearing swings and approximatively adjust the waveforms to the specific set points as proven in Fig. 13 which depicts the frequency responses of the machine under test using different PSSs in coordination of the LFC loop.

The frequency response yields the best damping performance using the proposed LFC-MPSS control design when compared to the remaining operation scenarios. Accordingly, the mixed PSS configuration demonstrates satisfactorily results revealing rapidly-damped out oscillations.

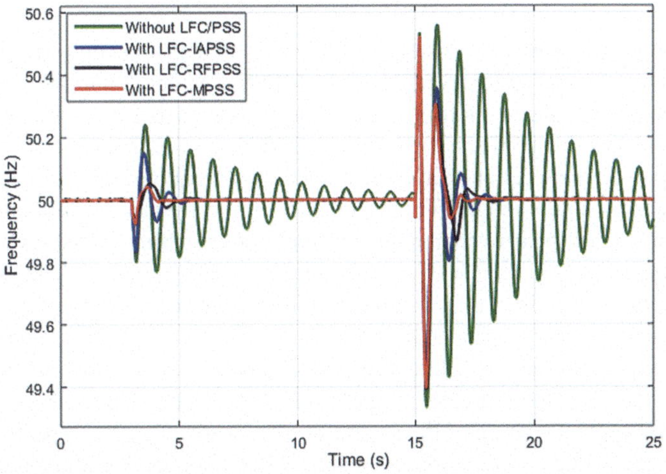

Fig. 13 Frequency response measured at generator 'Gen 2'

Table 6 Frequency responses for different scenarios considering DFIG WTs integration

	Normal operating conditions		Fault operating condition	
	Peak overshoot (Hz)	Settling time (s)	Peak overshoot (Hz)	Settling time (s)
Without LFC/PSS	0.23	>10	0.54	>10
With LFC-IAPSS	0.14	3.06	0.34	4.08
With LFC-RFPSS	0.05	2.55	0.29	3.5
With LFC-MPSS	0.03	1.27	0.27	2.55

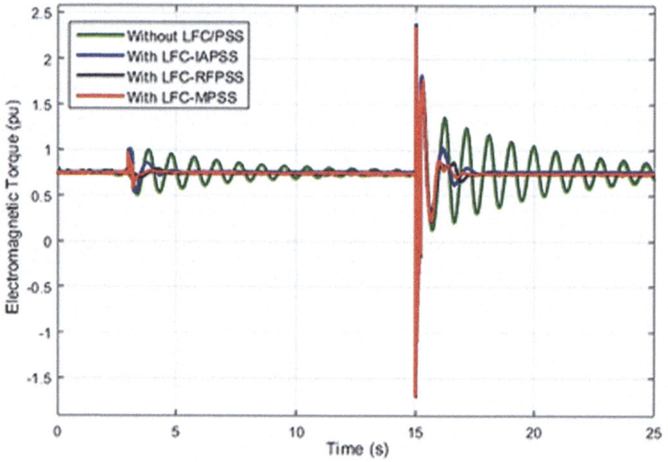

Fig. 14 Electromagnetic torque measured at generator 'Gen 2'

The numerical results are listed in Table 6 which proves that the frequency response improvement is achieved when the MPSS is integrated in the test system. Under normal operating conditions, the peak overshoot and the settling time are reduced significantly and reach the values of 0.03 Hz and 1.27 s, respectively. Likewise, comparison between the different simulated cases shows that combined LFC-MPSS design is able enough to damp effectively the frequency oscillations. The settling time and the peak overshoot are the least ones reaching the values 2.55 s and 0.27 Hz, respectively, when the system is subjected to a disturbance. Meanwhile, the dynamic stability margin of the system is kept within the permitted limits in presence of the intermittent energy of DFIGs WTs.

Further investigations on different operation scenarios prove the effectiveness of the proposal as depicted in Figs. 14 and 15 which show the electromagnetic torque and the rotor angle deviation of the generator of concern 'Gen 2', respectively.

It's noticeable that coordinating the designed LFC with the MPSS increased definitely the damping of system oscillations which take less time to stabilize and settle back to steady state. Along with an increase in load demand, the oscillatory swings are reasonably kept limited relatively at the scheduled values under fault conditions.

From Table 7, it is inferred that the settling times and amplitudes of electromagnetic torque oscillations are very high without adding any PSS. Yet, after the introduction of coordinated LFC-MPSS design in the system, the peak overshoot and settling times of oscillations are reduced at the values of 0.06 pu and 1.1 s, respectively, under normal operating conditions. Likewise, under fault condition, the simulation results show that the peak overshoot and the settling time are improved at the values of 0.12 pu and 1.76 s, respectively, compared to the remaining scenarios.

Equally, Table 8 proves that the coordination between the performed LFC and MPSS control designs is able to enhance the damping of the oscillatory response of rotor angle deviation. The peak overshoot is reduced and reaches the value of

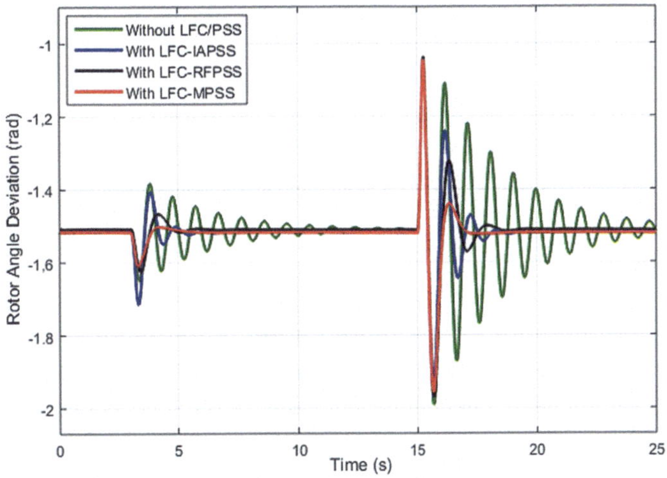

Fig. 15 Rotor angle deviation measured at generator 'Gen 2'

Table 7 Electromagnetic torque responses for different scenarios considering DFIG WTs integration

	Normal operating conditions		Fault operating condition	
	Peak overshoot (pu)	Settling time (s)	Peak overshoot (pu)	Settling time (s)
Without LFC/PSS	0.5	>10	1	>10
With LFC-IAPSS	0.25	1.76	0.5	3.23
With LFC-RFPSS	0.1	1.47	0.25	2.35
With LFC-MPSS	0.06	1.1	0.12	1.76

Table 8 Rotor angle deviation responses for different scenarios considering DFIG WTs integration

	Normal operating conditions		Fault operating condition	
	Peak overshoot (rad)	Settling time (s)	Peak overshoot (rad)	Settling time (s)
Without LFC/PSS	0.14	>10	0.4	>10
With LFC-IAPSS	0.12	2.77	0.28	3.6
With LFC-RFPSS	0.1	1.9	0.2	3.3
With LFC-MPSS	0.08	1.38	0.1	1.66

0.08 rad while the settling time is achieved at time of 1.38 s under normal operating conditions. These oscillations are more efficiently damped out reaching low overshoot and settling time with values of 0.1 rad and 1.66 s, respectively, when the system is subjected to a disturbance.

Accordingly, numerical and graphical results confirm the satisfactory performance of the proposed controllers in damping power system oscillations. In fact, compared to RFPSS and IAPSS performances, by using remote and local input signals, the MPSS shows improved and fast responses recovering the scheduled values in coordination with the optimized LFC loop under different operating conditions.

6 Conclusion

The present paper presents a novel control scheme that combines an optimized PI-based LFC with new designs of PSSs aiming mainly to increase the damping of power system oscillations. Great attention is given first to the LFC design which shows better responses whenever the loading conditions are unpredictably fluctuating. It is based on a tuned PI controller whose parameters are optimally set by using the H_∞ methodology. Additionally, a novel RFPSS regulator based on fuzzy system is investigated in coordination with the LFC design for system's damping improvement purposes. Furthermore, a review of wide-area damping control is discussed in this work. In fact, global measurements from PMUs are considered first as the input signals of the suggested IAPSS design. Later, a mixed configuration that combines single remote input with a local input is presented. The simulation studies on a modified IEEE 9-bus test system prove that the MPSS design yields the best and fast damping characteristics under different operation scenarios taking into account the impact of wind energy incorporation. In conclusion, the current work demonstrates that the application of real-time fuzzy logic PSS based PMU considering local inputs, whenever combined with an optimized LFC loop, exhibits high quality of the control signal that contributes in providing efficient dynamic system performances.

References

1. K. Tang, G.K. Venayagamoorthy, Adaptive inter-area oscillation damping controller for multi-machine power systems. Electric Power Syst. Res. **134**, 105–113 (2016)
2. G. Cai, D. Yang, C. Liu, adaptive wide-area damping control scheme for smart grids with consideration of signal time delay. Energies **6**, 4841–4858 (2013)
3. C. Sharma, B. Tyagi, Fuzzy type-2 controller design for small-signal stability considering time latencies and uncertainties in PMU measurements. IEEE Syst. J. **11**(2), 1149–1160 (2014)
4. S. Wivutbudsiri, K. Hongesombut, J. Rungrangpitayagon, Wide-area power system control using Thyristor Controlled Series Capacitor based fuzzy logic controller designed by observed signals. Int. Electr. Eng. Congr. (2014)
5. S. Ranjbar, M.R. Aghamohammadi, F. Haghjoo, A new scheme of WADC for damping inter-area oscillation based on CART technique and Thevenine impedance. Int. J. Electr. Power Energy Syst. **94**, 339–353 (2018)
6. B.P Padhy, S.C. Srivastava, N.K. Verma, A coherency-based approach for signal selection for wide area stabilizing control in power systems. IEEE Syst. J. **7**(4) (2013)
7. J. Zhang, C.Y. Chung, C. Lu, K. Men, L. Tu, A novel adaptive wide area PSS based on output-only modal analysis. IEEE Trans. Power Syst. **30**(5), 2633–2642 (2014)

8. M. Mokhtari, F. Aminifar, D. Nazarpour, S. Golshannavaz, Wide-area power oscillation damping with a fuzzy controller compensating the continuous communication delays. IEEE Trans. Power Syst. **28**(2), 1997–2005 (2013)
9. D. Murali, M. Rajaram, Comparison of damping performance of conventional and neuro–fuzzy based power system stabilizers applied in multi–machine power systems. J. Electr. Eng. **64**(6), 366–370 (2013)
10. A.B. Muljono, I.M. Ginarsa, I.A. Nrartha, Dynamic stability improvement of multimachine power systems using ANFIS-based power system stabilizer. TELKOMNIKA **13**(4), 1170–1178 (2015)
11. Y. Chompoobutrgool, L. Vanfrettiab, Using PMU signals from dominant paths in power system wide-area damping control. Sustain. Energy Grids Netw. **4**, 16–28 (2015)
12. I. Zenelis, X. Wang, Wide-area damping control for interarea oscillations in power grids based on PMU measurements. IEEE Control. Syst. Lett. **2**(4), 719–724 (2018)
13. S. Ranjbar, M.R. Aghamohammadi, F. Haghjoo, Damping inter-area oscillation in power system by using global control signals based on PSS devices, in *Iranian Conference on Electrical Engineering*, May 2017 (2017)
14. L.P. Kunjumuhammed, R. Singh, B.C. Pal, Robust signal selection for damping of inter-area oscillations. IET Gener. Transm. Distrib. **6**(5), 404–416 (2012)
15. T. Surinkaew, I. Ngamroo, Adaptive signal selection of wide area damping controllers under various operating conditions. IEEE Trans. Ind. Inf. **14**(2), 639–651 (2018)
16. P. McNabb, D. Wilson and J. Bialek, Classification of mode damping and amplitude in power systems using synchrophasor measurements and classification trees. IEEE Trans. Power Syst. **28**(2) (2013)
17. D. Molina, G.K. Venayagamoorthy, J. Liang, R.G. Harley, Intelligent local area signals based damping of power system oscillations using virtual generators and approximate dynamic programming. IEEE Trans. Smart Grid **4**(1) (2013)
18. D. Wang, M. Glavic, L. Wehenkel, Trajectory-based supplementary damping control for power system electromechanical oscillations. IEEE Trans. Power Syst. **29**(6) (2014)
19. H. Haes Alhelou, ME. Hamedani Golshan, M. Hajiakbari Fini, Wind driven optimization algorithm application to load frequency control in interconnected power systems considering GRC and GDB nonlinearities. Electr. Power Compon. Syst. **46**(11–12), 1223–1238 (2018)
20. H.H. Alhelou, ME. Golshan, ND. Hatziargyriou, A decentralized functional observer based optimal LFC considering unknown inputs, uncertainties, and cyber-attacks. IEEE Trans. Power Syst. **34**(6), 4408–4417 (2019)
21. H.H. Alhelou, M.E. Hamedani-Golshan, R. Zamani, E. Heydarian-Forushani, P. Siano, Challenges and opportunities of load frequency control in conventional, modern and future smart power systems: a comprehensive review. Energies **11**(10), 2497 (2018)
22. H.H. Alhelou, ME. Golshan, ND. Hatziargyriou, Deterministic dynamic state estimation-based optimal lfc for interconnected power systems using unknown input observer. IEEE Trans. Smart Grid (2019)
23. I. Nasiruddin, T.S. Bhatti, N. Hakimuddin, Automatic generation control in an interconnected power system incorporating diverse source power plants using bacteria foraging optimization technique. Electr. Power Compon. Syst. **43**(2), 189–199 (2014)
24. HH. Alhelou, ME. Hamedani-Golshan, E. Heydarian-Forushani, AS. Al-Sumaiti, P. Siano, Decentralized fractional order control scheme for LFC of deregulated nonlinear power systems in presence of EVs and RER, in *2018 International Conference on Smart Energy Systems and Technologies (SEST)*, 10 September 2018, pp. 1–6. (IEEE, 2018)
25. H.H. Alhelou, M.H. Golshan, J. Askari-Marnani, Robust sensor fault detection and isolation scheme for interconnected smart power systems in presence of RER and EVs using unknown input observer. Int. J. Electr. Power Energy Syst. **1**(99), 682–694 (2018)
26. T. Wang, A. Pal, James S. Thorp, Z. Wang, I. Liu, Y. Yang, Multi-polytope-based adaptive robust damping control in power systems using CART. IEEE Trans. Power Syste. **30**(4) (2015)
27. P.K. Ray, S.R. Paital, A. Mohanty, F.S. Eddy, H.B. Gooi, A robust power system stabilizer for enhancement of stability in power system using adaptive fuzzy sliding mode control. Appl. Soft Comput. **73**, 471–481 (2018)

28. R. Sedaghati, A. Rouhani, A. Habibi, A.R. Rajabi, A novel fuzzy-based power system stabilizer for damping power system enhancement. Indian J. Sci. Technol. **7**(11), 1729–1737 (2014)
29. J.M. Ramirez, R.E. Correa, D.C. Hernández, A strategy to simultaneously tune power system stabilizers. Int. J. Electr. Power Energy Syst. **43**(1), 818–829 (2012)
30. A. Hashmani, I. Erlich, Mode selective damping of power system electromechanical oscillations for large power systems using supplementary remote signals. Int. J. Electr. Power Energy Syst. **42**(1), 605–613 (2012)
31. P.R. Murty, *Power System Analysis* (Chap. 13), 2nd edn. (2017)

Wide-Area Monitoring of Large Power Systems Based on Simultaneous Processing of Spatio-Temporal Data

Emilio Barocio, Josue Romero, Ramon Betancourt, Petr Korba, and Felix Rafael Segundo Sevilla

Abstract Accurate identification of electromechanical oscillations on power systems and determination of its stability condition is a fundamental process in order to carry out an appropriate control action to prevent the partial loss or complete blackout of the system. However, the non-linear characteristics of measured variables often lead to incorrect information about the development of the electromechanical oscillations, making wide-area monitoring a challenging task. In addition, significant amount of information in extra large power systems is produced, which has to be stored on local servers requiring large amounts of central processing unit (CPU) storage. For these reasons, algorithms for Big Data problems in power systems are required and the methods presented on this chapter introduce some potential solutions. In this context, different data-driving methods based on spectral analysis of linear operator are presented for the analysis of electromechanical oscillations from a spatio-temporal perspective. These algorithms have the ability to process spatio-temporal data simultaneously, making possible to characterize inter-area and global oscillations (from 0.1 Hz to 1.0 Hz). To validate the effectiveness of the proposed approaches, two test systems with different structural and generation capacities are analysed: the Mexican Interconnected (MI) system and the initial dynamic model of Continental Europe from ENTSO-E. First, data collected from a transient stability study on the MI system are used to illustrate the ability of data-driving methods to characterize modal oscillations on longitudinal systems; where several inter-area modes produce interactions of different electrical areas. Then, simulation results from the initial dynamic model of ENTSO-E are analysed to characterize the propagation of its global electromechanical modes across Europe, which have been denominated as the North-South and East-West modes with frequencies of approximately 0.15 Hz

E. Barocio (✉) · J. Romero
Universidad de Guadalajara, Guadalajara, Mexico
e-mail: emilio.barocio@cucei.udg.mx

R. Betancourt
Universidad de Colima, Colima, Mexico

P. Korba · F. R. S. Sevilla
Zurich University of Applied Sciences, Winterthur, Switzerland

H. Haes Alhelou et al. (eds.), *Wide Area Power Systems Stability, Protection, and Security*, Power Systems, https://doi.org/10.1007/978-3-030-54275-7_7

and 0.25 Hz, respectively. The second analysis include the interconnection of Turkey (TR) to Continental Europe in December 2010, which derived on the grow of size and complexity of the original system having as result a decrease in the frequency value for the East-West mode and the introduction of a third inter-area mode on the system. The chapter concludes comparing the results of the proposed approaches against conventional methods available in the literature.

Keywords Central processing unit · WAMs · Mexican interconnected system · PMUs · Smart grid · Interconnected power systems

1 Introduction

Interconnections among different electrical power systems (EPS) offers significant technical, economic and environmental advantages. In the same way, energy exchange over distant regions provides flexibility in terms of maintaining the balance between generation and demand as result of the energy transfer condition [1, 2].

On the other hand, despite the advantages offered by these interconnections, there are several technical and economic limitations related to it. Particularly when the energy has to be transferred over long distances (generally over more than 100 km), which in power systems commonly represents spanning over one or more countries [3]. The interconnection of the EPS represents a complex problem for the system operators and is the main cause of low frequency oscillations when negative events such as trip of generation units, load variation or three-phase faults on transmission lines occur [4].

The presence of this type of oscillations, commonly referred as electromechanical oscillations, is a typical problem of interconnected systems around the world [3]. The electromechanical oscillations, which are the responsible of the low frequency oscillations can be classified as: local, inter-area and inter-continental oscillation modes, respectively [4]. These modes are characterized by its frequencies range, the number of participating and the location of generation units involved during the oscillatory process. Local modes oscillate between ~0.8 and 2 Hz and the participating machines are located in the same power station, which can accommodate up to 10 generation units. Inter-area modes range from ~0.25 to 0.7 Hz are characterized by oscillations between large groups of machines, which are located at different defined regions, especially when the interconnection in the system is weak. Intercontinental oscillation modes present a similar pattern as inter-area modes, however, the oscillation frequency is lower (between ~0.1 and 0.2 Hz) and involves large groups of machines during the oscillatory process, which are located on different countries [1, 2].

The effect and behaviour of low-frequency oscillations on interconnected power systems depends on different factors such as size of the areas conforming the interconnected system, the geographic distance between areas, the generation and power transfer capacity between the different areas, the demand of the loads, the synchronous machine type and the network topology. Identification of inter-area oscillations in large power systems represents an extra degree of complexity due to the processing capacity required from classical analysis tools, which are affected by the volume, speed and variety of the input data generated from numerical simulation based on models or monitoring systems [2, 5]. For these reasons, it is necessary to use new alternative approaches for the analysis of large dimensional power systems.

The recent development of wide area monitoring systems (WAMS) has open the opportunity to observe and track variables such as frequency and voltage magnitudes and angles of the voltage and current at strategic locations such as directly on the generation units, relevant loads and EPS compensation [6]. The fast recording (5–120 samples/s) achieved by phasor measurement units (PMUs), which is data including also the time stamp open new opportunities to develop tools for monitoring, analysis and control of electromechanical oscillations on EPS [7, 8]. However, is not straightforward to accomplish these tasks given the challenges related to these devices such as having partial observability of the system, the limitation for the detection and prediction of an instability condition in case of a disturbance and the capacity to process the data correlating the spatio-temporal information contained in the EPS.

The development and improvement of analytical tools [6–8] represents a difficult task due to the particular characteristics inherent within these datasets such as volume, variety and speed on which the data is generated, such as computer simulations or real measurement systems. These characteristics vary depending on the order (nodes, generators, among others), of the dynamic model of the EPS, the sampling and simulation time. Furthermore, the volume of data obtained from the measurement recordings varies depending on the number of PMUs placed in the network, as well as the sampling frequency and the window size collected. Therefore, one of the alternatives adopted in recent years to address this issue is the inclusion of data mining and data-driven techniques for evaluation of the security of the EPS, fault detection, as well as monitoring and analysis of electromechanical oscillations [8].

The aforementioned challenges motivate the development of algorithms with potential to process data that consider the spatio-temporal information available in the EPS. The proposed approach should include the ability to process volume, variety and speed within the data in order to capture the dynamic behaviour that occurs during an electromechanical oscillation. Similarly, the proposed methods should provide analytical information to understand the mechanism of propagation of the electromechanical oscillations. Upon this premise, this chapter presents different alternative approaches, to extract relevant modal characteristics of electromechanical oscillations that help to analyse this phenomenon on large interconnected electrical power systems.

2 Dynamic System Analysis Based on the Koopman Operator

The basis of the method referred as dynamic mode decomposition (DMD) is in fact the theory of the Koopman operator, which was introduced in 1931 for the analysis of Hamiltonian systems in discrete time [9]. DMD can be used as one algorithm for finding Koopman modes from spatio-temporal data. Each Koopman mode may be associated to a unique frequency and growth rate and interpreted as a nonlinear generalization of global eigenmodes of a linearized system. Based on the original definition of the Koopman operator in continuous time [10]:

Consider a continuous time dynamic system:

$$\frac{d\boldsymbol{x}}{dt} = f(\boldsymbol{x}) \tag{1}$$

where $\boldsymbol{x} \in \mathcal{M}$ is the state within a manifold $\mathcal{M}$ of dimension N. The Koopman operator $\mathcal{K}$ is a linear operator of infinite dimension, which operates on all observable functions $g : \mathcal{M} \rightarrow \mathbb{C}$ such that

$$\mathcal{K}g(x) = g(f(x)) \tag{2}$$

it is established that the Koopman operator performs a transformation from the representation in state space that considers a non-linear dynamic of finite dimension, towards the Koopman representation that considers a linear dynamic of infinite dimension.

In this case, the $f(*)$ term represents the dynamic of the system and the Koopman operator can be defined as a dynamic system in discrete time [11]. From (1), it can be induced a discrete system given by the flow map $\boldsymbol{F} : \mathcal{M} \rightarrow \mathcal{M}$ mapping the state $\boldsymbol{x}(t_0)$ to a future time $\boldsymbol{x}(t_0 + t)$:

$$\boldsymbol{F}(\boldsymbol{x}(t_0)) = \boldsymbol{x}(t_0 + t) = \boldsymbol{x}(t_0) + \int_{t_0}^{t_0+t} f(\boldsymbol{x}(\tau))d\tau \tag{3}$$

From the previous definition, the dynamic system is induced in discrete time as follows:

$$\boldsymbol{x}_{k+1} = \boldsymbol{F}(\boldsymbol{x}_k) \tag{4}$$

where the discrete time vector is defined as $\boldsymbol{x}_k = \boldsymbol{x}(kt)$ and $\boldsymbol{F}$ represents the flow map in discrete time. The analogue operation for the discrete time Koopman operator $\mathcal{K}$ for the observable function g is based in the continuous time Eq. (2) and it is formulated using the following expression:

$$\boldsymbol{K}g(\boldsymbol{x}_k) = g(\boldsymbol{F}(\boldsymbol{x}_k)) = g(\boldsymbol{x}_{k+1}) \tag{5}$$

where $\boldsymbol{K}$ denote the discrete time Koopman operator. By considering the spectral decomposition of the Koopman operator as an eigenvalue problem it is possible to represent the dynamic solution of the system:

$$\boldsymbol{K}\boldsymbol{\varphi}_k = \lambda_k \boldsymbol{\varphi}_k \tag{6}$$

where $\boldsymbol{\varphi}_k$ are the Koopman's *eigenfunctions*. Expanding these *eigenfunctions* $\boldsymbol{\varphi}_k$ based on the solution of the Koopman operator it is possible to represent the evolution of the dynamic of the system through expansion of the nonlinear observable functions g in terms of $\boldsymbol{\varphi}_k$:

$$\boldsymbol{g}(\boldsymbol{x}) = \begin{bmatrix} g_1(\boldsymbol{x}) \\ g_2(\boldsymbol{x}) \\ \vdots \\ g_m(\boldsymbol{x}) \end{bmatrix} = \sum_{k=1}^{\infty} \varphi_k(\boldsymbol{x})\boldsymbol{v}_k! \tag{7}$$

where $\boldsymbol{v}_k$ is the mode kth associated to $\boldsymbol{\varphi}_k$. Considering the eigenvalue problem described on (6), with the definition (7) it is possible to represent the dynamic evolution of the system using the following equation:

$$\boldsymbol{K}\boldsymbol{g}(\boldsymbol{x}) = \sum_{k=1}^{\infty} \boldsymbol{K}\varphi_k(\boldsymbol{x})\boldsymbol{v}_k = \sum_{k=1}^{\infty} \lambda_k\varphi_k(\boldsymbol{x})\boldsymbol{v}_k = \boldsymbol{g}(\boldsymbol{x}_{k+1}) \tag{8}$$

This expression represents the solution of the operator $\boldsymbol{K}$ in terms of the modes $\boldsymbol{v}_k$ and the eigenvalues λ_k of the system by means of an infinity sum. The dimensional problem can be tackled with a finite sum of modes that approximate the spectral solution of Koopman.

In the following section, a methodology to approximate a linear Koopman operator using considerations such as the measurements of the system under investigation is provided.

2.1 Schematic Visualization of Finite Dimensional Approximation of the Koopman Operator

To approximate the operator of infinite dimension $\mathcal{K}$, in [12] is considered one restriction in the group of nonlinear observable functions $\boldsymbol{g}$ in the form of an invariant subspace defined as $\mathbb{R}^N$, which includes *eigenfunctions* of the Koopman operator $\boldsymbol{K}$. With this restriction, the formation of a finite dimension operator $\boldsymbol{K}$ is induced and also the observation functions in the subspace $\mathbb{R}^N$ are mapped.

Figure 1a represents the transition of the flow map in state space $\boldsymbol{x}$ to the flow map of nonlinear observable functions $\boldsymbol{g} = \begin{bmatrix} g_1 \; g_2 \; \dots \; g_m \end{bmatrix}^T$ restricted by a finite subspace $\mathbb{R}^N$ through the Koopman operator $\boldsymbol{K}$. The space $\mathbb{R}^N$ can be interpreted as a delineation from which the mapping of the states of the system operate. On the other hand, Fig. 1b represents the sequence of the observation functions mapping $\boldsymbol{g}(\boldsymbol{x}_k) = \boldsymbol{y}_k$ conducted by the operator $\boldsymbol{K}$. In this case, the mapping of the operator $\boldsymbol{K}$ approximates the original trajectory on a linear space of infinite dimension. The mapping effect of the Koopman operator on dynamic systems has been numerically illustrated at Ref. [12].

Through this finite dimensional approximation, the DMD method was developed as an alternative to approximate the modes of the Koopman operator $\boldsymbol{K}$. The following section presents the relationship of the Koopman modal decomposition with the DMD model.

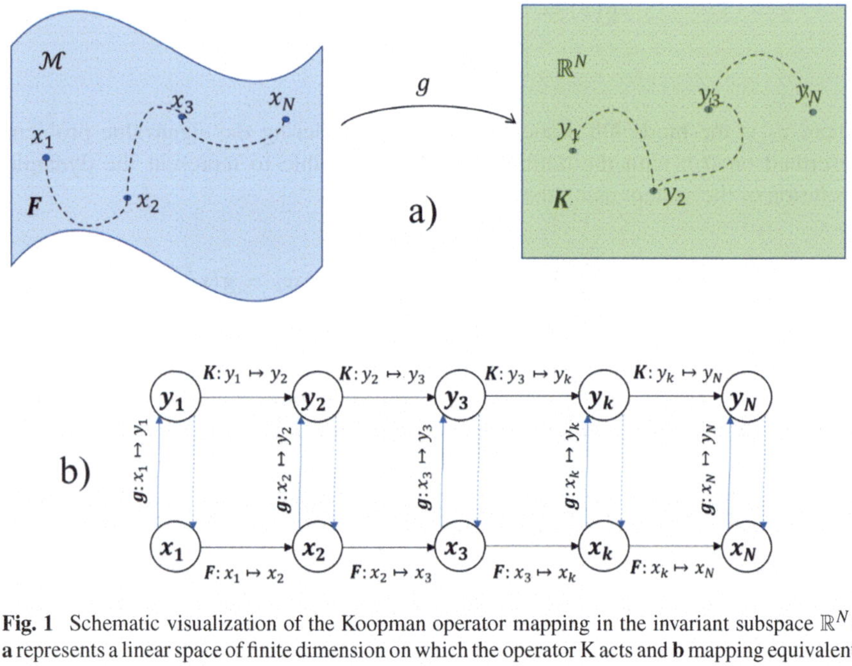

Fig. 1 Schematic visualization of the Koopman operator mapping in the invariant subspace $\mathbb{R}^N$: **a** represents a linear space of finite dimension on which the operator K acts and **b** mapping equivalent of the observable functions $\boldsymbol{y}$ and the states $\boldsymbol{x}$ [12]

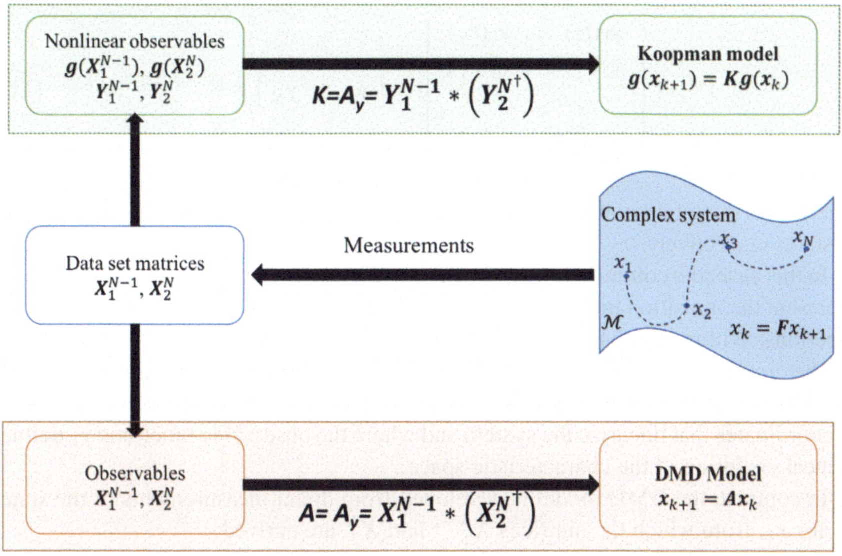

Fig. 2 Transition of DMD model approximation to Koopman model [12]

2.2 *Dynamic Mode Decomposition (DMD) as Approximation to the Koopman Operator*

As introduced on Sect. 2.1, DMD approach only requires the vector measurement of the states of the system, which is defined as $\boldsymbol{x}_k = \left[x_1(t_k)\ x_2(t_k)\ \ldots\ x_m(t_k)\right]^T \in \mathbb{R}^m, k = 1, 2, \ldots, N$, to approximate the dynamic of the system by means of the linear operator $\boldsymbol{A} \in \mathbb{R}^{mxm}$:

$$\boldsymbol{x}_{k+1} = \boldsymbol{A}\boldsymbol{x}_k \tag{9}$$

In (9), the operator $\boldsymbol{A}$ represents the approximation of the DMD model to the finite dimensional operator $\boldsymbol{K}$ in discrete time through a noise-free process [13]. The assembling of the DMD model as an approximation to the Koopman model is shown on Fig. 2.

From Fig. 2 it can be observed that the Koopman model is put together from nonlinear observable functions $\boldsymbol{y}_k = g(\boldsymbol{x}_k)$, from which the columns of the matrices $\boldsymbol{Y}_1^{N-1}$ and $\boldsymbol{Y}_2^N$ are formed as follows:

$$\boldsymbol{Y}_1^{N-1} = \begin{bmatrix} y_1(t_1) & \ldots & y_1(t_{N-1}) \\ y_2(t_1) & \ldots & y_2(t_{N-1}) \\ \vdots & \ddots & \vdots \\ y_m(t_1) & \ldots & y_m(t_{N-1}) \end{bmatrix} = \left[\boldsymbol{y}_1\ \ldots\ \boldsymbol{y}_{N-1}\right] \in \mathbb{R}^{mxN-1} \tag{10}$$

$$\boldsymbol{Y}_2^N = \begin{bmatrix} y_1(t_2) & \dots & y_1(t_N) \\ y_2(t_2) & \dots & y_2(t_N) \\ \vdots & \ddots & \vdots \\ y_m(t_2) & \dots & y_m(t_N) \end{bmatrix} = \begin{bmatrix} \boldsymbol{y}_2 \dots \boldsymbol{y}_N \end{bmatrix} \in \mathbb{R}^{mxN-1} \tag{11}$$

where N is the total number of *snapshots* and m is the total number of observations or states respectively.

In this case, the columns of $\boldsymbol{y}_k$ corresponding to the spectral analysis of Koopman represent the transition in the physical space of the mapping to a space of observable functions. Within this space, a representation of the dynamic of the system using (8) can be created. Using this representation, the intrinsic properties of the dynamic of the system represented in (1) are meet. Moreover, the *eigenfunctions* define a change of coordinates that linearize the system and where the observable functions $\boldsymbol{y}_k$ define a lineal evolution of the characteristic space.

In contrast, the DMD model is developed from direct measurements of the state vector $\boldsymbol{x}_k$, from which the matrices $\boldsymbol{X}_1^{N-1}$ and $\boldsymbol{X}_2^N$ are derived:

$$\boldsymbol{X}_1^{N-1} = \begin{bmatrix} x_1(t_1) & \cdots & x_1(t_{N-1}) \\ x_2(t_1) & \cdots & x_2(t_{N-1}) \\ \vdots & \ddots & \vdots \\ x_m(t_1) & \cdots & x_m(t_{N-1}) \end{bmatrix} = \begin{bmatrix} \boldsymbol{x}_1 \cdots \boldsymbol{x}_{N-1} \end{bmatrix} \in \mathbb{R}^{mxN-1} \tag{12}$$

$$\boldsymbol{X}_2^N = \begin{bmatrix} x_1(t_1) & \cdots & x_1(t_N) \\ x_2(t_1) & \cdots & x_2(t_N) \\ \vdots & \ddots & \vdots \\ x_m(t_1) & \cdots & x_m(t_N) \end{bmatrix} = \begin{bmatrix} \boldsymbol{x}_2 \cdots \boldsymbol{x}_N \end{bmatrix} \in \mathbb{R}^{mxN-1} \tag{13}$$

The relation between the models deduced with DMD and Koopman is based on the following two criteria:

1. The spectral decomposition of the operator $\boldsymbol{A}$ through the eigenvalue problem:

$$\boldsymbol{A}\boldsymbol{W} = \boldsymbol{W}\lambda_k \tag{14}$$

   where λ_k and $\boldsymbol{W}$ represent the *eigenvalues* and *eigenvectors*, respectively.

2. From the spectral decomposition of $\boldsymbol{A}$, the Theorem from [9] is presented: "Take φ_k as the *eigenfunctions* of $\boldsymbol{K}$ and eigenvalues λ_k, and assume $\varphi_k \in span\{g_j\}$, such that

$$\varphi_k(\boldsymbol{x}) = w_1 g_1(\boldsymbol{x}) + w_2 g_2(\boldsymbol{x}) + \cdots + w_m g_m(x) \tag{15}$$

For some $W = \begin{bmatrix} w_1 \; w_2 \; \ldots \; w_m \end{bmatrix}^T \in \mathbb{C}^m$. If $W \in R(X)$, where R is the rank of the matrix X, then W is a left *eigenvector* of the operator A with *eigenvalues* λ_k such that $W^* A = \lambda_k W^*$."

In this case, the associated *eigenvectors* W to the operator A are used to approximate the Koopman *eigenfunctions* φ_k and the *eigenvalues* λ_k represent an approximation to the *eigenvalues* λ_k associated to the operator K by means of the DMD algorithm.

After the relation between the DMD and the Koopman models have been presented, in the next section the different forms to compute the operator A and the potential applications of the DMD method are introduced.

2.3 Standard Approximation of the Operator A

Taking the formulation depicted on [13, 14] as reference, the development of the approximation of the operator DMD in a general form is introduced. From (9) and with the help of the Krylov sequence [15]:

$$X = \left[x_1 \; Ax_1 \; A^2 x_1 \; \ldots \; A^{N-1} x_1 \right] = \left[x_1 \; x_2 \; x_3 \ldots x_N \right] \tag{16}$$

$$X_1^{N-1} = \left[x_1 \; Ax_1 \; A^2 x_1 \; \ldots \; A^{N-2} x_1 \right] = \left[x_1 \; x_2 \; x_3 \cdots x_{N-1} \right] \tag{17}$$

$$X_2^{N} = \left[Ax_1 \; A^2 x_1 \; A^3 x_1 \; \ldots \; A^{N-1} x_1 \right] = \left[x_2 \; x_3 \; x_4 \ldots x_N \right] \tag{18}$$

in this case, Eq. (16) shows the succession from $x_2 = Ax_1, x_3 = Ax_2 = A(Ax_1) = A^2 x_1$ to $x_k = Ax_{k-1}$.

This technique is based on Arnoldi's method [15], which is related to the solution of a polynomial approximation problem. This technique assumes that the polynomial operator is invariant and the system measurements are linearly independent. With a sufficiently large number of *snapshots* it is considered that the last vector x_N can be represented as a linear combination of the previous *snapshots* [14], as indicated in the following expression:

$$x_N = c_1 x_1 + c_2 x_2 + \cdots + c_{N-1} x_{N-1} + r \tag{19}$$

From Eq. (19), a more compact form of the expansion of vector x_N by means of the coefficient vector $c = \left[c_1 c_2 \ldots c_{N-1} \right]^T \in \mathbb{R}^{N-1}$ is shown.

$$x_N = X_1^{N-1} c + r \tag{20}$$

The expansion of the sequence $\boldsymbol{X}_2^N$ can be represented using the following equation:

$$\boldsymbol{X}_2^N = \left[\boldsymbol{x}_2\,\boldsymbol{x}_3\,\boldsymbol{x}_4 \ldots \boldsymbol{X}_1^{N-1}\boldsymbol{c}\right] + \boldsymbol{r} \tag{21}$$

and in matrix form using the companion matrix $\boldsymbol{S}$:

$$\boldsymbol{X}_2^N = \boldsymbol{S}\boldsymbol{X}_1^{N-1} + \boldsymbol{r} \tag{22}$$

where the structure of the $\boldsymbol{S}$ matrix is shown in the following expression:

$$\boldsymbol{S} = \begin{bmatrix} 0\,0 \ldots 0 & c_1 \\ 1\,0 \ldots 0 & c_2 \\ 0\,1 \ldots 0 & c_3 \\ \vdots\,\vdots\,\ddots\,\vdots & \vdots \\ 0\,0 \ldots 1 & c_{N-1} \end{bmatrix} \in \mathbb{R}^{N-1xN-1} \tag{23}$$

On the other hand, by approximating the sequence $\boldsymbol{X}_2^N$ using the operator $\boldsymbol{A}$ through $\boldsymbol{X}_2^N = \boldsymbol{A}\boldsymbol{X}_1^{N-1}$ and taking (22) it can be shown that:

$$\boldsymbol{X}_2^N = \boldsymbol{A}\boldsymbol{X}_1^{N-1} = \boldsymbol{S}\boldsymbol{X}_1^{N-1} + \boldsymbol{r} \tag{24}$$

Conversely, two of the approximations for obtaining the companion matrix $\boldsymbol{S}$ are presented by means of the following algorithms [16]:

- Pseudoinverse

The solution through pseudoinverse matrix, also known as the Moore–Penrose matrix, is a generalization of the inverse matrix and represents the best approximation of the solution to the mean square error corresponding to the following optimization problem:

$$r = \left\|\boldsymbol{X}_2^N - \boldsymbol{X}_1^{N-1}\boldsymbol{S}\right\|_2 \tag{25}$$

The solution to this optimization problem focused on the Companion matrix is given by the following equation:

$$\boldsymbol{S} = \boldsymbol{X}_2^{N\dagger}\boldsymbol{X}_1^{N-1} \in \mathbb{R}^{N-1xN-1} \tag{26}$$

where † represents the Moore–Penrose matrix. With this expression, an approximation of the operator $\boldsymbol{A}$ is obtained through the Companion $\boldsymbol{S}$ matrix in an $N-1$ dimension.

- Orthogonal Projection Matrix

In this technique, the approximation of operator $\boldsymbol{A}$ is obtained from a low order model using a reduction technique such as proper orthogonal decomposition (POD). The basis of this technique is the singular value decomposition (SVD).

The SVD decomposition is based on a matrix representation $\boldsymbol{M} \in \mathbb{R}^{mxN}$ using two orthogonal unitary matrices $\boldsymbol{U} \in \mathbb{R}^{Nxm}$ and $\boldsymbol{V^*} \in \mathbb{R}^{NxN}$, which are denominated left and right singular vectors respectively, and a diagonal matrix $\boldsymbol{\Sigma} \in \mathbb{R}^{mxm}$, which contains the singular values of the matrix in descending order. It is possible to compute a reduced version of the SVD decomposition using the first r singular values of $\boldsymbol{\Sigma}$, where the dimension of $\boldsymbol{U}$, $\boldsymbol{V^*}$ and $\boldsymbol{\Sigma}$ is reduced to $\widetilde{\boldsymbol{U}} \in \mathbb{R}^{Nxr}$, $\widetilde{\boldsymbol{V}}^* \in \mathbb{R}^{rxr}$ and $\widetilde{\boldsymbol{\Sigma}} \in \mathbb{R}^{rxr}$.

From the SVD decomposition of the sequence $\boldsymbol{X}_1^{N-1}$:

$$\boldsymbol{X}_1^{N-1} = \boldsymbol{U\Sigma V^*} = \left[\widetilde{U}\right]\left[\widetilde{\Sigma}0\right]\begin{bmatrix}\widetilde{\boldsymbol{V}}^* \\ \widetilde{\boldsymbol{V}}_s^*\end{bmatrix} \tag{27}$$

The approximation of the sequence $\boldsymbol{X}_2^N$ is posed through the operator $\boldsymbol{A}$ using the following equation:

$$\widetilde{X}_2^N \approx \boldsymbol{A}\widetilde{\boldsymbol{U}}\widetilde{\boldsymbol{\Sigma}}\widetilde{\boldsymbol{V}}^* \tag{28}$$

As mentioned in [14], a representation of $\boldsymbol{A}$ is obtained in the base covered by the left singular vector's modes of the sequence $\boldsymbol{X}_1^{N-1}$ by means of the following expression

$$\widetilde{\boldsymbol{S}} \triangleq \widetilde{\boldsymbol{U}}^* \boldsymbol{A}\widetilde{\boldsymbol{U}}^* = \widetilde{\boldsymbol{U}}^* \boldsymbol{X}_2^N \widetilde{\boldsymbol{V}} \widetilde{\boldsymbol{\Sigma}}^{-1} \tag{29}$$

This approach seeks a reduced representation based on r dominant modes that capture the larger energy content in the dynamic of the system.

3 DMD Based Data-Driving Methods for Simultaneous Processing of Spatio-Temporal Data

Nowadays, the application of *data-driven* techniques in the modelling and control of physical systems is a field that has evolved rapidly due to the potential to work with measurements, either from historical data, numerical simulations or experimental data [12, 14]. DMD is one of the methods that has the potential to obtain the dynamics of complex and large systems. The DMD method was introduced by Schmid & Sesterhen for the analysis of dynamic fluids and was defined in [14] as a form to decompose complex flows into a representation based on coherent spatio-temporal structures. However, DMD has been widely used for the analysis of nonlinear dynamical systems, such as stock market [17], neuroscience [18], climate phenomena [19], thermodynamic process [20], foreground/background video separation [21] and more.

In the context of power system applications, DMD is one of the most recent post-processing tools for application on power system, where a large volume of data is collected from diverse monitoring systems (WAMS, SCADA, AMI), taking advantage of its ability to process simultaneously spatio-temporal data. Based on the pointed-out ability, DMD has been applied on ring-down modal identification analysis, [13, 22–26] state estimation and prediction and control [27], coherency identification [28–31], distortion harmonic identification [32], short-term electric load forecasting [33], voltage analysis [34] and various other power system applications [35].

In this chapter a comparison among DMD variants is carried out; its performance is evaluated through various experiments conducted on different power system scenarios with different interconnected system network. The effectiveness of the DMD based method is verified by comparing the results with conventional power system stability methods. The promising results suggest that the some of the DMD approach can be used as an efficient candidate for estimating the power system frequency and amplitude, damping rate, coherency groups identification on large interconnected power systems.

3.1 SVD Based DMD (SVD-DMD)

One form of interpreting the dynamic behaviour of a system can be obtained from the modal decomposition of the $\tilde{\mathbf{S}}$ operator by taking the orthogonal projection matrix as a basis

$$\tilde{\mathbf{S}}^{SVD} = \boldsymbol{\Upsilon}^{SVD}\boldsymbol{\Lambda}^{SVD}\boldsymbol{\Upsilon}^{SVD-1} \tag{30}$$

where $\boldsymbol{\Upsilon}^{SVD}$ represents the matrix of the left eigenvectors, $\boldsymbol{\Upsilon}^{SVD-1}$ is the matrix of the right eigenvectors and $\boldsymbol{\Lambda}^{SVD}$ is a diagonal matrix of eigenvectors and are defined by the following expressions:

$$\boldsymbol{\Lambda}^{SVD} \in \mathbb{R}^{mxm} = \begin{bmatrix} \lambda_1^{SVD} & \dots & 0 \\ \vdots & \ddots & \vdots \\ 0 & \dots & \lambda_m^{SVD} \end{bmatrix} \tag{31}$$

$$\boldsymbol{\Upsilon}^{SVD} \in \mathbb{R}^{mxm} = \begin{bmatrix} \vdots & \vdots & \vdots \\ \boldsymbol{v}_1^{SVD} & \dots & \boldsymbol{v}_m^{SVD} \\ \vdots & \vdots & \vdots \end{bmatrix} \tag{32}$$

With the modal decomposition of the operator $\widetilde{S}$ the analytical solution to the problem of reconstructing data is presented:

$$\widetilde{X}_2^N \approx \mathbf{\Phi}^{SVD}\mathbf{\Lambda}^{SVD}\mathbf{\Gamma}^{SVD}(t) \tag{33}$$

where the structure $\mathbf{\Phi}^{SVD}$ represents the spatial term of the dynamic of the system and is defined as follows:

$$\mathbf{\Phi}^{SVD} \in \mathbb{C}^{mxm} = \widetilde{\mathbf{U}}\mathbf{\Upsilon}^{SVD} = \left[\phi_1^{SVD} \dots \phi_m^{SVD}\right] \tag{34}$$

and the structure $\mathbf{\Gamma}^{SVD}(t)$ represents the temporal evolution of the modes of the system and is defined as [13]:

$$\mathbf{\Gamma}^{SVD}(t) \in \mathbb{C}^{mxN-1} = \mathbf{\Upsilon}^{SVD-1}\widetilde{\mathbf{\Sigma}}\widetilde{\mathbf{V}}^* = \begin{bmatrix} \sum_{k=1}^{m} \mathbf{\Upsilon}_{1k}^{SVD-1}\boldsymbol{a}_k^{SVD}(t) \\ \vdots \\ \sum_{k=1}^{m} \mathbf{\Upsilon}_{1m}^{SVD-1}\boldsymbol{a}_k^{SVD}(t) \end{bmatrix} = \begin{bmatrix} \widetilde{\boldsymbol{a}}_1^{SVD}(t) \\ \vdots \\ \widetilde{\boldsymbol{a}}_m^{SVD}(t) \end{bmatrix} \tag{35}$$

Figure 3 shows a schematic representation of the spatio-temporal structure associated to the dynamic of the system through the solution of the data reconstruction problem.

In this case, the spatio-temporal structure shown on Fig. 3 displays the principal components, a spatial component $\boldsymbol{\phi}$, a temporal component $\widetilde{\boldsymbol{a}}_k$ and a weighting factor λ. With these three components, the associated characteristics of the spatio-temporal structure of the dynamic of the system can be identified.

Identification of frequency and damping associate to mode $\boldsymbol{\phi}_j^{SVD}$ can be represented as follows [13]:

$$f_j^{SVD} = \frac{\Im\{\log(\lambda_i^{SVD})\}}{\frac{\Delta t}{2\pi}} \tag{36}$$

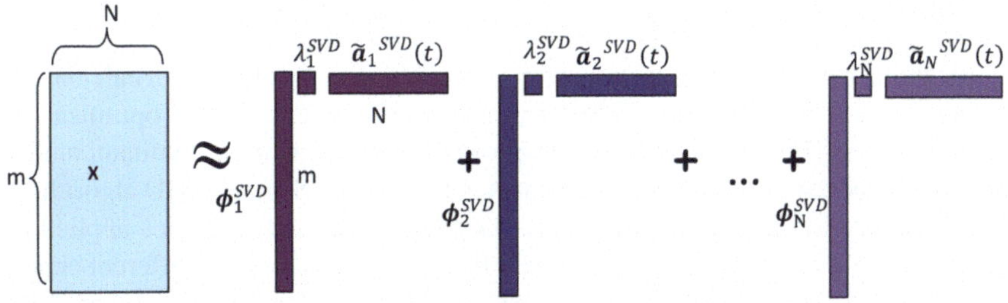

Fig. 3 Spatio-temporal modal decomposition structure of data by SVD-DMD

$$\zeta_j^{SVD} = \Re\{\log(\lambda_i^{SVD})\}/\Delta t \tag{37}$$

On the other hand, due to the spatial structure of the dynamic of the system associated to the modes of the system defined as $\boldsymbol{\phi}_j^{SVD}$ a participation factor related to each mode at instant t_0 can be calculated using the normalized magnitude of each mode $\|\boldsymbol{\phi}_j^{SVD}\|$. Similarly, the existing groups can be visualized with a similar dynamic behaviour using the phase $\angle\boldsymbol{\phi}_j^{SVD}$.

One form of visualizing the participation factors for each state of the system is through the time structure associated with the expression (35). When the energy resulting from each temporal term defined on Eq. (38) is considered

$$\widetilde{E}_k^{SVD} = \|\widetilde{\boldsymbol{a}}_k^{SVD}(t)\| \tag{38}$$

the relation mode-state is defined as in the Ref. [13]:

$$\widetilde{\boldsymbol{X}}_2^{NSVD} = \begin{bmatrix} \alpha_{11}^{SVD} & \alpha_{1j}^{SVD} & \alpha_{mm}^{SVD} \\ \vdots & \ddots & \vdots \\ \alpha_{m1}^{SVD} & \alpha_{mj}^{SVD} & \alpha_{mm}^{SVD} \end{bmatrix} \tag{39}$$

$$\alpha_{ij}^{SVD} = |\phi_{ij}^{SVD}\lambda_j^{SVD}\widetilde{E}_k^{SVD}(t)| \tag{40}$$

where the term α_{ij}^{SVD} is a measure of the participation factor of mode $\boldsymbol{\phi}_j^{SVD}$ in the states of the system.

It is important to note that the use of each variant depends on the characteristics of the dataset $\boldsymbol{X}$, whether $m < N$ or $m > N$, since the matrix structure in each case will be different. One of the application approaches to the $\widetilde{\boldsymbol{S}}$ variant allows a compact representation in the sense of the use of a certain amount of singular values, unlike the Companion $\boldsymbol{S}$ matrix, which focus its approximation on the number of available *snapshots*.

In the same way, in the modal identification approach, by means of the operator $\widetilde{\boldsymbol{S}}$ the modal analysis is based on the selection of the r singular values with the largest energy content considered in the SVD decomposition. In this case a full range matrix is assumed, i.e. $r = m$; on the other hand, the operator $\boldsymbol{S}$ considers $N - 1$ modal components. This part represents an advantage in the selection and display of a certain number of modal components for on-line applications.

Recently, one of the trends in the search for a solution to the problem of calculating the approximation to operator $\boldsymbol{A}$ based on (15) is through the use of optimization methodologies. This approach aims to improve the extraction of the dominant modal characteristics associated with the dynamics of the system. Several DMD algorithms have been developed with the inclusion of optimization methodologies applied to different fields of knowledge, with different analysis approaches and different objectives. In the following section seven approximations to the operator $\boldsymbol{A}$ based on optimization methodologies are presented.

As discussed before, the conventional approaches for approximating the operator $\boldsymbol{A}$ are mainly based on a polynomial variant $\boldsymbol{S}$ and a reduced variant $\widetilde{\boldsymbol{S}}$, which is stablished on an orthogonal projection. From these matrices, several works have been developed on diverse applications and proposing DMD algorithms with optimization methodologies that have converge to different results [36–38].

In the following sections, three methodologies based on optimization within the DMD method are presented. The analytical solution to the data reconstruction problem of each methodology is presented, and as an additional point, the mathematical formulation to obtain the characteristics of frequency, damping, mode shape, modal energy and participation factors in three of the methodologies is presented in the subsequent sections.

3.2 *Optimal Mode Decomposition (OMD)*

This methodology introduced by Goulart et al. in [36, 39] is a variant of the DMD technique that works with projections in low rank matrices. The mathematical formulation is proposed by means of an objective function that aims to identify a low dimensional subspace in a large dimensional system in which the trajectories of the system are optimally characterized. The formulation of the objective function is expressed by the following formulation:

$$\min_{O} \left\| \boldsymbol{X}_2^N - \boldsymbol{O}\boldsymbol{X}_1^{N-1} \right\|_2^2 \tag{41}$$

where the operator OMD $\boldsymbol{O}$ is defined as $\boldsymbol{O} = \boldsymbol{L}\boldsymbol{M}\boldsymbol{L}^T$, matrix $\boldsymbol{L}$ is a base of Stiefel type is defined as $\boldsymbol{L} \in \mathbb{R}^{rxr} | \boldsymbol{L}^T\boldsymbol{L} = \boldsymbol{I}, r \leq m$ and r is the rank of the matrix, in this case, a full rank is assumed such that $r = m$. The main objective is to maximize the base $\boldsymbol{L}$ through the following formulation:

$$\max_{L} \left\| \boldsymbol{L}^T \boldsymbol{X}_2^N \boldsymbol{Q}_L \right\|_F^2 \tag{42}$$

$$\boldsymbol{Q}_L = \boldsymbol{X}_1^{N-1^T} \boldsymbol{L} \left(\boldsymbol{L}^T \boldsymbol{X}_1^{N-1} \boldsymbol{X}_1^{N-1^T} \boldsymbol{L} \right)^{-1} \boldsymbol{L}^T \boldsymbol{X}_1^{N-1} \tag{43}$$

This optimization problem is solved using the ascending gradient-based algorithm described in detail in [39].

Alternatively, the matrix $\boldsymbol{M}$ represents an approximation of the operator $\widetilde{\boldsymbol{S}}$ and is dependant of the base $\boldsymbol{L}$. The assembling of this matrix is based on the solution to the following equation:

$$\boldsymbol{M}(\boldsymbol{L}) = \boldsymbol{L}^T \boldsymbol{X}_2^N \boldsymbol{X}_1^{N-1^T} \boldsymbol{L} \left(\boldsymbol{L}^T \boldsymbol{X}_1^{N-1} \boldsymbol{X}_1^{N-1^T} \boldsymbol{L} \right)^{-1} \tag{44}$$

This relationship is associated with the development of the optimization problem formulated in [36] where the dependence of the operator $\boldsymbol{M}$ with the orthogonal base $\boldsymbol{L}$ and the matrices $\boldsymbol{X}_1^{N-1}$ y $\boldsymbol{X}_2^N$ is stressed. Comparing this approach as an analogue formulation of the DMD method, substituting the base $\boldsymbol{L}$ with $\widetilde{\boldsymbol{U}}$ on Eq. (44) and considering the orthogonal characteristics of $\widetilde{\boldsymbol{U}}$, it can be proved that:

$$\boldsymbol{M}(\boldsymbol{U}) = \widetilde{\boldsymbol{U}}^T \boldsymbol{X}_2^N \boldsymbol{X}_1^{N-1^T} \widetilde{\boldsymbol{U}} \left(\widetilde{\boldsymbol{U}}^T \boldsymbol{X}_1^{N-1} \boldsymbol{X}_1^{N-1^T} \widetilde{\boldsymbol{U}} \right)^{-1} = \widetilde{\boldsymbol{U}}^* \boldsymbol{X}_2^N \widetilde{\boldsymbol{V}} \widetilde{\boldsymbol{\Sigma}}^{-1} = \widetilde{\boldsymbol{S}} \quad (45)$$

After finding the solution to the optimization problem, the analytic solution to the problem of data reconstruction is presented with the following expression:

$$\widetilde{\boldsymbol{X}}_2^N = \boldsymbol{O}\boldsymbol{X}_1^{N-1} = \boldsymbol{L}\boldsymbol{M}\boldsymbol{L}^T \boldsymbol{X}_1^{N-1} \quad (46)$$

Equation (46) can be represented through mode decomposition of the operator $\boldsymbol{M} = \boldsymbol{\Upsilon}^{OMD} \widetilde{\boldsymbol{\Lambda}}^{OMD} \boldsymbol{\Upsilon}^{OMD-1}$, which result on the following equation:

$$\widetilde{\boldsymbol{X}}_2^N = \boldsymbol{L}\boldsymbol{\Upsilon}^{OMD} \widetilde{\boldsymbol{\Lambda}}^{OMD} \boldsymbol{\Upsilon}^{OMD-1} \boldsymbol{L}^T \boldsymbol{X}_1^{N-1} = \boldsymbol{\Phi}^{OMD} \widetilde{\boldsymbol{\Lambda}}^{OMD} \boldsymbol{\Gamma}(t)^{OMD} \quad (47)$$

From Eq. (56) a spatial component can be defined using the mode matrix $\boldsymbol{\Phi}^{OMD}$, which is defined as follows:

$$\boldsymbol{\Phi}^{OMD} = \boldsymbol{L}\boldsymbol{\Upsilon}^{OMD} \quad (48)$$

Similarly, a temporal component can be defined as $\boldsymbol{\Gamma}(t)^{OMD}$ and represents the temporal evolution of the modes, which can be defined as follows:

$$\boldsymbol{\Gamma}^{OMD}(t) = \boldsymbol{\Upsilon}^{OMD-1} \boldsymbol{L}^T \boldsymbol{X}_1^{N-1} = \begin{bmatrix} \sum_{k=1}^{m} \boldsymbol{\Upsilon}_{1k}^{OMD-1} a_k^{OMD}(t) \\ \vdots \\ \sum_{k=1}^{m} \boldsymbol{\Upsilon}_{mk}^{OMD-1} \boldsymbol{a}_k^{OMD}(t) \end{bmatrix} = \begin{bmatrix} \widetilde{a}_1^{OMD}(t) \\ \vdots \\ \widetilde{a}_k^{OMD}(t) \end{bmatrix} \quad (49)$$

where the term $\boldsymbol{a}_k^{OMD}$ is defined through the relationship of $\boldsymbol{a}_k^{OMD} = \boldsymbol{L}_i^T \boldsymbol{x}_j$, $\boldsymbol{L}_i^T$ represent the columns of $\boldsymbol{L}^T$ and $\boldsymbol{x}_j$ are the rows of $\boldsymbol{X}_1^{N-1}$.

The variable $\boldsymbol{\phi}_j^{OMD}$ is the participation factor of the *jth* mode at time t_0 through the normalized magnitude of the same mode $\left\| \boldsymbol{\phi}_j^{OMD} \right\|$. Similarly, different groups existing in the time series that follow a similar dynamic behaviour can be observed with the phase $\angle \boldsymbol{\phi}_j^{OMD}$. Identification of particular characteristics such as frequency and damping associated to a particular mode $\boldsymbol{\phi}_j^{OMD}$ are defined by means of the following expressions [13]:

$$f_j^{OMD} = \frac{\Im\left\{\log\left(\lambda_j^{OMD}\right)\right\}}{\frac{\Delta t}{2\pi}} \tag{50}$$

$$\zeta_j^{OMD} = \Re\left\{\log\left(\lambda_j^{OMD}\right)\right\}/\Delta t \tag{51}$$

On the other hand, if we consider the energy extracted from each temporal term of Eq. (49) by means of the expression:

$$\widetilde{E}_k^{OMD} = \left\|\widetilde{\boldsymbol{a}}_k^{OMD}(t)\right\| \tag{52}$$

the relationship mode-state is defined:

$$\widetilde{\boldsymbol{X}}_2^{N^{OMD}} = \begin{bmatrix} \alpha_{11}^{OMD} & \alpha_{1j}^{OMD} & \alpha_{mm}^{OMD} \\ \vdots & \ddots & \vdots \\ \alpha_{m1}^{OMD} & \alpha_{mj}^{OMD} & \alpha_{mm}^{OMD} \end{bmatrix} \tag{53}$$

$$\alpha_{ij}^{OMD} = \left|\boldsymbol{\phi}_{ij}^{OMD} \lambda_j^{OMD} \widetilde{E}_k^{OMD}(t)\right| \tag{54}$$

where the term α_{ij}^{OMD} is a measure of the participation factor of mode $\boldsymbol{\phi}_j^{OMD}$ on the states of the system.

3.3 Nuclear Norm Regularised DMD (NNR-DMD)

This methodology, originally presented in Ref. [37], presents a DMD algorithm based on the use of the nuclear norm regularised. The goal of the objective function is to determine a low-rank representation of the $\widetilde{\boldsymbol{S}}$ matrix that captures the dynamics inherent in the data sequence through the following objective function:

$$\min_F \frac{1}{2}\left\|\widetilde{\boldsymbol{U}}^* \boldsymbol{X}_2^N - \widetilde{\boldsymbol{S}}^{NNR} \widetilde{\boldsymbol{\Sigma}} \widetilde{\boldsymbol{V}}^*\right\|_2^2 + \mu\left\|\widetilde{S}^{NNR}\right\|_* \tag{55}$$

where the constant μ represents a penalization term to the nuclear norm $\|\cdot\|_*$ in order to introduce a sparse methodology in the objective function.

The purpose of this formulation is to obtain an $\widetilde{\boldsymbol{S}}^{NNR}$ operator. By introducing the penalization term μ the problem becomes non-restrictive and the solution to the optimization problem is a system of equations obtained using the Split-Bregman method, which is described in detail in [37]. The solution of the S operator $\widetilde{\boldsymbol{S}}^{NNR}$ is shown below:

$$\widetilde{\boldsymbol{S}}^{NNR^{k+1}} = \left[\widetilde{\boldsymbol{U}}^* \boldsymbol{X}_2^N \left(\widetilde{\boldsymbol{\Sigma}} \widetilde{\boldsymbol{V}}^*\right)^* + \mu\left(\boldsymbol{H}^k - \boldsymbol{B}^k\right)\right]\left(\widetilde{\boldsymbol{\Sigma}} \widetilde{\boldsymbol{V}}^*\left(\widetilde{\boldsymbol{\Sigma}} \widetilde{\boldsymbol{V}}^*\right)^* + \eta \boldsymbol{I}\right) \tag{56}$$

where the terms μ, η, $\boldsymbol{H}$ and $\boldsymbol{B}$ are products resulting from the Split-Bregman method and the terms $\widetilde{\boldsymbol{U}}$, $\widetilde{\boldsymbol{\Sigma}}$ y $\widetilde{\boldsymbol{V}}^*$ are products of the SVD decomposition from Eq. (26). The solution of this iterative method is based on the Split-Bregman algorithm considering two different criteria: maximum number of iterations and an error criterion defined in Ref. [37].

By obtaining the solution to the optimization problem, the analytical solution to the data reconstruction problem is obtained by means of the following expression:

$$\widetilde{\boldsymbol{X}}_2^N \approx \boldsymbol{\Phi}^{NNR} \boldsymbol{B} \boldsymbol{\Gamma}(t)^{NNR} \tag{57}$$

where part of Eq. (57) can be defined as spatial through the modal matrix $\boldsymbol{\Phi}^{NNR}$ defined by the following equation:

$$\boldsymbol{\Phi}^{NNR} = \widetilde{\mathbf{U}} \boldsymbol{\Upsilon}^{NNR} \tag{58}$$

and where $\boldsymbol{\Upsilon}^{NNR}$ is calculated from the modal decomposition of the $\widetilde{\boldsymbol{S}}^{NNR} = \boldsymbol{\Upsilon}^{NNR} \widetilde{\boldsymbol{\Lambda}}^{NNR} \boldsymbol{\Upsilon}^{NNR-1}$ operator and the matrix $\boldsymbol{\Gamma}(t)^{NNR} \in \mathbb{C}^{mxN-1}$ represents the temporal evolution of the modes and is defined as follows:

$$\boldsymbol{\Gamma}^{NNR}(t) = \widetilde{\boldsymbol{T}}^{NNR} = \begin{bmatrix} \sum_{k=1}^{m} a_k^{NNR}(t) \\ \vdots \\ \sum_{k=1}^{m} \boldsymbol{a}_k^{NNR}(t) \end{bmatrix} = \begin{bmatrix} \widetilde{\boldsymbol{a}}_1^{NNR}(t) \\ \vdots \\ \widetilde{\boldsymbol{a}}_k^{NNR}(t) \end{bmatrix} \tag{59}$$

where the term $\boldsymbol{a}_k^{NNR}$ is described as $\boldsymbol{a}_k^{NNR} = \widetilde{\boldsymbol{T}}_j^{NNR}$ and the term $\widetilde{\boldsymbol{T}}_j^{NNR}$ corresponds to the rows of the matrix $\widetilde{\boldsymbol{T}}^{NNR}$. The weight of the temporal structure corresponds to the element b_i of matrix $\boldsymbol{B} := diag(\boldsymbol{b}) \in \mathbb{R}^{mxm}$, which represents the weight in descending order of relevance associated to each mode ϕ_j^{NNR} and is calculated using the sparse DMD approach. The mode frequency and damping rate is computed as:

$$f_j^{NNR} = \frac{\Im\left\{\log\left(\lambda_j^{NNR}\right)\right\}}{\frac{\Delta t}{2\pi}} \tag{60}$$

$$\zeta_j^{NNR} = \Re\{\log(\lambda_j^{NNR})\}/\Delta t \tag{61}$$

On the other hand, the energy extracted from each temporal term is considered by means of the expression:

$$\widetilde{E}_k^{NNR} = \left\| \widetilde{\boldsymbol{a}}_k^{NNR}(t) \right\| \tag{62}$$

and the relation mode-state is defined as:

$$\widetilde{X}_2^{NNNR} = \begin{bmatrix} \alpha_{11}^{NNR} & \alpha_{1j}^{NNR} & \alpha_{mm}^{NNR} \\ \vdots & \ddots & \vdots \\ \alpha_{m1}^{NNR} & \alpha_{mj}^{NNR} & \alpha_{mm}^{NNR} \end{bmatrix} \tag{63}$$

$$\alpha_{ij}^{NNR} = \left|\boldsymbol{\phi}_{ij}^{NNR} \widetilde{E}_k^{NNR}(t)\right| \tag{64}$$

where the term α_{ij}^{NNR} represents a measure of the participation factor of mode $\boldsymbol{\phi}_j^{NNR}$ on the states of the system.

3.4 Sparse-Promoting DMD (SP-DMD)

In the work developed in [38], a variant of the DMD algorithm is proposed. The alternative approach seeks to compensate the quality of the approximation in the formulation of the sequence of matrices and the number of modes used for the representation of the dynamic of the system. The process is carried out through exponential sequence of *eigenvalues* and *eigenvectors* based on a sparse methodology. In this case, the formulation of to the problem is based on finding the modal amplitude vector $\boldsymbol{b}$ such that weights the modal components in the most optimal form in the objective function $\boldsymbol{J}(\boldsymbol{b})$ described in the following equation:

$$\boldsymbol{J}(\boldsymbol{b}) = \boldsymbol{b}^*\boldsymbol{G}\boldsymbol{b} - \boldsymbol{q}^*\boldsymbol{b} - \boldsymbol{b}^*\boldsymbol{q} + s \tag{65}$$

where the terms $\boldsymbol{G}, \boldsymbol{q}, \boldsymbol{s}$ are defined by the following equations:

$$\boldsymbol{G} = (\widetilde{\boldsymbol{V}}^*\widetilde{\boldsymbol{V}}) \circ (\overline{\widetilde{\boldsymbol{T}}\widetilde{\boldsymbol{T}}^*}) \tag{66}$$

$$\boldsymbol{q} = \overline{diag\left(\widetilde{\boldsymbol{T}}\boldsymbol{\Upsilon}\widetilde{\boldsymbol{\Sigma}}^*\widetilde{\boldsymbol{V}}\right)} \tag{67}$$

$$s = tr\left(\widetilde{\boldsymbol{\Sigma}}^*\widetilde{\boldsymbol{\Sigma}}\right) \tag{68}$$

These terms correspond to the development of the problem presented in [38] and are based on the SVD decomposition presented in Eq. (26).

The algorithm for solving Eq. (65) is based on a sparse methodology by formulating a new objective function:

$$\min_b \boldsymbol{J}(\boldsymbol{b}) + \gamma \sum_{k=1}^{r} |b_k| \tag{69}$$

Previous equation represents a convex problem with the aim of finding a vector $\boldsymbol{b}$ and to determine the components that identify the modes with more influence on the system. The parameter γ has a direct impact on the number of components with a zero value that are obtained in the structure of vector $\boldsymbol{b}$. Thus, as the value of the parameter γ increases, the amount of zero components increases. The methodology proposes to solve this optimization problem is based on an ADMM method, and is presented in detail in the Ref. [38].

When the solution to the optimization problem is calculated, the analytical solution to the data reconstruction problem is obtained by means of the following expression:

$$\widetilde{\boldsymbol{X}}_2^N \approx \boldsymbol{\Phi}^{SP} \boldsymbol{B} \boldsymbol{\Gamma}(t)^{SP} \tag{70}$$

From (70) a spatial term is decoupled using the modal matrix $\boldsymbol{\Phi}^{SP}$ defined in the following equation:

$$\boldsymbol{\Phi}^{SP} = \widetilde{\mathbf{U}} \boldsymbol{\Upsilon}^{POD} \tag{71}$$

in this case, the modal matrices $\boldsymbol{\Phi}^{SP} = \boldsymbol{\Phi}^{POD}$ are equivalent. Matrix $\boldsymbol{\Gamma}(t)^{SP}$ represents a temporal structure during the process of reconstructing the signal and is defined as follows:

$$\boldsymbol{\Gamma}^{SP}(t) = \widetilde{\boldsymbol{T}} = \begin{bmatrix} \sum_{k=1}^{m} a_k^{SP}(t) \\ \vdots \\ \sum_{k=1}^{m} \boldsymbol{a}_k^{SP}(t) \end{bmatrix} = \begin{bmatrix} \widetilde{\boldsymbol{a}}_1^{SP}(t) \\ \vdots \\ \widetilde{\boldsymbol{a}}_k^{SP}(t) \end{bmatrix} \tag{72}$$

where the term $\boldsymbol{a}_k^{SP}$ is defined as $\boldsymbol{a}_k^{SP} = \widetilde{\boldsymbol{T}}_j$ and $\widetilde{\boldsymbol{T}}_j$ corresponds to the rows of the matrix $\widetilde{\boldsymbol{T}}$. The weight of this temporal structure corresponds to the element b_i from matrix $\boldsymbol{B} := diag(\boldsymbol{b}) \in \mathbb{R}^{mxm}$.

The identification of the frequency and damping characteristics associated to the mode ϕ_j^{SP}, following expressions (36) and (37), that is $f_j^{SP} = f_j^{POD}$ y $\zeta_j^{SP} = \zeta_j^{POD}$.

In this case, by considering the energy extracted from each temporal term by the expression:

$$\widetilde{E}_k^{SP} = \left\| \widetilde{\boldsymbol{a}}_k^{SP}(t) \right\| \tag{73}$$

The relation mode-state is defined:

$$\widetilde{\boldsymbol{X}}_2^{NSP} = \begin{bmatrix} \alpha_{11}^{SP} & \alpha_{1j}^{SP} & \alpha_{1m}^{SP} \\ \vdots & \ddots & \vdots \\ \alpha_{m1}^{SP} & \alpha_{mj}^{SP} & \alpha_{mm}^{SP} \end{bmatrix} \tag{74}$$

$$\alpha_{ij}^{SP} = \left| \boldsymbol{\phi}_{ij}^{SP} \widetilde{E}_k^{SP}(t) \right| \tag{75}$$

where the term α_{ij}^{SP} represents a measure of the degree of participation of $\boldsymbol{\phi}_j^{SP}$ mode in the system states.

3.5 Summary of Different DMD Approaches

Most of the DMD algorithms with optimization presented in this chapter consider the approximation of a low order operator as in the case of $\widetilde{S}$, with the exception of the Optimal Mode Decomposition (OMD) approach.

In Sects. 3.2 and 3.3 the main objective is the calculation of the approximation to the operator $\widetilde{S}$ directly. In the case of Sect. 3.4, the main objective is to obtain the vector of the amplitudes that adequately weights the modal components of the system. Different conditions are presented for the solution of the problem associated with the objective function of each section. In Sects. 3.3 and 3.4 regularization and penalty parameters are introduced for the formulation of non-restrictive optimization problems. In Sects. 3.3 and 3.4 a sparse parameter γ is introduced in the solution algorithm to identify the dominant modes of the system. In Sect. 3.3, two dimensionless parameters μ and η are introduced which are part of the solution of the optimization problem.

On the other hand, the analytical solution to the problem of reconstructing signals by means of modal components presents a different vision on different sections presented. In the case of the alternatives selected for evaluation, two main components are presented: a parameter associated to the spatial structure, in this case the modal matrix $\boldsymbol{\Phi}$, and a parameter associated to the temporal structure $\boldsymbol{\Gamma}(t)$. In Sects. 3.3 and 3.4 the temporal structure $\boldsymbol{\Gamma}(t)$ is constructed using the Vandermonde $\boldsymbol{T}$ matrix and the amplitude matrix $\boldsymbol{B}$ calculated from the disperse-based method. In Sect. 3.2 the time structure $\boldsymbol{\Gamma}(\mathrm{t})$ is formulated from the modal decomposition of the operator $\boldsymbol{M}$ and the base $\boldsymbol{L}$.

4 Wide Area Monitoring of Inter-area Oscillations Modes in a Longitudinal Interconnected System

The identification of inter-area oscillations presents an extra degree of complexity in large interconnected systems due to the volume and variety of information collected. Conventional tools for stability analysis based on mathematical models are limited by their accuracy and updating of their parameters. Similarly, the algorithms proposed to monitor spatio-temporal data from a wide area monitoring system are limited by the processing capacity of large volume of data. Therefore, it is necessary to propose new alternatives for the analysis of large interconnected electrical systems.

4.1 Description of Mexican Interconnected System

The test system used in this section, approximates the Mexican Interconnected (MI) system, which is distributed in seven electrical areas. The MI system has a longitudinal configuration characterized by long transmission lines and remote generation sources. As a consequence, SVC are the support voltage devices to improve the dynamic stability and voltage considerations. The network configuration requires implementation of several supplementary control schemes to meet the performance requirements. The overall generating capacity in the MIS is about 75.91 GW comprising 22.87 GW to renewable energy and the rest correspond to fossil energy. The bulk transmission system consists of 58,588 km of prevailing 400/230 kV lines, which is complemented by a network of 161 and 69 kV sub-transmission lines. The average demand of the MI system increases annually at a rate of about 7.1%. The MI system is characterized by a longitudinal infrastructure transmission system, supported in long transmission lines that help to import generation from neighbours areas; exciting undamped or poorly damped power oscillations when there are high power transfers from the areas I, II, III and VI, VII systems to areas IV and V. Based on this, the system studies carry out in this research assume that dynamic model of the generators are represented by a two-axis dynamic model, [5, 13] controlled with a simple excitation system. The loads of the system are assumed to be constant and SVCs are modelled to provide voltage support when required due to long transmission lines helping to interconnect all areas in the system. Figure 4 shows a simplified representation of the MI system control areas and illustrates the dynamic interaction of the different electrical areas.

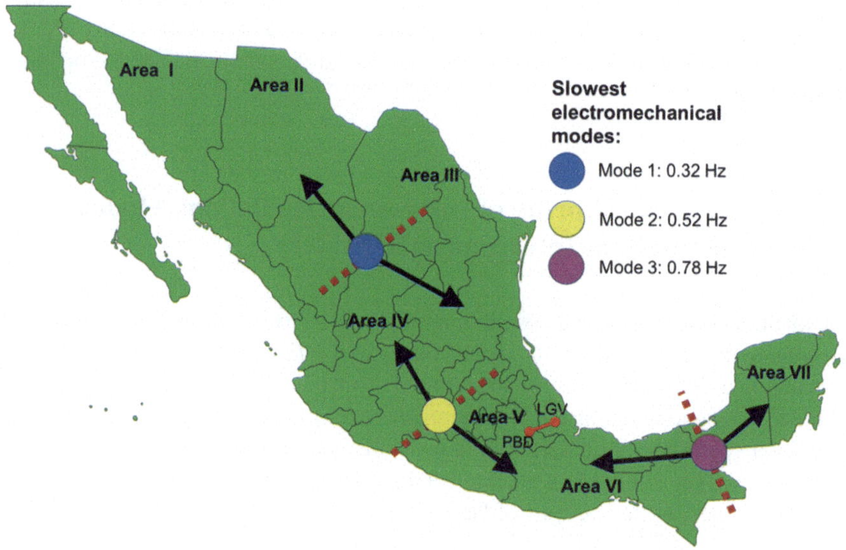

Fig. 4 Illustration of slowest electromechanical modes on the MI system [40]

Only 45 major generators distributed along the MI system were considered in this study. The interaction of the different electrical areas is caused by the presence of inter-area modes, which are excited by disturbances on the specific localization of the system [5, 41]. The equivalent of MI system is characterized by three inter-area modes involving the participation of different areas of the system [41].

4.2 *Power System Stability Analysis Using Conventional Tools*

The scenario for study is a three-phase fault event between the LGV-PBD transmission line located in area V at north-east of the country. This perturbation, stimulates the onset of the three inter-area modes, illustrated in Fig. 4. The dynamic responses are measured at the generator terminals and the angles signals $\Delta\delta_i$ are referenced to generator number 1 and collected. Figure 5 shows the transient responses of the referenced nodal angles $\boldsymbol{X}_{\Delta\delta} = \left[\Delta\delta_1 \; \Delta\delta_2 \; \ldots \; \Delta\delta_{41} \; \Delta\delta_{45} \right]^T \in \mathbb{R}^{45x2000}$, corresponding to 20 s of simulation, with an integration step of $\Delta t_i = 0.01$ s and a sampling frequency $f_s = 1/\Delta t_i$ of 100 Hz. The simulation was performed in Power System Toolbox (PST) open software [42]. From Fig. 5, the presence of inter-area oscillations can be clearly observed. In particular, a slow growing oscillation can be seen, suggesting an unstable condition of the system.

As a first approximation to the identification of the modal components associated with the oscillatory process observed in Fig. 5a; the fast Fourier transform (FFT) is applied to each of the signals $\Delta\delta_i$, to estimate their spectral content. Figure 5b shows the magnitude of each spectral component associated with each signal from $\Delta\delta_i$. Figure 5b shows three dominant frequencies obtained by the FFT: 0.39 Hz, 0.58 Hz and 0.782 Hz, respectively. The frequency components with the largest magnitude of the FFT are associated with the frequencies of 0.390 Hz and 0.781 Hz respectively.

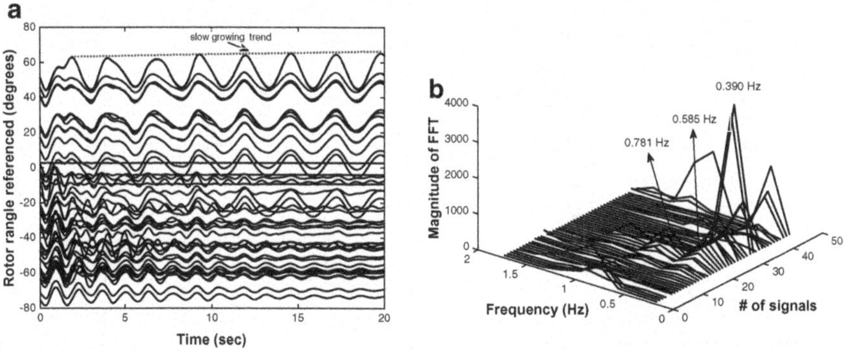

Fig. 5 **a** Dynamic response of the referenced nodal angle signals from 45 generators [42] and **b** spectral decomposition of the corresponding signals $\Delta\delta_i$ using the FFT

Generators #11 to 45 present a dominant frequency of 0.390 Hz. On the other hand, generators #1 to #10, #13 to #26 and #35 to #40 present a dominant frequency of 0.781 Hz. The results depicted on Fig. 5b suggest the presence of different coherent generators groups during the oscillatory process.

To gain more insight about the oscillation development on the MI system, a small signal stability analysis (SSSA) is performed. Figure 5b presents the frequency of the inter-area modes and their damping ratio coefficients resulted from the analysis.

From Fig. 5b, three modes of interest (0.385, 0.560 and 0.729 Hz) can be observed, which are indicated by red rectangular symbols. In particular, it can be seen that the 0.385 Hz mode has a damping coefficient of -0.0060, indicating the presence of an unstable electromechanical mode. The result agrees with the transient response depicted previously on Fig. 5a.

To visualize the separation mechanism and the interaction between electrical areas during an inter-area oscillation, mode shape is calculated using the right eigenvectors obtained from the decomposition by eigenvalues of the state matrix [42]. Figure 7 shows the mode-shape for the 3 modes of interest highlighted on Fig. 6.

Figures 7 shows the oscillation patterns involving different electrical areas of the system MI system under investigation. Figure 7a shows the oscillation pattern corresponding to the lowest frequency component and the interactions of the electric areas I, II and II against areas III, IV, VI and VII. While Fig. 7b depicts a different oscillation pattern, composed by the interaction of areas VII against IV and VI, respectively, which are located in the southeast of the country. Finally, Fig. 7c displays the interaction of areas IV against the area VI, involving a large number of hydraulic generators.

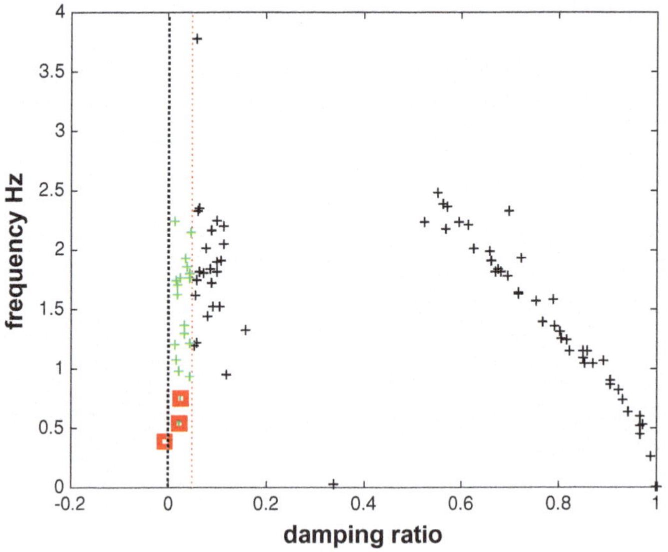

Fig. 6 Inter-area modes of MI system equivalent

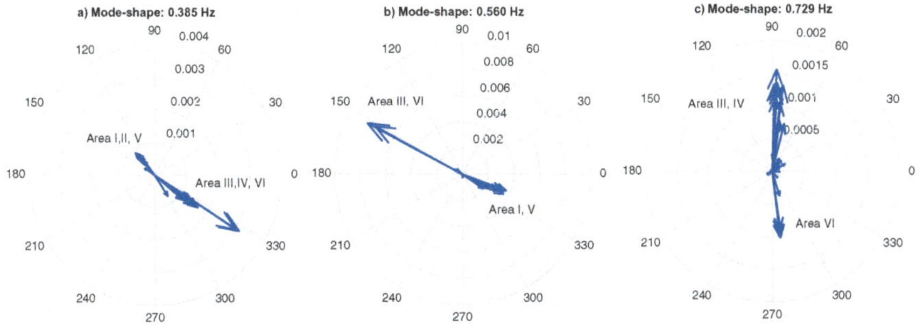

Fig. 7 Mode shape of modes inter-area showing the major coherent groups: **a** mode shape of 0.385 Hz, **b** mode shape of 0.560 Hz and **c** mode shape of 0.729 Hz

Table 1 Inter-area modes of MI system equivalent

# mode	Eigenvalue	Frequency (Hz)	Damping ratio ζ	Mode-shape
1	$-0.016 + j2.42$	0.385	−0.0060	I, II, V versus III, IV, VI, VII
2	$-0.080 + j3.51$	0.560	0.022	VII versus IV, VI
3	$-0.111 + j4.58$	0.729	0.023	IV versus VI

Table 1 shows a descriptive synthesis of the inter-area oscillation modes identified in the system according to previous studies.

The three-phase fault in area V excites the three inter-area oscillation modes. Mode #1 is the more complex mode because it involves participation of all electrical areas of the system and presents a dominant participation during the unstable oscillation process. In the following section a complementary analysis is performed using a space-time processing technique that allows to obtain a spectral analysis from the processing of the data collected from the simulation.

4.3 Spectral Analysis Based on SVD-DMD

Now, from the space-time decomposition simulation data given by $\boldsymbol{X}_{\Delta\delta}$; the SVD-DMD described at Sect. 3.1 is applied. The nature of system behaviour can be found by examining the empirical Ritz values, λ and their associated magnitudes [13, 14]. Figures 9 shows a plot of the empirical Ritz values, λ and their associated energy obtained from the norm of the time-dependent coefficients, $\left|\left|\widetilde{\boldsymbol{a}}_1^{SVD}(t)\right|\right|$, in (35).

As seen in Fig. 8a all the empirical Ritz values are on the unit circle $\lambda_j \approx 1.0$, indicating that the states of the dynamic system converge to a stable condition. Analysis of the relative energies in Fig. 8b, show that the modes with the largest energy contributions are those with frequencies of 0.384 Hz and 0.718 Hz, which are frequencies associated to the oscillation inter-area modes. The third identified mode

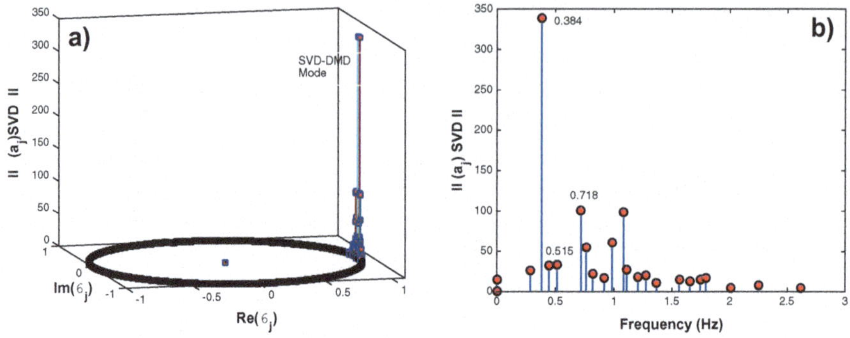

Fig. 8 Empirical Ritz values and their associated norms. **a** Empirical Ritz values estimated by SVD-DMD and **b** norm of dynamic modes and energy amplitudes associated with SVD-DMD modes

Table 2 Comparison of modal estimates for scenario: Time window 0–20 s and sampling frequency of 100 Hz

# of mode based on DMD nomenclature	FFT frequency	SSSA f ξ	SVD-DMD f $\rho/2\pi$
#27	0.390	0.385 –0.0060	0.384 –0.0040
#36	0.585	0.560 0.022	0.560 0.022
#31	0.781	0.729 0.024	0.718 0.019

with an approximate frequency of 0.515 Hz, represents the last inter-area mode and has a marginal impact during the oscillatory process.

Table 2 compares the modes estimation resulting from the application of SVD-DMD against conventional eigenvalue analysis. The results of modal estimated of frequencies obtained with SVD-DMD are in good agreement with the estimation from SSSA, however SVD-DMD estimation damping ratio are underestimated for mode #27 and #31 respectively.

In both methods, SSSA and SVD-DMD the, mode #1 produces a slow unstable oscillatory condition, while mode #2 and #3 are very well damped.

Clusters of coherent generators can be identified from the spatial signatures of SVD-DMD, contained in the modal vector $\Re\left\{\phi_j^{SVD}\right\}$. Figure 9a shows score plot for the three dominant modes obtained using SVD-DMD described at Sect. 3.1. SVD-DMD technique identifies three groups of coherent generators that involve all geographical areas.

From (39), a spatial (temporal) contribution factor that measure the contribution of each sensor to each state, can be defined. The strength of spatial contributions from each sensor to the observed data can be characterized and visualized. Figure 9b depicts a 2-D representation of the participation measures in (39) as a function of the sensor locations. Examination of Fig. 9b shows that mode #27 is strongly observable at sensors number #11 to #45, although the sensors #25 to #33 present the highest

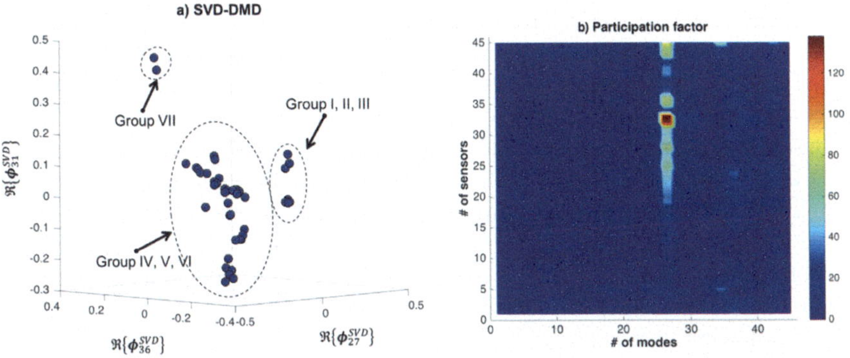

Fig. 9 Two ways to visualize the inter-area mode interaction: **a** coherency identification and **b** factor participation

participation factor in the system. This result has a strong relationship with the result depicted on Fig. 6.

4.4 Computational Effort and Time Window Simulation on Modal Parameter Estimation

Detailed simulations were conducted to assess the computational cost of SVD-DMD analysis for a study using a realistic dataset. Figure 10a shows the CPU time required

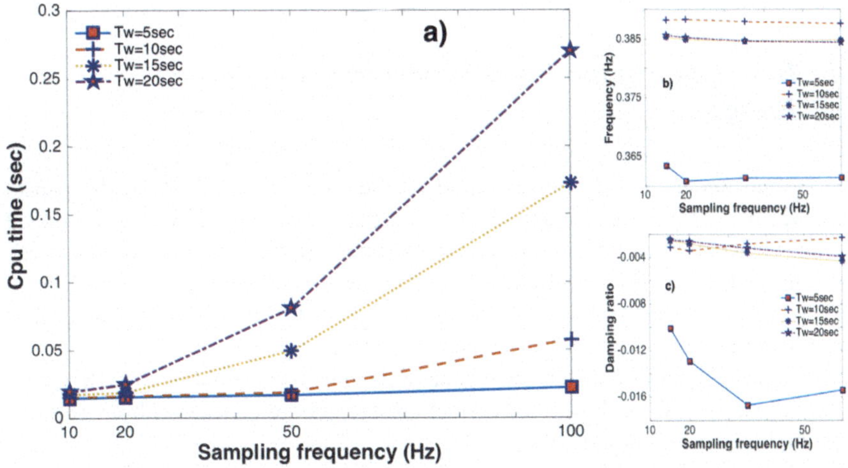

Fig. 10 **a** CPU time as a function of the sampling window, **b** time window effect on modal frequency, **c** time window effect on modal damping ratio estimation

to characterize the system behaviour for the scenario described before. Figure 10a shows a comparison of the CPU time for SVD-DMD analysis as a function of the size of the observation window. Previous results have illustrated that SVD-DMD is faster that Arnoldi-Koopman analysis for a similar sampling frequency [13, 43].

Based on the simulation results, it can be noted that the CPU time required by the SVD-DMD analysis is competitive in comparison with different modal estimation techniques [13]. In general, short time observation windows may degrade the quality of the estimation and result in various numerical problems, which is a common problem among other estimation techniques such as the Koopman mode analysis. Both observations are depicted at Fig. 10b, c respectively.

The following section presents a comparison of variants of the DMD technique in a larger interconnected continental system.

5 Wide-Area Monitoring of Global Oscillations Modes on Interconnected Continental System

As mentioned in previous sections, the presence of inter-area oscillations is a common problem around the world related to the interconnection of large and distant areas. This problem is more evident when groups of generators located on different geographical areas oscillate against each other. The objective in this section is to present a spectral analysis focused on identifying the modal characteristics associated with a disturbance taking place on an interconnected continental power system.

The following section presents a case study based on simulation data of an event in a given region of the European power system.

5.1 Description of the Power System from Continental Europe

The system under investigation is based on the studies carried out in the papers presented in [2, 44]. A representative schematic diagram of the test system is depicted on Fig. 11, where the selected regions are indicated; Spain (ES), Switzerland (CH), Germany (DE), Italy (IT) and Turkey (TK). The aforementioned countries have been selected based on the experience of the analysis described in [2, 44, 45] the level of detail in which these countries have been modelled.

Table 3 displays the distribution of generation units corresponding to each region.

Additionally, Table 4 shows the distribution of regions associated with each *cluster* identified in previous works [47].

In this subsection the exposure of the system to a three-phase fault of 100 sample/sec length in the region of France and applied at instant $t = 6$ s is presented. The fault leads to the disconnection of a 1.4 GW generation unit on its three phases.

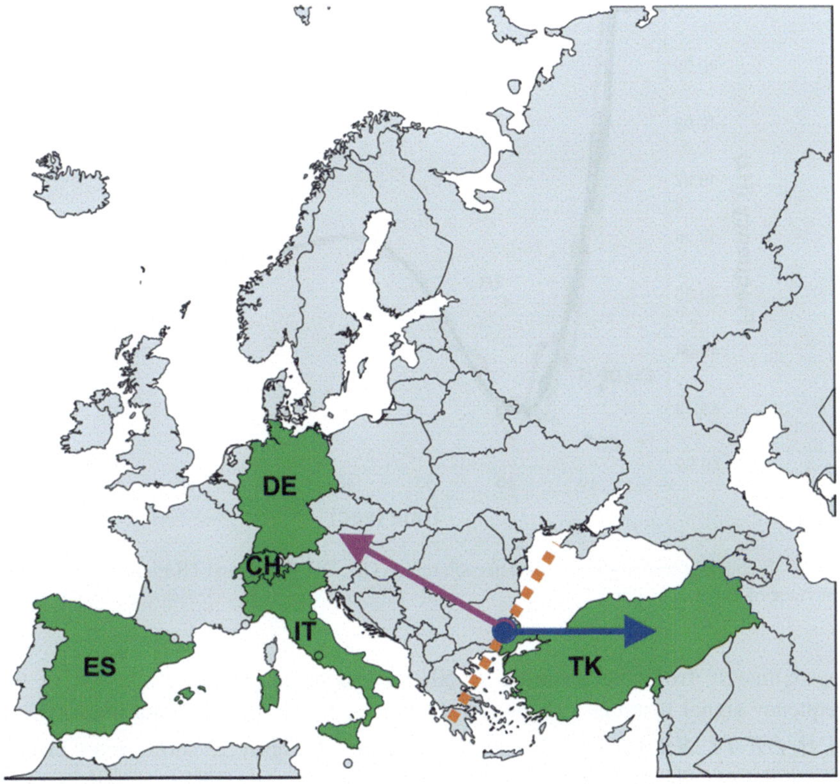

Fig. 11 Schematic illustration of ENTSO-E regions of continental Europe considered in this analysis [46]

Table 3 Number of generators per country

Country	CH	DE	IT	ES	TR	Total
# of generators	20	292	144	70	127	653

Table 4 Number of clusters and the associated countries

Group	c_1	c_2	c_3
Country	ES	DE, CH, IT	TR

The simulation was performed in the professional software DIgSILENT PowerFactory 2018 SP1with a sampling frequency $fs = 100$ Hz according to the recommendation of the IEEE standard for synchrophasor measurements in SEP C37.118.1-2011 [48]. The response of the frequency signals is displayed on Fig. 12.

Figure 12 shows the recording of 70 s of simulation at a frequency of 100 Hz, and it can be observed that the system converges to a new equilibrium point

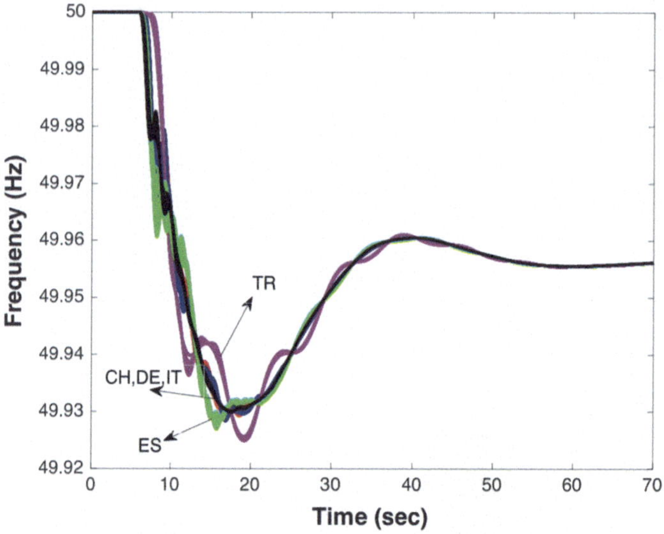

Fig. 12 Frequency responses associate at each region CH, DE, IT, ES and TR to the loss of a large generator of 1.4 GW in France

approximately from the instant $t = 60$ s. By considering the global set of the frequency signal response on each of the selected countries the data matrix $\boldsymbol{X} = \left[\boldsymbol{CH}\ \boldsymbol{DE}\ \boldsymbol{IT}\ \boldsymbol{ES}\ \boldsymbol{TR}\right]^T \in \mathbb{R}^{653x7000}$ is assembled. Figure 13 shows more clearly the oscillatory behaviour of the signals in the region of Turkey (TR) against the signals

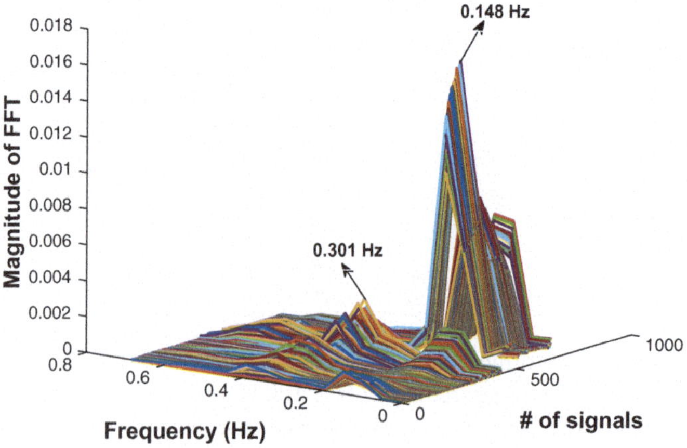

Fig. 13 Spectral decomposition by FFT of data set $\boldsymbol{X}$

on the other regions. This behaviour in the dynamic response of the frequency signals suggests the presence of low frequency oscillation modes.

5.2 Spectral Analysis Based on FFT Approach

The records from the WAMS system installed in the corresponding region of areas under investigation confirm the presence of two modes, which have frequency range of inter-area oscillation modes. These modes are the result of the power transfers between large geographical distances in continental Europe [2, 44]. Due to the structure of this network, two predominant low frequency global modes between 0.2 Hz (Global Mode #1) and 0.3 Hz (Global Mode #2) exist on the system. The interconnection with Turkey in December 2010 [45] increased the size and complexity of the original system, and as a consequence a new additional mode (Global Mode #3) of 0.15 Hz was incorporated.

As a first approach to identify the frequency components present in the system, the classical FFT tool is used. Figure 13 shows the calculation of the FFT applied to the 653 signals of the $\boldsymbol{X}$ data set:

Figure 13 illustrates the presence of three main low frequency components, in this case two of them corresponding to Global Modes #1 and #2. Figure 13 represents a first attempt to achieve modal identification on the response of frequency signals corresponding to the countries under analysis.

5.3 Spectral Analysis Based on Variants of DMD

One of the objectives in spectral analysis of PES is the identification of the dominant modes existing in the dynamics of the system. Figure 14 shows the result of the mode identification and their modal energy level calculated using Eqs. (38), (52), (62) and (70).

Figure 14 shows the 653 dynamic modes associated with the spatial structure m of the DMD operator. It can also be noted from the results of the different DMD variants that only a reduced number of them present a significant contribution to the dynamic behaviour of the system. In the case of the approaches OMD and SVD-DMD, approximately 25 modes that have different level of energy are identified. On the other hand, the results corresponding to approaches such as SP-DMD and NNR-DMD identify 20 dominant modes.

The difference between the number of identified modes on the different approaches correspond to the analytical solution of the reconstruction of data. Unlike the SVD-DMD and OMD methods, which have a temporal structure dependent on the modal decomposition of the DMD operator, sparsity-based approaches identify dominant modes in the system and assign a weight equal to zero to the remaining

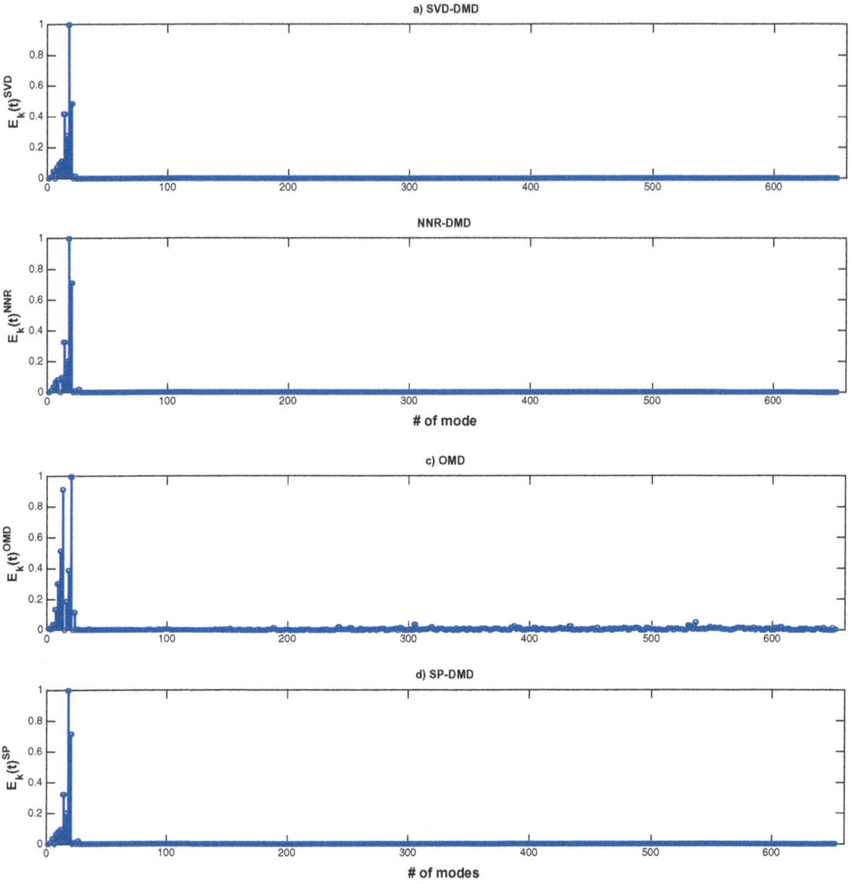

Fig. 14 Dynamic modes and energy amplitudes associated with the DMD variant modes: **a** SVD-DMD $\widetilde{E}_k^{SVD}$, **b** NNR-DMD $\widetilde{E}_k^{NNR}$, **c** OMD $\widetilde{E}_k^{OMD}$ and **d** SP-DMD $\widetilde{E}_k^{SP}$

modes. Figure 15 displays the evolution in time of the modes with largest energy content, which have been depicted on Fig. 14.

Figure 15 depicts the behaviour of the components $\widetilde{\boldsymbol{a}}_k^{NNR}(t)$ y $\widetilde{\boldsymbol{a}}_k^{SP}(t)$, which depend on the Vandermonde matrix $\widetilde{\boldsymbol{T}}$, while the components $\widetilde{\boldsymbol{a}}_k^{SVD}(t)$ and $\widetilde{\boldsymbol{a}}_k^{OMD}(t)$ are dependent of the right eigenvectors as result of the modal decomposition of the operators $\widetilde{\boldsymbol{S}}^{SVD}$ and $\boldsymbol{M}$ respectively.

In such a case, modes whose frequencies are approximate the same as Global Modes #2 and #3 are identified. Table 5 shows the mode number and its damping.

With the results presented in Table 5 it is possible to observe the temporal evolution of these modes on Fig. 16.

The temporal evolution of the modes corresponds to the oscillation that dissipates and reaches a new point of equilibrium, as shown in the response of the frequency

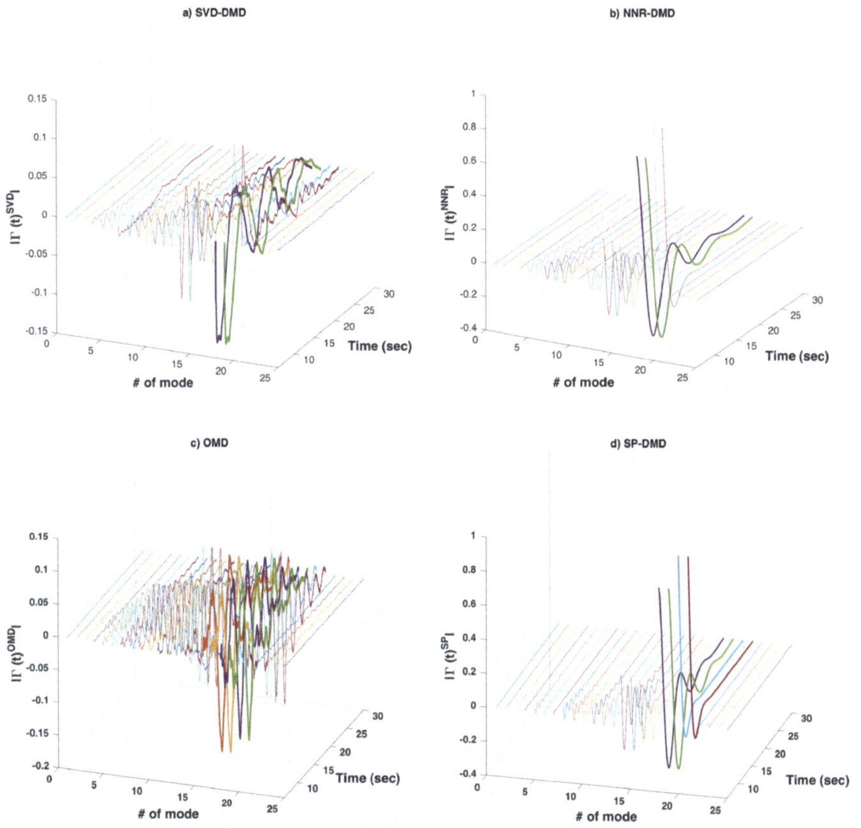

Fig. 15 Comparison of coherency identification DMD variants: **a** SVD-DMD $\boldsymbol{\Gamma}^{SVD}(t)$, **b** NNR-DMD $\boldsymbol{\Gamma}^{NNR}(t)$, **c** OMD $\boldsymbol{\Gamma}^{OMD}(t)$ and **d** SP-DMD $\boldsymbol{\Gamma}^{SP}(t)$

signals in Fig. 16. In this case, the components $\widetilde{\boldsymbol{a}}_k^{NNR}(t)$ and $\widetilde{\boldsymbol{a}}_k^{SP}(t)$ show an oscillation of lower amplitude corresponding to the structure of the Vandermonde matrix, where it is observed that the component corresponding to the frequency of the Global mode of 0. 15 Hz presents an abrupt increase and fast settling. Unlike the components $\widetilde{\boldsymbol{a}}_k^{SVD}(t)$ and $\widetilde{\boldsymbol{a}}_k^{OMD}(t)$ that base their response on the modal decomposition of the operators $\widetilde{\mathbf{S}}^{SVD}$ and $\mathbf{M}$, and that their oscillatory response has small amplitude and short oscillation time.

An important parameter in the modal analysis corresponds to the participation factors. Figure 17 shows the influence of the modes at each measurement point by means of the participation factors calculated from each different methodology.

As depicted on Fig. 17, the most influencing modes on the dynamic behaviour of the system, correspond to the modes identified first on the different variation of the analysis. The most susceptible areas are identified in a sensor range between #1 and #20, corresponding to the region of Switzerland (CH) and between the sensors #460 and #520, corresponding to Spain (ES) and Italy (IT) regions. Moreover, the modes

Table 5 Comparison of modal estimates of global modes

SVD-DMD			NNR-DMD			OMD			SP-DMD		
# mode	f (Hz)	ζ	# mode	f (Hz)	ζ	# mode	f (Hz)	ζ	# mode	f (Hz)	ζ
$\boldsymbol{\phi}_{20}^{SVD}$	0.1401	0.1063	$\boldsymbol{\phi}_{20}^{NNR}$	0.1401	0.1063	$\boldsymbol{\phi}_{18}^{OMD}$	0.1401	0.1063	$\boldsymbol{\phi}_{20}^{SP}$	0.1401	0.1063
$\boldsymbol{\phi}_{14}^{SVD}$	0.3738	0.0364	$\boldsymbol{\phi}_{14}^{NNR}$	0.3738	0.0364	$\boldsymbol{\phi}_{14}^{OMD}$	0.3738	0.0364	$\boldsymbol{\phi}_{14}^{SP}$	0.3738	0.0364

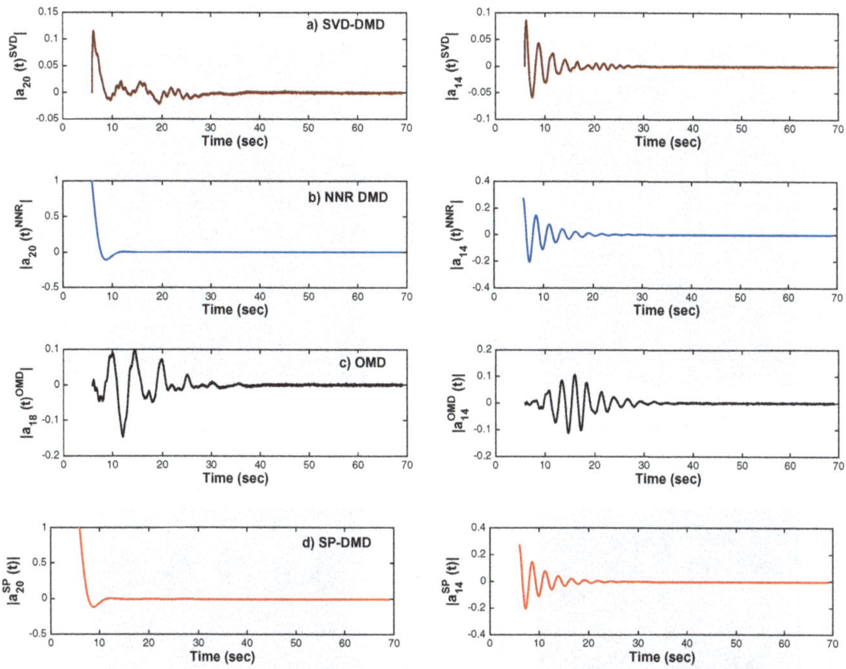

Fig. 16 Comparison of temporal amplitudes associated to dominant dynamic modes (0.15 and 0.39 Hz): **a** SVD-DMD $\tilde{\boldsymbol{a}}_K^{SVD}(t)$, **b** NNR-DMD $\tilde{\boldsymbol{a}}_k^{NNR}(t)$, **c** OMD $\tilde{\boldsymbol{a}}_K^{OMD}(t)$ and **d** SP-DMD $\tilde{\boldsymbol{a}}_k^{SP}(t)$

with the largest impact are in a range between #1 and #20. In this case, the SP-DMD and the NNR-DMD restrict the participation factors to a limited number of modes, identifying more clearly the critical areas in the system. The dominant participation factors are located among generators #480 to #510 corresponding to TR.

On the other hand, from the data presented in Table 5; Figs. 18 shows the formation of clusters by grouping the modal components associated with inter-area frequencies by means of the $\Re\{\boldsymbol{\phi}\}$ structure.

As shown in Fig. 18, three main clusters are identified: c_1, c_2 and c_3 as shown in Table 5. The distribution of the clusters corresponds to the response of the frequency signals observed in Fig. 18 where it is possible to identify three oscillation groups. In this case, a spatial distribution presenting three oscillation groups in the dynamic behaviour of the system is observed. However, the structure of the **M** operator corresponding to the OMD variant presents a different spatial structure due to the searching space within the gradient methodology. The formulation described in the rest of the variants presents a similar structure due to the characteristics of the $\widetilde{\mathbf{S}}$ operator, which depends on the SVD decomposition.

One approach to evaluate the performance of the different algorithms presented here, is the computational time associated to the calculation of the DMD operator

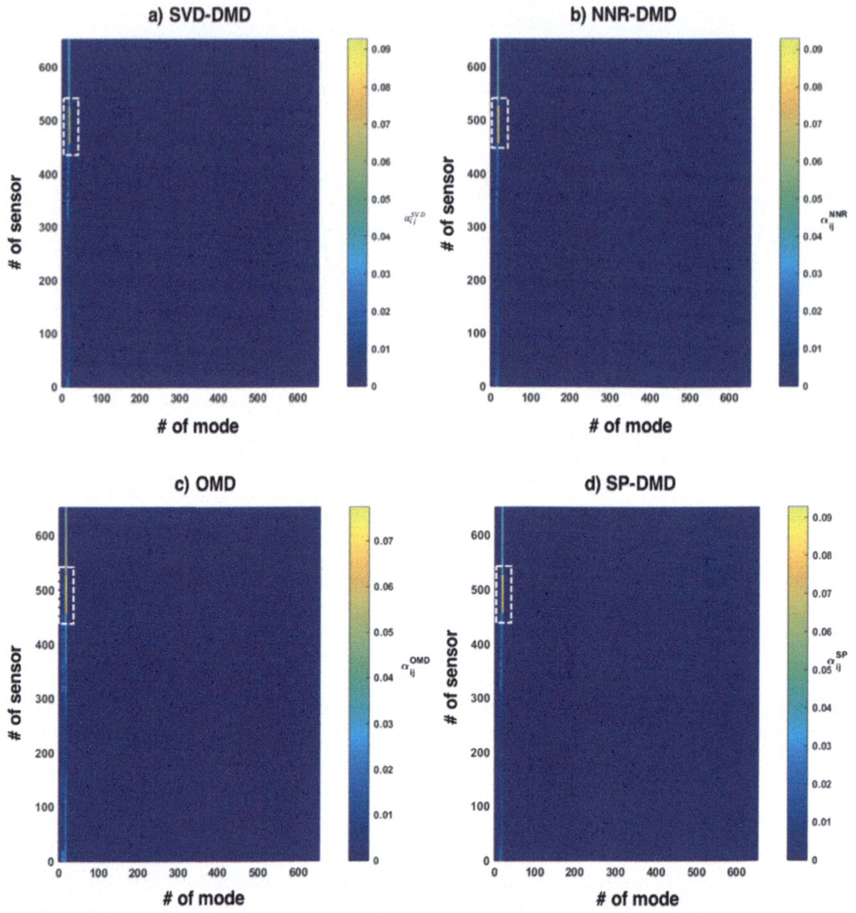

Fig. 17 Participation factors that relate the # of sensors and # of modes: **a** SVD-DMD α_{ij}^{SVD}, **b** NNR-DMD α_{ij}^{NNR}, **c** OMD α_{ij}^{OMD} and **d** SP-DMD α_{ij}^{SP}

and include the reconstruction signal process for each DMD variant. Table 6 shows the processing times for the dataset $\boldsymbol{X}$.

In this case, the method with the slowest processing time corresponds to the NNR-DMD method that integrates the weighting methodology corresponding to the sparse DMD variant as a solution to the problem of signal reconstruction. The computational time is an important parameter related to the processing capacity that must be considered when analysing large systems. Additionally, with the development of online algorithms the processing time is an important feature for the development of new tools for real-time application.

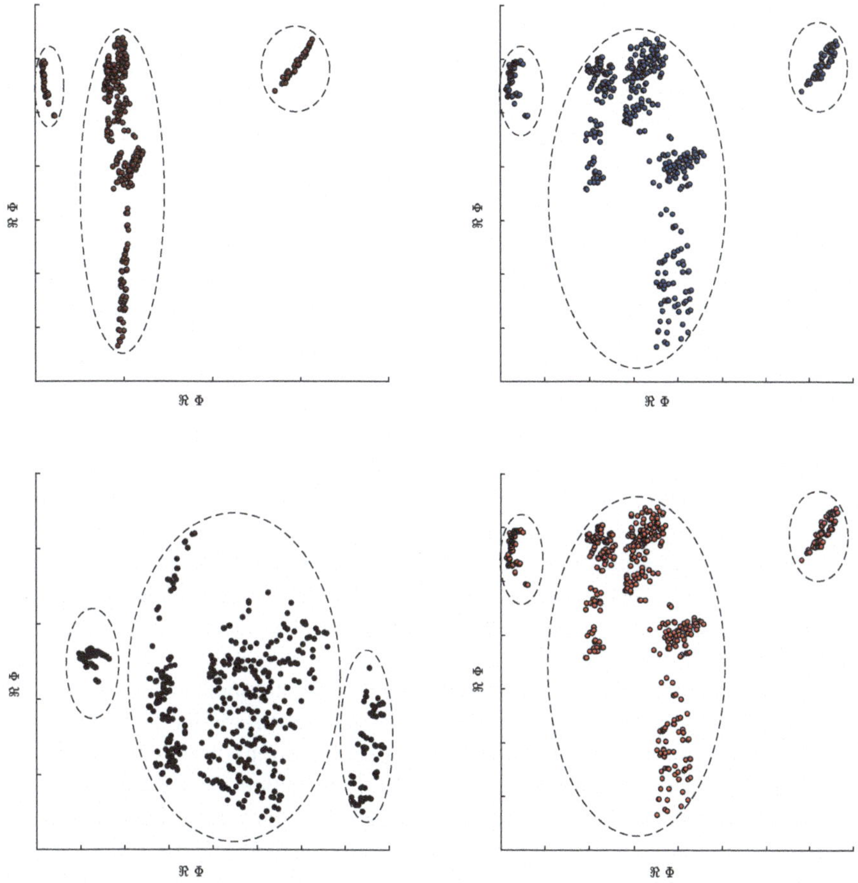

Fig. 18 Comparison of coherency identification DMD variants: **a** SVD-DMD, **b** NNR-DMD, **c** OMD (t) and **d** SP-DMD

Table 6 Comparison of DMD variant of CPU time simulated data

	SVD-DMD	NNR-DMD	OMD	SP-DMD
CPU time (s)	9.77	64.61	30.98	15.78

6 Conclusions

In this chapter, four variants of the DMD method were described. These alternative algorithms are based on a polynomial variant $\boldsymbol{S}$ and a reduced orthogonal projection matrix $\widetilde{\boldsymbol{S}}$, that are used to approximate the Koopman operator $\boldsymbol{A}$. In this form, DMD algorithms with optimization methodologies and their respective application were shown.

It has been demonstrated that the processing time depends mostly on the volume of the dataset under analysis. The variant representing the shortest processing time corresponds to the algorithm SVD-DMD. While the NNR-DMD variant resulted on the slowest algorithm among the others due to the double optimization process that is required.

One of the advantages observed when using a method based on sparsity to assign weights is the clear visualization of dominant components. This means, that the modal amplitude matrix shows only a limited number of components that have the largest effect on the dynamic of the system. This effect can be reflected in the visualization process of the participation factors. Modes that affect sensitive areas on the system are easily identified. However, the parameter γ immerse on the optimization problem must be correctly calibrated.

Participation factors represent a quantitative measure that is used to display the most affected areas on the system, the impact of the dominant modes on the states of the system and their geographical location. It has been observed that as the dimension of the system increase, the identification of these areas becomes more challenging and restrictive, because a smaller number of the dominant modes are visible.

The temporal evolution of the oscillation modes associated with the low frequency components in the system dynamics is an initial approximation within the system dynamics and affects the duration of the transient event. The behaviour resulting from the effect of the Vandermonde matrix, depends on the weighting of the eigenvalues raised to an exponential depending on the number of snapshots considered in the modal analysis, so the oscillatory behaviour depends partly on the location of the identified number of modes. On the other hand, the behaviour observed in the SVD-DMD and OMD variants depends on the modal decomposition of the $\widetilde{S}$ y $\mathbf{M}$ operators, respectively. In this case, through these variants, the duration of the oscillatory effects associated with characteristic related to the dominant modes can be more easily visualized.

References

1. Electricity european network of transmission system (2019), https://docstore.entsoe.eu/Documents/Publications/Statistics/Factsheet/entsoe_sfs2018_web.pdf
2. F. Segundo Sevilla, P. Korba, K. Uhlen, Evaluation of the ENTSO-E initial dynamic model of continental Europe subject to parameter variations, in *2017 IEEE Power and Energy Society Innovative Smart Grid Technologies Conference (ISGT 2017)*, Arlington, VA, USA (2017)
3. P. Kundur, Transient stability, *Power System Stability and Control* (McGraw-Hill Inc., Mexico City, 1993), pp. 827–828
4. G. Rogers, *Power System Oscillations* (Kluwer Academic Publishers, New York, 2000)
5. J.G. Calderón-Guizar, M. Ramirez-González, R. Castellanos-Bustamante, Low frequency oscillations in large power grids, in *2017 IEEE XXIV International Conference on Electronics, Electrical Engineering and Computing (INTERCON)*, Cusco, Perú (2017)
6. Y. Zhang, Wide-area frequency monitoring network (FNET) architecture and applications. IEEE Trans. Smart Grid **1**(2), 159–167 (2010)

7. J. Thambirajah, E. Barocio, N.F. Thornhill, Comparative review of methods for stability monitoring in electrical power systems and vibrating structures. IET Gener. Transm. Distrib. **4**(10), 1086–1103 (2010)
8. Task Force on Identification of Electromechanical Mode. IEEE Power and Energy Society, NA June 2012 (2012), http://resourcecenter.ieee-pes.org/pes/product/technical-publications/PESTR15. Accessed 01 Marzo 2019
9. B. Koopman, Hamiltonian systems and transformation in Hilbert space. Proc. Natl. Acad. Sci. USA **17**(5), 315–318 (1931)
10. C.W. Rowley, I. Mezic, S. Bagheri, P. Schlatter, D.S. Henningson, Spectral analysis of nonlinear flows. J. Fluid Mech. **641**, 115–127 (2009)
11. J.N. Kutz, J.L. Proctor, S.L. Brunton, Applied Koopman theory for partial differential equations and data-driven modeling of spatio-temporal systems. Hindawi Complex. **2018**, 1–17 (2018)
12. N.J. Kutz, S.L. Brunton, B.W. Brunton, J.L. Proctor, *Dynamic Mode Decomposition Data-Driven Modeling of Complex Systems* (SIAM, Philadelphia, 2016)
13. E. Barocio, B.C. Pal, N.F. Thornhill, A.R. Messina, A dynamic mode decomposition framework for global power system oscillation analysis. IEEE Trans. Power Syst. **30**(6), 2902–2912 (2015)
14. P.J. Schmid, Dynamic mode decomposition of numerical and experimental data. J. Fluid Mech. **656**, 5–28 (2010)
15. M. Hochbruck, C. Lubich, On Krylov subspace approximations to the matrix exponential operator. SIAM J. Numer. Anal. **24**(5), 12–17 (1997)
16. K.K. Chen, J.H. Tu, C.W. Rowley, Variants of dynamic mode decomposition: boundary conditions, Koopman, and Fourier analysis. J. Nonlinear Sci. 1–28 (2011)
17. H. Jia-Chen, R. Sukesh, J.L. McCauley, G.H. Gunaratne, Using dynamic mode decomposition to extract cyclic behavior in the stock market. Phys. A: Stat. Mech. Appl. **448**, 172–180 (2016)
18. B.W. Brunton, L. Johnson, J. Ojeman, J. Kutz, Extracting spatial-temporal coherent patterns in large-scale neural recordings using dynamic mode decomposition. J. Neurosci. Methods **258**, 1–15 (2016)
19. F.G. Georg, A. Gottwald, Detecting regime transitions in time series using dynamic mode decomposition. J. Stat. Phys. 1–23 (2019)
20. T. Grenga, J.F. MacArt, M.E. Mueller, Dynamic mode decomposition of a direct numerical simulation of a turbulent premixed planar jet flame: convergence of the modes. J. Combust. Theory Model. **22**(4), 795–811 (2018)
21. J.N. Kutz, X. Fu, S.L. Brunton, N.B. Erichson, Multi-resolution dynamic mode decomposition for foreground/background separation and object tracking, in *IEEE International Conference on Computer Vision Workshop*, Santiago, Chile (2015)
22. N. Mohan, K. Soman, K.S. Sachin, A data-driven approach for estimating power system frequency and amplitude using Dynamic Mode Decomposition, in *2018 International Conference and Utility Exhibition on Green Energy for Sustainable Development (ICUE)*, Phuket, Tailandia (2018)
23. S. Mohapatra, T.J. Overbye, Fast modal identification, monitoring, and visualization for large-scale power systems using Dynamic Mode Decomposition, in *2016 Power Systems Computation Conference (PSCC)*, Genova, Italia (2016)
24. J.J. Ramos, J.N. Kutz, Dynamic mode decomposition and sparse measurements for characterization and monitoring of power system disturbances. Nonlinear Sci. (2019)
25. Y. Suski, A. Chakrabortty, Introduction to Koopman mode decomposition for data-based technology of power system nonlinear dynamics. IFAC Pap. OnLine **51**(28), 327–332 (2018)
26. M.A. Hernandez, A.R. Messina, Nonlinear power system analysis using Koopman mode decomposition and perturbation theory. IEEE Trans. Power Syst. **33**(5), 5124–5134 (2018)
27. M. Netto, L. Mili, A robust data-driven koopman kalman filter for power systems dynamic state estimation. IEEE Trans. Power Syst. **33**(6), 7228–7237 (2018)
28. Y. Susuki, I. Mezic, Nonlinear Koopmnan modes and coherency identification of coupled swing dynamics. IEE Trans. Power Syst. **26**(4), 1904–1984 (2011)
29. Y. Susuki, I. Mezic, F. Raak, T. Hikihara, Applied Koopman operator theory for power systems technology. IECE Nonlinear Theory Appl. **7**(4), 430–459 (2016)

30. F. Raak, Y. Susuki, T. Hikihara, Data-driven partitioning of power networks via nonlinear koopman mode analysis. IEEE Trans. Power Syst. **48**(30), 1–10 (2015)
31. M. Netto, Y. Susuki, L. Mili, Data-driven participation factors for nonlinear systems based on koopman mode decomposition. IEEE Control. Syst. Lett. **3**(1), 198–203 (2018)
32. A. Saldaña, E. Barocio, A.R. Messina, J.J. Ramos, R.J. Segundo, G.A. Tinajero, Monitoring harmonic distortion in microgrids using dynamic mode decomposition, in *IEEE Power & Energy Society General Meeting*, Chicago, IL, USA (2017)
33. N. Mohan, K.P. Soman, S.S. Kumar, A data-driven strategy for short-term electric load forecasting using dynamic mode decomposition model. Appl. Energy **232**(15), 229–244 (2018)
34. Y. Susuki, K. Sako, Data-based voltage analysis of power systems via delay embedding and extended dynamic mode decomposition. IFAC-Pap. OnLine **51**(28), 221–226 (2018)
35. A. Alassaf, L. Fans, Dynamic mode decomposition in various power system applications, in *North American Power Symposium (NAPS)*, Wichita, KS, USA (2019)
36. P.J. Goulart, A. Wynn, D. Pearson, Optimal mode decomposition for high dimensional systems, in *IEEE 51st IEEE Conference on Decision and Control (CDC)*, Maui, HI (2012)
37. S. Wang, X. Qing, Nuclear norm regularised dynamic mode decomposition. IET Signal. Process. **10**(6), 626–632 (2016)
38. M.R. Jovanović, P.J. Schmid, J.W. Nichols, Sparsity promoting dynamic mode decomposition. Phys. Fluids **26**(2), 0234103(1–22) (2014)
39. A. Wynn, D.S. Pearson, B. Ganapathisubramani, P.J. Goulart, Optimal mode decomposition for unsteady flows. J. Fluid Mech. **773**(SN), 473–503 (2013)
40. Comisión Reguladora de Energía, 17 Diciembre 2019 (2019), https://www.gob.mx/, https://www.gob.mx/cre/documentos/reporte-de-confiabilidad-del-sistema-electrico-nacional-2016-2017. Accessed 17 Diciembre 2019
41. J. Ayon, E. Barocio, I. Cabrera, R. Betancourt, Identification of coherent areas using a power spectral density algorithm. Electr. Eng. **100**(2), 1–11 (2017)
42. C. Joe, R. Graham, Power system toolbox, version 3.0, Ontario, Canada (1991–2008)
43. M.A. Hernandez, A.R. Messina, An observability-based approach to extract spatiotemporal patterns from power system Koopman mode analysis. Electr. Power Compon. Syst. **45**(4), 355–365 (2016)
44. A. Semerow, S. Höhn, M. Luther, W. Sattinger, H. Abildgaard, A. Diaz Garcia, G. Giannuzzi, Dynamic study model for the interconnected power system of continental Europe in different simulation tools, in *IEEE Eindhoven PowerTech*, Eindhoven, Netherlands (2015)
45. ENTSO-E, Analysis of Continental Europe Inter-Area Oscillations on 1st December 2016 (2017)
46. European network of transmission system operators for electricity, 5 Junio 2019 (2019), https://docstore.entsoe.eu/Documents/Publications/Statistics/Factsheet/entsoe_sfs2018_web.pdf. Accessed 24 Diciembre 2019
47. E. Barocio, K. Petr, S. Walter, R. Felix, Segundo, Online coherency identification and stability condition for large interconnected power systems using an unsupervised data mining technique. IET Gener. Transm. Distrib. **13**(15), 3323–3333 (2019)
48. IEEE, Standard for synchrophasor measurements for power systems. Revision of IEEE Std. C37.118–2005 (2011)

Electromechanical Mode Estimation in Power System Using a Novel Nonstationary Approach

S. Rahul, Subin Koshy, and R. Sunitha

Abstract The modern power grid protection system should have considerable operational flexibility and resiliency to hedge the variability and uncertainty of high dimensional dependencies. The use of wide-area monitoring systems (WAMS) in the smart grid enables the real-time supervision of power system oscillations. With the help of advanced signal processing methods and big data analytics, time-synchronized phasor measurements can be used to extract valuable information concerning the electromechanical modal properties of power system oscillations. This chapter introduces a novel method for identifying electromechanical inter-area oscillation modes with the help of wide-area measurement data. Variational mode decomposition (VMD) can be considered as a flexible signal processing technique on the wide-area phasor measurements in power oscillation analysis. For the real-time operation, it is challenging to preset the value of the mode number in the VMD process. This issue has been addressed by improving the strategy for VMD, which is presented in this chapter. The first stage involves the use of Complete ensemble empirical mode decomposition with adaptive noise (CEEMDAN) technique to generate intrinsic mode functions and gives indexing based on the correlation factor. Depending on the indexing, the mode number is selected for the second stage VMD process. Techniques such as spectral analysis and Hilbert transform are quite suitable for the estimation of modal parameters. The study is based on significant features of power oscillations, such as determination of damping ratio, amplitude, and frequency. The identification and estimation of low-frequency modes have been performed using this improvised VMD technique, and the results have been compared with those obtained using empirical mode decomposition approaches. The proposed approach is also validated using

S. Rahul (✉) · S. Koshy · R. Sunitha
Electrical Engineering Department, National Institute of Technology Calicut, Calicut, India
e-mail: rahulsatheesh21@gmail.com

S. Koshy
e-mail: subin_loyolite@yahoo.com

R. Sunitha
e-mail: rsunitha@nitc.ac.in

H. Haes Alhelou et al. (eds.), *Wide Area Power Systems Stability, Protection, and Security*, Power Systems, https://doi.org/10.1007/978-3-030-54275-7_8

real-time data obtained from load dispatch centers. The results indicate the effectiveness of non-linear, nonstationary analysis methods for analysing the low-frequency modes and provide reliable validation of these algorithms in analysing real-time data patterns.

Keywords Electromechanical mode · Nonstationary method · Estimation · WAMS · PMUs · Variation mode decomposition · CEEMDAN · Power systems

1 Introduction

Increased power demand and deregulation policies have resulted in the power system being operated near its stability limit. In the modern scenario, maintaining the stability of a widely interconnected power system is highly essential, considering the increased integration of large-scale renewable resources. Even though the system is capable of damping most of the oscillations, a few undamped oscillations may lead to system collapse. Hence, the identification of low-frequency modes and estimation of mode parameters are a necessity in the modern power system context [1]. Conventionally, these modes are identified by applying Eigen value analysis on the linearized dynamic model of the system around an equilibrium point [2]. Nevertheless, this approach is not effective for large complex power systems and is computationally intensive. Hence it is mainly used for small signal stability analysis in offline conditions.

The estimation of inter-area oscillations using supervisory control and data acquisition (SCADA) is not possible on a real-time basis. Thus, the use of WAMS has to be utilized for the online estimation of oscillation parameters effectively through phasor measurement units (PMU) installations [3], which provides a real dynamic picture of the power system operating condition. With the widespread installation of PMUs, measurement-based methods and related WAMS techniques are employed extensively to identify and monitor the undamped modes in power systems accurately. The uneven inter-area oscillations affect the reliability and functioning of the modern power system profoundly, thereby restricting the power transfer through long transmission distances.

One of the intelligent aspects of smart real-time monitoring of the power system includes the online identification of coherency of generators and mode shape determination. The mode shape portrays the angular separation between system variables corresponding to a particular mode, which helps in the optimal determination of the generator or load tripping schemes, thus improving the damping of the less damped mode. The features, such as frequency, damping ratio and mode shape define the properties of the mode. Modal frequencies and damping ratio provide a picture of the power system stress or reduced grid capacity issues [4].

Recent advancements in signal processing techniques can be used for the identification of electromechanical oscillatory mode and its parameter estimation. Prony analysis deals with the determination of the polynomial coefficient as a solution for the matrix equation [5, 6]. However, Prony analysis is challenging to use in higher-order systems such as a complex power system. Matrix pencil method (MPM) has better statistical properties in the determination of the number of poles as a solution of eigenvalue analysis [7]. Advancements in MPM include the recursive MPM, as the MPM involves drawback with its pencil parameter and its selection. In [8], Fast Fourier Transform (FFT) is proposed to determine the low-frequency oscillatory mode. The major drawback of FFT is that it is only applicable to linear time invariant oscillation, and it also suffers from weaker frequency resolution.

The detection of the low-frequency oscillatory mode using a recursive Kalman filter is investigated in [9]. But it undergoes numerical instability. In [10], Wiener Hopf's linear prediction is used for the oscillation analysis. To estimate the electromechanical modes, Robust recursive least square has been used in [11]. The estimation of oscillation parameters using rotational invariance techniques (ESPRIT) is proposed in [12]. Transfer function methods and subspace methods can be used for the determination of mode shape and other properties. Transfer function methods include spectral analysis, frequency domain decomposition, and channel matching method [13]. Subspace methods have different variants of stochastic subspace identification (SSI) approach, robust SSI, etc. [14, 15]. Usually, the latter approach gives accurate results, but it takes more computation time. The transfer function method is suitable for real-time applications as it is computationally less complex.

The majority of the power system transients are non-linear and cannot be analysed using traditional linear analysis techniques. Multi-scale processes govern these transient phenomena and are fundamentally nonstationary due to non-linear dynamics and time-dependent control actions. Recently, non-linear nonstationary algorithms are found to adaptively analyse complex oscillatory properties in the power system [16]. They are capable of distinguishing nonstationary processes more thoroughly than other approaches and have the potential for online applications. Major nonstationary mode estimation algorithms are Hilbert Huang Transform (HHT), Wavelet Transform, Teager–Kaiser Operator, Koopman modes etc. The Wavelet transform adopted in [17] identifies the non-periodic and time-varying oscillations that work well for the sufficiently spaced modes, but it fails for closely spaced modes. HHT involves empirical mode decomposition (EMD), in which the primary signal is decomposed into its constituent intrinsic mode functions (IMF) [18]. After selecting the proper IMFs, the Hilbert transform can be employed for the instantaneous parameter estimation. Hilbert transform can be replaced by the Teager Kaiser operation, and it is implemented in [19]. The standard EMD does not have any mathematical validation and suffers from a mode mixing problem, which involves having more than one frequency component for a particular decomposed mode. Deering and Kaiser in [20], developed masking technologies for EMD to separate components that are similar in frequency using an additional masking signal. Mode mixing problem exhibited by traditional EMD can be eliminated up to a limit using some modified versions of EMD using averaging of modes and are known as ensemble empirical mode decomposition (EEMD) and complete ensemble empirical mode decomposition with adaptive noise (CEEMDAN) [21].

Of late, Dragomiretskiy and Zosso [22] proposed variational mode decomposition (VMD), an adaptive signal decomposition algorithm that can compute the center frequency and bandwidth of each component adaptively, by iteratively searching the optimal solution of variational modes. Over the last few years, it has been used by several researchers to analyse complex signals [23]. VMD is more effective compared to EMD in identifying the oscillatory modes and the estimation of instantaneous parameters. It is considered as a signal-dependent and non-linear multiresolution analysis technique to decompose a nonstationary signal into different components. The optimal formulation of these components is done through the Weiner filter solution. VMD technique requires predefined values for mode number and fidelity constraint and it is not an easy task to select the parameters adaptively [24].

With the help of appropriate models or techniques, the abnormality of the system can be accurately detected, and the reason for these failures can be identified in real-time. The proposed method in this work involves a novel method based on CEEMDAN and VMD, which can extract the electromechanical modes precisely from the non-linear and nonstationary PMU data. In the first processing stage, the collected data signal is processed using CEEMDAN into several IMFs. The suitable IMFs are then evaluated by the indexing of correlation parameters. The second stage employs VMD to decompose the selected IMF obtained from CEEMDAN effectively, and mode number for VMD is assigned based on the sensitive IMF index (SII) values. Finally, identification of frequency mode and Hilbert spectral analysis [25] is then performed to estimate the other instantaneous mode parameters.

The remaining sections of this chapter are organized as follows: Sect. 2 refers to the basic features of the CEEMDAN technique, VMD, sensitive IMF selection and parameter estimation through the Hilbert transform. The proposed oscillation monitoring system is described in Sect. 3. In Sect. 4, the effectiveness of the algorithms is tested using a simulated signal from the IEEE 16 machine 68 bus system. A case study using a real-time PMU data from POSOCO (Power system operation corporation) limited, is verified with the proposed method in Sect. 5 and is concluded in Sect. 6.

2 Methods Incorporated in the Proposed Technique

Various signal processing techniques are described in this section that is incorporated in the development of the proposed approach for identifying the low-frequency mode of oscillation in the power system.

2.1 Pre-processing

The quality of raw PMU data is low due to the high sampling rate. Hence proper data pre-processing methods are required. Vanfretti et al. suggested general techniques for pre-processing as defective data removal, data parcelling, outlier removal, interpolation, mean subtraction, down sampling etc. [26]. Defective data removal and mean subtraction are adopted for PMU data pre-processing in this work. A median filter is used for the POSOCO data.

2.2 Complete Ensemble Empirical Mode Decomposition with Adaptive Noise (CEEMDAN)

Improvements in empirical mode decomposition (EMD) include ensemble EMD (EEMD), complimentary EEMD, masking based EMD techniques and CEEMDAN. The improved CEEMDAN technique is highly robust from residual noise and free from fake modes during the decomposition stages. There are two operators needed for the CEEMDAN process, k^{th} mode operator $E_k(\cdot)$ and the local mean of the signal $M(\cdot)$. For a given signal $x(n)$; the CEEMDAN algorithm is described as follows.

1. Compute N number of realizations as, $x^i(n) = x(n) + E_1(\omega^i(n))$, where $i = 1, 2, 3, \ldots, N$ and $\omega^i(n)$ represents Gaussian white noise at different levels. The first residue is obtained as

$$r_1(n) = \frac{1}{N}\sum_{i=1}^{N} M(x^i(n)) \tag{1}$$

2. The first and the second modes are calculated using the formula

$$\widehat{IMF}_1(n) = x(n) - r(n) \tag{2}$$

$$\widehat{IMF}_2(n) = r_1(n) - r_2(n) = r_1(n) - \frac{1}{N}\sum_{i=1}^{N} M(r_1^i(n)) \tag{3}$$

3. For $i = 1, 2, 3, \ldots, P$ compute pth residue as

$$r_p(n) = \frac{1}{N}\sum_{i=1}^{N} M(r_p^i(n)) = \frac{1}{N}\sum_{i=1}^{N} M(r_{p-1}(n) + \delta_{p-1}E_p(\omega^i(n)) \tag{4}$$

   where P be the maximum number of IMFs.
4. Compute pth mode as

$$\widehat{IMF}_1(n) = r_{p-1}(n) - r_p(n) \tag{5}$$

Additional details of CEEMDAN is available in [27].

2.3 Sensitive IMF Selection

Proper IMF selection from the CEEMDAN process is most important for the effective determination of low-frequency mode of oscillation. Hence in this work, correlation analysis based sensitive IMF index (SII) is developed from the decomposed IMFs with the main signal, and the procedure is listed in the following steps.

1. The correlation coefficient is calculated with respect to the ith IMF and the main signal $x(n)$

$$\gamma_{c_i,x} = \frac{\sum_{i=1}^{N}(x(n) - \overline{x})(c_i(n) - \overline{c_i})}{\sigma_{ci}\sigma_x} \tag{6}$$

   where σ_{ci} and σ_x are the standard deviations; $c_i(n)$ is the ith decomposed IMF of CEEMDAN; $\overline{x}$ and $\overline{c_i}$ are the mean values.
2. Compute sensitive IMF index (SII)

$$I_{sii}(i) = \frac{\gamma_{c_i}}{\sum \gamma_{c_i}} \tag{7}$$

3. Rank these SII values in ascending order. IMFs with the highest value of SII value is selected from this second stage and are considered as the sensitive IMF to reconstruct the original signal most appropriately. The sensitive IMFs are considered as useful information, whereas the remaining IMFs are treated as noise components.

2.4 Variational Mode Decomposition (VMD) Technique

The decomposition of the real-valued non-linear nonstationary signal $f(t)$ using the VMD technique produces a combination of a discrete set of quasi-orthogonal IMFs represented as u_p, where p represents the mode number. This preset number of associated signals of IMF is considered as amplitude and frequency modulated signal, which has a center frequency of ω_p. VMD has the following main steps in the decomposition process:

(1) Hilbert Transform is applied to the one-sided frequency spectrum of individual mode
(2) To shift the frequency spectrum, an exponential multiplication factor $e^{-j\omega_p t}$ is considered.
(3) The gradient of the modulated signal is subjected to the L2 norm for the estimation of bandwidth.

VMD process can be considered as a constrained optimization or variational problem as in Eq. (8)

$$\min(\omega_p, u_p)\left\{\sum_{p=1}^{P}\left\|\delta_t\left[\left(\delta(t)+\frac{i}{\pi t}\right)*u_p(t)\right]e^{-j\omega_p t}\right\|_2\right\} \tag{8}$$

such that $\sum_{p=1}^{P} u_p(t) = f(t)$.

A quadratic penalty term (fidelity constraint) α and a Lagrangian multiplier λ are added to transform the problem to an unconstrained one, and the corresponding modification would be:

$$L(\{u_p\},\{\omega_p\},\{\lambda\} = \alpha\sum_{p=1}^{P}\left\|\delta_t\left[\left(\delta(t)+\frac{i}{\pi t}\right)*u_k(t)\right]e^{-j\omega_p t}\right\|_2^2 + \left\|f(t)-\sum_{p=1}^{P}u_p(t)\right\|_2^2 + \left\langle\lambda, f(t)-\sum_{p=1}^{P}u_p(t)\right\rangle \tag{9}$$

An optimization-based methodology named alternate direction method of multipliers [28] is used to compute the central frequencies and the IMFs on these frequencies simultaneously in VMD. The exact formulation of the optimization problem is continuous in the time domain. The different modes are calculated by the subsequent update of the previous mode and its center frequency using Eqs. (10) and (11).

$$\hat{u}_p^{n+1} = \frac{f - \sum_{i<p}\hat{u}_i^{n+1} - \sum_{i>p}\hat{u}_i^{n} + \frac{\lambda^n}{2}}{1+2\alpha(\omega-\omega_p^n)^2} \tag{10}$$

$$\hat{\omega}_p^{n+1} = \frac{\int_0^{\omega}\omega\left|\hat{u}_p^{n+1}(\omega)\right|^2 d\omega}{\int_0^{\omega}\left|\hat{u}_p^{n+1}(\omega)\right|^2 d\omega} \tag{11}$$

With the updation of modes and its center frequencies, simultaneous updation of the Lagrangian multiplier is also done and is represented in Eq. (12).

$$\hat{\lambda}^{n+1} = \hat{\lambda}^{n} + \left(f - \sum_{p}\hat{u}_p^{n+1}\right) \tag{12}$$

The updating process continues until it converges by the Eq. (13), where ε is the tolerance factor.

$$\sum_{p}\frac{\left\|\hat{u}_p^{n+1}-\hat{u}_p^{n}\right\|_2^2}{\left\|\hat{u}_p^{n}\right\|_2^2} < \varepsilon \tag{13}$$

Depending on the predefined mode number, IMFs are obtained. For the VMD process, parameters such as fidelity factor (α) and number of modes (p) are required for the optimal initialization. Usually, in a VMD process, these parameter values are assigned arbitrarily, which leads to unnecessary decomposition stages and hence requires more computation time. In this work, the mode number is assigned based on the count of sensitive IMFs after the indexing process. Hence, unnecessary decomposition stages can be avoided. Additional details regarding VMD is available in [22].

2.5 Power Spectral Density

The power spectral density P_{xx} of a stationary random signal $x(m)$, is mathematically connected to the autocorrelation sequence through Discrete-Time Fourier Transform (DTFT) and is expressed as:

$$P_{xx}(\omega) = \frac{1}{2\pi} \sum_{n=-\infty}^{\infty} R_{xx}(n) e^{-i\omega n} \tag{14}$$

where $R_{xx}(n)$ represents the autocorrelation sequence, and it is given by

$$\begin{aligned} &R_{xx}(n) = E\{x^*(m)x(m+n)\} \\ &-\infty < m < \infty \end{aligned} \tag{15}$$

where n represents the time delay, $E\{\cdot\}$ is the expectation value, and $*$ denotes complex conjugate operator.

2.6 Mode Parameter Estimation Through Hilbert Transform

The problem of LFO identification involves the estimation of the number of electromechanical modes of a power system and its main feature parameters such as modal frequency, equivalent damping ratio, amplitude, and the phase in the time or frequency domain. The Hilbert transform technique can be used to determine the instantaneous amplitude $a_p(t)$, frequency $f_p(t)$, and energy $e_p(t)$ values of the various IMFs. $H(.)$ denotes Hilbert Transform, and it is shown in the Eq. (16). For a selected mode, the instantaneous modal parameters are characterized by $\hat{u}_p(t)$ are estimated as in the Eqs. (17)–(19).

$$H(\hat{u}_p(t)) = \frac{1}{\pi} \int_{-\infty}^{\infty} \frac{\hat{u}_p(\tau)}{t - \tau} d\tau \tag{16}$$

$$a_p(t) = \sqrt{\hat{u}_p^2(t) + H(\hat{u}_p(t))^2} \tag{17}$$

$$f_p(t) = \frac{d(\arctan(H(\hat{u}_p(t))/\hat{u}_p(t))}{2\pi dt} \tag{18}$$

$$e_p(t) = 0.5a_p^2(2\pi f_p)^2 \tag{19}$$

The average frequency for the pth element can be considered as the slope of the straight line from the least-square fitting of the phase angle versus the time interval. The average damping coefficient σ_p for the pth element can be achieved as the slope of the straight line from the least-square fitting of the decaying amplitude $\ln(a_p(t))$ versus the time interval [29].

3 The Proposed Oscillation Monitoring System

In this section, a new systematic approach for oscillation monitoring of modes and its parameter estimation is discussed and is illustrated through a block diagram as shown in Fig. 1. The collected raw PMU data is properly pre-processed before its decomposition. CEEMDAN is utilized for the primary decomposition process, which produces several IMFs. Depends on the sensitive IMF selection, mode number is assigned for the VMD process, and the secondary decomposition of VMD starts. After attaining the required IMFs from VMD, the low-frequency points are identified using Power spectral densities. The estimation of modal parameters for the corresponding IMFs is obtained through the Hilbert transform. The proposed method is found to be computationally viable and accurate among the existing strategies of non-linear nonstationary approaches.

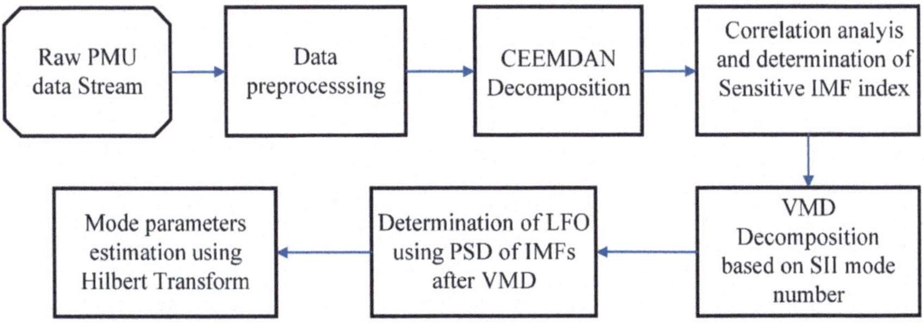

Fig. 1 Block diagram of the proposed approach

4 Analysis of Oscillation in Simulated Signals

In this section, the proposed method is employed on an IEEE 16 machine 68 bus system [30]. The power system toolbox (PST) in MATLAB Simulink is used to perform Non-linear simulations of the test system model. Synchronous generators in the test system are modeled as a classical model, and the input mechanical power is considered as constant. A ten percent increase in mechanical input torque to each generator was applied for a duration of 80 ms. The measurements were taken for 15 s at a sampling rate of 100 samples per second, and hence a total number of 1500 observations are considered. Data sets of rotor speed deviations are recorded and polluted with the white Gaussian noise of SNR 15 dB, as shown in Fig. 2.

For the analysis, machine 3 is considered as the test machine, and the rotor speed signal is extracted. The primary decomposition procedure involves the CEEMDAN process, which delivers ten IMFs, as shown in Fig. 3. In these IMFs, component 1 is a residual trend signal, which is avoided. Also, the last three IMFs behave as residues as per the observation of CEEMDAN output, which is also avoided. Hence SII for IMF2 to IMF7 can be calculated and are shown in Table 1. This elimination of IMFs reduces computational complexities. As per the correlational analysis, IMF 2 to IMF 4 has higher values of indexing compared to other IMFs, and it reconstructs the original signal to a high level of accuracy. Hence these three decomposed modes indicate the mode number in the next stage VMD process. IMF3 with higher indexing value is selected for the secondary decomposition VMD.

The next step consists of the VMD process, and its decomposition effect depends on proper initial parameter selection. As the incorrect parameter setting can affect the signal decomposition, the initial values of mode number and fidelity constraint are to be appropriately assigned for the VMD process. In the case of identification of low-frequency oscillations, fidelity constraint has a comparatively higher value, and hence it is assumed as 8000. For the mode number value, the three IMFs from the

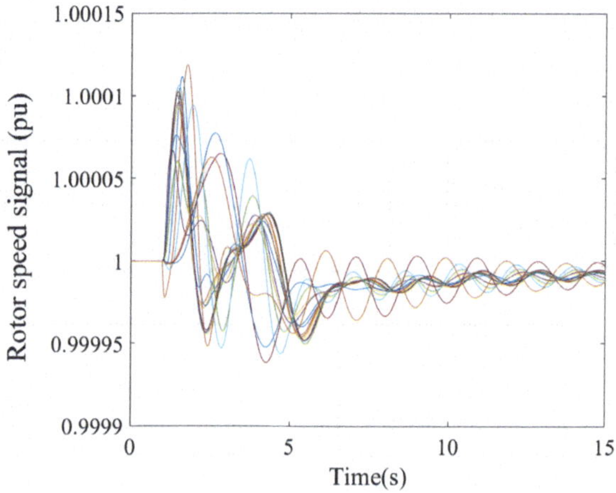

Fig. 2 Rotor speed signals of 16 machines

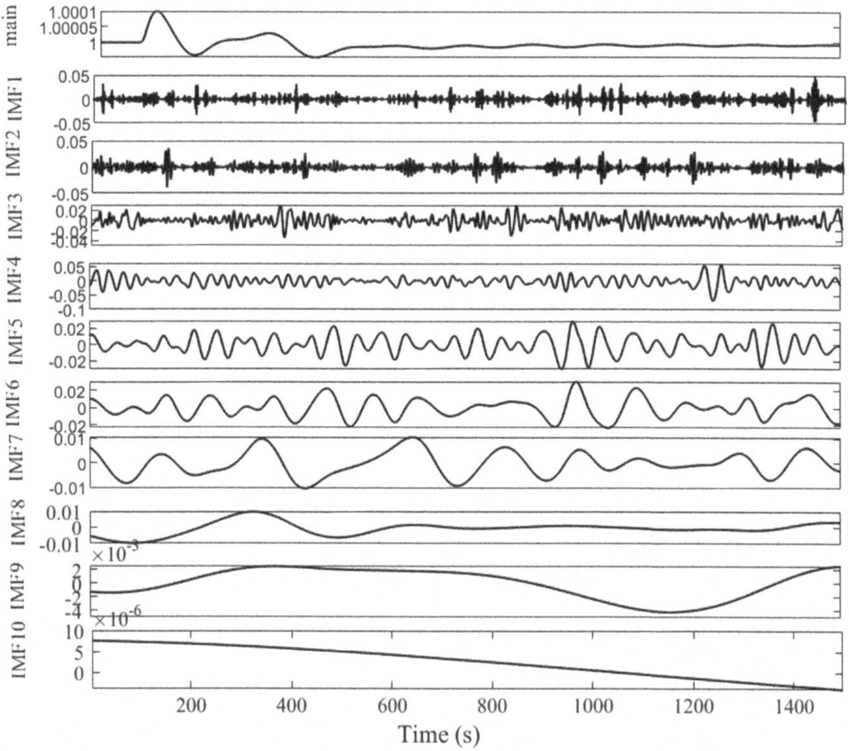

Fig. 3 Output of primary decomposition of CEEMDAN

Table 1 Sensitive IMF index values

IMF	2	3	4	5	6	7
SII values	0.3254	0.3845	0.3125	0.105	0.069	0.025

CEEMDAN process are recognized with adequate indexing, and the mode number is assigned as 3. The 3 IMFs obtained as the output of the VMD process are shown in Fig. 4, and the respective Power Spectral densities are shown in Fig. 5. From this It can be analysed that, three low-frequency oscillatory modes such as 0.523, 0.783, and 0.95 Hz are repeatedly occurring in the spectral densities of all the IMFs.

Hilbert transform is then applied to the low frequency oscillatory modes for the estimation of mode parameters. Figure 6 shows the instantaneous mode parameters obtained for low frequency mode of 0.523 Hz. Instantaneous amplitude and frequency are observed to indicate the disturbance occurring in the system. The damping ratio is calculated from the instantaneous amplitude by taking the slope of the least square fitting of $\ln a_p(t))$. From Fig. 6, it is observed that the signals comprise intricate non-linear patterns, and also some modes show fluctuating intervals, which could be replicated in the instantaneous amplitude and hence affect the

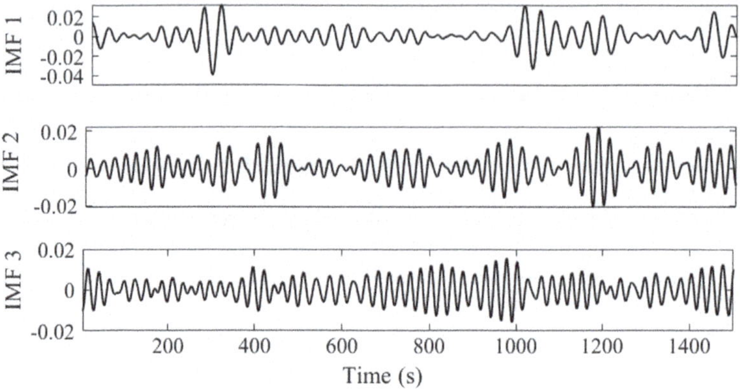

Fig. 4 Output of VMD process

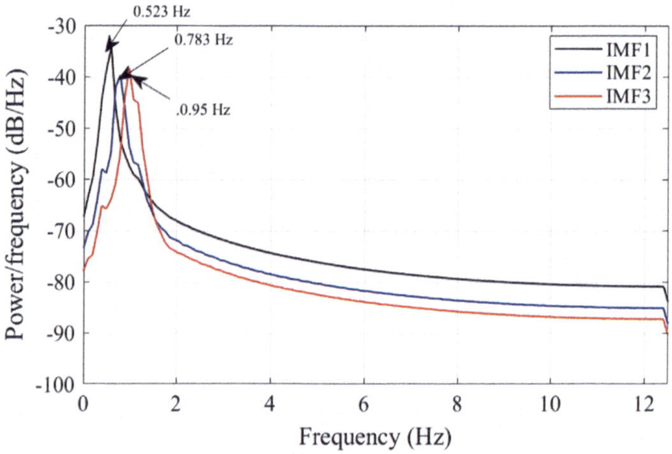

Fig. 5 Power spectral density for VMD output

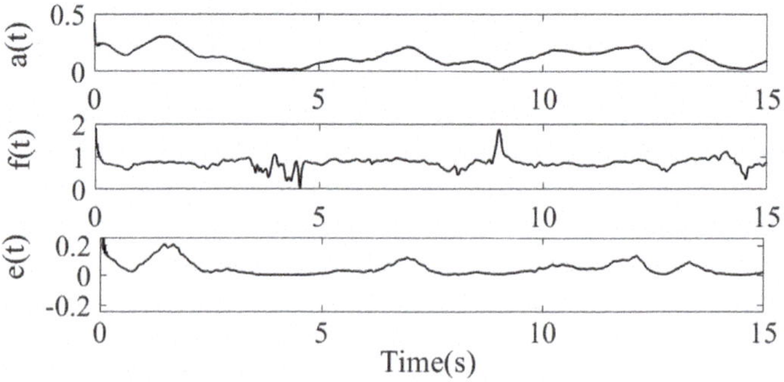

Fig. 6 Extracted features of low-frequency oscillatory mode of 0.523 Hz

Table 2 Decomposition error for different algorithms

Method	Frequency			Damping ratio		
	IMF 1	IMF 2	IMF 3	IMF 1	IMF 2	IMF 3
Proposed method	2.3561	2.342	2.4321	3.343	2.3451	3.5384
VMD	4.8732	2.3989	2.6731	4.7556	2.4521	5.6736
EEMD	5.4321	5.7812	6.3424	5.3440	3.5686	6.3441
EMD	12.6543	9.5647	8.3694	11.8576	12.8564	9.4561

Table 3 Oscillatory mode estimation using various algorithms

Mode	Proposed Method		VMD		EMD	
	Frequency (Hz)	Damping ratio (%)	Frequency (Hz)	Damping ratio (%)	Frequency (Hz)	Damping ratio (%)
1	0.523	4.26	0.52	6.48	0.47	5.98
2	0.783	2.13	0.74	5.124	0.86	3.68

accuracy in determining the damping ratio [25]. Interaction of these multiple modes results in the development of new modes, and this may cause inter-area mode with higher damping.

The effectiveness of the proposed method is analysed by comparing it with other mode estimation algorithms. For this, the simulated test signal is analysed using the nonstationary algorithms like VMD, EMD, EEMD, along with the proposed CEEMDAN-VMD method. For each method, the decomposition error or fitting error in frequency and damping ratio with respect to the original signal is calculated and is shown in Table 2. The fitting error can be calculated by using curve fitting application in MATLAB. From Table 2, it is clear that the decomposition error obtained is very less in the proposed method compared to other methods. For damping ratio, VMD and EEMD method are having almost nearer fitting errors in analysis. EMD method is the least performing algorithm in this analysis, mainly due to its drawbacks in identifying the frequency mode. The estimation of low-frequency oscillatory modes through different nonstationary mode estimation techniques is shown in Table 3. It can be observed that the proposed method and VMD give better identification of low-frequency mode than that of conventional EMD techniques.

5 Analysis of POSOCO's PMU Data

The proposed method is tested using real time–frequency data acquired from the PMU units located in the Indian grid, sampled at a frequency of 25 Hz. In the Indian power grid sector, the individual state grids are interconnected to form five regional grids (namely Northern, Eastern, Western, North Eastern, and Southern Grids), which enables the interstate electricity transmission in each region. Real-time frequency information is gathered from the Power System Operation Corporation Limited (POSOCO), Bangalore. Primary function of POSOCO is the supervision of all the regional load despatch centers (RLDC). Approximately 1000 samples of data from the PMU unit located at Ballia station on 17th March 2018 are considered here for the oscillation monitoring studies subjected to a variation in generation around 5% to 95% from Kahalgaon thermal power plant, and the frequency variations are recorded for 40 s. The massive change in the magnitude of generation must have introduced the low-frequency oscillation event.

There are frequency excursions observed within the data, and the enlarged view for a particular case is shown in Fig. 7 between 32 and 36 s. The proposed method of CEEMDAN-VMD is applied for identifying the low-frequency modes present in the signal. Proper pre-processing steps begin with the removal of outliers using the median filter. The mean value of test signals is calculated, and detrended value is assigned as the difference of the original signal with the mean value. The first decomposition using CEEMDAN process produces 12 IMFs. SII values are calculated for these IMFs and are presented in Table 4. The IMFs having very low SII values are not shown in the table. Based on the SII values, IMF2 to IMF5 are selected to reconstruct the main signal. So, these four IMFs represent the count for the mode number. Amongst this, IMF2 is selected for the VMD process, and finally, the IMFs obtained after the VMD is shown in Fig. 8. Four decomposed modes are produced, and power spectral densities are plotted, as shown in Fig. 9.

It can be observed that 2 low-frequency modes such as 0.785 and 1.23 Hz are repeatedly occurring in the spectral densities of the IMFs. The presence of these low frequency modes of frequencies 0.785 and 1.23 Hz in Indian grid, are verified from the earlier grid events as per the reports obtained from the POSOCO limited [31, 32].

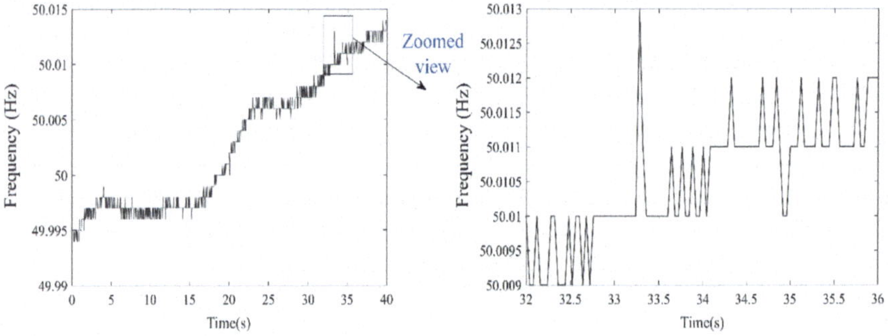

Fig. 7 Real-time PMU data from Ballia station

Table 4 Sensitive IMF index values for real time PMU data

IMF	2	3	4	5	6	7	8
SII values	0.456	0.342	0.3456	0.225	0.157	0.122	0.08

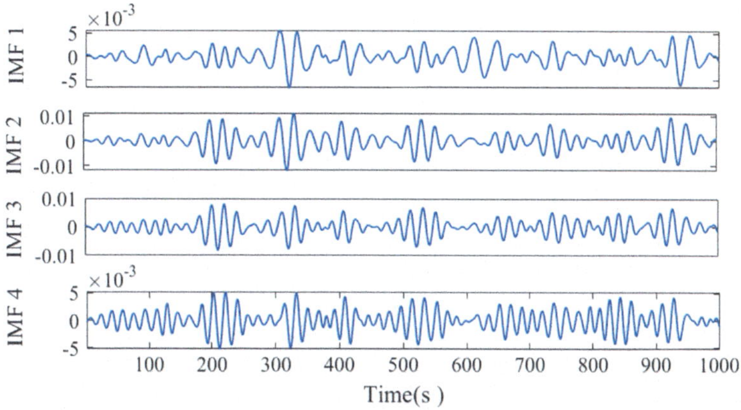

Fig. 8 Output IMFs after the proposed CEEMDAN-VMD process

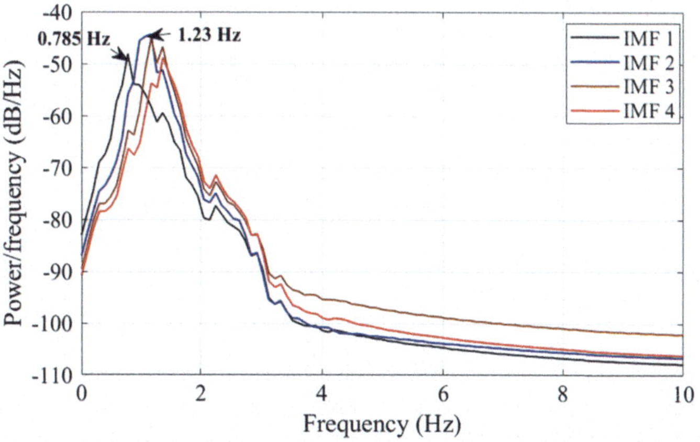

Fig. 9 Power spectral densities of IMFs

Hilbert transform technique performed for the estimation of instantaneous amplitude, frequency and energy for the selected modes. Figure 10 shows the parameter estimation for IMF3. Oscillatory mode estimation through different nonstationary algorithms like VMD and EMD are presented in Table 5. The damping ratio is calculated from the instantaneous amplitude, by taking the slope of the least square fitting of $\ln a_p(t)$).

From the observation, there are two significant disturbances in the instantaneous frequency (15 and 32 s) because of the sudden real power variation in the thermal power plant. These variations are predominant in the amplitude and energy curves. As the proposed method clearly identifies low frequency modes present time domain signal, it is possible to adapt this technique for real-time monitoring of the power system, employed with PMUs, thus providing support for the operator to identify the

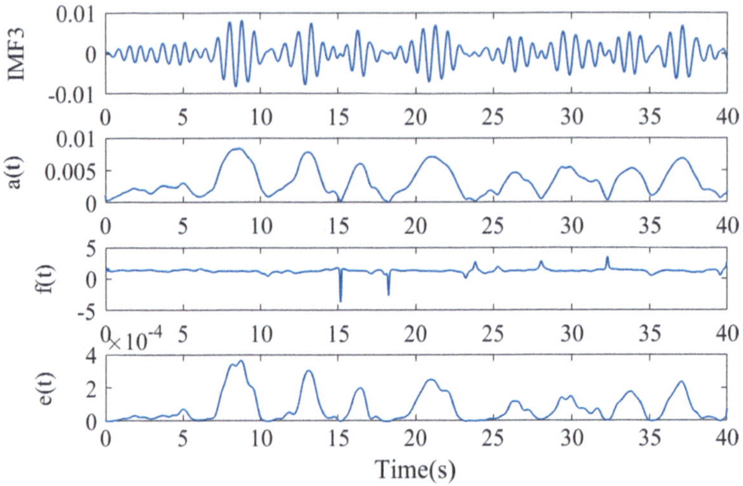

Fig. 10 Extracted oscillatory features of Ballia station

Table 5 Oscillatory mode estimation using various nonstationary algorithms for real-time data

Mode	Proposed Method		VMD		EMD	
	Frequency (Hz)	Damping ratio (%)	Frequency (Hz)	Damping ratio (%)	Frequency (Hz)	Damping ratio (%)
1	0.785	3.56	0.8	4.35	0.75	6.98
2	1.23	3.13	1.2	5.56	1.5	5.68

real events in the power system and enhance the overall performance in the power grid scenario. The proposed approach is compared with the various nonstationary algorithms, such as VMD, EEMD, EMD. Unnecessary decomposition stages arising due to manual prediction of mode number in VMD can also be eliminated through this method. The closer frequency modes are not an issue for the proposed approach. Non-linearity and non-stationarity of the signals increased with the complexities in the power system and reflected on the real-time PMU data. To cope up with this problem, the proposed CEEMDAN- VMD method provides a better solution for identifying the low frequency modes, than any other existing strategy.

As a future work and research gaps, the applications of PMU and wide-area monitoring systems in modern power systems operation, control, stability, protection and security should be well-investigated [33–42]. PMUs are the backbone of wide are monitoring system that can improve the overall security and stability of power systems. Therefore, it is suggested to make in-depth evaluation of the power system security under the new environment control based on PMUs data.

6 Conclusion

Undamped oscillations that arise in the power system due to the wide variety of events can lead to system collapse. Hence the identification of these low-frequency modes and estimation of mode parameters play an essential part in the monitoring of the power system. Conventional methods extract these modes using the linearized dynamic model and by applying Eigenvalue analysis around an equilibrium point, which are inaccurate and computationally hectic from a practical point of view. In this chapter, a dynamic approach for the identification of low-frequency oscillatory modes has been presented with a novel nonstationary approach. It has been found that existing mode estimation algorithms have so many disadvantages when it comes to nonstationary content. Hence the nonstationary algorithms like EMD and its modified versions face so many difficulties when implemented for a mode mixing problem. VMD is an efficient decomposition process but needs a manual operation in assigning the mode number. Using this proposed approach, an operator can purely pre-process the raw PMU data and apply it with the CEEMDAN-VMD process. The sensitive IMF index value obtained after the CEEMDAN process decides the mode number and aids in the capable selection of IMF through correlation analysis. Simulation results using the test signal from the IEEE standard bus system, and real-time PMU data obtained from POSOCO Limited are presented to highlight the superior feasibility and adaptability of the proposed work.

Acknowledgments The authors would like to thank Power System Operation Corporation Limited, Bangalore (India), for their technical assistance and providing PMU data.

References

1. G. Rogers, *Power System Oscillations* (Kluwer Academic Publishers, Boston, 2000)
2. P. Kundur, *Power System Stability and Control* (McGraw-Hill Inc., New York, 1994)
3. A.R. Messina, *Inter Area Oscillations in Power System-A Non-Linear and Nonstationary Perspective* Power Electronics and Power Systems. (Springer, US, 2009)
4. B. Wang, K. Sun, Location methods of oscillation sources in power systems: a survey. J. Mod. Power Syst. Clean Energy **5**(2), 151–159 (2017)
5. J.F. Hauer, C.J.Demeure, L.L.Scharf, Initial results in Prony analysis of power system response signals. IEEE Trans. Power Syst. **5**(1), 80–89 (1990)
6. M. Khodadadi Arpanahi, M. Kordi, R. Torkzadeh, H. Haes Alhelou, P. Siano, An augmented prony method for power system oscillation analysis using synchrophasor data. Energies **12**(7), 1267 (2019)
7. T.K. Sarkar, O. Pereira, Using the matrix pencil method to estimate the parameters of a sum of complex exponentials **37**(5), 48–55 (1994)
8. K.C. Lee, K.P Poon, Analysis of power system dynamic oscillations with heat phenomenon by Fourier transformation. IEEE Trans. Power Syst. **5**(1), 148–153 (1990)
9. M. Yazdanian, A. Mehrizi-Sani, M. Mojiri, Estimation of electromechanical oscillation parameters using an extended Kalman filter. IEEE Trans. Power Syst. **30**(6), 2994–3002 (2015)
10. J.W. Pierre, D.J. Trudnowski, M.K. Donnelly, Initial results in electromechanical mode identification from ambient data. IEEE Trans. Power Syst. **12**(3), 1245–1251 (1997)

11. N. Zhou, D.J. Trudnowski, J.W. Pierre, et al., Electromechanical mode online estimation using regularized robust RLS methods. IEEE Trans. Power Syst. **23**(4), 1670–1680 (2008)
12. I.Y.H. Gu, M.H.J. Bollen, Estimating inter harmonics by using sliding window ESPRIT. IEEE Trans. Power Deliv. **23**(1), 13–23 (2008)
13. L. Dosiek, N. Zhou, J.W. Pierre, Z. Huang, D.J. Trudnowski, Mode shape estimation algorithms under ambient conditions: a comparative review. IEEE Trans. Power Syst. **28**(2), 779–787 (2013)
14. T.Jiang, H. Yuan, H. Jia, et al. Stochastic subspace identification-based approach for tracking inter-area oscillatory modes in bulk power system utilizing synchrophasor measurements. IET Gener. Transm. Distrib. **9**(15), 2409–2418 (2015)
15. S.A. Nezam Sarmadi, V. Venkatasubramanian, Electromechanical mode estimation using recursive adaptive stochastic subspace identification. IEEE Trans. Power Syst. **29**(1), 349–358 (2014)
16. J.Sanchez-Gasca, D.Trudnowski, Identification of electromechanical modes in power system. IEEE Task Force on Identification of Electromechanical Modes of the Power System Stability, Power & Energy Society, Technical report (2012)
17. F. Shir, B. Mohammadi ivatloo, Identification of inter-area oscillations using wavelet transform and phasor measurement unit data. Int. Trans. Electr. Energy Syst. **25**(11), 2831–2846 (2015)
18. N.E. Huang, Z. Wu, A review on Hilbert-Huang transform: method and its applications to geophysical studies. Rev. Geophys. **46**(2), 1–23 (2008)
19. H. Xiao, J. Wei, H. Liu, Q. Li, Y. Shi, Identification method for power system low-frequency oscillations based on improved VMD and Teager–Kaiser energy operator. IET Gener. Transm. Distrib. **11**(16), 4096–4103 (2017)
20. R. Deering, J.F. Kaiser, The use of a masking signal to improve empirical mode decomposition, in *Proceedings. (ICASSP '05). IEEE International Conference on Acoustics, Speech, and Signal Processing* (2005)
21. H. Chen, P. Chen, W. Chen, C. Wu, J. Li, J. Wu, Wind turbine gearbox fault diagnosis based on improved EEMD and hilbert square demodulation. Appl. Sci. **7**(128) (2017)
22. K. Dragomiretskiy, D. Zosso, Variational mode decomposition. IEEE Trans. Signal Process. **62**(3), 531–544 (2014)
23. M.K. Jena, S.R. Samantaray, B.K. Panigrahi, Variational mode decomposition-based power system disturbance assessment to enhance WA situational awareness and post-mortem analysis. IET Gener. Transm. Distrib. **11**(13), 3287–3298 (2017)
24. R. Mario, A. Paterninaa, R.K. Tripathy, A. Zamora-Mendez, D. Dottad, Identification of electromechanical oscillatory modes based on variational mode decomposition . Electr. Power Syst. Res. **167**, 71–85 (2019)
25. D.S.Laila, A. Messina, B.C. Pal, A refined Hilbert–Huang transform with applications to inter-area oscillation monitoring. IEEE Trans. Power Syst. **24**(2), 610–619 (2009)
26. L. Vanfretti, S. Bengtsson, J.O. Gjerde, Preprocessing synchronized phasor measurement data for spectral analysis of electromechanical oscillations in the Nordic Grid. Int. Trans. Electr. Energy Syst. **25**, 348–358 (2015)
27. M.A. Colominas, G. Schlotthauer, M.E. Torres, Improved complete ensemble EMD: a suitable tool for biomedical signal processing. Biomed. Signal Proc. Control. **14**, 19–29 (2014)
28. C. Lu, S. Yan, Z. Lin, A unified alternating direction method of multipliers by majorization minimization. IEEE Trans. Pattern Anal. Mach. Intell. **40**(3), 527–541 (2018)
29. D. Lauria, C. Pisani, On Hilbert transform methods for low-frequency oscillations detection. IET Gener. Transm. Distrib. **8**(6), 1061–1074 (2014)
30. B. Pal, B.Chaudhuri, Robust Control in Power Systems (Springer, New York, 2005)
31. Posoco, Report on power system oscillations experienced in Indian Grid on 9th, 10th,11th and 12th August 2014, Task force report (Power System Operation Corporation Limited, New Delhi, 2014)
32. A. Singh, B.B. Singh, M.P. Reddy, et al., Report on low frequency oscillation in Indian power system, Task force report, March 2016 (Power System Operation Corporation Limited, New Delhi, 2016)

33. H.H. Alhelou, M.H. Golshan, J. Askari-Marnani, Robust sensor fault detection and isolation scheme for interconnected smart power systems in presence of RER and EVs using unknown input observer. Int. J. Electr. Power Energy Syst. **1**(99), 682–694 (2018)
34. H.H. Alhelou, M.E. Hamedani-Golshan, R. Zamani, E. Heydarian-Forushani, P. Siano, Challenges and opportunities of load frequency control in conventional, modern and future smart power systems: A comprehensive review. Energies **11**(10), 2497 (2018)
35. H. Haes Alhelou, M.E. Hamedani Golshan, F.M. Hajiakbari, Wind driven optimization algorithm application to load frequency control in interconnected power systems considering GRC and GDB nonlinearities. Electr. Power Compon. Syst. **46**(11–12), 1223–1238 (2018)
36. H.H. Alhelou, M.E. Hamedani-Golshan, E. Heydarian-Forushan, A.S. Al-Sumaiti, P. Siano, Decentralized fractional order control scheme for LFC of deregulated nonlinear power systems in presence of EVs and RER, in *2018 International Conference on Smart Energy Systems and Technologies (SEST)*, 10 Sept 2018, pp. 1–6 (IEEE, 2018)
37. H. Haes Alhelou, M.E. Hamedani-Golshan, T.C. Njenda, P. Siano, A survey on power system blackout and cascading events: research motivations and challenges. Energies **12**(4), 682 (2019)
38. H. Haes Alhelou, M.E. Hamedani Golshan, T.C. Njenda, P. Siano, Wams-based online disturbance estimation in interconnected power systems using disturbance observer. Appl. Sci. **9**(5), 990 (2019)
39. H.H. Alhelou, M.E. Golshan, N.D. Hatziargyriou, A decentralized functional observer based optimal LFC considering unknown inputs, uncertainties, and cyber-attacks. IEEE Trans. Power Syst. **34**(6), 4408–4417 (2019)
40. H.H. Alhelou, M.E. Golshan, N.D. Hatziargyriou, Deterministic dynamic state estimation-based optimal lfc for interconnected power systems using unknown input observer. IEEE Trans. Smart Grid (2019)
41. H.H. Alhelou, M.E. Golshan, T.C. Njenda, N.D. Hatziargyriou, An overview of UFLS in conventional, modern, and future smart power systems: challenges and opportunities. Electr. Power Syst. Res. **1**(179), 106054 (2020)
42. H.H. Alhelou, S.J. Mirjalili, R. Zamani, P. Siano, Assessing the optimal generation technology mix determination considering demand response and EVs. Int. J. Electr. Power Energy Syst. **1**(119), 105871 (2020)

Small Signal Stability Improvement of Pumped Storage Hydropower Using Wide Area Signal Considering Wind Farm

Mohsen Alizadeh Bidgoli, Davood Ganjali, Weijia Yang, and Saman Atrian

Abstract The electromechanical oscillations in the power system, known as local and inter-area modes, as well as power system oscillations in presence of wind turbines due to its inherent stochastic are two important cases in small signal stability study. Fixed speed (FS) pumped storage power plants (PSHP) similar to other power plants based on synchronous machine experience low frequency power oscillations. Therefore, a power system stabilizer (PSS) is developed for damping these oscillations. However, insufficient damping of these oscillations limited the capacity of energy transfer. On the other hand, state-of-the-art PSHP based on doubly fed induction machine (DFIM) known as variable speed (VS) have different effect on both small signal and transient stabilities of power system. Moreover, PSSs can be more important in multi-machine power grid to be tuned in a precise method. Nowadays, in smart power grids, PSS with wide area signal (WAS) instead of local signal is attended to decrease low frequency power oscillations, and therefore improve the small signal stability of the power system. This chapter intends to consider effect of DFIM and SM-based PSHP with different PSS tuning methods. Aiming at this purpose, a case of 343 MW hydro pump-turbine (HPT) coupled to DFIM with 381 MVA in comparison to the SM with same capacity, i.e., 381 MVA, as well as an aggregated wind farm are applied as the study case. Calculation and simulations are conducted in Digsilent 15.1 under diverse conditions. Also, modified New England test system, including10-machine and 39-bus system, is adopted as a large power network in presence of a wind farm. The results show using PSS with WAS can be a good option for FS-PSHP to improve damping low frequency oscillations.

M. Alizadeh Bidgoli (✉) · D. Ganjali · S. Atrian
Department of Electrical Engineering, Yadegar-e-Imam Khomeini (RAH) Share, Rey Branch, Islamic Azad University, Tehran, Iran
e-mail: m.alizadeh.b@iausr.ac.ir; m.alizadeh.b@gmail.com

W. Yang
State Key Laboratory of Water Resources and Hydropower Engineering Science, Wuhan University, Wuhan 430072, China

H. Haes Alhelou et al. (eds.), *Wide Area Power Systems Stability, Protection, and Security*, Power Systems, https://doi.org/10.1007/978-3-030-54275-7_9

Keywords Pumped storage hydropower plant (PSHP) · Wide area signal · power system stabilizer (PSS) · doubly fed induction machine (DFIM) · Small signal stability · Wind turbine

1 Introduction

In recent years, the dispatch power systems with high penetration of renewable energy sources (e.g., wind turbine and solar energy) has been a key issue due to stochastic inherent of wind speed and solar irradiation [1]. On the other side, the main goal of renewable resource owners in the market-based system is to attain maximum profit by injecting the active power into the grid. Therefore, in point of view power grids, using the flexibility of pumped storage hydropower plant (PSHP) is crucial to cope with unexpected rise or drop in the production. For example, a wind power ramp down event (1500–300 MW) occurred in the Electric Reliability Council of Texas (ERCOT) area of the United States in 2008 [2].

Several papers, such as those found in [3], have addressed only opportunities and barriers to pumped storage hydropower plant (PSHP) in different countries. A comparative study between fixed speed and variable speed PSHP has been performed for only in part of operating conditions in generating mode in [4], but the pumping mode has not been considered in this paper. The dynamic performance of two 320 MW PSHP have been reported using simplified converter model in [5]. Bidgoli set al. [6] propose nonlinear controller in fault ride-through (FRT) condition for both generating and motor modes, and Azbe and Mihalic [7] consider the transient stability of the large DFIM in a weak grid. The advantage of variable-speed pumped storage units for mitigating wind power variations is investigated based on a comparison with fixed-speed units [8]. Performance of a VS-PSHP during fault for pumping mode is considered in [9]. Also, Joseph et al. [10] consider the starting and braking of a 250 MW DFIM-based PSHP for pumping mode.

The small signal stability of rotor angle is assesed by low frequency electromechanical modes study of power systems. Insufficient damping of such oscillations is a major challenge in dynamic rotor angle stability problem. Among all the developed methods to improve damping of oscillation modes for large power systems, power system stabilizer (PSS) is the most cost-effective solution. However, not tuning of PSS may because of insufficient damping of oscillation modes of power system, even may be generate unstable modes into power system. In smart power system, wide area signals (WAS) are utilized and tuned using the eigenvalue or modal analysis to increase damping ratio of power system dynamic. Therefore, this chapter intends to explain briefly structure of variable speed PSHP versus fixed speed one. Then, effect of PSS parameters is considered to show significant of PSS parameters tuning. Also, a WAS from DFIM-based wind farm speed is used to improve performance of SM-based PSS.

This chapter is arranged as follows: Various structure of variable speed PSHP versus fixed speed one are introduced in Sect. 2. Also, mathematical model of DFIM-based PSHP is explained in Sect. 3. The implementation of the models and simulation results are presented in Sect. 4. Finally, in Sect. 5, conclusion and future works are summarized.

2 VS-PSHP Versus FS-PSHP

The list provided in Table 1 shows increase in tendency of various countries during recent years as for installation of variable-speed pumped-storage power plant. As reported in Table 1, the first-worldwide variable speed PSHP based on doubly fed induction machine (DFIM), installed in Japan in the 1990s, employed the cyclo-converter, while the first one in Europe was commissioned in 2004 (Goldisthal in Germany). This practical strategy has been recently used in many projects world-wide using the state-of-the-art power electronic converter, i.e., back to back voltage source converter, such as Avce in Slovenia or Linthal in Switzerland [11]. An interesting point is that the research and developments (R&D) part of companies still consider operational aspects of practical projects in different countries. In other words, although variable speed PSHPs are being preferred in sites where the water head variation is large, a designer may select each type of PSHP, i.e., either FS or VS, considering other features of a site.

2.1 FS-PSHP

The controller of FS-PSHP consists of two parts, as shown Fig. 1. First, the static excitation system which regulates the voltage of stator winding, and also an auxiliary input known PSS is added to excitation input for both generating and motor modes [12, 13]. However, HPT controller in motor and generating modes is different. The active power of the unit could be controlled by the turbine governor in generating mode, while the active power of the motor cannot be adjusted, and the guide vane opening is optimized to increase the efficiency of pump. Details of controllers can be found in [14].

2.2 VS-PSHP

As mentioned before, the speed of the state-of-the-art PSHP can be varied by different topologies. Three strategies may be used for medium or large-size PSHP to operate at variable speed.

Table 1 Variable-speed pumped-storage power plant installed in different locations during last years

Unit	Count	HPT (MW)	MG (MVA)	PEC (MVA)	PEC*	SVR (rpm)	Man**	Year
Narude	Japan	22	22	22	–	190–210	H	1987
Yagisawa	Japan	2 × 82	2 × 85	2 × 85	1	130–156	T	1990
Takami	Japan	105	140	24.3	2	209–254	M	1992
Ohkawachi	Japan	1 × 392 1 × 400	1 × 395 1 × 395	72	1	330–390 240–400	H	1993
Shiobara	Japan	330	360	51.1	1	356–394	T	1995
Okukiyotsu	Japan	340	345	31.5	2	408–450	T	1996
Yanbaru	Japan	30	31.5	3.96	2	423–477	T	1999
Goldisthal	Germ	2 × 330	2 × 331	–	1	300–347	V	2003
Kazunogawa	Japan		500	500		500		2005
Avce	Slove	185	195	195	3	576–636	ABB, M	2008
Omarugawa	Japan	330, 340	350, 370	350, 370	3, 1	576–624	M, H	2008
Kyogoku	Japan	228	230	230	2	–	T	2015
Linth-Limmern	Swit	4 × 250	–	–	3	470–530	A	2015
Nant de Drance	Swit	6 × 157	6 × 170	6 × 170	3	399–459	A	2017
Kazunogawa	Japan	2 × 475	2 × 475	2 × 475	2	480–520	T	2017
TEHRI	India	4 × 250	4 × 306	4 × 306	2	214–256	A	2017
Iowa Hill	USA	3 × 133	3 × 133	3 × 133	2	–	–	2018

*PEC: 1: Cyclo-converter, 2: GTO- back to back (VSI), 3: IGCT- back to back (VSI), 4: full converter (VSI)
**Manufacturer: H: HITACHI, T: TOSHIBA, M: MITSUBISHI, V: VATECH, A: ALSTOM

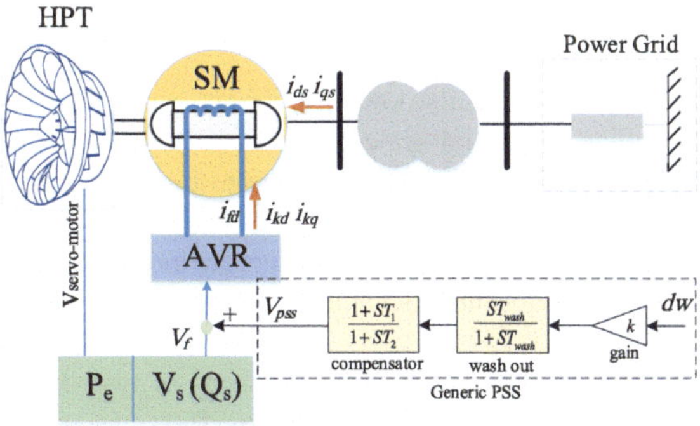

Fig. 1 Fixed speed PSHP connected to SMIB

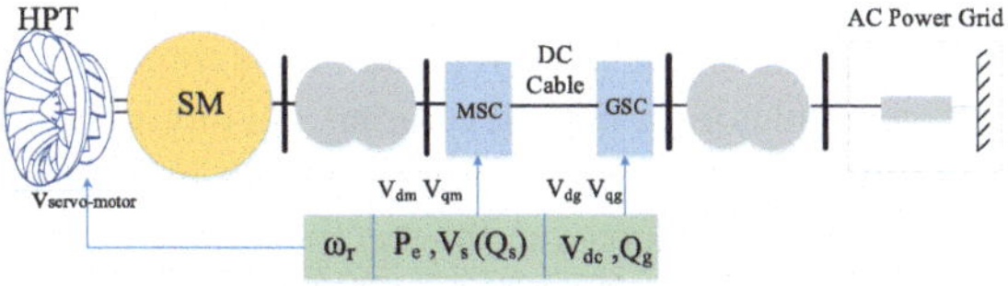

Fig. 2 Variable speed PSHP solution with HVDC transmission line

2.2.1 HVDC Solution

The synchronous machine, used in conventional applications, connecting to high voltage direct current (HVDC) link is the first solution as shown in Fig. 2. With this strategy, the range of speed variation may be within ±25% around the synchronous speed. This solution is attractive for power plants that are far away from the power grid [15]. In the short distance or AC transmission usage, two other schemes are available.

2.2.2 Full-Converter Solution

The second solution uses the synchronous machine which is connected to the grid through a full-scale converter as shown in Fig. 3. Although this scheme has been installed in Grimsel 2 by ABB in Switzerland [16], it is not widely-applied for other projects. It is noted that this method may be cost-effective for micro hydropower sites which use permanent magnetic synchronous machine (PMSM) [17, 18].

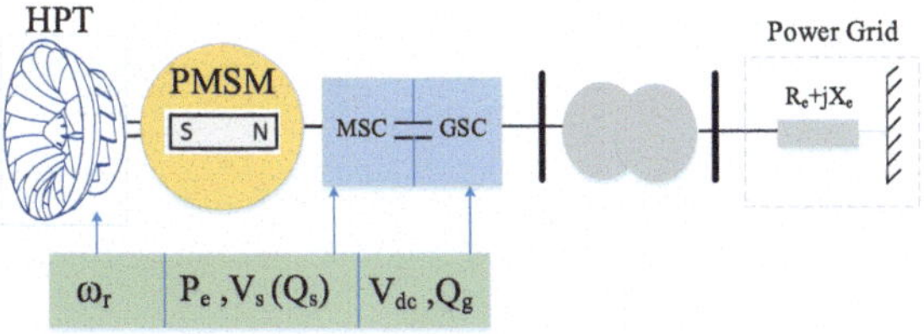

Fig. 3 VS-PSHP based with a full converter

2.2.3 Partial Converter Solution

In the last solution, also known as DFIM-based PSHP, wound rotor induction machine as well as the partial-scale converter is being used as shown in Fig. 4a, b. A DFIM, coupled to the HPT, uses the back-to-back converter, i.e., machine side converter (MSC) and grid side converter (GSC), whose power rating is partial scale of the DFIM rated power. These converters are located between the rotor and stator terminal that can transfer the mechanical power through both the stator and rotor as shown in Fig. 4b.

The controller of the MSC consists of two decoupled controllers, where each controller is comprised of two loops. First, the rotor current controller loop which is usually called fast or inner loop. Second, there are two outer or slow loops in the MSC controller. One of them is the electrical active power regulator in generating mode and is the rotor speed regulator in motor mode, while the other loop is the stator voltage regulator for both generating and motor modes. The main aim of the GSC is to maintain the voltage of dc-link constant in the nominal value regardless the rotor power direction. Moreover, the reactive power of the unit can be shared between the MSC and GSC to reduce the rated rotor current value and, therefore, design smaller DFIM. However, the extra reactive current for the GSC is normally higher than the reduction in the MSC reactive current rating due to the winding ratio. More details for designing controller of converters can be found in [6].

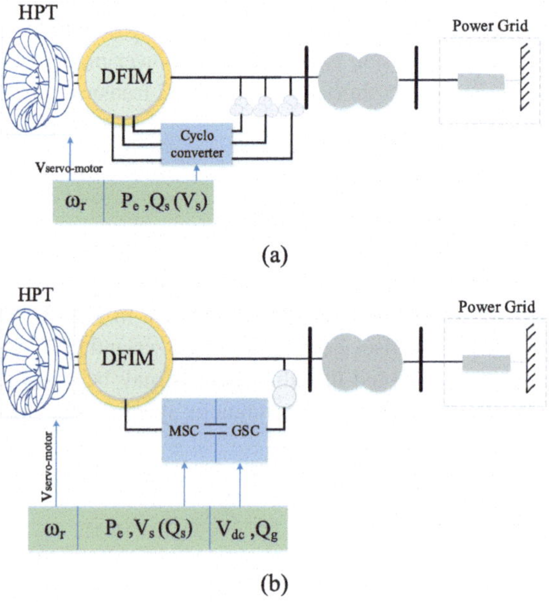

Fig. 4 DFIM-based PSHP with **a** cycloconverter **b** back to back converter

3 Modeling of DFIM-Based PSHP

In this section, a VS-PSHP equipped with DFIM coupling to hydro pump-turbine (HPT) is modelled with state space equations.

3.1 Hydraulic System

For the hydraulic system of this chapter, including a non-elastic water column with long length penstock without effecting the surge tank, a nonlinear hydraulic components model is considered [6]. The state space equations of the hydraulic subsystem in turbine mode are as Eqs. (1)–(2) [6]:

$$\begin{cases} \dot{q}_t = (h_s^{'} - h_t)/T_w \\ h_s^{'} = h_s - f_{eq}q_t^2 \\ h_t = (q_t/(g/(g_n - g_{nl})))^2 \\ T_w = Lq_n/(g'Ah_n) \end{cases} \tag{1}$$

$$\dot{g} = -k_a g + k_a v_{sm} \tag{2}$$

where $h_s^{'}$, h_s, and h_t are net static head, static head which is equal 1, and net head in the operating point of the turbine; Also, q is the water flow, g is the gate opening, v_{sm} is the servomotor voltage, T_w is water starting time of the pipe, k_a is the constant gain for servomotor modelling, and f_{eq} is equivalent frictional coefficient. Additionally, L is the length of the pipe, A is the area of the pipe, and g' is gravitational acceleration, which is equal to 9.81 m/s^2; note that subscripts n, nl, and t denote the nominal, no-load, and turbine mode values, respectively. Additionally, the mechanical power of the turbine is defined as follows:

$$\begin{cases} P_m = \eta p_r h_t(q_t - q_{nl}) \\ p_r = \frac{P_n(MW)}{S_n(MVA)} \end{cases} \tag{3}$$

where η is efficiency of the turbine and p_r is a constant value for per unit value conversion from the turbine base to DFIM; Also, P_n and S_n are nominal power of the turbine and nominal apparent power of the machine in generating mode, respectively.

3.2 DFIM and MSC

Assuming the positive direction for the stator and rotor currents to be motor conversion (into the machine), the fourth order electrical model of the DFIM (shown in

Fig. 4b), namely full order model, are written in the synchronous reference frame, d-q axis as Eqs. (4)–(7). Note that all parameters and equations are given in per unit except for time is s.

$$V_{ds} = R_s i_{ds} - \varphi_{qs} + \rho\varphi_{ds}/\omega_b \tag{4}$$

$$V_{qs} = R_s i_{qs} + \varphi_{ds} + \rho\varphi_{qs}/\omega_b \tag{5}$$

$$V_{dr} = R_r i_{dr} - s\varphi_{qsr} + \rho\varphi_{dr}/\omega_b \tag{6}$$

$$V_{qr} = R_r i_{qr} + s\varphi_{dr} + \rho\varphi_{qr}/\omega_b \tag{7}$$

where $i_{ds}, i_{qs}, v_{ds}, v_{qs}$ are the d-q axis stator currents and voltages; $i_{dr}, i_{qr}, v_{dr}, v_{qr}$ are the d-q axis rotor currents and voltages;ρ is derivative operator (*d/dt*); R_s and R_r are the stator and rotor resistances; *s* is the slip, and $\omega_b = 377$ (rad/s). Additionally,φ_{ds}, $\varphi_{qs}, \varphi_{dr}, \varphi_{qr}$ are the d-q axis stator and rotor flux magnitude which can be written as follows:

$$\begin{cases} \varphi_{ds} = L_s i_{ds} + L_m i_{dr} \\ \varphi_{qs} = L_s i_{qs} + L_m i_{qr} \\ \varphi_{dr} = L_r i_{dr} + L_m i_{ds} \\ \varphi_{qr} = L_r i_{qr} + L_m i_{qs} \end{cases} \tag{8}$$

where L_s, L_r are the stator and the rotor inductances; L_m is the magnetizing inductance; Additionally, swing equation is defined as follow.

$$\dot{\omega}_r = \frac{1}{2H}(T_m - T_e) \tag{9}$$

where T_m, T_e, and ω_r are the mechanical torque, electric torque, and rotor speed, respectively. Also, the electrical torque T_e, the stator reactive power Q_s, and the electrical real power P_s are defined as follows:

$$T_e = (L_m/L_s)\left(-\varphi_{ds} i_{qr} + \varphi_{qs} i_{dr}\right) \tag{10}$$

$$\begin{aligned} P_s &= -v_{ds} i_{ds} - v_{qs} i_{qs} \\ Q_s &= v_{ds} i_{qs} - v_{qs} i_{ds} \end{aligned} \tag{11}$$

3.3 TSC

Assuming the positive direction of the TSC is into the dc-link, and all parameters and equations in per unit except for time is s, the dc-link and the excitation transformer dynamics, shown in Fig. 4b, can be expressed in the synchronous reference frame, *d-q* axis, as follows [6]:

$$v_{ds} = R_{tr}i_{dtr} - L_{tr}i_{qtr} + (L_{tr}/\omega_b)\rho i_{dtr} + v_{dtr} \tag{12}$$

$$v_{qs} = R_{tr}i_{qtr} + L_{tr}i_{dtr} + (L_{tr}/\omega_b)\rho i_{qtr} + v_{qtr} \tag{13}$$

$$\rho V_{dc} = (P_{tr} - P_r)/(V_{dc}C) \tag{14}$$

where L_{tr} and R_{tr} are leakage inductance and resistance of the excitation transformer. Also P_r and P_{tr} in Eq. (14) are defined as $P_r = v_{dr}i_{dr} + v_{qr}i_{qr}$, $P_{tr} = v_{ds}i_{dtr} + v_{qs}i_{qtr}$.

4 Simulation Results

4.1 Case Study

New England 39-bus transmission network consists of 10 generators, 12 transformers, and 46 transmission lines. The technical parameters for the generation system and the transmission network are obtained from [19]. As shown in Fig. 5, the modified New England IEEE-39 bus system with a DFIG wind farm located in bus 33 instead of SM-based plant (G_4) is used as a study system. The power capacity of the wind farm is 100 MVA which is represented by an aggregated equivalent model. It should be mentioned G_1, located at bus 39, is interconnection of New England network to New York one. Type of all generators are reported in Table 2 which the last generator (G_{10}), located at bus 30, is hydro plant. In this study, G_{10} is substituted by either SM-based (salient pole rotor) named as *Sync Grid* or DFIM-based PSHP named as *DFIG Grid* in this study. Dc exciters of all generators are IEEE type 1 available in DIgSILENT PowerFactory, which are equipped with a PSS as shown in Fig. 5. Also, all generators are also equipped with generic governors, representing steam and hydro turbines. Moreover, DFIM is represented by a typical three order model neglecting the stator transients [6].

As mentioned, in this study, two grids, *DFIG Grid* and *Sync Grid*, are simulated independently. The former is the *DFIG Grid* which a 380 MVA DFIM-based PSHP is located in bus 30 instead of G_{10}, while the latter is *Sync Grid* which a 380 MVA SM-based PSHP is located in bus 30 instead of G_{10}. In this study, to demonstrate

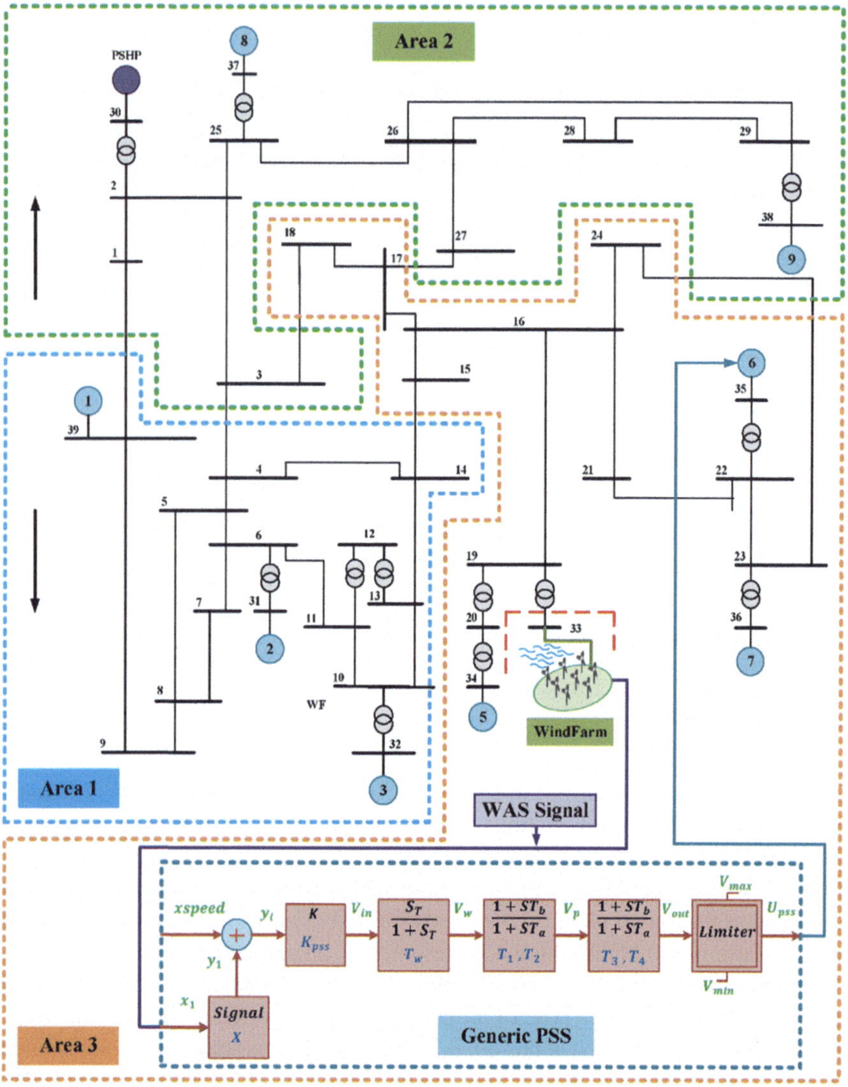

Fig. 5 Single line diagram of modified New England 39 bus test system [20]

the effectiveness of WAS method and well tuning of PSS parameters four cases are defined for both models *DFIG Grid* and *Sync Grid* as following.

- Case 1: base-case scenario with no PSS. It means all PSS signals for synchronous generators are inactive.
- Case 2: PSS without tuning. It means all PSS signals for synchronous generators are local and active with the local rotor speed feedback which parameters are in Table 3.

Table 2 Parameters of generator for under study 39 bus Test system

Name	Bus	P (MW)	Bus type	S (MVA)	V (p.u)	Rated voltage (KV)	Plant type
G1	39	1000	PV	10,000	1.03	345	Others
G2	31	–	SL	700	0.982	16.5	Nuclear
G3	32	650	PV	800	0.9831	16.5	Nuclear
G4	33	60	PV	100	1	16.5	Wind
G5	34	254	PV	300	1.0123	16.5	Coal
G6	35	650	PV	800	1.0493	16.5	Nuclear
G7	36	560	PV	700	1.0635	16.5	Coal
G8	37	540	PV	700	1.0278	16.5	Nuclear
G9	38	830	PV	1000	1.0265	16.5	Nuclear
G10	30	223	PV	381.111	1	18	Hydro

Table 3 PSS parameter values for all cases

Cases	Gen	K_{pss}	T_W	T_1	T_2	T_3	T_4	X	V_{Min}	V_{Max}
PSS without tuning	G02	41	10	1.2	0.02	1.1	0.04	0	–0.2	0.2
	G03	18	10	1.2	0.03	0.51	0.11	0	–0.2	0.2
	G05	31	10	0.72	0.02	1.2	0.13	0	–0.2	0.2
	G06	22	10	0.99	0.12	0.86	0.04	0	–0.2	0.2
	G07	27	10	0.94	0.11	1.1	0.03	0	–0.2	0.2
	G08	17	10	0.73	0.12	0.63	0.04	0	–0.2	0.2
	G09	37	10	0.85	0.03	0.59	0.04	0	–0.2	0.2
	G10	26	10	1.1	0.03	0.95	0.03	0	–0.2	0.2
PSS with tuning	G02	41	10	1.2	0.02	1.1	0.04	0	–0.2	0.2
	G03	18	10	1	0.02	0.51	0.11	0	–0.2	0.2
	G05	1	10	0.1	0.01	1.2	0.13	0	–0.2	0.2
	G06	12	10	0.99	0.12	0.86	0.04	0	–0.2	0.2
	G07	7	10	0.94	0.11	1.1	0.03	0	–0.2	0.2
	G08	17	10	0.95	0.13	0.63	0.04	0	–0.2	0.2
	G09	30	10	0.85	0.03	0.59	0.04	0	–0.2	0.2
	G10	30	10	2	0.02	0.95	0.03	0	–0.2	0.2
PSS with WAS	G02	41	10	1.2	0.02	1.1	0.04	0	–0.2	0.2
	G03	18	10	1	0.02	0.51	0.11	0	–0.2	0.2
	G05	1	10	0.1	0.01	1.2	0.13	0	–0.2	0.2
	G06	12	10	0.99	0.12	0.86	0.04	0	–0.2	0.2
	G07	7	10	0.94	0.11	1.1	0.03	1	–0.2	0.2
	G08	17	10	0.95	0.13	0.63	0.04	0	–0.2	0.2
	G09	30	10	0.85	0.03	0.59	0.04	0	–0.2	0.2
	G10	30	10	2	0.02	0.95	0.03	0	–0.2	0.2

- Case 3: PSS with tuning. It means all PSS signals for synchronous generators are active with the local rotor speed feedback which parameters are in Table 3.
- Case 4: PSS with WAS. As shown in Fig. 5, all PSS signals for synchronous generators (G_1–G_9) are active with the local rotor speed feedback except G_6 which an extra feedback signal comes from DFIG-based wind farm.

The simulations are conducted for all cases under a large disturbance as a three phase short-circuit fault at bus 19 for 50 ms.

4.2 *Modal Analysis*

4.2.1 Eigenvalues and Eigenvectors

For a matrix A, the λ is called eigenvalue of A if there exists a nonzero vector x that satisfies the following:

$$Ax_i = \lambda_i x_i \tag{15}$$

The vector $x_i \neq 0$ is known as the right eigenvector of A related with the eigenvalue λ_i (for $i = 1.2.3.\ldots n$).

The set of eigenvalues of the matrix A can be found readjust Eq. (15) as:

$$(A - \lambda I)x = 0 \tag{16}$$

Eigenvalues and their respective eigenvectors provide relevant information about the dynamics of the matrix A.

4.2.2 Participation Factors

From the sensitivity of an eigenvalue concept, the participation matrix ($P \in M_n$) combines right and left eigenvectors as a measure of association between the variables of a matrix and its eigenvalues. Defined as

$$P = \begin{bmatrix} P_1 & \ldots & P_n \end{bmatrix} \tag{17}$$

where

$$P_i = \begin{bmatrix} P_{1i} \\ P_{2i} \\ \vdots \\ P_{ni} \end{bmatrix} = \begin{bmatrix} x_{1i}\,y_{i1} \\ x_{2i}\,y_{i2} \\ \vdots \\ x_{ni}\,y_{in} \end{bmatrix} \tag{18}$$

where x_{ki} is the kith element of the the matrix X and y_{ik} is the ikth element of the matrix Y

$$P_{ki} = \frac{\partial \lambda_i}{\partial a_{kk}} \tag{19}$$

Therefore, the participation factors of λ_d will be all the elements of the diagonal of the sensitivity matrix of λ_d; this is generated by calculating $\partial \lambda_i$ with respect to all the elements of the matrix.

In a multi-machine power grid, two types of machine are used, i.e., synchronous machine and asynchronous machine. SMs are synchronously connected to power grid, they generate low frequency electromechanical oscillations. while asynchronous machine such as wind turbines and VS-PSHP have slip attitude, and therefore they do not participate in low frequency electromechanical oscillations [21]. These low frequency oscillations which are described in the range of 0.1–2 Hz can be divided to two modes including local and inter-area oscillation modes. Local oscillation mode is due to the rotating masses in SMs of specific area against each other in that area with associated frequency of 0.9–2 Hz. However, the swinging of group of SMs between areas (more than one area i.e., two or further areas) are named inter-area oscillation modes with associated frequency of 0.2–0.9 Hz. As mentioned earlier, PSHP and WT based on DFIM do not generate new oscillation modes into power networks account for the technologies of grid-connection. Note that low frequency of PSHP, generated by torque perturbations originated from the cavitation vortex rope in the turbine draft tube of hydro turbine operating at partial load, may be excited by oscillation modes of power grid.

Note that power systems have diverse dynamics involving different timescales such as those associated with the electromagnetic or the electromechanical phenomenon.Generally, there are two distinct processes that occur concurrently as follow [22].

- Electromagnetic dynamics: they are related to the magnetic flux linkages in generator which occur in the order of milliseconds.
- Electromechanical dynamics: they are related to the rotating masses in generators which occur in the order of seconds.

Since the first issue lasts in the timescale of milliseconds, it is not numerically efficient to model these dynamics in detail. Therefore, these oscillations in power system analysis are typically assumed to be fast and represented by algebraic equations. Two model including DFIM grid and SM grid are built in Digsilent version 15.1. The software automatically calculates the system equilibrium point, the system matrix, and the eigenvalues when the modal analysis procedure is enabled.

The location of the most dominant eigenvalues local and inter-area electromechanical oscillations are seen in FS-PSHP and other SM-based steam plants, i.e., *G2–G9*, are listed in Table 4 and Fig. 6 for four cases. Digsilent devotes randomly a tag number to every eigenvalue for identification purposes; using participation factors, it has been verified that when a different case is studied, although the numerical value

Table 4 Small signal simulation data in Digsilent

Modes				1st	2st	3st	4st	5st	6st	7st	8st	*9st*
Eigen value	Sync grid	Case 1	Re	−0.870	−0.700	−0.632	−0.781	−0.475	−0.559	−0.434	−0.366	−0.796
			Img	9.033	8.933	7.803	7.674	7.530	6.724	6.063	4.009	1.751
		Case 2	Re	*2.076**	*1.066*	−0.968	−0.599	−0.453	−0.555	−0.242	−1.986	−3.823
			Img	*17.709*	*15.118*	12.328	9.186	8.475	7.081	2.094	2.054	1.946
		Case 3	Re	−0.730	−0.559	−0.737	−0.594	−0.778	−0.442	−2.335	−2.819	−1.861
			Img	12.495	9.235	8.250	7.878	6.192	2.044	2.039	1.835	1.529
		Case 4	Re	−0.764	−0.554	−0.531	−0.761	−0.818	−2.312	−0.447	−2.818	−1.846
			Img	12.375	9.186	7.858	7.851	5.943	2.013	1.991	1.835	1.628
	DFIG grid	Case 1	Re	−0.700	−0.713	−0.790	−0.472	−0.561	−0.438	−0.379	−0.809	–
			Img	8.933	8.585	7.667	7.541	6.702	6.089	3.949	1.803	–
		Case 2	Re	*2.066*	*2.066*	*1.042*	−0.584	−0.453	−0.530	−0.234	−1.957	–
			Img	*17.691*	*−17.69*	*15.022*	9.270	8.527	7.169	2.136	2.014	–
		Case 3	Re	−0.726	−0.553	−0.737	−0.598	−0.787	−0.436	−2.355	−1.815	–
			Img	12.502	9.293	8.255	7.875	6.203	2.078	2.023	1.534	–
		Case 4	Re	−0.759	−0.549	−0.536	−0.761	−0.829	−0.445	−2.340	−1.784	–
			Img	12.385	9.251	7.857	7.850	5.949	2.025	2.000	1.629	–
Freq (Hz)	Sync rrid	Case 1		1.437	1.421	1.241	1.221	1.198	1.070	0.965	0.638	0.278
		Case 2		*2.818*	*2.406*	1.962	1.462	1.348	1.127	0.333	0.327	0.309
		Case 3		1.988	1.469	1.313	1.253	0.985	0.325	0.324	0.292	0.243
		Case 4		1.969	1.462	1.250	1.249	0.945	0.320	0.316	0.292	0.259
	DFIG grid	Case 1		1.421	1.366	1.220	1.200	1.066	0.969	0.628	0.287	–

(continued)

Table 4 (continued)

Modes			1st	2st	3st	4st	5st	6st	7st	8st	*9st*
		Case 2	*2.815*	*2.815*	*2.390*	1.475	1.357	1.141	0.340	0.320	–
		Case 3	1.989	1.479	1.313	1.253	0.987	0.330	0.322	0.244	–
		Case 4	1.971	1.472	1.250	1.249	0.946	0.322	0.318	0.259	–

*Hsighlighted row is shown the unstable modes

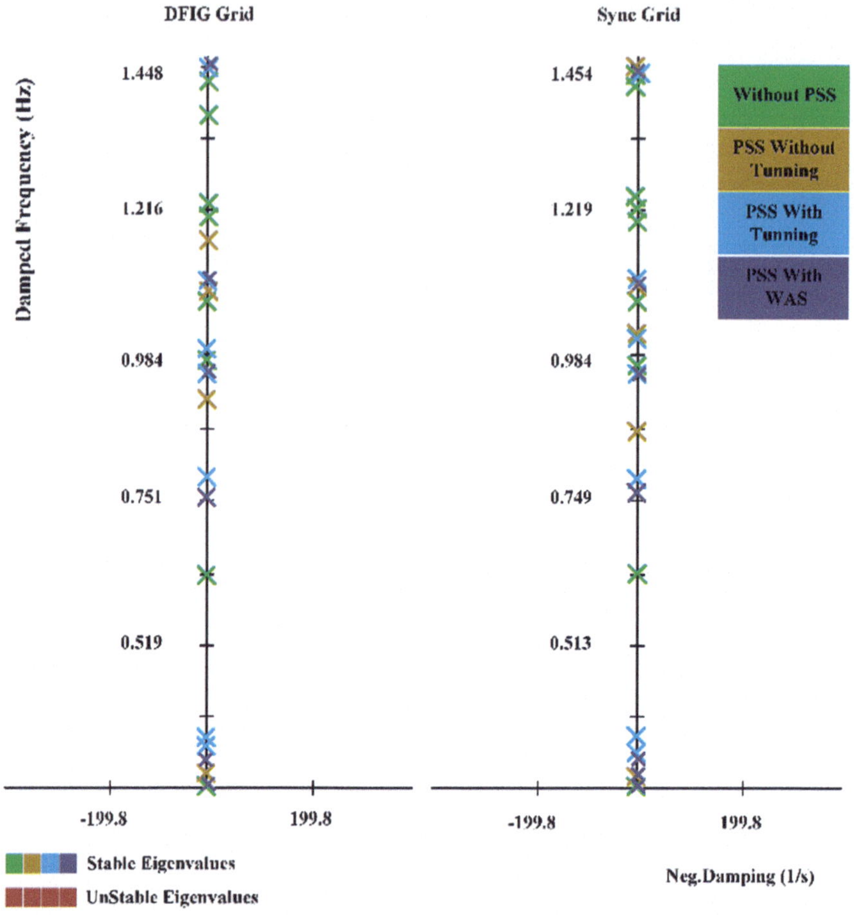

Fig. 6 Low frequency eigenvalue of four cases 1, 2, 3, and 4

of an eigenvalue may change, the assigned number corresponds to the same dynamic phenomenon.

As mentioned before, diverse dynamics in different timescales may be existed in electric power grids. Therefore, small signal stability analysis is a powerful method to determine electromechanical oscillations also known as modal analysis in order to anticipate the existence of electromechanical oscillations for different cases and take measures to avoid them. On the other hand, electromechanical oscillation modes are identified as those having the largest participation factors related to either the angular speed of SMs. As listed in the Table 4 eigenvalue analysis is conducted in the linearized system. Table 4 shows the eigenvalues corresponding to all oscillation modes and damping ratios at the normal operating condition for four cases in both *DFIG Grid* and *Sync Grid*. In this system there are six local modes, i.e. 1st, 2nd, 3rd, 4th, 5th, and 6th modes in the *Sync Grid*, five local modes for *DFIG Grid*, and three inter-area modes, i.e. modes 7th, 8th, and 9th modes for both *DFIG Grid* and *Sync Grid*.

4.3 RMS Simulation

Figures 7, 8, 9 and 10 compare the simulation results in term of *Sync Grid* and *DFIG Grid* for rotor angle of synchronous generators. Each figure shows result when a temporary three phase fault occurs at t = 50 s at bus 12 for 50 ms and it is cleared naturally. For the *DFIG Grid*, i.e., PSHP is DFIM-based, the deviation of rotor angle of G_1 is similar to *Sync Grid* except for case 3 (PSS with tuning) as shown in Fig. 9a, b. In case of G_1 when the PSHP is SM-based, cases 3 and 4 have similar oscillation results. But the WAS can improve the first swing stability of G_1 in case of SM-based PSHP as shown in Fig. 10a. Also, these explanations are accurate for G_3. Another important note in Fig. 7a–b is existing unstable mode for case 2 (PSS without tuning). This issue is extracted from modal analysis listed in Table 3. In summary, WAS signal (the rotor speed signal of wind farm) which is added to G_6 can be decrease deviation of the rotor angle of G_1 in the first swing. In other word, WAS not only does not negative effect of synchronous generators, but also can improve transient stability of DFIM-based power plants.

As a future work and research gaps, the applications of PMU and wide-area monitoring systems in modern power systems operation, control, stability, protection and security should be well-investigated [23–28]. PMUs are the backbone of wide are monitoring system that can improve the overall security and stability of power systems. Therefore, it is suggested to make in-depth evaluation of the power system security under the new environment control based one PMUs data. Likewise, the link between small signal stability stabilizers and automatic generation control loop can be suggested for improving the overall stability in future smart grids.

5 Conclusion

This chapter intends to present small signal modelling of DFIM-based and SM-based PSHP under generating operation mode using RMS-type simulation in Digsilent. Also, a wide area signal using the rotor speed of wind farm are added to PSS controller of a synchronous generator. The simulation results are conducted on a 10-machine 39-bus test system to assess effect of both types of PSHP and designed wide area signal on transient stability of large power system in presence of a wind farm. In summary, the following conclusions can be listed as follows.

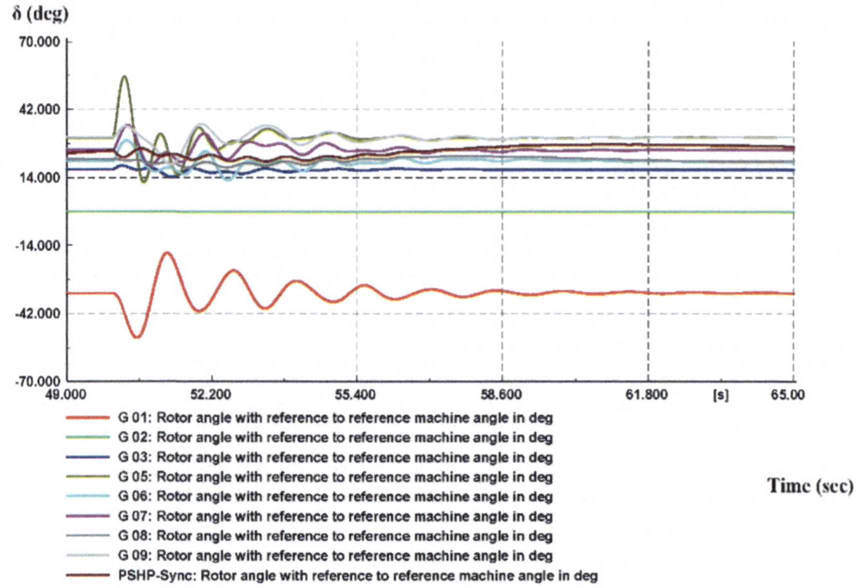

(a)

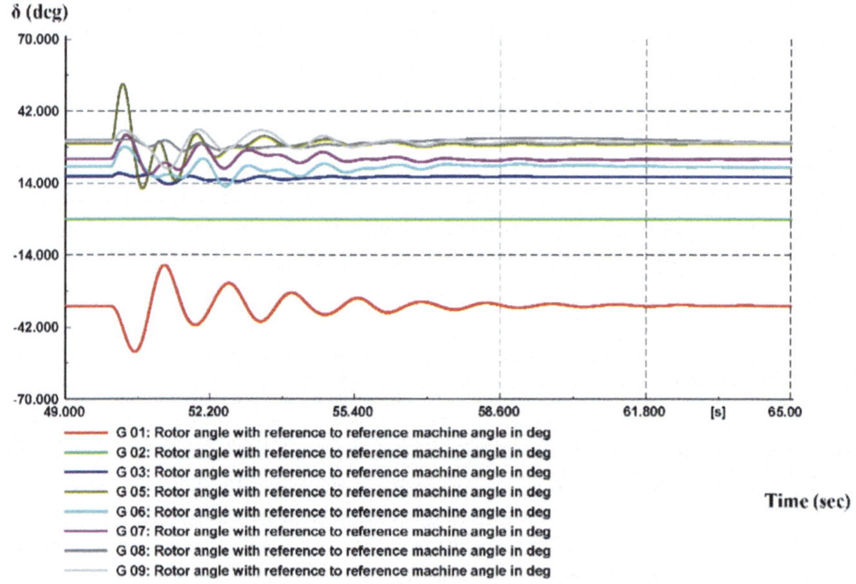

(b)

Fig. 7 Rotor angle of synchronous generators for case 1, **a** *Sync Grid*, **b** *DFIG Grid*

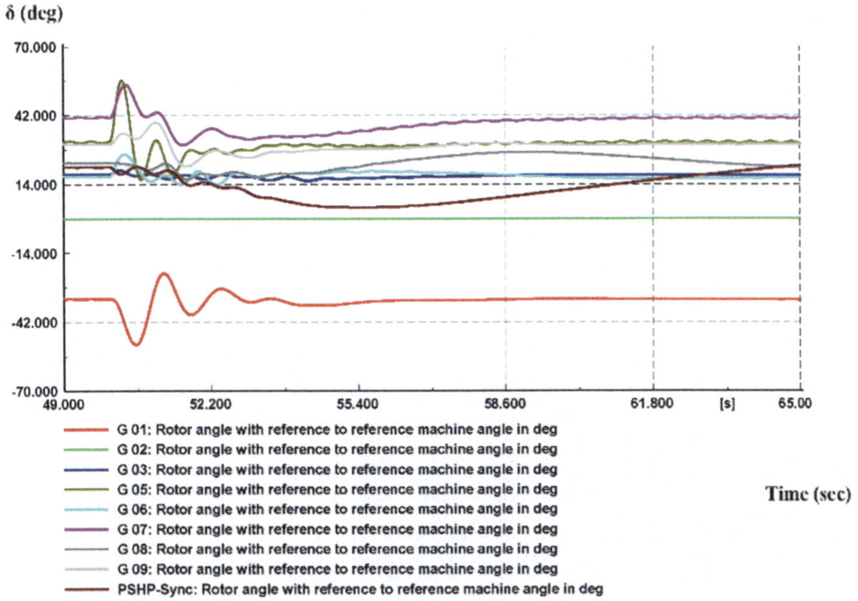

(a)

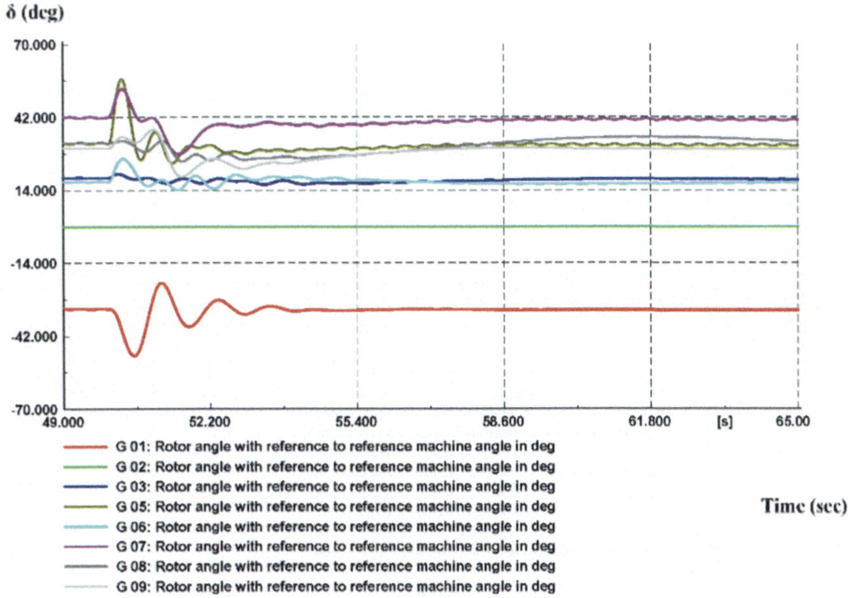

(b)

Fig. 8 Rotor angle of synchronous generators for case 2, **a** *Sync Grid*, **b** *DFIG Grid*

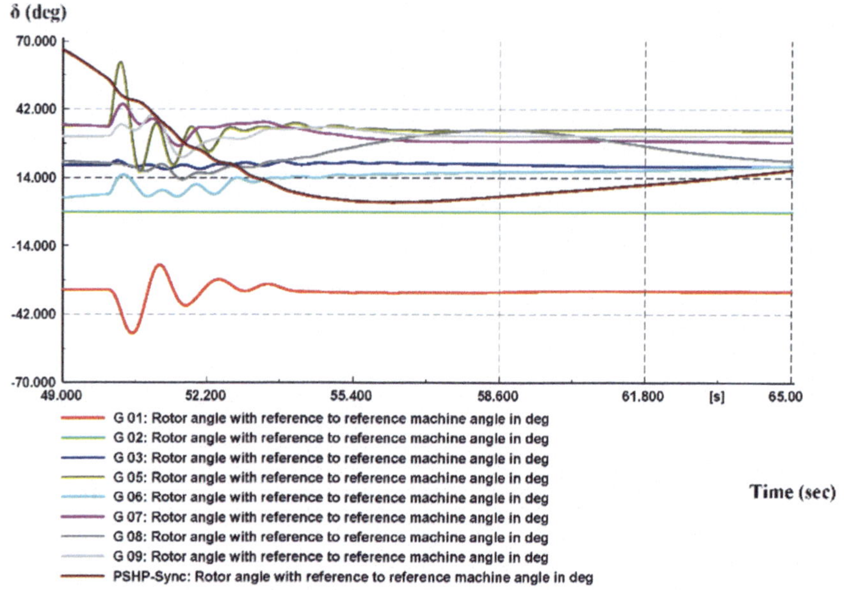

(a)

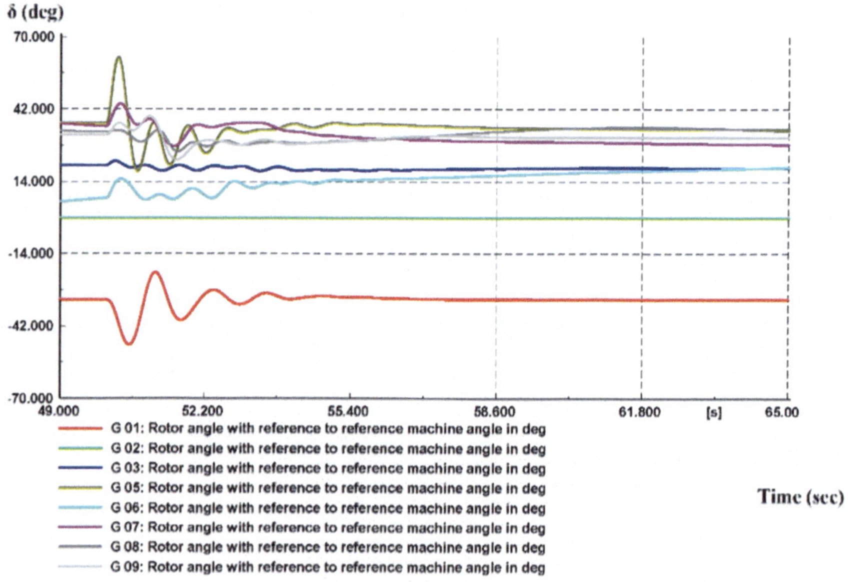

(b)

Fig. 9 Rotor angle of synchronous generators for case 3, **a** *Sync Grid*, **b** *DFIG Grid*

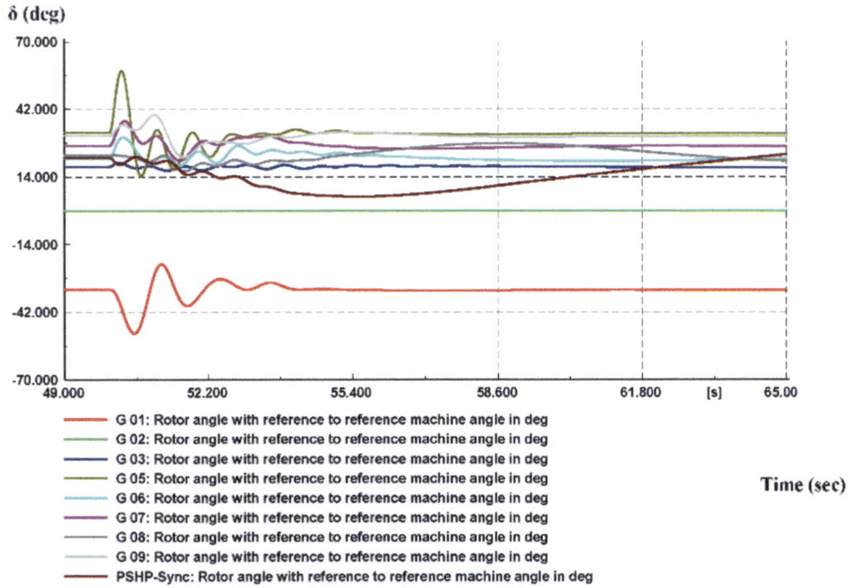

(a)

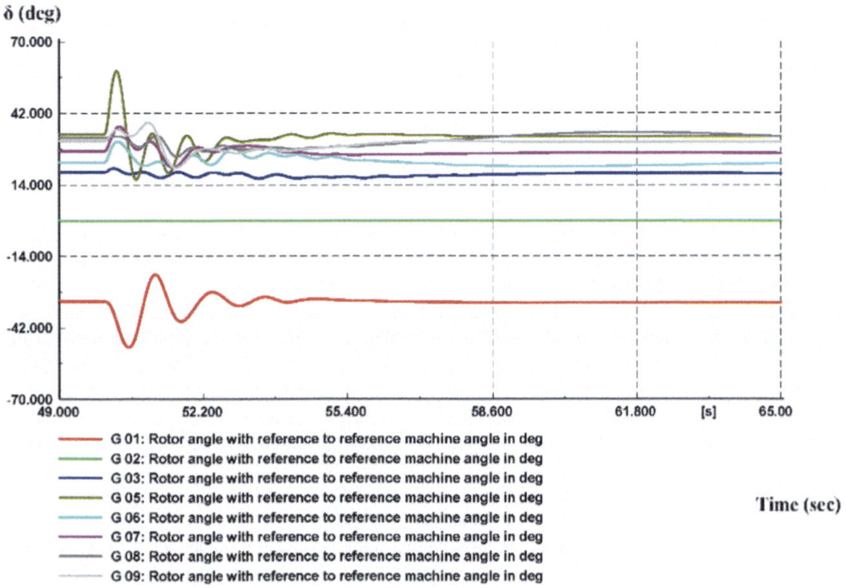

(b)

Fig. 10 Rotor angle of synchronous generators for case 4, **a** *Sync Grid*, **b** *DFIG Grid*

1. The speed of the state-of-the-art PSHP can be varied by different topologies which DFIM-based plants are mature rather than other technologies.
2. It is shown that a PSS can cause unstable modes, if it is not well tuned for low frequency oscillations.
3. It is shown that the wide area tuning of a PSS provides desired damping particularly for very low frequency oscillations.
4. Using DFIM-based PSHP in the interconnected power grids, not only the local oscillation modes of PSHP is eliminated, but also it can strongly damp inter-area oscillation modes and decrease the oscillation amplitude of other plants. This is because of the fast-injected power storing in the rotor of VS unit known flywheel effect.

Acknowledgments This chapter is the output of a research project supported by Yadegar -e-Imam Khomeini (RAH) Shahre Rey Branch, Islamic Azad University Iran. Also, the authors are thankful for the support from the National Natural Science Foundation of China (No.51809197, No.51879200).

References

1. M.A. Bidgoli, S.M.T. Bathaee, A. Shabani, Design a nonlinear auxiliary input for DFIG-based application using Lyapunov theory, in *The 5th Annual International Power Electronics, Drive Systems and Technologies Conference (PEDSTC 2014)*, pp. 102–107 (2014)
2. I.M.W.H. Americas et al., Technical analysis of pumped storage and integration with wind power in the Pacific Northwest (MWH, 2009)
3. N. Sivakumar, D. Das, N.P. Padhy, A.R. Senthil Kumar, N. Bisoyi, Status of pumped hydro-storage schemes and its future in India, Renew. Sustain. Energy Rev. **19**, 208–213 (2013). https://doi.org/10.1016/j.rser.2012.11.001
4. A.C. Padoan et al., Dynamical behavior comparison between variable speed and synchronous machines with PSS. IEEE Trans. Power Syst. **25**(3), 1555–1565 (2010). https://doi.org/10.1109/TPWRS.2009.2039586
5. Y. Pannatier, B. Kawkabani, C. Nicolet, J.-J. Simond, A. Schwery, P. Allenbach, Investigation of control strategies for variable-speed pump-turbine units by using a simplified model of the converters. IEEE Trans. Ind. Electron. **57**(9), 3039–3049 (2010) https://doi.org/10.1109/TIE.2009.2037101
6. M.A. Bidgoli, H.A. Mohammadpour, S.M.T. Bathaee, Advanced vector control design for DFIM-based hydropower storage for fault ride-through enhancement. IEEE Trans. Energy Convers. **30**(4), 1449–1459 (2015)
7. V. Azbe, R. Mihalic, Transient stability of a large doubly-fed induction machine in a pumped-storage plant. Electr. Power Syst. Res. **142**, 29–35 (2017)
8. W. Yang, J. Yang, Advantage of variable-speed pumped storage plants for mitigating wind power variations: integrated modelling and performance assessment. Appl. Energy **237**, 720–732 (2019)
9. A. Damdoum, I. Slama-Belkhodja, M. Pietrzak-David, M. Debbou, Low voltage ride-through strategies for doubly fed induction machine pumped storage system under grid faults. Renew. Energy **95**, 248–262 (2016). https://doi.org/10.1016/j.renene.2016.04.007
10. A. Joseph, R. Selvaraj, T.R. Chelliah, S.V.A. Sarma, Starting and braking of a large variable speed hydro-generating unit subjected to converter and sensor faults. IEEE Trans. Ind. Appl. 9994(i), 1–11 (2018)

11. S. Auber, Power on tap from variable speed pumped water storage scheme, Energize (2012)
12. M. Alizadeh Bidgoli, S.M.T. Bathaee, Full-state variables control of a grid-connected pumped storage power plant using non-linear controllers. Electr. Power Compon. Syst. **43**(3), 260–270 (2015). https://doi.org/10.1080/15325008.2014.980923
13. H.K. Pamsari, M.A. Bidgoli, M. Rajabzadeh, S.M.T. Bathaee, S. Ozgoli, Application of a new multivariable sliding mode controller for the single machine infinite bus systems, in *2011 2nd Power Electronics, Drive Systems and Technologies Conference*, pp. 211–216 (2011)
14. J. Liang, R.G. Harley, Pumped storage hydro-plant models for system transient and long-term dynamic studies. IEEE PES Gen. Meet. PES **2010**, 1–8 (2010). https://doi.org/10.1109/PES.2010.5589330
15. R.J. Kerkman, T.A. Lipo, W.G. Newman, J.E. Thirkell, An inquiry into adjustable speed operation of a pumped hydro plant part 1—machine design and performance. IEEE Trans. Power Appar. Syst. **PAS-99**(5), 1828–1837 (1980). https://doi.org/10.1109/TPAS.1980.319773
16. H. Schlunegger, A. Thöni, 100 MW full-size converter in the Grimsel 2 pumped-storage plant. Hydro (2013)
17. L. Belhadji, S. Bacha, I. Munteanu, A. Rumeau, D. Roye, Adaptive MPPT applied to variable-speed microhydropower plant. IEEE Trans. Energy Convers. **28**(1), 34–43 (2013). https://doi.org/10.1109/TEC.2012.2220776
18. A. Golshani, M.A. Bidgoli, S.M.T. Bathaee, Design of optimized sliding mode control to improve the dynamic behavior of PMSG wind turbine with NPC back-to-back converter. Int. Rev. Electr. Eng. **8**, 1170–1180 (2013)
19. M.A. Pai, *Energy Function Analysis for Power System Stability* (Springer Science & Business Media, 2012)
20. M. Khosravi-Charmi, T. Amraee, Wide area damping of electromechanical low frequency oscillations using phasor measurement data. Int. J. Electr. Power Energy Syst. **99**, 183–191 (2018)
21. A. Shabani, M.A. Bidgoli, A. Deihimi, Comparison of DDPMSG and DFIG concepts for wind turbines, in *The 5th Annual International Power Electronics, Drive Systems and Technologies Conference (PEDSTC 2014)*, pp. 512–517 (2014)
22. J.L. Rueda, J.C. Cepeda, I. Erlich, A.W. Korai, F.M. Gonzalez-Longatt, Probabilistic approach for risk evaluation of oscillatory stability in power systems, in *PowerFactory Applications for Power System Analysis*, pp. 249–266 (Springer, 2014)
23. H.H. Alhelou, M.E. Hamedani-Golshan, R. Zamani, E. Heydarian-Forushani, P. Siano, Challenges and opportunities of load frequency control in conventional, modern and future smart power systems: a comprehensive review. Energies **11**(10), 2497 (2018)
24. H.H. Alhelou, M.E. Golshan, J. Askari-Marnani, Robust sensor fault detection and isolation scheme for interconnected smart power systems in presence of RER and EVs using unknown input observer. Int. J. Electr. Power Energy Syst. **1**(99), 682–694 (2018)
25. H. Haes Alhelou, M.E. Hamedani Golshan, M. Hajiakbari Fini, Wind driven optimization algorithm application to load frequency control in interconnected power systems considering GRC and GDB nonlinearities. Electr. Power Compon. Syst. **46**(11–12), 1223–1238 (2018)
26. H.H. Alhelou, M.E. Golshan, N.D. Hatziargyriou, Deterministic dynamic state estimation-based optimal lfc for interconnected power systems using unknown input observer. IEEE Trans. Smart Grid (2019)
27. H.H. Alhelou, M.E. Golshan, N.D. Hatziargyriou, A decentralized functional observer based optimal LFC considering unknown inputs, uncertainties, and cyber-attacks. IEEE Trans. Power Syst. **34**(6), 4408–4417 (2019)
28. H.H. Alhelou, M.E. Golshan, T.C. Njenda, N.D. Hatziargyriou, An overview of UFLS in conventional, modern, and future smart power systems: challenges and opportunities. Electr. Power Syst. Res. **1**(179), 106054 (2020)

Impact Analysis and Robust Coordinated Control of Low Frequency Oscillations in Wind Integrated Power System

Abhilash Kumar Gupta, Akanksha Shukla, Kusum Verma, and K. R. Niazi

Abstract With rapid proliferation of wind generation in current generation mix, the issue of low frequency oscillations (LFOs) may get escalated in the modern power grids. The eigenvalue and dynamic sensitivity analysis have been employed to examine the effect of wind integration on system damping. Further, a wide area based robust damping improvement control is suggested. It involves the coordinated control of power system stabilizers (PSSs) of synchronous generators (SGs) and power oscillation dampers (PODs) of doubly fed induction generators (DFIGs). The robust control is attained by employing a new fitness function based on eigenvalue and damping ratio and optimized by Whale Optimization Algorithm (WOA). The wide area POD inputs are selected using modal observability criterion, obtained using phasor measurement units (PMUs) located optimally in the system. The results are verified on IEEE benchmark 68 bus NY-NE (New York- New England) test system. The simulation results highlights the robustness of proposed control to changing system conditions and shows its effectiveness in augmenting system damping and thus small signal stability with high level of wind penetration.

Keywords Oscillatory stability · Phasor measurement units · Doubly fed induction generator · Power oscillation dampers · Power system stabilizers · Small signal stability · Whale optimization algorithm · Wide area monitoring · Wide area control · Low frequency oscillations

---

A. K. Gupta (✉) · A. Shukla
Department of Electrical Engineering, GLA University, Mathura, India
e-mail: abhilash.gupta@gla.ac.in; abhidec91@gmail.com

A. Shukla
e-mail: akanksha.shukla@gla.ac.in

K. Verma · K. R. Niazi
Department of Electrical Engineering, Malaviya National Institute of Technology, Jaipur, India
e-mail: kverma.ee@mnit.ac.in

K. R. Niazi
e-mail: krniazi.ee@mnit.ac.in

H. Haes Alhelou et al. (eds.), *Wide Area Power Systems Stability, Protection, and Security*, Power Systems, https://doi.org/10.1007/978-3-030-54275-7_10

1 Introduction

The increasing environmental concerns to reduce generation from fossil fuels, leads to a sharp increase in renewables in the generation mix. The modern grid which already works closer to stability limits is further pushed to its limits with their integration as they usually have very less or no inertia [1]. The decrement in system inertia further convolute the system operation and stability issues. These issues needs to be settled to further improve the renewables penetration in the system. In recent times, wind proves to be the most beneficial renewable energy source for electricity generation [2]. The Doubly Fed Induction Generator (DFIG) based wind turbines are most popular due to their numerous benefits over other wind turbines [3]. The rapid shift towards green energy and their increased proliferation in the modern grid may lead to decrement in system low frequency oscillations (LFOs) damping causing small signal instability problems in the system [4]. The damping of LFOs should be 5–10% (for most systems) to maintain the small signal stability [5].

The DFIG penetration impacts the dynamic behavior of existing grids as their technology is quite different from the conventional synchronous generators (SGs) [6]. Their integration impacts the system load flow and changes the system configuration which affects the system damping [7]. Thus, the influence of DFIGs on system oscillatory stability need to be analyzed properly to facilitate higher wind integration in future.

The literature survey shows that many researchers carried out such type of analysis. Literature depicts the beneficial as well as detrimental effects of large scale integration of wind energy in the grid. The author in [8] shows that system damping could be improved by replacing SG which is actively participating in LFOs. The issues related to grid integration of DFIGs are reviewed in [1]. The eigenvalue sensitivity is utilized in [3] to study wind integration impact. The integration of Siemens wind turbine is investigated in [9] using modal analysis. The eigenvalue sensitivity to the Jacobian matrix is utilized to study the DFIG impact in [10]. A dynamic impact analysis using damping ratio (DR) sensitivity is presented in [11]. In [12], the tie line power is shown as an important parameter which effects LFOs with integration whereas [7] shows wind location as an important parameter. The DFIGs are converter based sources which makes them inertia less and thus their large integration may lead to escalation in LFO problems [6]. Normally, power system stabilizers (PSSs) are employed to boost the system damping. However, with high renewable integration, they might be not much effective and require support from supplementary controllers.

The literature suggests to employ power oscillation damper (POD) with DFIG to improve system oscillatory stability. The author in [13] improved damping using DFIG controller. Edrah et al. in [14] uses PSS inside voltage control loop of DFIG to augment system damping. Authors in [15] uses nonlinear controllers whereas wide area PSS is utilized in [16]. The recent literature suggest using PODs and PSSs together for damping LFOs. The simultaneous coordinated control of controllers is suggested in [17–19]. In [20], the plant dynamics is optimally controlled along with the coordinated control. The PSSs are tuned using the directional bat algorithm to improve system small-signal stability in [21]. The authors in [22] discusses the parameter setting strategy for the DFIG controller. The optimal PID controller has been utilized for LFO damping in [23]. The damping of LFOs has been enhanced using phasor measurement data in [24]. The choice of input signals for controllers is also significant and selected signals should have high observability of critical modes as suggested in [25]. The tuning of controller's parameters is usually obtained using analytical, numerical, meta-heuristic techniques, etc. However, out of them, metaheuristic techniques are preferred in recent literature.

Most of the authors in the literature have selected random location of DFIG for the impact analysis study and also for testing their control approach. However, it is quite evident that wind location does affect the system damping. Thus, this necessitates that a thorough analysis on the impact of DFIG location should be carried out. Also, most of the control strategies have been tested on small penetration of wind which might not be effective for higher penetrations. Literature also suggests that the actual impact of wind penetration can only be realized for higher wind penetrations (>15–20%). Thus, there is a need to carry out a robust control for higher wind penetration to improve system damping in critical scenarios.

In this work, a comprehensive analysis is performed and a control approach is proposed to improve LFOs damping in modern grids. Eigenvalue and dynamic trajectory sensitivity analysis are employed to analyze the effect of wind integration location on small signal stability of the system. This work also presents a robust coordinated tuning approach for PSSs and PODs in order to improve oscillatory stability of the system with high wind penetration. The coordinated tuning of controllers is carried out using a new meta-heuristic technique Whale optimization algorithm (WOA). The efficacy of the suggested tuning approach is investigated on IEEE 68-bus test system for a wide range of operating scenarios.

2 Wind Farm and Controllers Modelling

The DFIG wind farms have been modelled as single equivalent units at a particular bus. The MVA rating of wind farm has been assumed to be sum equivalent of all the wind turbines at that bus. The basic block diagram of DFIG employed in this work is shown in Fig. 1. GSC and RSC are grid side and rotor side converters respectively. The detailed modelling can be referred from [26]. In this work, reactive power is modulated using RSC controls to damp the LFOs.

A second-order lead-lag compensator has been utilized for the damping controllers modelling in this study. Both PSS and POD are similar in structure. The controllers have gain K (K_{PSS} or K_{POD}), wash-out time constant T_W (=10 s), and two phase compensators with T_1, T_2, T_3 & T_4 as time constants, shown in Fig. 2. The detailed modelling can be referred from [17].

The SG power output is the input signal for PSS while POD input signal is selected using observability criterion. The controller output is fed to automatic voltage regulator (AVR) in case of PSS and voltage controller in case of DFIG.

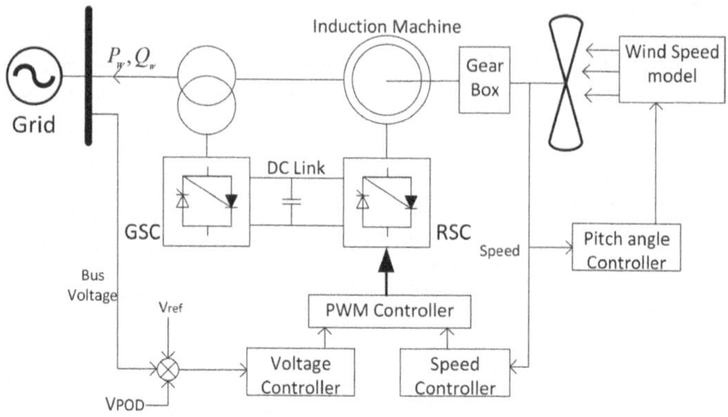

Fig. 1 DFIG block diagram

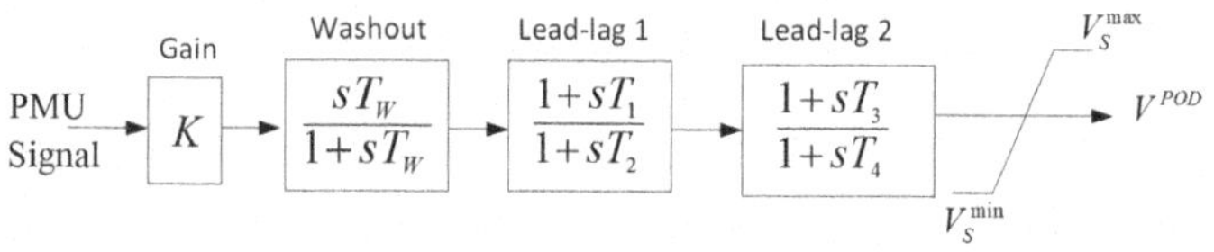

Fig. 2 Damping controller block diagram

3 Proposed Methodology

The goal of this study is to carry out the impact analysis of DFIG penetration on system damping and to propose a coordinated and robust control to improve system damping for high wind penetrated system.

3.1 Impact Analysis

Instead of integrating DFIGs randomly in the system, an impact analysis of DFIG location on the LFO damping is first carried out in this work. On that basis, suitable DFIG locations are finalized testing the control approach. The eigenvalue and dynamic sensitivity analysis have been employed to examine the effect of wind integration. The analysis involves following steps:

- In the base case, carry out eigenvalue analysis when there is no wind integrated in the system. Determine the damping ratio for all the LFO modes present in the system.
- As 68-bus system has 5 areas, seven DFIGs are integrated (2 each in Area 1 & 2 and 1 each in Area 3, 4 & 5) at random locations in the system. By keeping the

penetration same, locations are changed and eigenvalue analysis is performed for all such cases.

- All the possible placement combinations scenarios are tested for system damping in each case.
- Out of all cases, the best and worst locations according to LFO modes damping are computed.
- Next, the dynamic sensitivity analysis is carried out to verify the results obtained through eigenvalue analysis.

The eigenvalue analysis is an established approach for small signal stability analysis [27]. The LFO modes damping ratio and frequency are calculated using this method. The dynamic sensitivity analysis calculates the rotor angle sensitivity to changes in DFIG's active power [7]. It is determined using Eq. (1). An index called rotor angle sensitivity index (*RSI*) has been evaluated as given below.

$$RSI_i = \left| \sum_{t=1}^{T} w_t \left[\frac{\partial \delta_{i,COI}(t)}{\partial P_w} \right]_{t=t_n} \right| \tag{1}$$

where w_t depicts the weights attached to time instants. The *RSI* illustrate the DFIG integration impact using rotor angle values variation.

3.2 PODs Input

As discussed in introduction part, the input signal must possess high observability of critical LFO modes so as to provide a robust damping control over widely varying conditions [25]. As critical modes are global modes, wide area signals are required as they high observability of such LFOs. The modal observability is calculated for all possible candidate signals and the one having high value of this geometric measure is selected. It is calculated for *i*th mode as

$$OB_g = \frac{|C_g \phi_i|}{\|C_g\| \, \|\phi_i\|} \tag{2}$$

where C_g is the *g*th row of C matrix. In this study, the active power flow in the transmission lines are chosen as probable candidate input signals and observability is calculated for all such signals. These wide area signals can be obtained using the PMUs. The PMUs are optimally placed in the system as suggested in [28]. They are placed so that system is completely observable with minimum number of PMUs and ensure real-time critical buses monitoring even under single line or PMU outage in the system.

3.3 Coordinated Control Formulation

In this study, the robust control of controllers is suggested for system damping improvement. The main aim is to coordinate these supplementary damping controllers without using any complex methodologies. The proposed control is achieved by converting this problem into an optimization framework which calculates the optimal controller parameters.

The optimization formulation involves an improved multi-objective function which simultaneously maximize the system damping and move the critical modes to more stable left hand region of s-plane. These dual targets are achieved using a joined objective function defined as

$$f=min\left(\sum_{k=1}^{oc}\sum_{\xi_{i,k}\leq\xi_{spec}}\left(\frac{\xi_{spec}-\xi_{i,k}}{\xi_{spec}}\right)^2+\sum_{k=1}^{oc}\sum_{\sigma_{i,k}\geq\sigma_{spec}}\left(\frac{\sigma_{spec}-\sigma_{i,k}}{\sigma_{spec}}\right)^2\right)$$

$$\begin{aligned}&\text{subject to:}\\&K^{\min}\leq K\leq K^{\max}\\&T_{1,3}^{\min}\leq T_{1,3}\leq T_{1,3}^{\max}\\&T_{2,4}^{\min}\leq T_{2,4}\leq T_{2,4}^{\max}\end{aligned} \tag{3}$$

where oc represents operating conditions for which controllers were coordinated at the same time, $\sigma_{i,k}$ and $\xi_{i,k}$ are the real part and DR for the kth operating condition of the ith eigenvalue. Here, the values of threshold real part of eigenvalue and DR, σ_{spec} and ξ_{spec} are taken as -1.0 and $0.1(10\%)$ respectively. The gain K implies K_{PSS} and K_{POD}, and time constants T_1, T_2, T_3& T_4 implies values for controller's time constants ($T_1=T_3$, $T_2=T_4$). The values of $K^{\min}$, $K^{\max}$ are set as 0.1 and 50 respectively, $T_{1,3}^{\min}$, $T_{1,3}^{\max}$ are set as 0.1 and 1 respectively, and $T_{2,4}^{\min}$, $T_{2,4}^{\max}$ are set as 0.01 and 0.1 respectively. The above proposed fitness function will shift the critical modes to a stable D-shape region in the left-half of s-plane as shown in Fig. 3.

The proposed optimization of controller parameters is carried out using Whale Optimization Algorithm (WOA) proposed in [29]. The metaheuristic approaches are perfect techniques to solve such problems as they doesn't depend on system complicacy and volume, doesn't need past system information and provide quicker results. WOA is one such latest metaheuristic approach based on hunting behavior of humpback whales. It is derived from their bubble-net hunting strategy and utilized in this work because of its numerous advantages detailed in [29]. The proposed methodology is shown in Fig. 4. The design parameters of WOA and their details can be referred from [29].

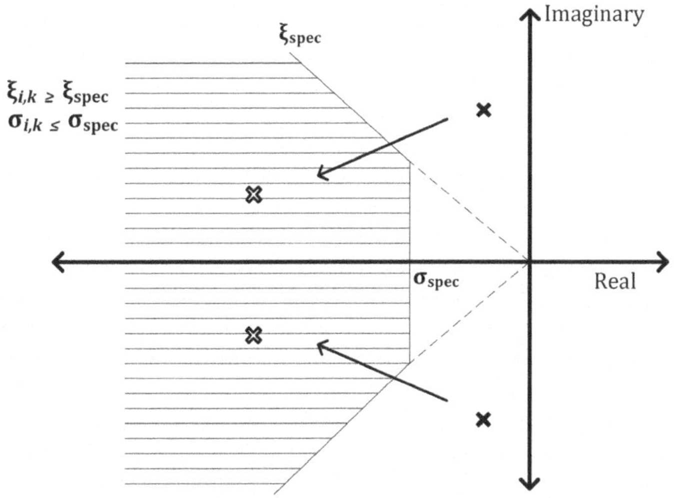

Fig. 3 D-shape stable region

4 Results and Discussion

The proposed analysis and coordinated control are implemented on the IEEE 68-bus system which is a benchmark system for such studies as suggested by [30]. According to report the system has high wind potential and suited for such study. The 68 bus system has 16 SGs, out of them 12 are equipped with AVRs and PSSs. The system is divided in 5 areas as shown in Fig. 5. The system shown has been modified and DFIGs have been located at suitable locations as identified by the impact analysis results.

The DFIG penetration in the system is calculated using Eq. (4) which is given below.

$$Wind\ Penetration\ (\%) = \frac{P_w}{P_G} \times 100 \tag{4}$$

where P_G represents the base case generation. The active power generation of the system in base case is 18,408.20 MW. The wind integration is kept at higher end (25%) in this work. A total of seven wind farms of equal size are integrated, as suggested in previous section, at non-generator buses. The load demand has also been ramped up accordingly to maintain the balance.

Initially, when there is no DFIG integration, the system excites 15 LFO modes. In this case the PSSs are also untuned. Out of these modes, there are 4 interarea modes. The base case results are shown in Table 1, showing the interarea modes present in the system. All four interarea modes has damping ratio less than 4%. In this work, the critical damping limit is taken as 10% for small signal stability. The next step is

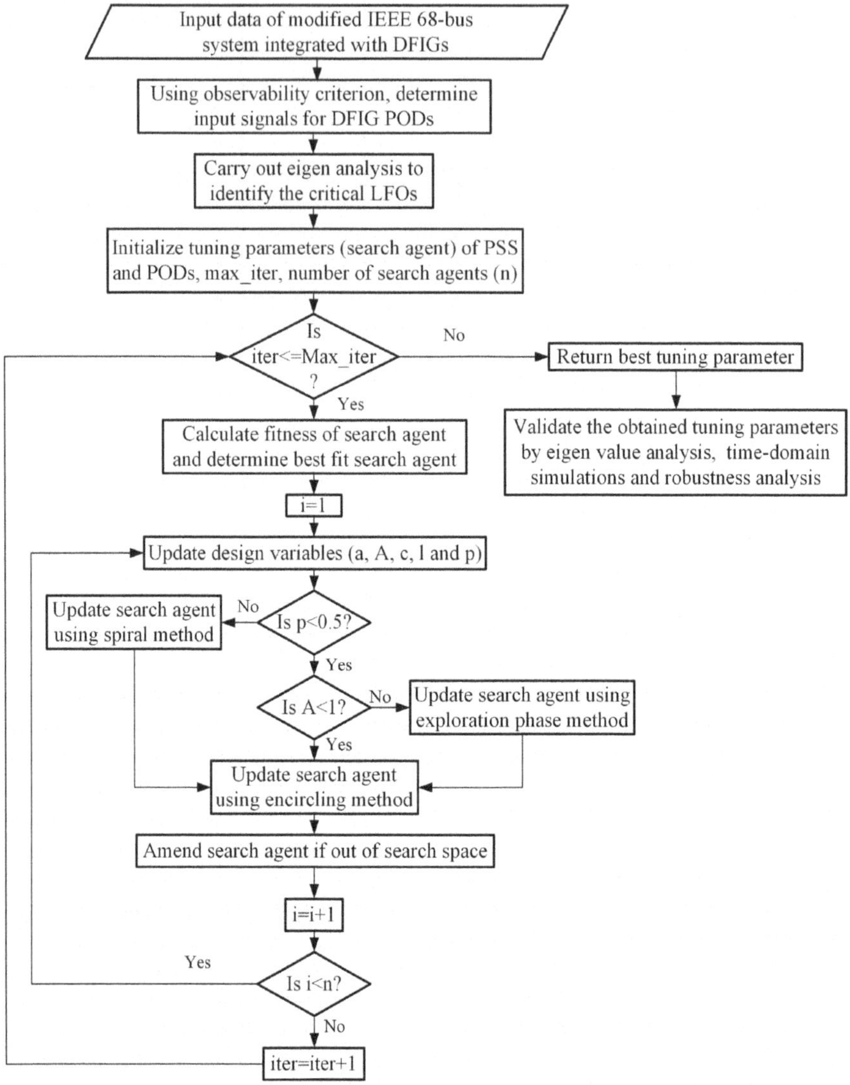

Fig. 4 Flowchart of proposed methodology

to study the effect of DFIG location on critical modes damping and thus oscillatory stability in the grid.

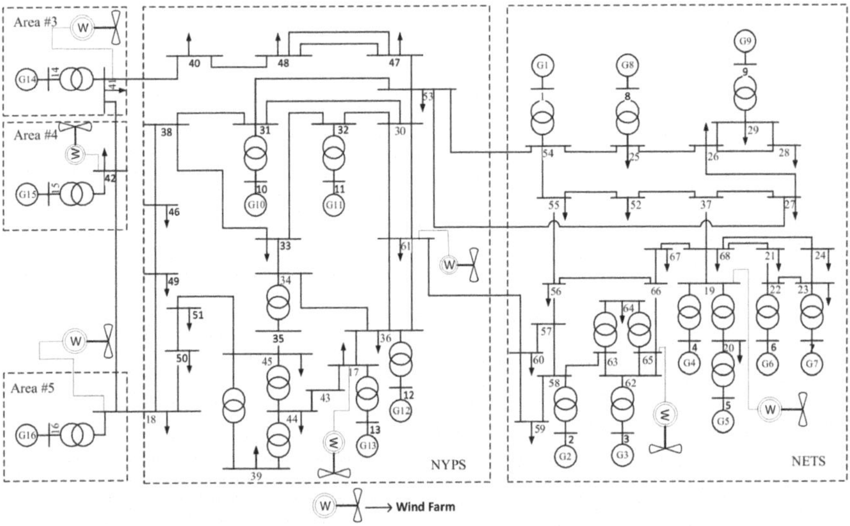

Fig. 5 IEEE 68 bus system

Table 1 Base Case Results

S.No	Frequency (Hz)	Damping ratio (%)
1	0.6821	03.14
2	0.7614	02.43
3	0.5233	02.30
4	0.4672	01.69

4.1 Impact Analysis Results

The DFIGs are located at seven non-generator buses as discussed earlier in the 68-bus system. All the possible placement combination scenarios are explored and studied. The eigenvalue analysis and dynamic trajectory sensitivity analysis have been carried out for all such cases. Out of all the cases studied, the case which shows the most improvement in the damping ratio of interarea modes and the case showing the most decrement in the damping ratio from the base case are picked out and shown in Table 2.

The findings of eigenvalue analysis for the two cases are presented here. When wind is integrated, five LFO modes are excited in the system. It means that due to DFIG integration a new LFO mode is excited in the system. If results are compared from Table 1, the difference in values of damping ratio is clearly visible. The worst DFIG placement positions for which the damping ratio of interarea modes decreases from the base case values are bus no. 17, 18, 19, 41, 42, 61, and 65. Also, for one mode (shown bold in the Table 2) the damping comes out to be negative which shows that the system becomes small signal unstable for that placement combination. The

Table 2 Eigenvalue analysis results for the best and worst case

S.No	Best case		Worst case	
	f (Hz)	DR (%)	f (Hz)	DR (%)
1	0.7916	04.39	0.7879	02.78
2	0.7895	03.88	0.6938	01.68
3	0.5638	03.01	0.5718	02.32
4	0.5144	03.41	0.4568	01.71
5	0.2895	02.92	0.2979	−00.34

best DFIG placement positions for which the damping ratio improves from the base case values are bus no. 18, 23, 35, 41, 42, 51, and 67. The results obtained shows that with change in the location of wind farms in the system, damping of system changes. Thus, it shows that location of DFIG impacts the small signal stability of the system and thus they should not be placed randomly while studying their impact on system stability. The results obtained from eigenvalue analysis are also verified with dynamic sensitivity analysis.

For the dynamic sensitivity analysis, a 3-phase fault is applied in the system at bus 53 at 1.0 s. The rotor angle values are recorded and the *RSI* is calculated using Eq. (1). The weights value is chosen as 1 for instants till 4 s after the disturbance, otherwise zero. For instants just after application of disturbance, higher weights are applied. The calculated values have been plotted for both the cases discussed above and shown for each SG. The plot is shown in Fig. 6. The bars indicate that the rotor angle variation is greater in worst case as compared to the best case, causing increase in oscillations and damping reduction.

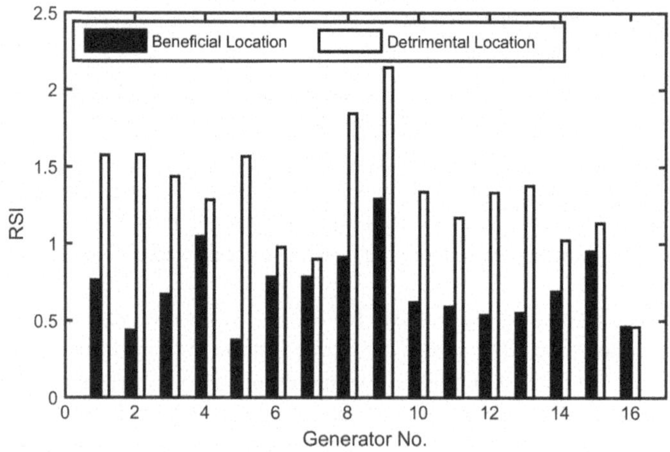

Fig. 6 RSI plot for the best and worst case

The results obtained using dynamic analysis further verifies the eigenvalue analysis results and reinstates that the location of DFIGs is an important factor affecting the small signal stability of the system. With wind integration, the active power flow in the system varies, which affects the rotor angles of SGs in the system. For some locations this rotor angle separation settles at smaller values and thus system has higher damping at such locations combination. Likewise, for some locations this angular separation increases to become higher than base case and thus leads to increment in oscillations and small signal instability in the system.

Practically, wind locations depend on the wind power resource and cannot be optimized. Thus, those locations could be either beneficial or detrimental to system damping which can be found out using impact analysis study carried out as above. The modes damping varies depending on the DFIG placement position. However, if reasoned practically, the LFO modes damping should not be the lone criteria to optimally place the wind farms. Moreover, the DFIG locations are practically governed by the abundance of wind resource available at a location. As the 68 bus system has multiple possible placement locations, so in place of arbitrary wind farm placement, the locations obtained in the impact analysis have been used for further analysis. The most detrimental locations obtained have been chosen as wind farm locations in this study. This is performed to test the effectiveness of proposed control approach. If proposed control improves the system damping for most detrimental case, it could be concluded that it will be equally effective for any other probable locations of DFIGs in the system.

4.2 PODs Input Signal Selection

The DFIGs are placed as per the worst case of impact analysis study, at bus no. 17, 18, 19, 41, 42, 61 and 65, as shown in Fig. 5. In this case system has 5 LFO modes all having damping ratio less than 10%. The damping of all these modes has to be ameliorated employing the proposed coordinated tuning method. The first step is to find out the input signals for the PODs employed with the DFIGs. The wide area input signals which have high observability of critical modes should be chosen. Also, based on the participation factor values of different DFIGs in the 5 critical modes, suitable DFIG has to be selected for damping a particular mode.

As DFIGs at 19, 42, 17, 41, 18 have high participation in modes 1, 2, 3, 4, 5 of Table 2 respectively, they are chosen for damping that particular mode. These five DFIGs need five wide area signals which has to be provided to their PODs input. As explained in previous section, the criterion of modal observability is used to select these signals. The observability of these five critical modes is evaluated for the active power flow in all the transmission lines in the system. The results are shown in Fig. 7 and also in Table 3. It can be seen from the figure that active power flow in line 53–27 is having highest observability of critical mode 1 and thus it has been selected as input for DFIG at bus 19 which we earlier chose for damping mode 1. The same results has been shown in the Table 3 also. Likewise, it has been shown for other

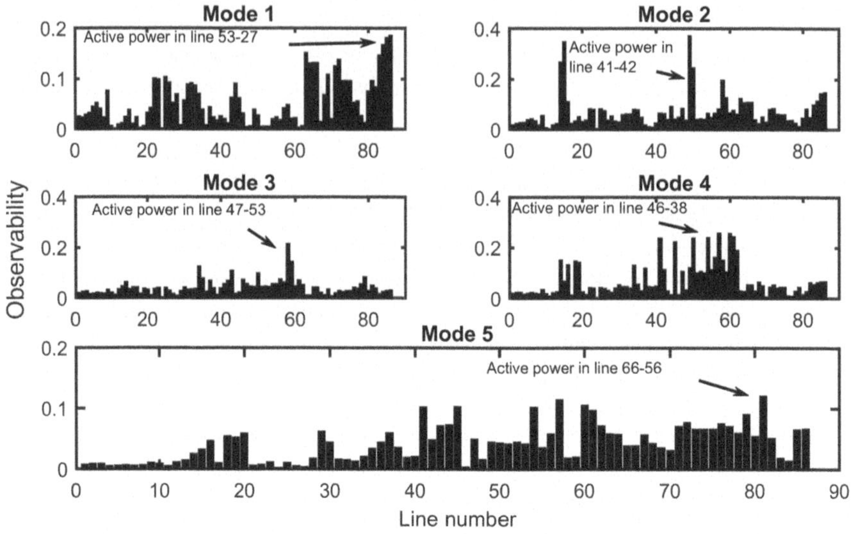

Fig. 7 Observability results for critical modes

Table 3 Observability Results

Mode No	DFIG selected (bus no.)	Input signal (from bus – to bus)
1	19	53–27
2	42	42–41
3	17	47–53
4	41	38–46
5	18	74–61

modes also. The active power signals selected are measured using PMUs placed in the system.

4.3 Results of Proposed Control Approach

As discussed in the methodology section, the next step is to design a coordinated tuning of controllers employed with SGs and DFIGs. The controllers are tuned using WOA with parameters set as: no of iterations = 100, population = 100, problem dimension would be as per no of controllers involved (three parameters are there for each controller). The parameters are simultaneously tuned for multiple operating conditions for achieving fairly robust results. These conditions are: base case condition with wind integrated, outage of line 30–31, outage of line 53–54, and outage of line 27–53. The efficacy of proposed control is verified by comparing the system damping results for the two cases:

Case 1—Base case when wind is integrated and all 12 PSSs are untuned, no PODs;

Case 2—12 PSSs and 5 PODs are tuned using proposed coordinated control approach.

Also, to assess the proposed control robustness to various faults and changes in system conditions, an index called Robustness Index (*RI*) is utilised. The value of *RI* is given as

$$RI = \sum_{m=1}^{n_g} \left| \frac{1}{T} \sum_{n=1}^{N} \left(\omega_{mn} - \omega_{COI,n} \right) \right| \quad (5)$$

where n_g represents the SGs number, T shows simulation time taken, N is number of data samples stored,ω_{mn} and $\omega_{COI,n}$ are the SGs rotor speed and rotor COI speed at the nth instant respectively. Lower value of *RI* shows system robustness to change in operating conditions.

The coordinated control is tested using eigenvalue analysis, time-domain simulations and robustness analysis. The results are detailed below.

4.3.1 Results of Eigenvalue Analysis

The eigenvalue analysis is conducted for the two cases studied and the results obtained are given in Table 4. The table shows the frequency, DR and real part of eigenvalue for the five LFO modes for the two cases. The results suggests that after the coordinated tuning of controllers one of the critical mode disappears. Also, the damping of all modes increases to become more than 40%. The modes are also shifted to more stable left half region, i.e., real part becomes more negative from the base case values. It basically shows that system damping improves leading to better small signal stability in the second case.

The eigenvalue plots for these two cases are also shown in Figs. 8 and 9. It can be observed from the figures that the modes which were on the left side of 10% damping

Table 4 Eigenvalue analysis results

Mode No	Case 1			Case 2		
	Real-part	f (Hz)	DR (%)	Real-part	f (Hz)	DR (%)
Mode 1	−0.22	0.7879	02.79	−2.29	0.6902	47.97
Mode 2	−0.17	0.6938	01.88	−1.89	0.5598	46.57
Mode 3	−0.12	0.5718	02.45	−1.61	0.4605	48.42
Mode 4	−0.13	0.4568	01.59	−1.79	0.5332	47.85
Mode 5	−0.05	0.2979	00.10	–	–	–

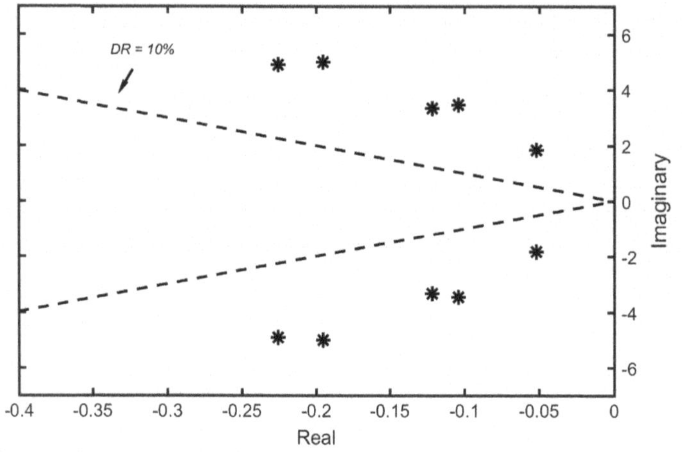

Fig. 8 Eigenvalue plot for Case 1

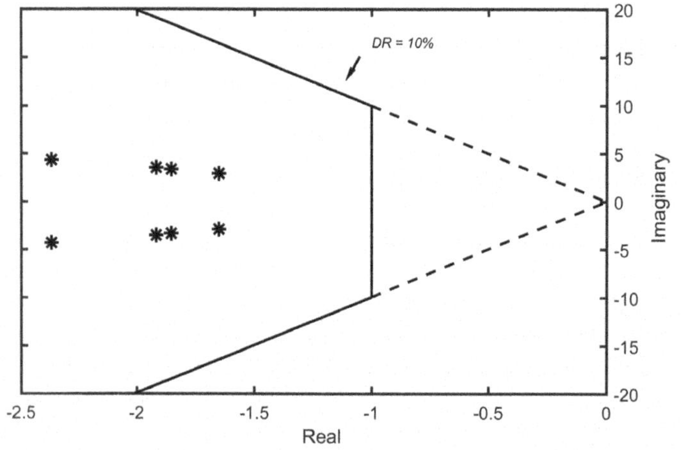

Fig. 9 Eigenvalue plot for Case 2

line in Case 1 are shifted to the right side in Case 2. Thus, the modes are shifted to a more stable region on application of the coordinated approach. Thus, from plots also it is evident that system damping improves with the proposed control approach. The results show that using the coordinated control approach as suggested will improve the damping in grids with high wind penetration. The proposed approach is also tested using time domain simulations.

4.3.2 Time Domain Simulations Result

In addition to above analysis, time domain simulations have also been executed on the modified 68-bus test system. One such case is discussed here. A 3-phase fault is applied at bus no 53 of six cycles at 1 s and following which the system's behaviour has been tracked for 30 s. The active power outputs of two generators are shown in Figs. 10 and 11 for comparing the two cases. Figure 10 shows the SG6 active power output and Fig. 11 shows the SG13 power output. It can be observed from the outputs of these two generators that the system has more damping in Case 2 and the oscillations settles down faster as compared to Case 1. Thus, with application of

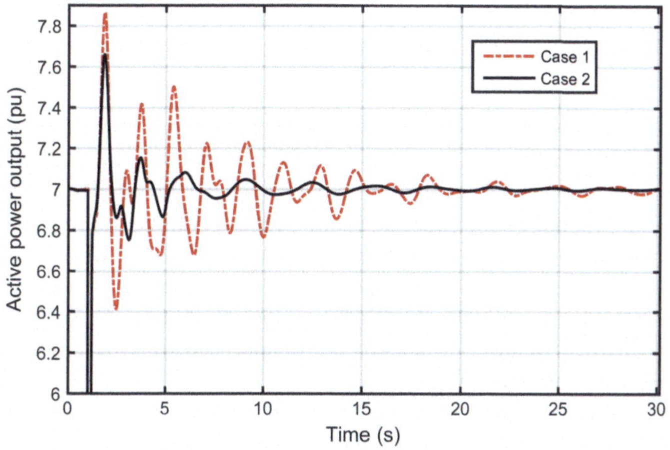

Fig. 10 SG6 active power plot

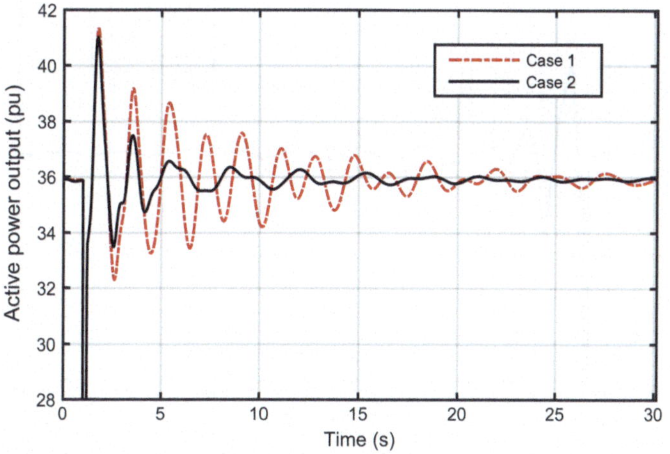

Fig. 11 SG13 active power plot

proposed coordinated control the oscillations die out early and thus makes system small signal stable. These results further verify the eigenvalue analysis results.

4.3.3 Robustness Analysis Results

In addition to the eigenvalue and time domain analysis, robustness analysis has also been carried out to test the effectiveness of proposed control strategy in critical scenarios. Various critical scenarios (CS) are designed to test the control approach, out of them two are presented here:

1. **CS1**: Outage of two important lines simultaneously (between buses 21–22 and 55–56) along with a 3-phase fault at bus 53 of 6 cycles at 1.0 s.
2. **CS2**: The decrement of system load by 10%, with same line outages and fault condition.

The comparison of two cases has been shown for these two critical scenarios in Figs. 12 and 13. The active power output of SG6 is shown in these figures for the two scenarios. It is clearly observable from the results that the oscillations are higher in Case 1 as compared to Case 2 in both the scenarios. In Case 2, oscillations are settled early as the proposed coordinated control is employed in this case which significantly improves the system damping. So, even in these critical contingencies the system stability is maintained with the proposed control. These results again proves the robustness of the coordinated control approach employed for small signal stability improvement in the system.

The values of *RI* are also calculated for the two cases and shown in Table 5. The sampling rate of rotor speed is considered 60 samples per second here and recorded for duration of 30 s to determine the value of *RI* from Eq. (5). These can be recorded

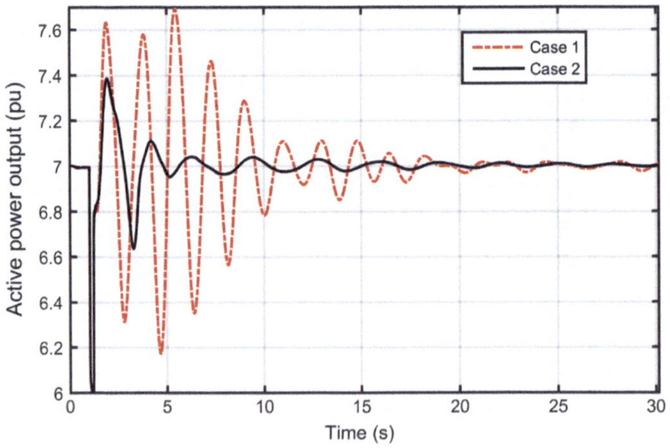

Fig. 12 CS1 result

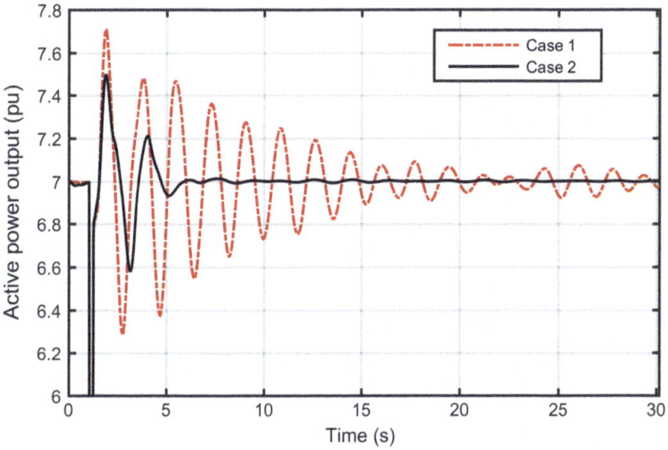

Fig. 13 CS2 result

Table 5 RI result

Critical scenario	Case 1	Case 2
CS1	0.0584	**0.0042**
CS2	0.8653	**0.0037**

using PMUs employed in the system. It is clear from the Table 5 that the value of RI is lowest for Case 2 for both the scenarios, which shows the controllers robustness to varying system conditions. The findings of robustness analysis further validate the findings obtained in earlier analysis.

The results obtained from different analysis shows that the controllers coordinated effectively in the proposed approach and significantly improves the oscillatory stability of the system.

As a future work and research gaps, the applications of PMU and wide-area monitoring systems in modern power systems operation, control, stability, protection and security should be well-investigated [31–36]. PMUs are the backbone of wide are monitoring system that can improve the overall security and stability of power systems. Therefore, it is suggested to make in-depth evaluation of the power system security under the new environment control based one PMUs data. Likewise, the link between small signal stability stabilizers and automatic generation control loop can be suggested for improving the overall stability in future smart grids.

5 Conclusion

The study presented in this work proposed a complete and methodical procedure which could be employed for damping improvement in large power systems having high penetrations of wind generation. The analysis carried out pointed the importance of DFIG location and its impact on system's oscillatory stability. A wide area based robust coordinated approach for controllers tuning is proposed for high wind penetrated systems using wide area signals. The controllers coordinated effectively and significantly improves the small signal stability of the system. The findings obtained from various analysis proves the efficacy of the control approach. The LFO modes are properly damped and oscillatory stability improves even in critical operating scenarios. The proposed control proves to be robust to changing system conditions and may facilitate future higher wind integration in modern grids while maintaining the small signal stability of the system.

References

1. H.T. Jadhav, R. Roy, A comprehensive review on the grid integration of doubly fed induction generator. Int. J. Electr. Power Energy Syst. **49**, 8–18 (2013)
2. A.K. Gupta, K. Verma, K.R. Niazi, Real-time low-frequency oscillations monitoring and coherency determination in a wind-integrated power system, in *Intelligent Computing Techniques for Smart Energy Systems*. Lecture Notes in Electrical Engineering, vol 607, eds. by A. Kalam, K. Niazi, A. Soni, S. Siddiqui, A. Mundra. (Springer, Singapore, 2020)
3. D. Gautam, V. Vittal, T. Harbour, Impact of increased penetration of DFIG-based wind turbine generators on transient and small signal stability of power systems. IEEE Trans. Power Syst. **24**(3), 1426–1434 (2009)
4. A.K. Gupta, K. Verma, K.R. Niazi, Robust coordinated control for damping low frequency oscillations in high wind penetration power system. Int. Trans. Electr. Energy Syst. **29**(5), e12006 (2019)
5. J. Paserba, J. Sanchez-Gasca, P. Kundur, Small signal stability and power system oscillations, in *Electric Power Engineering Handbook*, ed. by L.L. Grigsby, pp. 20–34 (CRC & IEEE Press, 2001)
6. J. Quintero, V. Vittal, G.T. Heydt, H. Zhang, The impact of increased penetration of converter control-based generators on power system modes of oscillation. IEEE Trans. Power Syst. **29**(5), 2248–2256 (2014)
7. A.K. Gupta, K. Verma, K.R. Niazi, (2017) Impact analysis of DFIG location on low-frequency oscillations in power system. J. Eng. **13**, 1413–1417 (2017)
8. J.G. Slootweg, W.L. Kling, The impact of large scale wind power generation on power system oscillations. Electr. Power Syst. Res. **67**(1), 9–20 (2003)
9. T. Knuppel, J.N. Nielsen, K.H. Jensen et al., Small-signal stability of wind power system with full-load converter interfaced wind turbines. IET Renew. Power Gener. **6**(2), 79–91 (2012)
10. M. Jafarian, A.M. Ranjbar, Interaction of the dynamics of doubly fed wind generators with power system electromechanical oscillations. IET Renew. Power Gener. **7**(2), 89–97 (2013)
11. A.K. Gupta, K. Verma, K.R. Niazi, Dynamic impact analysis of DFIG-based wind turbine generators on low-frequency oscillations in power system. IET Gener. Transm. Distrib. **11**(18), 4500–4510 (2017)
12. M. Garmroodi, D.J. Hill, G. Verbič, J. Ma, Impact of tie-line power on inter-area modes with increased penetration of wind power. IEEE Trans. Power Syst. **31**(4), 3051–3059 (2015)

13. Y. Mishra, S. Mishra, M. Tripathy et al., Improving stability of a DFIG-based wind power system with tuned damping controller. IEEE Trans. Energy Convers. **24**(3), 650–660 (2009)
14. M. Edrah, K.L. Lo, O. Anaya-Lara, Reactive power control of DFIG wind turbines for power oscillation damping under a wide range of operating conditions. IET Gener. Transm. Distrib. **10**(15), 3777–3785 (2016)
15. K. Elkington, M. Ghandhari, Non-linear power oscillation damping controllers for doubly fed induction generators in wind farms. IET Renew. Power Gener. **7**(2), 172–179 (2013)
16. D. Ke, C.Y. Chung, Design of probabilistically-robust wide-area power system stabilizers to suppress inter-area oscillations of wind integrated power systems. IEEE Trans. Power Syst. **31**(6), 4297–4309 (2016)
17. T. Surinkaew, I. Ngamroo, Hierarchical co-ordinated wide area and local controls of DFIG wind turbine and PSS for robust power oscillation damping. IEEE Trans. Sustain. Energy **7**(3), 943–955 (2016)
18. T. Surinkaew, I. Ngamroo, Two-level coordinated controllers for robust inter-area oscillation damping considering impact of local latency. IET Gener. Transm. Distrib. **11**(18), 4520–4530 (2017)
19. J. Zuo, Y. Li, D. Shi et al., Simultaneous robust coordinated damping control of power system stabilizers (PSSs), static var compensator (SVC) and doubly-fed induction generator power oscillation dampers (DFIG PODs) in multimachine power systems. Energies **10**(4), 565 (2017)
20. C. Zhang, D. Ke, Y. Sun, C.Y. Chung, J. Xu, F. Shen, Coordinated supplementary damping control of DFIG and PSS to suppress inter-area oscillations with optimally controlled plant dynamics. IEEE Trans. Sustain. Energy **9**(2), 780–791 (2018)
21. S. Gurung, J. Francisco, N. Sumate, S. Anawach, Comparative analysis of probabilistic and deterministic approach to tune the power system stabilizers using the directional bat algorithm to improve system small-signal stability. Electr. Power Syst. Res. **181**, 106176 (2020)
22. Y. Jia et al., Parameter setting strategy for the controller of the DFIG wind turbine considering the small-signal stability of power grids. IEEE Access **8**, 31287–31294 (2020)
23. M. Saadatmand, B. Mozafari, B. Gevork, S. Soleymani, Optimal PID controller of large scale PV farms for power systems LFO damping. Int. Trans. Electr. Energy Syst. e12372 (2020)
24. M. Khosravi-Charmi, A. Turaj, Wide area damping of electro-mechanical low frequency oscillations using phasor measurement data. Int. J. Electr. Power Energy Syst. **99**, 183–191 (2018)
25. L.P. Kunjumuhammed, B.C. Pal, Selection of feedback signals for controlling dynamics in future power transmission networks. IEEE Trans. Smart Grid **6**(3), 1493–1501 (2015)
26. F. Milano, Power System Modelling and Scripting. (Springer Science & Business Media, 2010)
27. P. Kundur, N.J. Balu, M.G. Lauby, *Power System Stability and Control*, vol. 7 (McGraw-hill, New York, 1994)
28. A.K. Gupta, K. Verma, K.R. Niazi, Contingency constrained optimal placement of PMUs for wide area low frequency oscillation monitoring, in *Proceedings of IEEE 7th International Conference on Power Systems*, Pune, India, pp. 1–6 (2017)
29. S. Mirjalili, A. Lewis, The whale optimization algorithm. Adv. Eng. Softw. **95**, 51–67 (2016)
30. C. Canizares, T. Fernandes, Benchmark systems for small-signal stability analysis and control. IEEE PES, 2019 (2015). https://resourcecenter.ieee-pes.org/pes/product/technical-reports/PES TR18PES-TR
31. H.H. Alhelou, M.E. Hamedani-Golshan, R. Zamani, E. Heydarian-Forushani, P. Siano, Challenges and opportunities of load frequency control in conventional, modern and future smart power systems: a comprehensive review. Energies **11**(10), 2497 (2018)
32. H.H. Alhelou, M.E. Golshan, J. Askari-Marnani, Robust sensor fault detection and isolation scheme for interconnected smart power systems in presence of RER and EVs using unknown input observer. Int. J. Electr. Power Energy Syst. **99**, 682–694 (2018)
33. H. Haes Alhelou, M.E. Hamedani Golshan, M. Hajiakbari Fini, Wind driven optimization algorithm application to load frequency control in interconnected power systems considering GRC and GDB nonlinearities. Electr. Power Compon. Syst. **46**(11–12), 1223–1238 (2018)

34. H.H. Alhelou, M.E. Golshan, N.D. Hatziargyriou, Deterministic dynamic state estimation-based optimal LFC for interconnected power systems using unknown input observer. IEEE Trans. Smart Grid (2019)
35. H.H. Alhelou, M.E. Golshan, N.D. Hatziargyriou, A decentralized functional observer based optimal LFC considering unknown inputs, uncertainties, and cyber-attacks. IEEE Trans. Power Syst. **34**(6):4408–4417 (2019)
36. H.H. Alhelou, M.E. Golshan, T.C. Njenda, N.D. Hatziargyriou, An overview of UFLS in conventional, modern, and future smart power systems: challenges and opportunities. Electr. Power Syst. Res. **179**, 106054 (2020)

Frequency Stability of Two-Area Interconnected Power System with Doubly Fed Induction Generator Based Wind Turbine

Ahmed Hamdy, Salah Kamel, Loai Nasrat, and Francisco Jurado

Abstract This chapter presents a comparison of the performance of integral (I) and proportional-integral-derivative (PID) controllers in frequency stabilization or load frequency control (LFC) of two-area interconnected power system considering generation rate constraints (GRCs) with Doubly fed induction generator (DFIG)-based wind energy. Two mathematically models are identified for investigations. Power system model 1 is two-area interconnected power system which contains two identical non-reheat thermal plants without DFIG participation. Whereas, power system model 2 contains two identical non-reheat thermal plants with dynamic participation of DFIG at both areas. Moreover, Harris Hawks Optimizer (HHO), Salp Swarm Algorithm (SSA), and Sine Cosine Algorithm (SCA) are applied to find the optimal values of the controller settings mentioned above. The effectiveness of the proposed controllers, which are optimally designed by several optimization techniques (i.e., HHO, SSA, and SCA) is tested and verified through an interconnected power system comprises two identical non-reheat thermal power plants with/without DFIG participation. Time-domain simulation results of the studied power system with all mentioned optimization techniques are carried out using Matlab/Simulink® software to validate the robustness of the proposed controllers.

Keywords WES · DFIG · LFC · HHO · SSA · SCA

A. Hamdy · S. Kamel · L. Nasrat
Department of Electrical Engineering, Faculty of Engineering, Aswan University, 81542 Aswan, Egypt
e-mail: ahmedhamdykasem2016@yahoo.com

S. Kamel
e-mail: skamel@aswu.edu.eg

L. Nasrat
e-mail: loainasrat@aswu.edu.eg

F. Jurado (✉)
Department of Electrical Engineering, University of Jaén, 23700 EPS Linares, Jaén, Spain
e-mail: fjurado@ujaen.es

H. Haes Alhelou et al. (eds.), *Wide Area Power Systems Stability, Protection, and Security*, Power Systems, https://doi.org/10.1007/978-3-030-54275-7_11

1 Introduction

The expanding of modern energy systems (MESs) which including varieties of conventional resources leads to depleting theses resources and will decrease over time. So the integration of renewable energy resources (RER) with the traditional power grid has become the optimal solution by the capacity expansion policy. Recently, wind energy has become one of the promising alternative RERs. The doubly fed induction generator (DFIG) is considered the widely renewable resource which has been depicted to participate in system frequency support [1]. Wind energy system (WES) technology has grown significantly in the last few decades. And also, the capacity of a single wind turbine has increased from a few kW ranges to machines producing 1–5 MW [2, 3]. This is because WES depends on natural source so that it can keep national income by reducing the expenditure on fuel that is used in traditional method of electricity production. The total installed capacity of WES is increased from 24 GW in 2011 to, 792 GW in 2020 as shown in Fig. 1 which illustrate these statistics [4].

According to the great and rapid progress in wind technology. The wind turbine rating can be taken as an example of this progress; where in 1980 wind turbine rating began with 50 KW and reached 10 MW by 2012 [5].

Accordingly, the rapid spread of wind energy; it was necessary to exploit it in electrical power systems. Frequency control represents one of the most important indexes which has gained a considerable attention due to its importance and wind energy able to enhance it. Frequency control is divided into four control levels, i.e., primary, secondary, tertiary, and emergency control levels. Primary frequency control loop summarized in performance of governor droop which attempt to recover system frequency before triggering over/under frequency protection relays. Secondary frequency control (LFC) or supplementary controller is responsible for

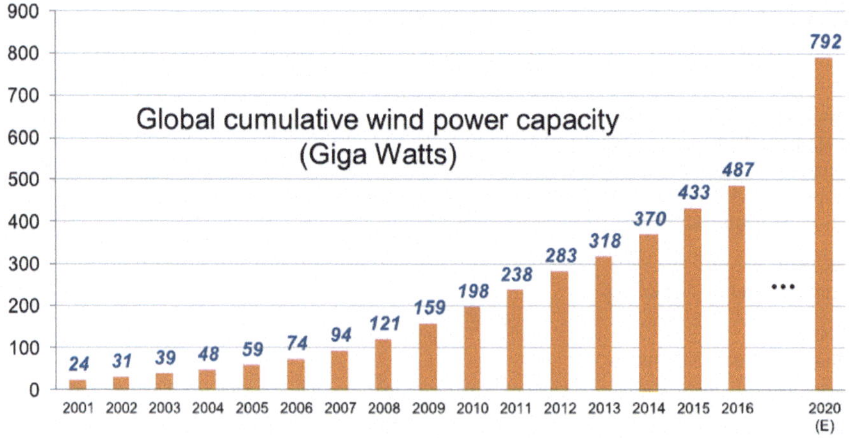

Fig. 1 Total capacity installed from 2011 till 2020 [4]

regulating system frequency and maintain it into desirable value with controlling also in interchanging power through tie-lines between different control areas. The re-dispatching of generation units is the main task of tertiary control due to continuing of oscillations. If the three mentioned controller types fail to overcome oscillations in frequency, emergency control type disconnect generation units or loads to secure the system from damage [6].

This chapter begins with discussing the construction of DFIG then discussing the dynamic modeling of the studied systems to solve LFC problem. One of them is two-area interconnected power system which contains two identical non-reheat thermal plants without DFIG participation. Whereas the another one contains two identical non-reheat thermal plants with dynamic participation of DFIG at both areas. Also, this chapter discuss the control model of DFIG when it is paralleling with conventional units to enhance and stabilize the system frequency and regulated it to normal value. Harris Hawks Optimizer (HHO), Salp Swarm Algorithm (SSA), and Sine Cosine Algorithm (SCA) are the proposed techniques which applied in this chapter to obtain Integral (I) and proportional-integral-derivative (PID) controllers. This chapter studies the frequency stabilizing of two-area interconnected power system with doubly fed induction generator based wind turbine.

2 Components of Doubly Fed Induction Generator (DFIG)

DFIG is basically a conventional wound-rotor induction machine in which the stator is directly connected to the grid through a transformer, and the connection of the rotor to the stator (and grid) is via a back-to-back voltage source convertor. The rotor converter system consists of a grid side converter (GSC) and rotor side converter (RSC) connected via a DC link [7–9]. A simplified schematic diagram of a DFIG based wind energy generation system is shown in Fig. 2.

The generator is called DFIG because the power is fed from both stator and the rotor circuits to the grid. The rotor circuit handles typically about 25–30% of the generator rated power, this percentage allows the DFIG to have about 30% operational speed range around the synchronous speed and reduces the rating and the cost of the rotor converter [10].

3 Dynamic Model of the Studied System to Solve LFC Problem

3.1 The Main Control Loops in Generator

Each generator in an interconnected power system has two basic equipment which known as Load Frequency Control (LFC) and Automatic Voltage Regulation (AVR)

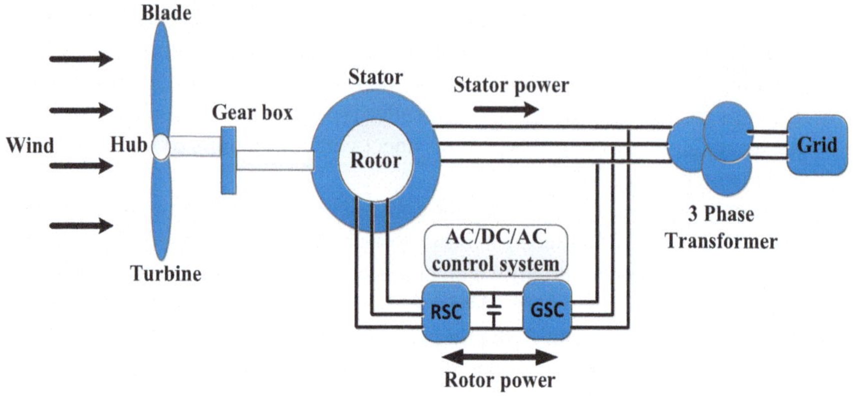

Fig. 2 Configuration of DFIG wind Energy conversion system (one single generator) using back-to-back converter

that must be installed. When there is a mismatch between generated active power and demand active power, it is necessary to adjust LFC equipment to maintain the frequency within a specified limit. AVR equipment must be installed at the moment of mismatching between generated reactive power and demand reactive power to maintain the voltage within a nominal value [11]. The frequency and voltage control can be named by the first control mechanism or AGC. For the stable operation in the interconnected power system, LFC is important to be adjusted to keep the system frequency and tie power exchange as close as possible the nominal values. In each area of any interconnected power system, the main target of LFC is to monitor the system frequency and the tie-line flows. After monitoring, LFC computes the net change in the generation required (which referred to as Area Control Error-ACE) and the set position of generators within the area is changed to achieve the time average of ACE at a low value (near to zero) [12]. ACE is defined as the linear combination of tie-line power exchanging and system frequency deviation and it represents as the controlled output of LFC. The main role of LFC is to drive ACE to zero which means that, both system frequency and tie-line power errors will be forced to zeros [13]. This chapter studies the enhancement of system frequency to achieve system stability and the schematic diagram of LFC loop is shown in Fig. 3.

3.2 Load Frequency Control (LFC)

The LFC strategy aims to balance between generated and demand active power, which is addressed in this chapter. Whenever the demand active power changes, a frequency deviation occurs. From Fig. 3, fg, f_{ref}, and fe represent; system frequency, reference signal of system frequency, and error signal of system frequency respectively. The relationship between fg, f_{ref} and fe can be calculated as follows:

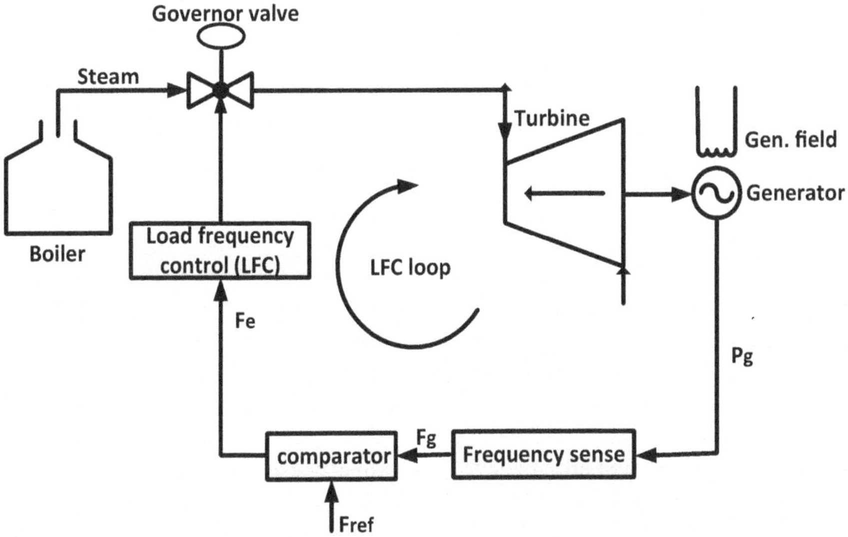

Fig. 3 Control methodology of LFC [14]

$$fe = fg - f_{ref} \tag{1}$$

If $fg = f_{ref}$, it means that there is no error signal and the governor valve is constant without any action. if $fg > f_{ref}$, it means that $fe > 0$ and the valve begin to close until $fg = f_{ref}$. If $fg < f_{ref}$, it means that $fe < 0$ and the valve begin to open until $fg = f_{ref}$.

3.3 The Mathematical Models of the Studied System

This chapter subjected to studying two models. The first model is two-area interconnected thermal-thermal power system without any penetration of DFIG at both areas. The second model is two-area interconnected power system with DFIG participation at both areas. The penetration percentage of DFIG at both areas is 30% which aid the system by supplying it additional inertia at step load perturbation conditions. The capacity of each thermal unit which studied in this chapter is equal to 1000 MW. Each of the individual area consists of turbine with an electrical governor system and generator. In this work, the GRC is taken into consideration for the two thermal units at both areas about 15% p.u.MW/min, as it has considerable effects on amplitude and settling time of oscillations. The governor, steam turbine, and generator transfer functions can be expressed respectively as follows [15]:

$$G(s)gov. = \frac{1}{Tgs + 1} \quad (2)$$

$$G(s)tur. = \frac{1}{Tts + 1} \quad (3)$$

$$G(s)gen. = \frac{Kp}{Tps + 1} \quad (4)$$

where; Tg is the governor time constant, Tt represents the turbine time constant,s is Laplace's operator, R represents the regulation droop constant and Tp, Kp is the time constant and gain of the generator model. To make sure that all deviations are vanished, the ACE must be reach to zero at steady state operation condition. All previous studies enhance the system performance by using an appropriate controller with a suitable optimization methodology.

Figure 4 shows the first investigated model which doesn't contain and DFIG penetration at both areas. DFIG-based wind turbine system contains a proposed controller for frequency support shown in Fig. 5. Also, the main target of this controller is to measure oscillations in frequency wave form and translate these oscillations into extra active power for enhancing system frequency.

To reduce the complexity of the two studied systems, the ring topology of control areas of the studied models are shown in Fig. 6. Figure 6a shows the control signals in the two-area thermal-thermal power system without any DFIG participation. Also, the control signals in the two-area thermal-thermal power system with DFIG participation at both areas are shown in Fig. 6b.

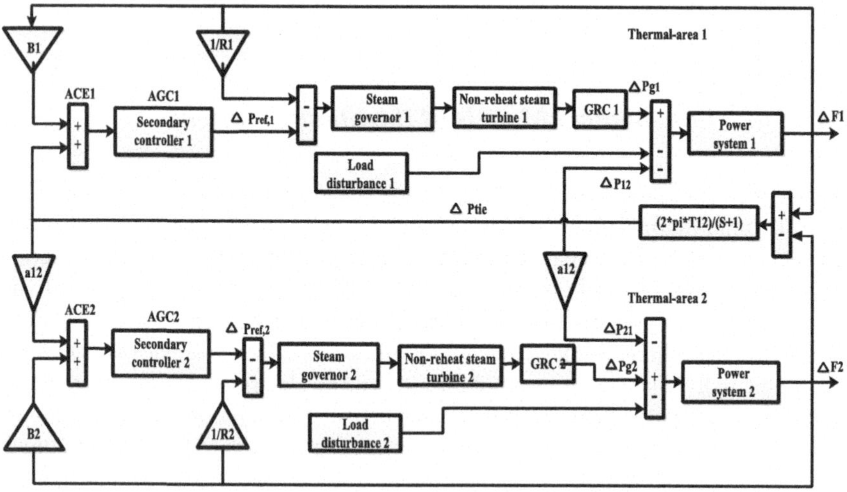

Fig. 4 Two-area interconnected power system without DFIG participation

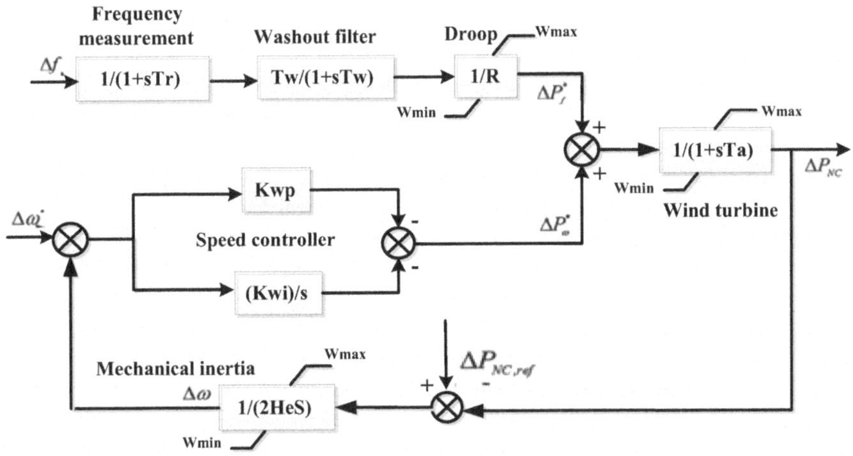

Fig. 5 Proposed controller for frequency support in DFIG based wind turbine

4 The Control Model of DFIG Based Wind Turbine

All over the world, power system planners design and implement power systems with an essential matter which is including of traditional generation units in this implementation. Traditional power plants e.g., thermal, gas, and hydro power plants, which represent the main source of frequency providing into power systems. Conventional power plants have the ability to enhance the power system performance under normal and abnormal operation conditions according to its inertia.

Also, the non-conventional power plants e.g., DFIG-based wind turbine, is able to enhance the power system performance by adding additional inertia through its speed controller [16, 17]. Figure 7 illustrates the control scheme between a conventional unit and non-conventional DFIG-based wind turbine. The relationship between the conventional and non-conventional units are shown in the following formulas [18]:

$$\frac{2HP_r}{f}\frac{d\Delta f}{\Delta t} = \Delta Pf - D\Delta f \tag{5}$$

$$\frac{2Hp_r}{f}\frac{d\Delta f}{dt} = \Delta P_g + \Delta P_{NC} - \Delta P_D - \Delta P_{tie} - D\Delta f \tag{6}$$

where, ΔP_D is represents the incremental demanded active power which it subtracted from the incremental active power generation of the conventional unit ΔP_g and non-conventional power plant ΔP_{NC}.Δf is the incremental change in frequency wave form. H and D is the inertia constant of the system and the damping constant factor, respectively. The concept of damping constant is illustrated as; when D equals to 1 value that means a 1% change in frequency would cause a 1% changing in the loads. The transferred power among neighbouring areas is ΔP_{tie} and P_r represents

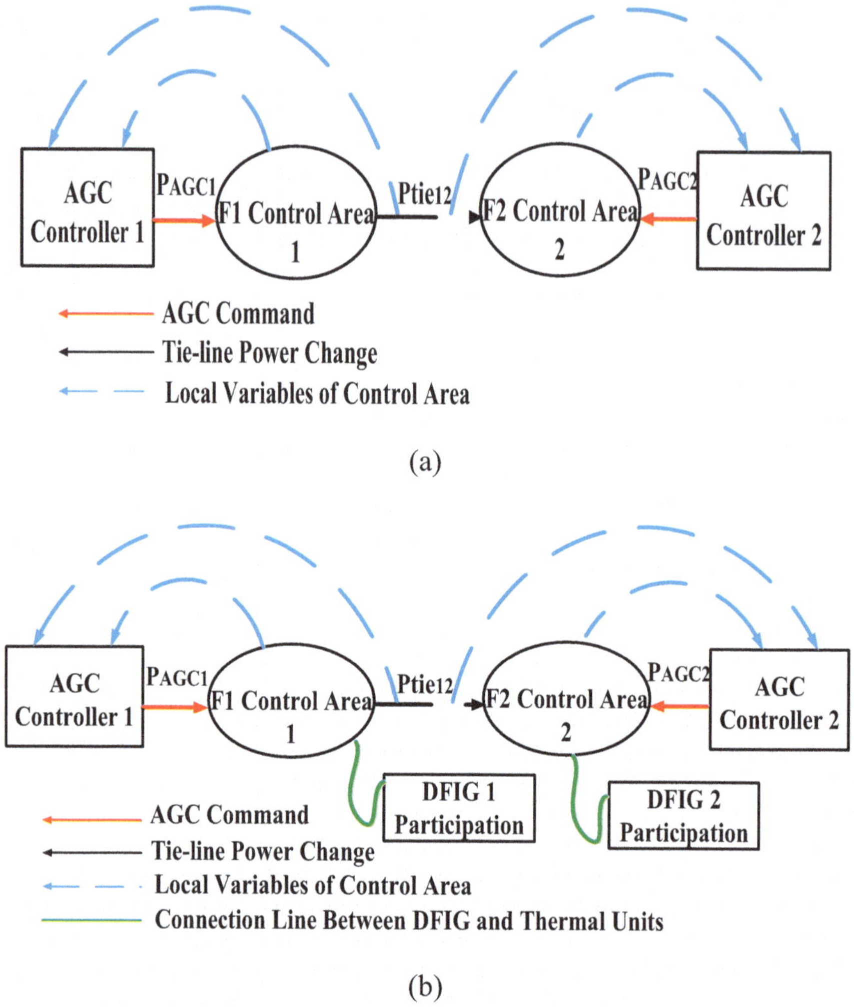

Fig. 6 Controllers between interconnected power system in a ring topology **a** without DFIG participation **b** with DFIG participation at both areas

the rated conventional generator capacity power. Dividing Eq. (6) by P_r and Eq. (7) is obtained as shown:

$$\frac{2H}{f}\frac{d\Delta f}{dt} = \frac{\Delta P_g}{P_r} + \frac{\Delta P_{NC}}{P_r} - \frac{\Delta P_D}{P_r} - \frac{\Delta P_{tie}}{P_r} - \frac{D\Delta f}{P_r} \tag{7}$$

Equation (7) is rewritten as in Eq. (8) by expressing it in per unit system.

$$\frac{2H}{f}\frac{d\Delta f}{dt} + D(p.u.)\Delta f = \Delta P_g(p.u.) + \Delta P_{NC}(p.u.)$$

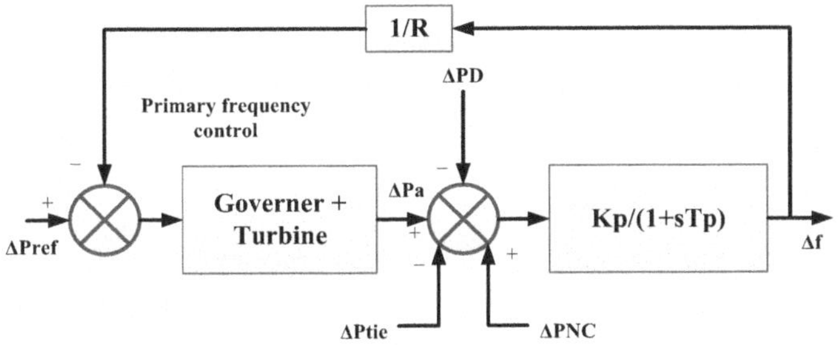

Fig. 7 Mixed generation units of power system model

$$-\Delta P_D(p.u.) - \Delta P_{tie}(p.u.) \tag{8}$$

The non-conventional power can be calculated as mentioned in (9) [19]:

$$\Delta P_{NC} = -Kdf\frac{d\Delta f}{dt} - Kpf\Delta f \tag{9}$$

where; Kpf and Kdf are the proportionality constants of the frequency deviation and its derivative respectively. The substitution from (9) into (8) then converting to Laplace's transform, Eq. (10) is obtained.

$$\Delta P_g(s) - \Delta P_D(s) - \Delta P_{tie}(s) = \hat{H}\, s\Delta f(s) + \check{D}\,\Delta f(s) \tag{10}$$

where;

$$\hat{H} = \frac{2H}{f} + Kdf, \check{D} = D + Kpf$$

The frequency response can be obtained by re-arranging Eq. (10) and formulate it in Eq. (11):

$$\Delta f(s) = [\Delta p_g(s) - \Delta p_D(s) - \Delta p_{tie}(s)]\frac{k_p}{1 + sT_p} \tag{11}$$

where;

$$T_p = \frac{1}{\check{D}}, k_p = \frac{\hat{H}}{\check{D}}$$

Equation (11) illustrates the impact of DFIG in changing the damping ratio of the conventional power system and also in changing the power system area inertia constant. DFIG mechanical equation can be expressed as shown in Eq. (12). Figure 8

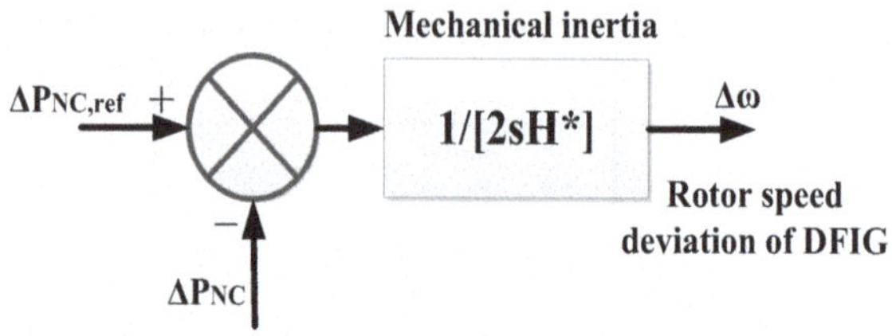

Fig. 8 DFIG mechanical inertia block diagram

shows the mechanical inertia control block diagram of DFIG based wind turbine. The output of mechanical inertia block diagram represents an indication of rotor speed variations.

$$2H^{*}\frac{d\Delta\omega}{dt} = \Delta p_{NC,ref} - \Delta p_{NC} \tag{12}$$

Figures 7 and 8 are modified such as shown in Fig. 9 by adding the two controller units in the power system.

The first controller is the supplementary frequency control unit G_1 which supplying active power from traditional unit into power system. The second controller is the speed controller unit G_2 which allow fast speed recovery of DFIG with shorter period of transient speed variation until, the machine not enter in the stalling phase. Designing of these two controller parameters helps in enhancement of power system performance. The hybrid system of synchronous generator with DFIG based wind turbine is illustrated in Fig. 9. The transfer function of the wind turbine is expressed as given:

$$\frac{1}{T_a s + 1} \tag{13}$$

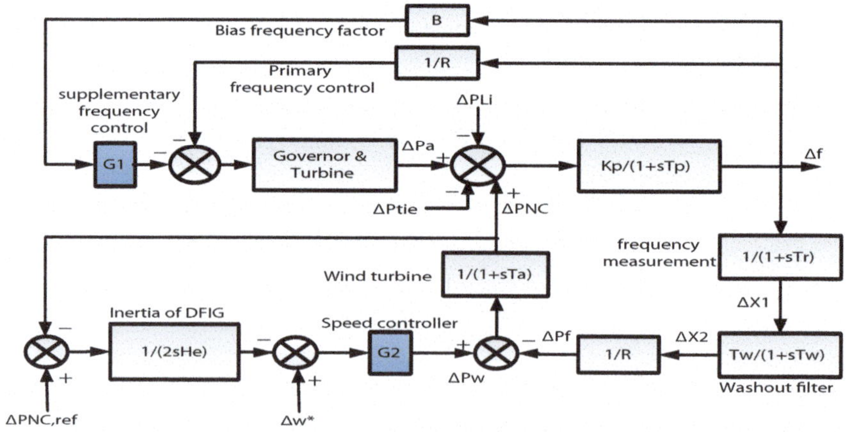

Fig. 9 Block diagram of frequency control via connection between traditional unit and DFIG

where; T_a represents the wind turbine time constant. The DFIG speed controller consists of a proportional plus integral controller parameter. According to different participation level of DFIG, the speed control parameters will be changed. The speed deviations should be regulated with the speed controller design $T_{wt}(s)$ as given:

$$T_{wt}(s) = k_{wp} + \frac{k_{wi}}{s} \tag{14}$$

where; $T_{wt}(s)$ represents the speed controller that control in the magnitude of extra active power from DFIG. System frequency has been sensed and measured through a filter which have the time constant (T_r). There is a washout filter after the first filter which have the time constant (T_w). To overcome the frequency oscillation and achieving steady state case, the output power setting reference (Δp_f) must be developed. The power setting point acts an important factor which in a similar manner of the primary frequency control related with the conventional power plants using the droop constant index (R) as:

$$\Delta p_f = \frac{1}{R}\Delta x \tag{15}$$

where; Δx is the sensing signal of load frequency in the studied power system.

5 Optimization Techniques Applied with Studied Models and Problem Statement Issue

5.1 *Problem Statement Issue*

Integral (I) and PID controllers are implemented in the investigated studied model to obtain more system stability by solving LFC problem. The transfer functions of I and PID controllers can be obtained as follows [20]:

$$Gi, I(s) = \frac{k_{i,n}}{s} \tag{16}$$

$$Gi, PID(s) = kp, n + \frac{ki,n}{s} + kd, ns \tag{17}$$

where; kp, n, ki, n, kd, n are the implemented proportional, integral, derivative controller gains of nth area, respectively. The control signal of ith area can be expressed as mentioned as follows [21, 22]:

$$Ui(s) = Gi(s)ACEi(s) \tag{18}$$

where;

$$ACE_1 = B_1 \Delta f_1 + \Delta p_{tie} \tag{19}$$

$$ACE_2 = B_2 \Delta f_2 + a_{12} \Delta p_{tie} \tag{20}$$

$$a_{12} = \left(-\frac{p_{r1}}{p_{r2}} \right) \tag{21}$$

where; ACE represents the area control error index, B is the frequency bias constants and a_{12} represents the product of rated power at the first area p_{r1} division by rated power at the second area p_{r2}. The objective function can be expressed in time domain manner as shown in Eq. (22). The Integral Square Error criterion (ISE) is used for minimizing the objective function with achieving the studied system stability, which can be expressed as follows [23]:

$$J = ISE = \int_0^{T_{sim}} (\Delta f_1^2 + \Delta f_2^2 + \Delta p_{tie}^2).dt \tag{22}$$

where; J represents the objective function expression and dt is a given time interval for taking samples along the simulation process. The main target of implementing the controller is to minimize J for achieving the system stability without any oscillation. The constraint of kp, n,ki, n and kd, n gains along the run simulation time is expressed as follows:

$$kp, n^{\max} \geq kp, n \geq kp, n^{\min} \tag{23}$$

$$ki, n^{\max} \geq ki, n \geq ki, n^{\min} \tag{24}$$

$$kd, n^{\max} \geq kd, n \geq kd, n^{\min} \tag{25}$$

where; $kp, n^{\min}$, $ki, n^{\min}$, $kd, n^{\min}$, $kp, n^{\max}$, $ki, n^{\max}$ and $kd, n^{\max}$ are the minimum and the maximum values of the proportional, integral and derivative controller parameters respectively and also, the constrain of parameters be in [−5, 5] period. The constrains of the PI speed controller of DFIG (kwp_1, kwi_1) are given below:

$$kwp_1^{\max} \geq kwp_1 \geq kwp_1^{\min} \tag{26}$$

$$kwi_1^{\max} \geq kwi_1 \geq kwi_1^{\min} \tag{27}$$

where; $kwp_1^{\min}$, $kwi_1^{\min}$, $kwp_1^{\max}$ and $kwi_1^{\max}$ are the minimum and the maximum values of the PI speed controller respectively and also, the constraints of parameters be in [0, 5] period.

5.2 Harris Hawks Optimizer (HHO)

The HHO algorithm is a new developed method that was invented in 2019 by Seyedali Mirjalili and inspired by nature [24]. It represents one of the recent metaheuristic intelligence methods. The methodology of this intelligent method is summarized in, several hawks attack the prey from different locations in an attempt to surprise it. According to escape scenarios of the prey, hawks can reveal a variety of stalking patterns. HHO algorithm can be discussed briefly with following steps:

(i) Exploration phase

This phase determines the mathematically wait, search and discover the desired prey. The updated location of hawks can be calculated as follows:

$$x(t+1) = \begin{cases} x_{rand}(t) - r_1|x_{rand}(t) - 2r_2x(t)|\, q \geq 0.5 \\ (x_{rabbit}(t) - x_m(t)) - r_3(LB + r_4(UB - LB))\, q < 0.5 \end{cases} \tag{28}$$

where; $x\ (t+1)$ represents the position vector of hawks in the next iteration t, $x_{rabbit}(t)$ is the rabbit position, x(t) is the current position vector of hawks, r_1, r_2, r_3, r_4 and q are random numbers inside (0, 1), which are updated for each iteration, $x_{rand}(t)$ represents the random selected hawks from current population and x_m refers to the average hawks position of the current population. The average hawks position can be expressed as follows:

$$x_m(t) = \frac{1}{N}\sum_{i=1}^{N} x_i(t) \tag{29}$$

where; $x_i(t)$ indicates the location of each hawk in each iteration t and N refers to the total number of hawks.

(ii) Transition from Exploration to Exploitation

Let consider that T and E_0 (−1, 1) be the maximum size about the repetitions and the initial energy during each step respectively. The escaping energy of rabbit E can be detected through HHO as follows:

$$E = 2E_0\left(1 - \frac{iter}{T}\right) \tag{30}$$

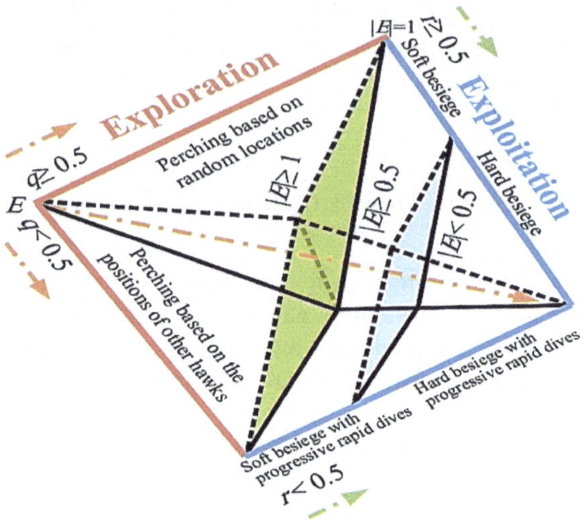

Fig. 10 Various phases of Harris Hawks optimization [24]

According to this value, the exploration and exploitation processes can be changed. If $|E| \geq 1$, the exploration phase gets started; otherwise, the rest solutions are aimed to be exploited.

(iii) Exploitation phase

This phase represents attacking on the prey. For $|E| < 1$ and $r = rand()$, the hawks move to pounce on the prey. A so-called parameter "r" is defined to measure the escaping chance of the prey. Accordingly, $r < 0.5$ represents a successful escape. In addition, when $|E| \geq 0.5$, HHO takes soft surround and when $|E| < 0.5$, hard surround is applied. It is worth noting that even if the prey is able to escape (i.e., $|E| \geq 0.5$), its success also depends on r. The attack procedure is influenced by the escaping and pursuing strategy of the prey and hawks, respectively. In this sense, four major steps are considered which are broadly explained in [25, 26] (Fig. 10).

5.3 Salp Swarm Algorithm (SSA)

This algorithm was invented by Seyedali Mirjalili in 2017 [27]. SSA emulate the behaviour of salps in swimming process in oceans. Also, it is mimicking their social interaction such as motion and foraging. Salps are quite similar to jelly fishes in their tissues and in their movements with an arranged chain. Salps is driven forward by water which is considered as a force for searching and eating food. The population of salp chains is divided into two groups; leader which be in the front and followers which follow the movement of leader. The updated location of leaders can be calculated as follows:

$$s_j^1 = \begin{cases} F_j + c_1((ub_j - lb_j))c_2 + lb_j)\, c_3 \geq 0 \\ F_j - c_1((ub_j - lb_j))c_2 + lb_j)\, c_3 < 0 \end{cases} \tag{31}$$

where; s_j^1 represents the leader position in jth dimension, F_j is the food source position in jth dimension ub_j, and lb_j indicate to the upper and lower bound of jth dimension, c_1, c_2 and c_3 are random numbers.c_1 it represents the most important parameter in SSA technique as it balances between exploration and exploitation process that formulated as follows:

$$c_1 = 2e^{-\left(\frac{4l}{L}\right)^2} \tag{32}$$

where; l is the current iteration and L represents the maximum number of iterations.

The updated location of followers can be calculated as follows:

$$s_j^i = \frac{1}{2} a\, t^2 + v_o t \tag{33}$$

where; $i \geq 2$, s_j^i represents the position of ith followers in jth dimension. t is the time,v_o is the initial speed, and $a = \frac{v_{final}}{v_Q}$ where $v = \frac{x - x_Q}{t}$. The discrepancy between iterations is equal to 1 as the time in optimization is iteration, and considering $v_o = 0$, the equation of updating position of follower slaps can be written as follows:

$$s_j^i = \frac{1}{2}\left(s_j^i - s_j^{i-1}\right) \tag{34}$$

where; $i \geq 2$, s_j^i represents the position of ith followers in jth dimension (Fig. 11).

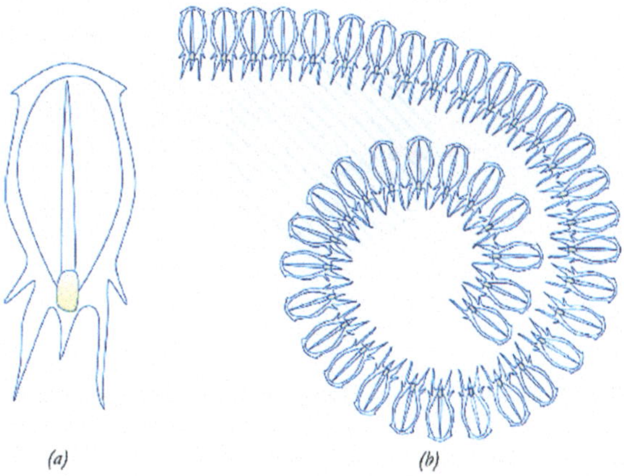

Fig. 11 **a** Individual salp, **b** swarm of salps (salps chain) [27]

5.4 Sine Cosine Algorithm (SCA)

The sine cosine algorithm (SCA) behavior represents as achieving the optimal solutions that based on the sine and cosine functions [28]. Multiple initial random solutions are created through SCA to fluctuate towards or outwards the best solutions. Steps of SCA to find the optimum solution can be discussed as follows:

The following position updating equations are proposed for sine and cosine phases:

$$x_i^{t+1} = x_i^t + r_1 \times \sin(r_2) \times \left|r_3 p_i^t - x_i^t\right| \tag{35}$$

$$x_i^{t+1} = x_i^t + r_1 \times \cos(r_2) \times \left|r_3 p_i^t - x_{ii}^t\right| \tag{36}$$

where; x_i^t is the position of the current position in ith dimension at tth iteration, $r_1/r_2/r_3$ are random numbers, p_i is the position of the destination point in ith dimension, and $\|$ represents the absolute value. These two equations are combined as follows:

$$x_i^{t+1} = \begin{cases} x_i^t + r_1 \times \cos(r_2) \times \left|r_3 p_i^t - x_{ii}^t\right| r_4 \geq 0.5 \\ x_i^t + r_1 \times \sin(r_2) \times \left|r_3 p_i^t - x_i^t\right| r_4 < 0.5 \end{cases} \tag{37}$$

where; r_4 is a random number in [0, 1]. The four main parameters in SCA are r_1, r_2, r_3 and r_4. r_1 indicates the next position regions, r_2 defines how far the movement should be towards or outwards the destination. r_3 gives random weights for destination. r_4 switches between the sine and cosine components as mentioned in Eq. (37) (Fig. 12).

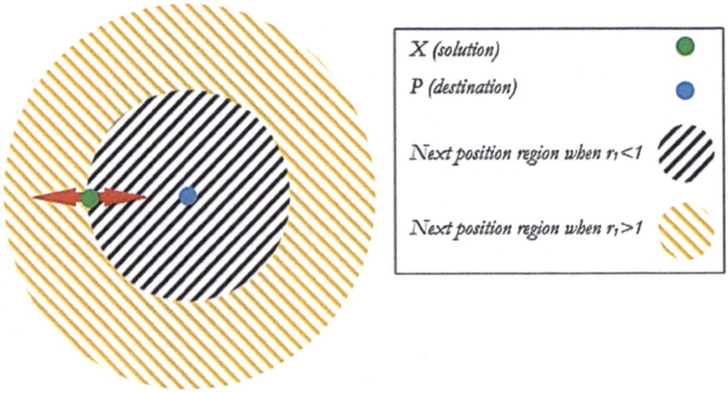

Fig. 12 Effects of Sine and Cosine in Eq. (37) on the next position [28]

6 Performance of System Frequency Interconnected DFIG

In this section, frequency response enhancement of two-area interconnected power system with/without DFIG participation has been presented by using different controller schemes that tuned by different optimization techniques. These studies have been made to validate and verify the efficiency of the proposed techniques and ensure that, when DFIG is participate with the interconnected power system, the optimum performance of power system is achieved. As the higher degree of stability and the quick response are desirable for any control process in the power system. Also, at any interconnected power system, the performance indices and the degree of relative stability must be focused on them. According to the above, investigation study of power system dynamics has been done and the acceptability of the designed controller is established by computing peak overshoot, peak undershoot and settling time of the transient responses. The small step load perturbation (SLP) term is employed to examine the dynamic system stability of different power systems.

In this manuscript, two investigated interconnected power system are presented. The first studied model summarized as a two-area thermal-thermal power system considering GRCs with a neighbouring tie-line between them with 1% SLP and without any DFIG penetration at both areas. The second investigated model is quite like the first studied model, but the only difference is that DFIG participation at both areas. The I and PID controllers are installed and adjusting to solve LFC problem at both studied models. The I and PID controller parameters are optimized by Harris Hawks Optimizer (HHO), Salp Swarm Algorithm (SSA) and Sine Cosine Algorithm (SCA) and selecting the best technique with the best controller to achieve more system stability. MATLAB platform is the program which used to do simulations.

6.1 Impact of Integral (I) Controller for Enhancing the Studied System Frequency

This scenario presents two studied model via applying HHO, SSA and SCA algorithms to obtain the I controller parameters for achieving the system stability. Comparisons have been made between these techniques to obtain the optimal controller parameters for studied models. Whereas, the first model does not contain any participation of DFIG at both areas, and it is the main different in the second model which have DFIG penetration at both areas. Simulink results ensure that, the performance of system frequency has been enhanced with DFIG participation in frequency control and adjust. However, without DFIG participating in power system, the overshoots (OS) and the undershoots (US) have high values and the settling time become large. The prime movers and governors in power system have detected values for normal operation condition while, for enhancing in the frequency response that subjected to distortion after load perturbation, prime movers and governors are added to the system with a new value.

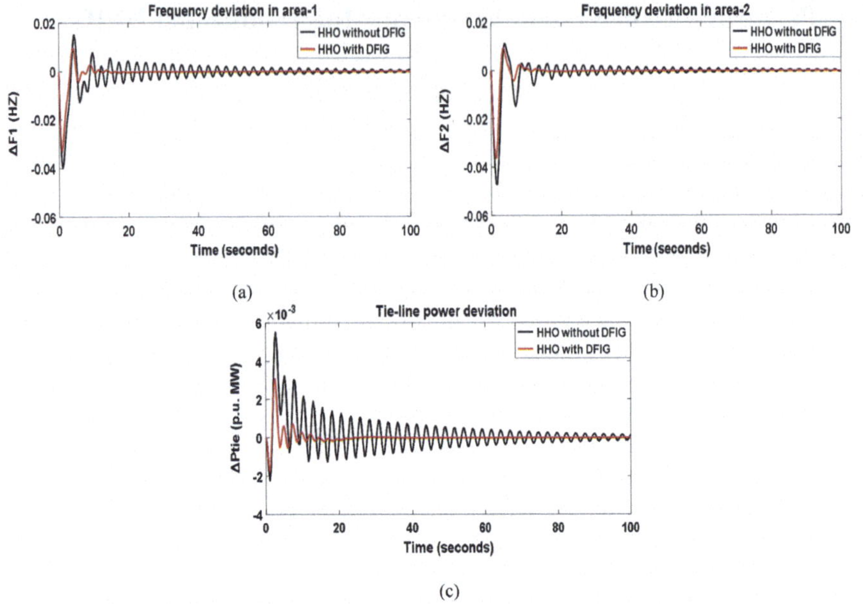

Fig. 13 Dynamic performance of two-area interconnected power system considering GRCs using the optimal I controller-based HHO, **a** frequency deviation in area 1, **b** frequency deviation in area 2, tie-line power deviation

6.2 Impact of HHO Using I Controller to Enhance System Frequency with/Without DFIG

Figure 13 shows frequency responses of two-area deregulated power system considering GRCs and the exchanging power between both areas using HHO.

6.3 Impact of SSA Using I Controller to Enhance System Frequency with/Without DFIG

Figure 14 shows frequency responses of two-area deregulated power system considering GRCs and the exchanging power between both areas using SSA.

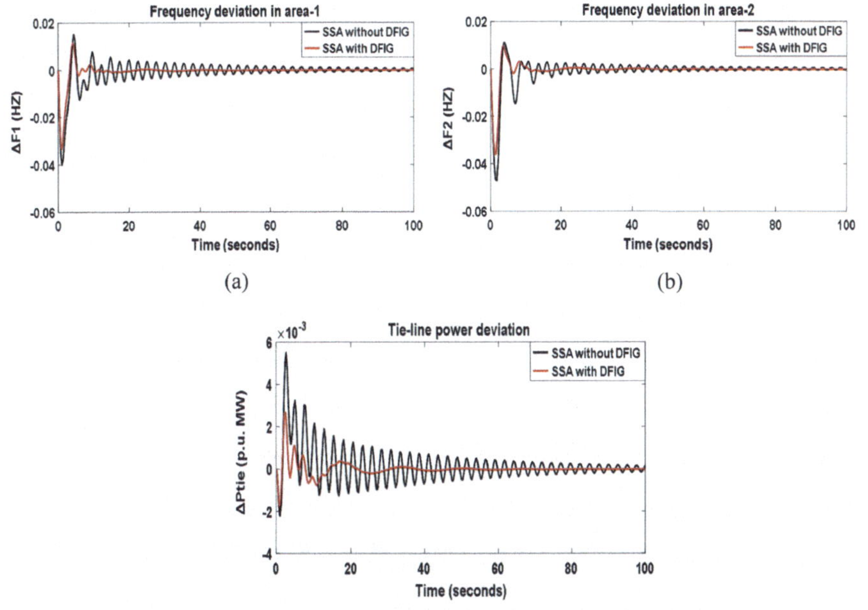

(a) (b) (c)

Fig. 14 Dynamic performance of two-area interconnected power system considering GRCs using the optimal I controller- based SSA, **a** frequency deviation in area 1 **b** frequency deviation in area 2, tie-line power deviation

6.4 Impact of SCA Using I Controller to Enhance System Frequency with/Without DFIG

Figure 15 shows frequency responses of two-area deregulated power system considering GRCs and the exchanging power between both areas using SCA.

It is clear that, there is a significant effect by adding DFIG at both areas via applying all mentioned optimization techniques. There is a noticeable difference in the frequency deviation of the first area in the case of with/without DFIG by applying HHO, SSA and SCA. This difference represents in reducing both oscillations and settling time and achieving more system stability after load perturbations. Table 1 presents the optimal parameters of I controller based mentioned optimization techniques of this studied system model without the DFIG participation, as well as the performance specifications; OS and US, of the studied system are shown in Table 2. Similarly, Table 3 gives the optimal parameters of I controller based mentioned optimization techniques of the studied system with the DFIG participation, and the performance specifications; OS and US, of this studied system are shown in Table 4.

It is remarkable that in the case of the studied system considering GRCs without DFIG participation, different optimization techniques gained the same objective function which equals 0.033268 and the fastest optimization technique reaches

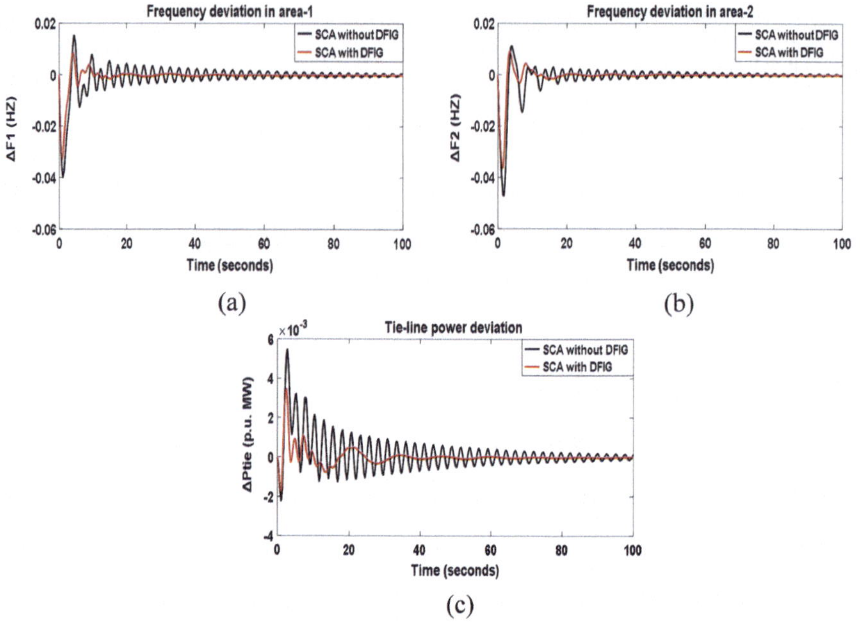

Fig. 15 Dynamic performance of two-area interconnected power system considering GRCs using the optimal I controller- based SCA **a** frequency deviation in area 1, **b** frequency deviation in area 2, tie-line power deviation

Table 1 Obtaining I controller parameters of the studied system without DFIG participation via different optimization techniques

Controller parameters	HHO	SSA	SCA
$Ki1$	0.4183	0.4174	0.4145
$Ki2$	0.174	0.1744	0.1761

Table 2 Optimized OS and US of area frequencies, tie power exchange and performance indices for the studied system without DFIG participation via different optimization techniques via I controller

Different dynamic responses		HHO	SSA	SCA
Dynamic response of (F1)* 10^{-3}	OS	15.286	15.291	15.329
	US	−40.171	−40.174	−40.185
Dynamic response of (F2)* 10^{-3}	OS	11.05	11.042	11.044
	US	−47.217	−47.216	−47.209
Dynamic response of (Ptie)* 10^{-3}	OS	5.54	5.527	5.485
	US	−2.253	−2.253	−2.254
Objective function minimizing* 10^{-3}	J	33.268	33.268	33.268

Table 3 Obtaining I controller parameters of the studied system with DFIG participation via different optimization techniques

Controller parameters	HHO	SSA	SCA
$Ki1$	0.5493	0.5863	0.6087
$Ki2$	0.4433	0.498	0.4305
$Kwp1$	0.5429	2.179	0.2728
$Kwi1$	0.571	1.6469	1.432
$Kwp2$	0.2	0.1586	0.1
$Kwi2$	0.1963	0.8752	0.6083

Table 4 Optimized OS and US of area frequencies, tie power exchange and performance indices for the studied system with DFIG participation via different optimization techniques via I controller

Different dynamic responses		HHO	SSA	SCA
Dynamic response of (F1)* 10^{-3}	OS	9.929	11.867	8.974
	US	−33.291	−33.396	−33.172
Dynamic response of (F2)* 10^{-3}	OS	9.478	9.666	9.161
	US	−36.049	−35.877	−36.049
Dynamic response of (Ptie)* 10^{-3}	OS	3.124	2.708	3.504
	US	−1.762	−1.772	−1.752
Objective function minimizing* 10^{-3}	J	3.15	3.2	3.10

firstly is HHO, then SCA, and finally SSA as shown in Fig. 16 which represents the convergence curves of techniques.

Figure 17 shows the convergence curves of those optimization techniques with the studied system considering GRCs with DFIG participation at both areas. It can be

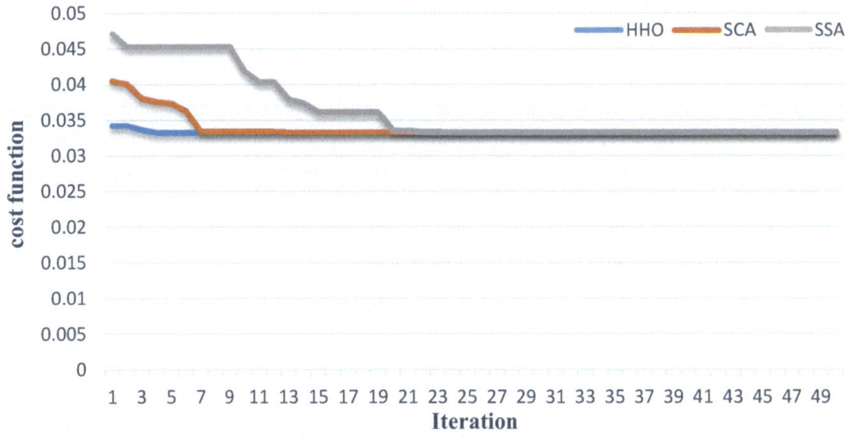

Fig. 16 Convergence curves of HHO, SCA and SSA in the case of the studied system considering GRCs without DFIG participation via I controller

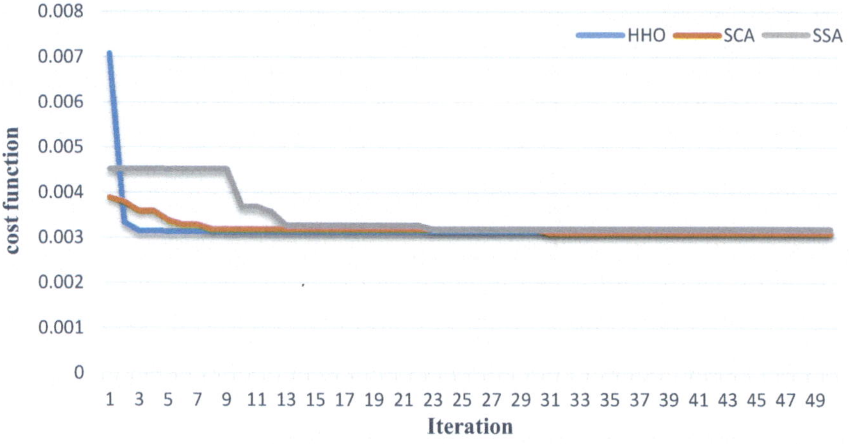

Fig. 17 Convergence curves of HHO, SCA and SSA in the case of the studied system considering GRCs with DFIG participation via I controller

noticeable that, all OS and US values have less magnitude compared with the state in the first model. The comparison between the convergence curves of the studied system with/without DFIG participation ensures that the performance indices be better with DFIG participation as noticeable in Table 4. In Fig. 17, HHO technique achieve the objective function faster than SCA and SSA techniques. Whereas, objective function value with SSA is 0.0032 but HHO and SCA objective functions are 0.00315 and 0.00310 respectively.

6.5 *Effect of Applying Proportional-Integral-Derivative (PID) Controller to Solve LFC Problem*

This scenario presents two studied model via applying HHO, SSA and SCA algorithms to obtain the PID controller parameters for achieving the system stability. Comparisons have been made between these techniques to obtain the optimal controller parameters for studied models. Whereas, the first model does not contain any participation of DFIG at both areas, and it is the main different in the second model which have DFIG penetration at both areas.

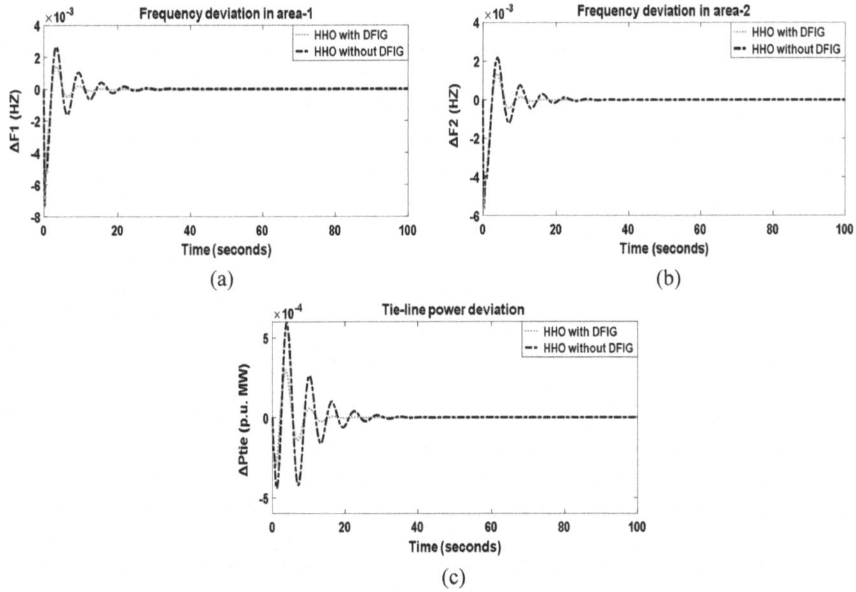

Fig. 18 Dynamic performance of two-area interconnected power system considering GRCs using the optimal PID controller-based HHO **a** frequency deviation in area 1, **b** frequency deviation in area 2, tie-line power deviation

6.6 Effect of Applying HHO Using PID Controller to Solve LFC Problem with/Without DFIG

Figure 18 shows frequency responses of two-area deregulated power system considering GRCs and the exchanging power between both areas using HHO.

6.7 Effect of Applying SSA Using PID Controller to Solve LFC Problem with/Without DFIG

Figure 19 shows frequency responses of two-area deregulated power system considering GRCs and the exchanging power between both areas using SSA.

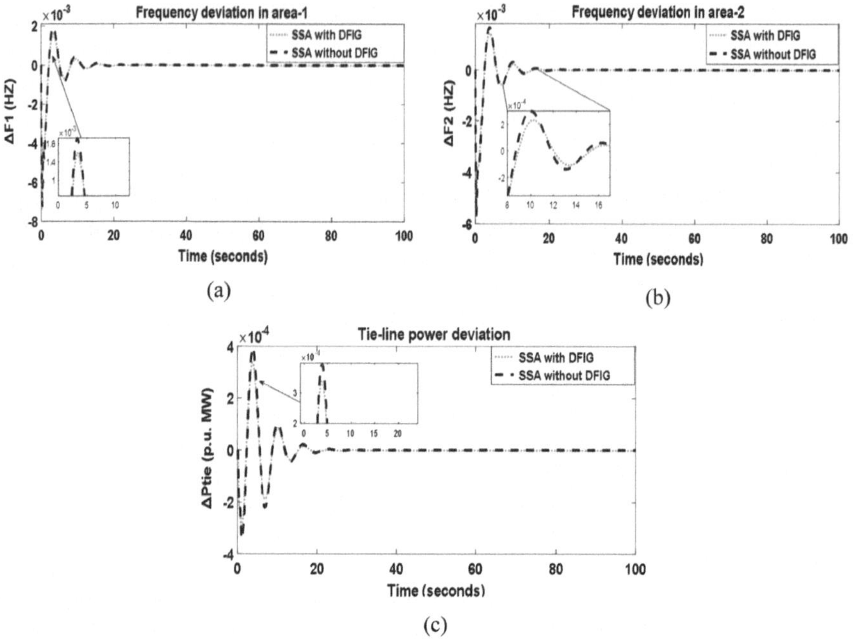

Fig. 19 Dynamic performance of two-area interconnected power system considering GRCs using the optimal PID controller-based SSA **a** frequency deviation in area 1, **b** frequency deviation in area 2, tie-line power deviation

6.8 *Effect of Applying SCA Using PID Controller to Solve LFC Problem with/Without DFIG*

Figure 20 shows frequency responses of two-area deregulated power system considering GRCs and the exchanging power between both areas using SCA.

It is clear that, there is a significant effect by adding DFIG at both areas via applying all mentioned optimization techniques. There is a noticeable difference in the frequency deviation of the first area in the case of with/without DFIG by applying HHO, SSA and SCA. This difference represents in reducing both oscillations and settling time and achieving more system stability after load perturbations. Table 5 presents the optimal parameters of PID controller based mentioned optimization techniques of this studied system model without the DFIG participation, as well as the performance specifications; OS and US, of the studied system are shown in Table 6. Similarly, Table 7 gives the optimal parameters of PID controller based mentioned optimization techniques of the studied system with the DFIG participation, and the performance specifications; OS and US, of this studied system are shown in Table 8.

In the studied case without DFIG participation, Fig. 21 shows the convergence curves of mentioned techniques to obtain the objective function value. The faster

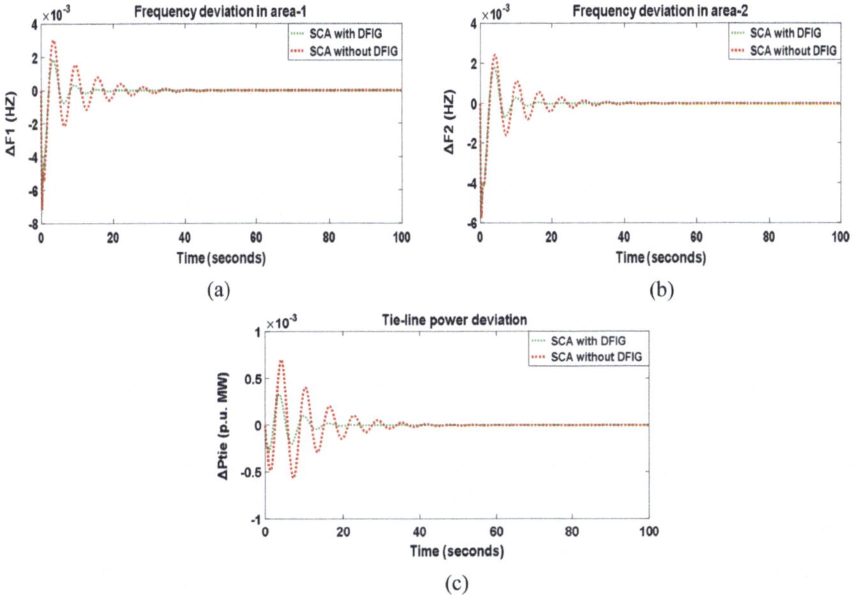

Fig. 20 Dynamic performance of two-area interconnected power system considering GRCs using the optimal PID controller-based SCA **a** frequency deviation in area 1, **b** frequency deviation in area 2, tie-line power deviation

Table 5 Obtaining PID controller parameters of the studied system without DFIG participation via different optimization techniques

Controller parameters	HHO	SSA	SCA
Kp1	0.5312	1.3525	0.135
Ki1	5	5	5
Kd1	4.1252	4.0511	4.2915
Kp2	5	4.9797	5
Ki2	5	5	5
Kd2	5	5	5

technique which obtained the objective function is SSA with 0.00011816 value, then HHO technique with 0.00011875 and finally SCA technique with 0.00012294 value.

Figure 22 shows the convergence curves of those optimization techniques with the studied system considering GRCs with DFIG participation at both areas. It can be noticeable that, all OS and US values have less magnitude compared with the state in the case of without DFIG participation. The comparison between the convergence curves of two studied system ensure that, the performance indices be better with DFIG participation as noticeable in Table 8. The best objective function obtained by SSA technique with 0.00002169 value, then HHO technique with 0.000021748 value and finally SCA with 0.000022686 value. These results ensure that, the best optimization technique which applied for obtaining more system stability is SSA technique. To

Table 6 Optimized OS and US of area frequencies, tie power exchange and performance indices for the studied system without DFIG participation via different optimization techniques via PID controller

Different dynamic responses		HHO	SSA	SCA
Dynamic response of (F1)* 10^{-3}	OS	2.649	1.91	3.036
	US	−7.329	−7.343	−7.177
Dynamic response of (F2)* 10^{-3}	OS	2.165	1.649	2.432
	US	−5.734	−5.732	−5.735
Dynamic response of (Ptie)* 10^{-3}	OS	0.594657	0.395146	0.703034
	US	−0.437774	−0.344077	−0.562747
Objective function minimizing* 10^{-4}	J	1.1875	1.1816	1.2294

Table 7 Obtaining PID controller parameters of the studied system with DFIG participation via different optimization techniques

Controller parameters	HHO	SSA	SCA
$Kp1$	1.7817	1.8154	1.3604
$Ki1$	5	5	5
$Kd1$	3.8809	3.9013	3.8021
$Kp2$	5	5	3.7725
$Ki2$	5	5	4.9925
$Kd2$	5	5	5
$Kwp1$	0.6505	4.9875	0.6015
$Kwi1$	0.8492	4.9796	0.3103
$Kwp2$	3.8236	3.6131	1.6409
$Kwi2$	1.4185	0.1715	0.1334

Table 8 Optimized OS and US of area frequencies, tie power exchange and performance indices for the studied system with DFIG participation via different optimization techniques via PID controller

Different dynamic responses		HHO	SSA	SCA
Dynamic response of (F1)* 10^{-3}	OS	1.44	1.602	1.849
	US	−7.42	−7.394	−7.555
Dynamic response of (F2)* 10^{-3}	OS	1.354	1.443	1.813
	US	−5.66	−5.66	−5.748
Dynamic response of (Ptie)* 10^{-3}	OS	0.299986	0.339562	0.33154
	US	−0.297823	−0.305615	−0.28545
Objective function minimizing* 10^{-4}	J	0.21748	0.2169	0.22686

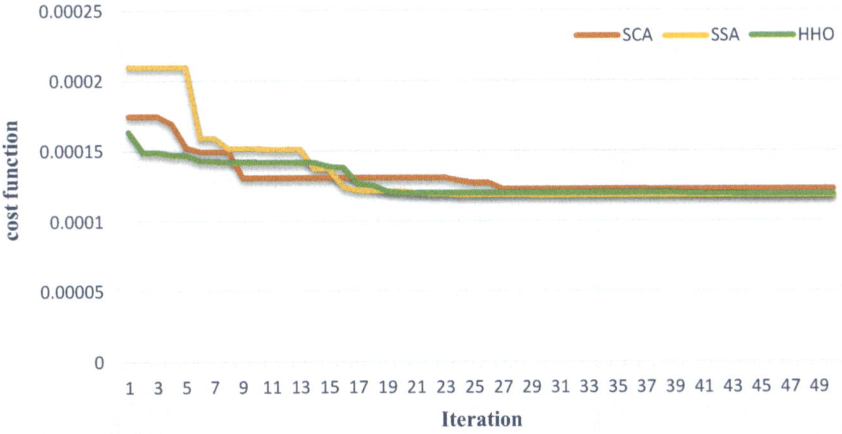

Fig. 21 Convergence curves of HHO, SCA and SSA with the studied system considering GRCs without DFIG participation via PID controller

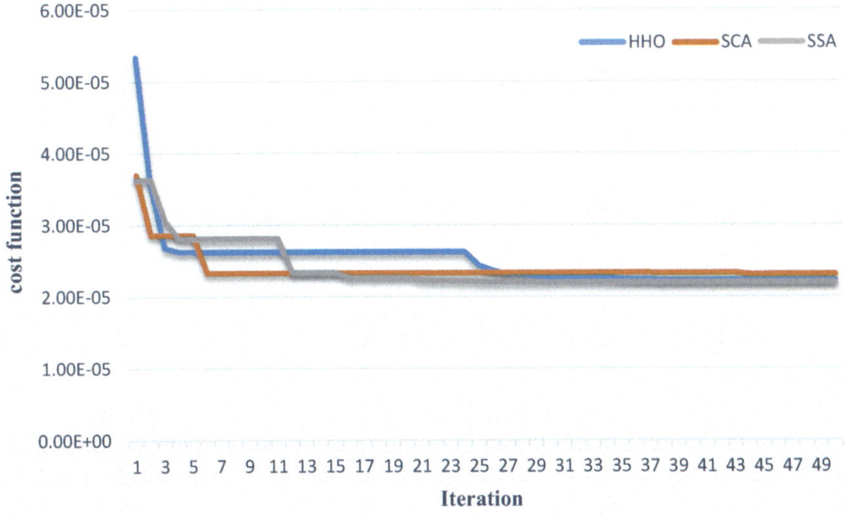

Fig. 22 Convergence curves of HHO, SCA and SSA with the studied system considering GRCs with DFIG participation via PID controller

make the comparison more obviously, Fig. 23 contain that; (a) cost function of the applied optimization techniques in the studied system without DFIG participation using I controller, (b) cost function of the applied optimization techniques in the studied system with DFIG participation using I controller, (c) cost function of the applied optimization techniques in the studied system without DFIG participation using PID controller, and (d) cost function of the applied optimization techniques in

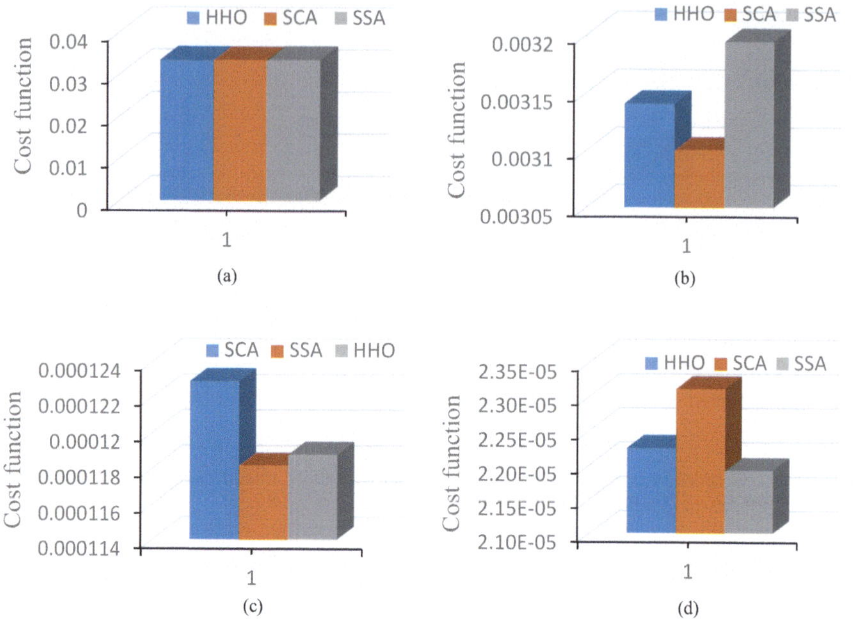

Fig. 23 Comparisons between optimization techniques performance for minimizing the objective function **a** first model with I controller, **b** second model with I controller, **c** first model with PID controller, **d** second model with PID controller

the studied system with DFIG participation using PID controller. Figure 23a shows that all optimization techniques achieve the same objective function and it refers into there is no method is better than the other but, Fig. 23b shows that SCA achieve the least objective function means that more system stability has been obtained by using the optimal I controller-based SCA. It can be noticeable that all cost functions in Fig. 23b gained less values compared to which gained in Fig. 23a because of the contribution of DFIG participation. Figure 23c shows that SSA achieve the least objective function means that more system stability has been obtained by using the optimal PID controller-based SSA. It is clear that, the obtained cost functions in Fig. 23c are more little than objective functions in Fig. 23b, which using DFIG. Hence, the performance of the PID controller is more effective than I controller even though, DFIG has been penetrating with the model which contain supplementary I controllers. Figure 23d shows the least obtained cost functions according to adjusting PID controller parameters with DFIG participation. SSA achieve the least cost function by 0.00002169 value compared to SCA and HHO. It can be said that, DFIG make a significant effect by paralleling with traditional units and using PID secondary controllers to adjust the LFC and achieve more system stability with less oscillation.

As a future work and research gaps, the applications of PMU and wide-area monitoring systems in modern power systems operation, control, stability, protection and security should be well-investigated [29–33]. PMUs are the backbone of wide are monitoring system that can improve the overall security and stability of power systems. Therefore, it is suggested to make in-depth evaluation of the power system security under the new environment control based one PMUs data. Likewise, the link between small signal stability stabilizers and automatic generation control loop can be suggested for improving the overall stability in future smart grids.

7 Conclusions

This chapter provides a brief introduction to the DFIG that used widely in wind farms. The mathematical modelling of two-area interconnected power system considering GRCs was presented in this chapter. The first proposed model consists of two-area interconnected power system without DFIG penetration. The second investigated model looks like the first one except the penetration of DFIG at both areas. Also, this chapter presented the control model of DFIG based wind turbine paralleling with conventional generators. HHO, SSA and SCA optimization techniques have been applied in the deregulated power systems for obtaining optimal I and PID controller parameters to solve LFC problem. Many comparisons were made for selecting the best optimization techniques which achieve optimal controller parameters at both studied models to enhance the system frequency. From results, it can be concluded that, the system performance has been enhanced by participating DFIG with thermal power plants. System performance enhancement has been summarized in minimizing OS and US of two-area frequencies and tie-line power exchange signals. The robustness of the PID controller has been confirmed by making several studying cases. It is noteworthy, the PID controller has the ability to enhance system stability more than I controller even though using DFIG with these I secondary controller at both areas. Briefly, the least objective function has been achieved via SSA which has become dominated by obtaining the optimal PID controller parameters with the second studied model which contains DFIG. Also, it can be said that, when planning engineers intends to establish two-area thermal-thermal power system, the presence of DFIG at both areas is necessary to enhance the performance. Also, system stability has been achieved using SSA to adjust the secondary PID controllers at both areas.

Appendix

Parameter (Notation)	Value (Unit)
DFIG inertia constant at first area (H_{e1})	3.5 (p.u.M.W)
DFIG inertia constant at second area (H_{e2})	3.5 (p.u.M.W)

(continued)

(continued)

Parameter (Notation)	Value (Unit)
Power system gain at first area (Kp1)	62 Hz/(p.u.MW)
Power system gain at second area (Kp2)	62 Hz/(p.u.MW)
Power system time constant at first area (Tp1)	10 s
Power system time constant at second area (Tp2)	15 s
Rated power at the first area ($pr1$)	1000 MW
Rated power at the second area ($pr2$)	1000 MW
Generation rate constraint (GRC) at non-reheat thermal unit 1	15% (p.u.MW)/min
Generation rate constraint (GRC) at non-reheat thermal unit 2	15% (p.u.MW)/min
Coefficient of synchronizing (T12)	0.07p.u.MW/HZ
Constant of speed regulation at first area (R1)	3 Hz/(p.u.MW)
Constant of speed regulation at second area (R2)	3 Hz/(p.u.MW)
Wind turbine time constant at first area (Ta1)	0.2 s
Wind turbine time constant at second area (Ta2)	0.2 s
Speed governor time constant at first area (Tg1)	0.1 s
Speed governor time constant at second area (Tg2)	0.1 s
Steam turbine time constant at first area (Tt1)	1 s
Steam turbine time constant at second area (Tt2)	1 s
Frequency measurement starting time at first area (Tr1)	0.1 s
Frequency measurement starting time at second area (Tr2)	0.1 s
Washout filter starting time at first area (Tw1)	6 s
Washout filter starting time at second area (Tw2)	6 s

References

1. F. Yao, R.C. Bansal, Z.Y. Dong, R.K. Saket, J.S. Shakya, Wind energy resources: theory, design and applications, in *Handbook of Renewable Energy Technology* (World Scientific, 2011), pp. 3–20
2. E. Muljadi, C. Butterfield, R. Yinger, H. Romanowitz, Energy storage and reactive power compensator in a large wind farm, in *42nd AIAA Aerospace Sciences Meeting and Exhibit* (2004), p. 352
3. E. Muljadi, V. Gevorgian, Short-circuit modeling of a wind power plant, in *2011 IEEE Power and Energy Society General Meeting* (2011), pp. 1–9
4. F. Blaabjerg, K. Ma, Wind energy systems. Proc. IEEE **105**, 2116–2131 (2017)
5. B. Wu, Y. Lang, N. Zargari, S. Kouro, *Power Conversion and Control of Wind Energy Systems* (Wiley, 2011)
6. H.H. Alhelou, M.E. Hamedani-Golshan, R. Zamani, E. Heydarian-Forushani, P. Siano, Challenges and opportunities of load frequency control in conventional, modern and future smart power systems: a comprehensive review. Energies **11**(10), 2497 (2018)
7. A.H. Elkasem, S. Kamel, A. Rashad, F.J. Melguizo, Optimal performance of doubly fed induction generator wind farm using multi-objective genetic algorithm. IJIMAI **5**, 48–53 (2019)

8. A.H. Elkasem, S. Kamel, A. Rashad, F. Jurado, Optimal performance of DFIG integrated with different power system areas using multi-objective genetic algorithm, in *2018 Twentieth International Middle East Power Systems Conference (MEPCON)* (2018), pp. 672–678
9. A.M. Rshad, S. Kamel, A.H. Elkasem, Enhancement of combined wind farm performance using crowbar, in *2018 International Conference on Innovative Trends in Computer Engineering (ITCE)* (2018), pp. 283–288
10. E. Muljadi, M. Singh, V. Gevorgian, Doubly fed induction generator in an offshore wind power plant operated at rated V/Hz. IEEE Trans. Ind. Appl. **49**, 2197–2205 (2013)
11. H. Saadat, *Power System Analysis* (1999)
12. P. Kundur, N.J. Balu, M.G. Lauby, *Power System Stability and Control*, vol. 7 (McGraw-hill New York, 1994)
13. O. Elgerd, Energy Systems Theory: an Introduction (New Delhi, McGraw-Hill, 1983)
14. T. Shyama, R.S. Kumar, V. Shanmugasundaram, Design of FGSPI controller based combined LFC and AVR of two area interconnected power generating system. Int. J. Eng. Adv. Technol. **1**, 135–139 (2012)
15. F.K. Abo-Elyousr, Load frequency controller design for two area interconnected power system with DFIG based wind turbine via ant colony algorithm, in *Power Systems Conference (MEPCON), 2016 Eighteenth International Middle East* (2016), pp. 253–260
16. F.M. Hughes, O. Anaya-Lara, N. Jenkins, G. Strbac, Control of DFIG-based wind generation for power network support. IEEE Trans. Power Syst. **20**, 1958–1966 (2005)
17. R.G. De Almeida, E.D. Castronuovo, J.P. Lopes, Optimum generation control in wind parks when carrying out system operator requests. IEEE Trans. Power Syst. **21**, 718–725 (2006)
18. J.M. Mauricio, A. Marano, A. Gómez-Expósito, J.L.M. Ramos, Frequency regulation contribution through variable-speed wind energy conversion systems. IEEE Trans. Power Syst. **24**, 173–180 (2009)
19. M. Jalali, DFIG based wind turbine contribution to system frequency control (University of Waterloo, 2011)
20. B. Sahu, P. Mohanty, S. Panda, S. Kar, N. Mishra, Design and comparative performance analysis of PID controlled automatic voltage regulator tuned by many optimizing liaisons, in *2012 International Conference on Advances in Power Conversion and Energy Technologies (APCET)* (2012), pp. 1–6
21. P.C. Pradhan, R.K. Sahu, S. Panda, Firefly algorithm optimized fuzzy PID controller for AGC of multi-area multi-source power systems with UPFC and SMES. Eng. Sci. Technol. Int. J. **19**, 338–354 (2016)
22. M. Elsisi, M. Soliman, M. Aboelela, W. Mansour, ABC based design of PID controller for two area load frequency control with nonlinearities. TELKOMNIKA Indones J. Electr. Eng. **16**, 58–64 (2015)
23. P. Bhatt, R. Roy, S. Ghoshal, Dynamic participation of doubly fed induction generator in automatic generation control. Renew. Energy **36**, 1203–1213 (2011)
24. A.A. Heidari, S. Mirjalili, H. Faris, I. Aljarah, M. Mafarja, H. Chen, Harris hawks optimization: algorithm and applications, in *Future Generation Computer Systems* (2019)
25. H. Moayedi, A. Osouli, H. Nguyen, A.S.A. Rashid, A novel Harris hawks' optimization and k-fold cross-validation predicting slope stability, Eng. Comput. 1–11 (2019)
26. D.T. Bui, H. Moayedi, B. Kalantar, A. Osouli, B. Pradhan, H. Nguyen et al., A novel swarm intelligence—Harris Hawks optimization for spatial assessment of landslide susceptibility. Sensors **19**, 3590 (2019)
27. S. Mirjalili, A.H. Gandomi, S.Z. Mirjalili, S. Saremi, H. Faris, S.M. Mirjalili, Salp swarm algorithm: a bio-inspired optimizer for engineering design problems. Adv. Eng. Softw. **114**, 163–191 (2017)
28. S. Mirjalili, SCA: a sine cosine algorithm for solving optimization problems. Knowl.-Based Syst. **96**, 120–133 (2016)
29. H.H. Alhelou, M.E. Golshan, J. Askari-Marnani, Robust sensor fault detection and isolation scheme for interconnected smart power systems in presence of RER and EVs using unknown input observer. Int. J. Electr. Power Energy Syst. **99**, 682–694 (2018)

30. H. Haes Alhelou, M.E. Hamedani Golshan, M. Hajiakbari Fini, Wind driven optimization algorithm application to load frequency control in interconnected power systems considering GRC and GDB nonlinearities. Electr. Power Compon. Syst. **46**(11–12), 1223–1238 (2018)
31. H.H. Alhelou, M.E. Golshan, N.D. Hatziargyriou, Deterministic dynamic state estimation-based optimal lfc for interconnected power systems using unknown input observer. IEEE Trans. Smart Grid (2019)
32. H.H. Alhelou, M.E. Golshan, N.D. Hatziargyriou, A decentralized functional observer based optimal LFC considering unknown inputs, uncertainties, and cyber-attacks. IEEE Trans. Power Syst. **34**(6), 4408–5517 (2019)
33. H.H. Alhelou, M.E. Golshan, T.C. Njenda, N.D. Hatziargyriou, An overview of UFLS in conventional, modern, and future smart power systems: challenges and opportunities. Electr. Power Syst. Res. **179**, 106054 (2020)

Wide-Area Measurement-Based Voltage Stability Assessment by Coupled Single-Port Models

Jian-Hong Liu, Heng-Yi Su, and Chia-Chi Chu

Abstract As the power system becomes more stressed and the penetration of intermittent renewable energies increase, voltage stability assessment (VSA) becomes a key concern for maintaining and enhancing the security of bulk power systems. Physically, the phenomenon of voltage instability is indeed caused by an uncontrollable drop in system voltage after being subjected to a disturbance. This deterioration may ultimately result in voltage collapse that has been responsible for several blackout incidents. So far, a vast number of methods ranging from simple static techniques to complex dynamic methods have been proposed for performing VSA. More recently, with wide deployment of synchronized phasor measurement units (PMUs), PMU-based wide area measurement system (WAMS) has attracted lots of interests from both academia and industry. In this chapter, recent developments of measurement-based coupled single-port models will be presented for VSA. Generally speaking, the concept of the coupled single-port model is to decouple a mesh power grid into several single-port local equivalent models with considering extra coupling impedances. By collecting real-time PMU measurements in each individual load bus, the reactive power response derived from the extended Ward-type equivalent model can be applied to eliminate the reactive power mismatch of the existing single-port model. Meanwhile, these parameters of the Thevenin equivalent circuit in the existing single-port model will be modified by a mitigation factor to improve the model accuracy of VSA. Since the proposed method is simple, several voltage stability indicators can be easily extended with slight modifications. Simulations are conducted on two

J.-H. Liu (✉)
Department of Electrical Engineering, Yuan Ze University,
Taoyuan Taiwan 32003, Republic of China
e-mail: jhliu727@saturn.yzu.edu.tw

H.-Y. Su
Department of Electrical Engineering, Feng Chia University,
Taiwan 40724, Taichung, Republic of China
e-mail: hengyisu@fcu.edu.tw

C.-C. Chu
Department of Electrical Engineering, National Tsing Hua University,
Taiwan 30013, Hsinchu, Republic of China
e-mail: ccchu@ee.nthu.edu.tw

H. Haes Alhelou et al. (eds.), *Wide Area Power Systems Stability, Protection, and Security*, Power Systems, https://doi.org/10.1007/978-3-030-54275-7_12

test systems, including IEEE 57-bus and IEEE 118-bus test systems, to validate the accuracy of the proposed method.

Keywords Voltage stability assessment · Secondary voltage control · Coupled single-port models · WAMs · Voltage control · Reactive power control · Power system stability

1 Introduction

As the increasing penetration level of intermittent renewable energies becomes the recent trend of energy utilizations in modern power systems, the voltage stability assessment (VSA) has raised a significant concern for the maintenance and enhancement of the security in the large-scale power grid [1]. Nowadays, a vast number of approaches regarding to VSA, ranging from static methods to complex dynamic techniques, have been addressed.

In the past few decades, two major approaches, model-based methods and measurement-based methods, have been developed to solve this VSA. Usually, accurate system parameters are required in model-based approaches in order to achieve precise VSA estimations. Several methods, including power flow approaches (CPFLOW) [2–4], direct methods [5], and optimal power flow appraoches [6, 7], have been developed along this direction. In these model-based approaches, physical constraints, such as line thermal limits and reactive power limits of generators, can be included in the analysis of VSA. However, theses approaches inherently exist computational complexities when applied for practical large-scale power grids under real-time environment.

With rapid advances of the technologies in Phasor Measurement Unit (PMUs), new advanced technologies of measurement-based approaches are very suitable to perform the monitoring of the voltage stability by exploring their characteristics from vast measurement data. In early advance, measurement-based approaches are based on a single-port model which is constructed by local measurements and used for VSA on the basis of the maximum power transfer theorem. Similarly, several voltage stability indicators (VSIs) have also been proposed [8–12]. Moreover, various emergency control methods have also been recommended [13–15]. The superior advantage of measurement-based approaches is the computational simplicity and make real-time applications possible. On the contrary, these methods may have inaccuracies due to the limited measurement information from a single location. Recognizing the demand of unifying measurements from different locations, relying by a trustworthy communication network, various approaches have been proposed. Along this direction, the monitoring systems have been developed for wide-area voltage stability prediction. In recent years, the trend for wide-area VSA focuses on the development of the voltage stability estimation systems in real-time environment [16, 17]. The core technology in wide-area VSA is based on multi-port equivalent Thevenin models. Based on this circuit topology, coupled single-port models were proposed to represent equivalent Thevenin parameters measured by wide-area measurements [18–21]. The fundamental idea of coupled single-port models is to decouple a complicated power system into several individual equivalent single-port branch with an extra coupling impedance. Under the scenario of proportional increasing loads, the accurate VSI can be achieved from this individual equivalent branch. However, from extensive simulations using the existing coupled single-port models, underestimations have been observed for VSA when the non-proportional load-increase scenario is considered.

In order to effectively eliminate inaccuracies, we have proposed the modified coupled single-port models to improve its accuracy [22]. In the proposed model, reactive power response factor (RPRF) [23, 24], defined by the voltage ratio of the reactive power and the voltage variations for load buses, is adopted. The core technique is to eliminate the RPRF mismatch between the power network and the measurement-based model. Therefore, for each load bus, the accuracy of voltage profile is able to greatly ameliorated. The extra reactive power support carried out by inserting the shunt admittance is provided to modulate the RPRF on the coupled single-port model. Specifically, the associated procedure of this modulation can be described as the parameter modifications of the existing model by using a mitigation factor. Under load scenarios, the proposed modified model is able to report acceptable voltage profiles. Therefore, the accuracy can be significantly improved for load margin estimations in VSA. Other salient features of the proposed modified coupled single-port model can be summarized as follows:

1. Any load variation scenarios can be utilized and have no limitation.
2. The extended Ward-type equivalent model can be converted to a special version to represent our proposed model. Therefore, the accurate reactive power response can also be expected.
3. By using the proposed mitigation factor, the existing single-port model will be modified to construct the proposed modified model.
4. Based on two consecutive PMU measurements, the mitigation factor is calculated.
5. These existing VSIs such as the L-index and the maximal loading parameter, can be slightly modified for performing more accurate VSA.

This chapter will report recent advances in wide-area measurement-based VSA by coupled single-port models. The rest of the chapter is organized as follows. Sect. 2 will provide an overview of existing coupled single-port models. Technical details of the proposed modified coupled single-port model will be addressed in Sect. 3. Simulation studies of IEEE 57-bus and IEEE 118-bus test systems are conducted in Sect. 4 to validate the accuracy of the proposed method. Finally, some conclusions are made in Sect. 5.

2 Existing Coupled Single-Port Models

The fundamental idea of the measurement-based VSA is based on the so-called *impedance matching theorem* used in the single-port Thevenin equivalent model.

Current and voltage phasors for local load buses are measured by PMUs, and the associated single-port Thevenin equivalent branch model of the rest power system, seen from the local bus, can be easily obtained according to the measured voltage and current phasors. When the impedance matching between the equivalent load impedance and the system Thevenin equivalent impedance occurs, the voltage instability will appear. Based on this fundamental of single-port equivalent model, several works, such as the conventional L-index [25] and the improved L-index [26], have been proposed for VSIs in the past two decades. Nevertheless, when dynamic loads with nonlinear complexities get involved, variations in Thevenin equivalent impedances cannot be well addressed in such single-port model. In order to overcome such predicament, the multi-port equivalent models have been proposed in recent years [18–20].

2.1 Fundamental of Coupled Single-Port Models

In the framework of the multi-port equivalent models, several coupled single-port branch models are organized together to form a coupled single-port network [21]. In the inter-connected power system, all buses can be categorized by three types: generator buses, load buses, and zero-injection buses. It is noted that the injection current at zero-injection buses is zero. Accordingly, the relationship between the injection current and the terminal voltage for these three types of buses can be formulated as [20]

$$\begin{bmatrix} \bar{I}_L \\ 0 \\ I_G \end{bmatrix} = \begin{bmatrix} Y_{LL} & Y_{LT} & Y_{LG} \\ Y_{TL} & Y_{TT} & Y_{TG} \\ Y_{GL} & Y_{GT} & Y_{GG} \end{bmatrix} \begin{bmatrix} V_L \\ V_T \\ V_G \end{bmatrix}, \tag{1}$$

where Y denotes the system admittance matrix. Voltage and current phasor for buses can be expressed by V and I. Load buses, zero-injection buses and generator buses can be presented in the subscripts L, T, G. When suitable PMU placements are assumed, the power system can be completely full observable. Thus, for load buses, voltage and current phasor can be measured properly by real-time PMU [27]. As the first row, corresponding to the load current in (1), is arranged, the circuit equation of the Thevenin equivalent model can be expressed by [20]:

$$V_L = Z_{LL}(Y_{LT}Y_{TT}^{-1}Y_{TG} - Y_{LG})V_G + Z_{LL}\bar{I}_L = E_{eq} - Z_{LL}I_L, \tag{2}$$

where the reversal load current is denoted by $\bar{I}_L = -I_L$. The equivalent voltage source is expressed by $E_{eq} = Y_{LL}^{-1}(Y_{LT}Y_{TT}^{-1}Y_{TG} - Y_{LG})V_G$. The system equivalent impedance is shown by $Z_{LL} = (Y_{LL} - Y_{LT}Y_{TT}^{-1}Y_{TL})^{-1}$ [20]. When (2) considers the coupling term, for the i-th load bus, the equivalent circuit model can be represented as [20]:

$$V_{Li} = E_{eq,i} - Z_{LLii} I_{Li} - \sum_{j=1,i\neq j}^{n} Z_{LLij} I_{Lj} = E_{eq,i} - Z_{LLii} I_{Li} - E_{coupled,i}, \quad (3)$$

for $i = 1, 2, 3, \cdots, n$, where the diagonal entry in Z_{LL} can be denoted by Z_{LLii}, and the $i - j$ non-diagonal entry is expressed by Z_{LLij}. For the i-th equivalent branch model, the equivalent voltage source can be denoted by $E_{eq,i}$. The voltage and current phasors are represented by V_{Li} and I_{Li} for the i-th load bus. For the ith equivalent branch model, the coupling effect from other loads is modelled as the coupling voltage source and expressed by $E_{coupled,i}$. Several previous methods have been mentioned on modelling the coupling voltage source $E_{coupled,i}$ when the single-port topology is maintained [20]. If such coupling effect is modelled as the coupled impedance, the coupled impedance will remain approximately constant as loads increase. It means that the coupling effect is more suitable to be represented as the coupled impedance for VSA. Specifically, the coupled impedance $Z_{coupled,i}$ is expressed by [20]

$$\begin{aligned} Z_{coupled,i} = E_{coupled,i}/I_{Li} &= \sum_{j=1,i\neq j}^{n} Z_{LLij}(I_{Lj}/I_{Li}) \\ &= \sum_{j=1,i\neq j}^{n} Z_{LLij}(S^*_{Lj}/S^*_{Li}) \times (V^*_{Li}/V^*_{Lj}), \end{aligned} \quad (4)$$

where the complex power load at the i-th load bus is denoted by S_{Li}. It is interesting to note that the complex power ratio S_{Lj}/S_{Li} remains almost constant when loads are increasing proportionally. Moreover, the voltage ratio V_{Li}/V_{Lj} also keeps approximately constant. Two constant ratios will lead to the fact of the nearly constant coupled impedance $Z_{coupled,i}$ [20]. It also reveals that the equivalent impedance $Z_{eq,i}$ is also nearly constant, which can be modelled as the series impedance connection of Z_{LLii} and $Z_{coupled,i}$, as expressed by

$$Z_{eq,i} = Z_{LLii} + Z_{coupled,i}. \quad (5)$$

Now, the coupled single-port models with n equivalent branch models are defined and depicted in Fig. 1. All parameters in coupled single-port models shown in Fig. 1 can be calculated if current and voltage phasors, I_{Li} and V_{Li}, are measured by PMUs.

2.2 *Voltage Stability Indicators*

In this chapter, two VSIs will be discussed. According to its simplicity, the great potential of these indicators make them possible for the application of real-time VSA [11, 25, 26]. Due to the simplicity of the result expressions and the numerical computations, these indicators have prominent potential to be applied in real-time voltage stability assessment.

2.2.1 Maximum Loading Parameter

According to the voltage law in the Kirchhoff circuit theorem, for the ith equivalent branch, the corresponding circuit equation regarding to the load power $P_i + jQ_i$ and the load voltage V_{Li} can be expressed by

$$|Z_{eq,i}|^2\left[|V_{Li}|^4+(2P_iR_{eq,i}+2Q_iX_{eq,i}-|E_{eq,i}|^2)|V_{Li}|^2\right]+|Z_{eq,i}|^4(P_i^2+Q_i^2)=0, \tag{6}$$

where the equivalent Thevenin impedance is expressed by $Z_{eq,i} = R_{eq,i} + jX_{eq,i}$. When the load power is parameterized as $\lambda_i(P_i + jQ_i)$ by the loading parameter λ_i, (6) can be reformed as a quadratic function of $|V_{Li}|^2$. When the loading parameter λ_i arrives at the maximal loading parameter λ_i^*, there exists the unique solution for the quadratic function [13], where

$$\lambda_i^* = \frac{|E_{eq,i}|^2(-P_iR_{eq,i}-Q_iX_{eq,i}+\sqrt{(P_i^2+Q_i^2)|Z_{eq,i}|^2})}{2(R_{eq,i}Q_i-X_{eq,i}P_i)^2} - 1. \tag{7}$$

Then, if all coupled single-port models have collected the corresponding maximal loading parameters λ_i^*, the smallest one is selected as the system maximal loading parameter, as expressed by

$$\lambda_{sys}^* = \min_{i=1,\ldots,n} \lambda_i^*. \tag{8}$$

2.2.2 *L*-Index

The fundamental of the L-index comes from the impedance matching theorem. Thus, it is defined by the ratio of the equivalent impedance $Z_{eq,i}$ to the equivalent load impedance Z_{Li}. For the i-th load bus, the L-index is defined as

$$L_i = \left|Z_{eq,i}/Z_{Li}\right| = \left|(Z_{LLii}I_{Li} + \sum_{j\neq i} Z_{LLij}I_{Lj})/V_{Li}\right|$$

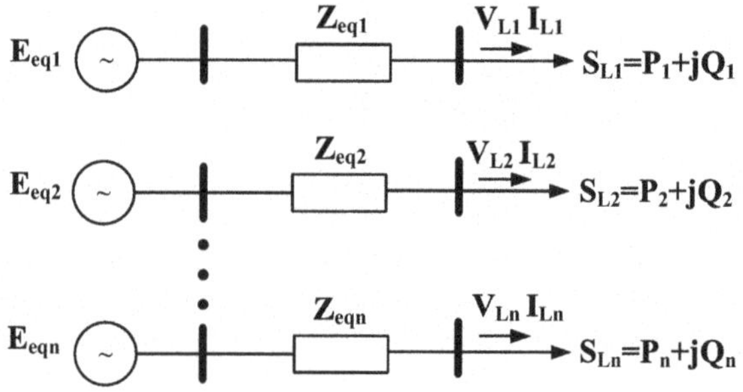

Fig. 1 The coupled single-port equivalent multi-port model

$$= \left| (Z_{LL} I_L)_i / V_{Li} \right| = \left| 1 - (Z_{LL} \bar{Y}_{LG} V_G)_i / V_{Li} \right|, \tag{9}$$

where the composite admittance is denoted by $\bar{Y}_{LG} = Y_{LT} Y_{TT}^{-1} Y_{TG} - Y_{LG}$. When the power system become stressed and is close to the voltage collapse point, L_i at the critical load bus will approach to 1.

2.3 Limitations of The Current Model

When the proportional-increase load scenario is considered, the equivalent Thevenin impedance $Z_{eq,i}$ and the coupling impedance $Z_{coupled,i}$ keep constant in the existing coupled single-port model. However, we have observed that the coupled impedance $Z_{coupled,i}$ fails to maintain constant while all loads are not increased proportionally. Therefore, the nonlinear nature of power flow variables is reflected in the nonlinearity of power flow equations and furthermore results in the inaccuracy for VSA. In order to concretely illustrate such inaccuracy, the simulation in IEEE 14-bus system is considered. In IEEE 14-bus system, eight loads are utilized. By PMU measurements, the variations on real and reactive power for load buses are given by $\Delta P = [0.1,\ 0.2,\ 0.1,\ 0.2,\ 0.5,\ 0.2,\ 0.1,\ 0.2]$ and $\Delta Q = [0.1,\ 0.2,\ 0.1,\ 0.2,\ 0.25,\ 0.2,\ 0.1,\ 0.2]$. Obviously, all loads are assigned non-proportional-increase load scenarios.

In Fig. 2, the voltage ratio V_{Li}/V_{L1} with respect to all load buses at the 1-th load bus is illustrated under various loading parameters. Figure 3 depicts the waveform of the couple impedance $Z_{coupled,i}$ for all equivalent branches under various loading parameters. The number showing in figures indicates the label of the analysed eight loads. By using the existing model, all dispersed maximum loading parameters for all

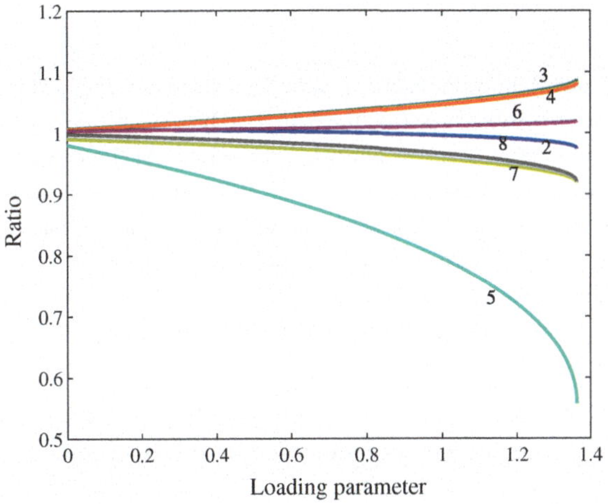

Fig. 2 The voltage ratio V_{Li}/V_{L1} with respect to all load buses at the 1-th load bus

Fig. 3 The waveform of the coupled impedance $Z_{coupled,i}$ for all the equivalent branches model under various loading parameters

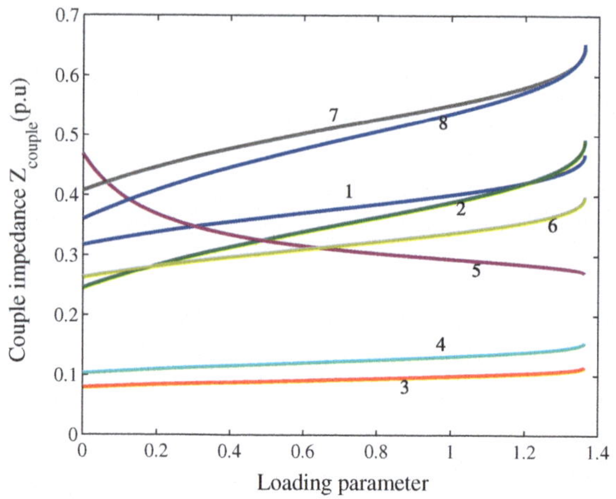

Fig. 4 The estimated maximum loading parameters by the existing model in IEEE 14-bus test system

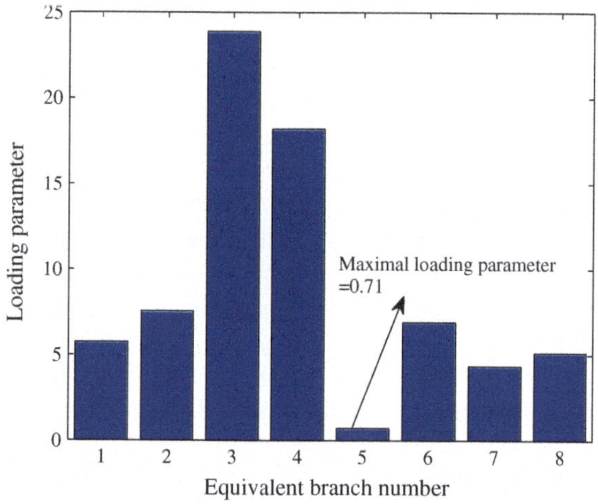

equivalent models are sketched, as shown in Fig. 4. Exploring these eight load buses, it can be observed that the 5-th load bus has the smallest maximum loading parameter and is regarded as the critical load bus. At the critical load bus, by using CPFLOW method and the existing model, P-V curves can be analysed and will be compared, as shown in Fig. 5. Clearly, the larger mismatch exists between estimated load margin $\lambda^*_{sys} = 0.71$ from existing model and actual load margin $\lambda^*_{CPFLOW} = 1.363$ from CPFLOW.

Through examining theories and simulations, three observations can be summarized:

1. At each branch of existing coupled single-port model, the coupled impedance variations are not ignorable if non-proportional load-increase scenarios are con-

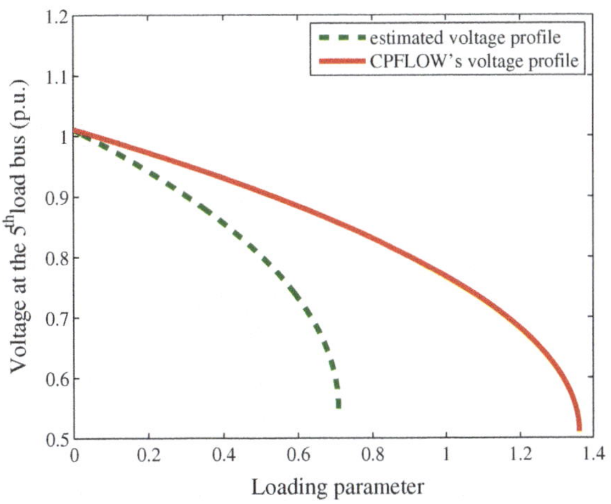

Fig. 5 P-V curves depicted by CPFLOW method and analysed by the existing model for the critical load bus

sidered. If loads are not increased proportionally in the existing model, the variations in the coupled impedance are significant and cannot be neglected.
2. Due to inconstant coupling impedance, the voltage profile in the existing model may be quite worse.
3. If real-time PMU measurements are utilized to modify the coupled impedance, the voltage profile in the modified model can be greatly improved and has more accurate VSA.

In order to overcome these difficulties, the notion of the reactive power response will be introduced in the next section for constructing the modified coupled single-port model to properly address the voltage profile from load variations.

3 The Modified Coupled Single-Port Model

3.1 Extended Ward Equivalent

The basic in the Ward-type equivalent leads to constructing equivalent networks for the external power system [23, 28]. By the initialization of the Ward-type equivalent, real power flows can be accurately addressed. Nevertheless, the poor accuracy appears in reactive power flows. In order to improve the Ward-type equivalent against such poor accuracy, the extended Ward-type equivalent is adopted later to retain the reactive power response of the power system. As a matter of fact, the coupled single-port model can be treated as a particular Ward-type equivalent in which each load bus is considered as the boundary bus, as shown in Fig. 6. Local load models and coupled single-port models can be modelled as the internal and external systems from the particular Ward-type equivalent. In order to improve the poor reactive power

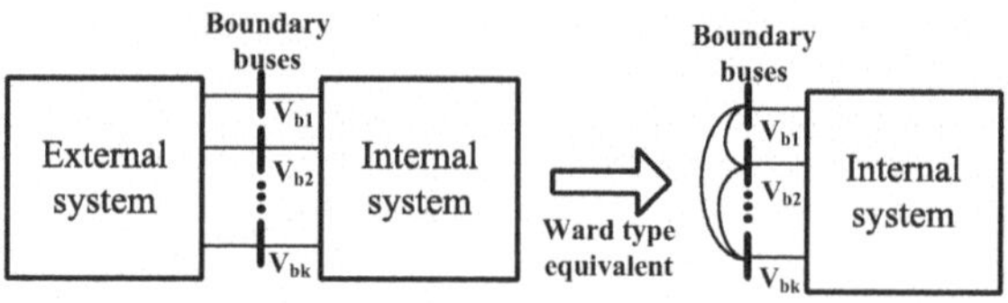

Fig. 6 Ward-type equivalent model

response in the current model, the modified coupled single-port model is built up by utilizing the extended Ward-type equivalent. Decoupled reactive power-flow equations can be utilized to describe the extended Ward-type equivalent models [23, 28]. The conventional decoupled reactive power-flow equations can be utilized to express the extended Ward-type equivalent as follows:

$$B_w \begin{bmatrix} \Delta|V_b| \\ \Delta|V_t| \end{bmatrix} = \begin{bmatrix} B'_w & B'_{bt} \\ B'_{tb} & B'_{tt} \end{bmatrix} \begin{bmatrix} \Delta|V_b| \\ \Delta|V_t| \end{bmatrix} = \begin{bmatrix} \frac{\Delta Q_b}{|V_b|} \\ \frac{\Delta Q_t}{|V_t|} \end{bmatrix}, \tag{10}$$

where B_w matrix comes from the Gaussian elimination of the admittance matrix Y. The voltage regarding to the internal system and the boundary buses can be expressed by V_t and V_b. $\Delta|V_t|$ and $\Delta|V_b|$ represent the voltage deviations at the internal system and the boundary buses. ΔQ_t and ΔQ_b are expressed as the deviations of the reactive power for the internal system and the boundary buses.

By rearranging (10), for boundary buses, the relationship between the voltage and the reactive power deviation is denoted as [23, 24]

$$\Delta Q_{wb} = |V_b| B'_w \Delta|V_b|. \tag{11}$$

Due to measurable deviations ΔQ_{wb} and $\Delta|V_b|$ estimated by two consecutive measurements, can be calculated as follows:

$$B'_w = \Delta Q_{wb}/(|V_b|\Delta|V_b|). \tag{12}$$

Here, the *reactive power response factor* (RPRF) can be defined by the term B'_w, representing the ratio of the two deviation, $\Delta|V_b|$ and ΔQ_{wb}, for boundary buses. Since the existing model has poor voltage approximations, RPRF analysed in the formula (11) has mismatch with the one calculated in (12) by PMU measurements. Hence, minimizing the mismatch is necessary.

3.2 Modifications of Coupled Single-Port Model

In this subsection, the major aim is to assure that RPRF calculated from equivalent branch models can be much closer to that measured from power grid model. By the case, the mismatch of the voltage profile between power system and equivalent branch

Fig. 7 RPRF estimations for eight loads on IEEE 14-bus test system

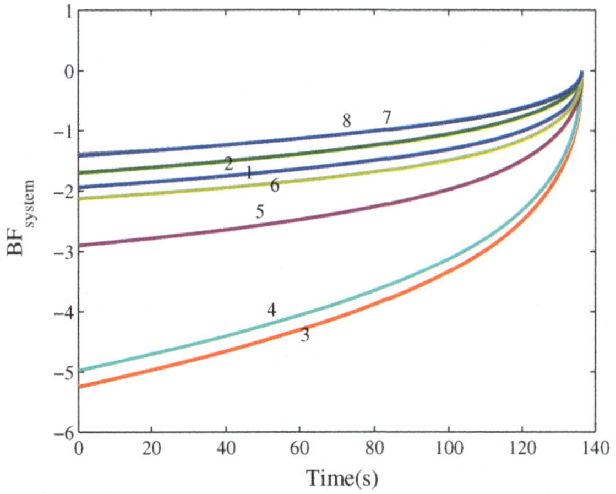

models can be eliminated. At the i-th load bus, by using two available consecutive PMU measurements, the load variation direction $\gamma_i(k)$ at a time stamp k can be denoted by

$$\gamma_i(k) = \frac{\Delta P_i(k)}{\Delta Q_i(k)} = \frac{P_i(k) - P_i(k-1)}{Q_i(k) - Q_i(k-1)}. \tag{13}$$

The current measurement technology has been addressed the ranged sampling time from about 8 ms to 30 ms [29]. In our application, 10 ms is adopted. The sensitivity formula is utilized to represent the relationship between the load voltage variation $\Delta|V_{Li}|$ and the reactive power variation ΔQ_i. Then, (6) is performed the partial derivative to $\Delta|V_{Li}|$, ΔQ_i and ΔP_i, and the sensitivity formula can be obtained. Due to the fact of $\Delta P_i(k) = \gamma_i(k)\Delta Q_i(k)$, the sensitivity formula can be expressed by

$$4|Z_{eq,i}|^2|V_{Li}|^3\Delta|V_{Li}| + 2|Z_{eq,i}|^2\Big[2P_i R_{eq,i} + 2Q_i X_{eq,i} - |E_{eq,i}|^2\Big]|V_{Li}|\Delta|V_{Li}|$$

$$+ 2|Z_{eq,i}|^2\Big[P_i\gamma_i(k)|Z_{eq,i}|^2 + 2Q_i|Z_{eq,i}|^2 + X_{eq,i}|V_{Li}|^2 + R_{eq,i}|V_{Li}|^2\gamma_i(k)\Big]\Delta Q_i = 0, \tag{14}$$

where the high-order terms can be neglected and the ratio $\Delta Q_i/\Delta|V_{Li}|$ can be derived. Hence, the ratio $\Delta Q_i/\Delta|V_{Li}|$ can be calculated by

$$\Delta Q_i/\Delta|V_{Li}| = d/e,$$

where

$$d = -4|Z_{eq,i}|^2|V_{Li}|^3 - 2|V_{Li}||Z_{eq,i}|^2(2P_i R_{eq,i} + 2Q_i X_{eq,i} - |E_{eq,i}|^2), \text{ and}$$
$$e = 2|Z_{eq,i}|^2|V_{Li}|^2[X_{eq,i} + R_{eq,i}\gamma_i(k)] + 2|Z_{eq,i}|^4(P_i\gamma_i(k) + Q_i).$$

Now, at the i-th equivalent branch model, the associated RPRF $BF_{eq,i}(k)$ can be expressed by

$$BF_{eq,i}(k) = \Delta Q_i/(|V_{Li}|\Delta|V_{Li}|) = d/(|V_{Li}|e). \tag{15}$$

On the other aspect, through two consecutive PMU measurements, wide-area RPRF $BF_{system,i}$ at the i-th load bus is assessed as follows:

$$BF_{system,i}(k) = \frac{Q_i(k) - Q_i(k-1)}{[V_{Li}(k) - V_{Li}(k-1)]V_{Li}(k)}. \tag{16}$$

Due to the rapid PMU measurements, $BF_{system,i}(k)$ will vary significantly with various loading conditions. As the voltage collapse point approaches, $BF_{system,i}(k)$ is close to zero, as depicted in Fig. 7. The poor and inaccurate voltage profile in the existing model reveals that $BF_{system,i}(k)$ for power system is larger than $BF_{eq,i}(k)$ in the equivalent branch. It also leads to the fact that the equivalent impedance $Z_{eq,i}$ has the higher mismatch with the one measured by PMU. Thus, the equivalent impedance $Z_{eq,i}$ must be scaled down. In order to effectively perform the reduction in the equivalent impedance, an additional shunt admittance Y_{Ci} is connected in the existing model, as shown in Fig. 8. The associated formula in the modified equivalent branch is expressed by

$$V_{Li} = E_{eq,i} - Z_{eq,i}I'_{Li}, \tag{17}$$

where $I'_{Li} = (1 - Y_{Ci}V_{Li}/I_{Li})I_{Li}$ denotes the modified load current. If the modified load current is introduced to (17), the load voltage V_{Li} is rewritten as

$$V_{Li} = (\alpha_i Z_{eq,i}I_{Li} + V_{Li}) - \alpha_i Z_{eq,i}I_{Li}, \tag{18}$$

where the mitigation factor is defined by $\alpha_i = 1 - Y_{Ci}V_{Li}/I_{Li}$. Therefore, the equivalent impedance $Z_{eq,i}$ can be modified through the multiplied mitigation factor α_i. When forcing RPRF of each modified model to be identical to the one measured from wide-area PMUs, the mitigation factor α_i can be obtained. It can be expressed as

$$BF'_{eq,i}(k) = BF_{system,i}(k), \tag{19}$$

where the RPRF estimated from the modified model can be represented by $BF'_{eq,i}(k)$ with the mitigation factor α_i. For the purpose of the effective equivalent impedance reduction, the range $0 \le \alpha_i \le 1$ is the constrains for the mitigation factor α_i. Now, at the i-th equivalent branch, the modified coupled single-port model is utilized to represent the load voltage V_{Li}.

$$V_{Li} = E'_{eq,i} - Z'_{eq,i}I_{Li}, \tag{20}$$

where the modified equivalent voltage source and the equivalent impedance can be expressed by $E'_{eq,i} = \alpha_i Z_{eq,i}I_{Li} + V_{Li}$ and $Z'_{eq,i} = \alpha_i Z_{eq,i}$. Then, the i-th modified

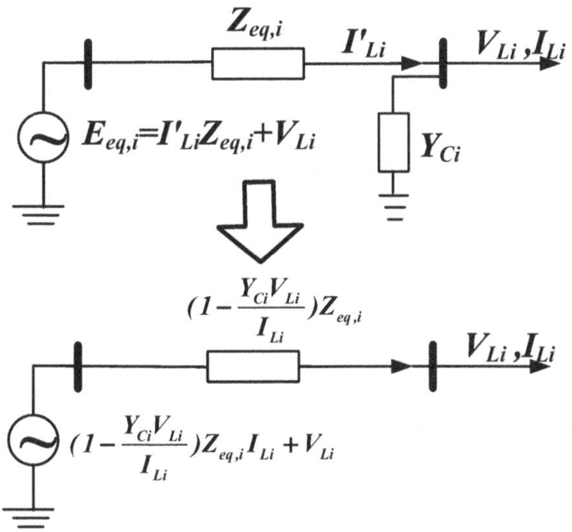

Fig. 8 Equivalent series compensation of the i-th coupled single-port model from the shunt admittance Y_{Ci}

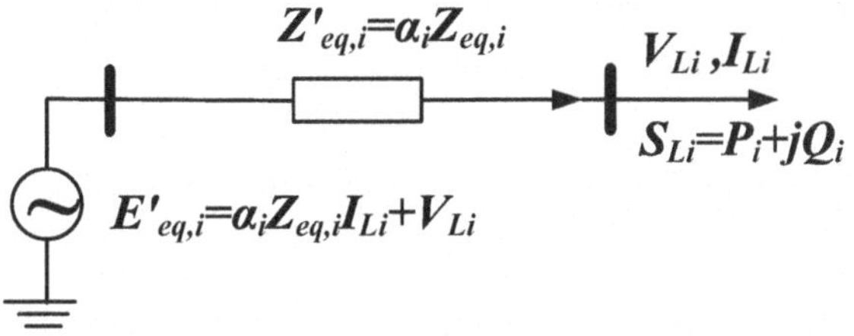

Fig. 9 The the i-th modified coupled single-port model

coupled single-port model can come from the modification of the i-th coupled single-port model, as depicted in Fig. 9.

Now, for the ith modified model, the equivalent impedance is replaced by $Z'_{eq,i}$ in (15) to obtain the modified RPRF $BF'_{eq,i}$. The modified equivalent voltage $E'_{eq,i}$ can be separated into the real and imaginary components, as written by

$$E'_{eq,i} = V_{Li} + Z'_{eq,i} I_{Li} = V_{Li} + \alpha_i V_{Line,i} = (V_{LRi} + jV_{LMi}) + \alpha_i (V_{Ri} + jV_{Mi}), \tag{21}$$

where the two-axis representation of the load voltage is $V_{Li} = V_{LRi} + jV_{LMi}$. The cross voltage at the equivalent impedance is denoted by $V_{Line,i} = V_{Ri} + jV_{Mi}$. The formula (19) can be reformulated through the replacement of (15) and (21) and modified by

$$BF_{system,i}(k) = \frac{\alpha_i^2 \kappa_1 + \alpha_i \kappa_2 + \kappa_3}{\alpha_i^2 \kappa_4 + \alpha_i \kappa_5}, \tag{22}$$

where all coefficients can be expressed by

$$
\begin{aligned}
\kappa_1 &= |Z_{eq,i}|^2(V_{Ri}^2 + V_{Mi}^2),\\
\kappa_2 &= |Z_{eq,i}|^2(2V_{Ri}V_{LRi} + 2V_{Mi}V_{LMi} - 2P_iR_{eq,i} - 2Q_iX_{eq,i}),\\
\kappa_3 &= |Z_{eq,i}|^2(V_{LRi}^2 + V_{LMi}^2 - 2|V_{Li}|^2,)\\
\kappa_4 &= |Z_{eq,i}|^4(P_i\gamma_i(k) + Q_i), \text{ and}\\
\kappa_5 &= |Z_{eq,i}|^2|V_{Li}|^2(X_{eq,i} + R_{eq,i}\gamma_i(k)).
\end{aligned} \tag{23}
$$

It can be noticed that κ_2 is equal to zero since we can formulate each term of κ_2 as

$$
\begin{aligned}
V_{Ri}V_{LRi} &= (R_{eq,i}I_{LRi} - X_{eq,i}I_{LMi})V_{LRi} = R_{eq,i}I_{LRi}V_{LRi} - X_{eq,i}I_{LMi}V_{LRi},\\
V_{Mi}V_{LMi} &= (R_{eq,i}I_{LMi} + X_{eq,i}I_{LRi})V_{LMi} = R_{eq,i}I_{LMi}V_{LMi} + X_{eq,i}I_{LRi}V_{LMi},\\
P_iR_{eq,i} &= (V_{LRi}I_{LRi} + V_{LMi}I_{LMi})R_{eq,i} = V_{LRi}I_{LRi}R_{eq,i} + V_{LMi}I_{LMi}R_{eq,i}, \text{ and}\\
Q_iX_{eq,i} &= (V_{LMi}I_{LRi} + V_{LRi}I_{LMi})X_{eq,i} = V_{LMi}I_{LRi}X_{eq,i} - V_{LRi}I_{LMi}X_{eq,i},
\end{aligned} \tag{24}
$$

where the load current can be denoted by $I_{Li} = I_{LRi} + jI_{LMi}$. Clearly, it can be found that combining terms in (24) leads to $\kappa_2 = 0$. Furthermore, by simplifying the formula (22), the mitigation factor α_i can be obtained by solving the following quadratic equation:

$$
a\alpha_i^2 + b\alpha_i + c = 0, \tag{25}
$$

where coefficients a, b and c can be represented by

$$
\begin{aligned}
a &= BF_{system,i}(k)|Z_{eq,i}|^2(P_i\gamma_i(k) + Q_i) - |V_{Line,i}|^2,\\
b &= BF_{system,i}(k)|V_{Li}|^2(X_{eq,i} + R_{eq,i}\gamma_i(k)), \text{ and}\\
c &= |V_{Li}|^2.
\end{aligned} \tag{26}
$$

Coefficients a, b, and c possess the following characteristics:

- If the positive reactive power variation is considered for $\Delta Q_i > 0$, the nature of loads leads to the negative load voltage variation $\Delta V_{Li} < 0$. It can be concluded that $BF_{system,i}(k) \le 0$.
- Since ZIP load model or voltage dependent load model [30, 31] can be considered in the modelling of the studied loads, at each load bus, real and reactive power will increase together, indicating the consequence $\gamma_i(k) \ge 0$ by (13).

If the conditions $\gamma_i(k) \ge 0$ and $BF_{system,i}(k) \le 0$ are held, these coefficients a and b can be identified as $a < 0$ and $b \le 0$. Moreover, the coefficient c is ranged by $c > 0$ due to $|V_{Li}| \in [0, 2)$. Thus, the mitigation factor α_i can be solved for the ith modified coupled single-port model by

$$
\alpha_i = \frac{-b - \sqrt{b^2 - 4ac}}{2a}. \tag{27}
$$

Due to $\alpha_i \in [0, 1]$, we choose the smaller solution.

Thus, the generated mitigation factor α_i leads to the construction of the proposed modified coupled single-port models. Obviously, it can be clearly observed in (20) that the modified equivalent voltage source $E'_{eq,i}$ and the modified equivalent impedance $Z'_{eq,i}$ can be analysed as functions of α_i, $Z_{eq,i}$, V_{Li} and I_{Li}. It is worthy to know that all these terms, α_i, $Z_{eq,i}$, V_{Li}, I_{Li}, $BF_{system,i}(k)$ and $\gamma_i(k)$, can be measured by PMUs. Thus, following such framework, the conventional L-index and the maximal loading parameters can be accordingly modified to perform the enhancement of VSA.

3.3 Computational Algorithm

The explicit procedures of establishing the modified coupled single-port model are addressed as follows:

- **Step 0**: Read the current power system data.
- **Step 1**: Use (3) to construct the existing multi-port equivalent model. The equivalent voltage source and the equivalent impedance can be calculated in (5) by PMU measurements.
- **Step 2**: Use two consecutive PMU measurements to compute γ_i and $BF_{system,i}$ in (13) and (16).
- **Step 3**: Solve the quadratic equation (25) to obtain the mitigation factor α_i.
- **Step 4**: By using the modified equivalent impedance $Z'_{eq,i}$ and the modified equivalent voltage source $E'_{eq,i}$, the modified coupled single-port models are constructed.
- **Step 5**: Perform VSA by computing L-index and the maximum loading parameters in modified coupled single-port models.

IEEE 14-bus test system is utilized to verify the feasibility of the proposed modified coupled single-port model. All maximum loading parameters in modified coupled single-port models are depicted in Fig. 10. CPFLOW method and the modified coupled single-port model perform P-V curve analysis at the critical load bus, as shown in Fig. 11. From the simulation results, the load margin estimation in the proposed model is reported by $\lambda^*_{sys} = 1.3544$ while the actual one by CPFLOW method is $\lambda^* = 1.363$. Obviously, the poor voltage profile in the proposed model is greatly improved. Besides, the L indices derived from the existing model and the proposed model are also analysed at the critical load bus and depicted in Fig. 12. Apparently, it can be found that the modified L-index is more accurate on VSA.

3.4 Extension to Generator Reactive Limits

In the framework of the proposed model, VSA still meets the limitation of the accuracy since physical constraints of power system cannot be included in the proposed model. In order to illustrate the feasibility, the case study of considering generator

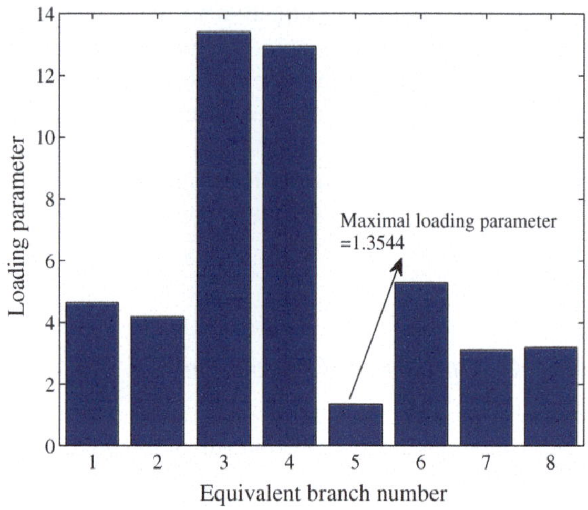

Fig. 10 The all maximal loading parameters calculated from eight modified coupled single-port models in IEEE 14-bus test system

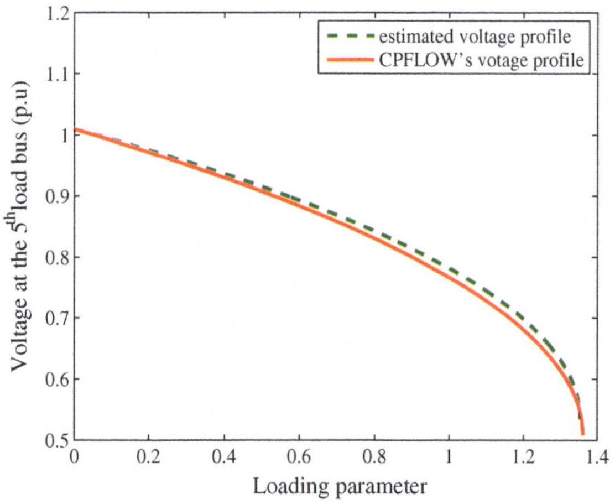

Fig. 11 P-V curves analysed by the modified coupled single-port model and CPFLOW method at the critical load bus on IEEE 14-bus test system

reactive power limits is conducted in IEEE 14-bus test system. If the load-increase scenarios are given the same load variations as Fig. 2, reactive power limits will be encountered at generators 1, 2, and 3 one by one. According to the nature of synchronous generators, the violation of reactive power limits leads to the conversion of the operation in which the generator buses of synchronous generators are changed to negative load buses. As a result, optimistic load margin estimation results may be reported in the proposed model, as depicted in Fig. 13. The most effective strategy to improve the model deficiency is re-partitioning the multi-port framework of the proposed model when the reactive power limits are encountered. Therefore, the re-partitioning is implemented by changing PV bus mode, also denoted by the generator bus mode, into PQ bus mode, expressed by load bus mode. Next, the transition on bus modes will be illustrated in the following simulation example.

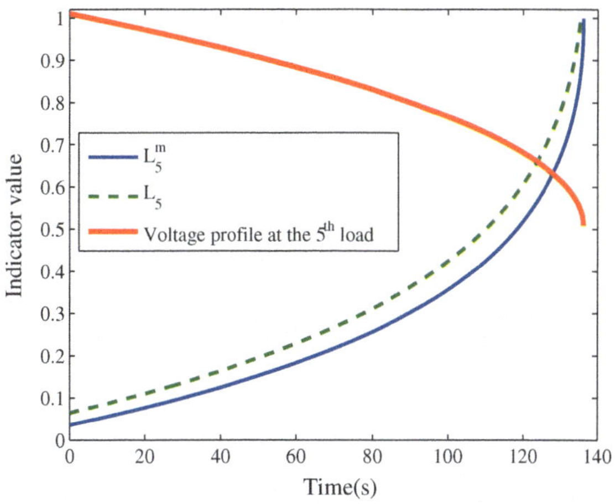

Fig. 12 Comparisons between the existing L_5 and the proposed modified L_5^m on IEEE 14-bus test system

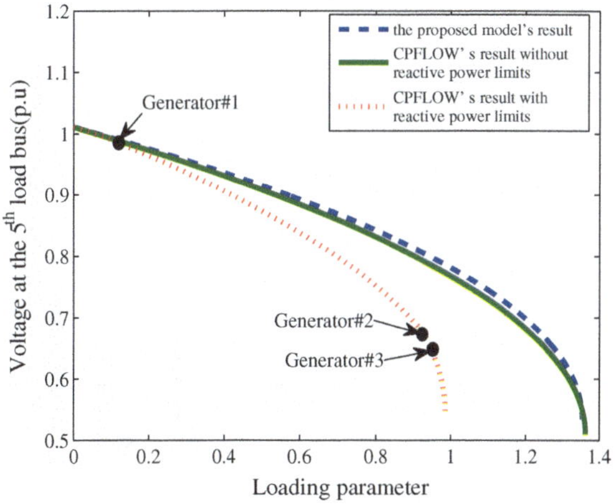

Fig. 13 P-V curve analysis results under considering generator reactive power limits, not considering generator reactive power limits and the modified coupled single-port model on IEEE 14-bus test system

Here, generator reactive power limit is considered in the case study of IEEE 14-bus system. In this case study, only generator at the bus 6 considers its reactive power limit with $-1 < Q_{limit} < 1$ such that its reactive power may meet the reactive power limit ($Q_6 = 1$) when its output real power poses $P_6 = 0.5$. As a result, the generator increases its output real power and keeps its output reactive power constant. The critical load bus is also recognized at the 5-th load. For IEEE 14-bus test system, Fig. 14 shows P-V curves of CPFLOW, PV bus mode and PQ bus mode with considering generator reactive power limit. The maximal loading parameters of PV bus mode and PQ bus mode obtained from the modified model on IEEE 14-bus system can be depicted in Fig. 15. PV bus mode represents that the generator keeps its PV bus mode in the proposed modified couple single-port models when it meets the reactive power limit. On the contrary, PQ bus mode denotes that the original PV bus mode is changed into PQ bus mode in the proposed modified models. As depicted in Fig. 14,

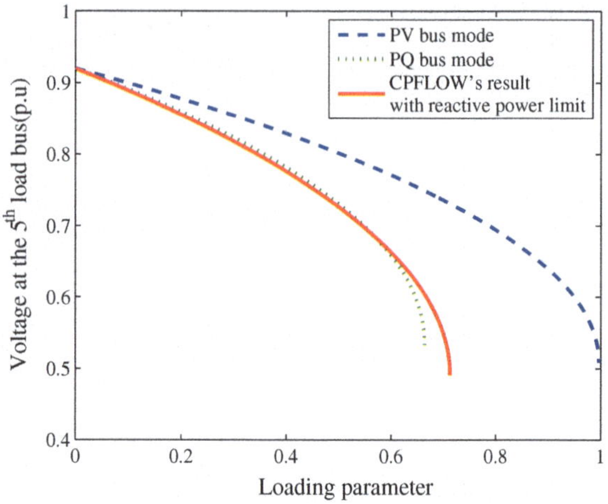

Fig. 14 P-V curves of CPFLOW, PV bus mode, and PQ bus mode with considering generator reactive power limit in IEEE 14-bus system

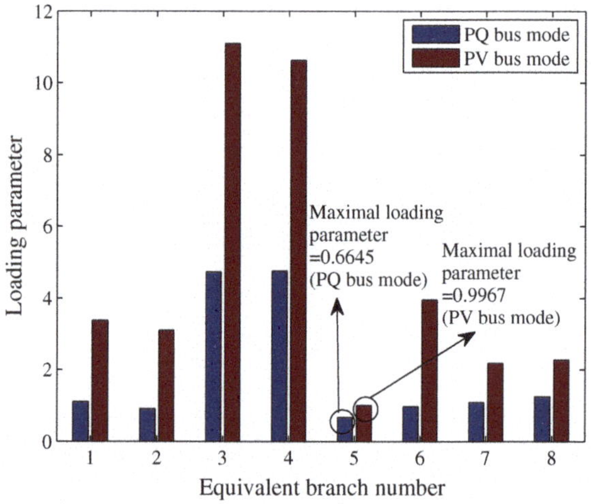

Fig. 15 The maximal loading parameters of PV bus mode and PQ bus mode obtained from the modified coupled single-port model in IEEE 14-bus system

the load margin from CPFLOW is $\lambda^* = 0.7126$ and the result from PQ bus mode has a slight underestimation result $\lambda^*_{sys} = 0.6645$. On the other hand, the result from PV bus mode has a larger overestimation $\lambda^*_{sys} = 0.9967$. Results can be concluded that the generator has been changed as a negative load when it meets the reactive power limit such that the original PV bus mode of the generator is necessary to be modified to PQ bus mode.

According to theoretical developments and numerical explorations in simple system mentioned earlier, now we are in the position to extend the proposed computational algorithm of the modified coupled single-port model to develop the modified algorithm with considering reactive power limits of generators as follows:

- **Step 0**: Read the current power system data.
- **Step 1**: Use (3) to construct the existing multi-port equivalent model. The equivalent voltage source and equivalent impedance can be determined in (5) by PMU measurements at each load buses.
- **Step 2**: Use two consecutive PMU measurements to compute γ_i and $BF_{system,i}$ described in (13) and (16).
- **Step 3**: Solve the quadratic equation (25) to obtain the mitigation factor α_i.
- **Step 4**: Construct the modified coupled single-port model by using the modified equivalent impedance $Z'_{eq,i}$ and the modified equivalent voltage source $E'_{eq,i}$.
- **Step 5**: Perform VSA by computing L-index and the maximum loading parameters in modified coupled single-port models.
- **Step 6**: If reactive power limits of some generators are encountered, the re-partition of the multi-port equivalent models is performed by changing these generator buses into load buses; Otherwise, stop the algorithm.
- **Step 7**: Go to Step 1 by fixing the reactive loading power at those new load buses as the reactive power limits of these generators.

To validate the performance of the above the modified algorithm with considering generator reactive limits, simulation studies on two IEEE test systems will be reported in the next section.

4 Case Studies

In this section, simulation studies will be conducted on IEEE 57-bus and IEEE 118-bus test systems to verify the accuracy of the proposed model. Simulation studies are designed to perform VSA under conditions, with and without considering reactive power limits of generators, on IEEE 57-bus and IEEE 118-bus test systems. In the simulations, all load variations are assumed by $\Delta P_i(k) = \Delta Q_i(k) = 0.2$ for even number i and $\Delta P_i(k) = \Delta Q_i(k) = 0.1$ with odd number i.

4.1 IEEE 57-Bus System without Considering Generator Reactive Power Limit

In the subsection, reactive power limit of generators is not included in the the simulation on IEEE 57-bus test system. Total load consumption is assigned $40 + j19.37$ MVA with a 0.9 lagging power factor for each load bus. At the beginning, by the analysis of CPFLOW method, the critical load bus is identified at the 5-th load bus.

The system maximum loading parameter, determined by CPFLOW method, reports $\lambda^*_{CPFLOW} = 1.955$. However, the underestimation result of VSA, $\lambda^*_{sys_o} = \min_i\{\lambda^*_i\} = \lambda^*_5 = 0.3354$, is reported in the existing model, as depicted in Fig. 16. On the contrary, more accurate estimation of the maximum loading parameter,

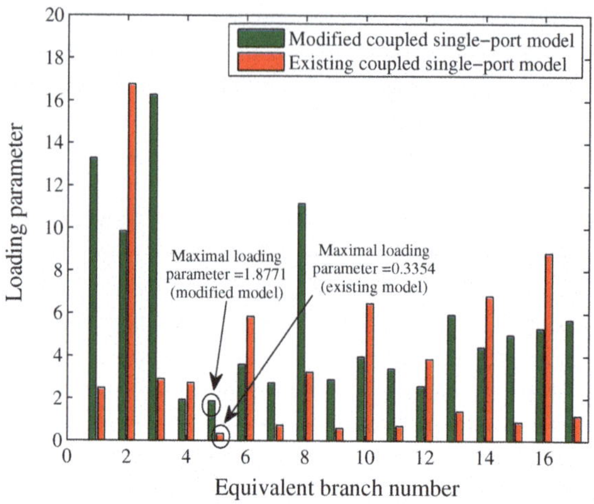

Fig. 16 Maximum loading parameters estimations without considering generator reactive power limits in IEEE 57-bus system

Table 1 Comparisons of maximal loading parameters without generator reactive power limit

Test systems	Models	Estimated λ^*_{sys}	λ^*_{CPFLOW}	Mismatch(%)
(a) IEEE 14-bus system	Existing model	0.71	1.363	−47.9
	Modified Model	1.3544	1.363	−0.63
(b) IEEE 57-bus system	Existing model	0.3354	1.955	−82.84
	Modified Model	1.8771	1.955	−3.984
(c) IEEE 118-bus system	Existing model	2.8225	5.0746	−44.379
	Modified Model	4.9856	5.0746	−1.753

$\lambda^*_{sys_m} = 1.8771$, can be obtained in the proposed modified model. The two models can be compared and list the comparison in Table 1b. In Table 1b, it clearly shows that the proposed model is more accurate than the existing model. The mismatch between the existing and the proposed models can be depicted in Fig. 17 under various loading conditions. Obviously, the VSA estimation in the proposed model is more accurate than the existing model even for the light load case. Besides, the modified version of the L-index is also used to perform VSA. In Fig. 18, the comparison between the modified and the existing L-indices are illustrated for the 5-th load bus. Results clearly address that the modified L-index by the proposed model can produce more accurate VSA to exactly detect the voltage collapse point.

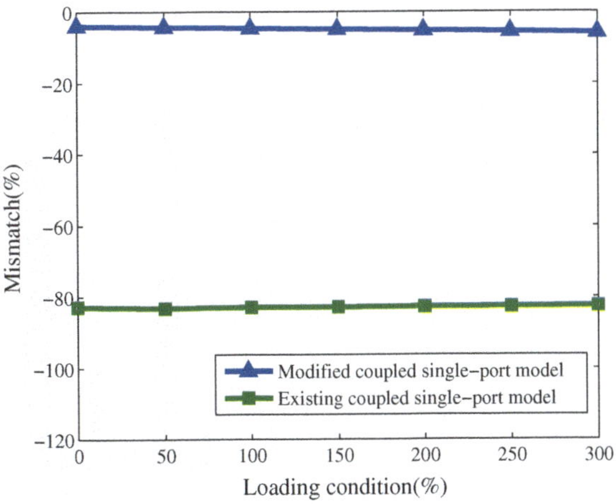

Fig. 17 Mismatch of the system maximal loading parameters estimation λ^*_{sys} under various loading conditions without considering generator reactive power limits on IEEE 57-bus test system

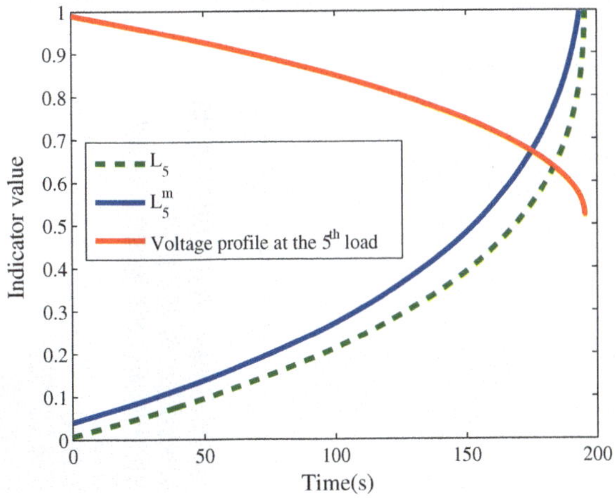

Fig. 18 Comparisons of the existing L_5 and the proposed modified L_5^m without considering generator reactive power limits on IEEE 57-bus test system

4.2 *IEEE 57-Bus System with Considering Generator Reactive Power Limit*

In this subsection, three generators at buses 3, 6 and 8 consider their reactive power limits given by $-1 < Q_{limit} < 1$, and three generators are designed to hit their reactive power limits in which $Q_{G3} = -1$, $Q_{G6} = -1$ and $Q_{G8} = 1$ are addressed respectively under the specific output real power $P_{G3} = 1.4$, $P_{G6} = 1.3$ and $P_{G8} = 2.5$.

All loads are also assigned with $40 + j19.37$ MVA with the power factor 0.9 lagging. As generators encounter the reactive power limits, the generator reactive power tends to keep constant with $Q_{G3} = -1$, $Q_{G6} = -1$ and $Q_{G8} = 1$ while the output real power keeps increasing. CPFLOW method reports the maximal loading parameter $\lambda^*_{CPFLOW} = 1.8868$. As the existing couple single-port model is utilized,

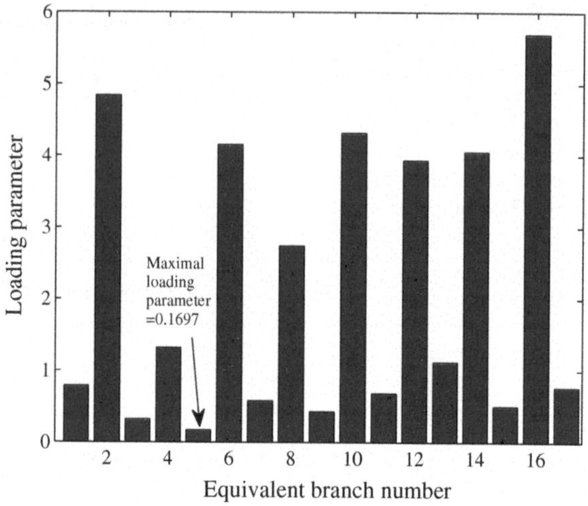

Fig. 19 The maximal loading parameters of the existing model with considering generator reactive power limit in IEEE 57-bus system

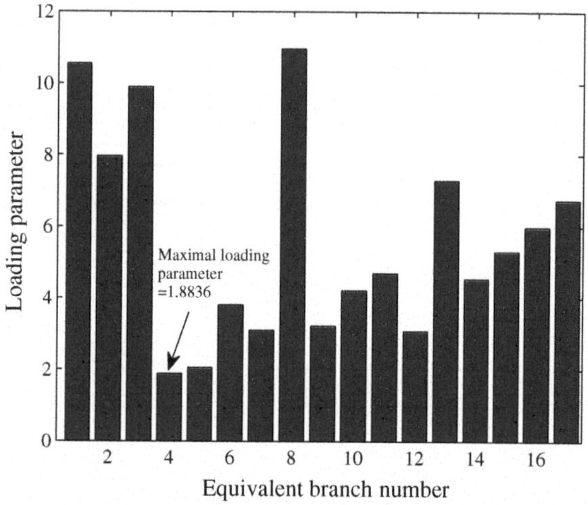

Fig. 20 The maximal loading parameters of the modified model with considering generator reactive power limit in IEEE 57-bus system

the critical load is addressed at the 5-th load bus. As shown in Fig. 19, the result $\lambda^*_{sys_o} = \min_i\{\lambda^*_i\} = \lambda^*_5 = 0.1697$ reveals the underestimations of load margin in the existing model. If the proposed modified model is employed, the estimated result $\lambda^*_{sys_m} = 1.8836$, addressed at the critical 4-th load bus, represents the accurate result, as illustrated in Fig. 20.

The comparison is conducted on both models, as appeared in Table 2a. It reveals that the proposed modified model is more accurate. Fig. 21 shows the waveform of the modified L^m_5 and the existing L_5. Results clearly demonstrate that the associated modified L-index is also accurate for detecting the voltage collapse point.

Table 2 Comparisons of maximal loading parameters with generator reactive power limit

Test systems	Models	Estimated λ^*_{sys}	λ^*_{CPFLOW}	Mismatch(%)
(a) IEEE 57-bus system	Existing model	0.1697	1.8868	−91.005
	Modified Model	1.8836	1.8868	−0.1696
(b) IEEE 118-bus system	Existing model	2.7444	4.935	−44.3891
	Modified model	4.8443	4.935	−1.8379

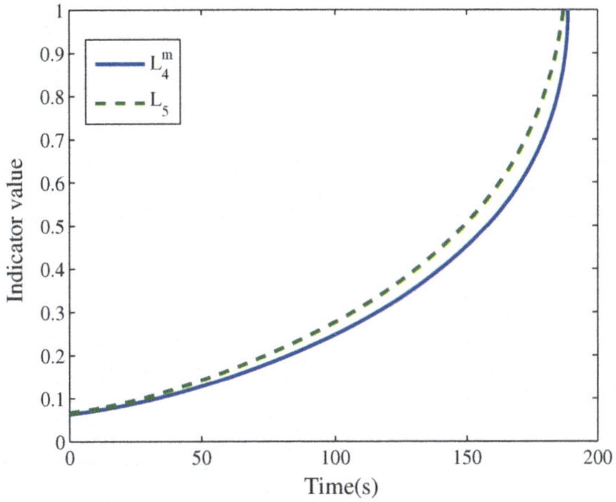

Fig. 21 Comparisons of the existing L-index L_5 and the modified L-index L_4^m with considering generator reactive power limit in IEEE 57-bus system

4.3 *IEEE 118-Bus System without Considering Generator Reactive Power Limit*

The numerical simulation are performed without considering generator reactive power limits on IEEE 118-bus test system. In IEEE 118-bus test system, 53 loads are amounted to the total complex power loading of $530 + j256.69$ MVA with the 0.9 lagging power factor for each load bus. In the beginning, the system maximum loading parameter, analysed by CPFLOW method, is reported by $\lambda^*_{CPFLOW} = 5.0746$ and also utilized to address the critical load bus at the 10-th load bus. When the system maximum loading parameter is determined in the existing model, underestimation results present with $\lambda^*_{sys_o} = 2.8225$. On the contrary, more accurate estimation result, $\lambda^*_{sys_m} = 4.9856$, appears in the proposed model. All the maximum loading parameters of all equivalent branch models for the existing and the proposed models are integrated in Fig. 22. The comparison of the mismatches between the existing and the proposed models are summarized in Table 1c. Under various loading conditions, the mismatches of the maximum loading parameter estimation between the existing and the proposed models are shown in Fig. 23. It can be concluded that the significant improvement of the accuracy appears in the proposed model.

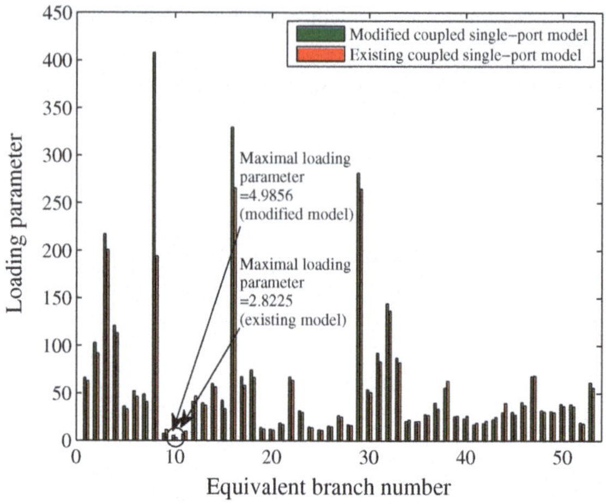

Fig. 22 Maximal loading parameter estimations without considering generator reactive power limits on IEEE 118-bus system

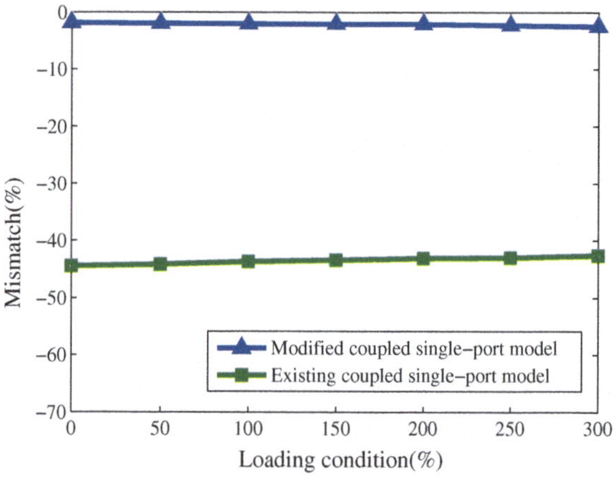

Fig. 23 Mismatch of maximal loading parameter estimation λ^*_{sys} under various loading conditions without considering generator reactive power limits on IEEE 118-bus system

In addition, the L-index is also utilized to verify the feasibility for VSA. For that, Fig. 24 clearly shows that the L-index L^m_{10} in the proposed model is very accurate to locate the voltage collapse point.

4.4 *IEEE 118-Bus System with Considering Generator Reactive Power Limit*

In this subsection, the reactive power limit ($-0.5 < Q_{limit} < 0.5$) of five generators, equipped at buses 72, 26, 34, 73 and 103 , is considered in the simulation. In the simulation, five generators hit the reactive power limits of $Q_{G72} = Q_{G26} = Q_{G34} =$

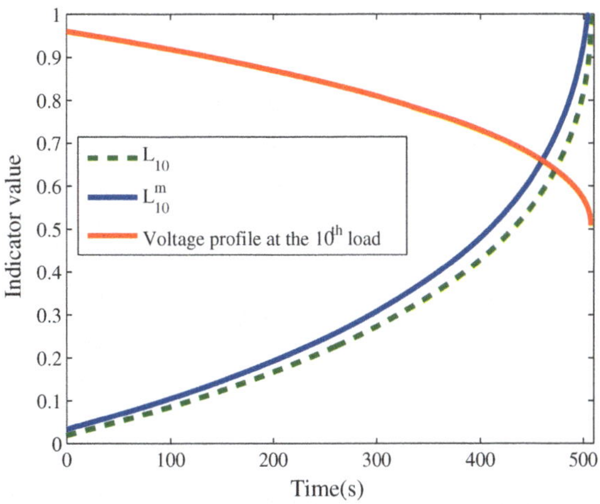

Fig. 24 Comparisons of the existing L_{10} and the modified L_{10}^{m} without considering generator reactive power limits on IEEE 118-bus system

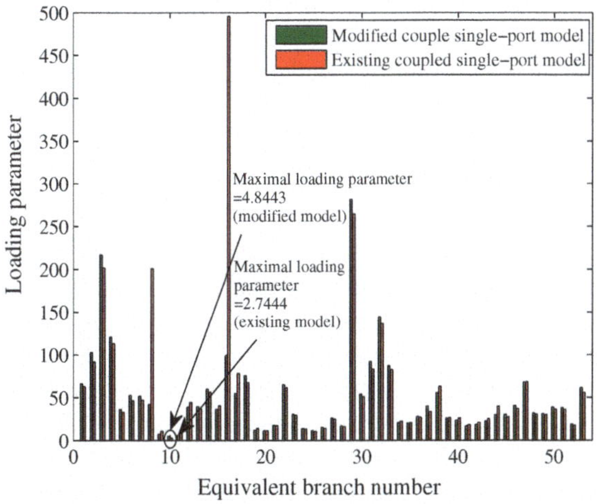

Fig. 25 Estimations of the maximal loading parameters with considering generator reactive power limit in IEEE 118-bus system

$Q_{G73} = Q_{G103} = 0.5$ while their output real power are given by $P_{72} = 0.1$, $P_{26} = 3.14$, $P_{34} = 0.1$, $P_{73} = 0.1$ and $P_{34} = 0.4$ respectively.

Load consumption $530 + j256.69$ MVA with the power factor 0.9 lagging is assigned in all loads. When encountering reactive power limit, the generator reactive power keeps constant and their output real power keeps increasing. CPFLOW method reports the maximal loading parameter $\lambda^*_{CPFLOW} = 4.935$. The critical load bus is identified at the 10-th load bus. As shown in Fig. 25, if the existing model is utilized, the estimated result $\lambda^*_{sys_o} = \min_i\{\lambda^*_i\} = \lambda^*_5 = 2.7444$ shows the underestimations. On the contrary, the proposed modified model is accurate by the estimated result $\lambda^*_{sys_m} = 4.8443$. Comparison studies, as illustrated in Table 2b, also reveal that the proposed modified model is more accurate by the much smaller mismatch. Fig. 26

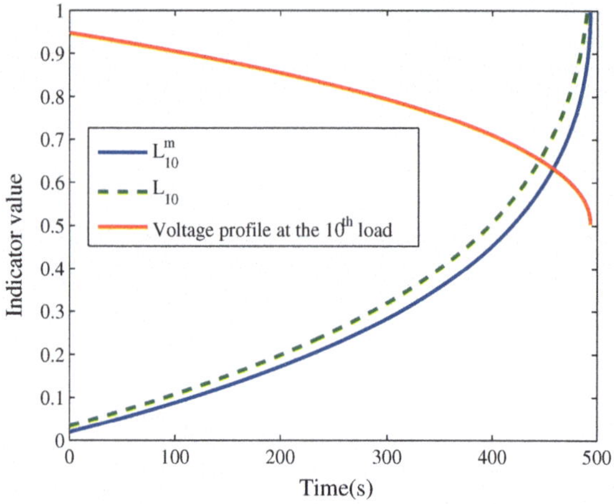

Fig. 26 Comparisons of the existing L-index L_{10} and the modified L-index L^m_{10} with considering generator reactive power limit in IEEE 118-bus system

shows the waveform of the existing L_{10} and the modified L^m_{10}. It can be observed that the accurate modified L-index is also accurate to detect the voltage collapse point.

5 Conclusion

In this work, the modified coupled single-port model is proposed to perform the high accurate measurement-based long-term VSA. By using two consecutive PMU measurements, the mitigation factor can be calculated and utilized to modify the existing coupled single-port model. Accordingly, the proposed modified coupled single-port model can be constructed. The additional reactive power support corresponds to the reactive power response from the extended Ward-type network equivalent, also indicating to the basic of the mitigation factor. Due to the measurement-based calculation of the mitigation factor, it is suitable to be applied in real-time environment. Moreover, the proposed model can be slightly modified to perform VSA considering generator reactive power limits. In the proposed work, both L-index and maximum loading parameters are examined for VSA. Case studies are conducted on IEEE 57-bus and IEEE 118-bus test systems to verify the accuracy of the proposed modified coupled single-port model in VSA.

References

1. IEEE/CIGRE Joint Task Force on Stability Terms and Definitions, Definition and classification of power system stabilit. IEEE Trans. Power Syst. **19**(2), 1387–1401 (2004)
2. C.A. Canizares, F.L. Alvarado, C.L. DeMarco, I. Dobson, W.F. Long, Point of collapse methods applied to AC/DC power systems. IEEE Trans. Power Syst. **7**(2), 673–683 (1992)
3. V. Ajjarapu, C. Christy, The continuation power flow: a tool for steady state voltage stability analysis. IEEE Trans. Power Syst. **7**(1), 416–422 (1992)

4. H.D. Chiang, A.J. Flueck, K.S. Shah, N. Balu, CPFLOW: a practical tool for tracing power system steady-state stationary behaviour due to load and generation variation. IEEE Trans. Power Syst. **10**(2), 623–633 (1995)
5. I. Dobson, Computing a closest bifurcation instability in multidimensional parameter space. J. Nonlinear Sci. **3**(1), 307–327 (1993)
6. C.A. Canizares, Calculating optimal system parameters to maximize the distance to saddle-node bifurcations. IEEE Trans. Circuit Syst.-I Fundam. Theory Appl. **45**(3), 225–237 (1998)
7. G.D. Irisarri, X. Wang, J. Tong, S. Mokhtari, Maximum loadability of power systems using interior point nonlinear optimization method. IEEE Trans. Power Syst. **12**(1), 6162–6172 (1997)
8. G. Verbic, F. Gubina, A new concept of voltage-collapse protection based on local phasors. IEEE Trans. Power Deliv. **19**(2), 567–581 (2004)
9. G. Verbic, F. Gubina, Fast voltage-collapse line-protection algorithm based on local phasors. IEEE Proc. Gener. Transm. Distrib. **150**(4), 482–486 (2003)
10. M. Glavic T.V. Cutsem, A short survey of methods for voltage instability detection, *Proceeding of the 2011 IEEE/PES Winter Meeting* (2011), pp. 1–8
11. K. Vu, M.M. Bagovic, D. Novosel, M.M. Saha, Use of local measurements to estimate voltage stability margin. IEEE Trans. Power Syst. **14**(3), 1029–1035 (1999)
12. M. Parniani, M. Vanouni, A fast local index for online estimation of closeness to load limit. IEEE Trans. Power Syst. **25**(1), 584–585 (2010)
13. B. Venjatesh, A. Rost, L. Chang, Dynamic voltage collapse index-wind generation application. IEEE Trans. Power Deliv. **22**(1), 90–94 (2007)
14. A. Wiszniewski, New criteria of voltage stability margin for the purposed of load shedding. IEEE Trans. Power Deliv. **22**(3), 1367–1371 (2007)
15. K. Seethalekshimi, S.N. Singh, S.C. Srivastava, A synchrophasor assisted frequency and voltage stability based load shedding scheme. IEEE Trans. Smart Grid **2**(2), 221–230 (2011)
16. Y.V. Makarov, P. Du, S. Lu, T.B. Nguyen, X. Guo, J.W. Burns, J.F. Gronquist, M.A. Pai, PMU-based wide-area security assessment: concept, method, and implementation. IEEE Trans. Smart Grid **3**(3), 1325–1332 (2012)
17. C.W. Taylor, The future in on-line security assessment and wide-area stability control, *Proceeding of the 2000 IEEE/PES Winter Meeting*, vol. 1 (2000)
18. W. Li, Y. Wang, T. Chen, Investigation on the Thevenin equivalent parameters for online estimation of maximum power tansfer limits. IET Gener. Transm. **18**(1), 121–127 (2010)
19. Y. Wang, I.R. Pordanjani, W. Li, W. Xu, E. Vaahedi, Strategy to minimise the load shedding amount for voltage collapse prevention. IET Gener. Transm. **5**(3), 307–313 (2010)
20. Y. Wang, I.R. Pordanjani, W. Li, W. Xu, T. Chen, E. Vaahedi, J. Gurney, Voltage stability monitoring based on the concept of couple single-port circuit. IEEE Trans. Power Syst. **26**(4), 2154–2163 (2011)
21. W. Xu, I. Pordanjani, Y. Wang, E. Vaahedi, A network decoupling transform for phasor data based voltage stability analysis and monitoring. IEEE Trans. Smart Grid **3**(1), 261–270 (2012)
22. J.H. Liu, C.C. Chu, Wide-area measurement-based voltage stability indicators by modified coupled single-port models. IEEE Trans. Power Syst. **29**(2), 756–764 (2013)
23. F.F. Wu, A. Monticelli, Critical review of external network modelling for online security analysis. Int. J. Electr. Energy Syst. **5**(4), 222–235 (1983)
24. P.M. De Oliveira-De Jesus, E.D. Castronuovo, M.T. Ponce de Leao, Reactive power response of wind generators under an incremental network-loss allocation approach. IEEE Trans. Energy Convers. **23**(2), 612–621 (2008)
25. P. Kessel, H. Glavitsch, Estimating the voltage stability of a power system. IEEE Trans. Power Deliv. **1**(3), 346–352 (1986)
26. H. Jia, X. Yu, Y. Yu, An improved voltage stability index and its application. Int. J. Electr. Power Energy Syst. **27**(8), 567–574 (2005)
27. T.L. Baldwin, L. Mili, M.B. Boisen, R. Adapa, Power system observability with minimal phasor measurement placement. IEEE Trans. Power Syst. **8**(2), 707–715 (1993)
28. A. Monticelli, S. Deckmann, A. Garcia, B. Sccot, Real-time external equivalents for static security analysis. IEEE Trans. Power Appar. Syst. **PAS-98**(2), 498–503 (1979)

29. P. Kansal, A. Bose, Bandwidth and latency requirements for smart transmission grid applications. IEEE Trans. Smart Grid **3**(3), 1344–1352 (2012)
30. B. Milosevic, M. Begovic, Voltage-stability protection and control using a wide-area network of phasor measurements. IEEE Trans. Power Syst. **18**(1), 121–127 (2003)
31. D.J. Hill, Nonlinear dynamic load models with recovery for voltage stability studies. IEEE Trans. Power Syst. **8**(1), 166–176 (1993)

Adaptive WAMS-Based Secondary Voltage Control

Heng-Yi Su, Jian-Hong Liu, and Chia-Chi Chu

Abstract Voltage instability is a growing threat to the security and the reliability of power grids, especially as the penetration level of intermittent renewable energies increase significantly in recent years. Voltage instability and even voltage collapse will take place as the loss of control of the voltage profiles in a power system. To achieve more efficient voltage regulation in power systems, the hierarchical three-level coordinated voltage control mechanism has been developed recently to prevent voltage collapse through the appropriate management of reactive power sources. This chapter presents recent developments in adaptive secondary voltage control (SVC) by utilizing real-time measurements of power systems obtained from the wide-area measurement system (WAMS). These methods are adaptive in the sense that load disturbances are estimated from synchronized phasors of WAMS in nearly real-time. Thus, these control inputs of SVC can be synthesized to minimize deviations in load voltage profiles under the worst-case scenario. Uncertainties in measurement are also taken into considerations by exploring the maximum likelihood (ML) method to further improve SVC performance. Comprehensive simulations on a variety of IEEE benchmark systems have been performed to verify the feasibility and the effectiveness of these schemes.

Keywords Phasor measurement unit (PMU) · Maximum likelohood · Secondary voltage control (SVC) · Wide area measurement system (WAMS) · Worst-case design

---

H.-Y. Su
Department of Electrical Eng, Feng Chia University, Taichung 40724, Taiwan
e-mail: hengyisu@fcu.edu.tw

J.-H. Liu
Department of Electrical Eng, Yuan Ze University, Taoyuan 32003, Taiwan
e-mail: jhliu727@saturn.yzu.edu.tw

C.-C. Chu (✉)
Department of Electrical Eng, National Tsing Hua University, Hsinchu 30013, Taiwan
e-mail: ccchu@ee.nthu.edu.tw

H. Haes Alhelou et al. (eds.), *Wide Area Power Systems Stability, Protection, and Security*, Power Systems, https://doi.org/10.1007/978-3-030-54275-7_13

1 Introduction

Voltage instability has recently become one of the primary threats to security and reliable operation of power grids, especially as the penetration level of intermittent renewable energies increase significantly in recent years [1]. Voltage instability and even voltage collapse will take place as the loss of control of the voltage profiles in a power system. In history, a number of severe blackout events have been reported to be associated with this class of voltage instability problems [2].

To achieve more efficient voltage regulation in power systems, the Automatic voltage control (AVC) scheme has been developed in the past few decades as a countermeasure to reduce the risk of voltage collapse. As shown in Fig. 1, the hierarchical architecture of AVC consists of three levels: tertiary, secondary, and primary voltage regulation. Tertiary voltage control (TVC) is tasked with determining the optimal pilot-bus set points for secondary voltage control (SVC). SVC is tasked with regulating voltage and reactive power in the power system. Primary voltage control (PVC) is tasked with handling random short-term deviations in voltage via automated voltage regulators. Numerous research studies have been achieved in developing SVCs. For example, the pioneer work was initialized in France [3]. Later works have been developed in other European countries [3–7], Brazil [8], South Africa [9], the United States [10, 11], and China [12].

In the past two decades, various methods have been presented to design SVCs. Some of them have been devoted to the pilot-bus selection problem [3–6, 16, 17]. Once the locations of pilot nodes are determined, two available control laws can be

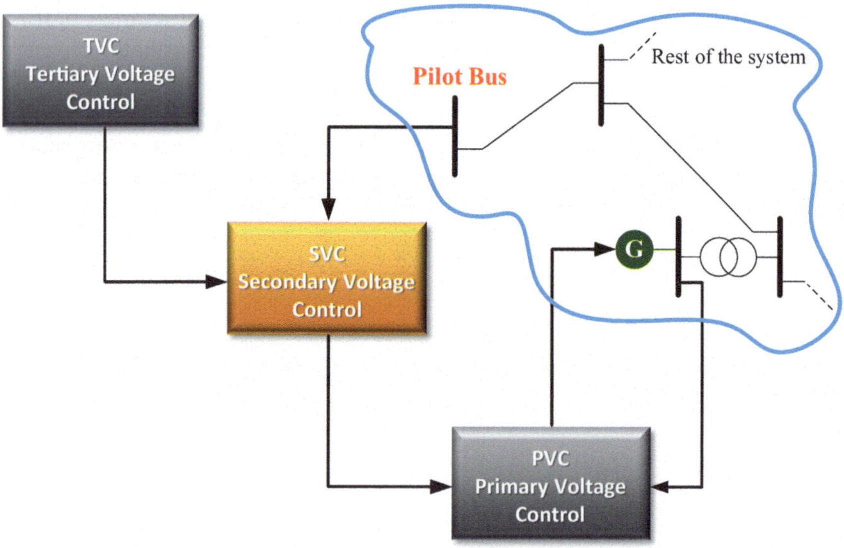

Fig. 1 Hierarchical architecture of AVC

applicable in SVC design. Typically, the control method focuses only on regulating voltages at the given pilot buses, but this kind of method is easily influenced by system topological variations. However, this approach is susceptible to variations in system topology. Recent advances in SVC consider voltage variations in pilot buses as well as other load buses to ensure good performance despite variations in topology [17–25]. Researchers have also developed a system referred to as robust secondary voltage control (RSVC) based on the worst-case design (WCD) scheme [21, 22]; however, that approach depends heavily on a given set of assumptions and is not easily adapted to prevailing system conditions.

To overcome the drawback of the RSVC method, our recent developments in adaptive SVC will be presented in this chapter [23–25]. These proposed methods are adaptive in the sense that load disturbances are estimated in real-time from the wide-area measurement systems (WAMS) [26, 27]. Wide-area measurement systems (WAMS) comprise a variety of phasor measurement units (PMUs), which give precise time-synchronized measurements from voltage phasors and current phasors covering a wide area. Thus, this coherent and time-synchronized WAMS can be considered in synthesizing more accurate control schemes for wide-area power system voltage control. In this study, it was assumed that PMUs were placed at target pilot nodes. Changes in pilot-bus voltage at a given location are re-formulated as an equality constraint. Control inputs are used to ensure that the most extreme changes in load voltage are constrained within a predefined range. Voltage magnitudes at pilot buses are used as inputs, which generate control actions as outputs.

In our first study [23], we presented a two-phase algorithm. The objective in the first phase was to minimize the total reactive power under the equality constraint generated by the pilot bus. The objective in the second phase was to identify an optimal set of control solutions capable of minimizing the worst-case deviations in load voltage. The fact that the objective function involved minimizing the worst-case change in load voltage made it possible to ensure precise and highly-responsive secondary voltage control. That scheme was shown to outperform the conventional robust method in terms of control accuracy and computation time.

In our second study [24], we presented a WAMS-based SVC method in which the successive approximate vector ∞-norm minimization algorithm was used to solve the nonlinear constrained optimization problem, while taking into account power flow equality constraints as well as control variable inequality constraints. The basic strategy involved the iterative linearization of the nonlinear constrained equation as fast-decoupled power flow (FDPF) equations, wherein deviations in load bus voltage were expressed in terms of load disturbance and control variables. Extensive simulations on various IEEE benchmark systems demonstrated the feasibility and effectiveness of this approach.

In our most recent study [25], we reformulated the SVC problem as a robust optimization problem using the maximum likelihood (ML) approach to overcome errors in the measurement data. An interior-point technique was employed to solve the ML problem. This approach greatly outperforms other worst-case-based SVC methods in terms of advantages and effectiveness. Numerical studies on several IEEE benchmark systems have confirmed these claims.

The remainder of this chapter is organized as follows. In Sect. 2, an overview of WAMS-based SVC will be provided. A two-phase adaptive WAMS-based SVC Scheme will be proposed in Sect. 3. The successive approximate vector ∞-norm minimization algorithm to solve the nonlinear constrained optimization problem with considering both power flow equality constraints and control variable inequality constraints will be studied in Sect. 4. The ML-based robust SVC will be investigated in Sect. 5. Finally, some comparison studies and conclusions will be made in Sects. 6 and 7 respectively.

2 Overview of WAMS-Based SVC

2.1 Problem Statement

Consider the following voltage-var control model

$$[\Delta \mathbf{Q}] = B[\boldsymbol{\Delta}|\mathbf{V}|], \tag{1}$$

where $\boldsymbol{\Delta}\mathbf{Q}$, $\boldsymbol{\Delta}|\mathbf{V}|$, and B are injected reactive power deviations, bus voltage magnitude deviations, and system susceptance, respectively. We rewrite (1) as follows:

$$\begin{bmatrix} \Delta \mathbf{Q}_{\mathrm{G}} \\ \Delta \mathbf{Q}_{\mathrm{L}} \end{bmatrix} = \begin{bmatrix} B_{GG} & B_{GL} \\ B_{LG} & B_{LL} \end{bmatrix} \begin{bmatrix} \boldsymbol{\Delta}|\mathbf{V}_{\mathrm{G}}| \\ \boldsymbol{\Delta}|\mathbf{V}_{\mathrm{L}}| \end{bmatrix}, \tag{2}$$

where subscripts L and G are utilized to denote load buses and generator buses. Furthermore, let

$$\mathbf{q} = \Delta \mathbf{Q}_{\mathrm{L}} \text{ and} \tag{3}$$

$$\mathbf{u} = \boldsymbol{\Delta}|\mathbf{V}_{\mathrm{G}}| \tag{4}$$

respectively refer to reactive power load disturbances and control variables. Therefore, $\boldsymbol{\Delta}|\mathbf{V}_{\mathrm{L}}|$ is expressed in terms of $\mathbf{q}$ and $\mathbf{u}$ as

$$\boldsymbol{\Delta}|\mathbf{V}_{\mathrm{L}}| = J_1\mathbf{q} - J_2\mathbf{u}, \tag{5}$$

where J_1 and J_2 are matrices in the following forms:

$$J_1 = B_{LL}^{-1}, \tag{6}$$

$$J_2 = J_1 B_{LG}. \tag{7}$$

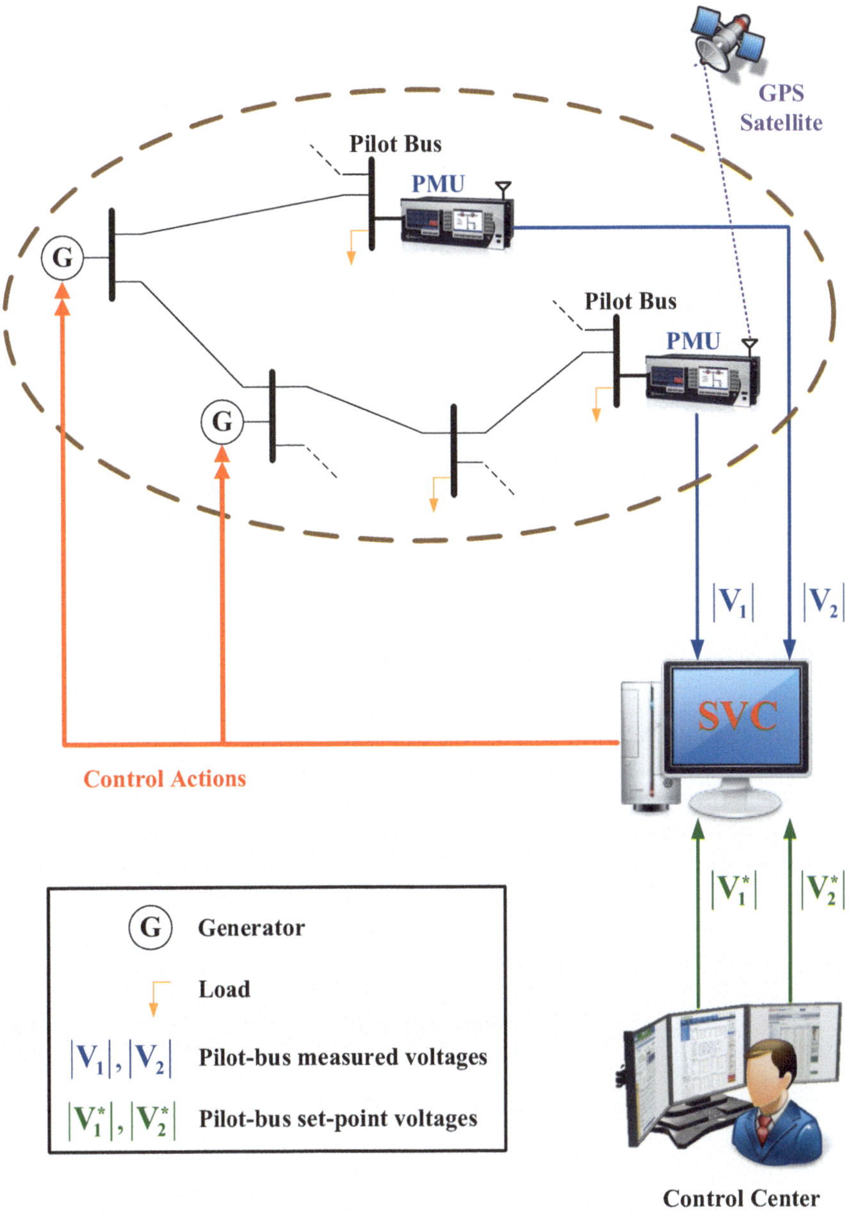

Fig. 2 Typical structure of WAMS-based SVC

2.2 WAMS-Based SVC

Figure 2 presents the feedback control structure of SVC, where monitored deviations in pilot-bus voltage are used as control inputs to generate VC actions. Suppose that PMUs are installed to monitor voltage variations at all of the target pilot nodes, which can be defined as follows:

$$\Delta|\mathbf{V_P}| = \left|\mathbf{V_P^*}\right| - |\mathbf{V_P}|, \tag{8}$$

where $\left|\mathbf{V_P^*}\right|$ and $|\mathbf{V_P}|$ respectively denote pilot-bus set-point voltages and pilot-bus measured voltages.

On the other hand, since $\mathbf{\Delta}|\mathbf{V_P}|$ are only some elements of $J_l\mathbf{q}$, $\mathbf{\Delta}|\mathbf{V_P}|$ can be rewritten as

$$\mathbf{\Delta}\left|\mathbf{V_p}\right| = J_p\mathbf{q}, \tag{9}$$

where J_p is the matrix with the rows of J_l associated with the buses selected.

3 Adaptive WAMS-Based SVC Scheme

In our first study on WAMS-based SVC, the objective function of our optimal control model was to minimize the worst-case variations in $\mathbf{\Delta}|\mathbf{V_L}|$. In mathematical terms, the problem involved minimizing $\| \mathbf{\Delta}|\mathbf{V_L}| \|_\infty$. Figure 3 details the computational procedure of the proposed adaptive SVC (ASVC) scheme, in which $\mathbf{\Delta}|\mathbf{V_P}|$ are used as control inputs. The task in the first phase is to identify a value for $\mathbf{q}$, such that $J_p\mathbf{q} = \mathbf{\Delta}\left|\mathbf{V_p}\right|$. In this research, we use the least-norm technique to approximate parameter vector $\mathbf{q}$ without the need to take further measurements. Hence, we can state the problem of optimizing $\mathbf{q}$ as follows:

$$\text{minimize} \| \mathbf{q} \|_2 \tag{10}$$

$$\text{subject to } J_p\mathbf{q} = \mathbf{\Delta}\left|\mathbf{V_p}\right|. \tag{11}$$

The optimal $\mathbf{q}^*$ can be obtained by

$$\mathbf{q}^* = J_p^T\left(J_pJ_p^T\right)^{-1}\mathbf{\Delta}\left|\mathbf{V_p}\right|. \tag{12}$$

The second phase of the computation process involves identifying the optimal set of control solutions by minimizing $\| \mathbf{\Delta}|\mathbf{V_L}| \|_\infty$ based on $\mathbf{q}^*$. Thus, the voltage control strategy is formulated as the following search problem:

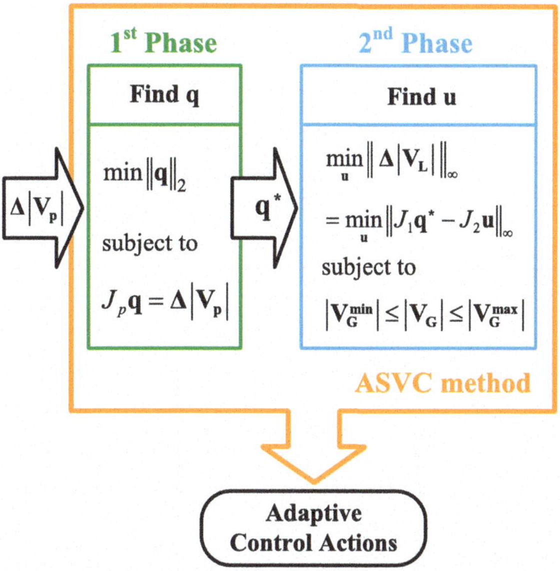

Fig. 3 Computational procedure used in ASVC

$$\min_{\mathbf{u}} \quad \| \Delta|\mathbf{V}_L| \|_\infty = \min_{\mathbf{u}} \left\| J_1\mathbf{q}^* - J_2\mathbf{u} \right\|_\infty \tag{13}$$

$$\text{subject to } \mathbf{u}^{\min} \le \mathbf{u} \le \mathbf{u}^{\max} \tag{14}$$

where $\mathbf{u}^{\min}$ and $\mathbf{u}^{\max}$ represent the lower and upper limits of $\mathbf{u}$. The expression (13) can be cast as a linear programming (LP) problem. Such LP problem can be solved by interior-point algorithm [28].

4 Successive Approximation Algorithm for ASVC

Researchers have developed a wide range of SVC methods [21–23] based on WCD. A review of the literature revealed that WCD schemes are based on the linearized model of FDPF equations. Note however that linearized Q-V models do not necessarily address all of the voltage stability issues in the electrical power systems encountered in the real world.

In our second study, the VC problem was formulated as a nonlinear constrained optimization problem:

$$\min_{\mathbf{u}} \quad f = \| \Delta|\mathbf{V}_L| \|_\infty \tag{15}$$

$$g(\mathbf{x}, \mathbf{u}, \mathbf{P}, \mathbf{Q}) = \mathbf{0}, \text{ and} \tag{16}$$

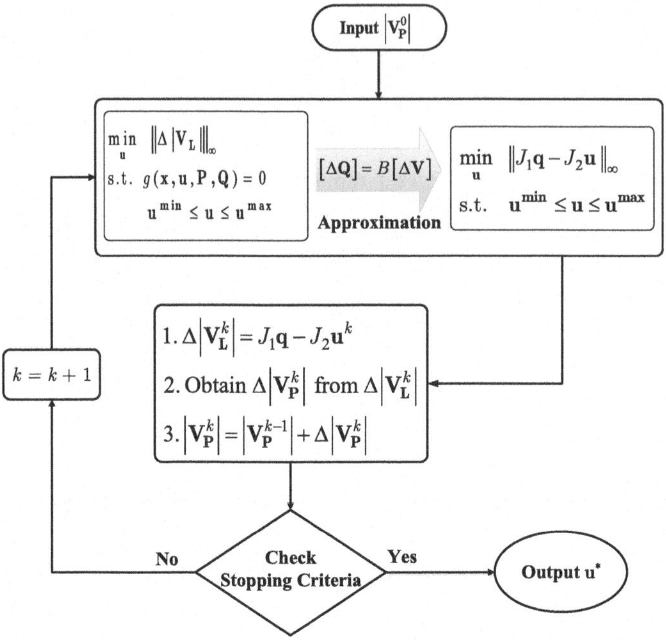

Fig. 4 Solution process of the proposed successive approximation algorithm

$$\mathbf{u}^{\min} \leq \mathbf{u} \leq \mathbf{u}^{\max}, \tag{17}$$

where (16) denotes the nonlinear power flow equation, in which $\mathbf{x}$, $\mathbf{P}$, and $\mathbf{Q}$ are vector of bus voltage phasors, vector of active bus powers, and vector of reactive bus powers, respectively.

Figure 4 illustrates the solution process of the proposed successive approximation algorithm for solving the problem (15)–(17). The idea of this algorithm is that the nonlinear constrained Eq. (16) is iteratively linearized as the FDPF equations which approximate (16). In the FDPF equations, $\Delta|\mathbf{V}_L|$ is expressed in terms of $\mathbf{q}$ and $\mathbf{u}$, where $\mathbf{q}$ can be linearly estimated by $|\mathbf{V}_P|$. Then the problem (15)–(17) was resolved iteratively by successively solving the approximate LP problem. The convergence of FDPF equations is well-established in the literature [29]; therefore, it is reasonable to expect that this algorithm should always reach convergence.

5 ML-Based Robust SVC

Traditional WAMS-based SVC methods [22–24] do not consider cases in which measured data from WAMS may contain errors. In this research, we consider the following problem

$$v_i = \tilde{\mathbf{b}}_i^T \mathbf{u} + \varepsilon_i, \quad i = 1, \ldots, n_L \tag{18}$$

where v_i are load bus voltage data measured from PMUs; ε_i are measurement errors with density $p_u(\varepsilon)$; $\tilde{\mathbf{b}}_i^T$ are the rows of J_2; and n_L is the number of load buses.

The log-likelihood function for (18) is defined by

$$\begin{aligned} L(\mathbf{u}) &\triangleq \log\left(\prod_{i=1}^{n} p_u(\varepsilon_i)\right) \\ &= \log\left(\prod_{i=1}^{n} p_u(v_i - \tilde{\mathbf{b}}_i^T \mathbf{u})\right) \\ &= \sum_{i=1}^{n} \log\left(p_u(v_i - \tilde{\mathbf{b}}_i^T \mathbf{u})\right) \end{aligned} \tag{19}$$

Due to the fact that ε in field PMU data has a Laplace distribution [30], $p_u(\varepsilon) = (1/2\sigma)\exp\left(-|\varepsilon|/\sigma\right)$ for some $\sigma > 0$. Hence, we rewrite (19) as

$$\begin{aligned} L(\mathbf{u}) &= \sum_{i=1}^{n} \log\left(\frac{1}{2\sigma}\exp\left(-\frac{\left|v_i - \tilde{\mathbf{b}}_i^T \mathbf{u}\right|}{\sigma}\right)\right) \\ &= \sum_{i=1}^{n}\left(-\log(2\sigma) - \frac{\left|v_i - \tilde{\mathbf{b}}_i^T \mathbf{u}\right|}{\sigma}\right) \\ &= -n\log(2\sigma) - \frac{1}{\sigma}\sum_{i=1}^{n}\left|v_i - \tilde{\mathbf{b}}_i^T \mathbf{u}\right| \end{aligned} \tag{20}$$

The concept underlying the maximum likelihood method [31] is to estimate $\mathbf{u}$ capable of maximizing $L(\mathbf{u})$, i.e.,

$$\text{minimize} \quad -L(\mathbf{u}) = n\log(2\sigma) + \frac{1}{\sigma}\sum_{i=1}^{n}\left|v_i - \tilde{\mathbf{b}}_i^T \mathbf{u}\right| \tag{21}$$

The solution to the problem (21) is independent of σ; therefore, the problem (21) can be cast as a LP problem. Such problem can be solved by interior-point algorithm [28]. The illustration of the proposed ML-based robust SVC is given in Fig. 5.

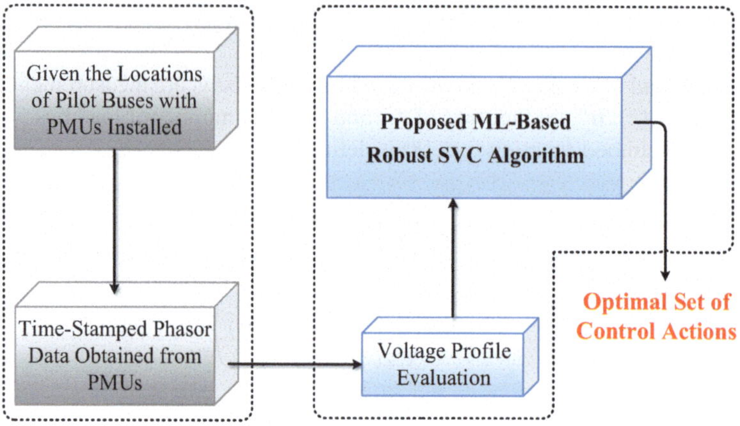

Fig. 5 Overview of the proposed ML-based robust SVC

6 Case Studies and Results

In this section, we present numerical examples of the control methods mentioned above. Analysis was performed using the IEEE 14-bus, IEEE 30-bus, IEEE 118-bus benchmark test systems. All of the data are provided in [32]. All of the control methods were implemented in MATLAB. Note that all numerical simulations were performed on a personal computer with an Intel Core i7 CPU (3.4 GHz) with 8-GB RAM.

To evaluate SVC performance, voltage profile improvement (VPI) and loss ratio (LR) are considered. Both of them are defined by

$$y^{\mathrm{rms}} = \sqrt{\frac{1}{n_L}\sum_{j=1}^{n_L} \left\| y_j \right\|_2^2} \tag{22}$$

$$\mathrm{LR}(\%) = \frac{\text{transmission loss}}{\text{total system load}} \times 100 \tag{23}$$

where y_j is the resulting load bus voltage change. Notice that lower VPI and LR values indicate better performance.

Figure 6 presents a typical case involving the IEEE 118-bus, before and after application of our SVC scheme. It is clear that $|\mathbf{V}_\mathrm{L}|$ improved considerably following implementation of the control actions.

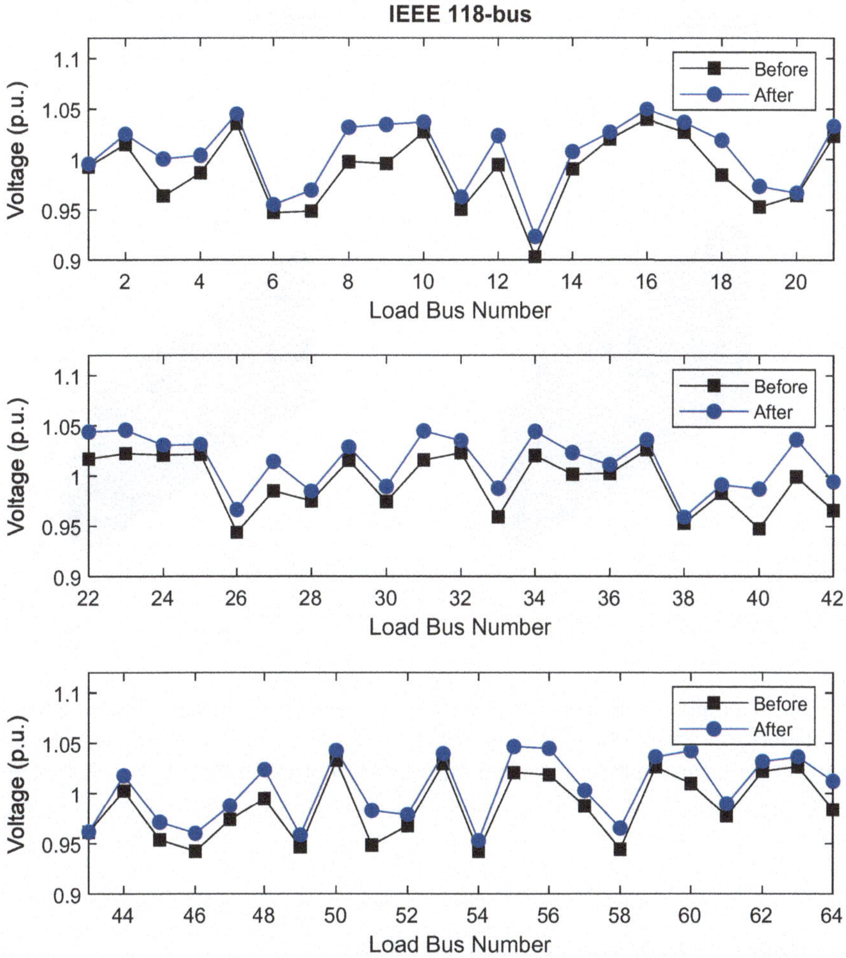

Fig. 6 Typical case for IEEE 118-bus

6.1 *Effects of the Number of Pilot Buses*

We performed a performance comparison of the proposed ASVC with RSVC when applied to the IEEE 14-bus and IEEE 30-bus test systems. The measure used to evaluate the performance can be expressed as

$$\bar{y}_{n_p}^{rms} = \frac{1}{n_r} \sum_{j=1}^{n_r} y_j^{\mathrm{rms}} \tag{24}$$

where n_p is the number of pilot buses and n_r is the number of simulation runs.

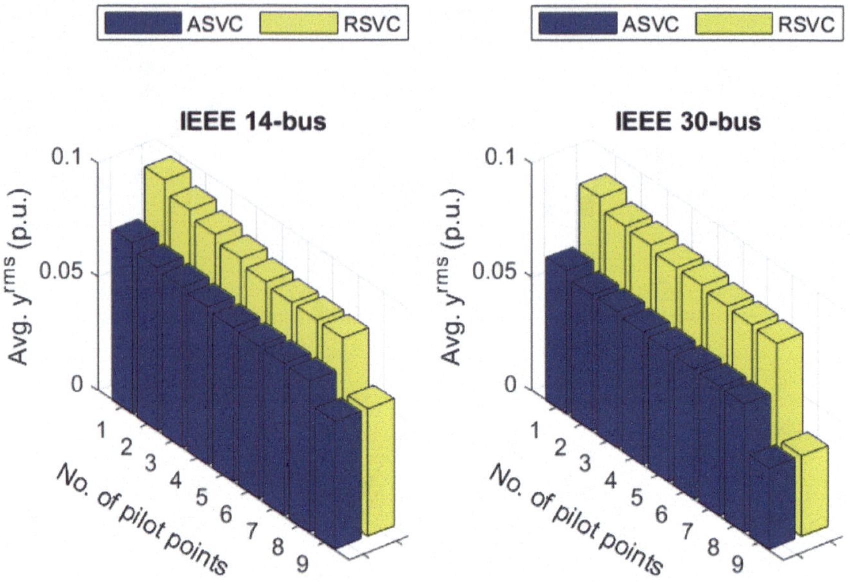

Fig. 7 Comparison results in terms of $\bar{y}_{n_p}^{rms}$ for each n_p

Figure 7 compares the performance of ASVC and RSVC under selected cases. Clearly the $\bar{y}_{n_p}^{rms}$ values obtained using ASVC are far lower than those obtained using RSVC. This means that for a limited number of pilot buses, ASVC is superior to RSVC. As shown in Fig. 8, ASVC also outperformed RSVC in terms of computation time.

6.2 *Impact of Incoherent Dataset of* $|\mathbf{V_p}|$

We examined the impact of an incoherent dataset of $|\mathbf{V_p}|$ on SVC performance by performing multiple simulations under a variety of load disturbances with various n_p. Let

$$\left|\mathbf{V_p^0}\right| = \begin{bmatrix} \left|V_1^0\right| \\ \left|V_2^0\right| \\ \vdots \\ \left|V_{n_p}^0\right| \end{bmatrix} \text{and} \left|\mathbf{V_p^1}\right| = \begin{bmatrix} \left|V_1^0\right| + \alpha \\ \left|V_2^0\right| + \alpha \\ \vdots \\ \left|V_{n_p}^0\right| + \alpha \end{bmatrix} \quad (25)$$

respectively represent a coherent dataset and an incoherent dataset. In (25), $\{\left|V_i^0\right| \mid i = 1, 2, \ldots, n_p\}$ denote $|\mathbf{V_p}|$ with the same time-stamp and α is a random number uniformly distributed between −0.03 and 0.03 p.u.

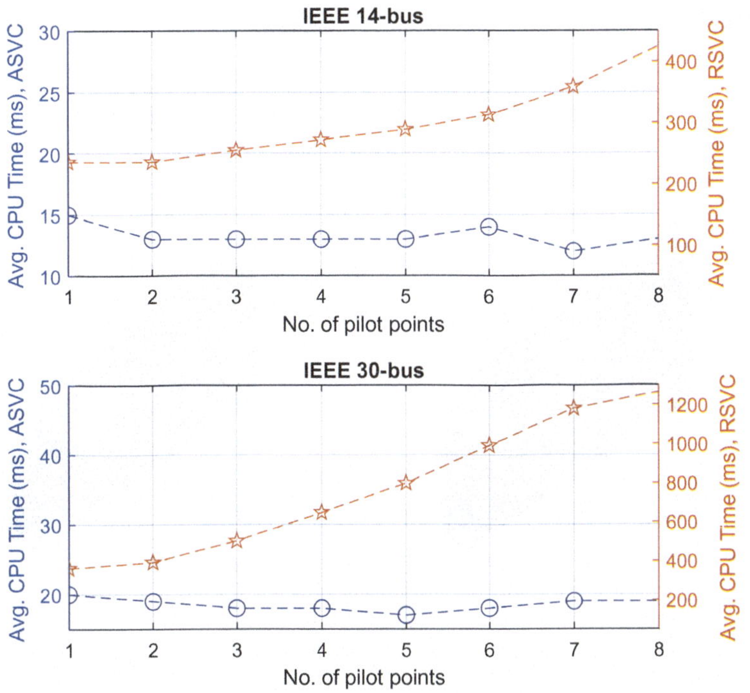

Fig. 8 Comparison results in terms of average computation time for each n_p

Figure 9 demonstrates the test results in which a smaller $\bar{y}_{n_p}^{rms}$ value is indicative of better performance. This figure clearly shows performance degradation when the control method was applied to an incoherent dataset.

6.3 *Impact of Load Disturbances*

The robustness and performance of the WCD-based methods were verified in a wide range of experiments using different types of load patterns. The variations in load included active and reactive powers ranging from roughly 10% to 40% of the initial load level.

Figure 10 presents the results of simulations performed on the various test systems under different load levels using an assortment of load variations and load patterns. Clearly, the VPI values computed using the method [22] and method [23] are far higher than the proposed method. Conversely, the VPI points obtained using the proposed method are smaller.

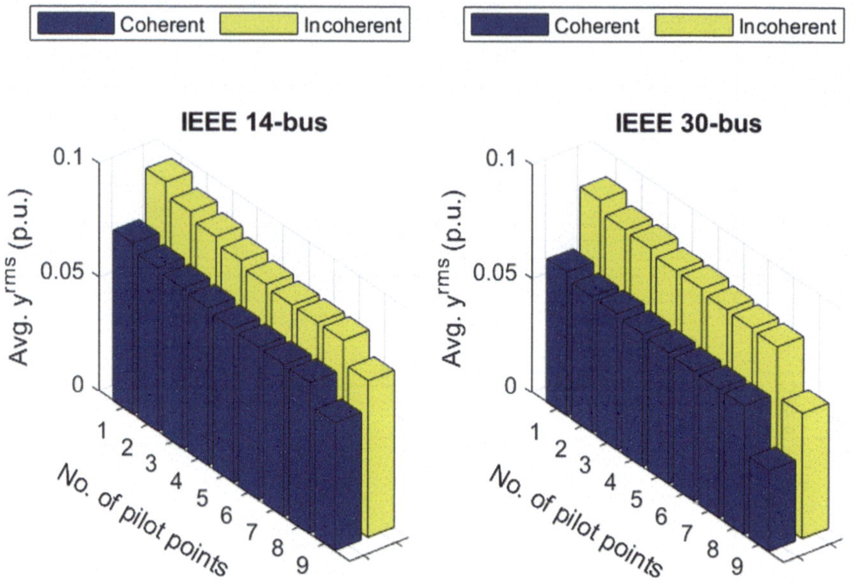

Fig. 9 Impact of an incoherent dataset of $|\mathbf{V_P}|$ on SVC performance

In the above-mentioned simulations using IEEE test platforms (ranging from small to bulk systems), the proposed method achieved acceptable performance regardless of system size.

6.4 Impact of Topology Changes

We studied a large number of test cases before assessing the performance of the WCD-based methods. We conducted these simulations using a variety of test systems under a variety of line outage situations. Figure 11 presents VPI index results obtained under various line outage conditions. Clearly, the proposed method outperformed the method [22] and the method [23], regardless of the line outage conditions.

6.5 Impact of Measurement Errors

In this study, we considered a total vector error of 1% in our simulations of noisy PMU data, in accordance with IEEE standards [33]. The results obtained using WCD-based methods [22, 23] are compared with those obtained using the proposed robust ML method. Figures 12 and 13 respectively present the compared results under different

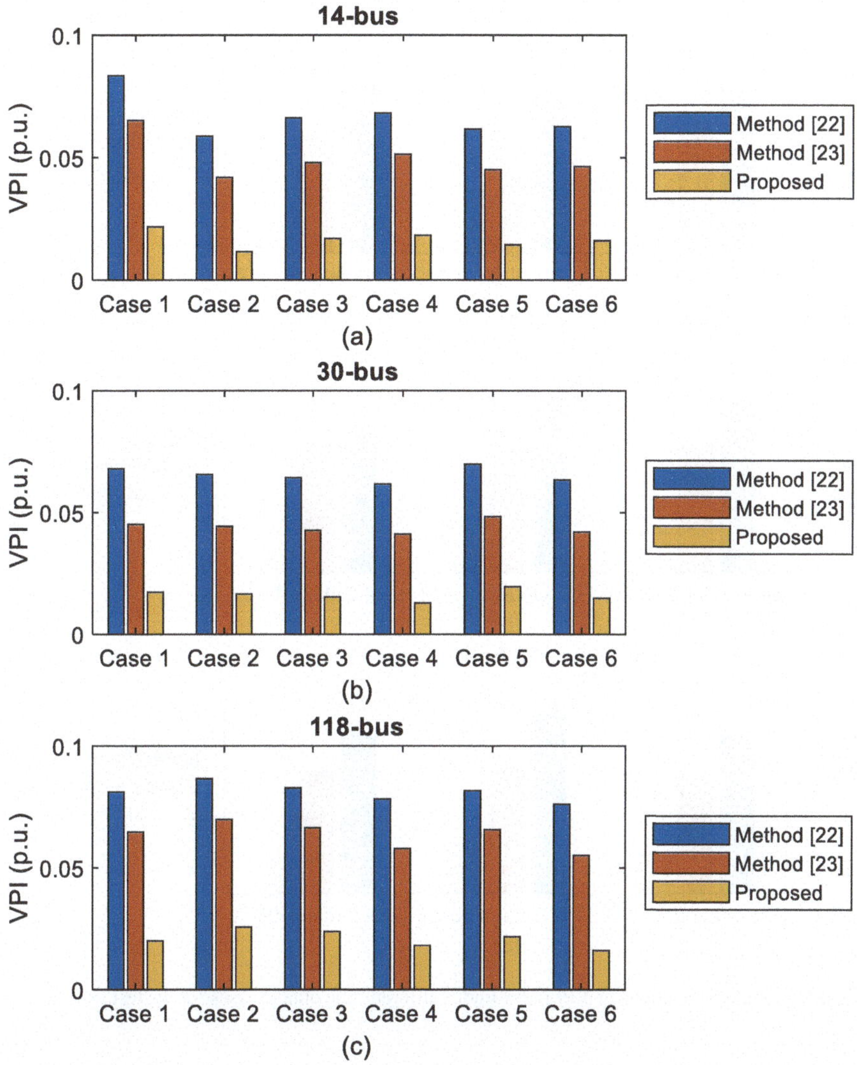

Fig. 10 Comparison results of VPI index under different load disturbance cases for the test system

measurement error cases. Form the show figures, it is clear that the proposed robust ML method performed well even in cases of 1% measurement error.

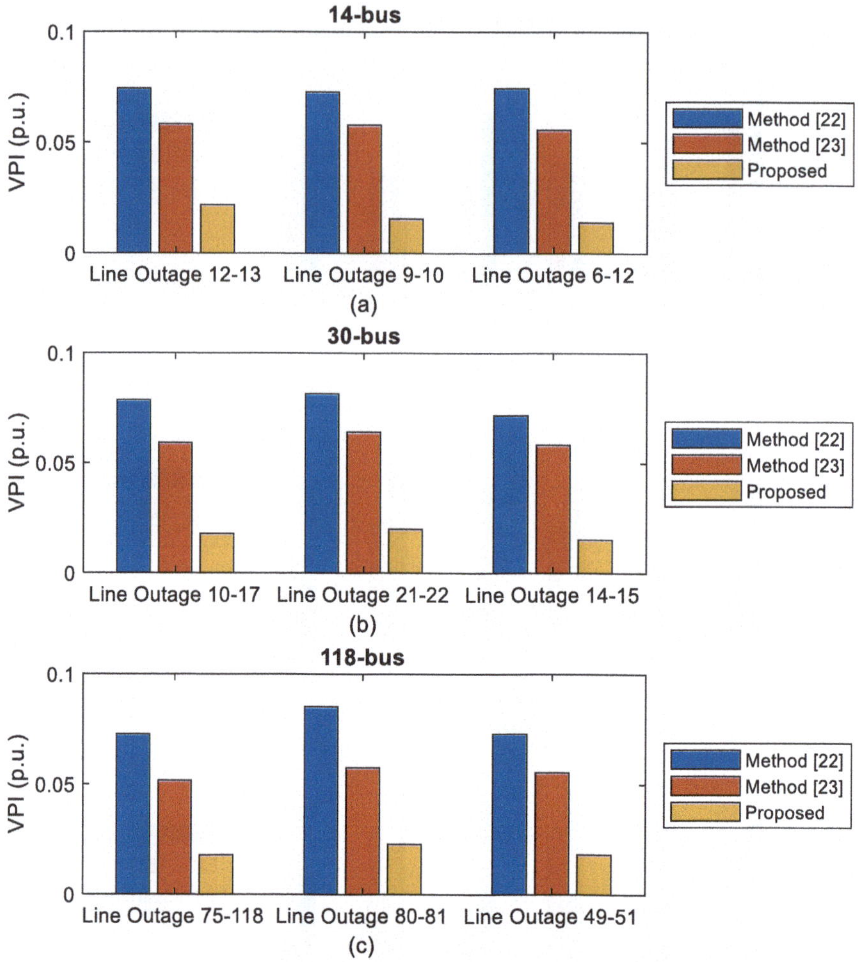

Fig. 11 Comparison results of VPI index under different line outage cases for the test system

7 Conclusion

To achieve better voltage control in power grids, this chapter provide an overview of our recent developments about adaptive WAMS-based SVC schemes. The optimal control actions are determined by using WCD. The proposed schemes aim to use voltage magnitudes at pilot buses as control feedback to optimally manage reactive power sources in the enhancement of system voltage profile. With the utilization of synchrophasors from WAMS as trigger signals, an efficient and accurate SVC can be achieved. Extensive simulations on various IEEE benchmark systems have been investigated to demonstrate the advantage and effectiveness of the proposed schemes.

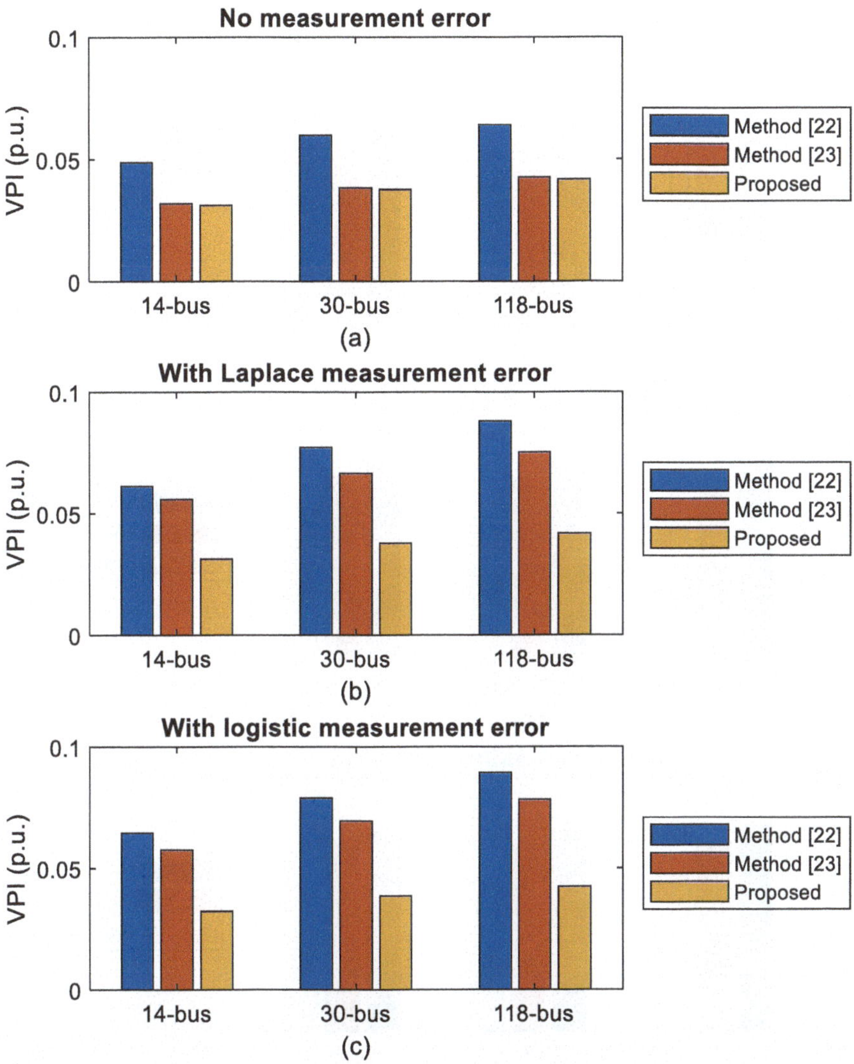

Fig. 12 Comparison results of VPI index under different measurement error cases for the test system

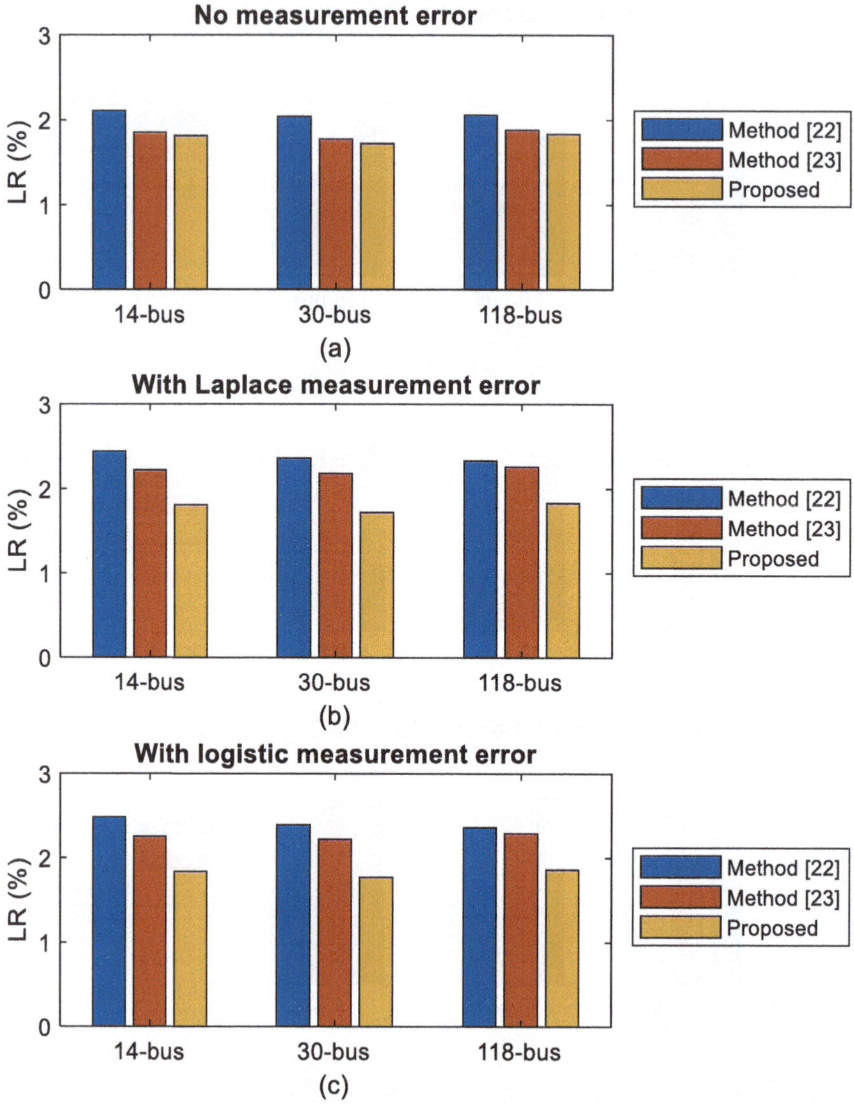

Fig. 13 Comparison results of LR index under different measurement error cases for the test system

References

1. P. Kundur et al., Definition and classification of power system stability IEEE/CIGRE joint task force on stability terms and definitions. IEEE Trans. Power Syst. **19**(3), 1387–1401 (2004)
2. V. Ajjarapu, *Computational Techniques for Voltage Stability Assessment and Control* (Springer, New York, NY, USA, 2006)
3. J.P. Paul, J.T. Leost, J.M. Tesseron, Survey of the secondary voltage control in France: present realization and investigations. IEEE Trans. Power Syst. **2**(2), 505–511 (1987)

4. S. Corsi, P. Marannino, N. Losignore, G. Moreschini, G. Piccini, Coordination between the reactive power scheduling function and the hierarchical voltage control of the EHV ENEL system. IEEE Trans. Power Syst. **10**(2), 686–694 (1995)
5. H. Vu, P. Pruvot, C. Launay, Y. Harmand, An improved voltage control on large scale power system. IEEE Trans. Power Syst. **11**(3), 1295–1303 (1996)
6. J.L. Sancha, J.L. Fernandez, A. Cortes, J.T. Abarca, Secondary voltage control: analysis, solutions, simulation results for the Spanish transmission system. IEEE Trans. Power Syst. **11**(2), 630–638 (1996)
7. S. Corsi, M. Pozzi, C. Sabelli, A. Serrani, The coordinated automatic voltage control of the Italian transmission grid-Part I: reasons of the choice and overview of the consolidated hierarchical system. IEEE Trans. Power Syst. **19**(4), 1723–1732 (2004)
8. G.N. Taranto, N. Martins, D.M. Falcao, A.C.B. Martins, M.G. dos Santos, Benefits of applying secondary voltage control schemes to the Brazilian system, in *Proceedings of the IEEE Power Engineering Society Summer Meeting*, vol. 2 (2000), pp. 937–942
9. S. Corsi, F. De Villiers, Vajeth, Secondary voltage regulation applied to the South Africa transmission grid, in *Proceedings of the IEEE Power and Energy Society General Meeting* (2010), pp. 1–8
10. Q. Guo, H. Sun, J. Tong, M. Zhang, B. Wang, B. Zhang, Study of system-wide automatic voltage control on PJM system, in *Proceedings of the IEEE Power and Energy Society General Meeting* (2010), pp. 1–6
11. J. Tong, D.W. Souder, C. Pilong, M. Zhang, Q. Guo, H. Sun, B. Zhang, Voltage control practices and tools used for system voltage control of PJM, in *Proceedings of the IEEE Power and Energy Society General Meeting* (2011), pp. 1–5
12. H. Sun, Q. Guo, B. Zhang, W. Wu, B. Wang, An adaptive zone-division-based automatic voltage control system with applications in China. IEEE Trans. Power Syst. **28**(2), 1816–1828 (2013)
13. P. Lagonotte, J.C. Sabonnadiere, J.Y. Leost, J.P. Paul, Structural analysis of the electrical system: application to the secondary voltage control in France. IEEE Trans. Power Syst. **4**(2), 1477–1484 (1989)
14. H. Mehrjerdi, S. Lefebvre, M. Saad, D. Asber, A decentralized control of partitioned power networks for voltage regulation and prevention against disturbance propagation. IEEE Trans. Power Syst. **28**(2), 1461–1469 (2013)
15. V. Alimisis, P.C. Taylor, Zoning evaluation for improved coordinated automatic voltage control. IEEE Trans. Power Syst. **30**(5), 2736–2746 (2015)
16. A. Conejo, M.J. Aguilar, Secondary voltage control: nonlinear selection of pilot buses, design of an optimal control law, and simulation results. IEE Proc.-Gener. Trans. Distrib. **145**(1), 77–81 (1998)
17. M. Ilic, J. Christensen, K.L. Eichorn, Secondary voltage control using pilot point information. IEEE Trans. Power Syst. **3**(2), 660–668 (1988)
18. A. Zobian, M.D. Ilic, A steady state voltage monitoring and control algorithm using localized least square minimization of load voltage deviations. IEEE Trans. Power Syst. **11**(2), 929–938 (1996)
19. C. Yu, Y.T. Yoon, M.D. Ilic, A. Catelli, On-line voltage regulation: the case of New England. IEEE Trans. Power Syst. **14**(4), 1477–1484 (1999)
20. A. Stankovic, M. Ilic, D. Maratukulam, Recent results in secondary voltage control of power systems. IEEE Trans. Power Syst. **6**(1), 94–101 (1991)
21. J.S. Thorp, M. Ilic-Spong, M. Varghese, An optimal secondary voltage-var control technique. Automatica **22**(2), 217–222 (1986)
22. Z. Liu, M.D. Ilic, Toward PMU-based robust automatic voltage control (AVC) and automatic flow control (AFC), in *Proceedings of the IEEE Power and Energy Society General Meeting* (2010), pp. 1–8
23. H.Y. Su, C.W. Liu, An adaptive PMU-based secondary voltage control scheme. IEEE Trans. Smart Grid **4**(3), 1514–1522 (2013)
24. H.Y. Su, F.M. Kang, C.W. Liu, Transmission grid secondary voltage control method using PMU data. IEEE Trans. Smart Grid **9**(4), 2908–2917 (2018)

25. H.Y. Su, T.Y. Liu, Enhanced worst-case design for robust secondary voltage control using maximum likelihood approach. IEEE Trans. Power Syst. **33**(6), 7324–7326 (2018)
26. F. Aminifar et al., Synchrophasor measurement technology in power systems: panorama and state-of-the-art. IEEE Access **2**, 1607–1628 (2014)
27. E. Schweitzer, D. Whitehead, G. Zweigle, K.G. Ravikumar, Synchrophasor-based power system protection and control applications, in *Proceedings of the Modern Electric Power System* (2010), pp. 1–10
28. S. Boyd, L. Vandenberghe, *Convex Optimization* (Cambridge University Press, 2004)
29. B. Stott, O. Alsac, Fast decoupled load flow. IEEE Trans. Power Appar. Syst. **PAS-93**(3), 859–69 (1974)
30. S. Wang, J.B. Zhao, Z. Huang, R. Diao, Assessing Gaussian assumption of PMU measurement error using field data. IEEE Trans. Power Del. **33**(6), 3233–3236 (2018)
31. K.J. Astrom, Maximum likelihood and prediction error methods. Automatica **16**, 551–574 (1980)
32. Power Systems Test Case Archive, University of Washington College of Engineering. http://www.ee.washington.edu/re-serach/pstcal/
33. Power System Relaying Committee, IEEE Stand. Synchrophasors Power Syst. IEEE PES, C37.118.1-2011 (2011)

Applications of Decision Tree and Random Forest Methods for Real-Time Voltage Stability Assessment Using Wide Area Measurements

Heng-Yi Su, Yu-Jen Lin, and Chia-Chi Chu

Abstract Traditionally, voltage stability assessment (VSA) are widely investigated by model-based approaches. Several achievements have been developed along this direction, including continuation power flow methods (CPFLOW), direct methods, and optimal power flow methods. Since precise model information are required and their computations are very demanding, their applications to real-time VSA are challenging, especially when network operating conditions and/or network topology may be always changed. In recent years, with wide deployment of synchronized phasor measurement unit (PMUs), PMU-based wide area measurement system (WAMS) has already attracted lots of interests in investigating VSA in advanced artificial intelligence approaches. By collecting real-time big data from power grids and studying these historical data through statics analytics, some prediction models can be constructed for VSA of the current operation conditions. This chapter presents some recent advances in data mining framework for power system VSA under real-time environments. The proposed framework adapts a new enhanced online random forest (EORF) algorithm to update decision trees (DTs), such as tree growth and replacement. By means of weighted majority voting, one of the ensemble learning skills, DTs in the random forest are able to reach consesus to deal with power system changes. The proposed EORF framework is first tested on IEEE 57-bus power systems, and then is applied to Taiwan 1821-bus power system. Through comprehensive computer simulations, the robustness, the computation speed, as well as the assessment accuracy, of the proposed EORF framework are justified for assessing the power system voltage stability in real-time.

H.-Y. Su
Department of Electrical Engineering, Feng Chia University, Taichung 40724, Taiwan
e-mail: hengyisu@fcu.edu.tw

Y.-J. Lin
Department of Electrical Engineering, I-Shou University, Kaohsiung 84001, Taiwan
e-mail: yjlin@isu.edu.tw

C.-C. Chu (✉)
Department of Electrical Engineering, National Tsing Hua University, Hsinchu 30013, Taiwan
e-mail: ccchu@ee.nthu.edu.tw

H. Haes Alhelou et al. (eds.), *Wide Area Power Systems Stability, Protection, and Security*, Power Systems, https://doi.org/10.1007/978-3-030-54275-7_14

Keywords Voltage stability assessment · Data mining · Decision tree · Random forest · Online learning · Wide area measurements

1 Introduction

Voltage collapse is a crucial issue in operations and control of modern power systems, as many major power outage incidents around the world have been associated with this class of instability mechanism [1, 2]. This issue not only results in significant economic losses but also exerts unpredictable impacts on people's lives and industrial productions. In view of this, an accurate real-time voltage stability assessment (VSA) tool to quickly identify whether a current operating state is prone to voltage collapse or not becomes a necessity for system operators to take corresponding control actions for preventing possible voltage collapse.

Traditionally, VSA are widely investigated by model-based approaches. Several achievements have been developed along this direction including continuation power flow methods (CPFLOW), direct methods, and optimal power flow methods [3, 4]. Since precise model information are required and their computations are very demanding, their applications to real-time VSA are challenging especially when network operating conditions and/or network topology may be always changed. In recent years, with wide deployment of synchronized phasor measurement unit (PMUs), PMU-based wide area measurement system (WAMS) has already attracted lots of interests in investigating VSA in advanced artificial intelligence (AI) approaches [5, 6]. By collecting real-time big data from power grids and studying these historical data through statics analytics, some prediction models can be constructed for VSA of the current operation conditions. Several recent studies have been demonstrated that VSA by advanced AI approaches have better performance in terms of the accuracy and the computational speed in comparison with conventional model-based approaches [7–12].

Among diverse branches of AI, data mining (DM) approaches have attracted much attention since they exhibit some ensemble learning ability. Classification and regression are the two main streams in data mining technology. Generally speaking, data mining approaches aim to extract models and patterns of interest from large-scale data sets. Among them, a decision-tree (DT), a non-parametric supervised learning method, is the simplest representation for classification and regression. The goal is to create a model that predicts the value of a target variable by learning simple decision rules inferred from the data features. Tree learning comes closest to meeting the requirements for serving as an off-the-shelf procedure for data mining, however, they are seldom accurate: trees that are grown very deep tend to learn highly irregular patterns: they over-fit their training sets. In order to overcome this difficulty, random forest (RF) methods have been proposed later to explore the ensemble learning of DT techniques by averaging multiple deep DTs, trained on different parts of the same training set, with the goal of reducing the variance. A RF classifier is constructed from random sample selections and random variable selections, where each tree classifier in the forest relies on the values of a random vector sampled independently and with the same distribution for all trees. Therefore, the RF classifier can be treated as an improved composite classification model.

In recent modern power grid applications, some of these advanced AI methods have been devoted to transient stability problems as well as voltage stability problems. For example, techniques with considering local regression and adaptive database in DM approaches have been developed for achieving the seamless adaptation among different operating conditions and network configurations [13]. It has been demonstrated that this method is more applicable in the system with high renewable penetration.

In general, collections of new data points occurs offline, and data points are transmitted periodically. In this sense, the prediction model for online voltage stability assessment may need to be retrained based on the updated training database only. To the author's best knowledge, most of the existing data mining based methods retrain their voltage stability prediction models using the whole training database when new data is appended. Such retraining process would result in a lot of repetitive computation, and it can only be done in offline mode. Thus, the algorithm for online learning is much more useful.

In this context, this chapter is concerned with developing a new online learning algorithm called enhanced online random forest (EORF) [14]. It enables to online update these trees involving tree growth and/or tree replacement. These trees in the forest are then combined via a weighted majority voting, which makes the decision model better adapted to system changes. Thus, it can effectively and rapidly predict voltage stability in power grids not only under operating condition variations but also under system topology changes. In the proposed framework, the permutation-based method is employed to identify the critical variables for the final model. The synchronized phasor measurements of the critical variables from PMUs are then utilized as inputs to the trained model which predicts the system voltage stability status. To continuously keep the online prediction model which is better adapted to changing environments, an EORF algorithm, together with a weighted majority voting strategy is developed. The developed algorithm is able to update the prediction model in an online fashion.

This chapter is organized as follows. An overview of DM techniques, including both DT and RT, and their applications to VSA are presented in Sect. 2. Section 3 explains the details of the proposed EORF framework for real-time VSA. Section 4 presents experimental results of case studies with the IEEE 57-bus system and a practical Taiwan power system. Finally, Sect. 5 concludes the paper

2 Overview

Data mining can be understood literally. It is a process of exploring data and extracting knowledge from data. In general, the scale of mined data is too large to extract knowledge at a glance. Data mining refers to the systematic process to acquire useful knowledge from a large scale dataset. It covers a wide range of issues, including data pre-processing, data warehousing, analytical processing, data visualization in multi-dimensional space, outlier detection, and knowledge expression. This chapter briefs some frequently-used but not all data mining tools, and also introduces their applications to VSA in power systems and related topics.

Statistics can be considered a tool in the realm of data mining as well. It collects, organizes, and analyses numerical data. Statistics can provide information of numerical data, such as the average, median value, standard deviation, variance and distribution. A popular statistics application is regression, which guides us to interpolate data, infer a regression model, and use the regression model for further extrapolation. Statistics is a useful tool indeed, but unfortunately it is unable to provide solutions for extracting knowledge and discovering patterns, which are the main interests of data mining.

2.1 Artificial Neural Networks

Artificial neural networks (ANNs) are mathematical models which are able to mimic human brain reasoning function. An ANN comprises a number of numerical processing units, also called neurons as they are similar to biological brain neurons. Neurons are connected by links associated with numerical weights. Those numerical weights are used to emphasize the importance of the links between neurons. In ANNs, links actually store long-term memory. These links are continuously adjusted. Adjusting these links called training or learning. Link adjustment in an ANN is based on the dataset as well as learning algorithms.

Using ANN to mine a dataset, one has to decide the input attributes and output attributes of the dataset. Input attributes and output attributes constitute training samples. Various learning algorithms have been developed and trained ANNs successfully. Well-trained ANNs can learn the relationship between input attributes and output attributes of the dataset. Therefore, we may say that an ANN is trained by data, or an ANN learns from data.

Kamalasadan et al. used artificial feed-forward neural networks to assess power system voltage stability [15]. They constituted a dataset by off-line simulations of various loading conditions. They considered real and reactive power, and voltages of generator and load buses for neural network's input, and L-index for neural network's output. They showed their proposal fast, robust, and accurate.

Devaraj and Roselyn presented an ANN based approach for on-line voltage security assessment [16]. They considered radius basis function neural networks. Inputs of that neural networks were power flow states before contingency, and the putout was the voltage stability indicated by maximum L-index of load bus. They found the voltage index predicted by radius basis function neural network was more accurate than other neural networks, based on the simulation results conducted by IEEE 30-bus power system and an Indian practical 76-bus power system.

Bahmanyar and Karami used ANNs to estimate voltage stability margin [17]. They used Gram-Schmidt orthogonalization process and sensitivity technique to find the fewest inputs that can estimate voltage stability margin with very good accuracy. Their study also considered the change of the power system's configuration. They estimated voltage stability margin of the New England 39-bus and the southern and eastern Australian power system successfully and accurately, even if the system configuration has changed.

A hierarchical, self-adaptive method for assessing short-time voltage stability has been proposed by Zhang et al. [18].The proposed method used strategically designed ensemble-based randomized learning model and extreme learning machine. The proposed method can assess voltage instability and fault-induced delayed voltage recovery phenomenon. The assessment accuracy, speed are tested on a benchmark power system.

Ren et al. proposed an intelligent system containing multiple randomized algorithms in an ensemble learning manner to assess short-term voltage stability [19]. The proposed intelligent system included random vector functional link networks and extreme learning machine. The authors also developed a multi-objective optimization programming framework that balanced computation speed and accuracy. To verify the proposal's superiority over the others, the authors applied their intelligent system to New England 39-bus power system and Nordic test system.

2.2 *Support Vector Machines*

A support vector machine (SVM) can perform classification, regression and even outlier detection very well. An SVM classifier attempts to widen the boundary between classes in a dataset. An SVM classifier is supported by data located on the edge of classes. These data are called the support vector. A way that an SVM handles nonlinear data is to add features or to transform. As a result, a nonlinear dataset could be transformed into a linear separable dataset. Another commonly-used kernel technique is Gaussian radius bias function. Unlike SVM classifiers which find the boundary between different classes of data as wide as possible, SVM regression tries to approximate one class of data as close as possible. The same kernel concept can be applied to SVM regression problems in order to deal with nonlinearity.

SVM was used in [20] to predict dynamic voltage collapse on an actual 87-bus power system. Time domain simulation results considering load increase were the inputs of the dataset, and output was power transfer stability index (PTSI). PTSI was in the range of 0 and 1 in which voltage collapsed. SVM was used as a regression machine.

Sajan et al. proposed a SVM approach to monitoring power system long-term voltage stability [21]. To develop an efficient SVM at computing accuracy and speed, the authors utilized genetic algorithms (GA). Voltage magnitude and phase angle measured from PMU were fed into the GA based SVM and the output was voltage stability margin index. The test results conducted from New England 39 bus power system and 246-bus Indian Northern Region Power Grid have shown the feasibility of their approach.

Yang et al. proposed a lease square SVM to predict short-term voltage instability [22]. In their study, they considered composite load model, which comprised a static impedance-current-power load model and a dynamic induction machine load model. They used least square SVM to predict induction machine trajectory.

Mohammadi et al. presented a voltage stability assessment method using an SVM [23]. They used multi-objective optimization to reduce input futures and to train the SVM. They also reduced data size and misclassification of the SVM by means of multi-objective biogeography-based optimization. They applied their method to 39-bus bench mark power system and 66-bus real power grid.

2.3 Decision Trees

A decision tree (DT) consists of nodes, branches and leaves, just like a natural tree in real life. A grown DT virtually illustrates a series of reasoning processes of a dataset, therefore it can be used to mine useful information contained in the dataset. At first, all data are stored at root node of a DT, whose impurity is large currently since all data have been mixed up. Next, the root node starts to be split into nodes according to the greatest reduction in impurity of the whole DT. Such a split indicates the best reasoning process of dataset because it can purify the DT the most. Keeping split and the DT is growing up continuously. Finally, the grown up DT becomes a map that is able to reason the dataset. DTs that are applied to classification problems are called classification trees (CTs). Likewise, DTs that are applied to regression problems are called regression trees (RTs).

Among other data mining tools, DTs have been given many thumbs up due to following reasons. First, growing DTs is straightforward. Unlike other data mining tools which require many parameters to set up mathematical models, DTs are non-parametric. Second, DTs can reach the conclusion immediately so that they are found in many on-line, real-time applications. Third, DTs are explicable. Researchers can extract knowledge from rules inferred in grown DTs. Above reasons have significantly prompted DTs' success in voltage stability research, particularly in assessment, classification, prediction and control aspects [24–35].

Issues on assessment, classification and prediction problem demands high accuracy; while control actions design requires the knowledge of what variables shall be controlled, and how to set the values of those control variables. It has been widely accepted that DTs are explicable, but the manner of transforming the acquired knowledge into control action designs is not unique. Control actions that can alleviate threats brought by contingency are termed enhancement control, and control actions that can save power system from collapsing post-contingency are termed preventive control. In [26, 29], preventive control rules were derived from decision trees. Control variables had been selected but their values were not specified further. In [29], decision rule derived from a regression tree was used to determine the number of generator unit tripping in emergency control. Ref. [32] assessed security by a path-based method. Preventive control was derived along the path and was weighted by several important factors. In [34], the security boundary obtained from the grown DT was used as a guide to design preventive control. To ensure conservative control actions, a small margin was considered to keep away from the determined security boundary. The preventive control resulted from a linear combination of input variables of the classification tree. Ref. [35] expressed the decision rules in terms of linear combination and determined the security region. The linear combination associated with optimal scheduling of generation to constitute final preventive control.

Beiraghi and Ranjbar presented a method based on DT algorithm to assess voltage stability online [36]. They produced a database which contained one day ahead load and generation variation scenarios, and labeled secure or insecure decided based on voltage stability criteria. They considered ensemble learning concept, bagging and adaptive boosting, to constitute combined models in attempt to increase accuracy and to eliminate model parameters dependence. The authors have tested their method on IEEE 118-bus power system.

Meng et al. used C4.5 algorithm to construct decision tree for online voltage stability assessment [37]. They first generated samples by using P-V curve analysis. Then they considered participation factor analysis and relief algorithm to select attributes for decision tree. Finally they used C4.5 algorithm to construct decision tree, and to extract rules. The authors applied their decision tree to a practical power system to assess the power system voltage stability status, and to extract rules for power system operating guidelines.

3 Proposed EORF Scheme

In this section, we outline a novel algorithm for online voltage stability (VS) monitoring using the measured data from WAMS. Support for online model updating allows the proposed scheme to be highly adaptive to system information changes. More details are presented as follows.

3.1 Drift Detection

Indeed, system information must be checked and updated in a timely manner to remain current with changes in operating conditions. In this study, we adopted the least confidence strategy for the detection of such changes.

The uncertainty of the online classifier can be assessed in terms of posterior probability, which reflects the uncertainty of measuring an observation to a particular class [38]. Generally, examples are flagged if the degree of certainty falls below a required threshold. During periods without change, the uncertainty degree pertaining to recent examples should be relatively low. Any significant increase in classification uncertainty is an indication that the concept is undergoing change.

In this research, we denote e as the percentage of data that are uncertain. In monitoring the evolution of e, two indicators, l_w and l_d, are employed, where l_w is the warning level and l_d is the drift level. If $e \times l_w$, then a warning message is issued. If $e \geq l_d$, it can be assumed that there has been changes in data distribution.

3.2 EORF Algorithm

Most existing methods handle significant changes in the network topology or operating conditions by rebuilding their voltage stability prediction model using added training examples in offline fashion. Unfortunately, these reconstructing procedures are time-consuming. In this study, we developed a novel EORF algorithm, which makes it possible to update the model online. In order to prevent overfitting, online bagging [39] with control over leaf node splitting is applied to update the model. Note also that the proposed model is highly randomized [40], through the random generation of split functions and thresholds. The following summarizes the steps in the overall process.

Step (1) Initialization:

(1) In this research, we denote the number of new trees and the total number of trees as n_u and n_t, respectively. Randomly choose n_u trees from $T = \{t_k\}_{k=1}^{n_t}$ to form $T_{new} = \left\{t_k^{new}\right\}_{k=1}^{n_u}$
(2) Let T_{old} represent the trees exempted from updating.
(3) Let g_s represent the minimum gain a split must achieve, n_s represent the minimum number of samples a leaf node must see, and $h^{\max}$ represent the maximum tree height.
(4) Let D_{learn}^* represent the learning set for online processing and initialize $D_{learn}^* = \emptyset$; let c_d represent the concept drift index and initialize $c_d = 0$.

Step (2) Detection of changes:

(1) Calculate e of $T = \{t_k\}_{k=1}^{n_t}$ using the drift detection approach outlined in Sect. 3.1.
(2) Set $c_d = 1$ when $e \geq l_w$. Store the sample ($\boldsymbol{x}$, y) via

$$D_{learn}^* = D_{learn}^* \cup (\mathbf{x}, y). \tag{5}$$

D_{learn}^* is utilized for tree reconstruction. When $e \geq l_d$, set $c_d = 2$.
(3) Proceed to Step 3 when $c_d \neq 0$.

Step (3) Tree updating strategies:

(1) $T_{new} = \{t_k^{new}\}_{k=1}^{n_u}$ is updated on each sample n_r times, where n_r is a random number produced by a Poisson ($\boldsymbol{\Gamma}$) distribution.
(2) Proceed to 3 when $n_r > 0$ and $c_d = 1$. Otherwise, proceed to 4 when $n_r = 0$ and $c_d = 2$.
(3) *Tree Growth:* Proceed to 4 when $h_k = h^{\max}$. Otherwise, proceed to the following process.
   (a) The candidμate splits is defined by

$$\mu_s = \{(f_m(\mathbf{x}), \theta_m)\}_{m=1}^{M} \tag{6}$$

   where θ_m represents a scalar threshold and $f_m(\mathbf{x})$ represents a split function.
   (b) Calculate the gain function

$$\Delta G_j^s = G(D_j) - \frac{|D_j^{sl}|}{|D_j|} \cdot G(D_j^{sl}) - \frac{|D_j^{sr}|}{|D_j|} \cdot G(D_j^{sr}) \tag{7}$$

   where $|\cdot|$ denotes the number of samples it contains, $G(\cdot)$ denotes the entropy, D_j^{sr} and D_j^{sl} denote the right and left partitions made by μ_s, and D_j indicates the samples arriving at node j.
   (c) Split leaf node j utilizing μ_s if $|D_j| > n_s$

$$\mu_s = \arg\max_s (\Delta G_j^s) \tag{8}$$

(4) Replace t_k^{new} with a new one built by D_{learn}^*.
(5) Increase the value of k. Once $T_{new} = \{t_k^{new}\}_{k=1}^{n_u}$ is achieved.

Step (4) Hybrid model creation:

After obtaining T_{new}, the proposed online classifier is built as follows:

$$T = T_{old} \cup T_{new} \tag{9}$$

A new online classification model is created by following the above steps. Note that many of the trees in the forest may differ in terms of age after the EORF algorithm has been running for an extended period. In this situation, older trees may be given priority over younger trees in terms of modification.

3.3 Weighted Majority Vote

This section describes the process of estimating class-membership probabilities $\hat{\eta}_{yk}$ from t_k and the use of weighted majority voting to obtain the final prediction.

In this study, the Laplace estimate [41] is used to smooth out the estimated probabilities, such that η_{yk} $(y/\mathbf{x})$ is given by

$$\hat{\eta}_{yk} = \eta_{yk}(y/\mathbf{x}) = \frac{1}{N + n_y} \sum_{i=1}^{N} I(y_i, y) + 1 \tag{10}$$

where $I(y_i, y)$ denotes a zero-one loss function; n_y denotes the number of classes; y_i denotes the class label for the i th training data; N indicates the number of examples associated with y that fall within the same leaf.

To ensure that $T = T_{old} \cup T_{new}$ is well-adapted to new system information, T is reweighted according to age. Specifically, T_{new} probably outperform T_{old} and are therefore assigned greater weights. In other words, trees designated as T_{old} are given less voting power. Hence, $\hat{T}(\mathbf{x})$ is given by

$$\hat{T}(\mathbf{x}) = \arg\max_{y} \left(\sum_{k=1}^{n_u} \frac{w_{new}}{n_u} \hat{\eta}_{yk}^{new} + \sum_{k=1}^{n_t - n_u} \frac{w_{old}}{n_t - n_u} \hat{\eta}_{yk}^{old} \right) \tag{11}$$

where $\hat{\eta}_{yk}^{new}$ and $\hat{\eta}_{yk}^{old}$ respectively indicate the estimated probabilities from t_k in T_{new} and T_{old}, whereas $w_{new} \in (0, 1)$ end $w_{old} \in (0, 1)$ respectively indicate the weights assigned to T_{new} and T_{old}.

3.4 Overall Scheme

Figure 1 presents an overall scheme of the proposed approach. Note that the designed scheme includes two major steps: (1) offline training and (2) online prediction and updating. The steps are described below.

(1) *Offline training*: The dataset involves system operating points (OPs) and their corresponding VS states. In this research, we selected $X = \{(|V_{j|}, \delta_j)\}|\ j \in$ selected load buses} as input variable and $Y = \{I, S\}$ as output variable, in which "I" and "S" respectively refer to insecure and secure cases. Given $D_{train} = \{(\mathbf{x}_i, y_i)|\mathbf{x}_i \in X, y_i \in Y\}_{i=1}^{N}$, $T = \{t_k\}_{k=1}^{n_t}$ is then trained to achieve online VS monitoring.

(2) *Online prediction and updating:* The trained $T = \{t_k\}_{k=1}^{n_t}$ is utilized to predict VS under the current OPs based on new $\mathbf{x}$ from WAMS. The prediction model is updated using the proposed EORF algorithm with the aim of making the online VS monitoring scheme adaptive to changing operating conditions. When the concept drift is determined via the drift detection method, updating and reweighting strategies are triggered, and a new prediction model with weighted voting is constructed. The newly-constructed model can make decisions for classifying the voltage stability for a range of operating points under variable operating conditions and system topologies.

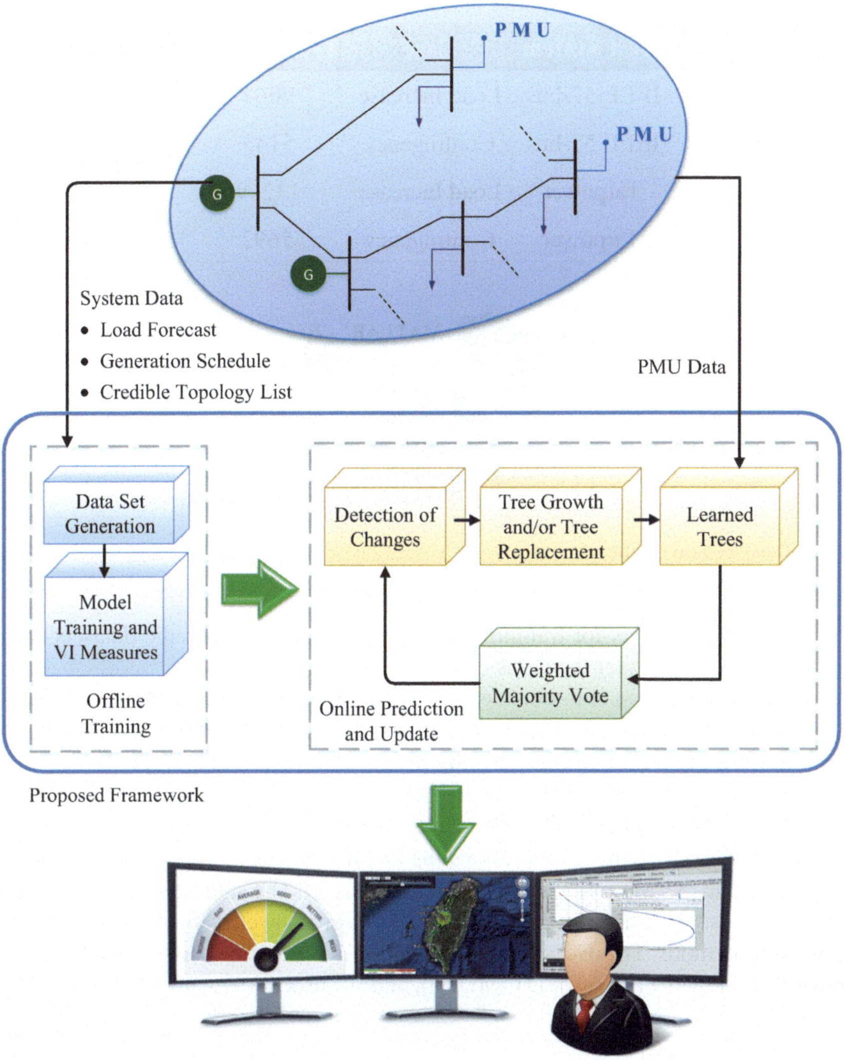

Fig. 1 Overall scheme of the EORF algorithm for VS monitoring

4 Simulation Results

The proposed scheme was tested on two power system models of different size: the IEEE 57-bus and Taiwan Power (Taipower) system (TPS) comprising 1821 buses. Moreover, all experiments were run within the MATLAB® environment.

In our research, the dataset comprises known OPs and their associated VS conditions. All OPs and VS conditions were produced utilizing PSAT [42]. Furthermore, two cases are considered: continuous load increase and contingency cases.

Dataset	Test System	Case	Training Set	Test Set
I	IEEE 57-bus	Load Increase	9064	4215
II	IEEE 57-bus	Contingency	5143	2678
III	Taipower	Load Increase	13269	5687
IV	Taipower	Contingency	5692	2440

Fig. 2 Datasets used for the test systems

We compiled 4 datasets for training and testing. Figure 2 details the datasets used for the test systems under the cases considered in these simulations.

Accuracy is the most common measure of classifier performance; however, security and reliability are also used as measures to account for the output class imbalance problem [38]. Figure 3 demonstrates the results obtained by the proposed scheme. As shown in Fig. 3, the proposed scheme achieved promising results.

We also compared the simulation results with those obtained using four classification methods, involving adaptive boosting (AdaBoost), decision tree (DT), support vector machine (SVM), and k-nearest neighbor (kNN). The same datasets were used to train and test all of the algorithms. Figure 4 presents the classification accuracy of each of the methods. The overall accuracy was as follows: proposed scheme (98.4%), kNN (95.2%), SVM (96.6%), DT (96.9%), and AdaBoost (97.4%).

4.1 Robustness Evaluation

We also evaluated the robustness of the proposed scheme using simulations based on the datasets presented in Fig. 2. In this research, noisy data and missing data are considered.

Measurements from PMUs often include errors. Numerous experiments have been conducted to characterize the effects of noisy PMU data. The current test simply involved the addition of random noise to the original PMU data. Figure 5 presents the results with and without noisy PMU data, illustrating that measurement inaccuracy can indeed degrade the accuracy of the classification method.

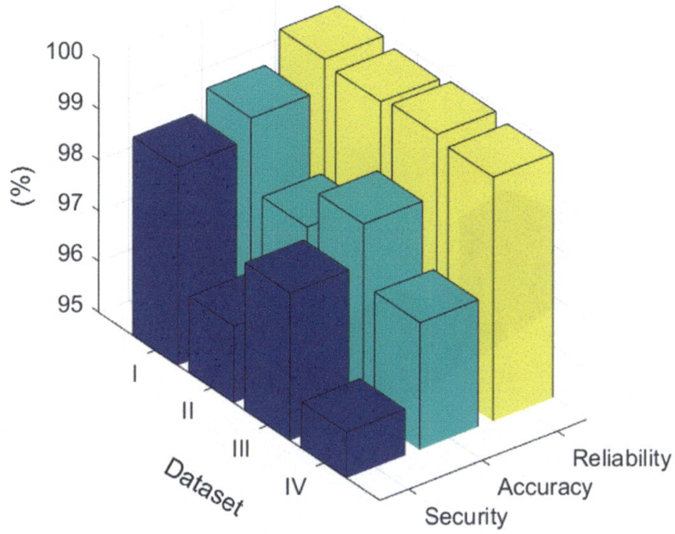

Fig. 3 Results in terms of security, reliability, and accuracy for EORF

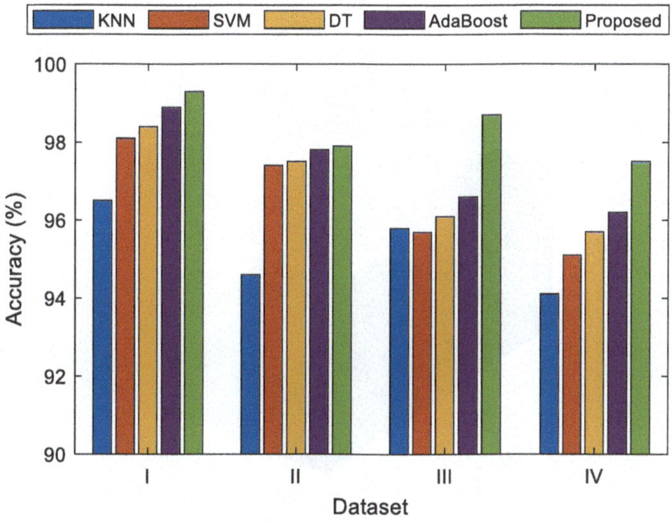

Fig. 4 Results in terms of accuracy for the compared methods

We performed a series of simulations to show the impact of missing data on EORF performance under various system operating conditions. In these simulations, ρ was used to indicate the percentage of the missing data. The simulation results are shown in Fig. 6. As indicated by the test results in Figs. 5 and 6, the proposed scheme is

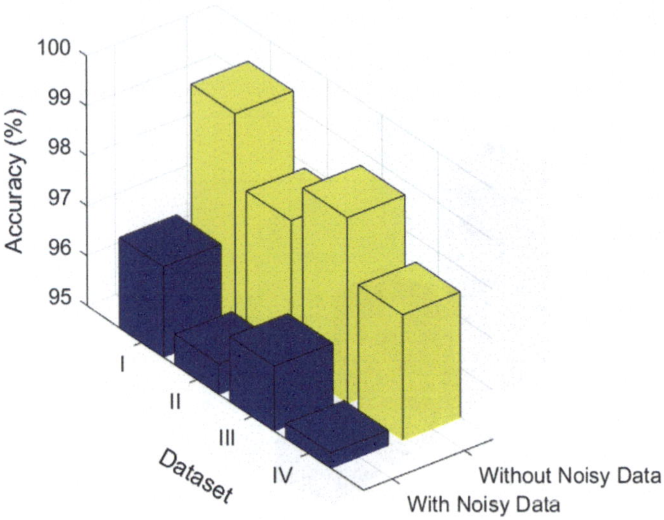

Fig. 5 Robustness against noisy PMU data

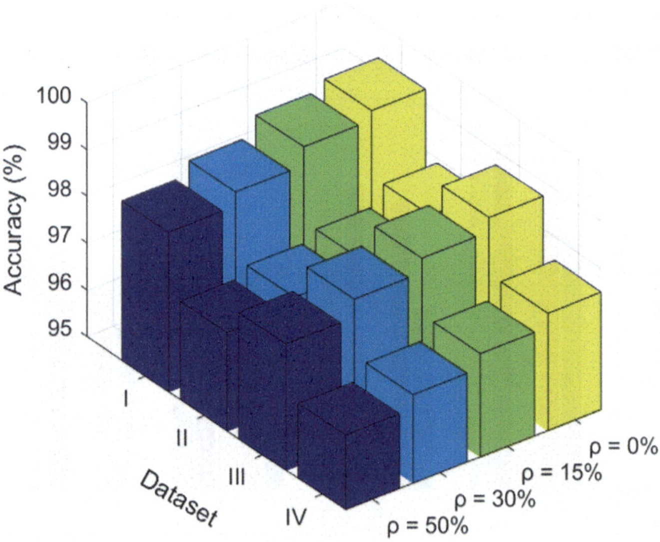

Fig. 6 Robustness against missing PMU data

robust to noisy data and missing data.

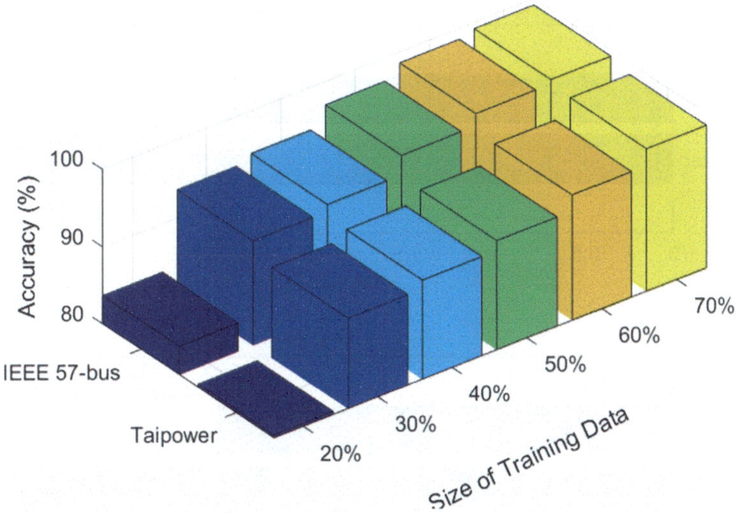

Fig. 7 Effects of training set size

4.2 Effects of Training Set Size

We also sought to determine the degree to which the size of the training data affects classification accuracy. This was achieved by creating a series of training sets using 20, 30, 40, 50, 60, or 70% of the original dataset via random sampling. Figure 7 presents the overall accuracy of each test system in simulations that were repeated 10 times. As shown in the figure, classification accuracy was proportional to the size of the training set.

4.3 CPU Time Evaluation

CPU time can be used as a proxy to estimate computational costs of the EORF. Figure 8 summarizes the training and testing used to evaluate each test system in terms of CPU time. As shown in Fig. 8, the CPU times for IEEE 57-bus and Taipower under the test set were 9 and 13 s, respectively. This indicates that regardless of the test system, a new OP can be assessed using the trained classifier in less than 0.002 s.

4.4 Effects of Concept Drift

We performed a large number of experiments on the test systems to elucidate the impact of concept drift on EORF performance. We generated 21,100 examples for

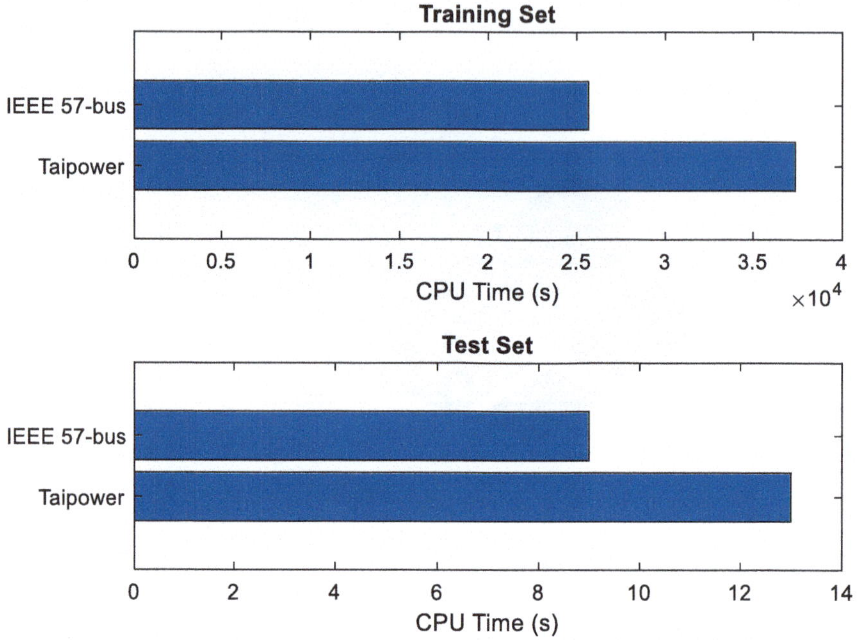

Fig. 8 CPU time evaluation

IEEE 57-bus and 27,088 for TPS. Among these, 70% were used for training and the remaining 30% were used for testing.

We then compared our simulation results with those obtained using two online learning schemes: online AdaBoost (OAB) and online Bagging (OB). The overall results of which are presented in terms of prequential error in Fig. 9. As shown in Fig. 9, the proposed method outperformed both of the other methods in terms of prequential errors. Furthermore, our results demonstrate the benefits of drift detection in making the proposed scheme adaptive to new system information.

5 Conclusions

A novel online real-time power system voltage stability assessment framework is developed in this chapter. The framework is relied on wide area measurement system (WAMS) data using phasor measurement units (PMUs), and is based on enhanced-online-random-forest (EORF) algorithm. The framework is able to process noisy or missing data, update EORF model online, and adapt to various power system operating conditions. As a consequence, the proposed framework assigns a robust voltage stability assessment to power utilities. To verify the proposed framework, this chapter applies the proposed framework to IEEE 57-bus and 1821-bus Taiwan

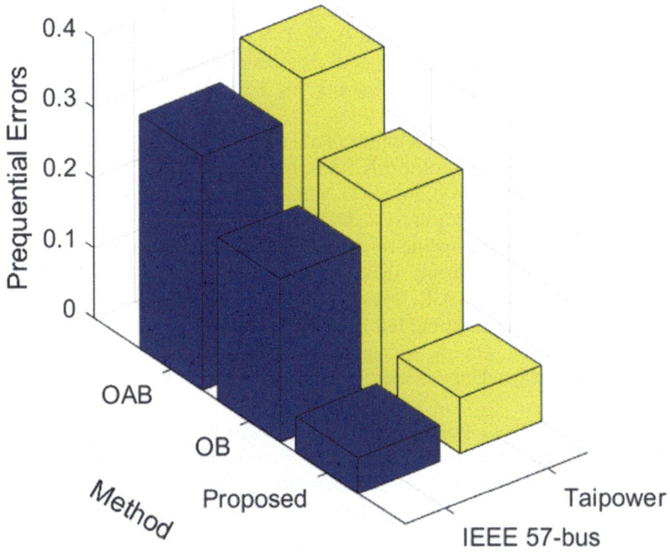

Fig. 9 Comparison results in terms of prequential errors

power systems. Comprehensive computer simulation results justify the feasibility of the proposed EORF framework.

References

1. P. Pourbeik, P.S. Kundur, C.W. Taylor, The anatomy of a power grid blackout. IEEE Power & Energy Mag. 22–29 (2006)
2. C.J. Mozina, A shot in the dark. IEEE Ind. Appl. Mag. (2008), 45–52
3. T. Van Cutsem, C. Vournas, *Voltage Stability of Electric Power Systems* (Kluwer Academic Publishers, 1998)
4. V. Ajjarrapu, *Computational Techniques for Voltage Stability Assessment and Control* (Springer, 2006)
5. J.-H. Liu, C.-C. Chu, Wide-area measurement-based voltage stability indicators by modified coupled single-point models. IEEE Trans. Power Syst. **29**(2), 756–764 (2014)
6. H.-Y. Su, C.-W. Liu, Estimating the voltage stability margin using PMU measurements. IEEE Trans. Power Syst. **31**(4), 3221–3229 (2016)
7. V.K. Singh, M. Govindarasu, Decision tree based anomaly detection for remedial action scheme in smart grids using PMU data, in *IEEE Power & Energy Society General Meeting (PESGM)* (2018), pp. 1–5
8. S.S. Maaji, G. Cosma, A. Taherkhani, A.A. Alani, T.M. McGinnity, Online voltage stability monitoring using an ensemble Adaboost classifier, in *4th International Conference on Information Management (ICIM)* (2018), pp. 253–259
9. Y. Wang, H. Pulgar-Painemal, K. Sun, Online analysis of voltage security in a microgrid using Convolutional neural networks, in *IEEE Power & Energy Society General Meeting (PESGM)* (2017) pp. 1–5

10. Z. Nie, D. Yang, V. Centeno, K.D. Jones, A PMU-based voltage security assessment framework using heffding-tree-based learning, in *19th International Conference on Intelligent System Application to Power Systems (ISAP)* (2017), pp. 1–6
11. L.D. Arya, L.S. Titare, D.P. Kothari, Probabilistic assessment and preventive control of voltage security margins using artificial neural network, Int. J. Electr. Power Energy Syst. **29**, 99–105 (2007)
12. H.M. Ma, K.-T. Ng, K.F. Man, A multiple criteria decision-making knowledge-based scheme for real-time power voltage control. IEEE Trans. Ind. Inform. **4**(1), 58–66 (2008)
13. S. Li, V. Ajjarapu, M. Djukanovic, Adaptive online monotiring of voltage stability margin via local regression. IEEE Trans. Power Syst. **33**(1), 701–713 (2018)
14. H.-Y. Su, T.-Y. Liu, Enhanced-online-random-forest model for static voltage stability assessment using wide area measurements. IEEE Trans. Power Syst. **33**(6), 6696–6704 (2018)
15. S. Kamalasadan, D. Thukaram, A.K. Srivastava, A new intelligent algorithm for online voltage stability assessment and monitoring, Int. J. Electr. Power Energy Syst. (31), 100–110 (2009)
16. D. Devaraj, J.P. Roselyn, On-line voltage stability assessment using radius basis function network model with reduced input features. Int. J. Electr. Power Energy Syst. (33), 1550–1555 (2011)
17. A.R. Bahmanyar, A. Karami, Power system voltage stability monitoring using artificial neural networks with reduced set of inputs. Int. J. Electr. Power Energy Syst. (58), 246–256 (2014)
18. Y. Zhang, Y. Xu, Z.Y. Dong, R. Zhang, A hierarchical self-adaptive data-analytics method for real-time power system short-term voltage stability assessment. IEEE Trans. Ind. Inform. **15**(1), 74–84 (2019)
19. C. Ren, Y. Xu, Y. Zhang, R. Zhang, A hybrid randomized learning system for temporal-adaptive voltage stability assessment of power systems. IEEE Trans. Ind. Inform. **16**(6), 3672–3684 (2020)
20. M. Nizam, A. Mohamed, A. Hussian, Dynamic voltage collapse prediction in power systems using support vector regression. Expert Syst. Appl. **37**, 3730–3736 (2010)
21. A.R. Bahmanyar, A. Karami, Genetic algorithm based support vector machine for on-line voltage stability monitoring. Int. J. Electr. Power Energy Syst. (73), 200–208 (2015)
22. H. Yang, W. Zhang, J. Chen, L. Wang, PMU-based voltage stability prediction using least square support vector machine with online learning. Electr. Power Syst. Res. **160**, 234–242 (2018)
23. H. Mohammadi, G. Khademi, M. Dehghani, D. Simon, Voltage stability assessment using multi-objective biogeography-based subset selection. Int. J. Electr. Power Energy Syst. (103), 525–536 (2018)
24. T. Van Cutsem, L. Wehenkel, M. Pavella, B. Heibronn, M. Goubin, Decision tree approaches to voltage security assessment. IEE Proc-C **140**(3), 189–198 (1993)
25. L. Wehenkel, Contingency severity assessment for voltage security using non-parametric regression techniques. IEEE Trans. Power Syst. **11**(1), 101–111 (1996)
26. E.S. Karapidakis, N.D. Hatziargyriou, Online preventive dynamic security of isolated power systems using decision trees. IEEE Trans. Power Syst. **17**(2), 297–304 (2002)
27. S.R. Nandanwar, S.B. Warkad, Voltage security assessment with application of PMUs using decision tree, in *8th International Conference on Computational Intelligence and Communication Networks (CICN)* (2016), pp. 365–369
28. J.A. Huang, S. Harrison, G. Vanier, A. Valette, L. Wehenkel, Application of data mining to optimize settings for generator tripping and load shedding system in emergency control at Hydro-Québec. Int. J. Comput. Math. Electr. Electron. Eng. (COMPEL) **23**(1), 21–34 (2004)
29. J.-P. Paul, K.R.W. Bell, A flexible and comprehensive approach to the assessment of large-scale power system security under uncertainty. Int. J. Electr. Power Energy Syst. **26**, 265–272 (2004)
30. E.A. Leonidaki, D.P. Georgiadis, N.D. Hatziargyriou, Decision tree for determination of optimal location and rate of series compensation to increase power system loading margin. IEEE Trans. Power Syst. **21**(3), 1303–1310 (2006)
31. N. Senroy, G.T. Heydt, V. Vittal, Decision tree assisted controlled Islanding. IEEE Trans. Power Syst. **21**(4), 1790–1797 (2006)

32. K. Sun, S. Likhate, V. Vittal, V.S. Kolluri, S. Mandal, An online dynamic security assessment scheme using phasor measurements and decision trees. IEEE Trans. Power Syst. **22**(4), 1935–1943 (2007)
33. R. Diao, K. Sun, V. Vittal, R.J. O'Keefe, M.R. Richardson, N. Bhatt, D. Stradford, S.K. Sarawgi, Decision tree-based online voltage security assessment using PMU measurements. IEEE Trans. Power Syst. **24**(2), 832–839 (2009)
34. R. Diao, V. Vittal, N. Logic, Design of a real-time security assessment tool of situational awareness enhancement in modern power systems. IEEE Trans. Power Syst. **25**(2), 957–965 (2010)
35. I. Genc, R. Diao, V. Vittal, S. Kolluri, S. Mandal, Decision tre-based preventive and corrective control applications for dynamic security enhancement in power systems. IEEE Trans. Power Syst. **25**(3), 1611–1619 (2010)
36. M. Beiraghi, A.M. Ranjbar, Online voltage security assessment based on wide-area measurements. IEEE Trans. Power Deliv. **28**(2), 989–997 (2013)
37. X. Meng, P. Zhang, Y. Xu, H. Xie, Construction of decision tree based on C4.5 algorithm for online voltage stability assessment. Int. J. Electr. Power Energy Syst. (58), 105793 (2020)
38. J.H, M.K, *Data Mining: Concepts and Techniques* (Morgan Kaufmann, 2012)
39. N.C. Oza, S. Russell, Online bagging and boosting. Proc. Artif. Intell. Stat. 105–112 (2001)
40. P. Geurts, D. Ernst, L. Wehenkel, Extremely randomized trees. Mach. Learn. **63**(1), 3–42 (2006)
41. F. Provost, P. Domingos, Tree induction for probability-based ranking. Mach. Learn. **52**(3), 199–215 (2003)
42. F. Milano, *Power System Analysis Toolbox (PSAT)*, ver.2.1.10 (2016)

Superseding Mal-Operation of Distance Relay Under Stressed System Conditions

Nilesh Kumar Rajalwal and Debomita Ghosh

Abstract Mal-operation of distance relay imposes serious threats to system stability and a big reason for large scale blackouts. These relays operate in its third zone due to the inability of detecting fault during stressed system conditions. These stressed conditions are load encroachment, power swing, voltage instability conditions, extreme contingencies, etc. Conventional distance relay operates on the basis of local measurements. It calculates the impedance from the relay to the fault point for its operation. Load encroachment and power swing are very similar to the symmetrical fault condition and it is difficult for these conventional relays to distinguish these stressed conditions from symmetrical faults. It is therefore important to make the distance relay intelligent enough so that it will be able to discriminate between a fault and stressed system condition. With the advancement in synchro-phasor technology, the drawbacks of conventional relays have been overcome. The wide-area monitoring system (WAMS) is capable of development of online intelligent techniques that can segregate the stressed system condition from any fault. With these advanced techniques, the mal-operation of distance relays can be avoided and thus wide-area blackouts can be stopped. In this chapter, a new scheme for detecting the zone-III operation of distance relay is proposed to discriminate the stressed system conditions such as voltage instability, power swing, or load encroachment from fault. The proposed scheme is based on the monitoring of active and reactive power of the load buses using WAMS. Various cases are created on WSCC-9, IEEE-14 and IEEE-30 bus system to test the performance of the proposed algorithm. The simulations have been done on the MATLAB Simulink platform. Results shows that the proposed method is helpful to avoid the unwanted distance relay operation under stressed system conditions.

Keywords Wide-area monitoring system · Phasor measurement units · Distance relays · Power system protection

---

N. K. Rajalwal (✉) · D. Ghosh
Department of Electrical and Electronics Engineering, Birla Institute of Technology, Mesra, Ranchi, India
e-mail: nileshrajalwal@bitmesra.ac.in

H. Haes Alhelou et al. (eds.), *Wide Area Power Systems Stability, Protection, and Security*, Power Systems, https://doi.org/10.1007/978-3-030-54275-7_15

Abbreviations

WAMS	Wide-area monitoring system
PMU	Phasor measurement unit
OOS	Out of step
NERC	North American electric reliability corporation
SVM	Support vector machine

Symbols

Z_{seen}	Impedance seen by the relay
P_{iload}	Active power consumption of load bus i
Z_{set}	Impedance setting of distance relay
Q_{iload}	Reactive power consumption of load bus i
Z_{k}	Impedance of element k
P_{set}	Active power setting for distance relay
V_{i}	Voltage phasor of bus i
Q_{set}	Reactive power setting for distance relay
I_{i}	Current phasor of bus i
P_{AB}	Active power flow between bus A and B
δ_i	Phase angle of bus i
Q_{AB}	Reactive power flow between bus A and B

1 Introduction

The unwanted operation of the distance relay in its third zone due to stressed system conditions such as load encroachment, voltage instability or power swing enhances the chances of cascaded system outage of the power system. The traditional distance protection calculates the impedance at relay location on fundamental frequency based on the local information. The measured local impedance reduces in faulty condition and the distance relay operates. Apart from the fault conditions, a stressed system condition may also reduce the measured impedance and the conventional distance relay may mal-operate in its zone-III.

Many researchers have suggested different methods to avoid this unwanted distance relay operation. In some literature, the discrimination between stressed system conditions and a faulty condition is done using the local measurements. In [1], blinders are utilized to avoid the unwanted operation of distance relay during power swing. In [2], a cross blocking criterion is proposed for blocking of the distance relay during power swing by measuring active and reactive power of the system. Another method that discriminate between fault and power swing is proposed based on the rate of change of voltage intend of taking only the voltage magnitude [3]. Along with the rate of change of voltage, the rate of change of current is also used to avoid unwanted distance relay operation in [4]. Computational intelligent methods such as support vector machine (SVM) is also used for segregation between the fault and stressed system condition and an adaptive neuro-fuzzy inference system is used for power swing blocking in [5]. In [6], a decision-tree based approach is proposed for fault detection by taking the positive sequence voltage, current and zero sequences current as input. With the advancement in the wide-area monitoring system, advance methods based on the PMUs data is also proposed by the researchers. In [7], a method is proposed for the operation of zone-III of a relay, based on adjacent relays operation in their zone-I and zone-II. Various integrity protection schemes are also proposed for avoiding the unwanted operation of distance relay. In [8], the relay boundary setting is changed online based on the data obtained from WAMS. Additionally, a load rejection scheme is proposed for the formation of integrity protection scheme. A relay security index is proposed in [9] to avoid the unwanted distance relay operation. In the same work a system stability index is also proposed for system instability prediction. Another method proposed in [10] to distinguish the faulty condition from the stressed system condition based on the change in active power and angle difference between load and source. A differential power coefficient is proposed in [11] to determine the difference between the stable and unstable power swing using WAMS. In [12], an online sequential extreme learning machine is proposed for discrimination between voltage instability, power swing, and faulty condition. The proposed method consists of a two stage classifier that utilizes the WAMS for the computation.

Each method discussed above has some limitations. Methods based on the local information may suffer in discriminating the fault condition from the stressed condition when system complexity increase. Computational intelligent methods require large training data sets for the operation. Methods based on Thevenin's equivalent need to compute online Thevenin's impedance that requires fast computing efforts. In this chapter, a method based on the monitoring of real-time active & reactive power of the load is proposed for blocking the operation of distance relay. The active & reactive power flow at loads are significantly different during a fault and stressed system conditions. This significant difference is utilized here for proposing a new algorithm for easy discrimination between a fault and the stressed system condition.

2 Distance Relay Mal-Operation Scenarios

In this section load encroachment, power swing, and voltage instability conditions are described and the effect of these scenarios on distance relay characteristic have been discussed.

Figure 1 shows a two bus system having two sources connected at bus A and B with a transmission line between them having an impedance Z_L. The generator 1 connected to bus A has induced EMF $E_1\angle\delta$ and internal impedance Z_1.

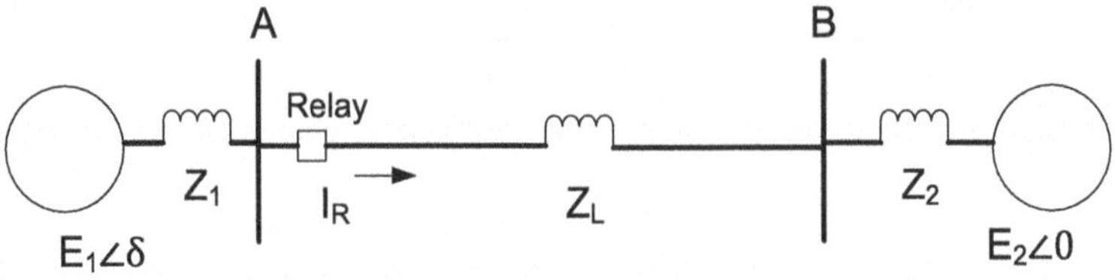

Fig. 1 Two bus system

The generator 2 connected to bus *B* has an induced EMF of $E_2\angle 0$ and the internal impedance of Z_2. A distance relay is connected in the line near bus *A* to detect any faulty condition in the transmission line. If V_R and I_{R} is the voltage and current measured by the relay, then the impedance seen by the relay Z_{seen} will be:

$$Z_{seen} = \frac{V_R}{I_R} \tag{1}$$

A fault in line *AB* will increase the value of current I_R and will decrease the voltage V_R at relay location. These changes further reduces the value of Z_{seen} [13].

2.1 Distance Relay Performance During Power Swing

Due to a large disturbance such as switching of large loads, line contingencies or faults, the power flow in the transmission lines oscillates. These oscillations are known as power swings. A distance relay realizes these swings as changed in impedance and thus treat the power swing as a faulty condition [14]. The power swing phenomenon can be understood using the system shown in Fig. 1 and Eq. (1). The voltage seen by the relay V_{R} and the current seen by the relay I_{R} can be expressed by:

$$V_R = E_1\angle\delta - I_R Z_1 \text{ and } I_R = \frac{E_1\angle\delta - E_2\angle 0}{Z_T} \tag{2}$$

The value of V_R and I_R replaced in Eq. (1) and rearranging the equation for Z_{seen}:

$$Z_{seen} = \left(\frac{E_1\angle\delta}{E_1\angle\delta - E_2\angle 0} \times Z_T\right) - Z_1 = \left(\frac{1}{1 - \frac{E_2\angle -\delta}{E_1}} \times Z_T\right) - Z_1 \tag{3}$$

Here, $Z_T = Z_1 + Z_L + Z_2$ is the total impedance. Also in Eq. (3), assuming that the voltage ratio $E_2/E_1 = k$, thus:

$$Z_{seen} = \left(\frac{1}{1 - k(\cos\delta + j\sin\delta)} \times Z_T\right) - Z_1 \tag{4}$$

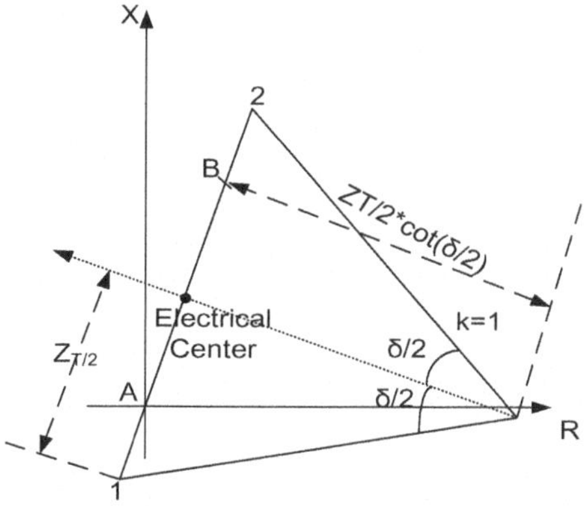

Fig. 2 Representation of Eq. (5) on R-X plane

In Eq. (4) if voltages E_1 and E_2 are equal so $k = 1$, then Eq. (4) will be modified to

$$Z_{seen} = \left(\frac{1}{1 - (\cos\delta + \mathrm{j}\,\sin\delta)} \times Z_T\right) - Z_1$$

or

$$Z_{seen} = \left(-Z_1 + \frac{Z_T}{2}\right) - j\left(j\frac{Z_T}{2}\cot\frac{\delta}{2}\right) \tag{5}$$

On the R-X plane, the Eq. (5) can be represented as shown in Fig. 2. In Eq. (5), the term $\left(-Z_1 + \frac{Z_T}{2}\right)$ represents a constant offset and $\left(j\frac{Z_T}{2}\cot\frac{\delta}{2}\right)$ represents a perpendicular line segment on the constant offset. As $k = 1$, the swing locus is a straight line that cuts the impedance line at the electrical center.

If voltages E_1 and E_2 are not equal, then Eq. (4) will be represented as:

$$Z_{seen} = -Z_1 + \frac{k[(k - \cos\delta) - j\sin\delta]}{(k - \cos\delta)^2 + (\sin\delta)^2} \times Z_T \tag{6}$$

Here, Eq. (6) represents the family of the circles with δ as a variable and k as a parameter as represented in Fig. 3. If the impedance swing remains in the distance relay characteristic, the distance relay measures it as a fault and operates.

2.2 Distance Relay Performance During Load Encroachment

During load encroachment condition, the impedance measured by a distance relay determined by the loadability limit at a specified power factor exceeds due to load voltage and current. The loadability limit is defined in VA for a distance relay at

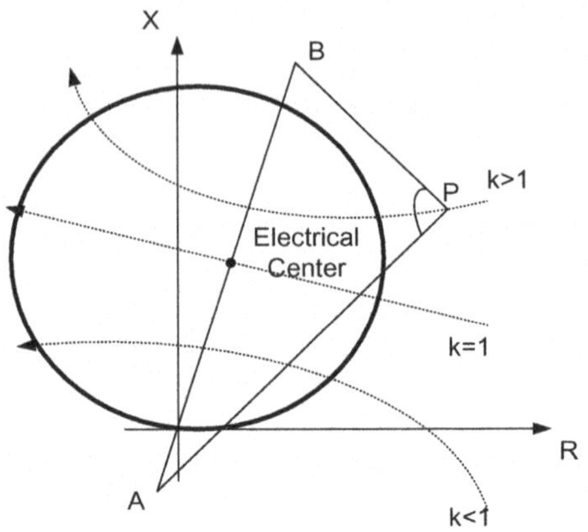

Fig. 3 MHO relay characteristic for Fig. 1

nominal voltage and a specified power factor. The load encroachment is prevalent in long heavily loaded transmission lines and in zone-III of distance relay [15]. For understanding the load encroachment phenomenon, consider the voltage of bus A in Fig. 1 is represented by V_R, the active power flow from bus A to B is P_{AB} and the reactive power flow from bus A to B is Q_{AB}. If the current flowing from bus A to B is I_R, then Eq. (7) defines the relation between P_{AB}, Q_{AB}, $V_{R,}$ and I_R.

$$P_{AB} - jQ_{AB} = V_R * I_R \tag{7}$$

The value of I_R from Eq. (1) is replaced in Eq. (7) we have

$$P_{AB} - jQ_{AB} = V_R * \frac{V_R}{Z_{seen}}$$

$$Z_{seen} = \frac{V_R^2(P_{AB} + jQ_{AB})}{P_{AB}^2 + Q_{AB}^2} \tag{8}$$

From Eq. (8) it is clear that the apparent power flows through the line (i.e. increase in the loading condition) is inversely proportional to the impedance seen by the relay. Due to high loading, the impedance seen by the relay may reduce and fall under the zone-III characteristic and creates unnecessary tripping.

2.3 Distance Relay Performance During Voltage Instability

From Eq. (8) it is observed that the impedance seen by the distance relay is proportional to the square of the bus voltage magnitude. Thus a reduction in voltage also decreases the impedance seen by the relay. There is a possibility for the unwanted operation of distance relay in voltage instability scenarios. In a power system, the voltage instability conditions arise due to a reduction in voltage at several locations or the increased in reactive power of the loads connected at different locations. In a modern power system, the stressed voltage condition may occur due to other reasons that include generator reactive power limit and voltage limits, characteristic of FACTS devices, the action of tap changing transformers, voltage sensitivity of loads, etc. [16]. In such situations, the impedance seen by relay may enter into the third zone of distance relay and lead towards unnecessary tripping [17].

3 Conventional Methods to Avoid Distance Relay Mal-Operation

3.1 Distance Relay Blocking During Power Swing

The power swing is detected by measuring the change in the apparent impedance seen by the distance relay. This change is gradual during power swing compared to a faulty condition and also limited by generators inertia. For the identification of the power swing, an out-of-step (OOS) blocking unit is used. This OOS unit has a characteristic similar to the MHO relay with a larger radius. The power swing will first enter in the OOS blocking relay and then enters in the tripping relay characteristic. If the time for traveling impedance locus from OOS relay characteristic to entering tripping relay characteristic is larger than a pre-set value, the OOS relay will block the operation of distance relay [18]. The decrease in the area of third zone characteristic also reduces the chances of relay operation during power swing. The relay characteristic area can be changed with the method discussed in the next section.

3.2 Distance Relay Blocking During Load Encroachment

The distance relay operation under load encroachment is undesirable and methods discussed below are normally used to avoid this undesirable operation [19]:

- In MHO relays, the reach is maximum at an angle known as maximum torque angle. Increase in the maximum torque angle shifts the relay characteristic, decreases the load angle and increase the loadability limit of the relay [20].
- To reduce the large area of third zone of MHO relay, blinders are provided in series with the MHO relay along the R axis. As the blinders move towards the origin, the area of operation of relay reduces and loadability limit increases [20].
- The MHO relay's circular characteristic could easily be changed to a lens by adjusting the coincident timer. A lens structure has less chances of tripping during power swing due to less area on R-X plane. Also the loadability limit will increase by the using the lens structure [20].

- In a faulty condition, the voltage, current and power angles changes instantaneously, when compared to a stressed system condition. This property is utilized for OOS blocking using offset MHO relays. Fully offset MHO relays are suitable for the remote backup or for third zone protection. With the implementation of the offset backup, the loadability limit also increases [20].

3.3 Distance Relay Blocking During Stressed Voltage Conditions

For a distance relay, it is difficult to discriminate a symmetrical fault and stressed voltage condition and sometimes relay operates in its third zone due to voltage instability. During a faulty scenario, the rate of change of voltage is high compared to stressed voltage situation. Thus rate of change of voltage is considered as a discriminating element between fault and voltage instability. But the change in voltage may also takes place due to switching of capacitors, line and reactors thus these elements are needed to be considered while looking for voltage change [3]. Voltage stability indices are used to predict the voltage instability in the system. Thus, voltage stability index is also considered for discrimination in fault and voltage instability [21].

4 Proposed Method

With the advancement in PMU technology, real-time state estimation of the bulk power system is now possible. PMU provides the time-synchronized data of voltage, currents, and phase angles. Optimally located PMUs may provide the real-time power flow of a bulk power network. In the proposed method, PMUs data is utilized for calculating the real-time active and reactive powers of load buses. PMUs are the back-bone of WAMS and failure of PMUs may cause large scale blackouts. Thus the availability of PMU is an important factor while designing a WAMS.

For an illustration of the proposed method, IEEE 3 bus system is used as shown in Fig. 4. PMUs connected to buses 1 and 3 will provides the voltage and current phasors for the calculation of the real-time load active and reactive power measurement.

4.1 Understanding Scenarios for the Proposed Algorithm

4.1.1 Case-I: Normal System Condition

At rated generation and loading conditions, the system shown in Fig. 4 is simulated on MATLAB Simulink. The impedance seen by the relay R (Z_{seen}) is recorded and shown in Fig. 5a. From Fig. 5a, it is clear that Z_{seen} is constant and high (330) during

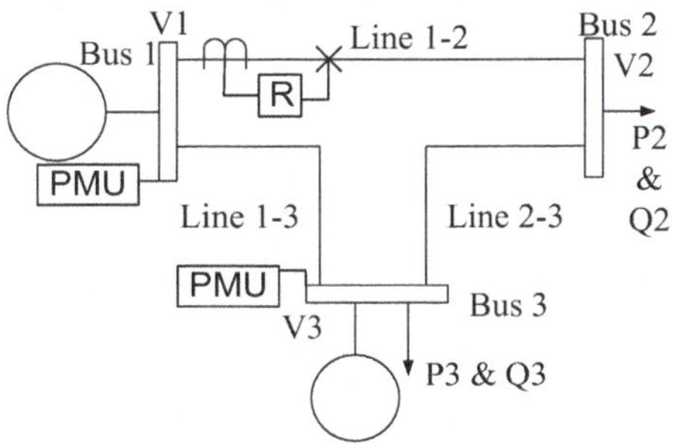

Fig. 4 IEEE-3 bus system

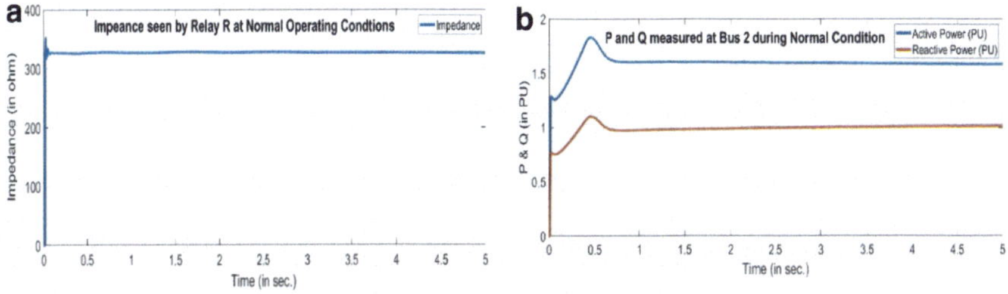

Fig. 5 **a** Measured Z_{seen} for relay R. **b** Active and reactive power at load bus 2

normal operation and it will not enter the operating zone of relay R. The active and reactive power supplied to the load at bus 2 is shown in Fig. 5b.

4.1.2 Case-II: Fault in Line 1–2 at 70% of Line Length

The system of Fig. 4 is simulated again and a three-phase fault is created in line 1–2 at 70% of line length from bus 1 at 1 s. Due to the fault, Z_{seen} is decreased to a low value as shown in Fig. 6a. This low value of impedance normally comes under the zone-I characteristic of relay R. If the relay is set to trip for such a scenario, it will

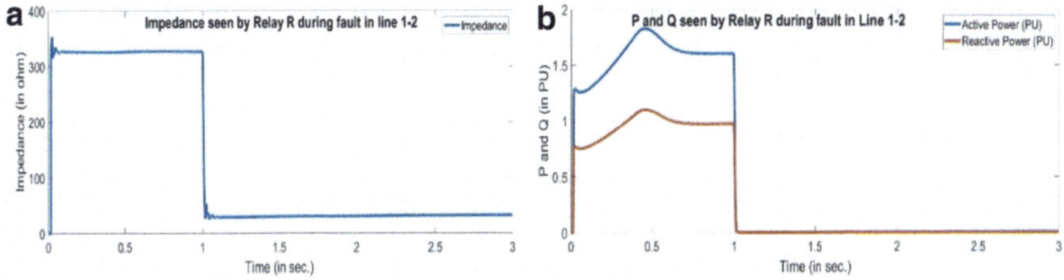

Fig. 6 **a** Measured Z_{seen} for relay R. **b** Active and reactive power at load bus 2

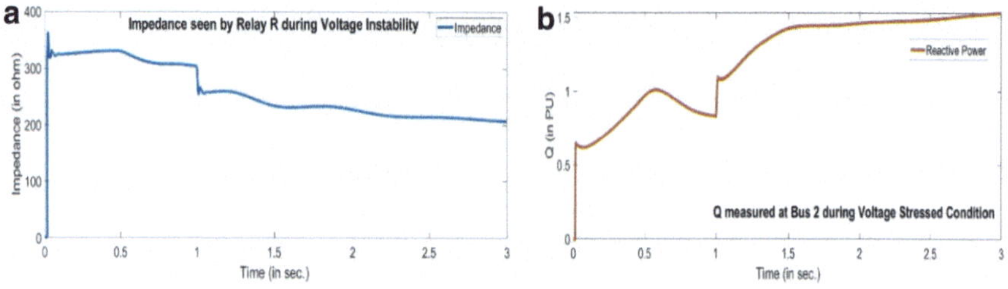

Fig. 7 **a** Measured Z_{seen} for relay R. **b** Reactive power at bus

trip in its zone-I. During this period, the active and reactive powers at the load end will also reduce as seen from Fig. 6b.

4.1.3 Case-III: Creating Voltage Instability

The voltage instability condition is created by increasing the amount of reactive power consumption at bus 2 with line 2–3 switch off. At 1 s., the reactive power of bus 2 is doubled that makes the Z_{seen} reduce from its normal value as shown in Fig. 7a. This reduced value of Z_{seen} may insert in the relay operating characteristic, especially in zone-III. During this period the active power at bus 2 remains constant. The reactive power at bus 2 as shown in Fig. 7b is increased significantly. It is visible that the impedance is reduced without any fault in the system but the measured reactive power at load end clearly reflecting a stressed system condition.

4.1.4 Case-IV: Creating Load Encroachment Condition

For this condition, the bus 2 load is increased to double of its rated value at 1 s. Figure 8a shows that the Z_{seen} is reduced from its normal value and may enter in zone-III relay characteristic. While, the increased amount of active and reactive power at bus 2 shown in Fig. 8b differentiating this scenario from the faulty condition.

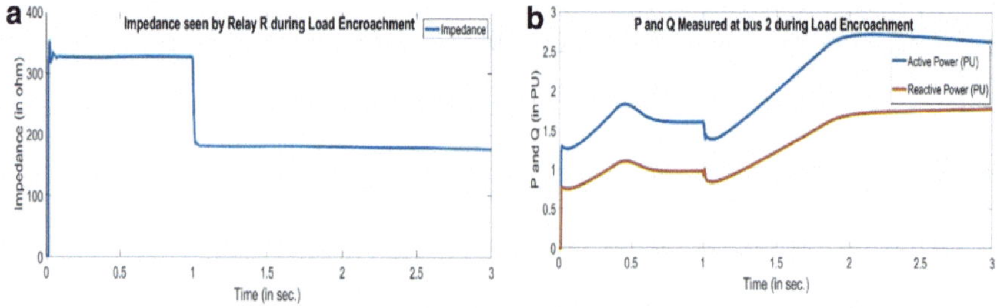

Fig. 8 **a** Measured Z_{seen} for relay. **b** Active & reactive power at bus 2

Table 1 Comparison of Z_{seen} by relay R and active & reactive powers at bus-2 during various cases

Cases	Conditions on which impedances measured	Z_{seen} (Ω)	P (PU)	Q (PU)
I	Normal operating conditions	330	1.6	1.05
II	After fault at 1 s	30	0	0
III	Creating voltage stressed condition at 1 s	210	1.6	1.6
IV	Creating load encroachment condition at 1 s	180	2.6	1.8

4.2 Comparison of the Discussed Cases

The impedances seen by relay *R* and the active & reactive powers in above simulated cases are shown in Table 1.

Following are the comparative analysis of above discussed cases.

- During normal condition (Case-I), the impedance seen by the relay is 330 Ω with the rated active and reactive power load at bus 2 and 3.
- In Case-II, before the fault in line 1–2, the value of Z_{seen} is 330 Ω and the active and reactive powers at bus-2 is at rated values. A fault at 1 s. in Case-II reduces the Z_{seen} to 30 Ω and the amount of active and reactive power flow to the load connected to bus 2 is reduced to 0 PU as shown in Table 1.
- It is visible from Fig. 7, that in Case-III, initially, the Z_{seen} is 330 Ω and the active & reactive powers for load connected to bus 2 are at their rated values. The stressed condition is created at 1 s. by increasing the reactive power of bus 2 at double of its rated value that reduces the impedance seen by relay to 210 Ω.
- In Case-IV, initially, the Z_{seen} is 330 Ω and the active and reactive powers for bus 2 are their rated values as shown in Fig. 8. After creating the load encroachment condition by increasing the active and reactive powers of bus-2 at 1 s. in Case-IV, the Z_{seen} is reduced to 180 Ω.

From the above discussed cases it is clear that a fault in the system reduces the active & reactive power flow into the system and also reduce the Z_{seen}. While in a stressed system condition the impedance seen by the relay reduces while the active and/or reactive powers flow at load buses increases. This difference in the active and reactive power flow at loads is utilized here to propose a new algorithm.

4.3 Proposed Algorithm

To avoid the chances of the unwanted distance relay operation, the algorithm of distance relay is modified using the data obtained from WAMS. In WAMS, PMUs are used to collect time-synchronized voltage, currents, and phase angles. These voltage and current signals of load buses are used for calculation of real-time active and reactive powers connected to load buses. The flow chart of the proposed method is shown in Fig. 9.

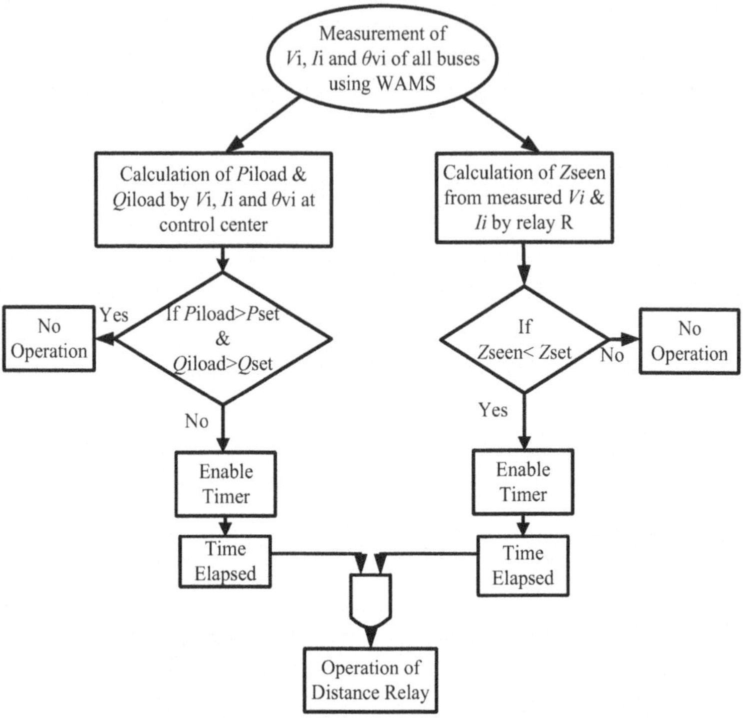

Fig. 9 Flow chart of the proposed algorithm

The first condition is the conventional condition in which if the value of impedance seen by the relay Z_{seen} is less than the set value of relay impedance Z_{set}, the relay will operate. The first condition is expressed in Eq. (9).

$$Z_{seen} < Z_{set} \tag{9}$$

The scenario of Eq. (9) may come either due to faulty condition or due to stressed system conditions discussed in Sect. 2. To avoid the operation of distance relay in stressed situations, a second operating condition is added here which is based on the parameters obtained from WAMS. The voltage and currents obtained from WAMS is used for calculation of real-time active P_{iload} and reactive power flow of load buses Q_{iload}. The load encroachment and voltage instability condition occur when the value of P_{iload} and/or Q_{iload} increased from its rated value. The power swing condition may occur in the system when there is a fault in other transmission lines or a large generator may trip. In all such conditions, the value of P_{iload} and Q_{iload} will not reduce to zero. The P_{iload} and Q_{iload} will be reduced to zero only when there is a fault in the transmission line that is supplying the load P_{iload} and Q_{iload}. Based on the logic discussed here, another condition is proposed for the operation of the distance relay. Equation (10) describe the operating condition of the distance relay. If the value of real-time load bus active and reactive powers reduces below a set value P_{set}, the distance relay will operate.

$$P_{iload} < P_{set} \& Q_{iload} < Q_{set} \tag{10}$$

Here, P_{set} and Q_{set} are predefined settings of active & reactive power. It must be noted that distance relay will operate only when both Eqs. (9) and (10) will be satisfied.

5 Case Study

5.1 WSCC-9 Bus System

In WSCC-9 bus system shown in Fig. 10, a 3-stepped distance relay R is placed near bus 7 in line 7–8. The setting of distance relay is such that zone-I consist of 80% of the line 7–8 length, zone-II covers full length of line 7–8 and 50% of the adjoining line 8–9 and zone-III have the full length of line 8–9. The active and reactive powers at load buses is determined by using PMUs placed at optimum bus locations i.e. bus-4, bus-7 and bus-9.

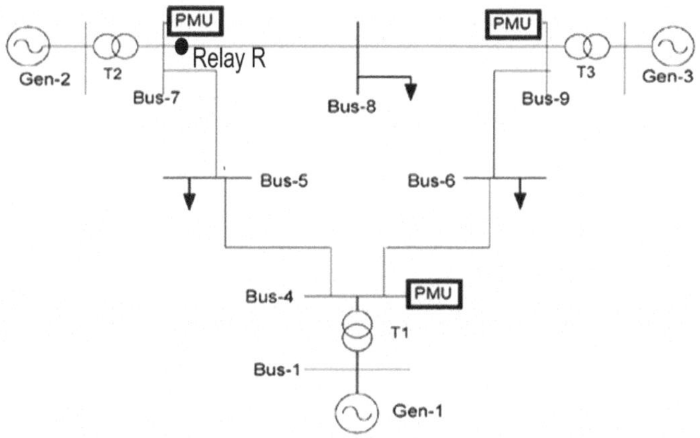

Fig. 10 WSCC-9 bus Test System

5.1.1 Calculate Active & Reactive Power of Bus-8 with Optimum PMU Location

The PMUs connected to bus 4, 7 and 9 will provide the time stamped voltage, current magnitude and phase angles of respective buses. Here, the data obtained from PMUs are utilized to calculate the real time active and reactive powers at bus 8.

Assumed that the voltage vector of bus 8 is V_8 then

$$V_8 = I_8 \times Z_{L8} \tag{11}$$

Here, I_8 is the current flowing through load 8 and Z_{L8} is the load impedance.

By implementing the nodal analysis at bus 8, the current I_8 can be rewrite as:

$$I_8 = I_{78} + I_{89} \tag{12}$$

Equation (11) is modified by putting the value of I_8 from Eq. (12) as:

$$V_8 = (I_{78} + I_{89}) \times Z_{L8}$$

Here, I_{78} is the current in line 7–8, and I_{89} is the current vector of line 8–9.

$$\text{Also, } V_8 = V_7 - I_{78} \times Z_{78} \tag{13}$$

Here, V_7 is the voltage vector of the bus 7, and Z_{78} is the impedance of line 7–8.

From Eqs. (12) to (13): $(I_{78} + I_{89}) \times Z_{L8} = V_7 - I_{78} \times Z_{78}$

$$\text{or } V_7 = (Z_{78} + Z_{L8}) \times I_{78} + I_{98} \times Z_{L8} \tag{14}$$

$$\text{Similarly, } V_9 = (Z_{98} + Z_{L8}) \times I_{98} + I_{78} \times Z_{L8} \tag{15}$$

On solving Eqs. (14) and (15)

$$I_{78} = \frac{V_9 Z_{L8} - V_7(Z_{89} + Z_{L8})}{Z_{L8}^2 - (Z_{L8} + Z_{78})^2} \;\&\; I_{89} = \frac{V_9 Z_{L8} - V_7(Z_{89} + Z_{L8})}{Z_{L8}^2 - (Z_{L8} + Z_{89})^2} \tag{16}$$

Putting the values of I_{78} and I_{89} from Eq. (16) and (17) to Eq. (12)

$$I_8 = \frac{Z_{89}(V_7 + V_9)}{(Z_{L8} + Z_{89})^2 - Z_{L8}^2}$$

Once the I_8 is determined from Eq. (12), V_8 can be also be determined by Eq. (11).

The value of voltage and current phasors of bus 8 i.e. V_8 and I_8 is used to determine the bus 8 active and reactive powers. These real time active and reactive powers are fed to the distance relay for implementing the proposed algorithm.

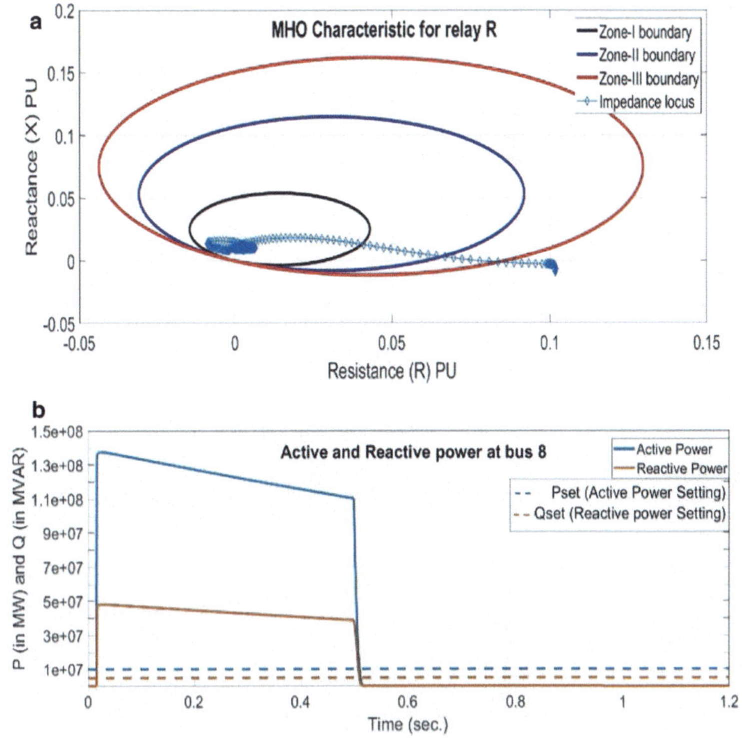

Fig. 11 **a** Relay R characteristic near bus 7 during fault condition. **b** Active and reactive power at bus 8

5.1.2 Simulated Cased on WSCC-9 Bus System

Following cases are simulated on for verification of the proposed algorithm.

Case A—During Faulty Condition Near Bus 8

In this case a fault is initiated in line 7–8 at 80% of its line length from bus-7 at 0.5 s. Figure 11a shows that due to fault, the impedance locus entered in zone-I of the distance relay. The Eq. (9) operating criteria is satisfied as the impedance locus is entering in the distance relay operating zone.

Figure 11b shows the active and reactive power of bus 8 for the simulated fault. It is clear that bus 8 active and reactive powers are reduced to zero which is below the set active power and reactive power limit i.e. P_{set} and Q_{set} respectively (10% of the rated value of bus-8).

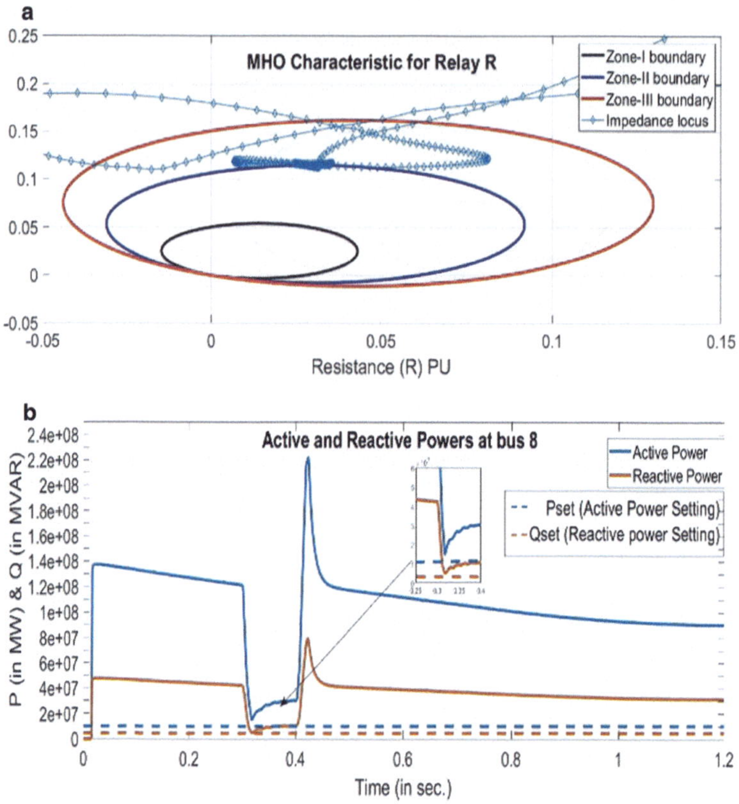

Fig. 12 **a** Relay R characteristic during power swing. **b** Active and reactive power at bus 8

The operating condition set in Eq. (10) is also confirming the operation of the relay. As both Eqs. (9) and (10) are satisfying with the given condition, the relay will operate for the simulated fault.

Case B—During Stable Power Swing at Bus 8

The stable power swing condition is initiated on line 7–8 by creating a 3-phase fault at mid-section of line 5–7 from 0.3 to 0.4 s. The fault is cleared by opening the line circuit breakers. Due to fault, the impedance locus entered in zone-III of the distance relay as shown in Fig. 12a.

Figure 12b, shows the active & reactive power flow for bus-8. It is clear that the active and reactive powers of bus-8 is not crossing the P_{set} and Q_{set} values for load bus 8. It is clear that this scenario is not satisfying the Eq. (10) operating condition. From the two of the operating condition, only one Eq. (9) is satisfying thus relay will block its operation.

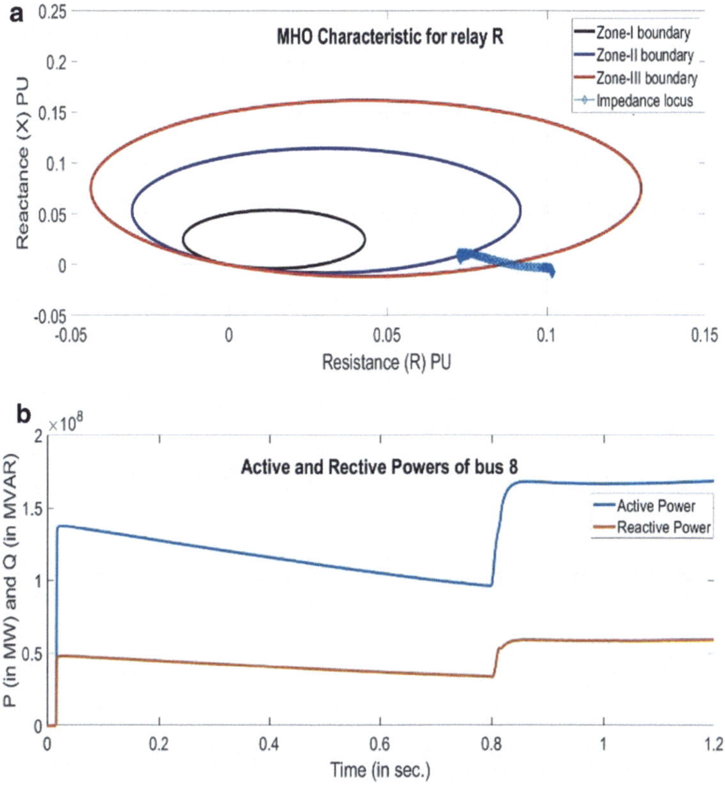

Fig. 13 **a** Relay R characteristic near bus 7 during Case-C. **b** Active and reactive power at bus-8 for Case-C

Case C—During Load Encroachment at Bus 8

For load encroachment, extreme loading condition is created at bus 8 by doubling its load as per the NERC standard of extreme loading criteria. Figure 13a shows that the relay R characteristic with impedance locus tending towards zone-III due to load encroachment condition. As the Z_{seen} reduced from the value of Z_{set}, it will operate a conventional distance relay as it satisfies the operation criteria mentioned in Eq. (9). It is clear that without any fault, the impedance is inserting in zone-III of the distance relay. However, Fig. 13b shows the active and reactive powers at bus 8. It is clear that the real and reactive power of bus 8 is higher than the set values and thus Eq. (10) is not satisfied that blocks the distance relay operation.

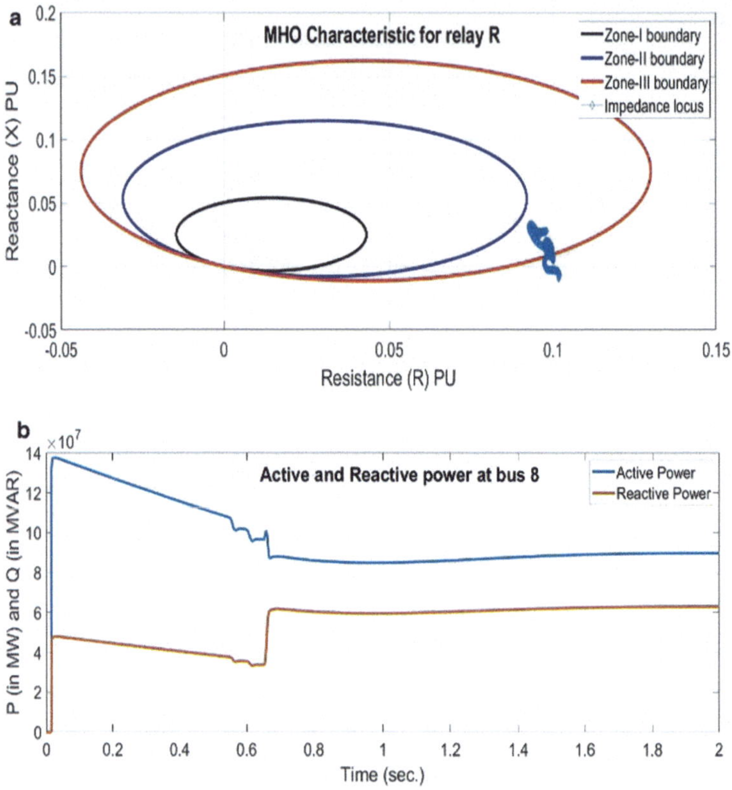

Fig. 14 **a** Relay R characteristic for Case-D. **b** Active and reactive power at bus 8

Case D—During Voltage Instability at Bus 8

The voltage instability situation is imposed by increasing the amount of reactive powers of load connected to bus 5, 6 and 8 such that the impedance locus enters in zone-III of relay R as shown in Fig. 14a. Figure 14b shows the active and reactive powers of bus 8 indicating a no-fault condition. From the power flow, it is clear the locus entering the zone-III is due to voltage instability condition and thus proposed method prevents the relay operation. The active and reactive powers of bus-8 is only shown in the Fig. 14b because the relay placed in line 7–8 will have the active and reactive power data of bus-8 for its operation.

5.2 *IEEE 14 Bus System*

The proposed scheme is also tested on IEEE 14- bus system shown in Fig. 15. The three stepped distance relay R_1 is placed in line 9–14 on bus 9. The settings of

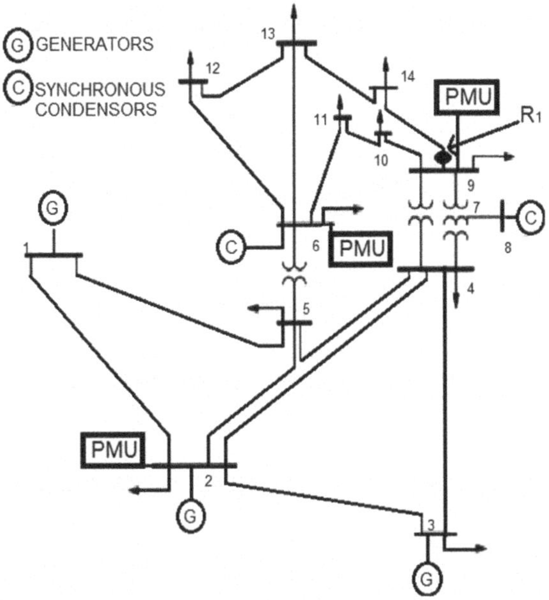

Fig. 15 IEEE 14-bus test system [22]

distance relay are such that zone-I consist of 80% of line length 9–14, the second zone includes 50% of line 14–13 and third zone covers full length of line 14–13.

The proposed scheme requires real-time active & reactive power of load bus 13.

Optimum PMU placed at bus 2, 6 and 9 will provide the load bus-8 voltage and currents as discussed in Sect. 5.1.1 for WSCC-9 bus system. The real-time load powers are fed to the relay for implementing the proposed algorithm.

5.2.1 Simulated Cased on IEEE-14 Bus System

Following cases are conducted for verification of the proposed algorithm.

Case A—During Faulty Condition in Line 13–14

Figure 16a shows that the impedance is approaching in Zone-II during the faulty condition. This condition is fulfilling the operating condition of the distance relay mentioned in Eq. (9). The second condition of the proposed approach will look towards the active power P_{13load} and reactive powers Q_{13load} of bus 13 for the relay operation. The fault is reducing the active and reactive powers of bus 13 to zero as shown in Fig. 16b. The P_{set} and Q_{set} are considered here to 10% of the rated value of load connected to bus 13. It is clear here that

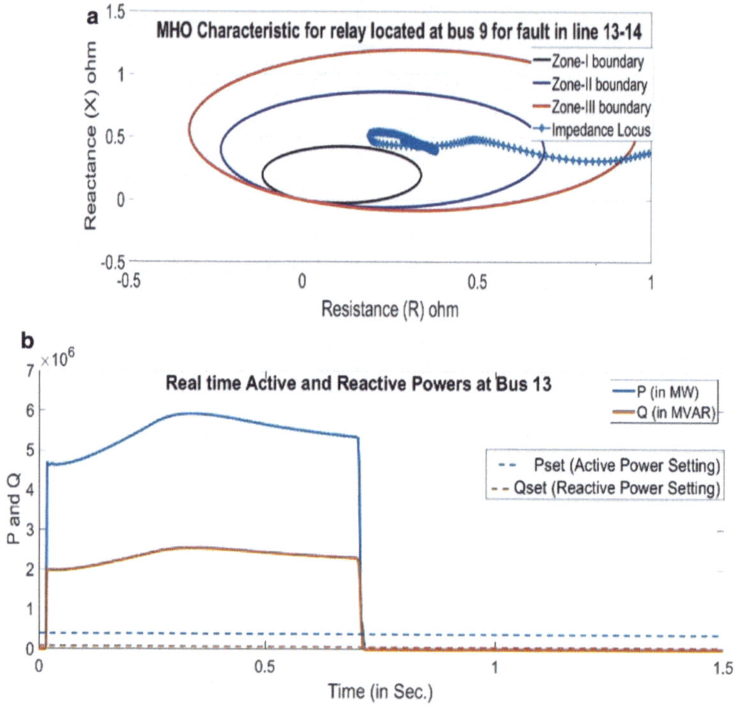

Fig. 16 **a** Relay R_1 characteristic during fault. **b** Active and reactive powers at bus 13 during fault

$$P_{13load} < P_{set}$$

$$Q_{13load} < Q_{set}$$

This scenario is satisfying the Eq. (10). As both Eqs. (9) and (10) are true for the given fault, the proposed algorithm will trip the relay.

Case B—During Stable Power Swing at Bus 13

For stable power swing, a temporary 3-phase fault is initiated on line 9–10 from 0.7 to 1.2 s. The Z_{seen} by relay placed at bus 9 observe this condition as a fault in its zone-II. The impedance trajectory is shown in Fig. 17a. It is clear that Z_{seen} is reduced and satisfies the operating condition discussed in Eq. (9). A conventional distance relay will operate in this a scenario. In the proposed algorithm, the relay will also look for active & reactive powers of bus 13 shown in Fig. 17b. It is clear that the active & reactive power of bus 13 is reducing due to fault but not going beyond the relay setting. P_{set} and Q_{set}. Due to this, the criteria for the operation of the relay shown in Eq. (10) is not fulfilled and the relay avoids the tripping.

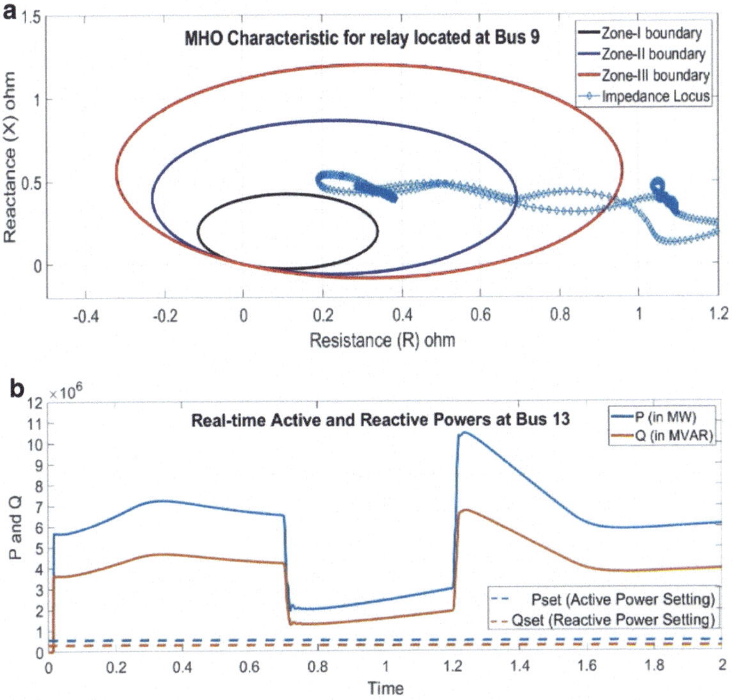

Fig. 17 **a** Relay R_1 characteristic during power swing. **b** Active & reactive powers at bus 13 during Case B

Case C—During Load Encroachment at Bus 13

For load encroachment, extreme loading condition is created at bus 13 and 14 by doubling its load as per the NERC standard of extreme loading condition. Figure 18a shows that the relay R_1 characteristic with impedance locus tending towards zone-III due to load encroachment.

As the Z_{seen} reduced, it may operate a conventional distance relay as it satisfies the operation criteria of Eq. (9). Figure 18b shows the active and reactive powers at bus 13. It is clear that the real and reactive power of bus 13 i.e. P_{13load} and Q_{13load} is higher than the set values P_{set} and Q_{set}. As the condition of Eq. (10) is not satisfied, the proposed algorithm avoids the operation of the distance relay.

Case D—During Voltage Instability at Bus 13

The voltage instability conditions in a power system may also operate a distance relay as the impedance trajectory enters in the zone-III of distance relay. The voltage instability the situation is created by increasing the reactive powers of bus 10, 13 and 14 such that the impedance locus enters in zone-III of relay R_1 as shown in Fig. 19a.

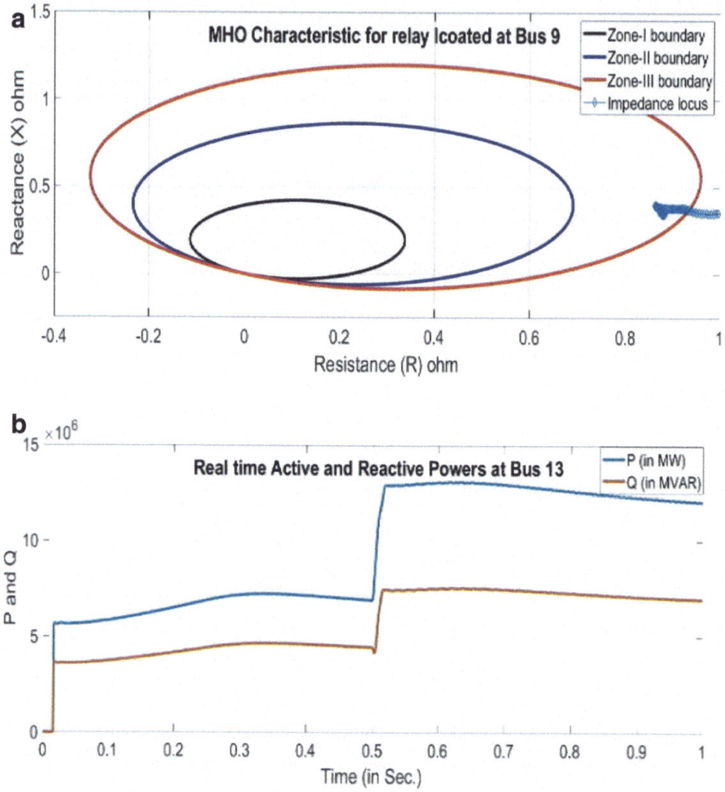

Fig. 18 **a** Relay R_1 characteristic during load encroachment. **b** Active & reactive powers at bus 13 during load encroachment

Figure 19b shows the active and reactive powers of bus 10, 13 and 14 indicating a no-fault condition. From the power flow, it is clear the locus entering the zone-III is due to voltage instability condition and thus proposed method prevents the relay operation.

5.3 *IEEE 30 Bus System*

Proposed scheme is also tested on IEEE 30 bus system discussed in [23]. The three stepped distance relay R_1 is placed in line 15–23 on bus 15. The settings of distance relay are such that zone-I consist of 80% of line 15–23, the second zone includes 50% of line 23–24 and third zone covers full length of line 23–24. As the system complexity increased here, real time active and reactive power of two load buses (bus 23 and 24) are fed to the relay for its operation.

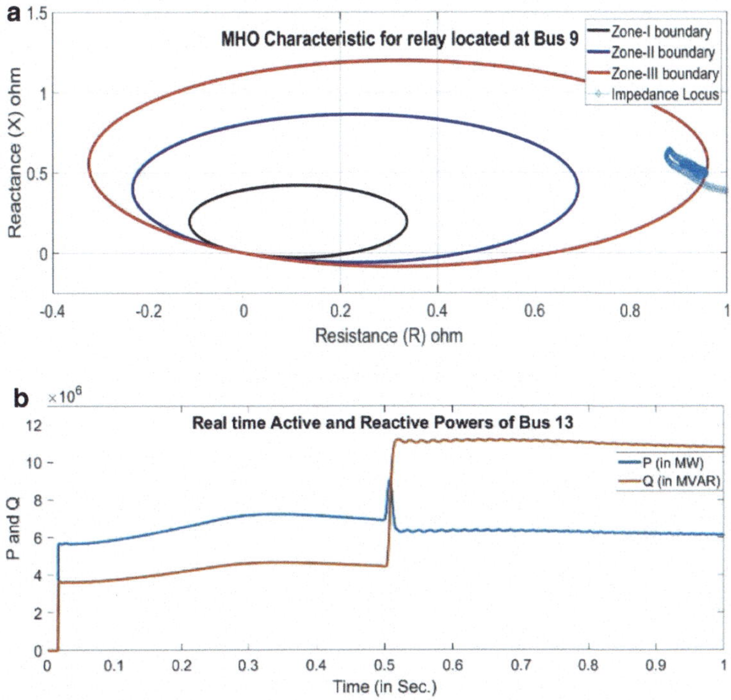

Fig. 19 **a** Relay R_1 characteristic during voltage instability. **b** Active & reactive powers during voltage instability

5.3.1 Simulated Cases on IEEE-30 Bus System

Following cases are conducted for verification of the proposed algorithm on IEEE 30 bus system.

Case A—During Normal Operating Condition

During normal operating condition, the impedance seen by the relay is high enough represented in Fig. 20a. Figure 20b shows the flow of active and reactive power in load bus 24. Figure 20 clearly shows that the relay will not operate during normal operating condition because the operating criteria set by Eqs. (9) and (10) will not be satisfied.

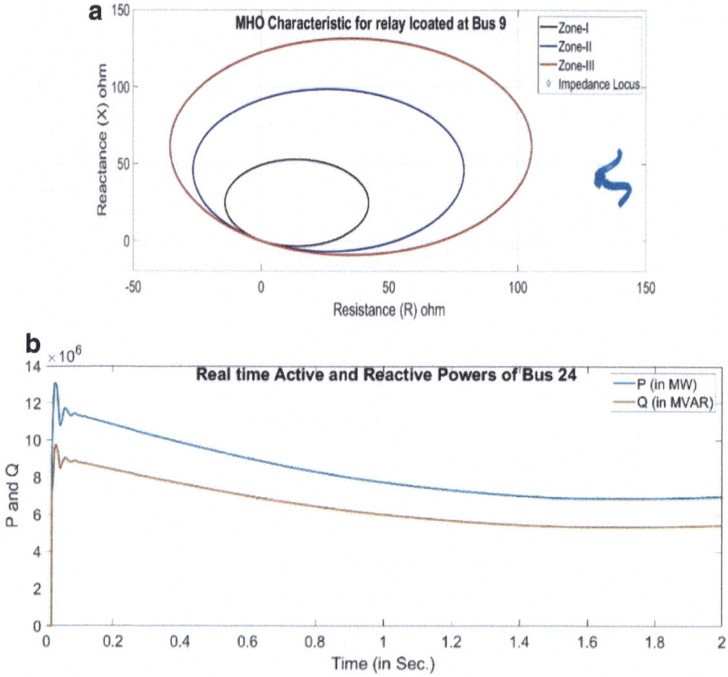

Fig. 20 **a** Relay R_1 characteristic during normal condition. **b** Active and reactive powers at Bus 24

Case B—During Fault Condition

To check the relay performance during a faulty condition, a fault is initiated in line 15–23. Figure 21a shows that the impedance is approaching in Zone-I during the faulty condition. This condition is fulfilling the operating criteria of the distance relay mentioned in Eq. (9).

The second condition of the proposed approach will look towards the active power and reactive powers of load buses for the relay operation. The fault is reducing the active and reactive powers of bus 23 to zero as shown in Fig. 21b. This scenario is satisfying the Eq. (10).

As operating criteria of both Eqs. (9) and (10) are satisfying for the given fault, the proposed algorithm will trip the relay.

Case C—During Stable Power Swing

For stable power swing, a temporary 3-phase fault is initiated on line 24-25. The Z_{seen} by relay placed at bus 15 observe this condition as a fault in its zone-III. The impedance trajectory is shown in Fig. 22a. It is clear that Z_{seen} is reduced and

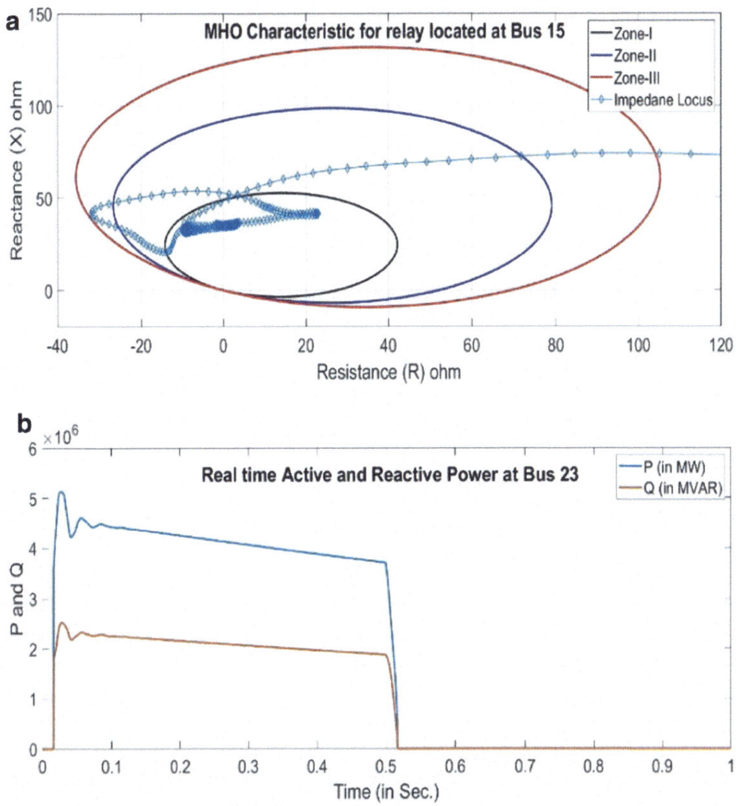

Fig. 21 **a** Relay R_1 characteristic during power swing. **b** Active & reactive powers at bus 24 during Case B

satisfies the operating condition discussed in Eq. (9). A conventional distance relay will operate in this a scenario.

In the proposed algorithm, the relay will also look for active & reactive powers of bus 23 and 24. The active and reactive powers of bus 24 is shown in Fig. 22b. It is clear from the Fig. 22b that the active and reactive powers are not reducing to zero as in previous case. A suitable value of P_{set} and Q_{set} discussed in Eq. (10) may block the unwanted operation of distance relay.

Case D—During Load Encroachment

For load encroachment, extreme loading condition is created at bus 23 and 24 by doubling its load. Figure 23a shows that the relay R_1 characteristic with impedance locus tending towards zone-III due to load encroachment. As the Z_{seen} reduced, it may operate a conventional distance relay as it satisfies the operation criteria of Eq. (9).

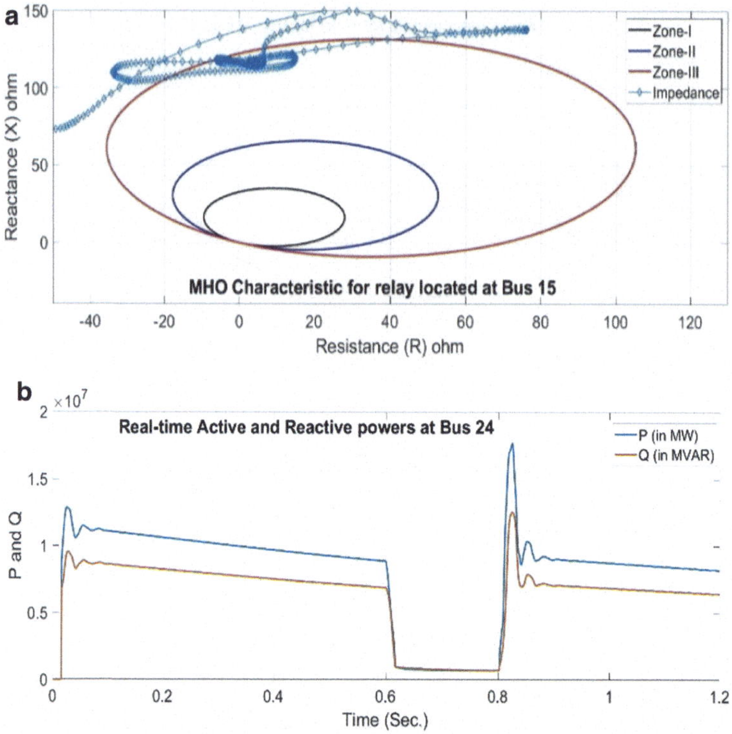

Fig. 22 **a** Relay R_1 characteristic during power swing. **b** Active & reactive powers at bus 24 during Case B

Figure 23b shows the active and reactive powers at bus 24. It is clear that the real and reactive power of bus 24 is higher than the rated values. It is not satisfying the operating criteria discussed in Eq. (10). The proposed algorithm avoids the operation of the distance relay.

5.4 Comparison with Existing Methods

The proposed algorithm is tested on the WSCC-9, IEEE-14 and IEEE-30 bus test systems for various cases such as stable load encroachment, power swing, and voltage instability condition. It is found that the active & reactive power at load buses have significant difference during faults and abnormal operating conditions. The load bus powers are utilized for blocking the relay operation with the proposed algorithm. The proposed method is compared with the existing methods discussed in literature based on the following points:

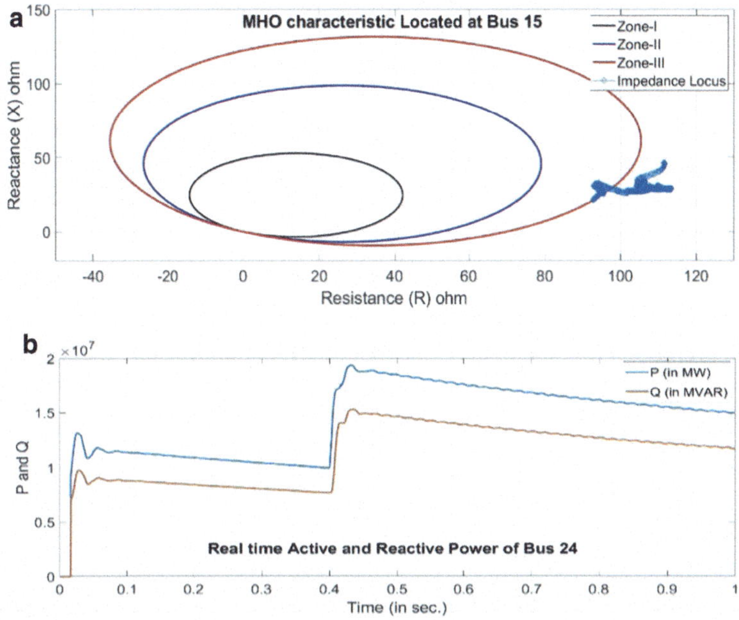

Fig. 23 **a** Relay R_1 characteristic during load encroachment. **b** Active & reactive powers during load encroachment

1. PMUs Location: The methods represented in [7–9, 12] need the PMU at all buses of a network. For practical applications, PMUs are placed only at optimal locations for the data collection. The proposed method does not require the PMU at all locations and utilizes the optimal PMU data for decision making. This also makes the approach more economical in practical perspective.
2. Training Requirement: The methods discussed in [5, 6, 12] needs to train the model before implementation. This training process can be done offline but needs large data sets for effective results. While the proposed method does not require any data training model.
3. Network conversion: The method discussed in [8, 10] required to convert the complete network in an equivalent TE model or SMIB system. The conversion of a real-time network into another equivalent model needs additional computational time. In the proposed method, there is no need to convert the network into another equivalent network, thus reduces the computational burden.
4. Utilization of wide-area information: Protection scheme discussed in [5, 6] makes the operating decision based on the local information. In a large system decision making using only the local information may lead to unwanted relay operation. The proposed method utilizes the WAMS and has high accuracy in the prediction of the power system state.

Table 2 Comparison of different methods

Methods→	[5]	[6]	[7]	[8]	[9]	[10]	[12]	Proposed
PMUs Locations	–	–	All	All	All	Optimum	All	Optimum
Training required	✓	✓	✓	✗	✗	✗	✓	✗
Network Conversion	✗	✗	✗	✓	✗	✓	✗	✗
Wide area information	✗	✗	✓	✓	✓	✓	✓	✓
Applicable for Load Encroachment	✗	✗	✓	✓	✓	✓	✗	✓

Table 2 represents a comparison of the proposed method with the existing methods. The Load encroachment phenomenon which is not considered by few authors is also incorporated in the existing method to avoid distance relay mal-operation.

6 Conclusion

Advancement in the WAMS technology opens the door for researchers to propose new algorithms for power system operation, control, and protection. The protection schemes implemented using WAMS reduces the risks of large area blackouts. Unwanted distance relay mal-operation may create cascaded system tripping and it is one of the reasons for large area blackouts. This chapter utilizes the data from WAMS to address the issues of unwanted distance relay operation due to stressed system conditions. The scenarios in which a distance relay mal-operates are thoroughly discussed in this work. Based on the observations from the simulation on IEEE 3 bus system, a method has been proposed to avoid the unwanted operation of distance relay using monitored changes in load active and reactive power. Real-time current and voltage phasors obtained from optimally located PMUs are utilized for calculation of load active and reactive power. To check the effectiveness of the proposed algorithm, various cases are simulated on WSCC-9 bus, IEEE-14 bus and IEEE-30 bus test systems. It is observed that the proposed method successfully blocks the distance relay operation under stressed system conditions.

References

1. G. Benmouyal, D. Hou, D. Tziouvatas, Zero-setting power-swing blocking protection, in *31st Annual Western Protective Relay Conference* (WA, USA, 2004)
2. X. Lin, Y. Gao, P. Liu, A novel scheme to identify symmetrical faults occurring during power swings. IEEE Trans. Power Deliv. **23**, 73–78 (2008)
3. M. Jonsson, J. Daalder, An adaptive scheme to prevent undesirable distance protection operation during voltage instability. IEEE Trans. Power Deliv. **18**(4), 1174–1180 (2003)

4. M. Sharifzadeh, H. Lesani, M. Sanaye-Pasand, A new algorithm to stabilize distance relay operation during voltage degraded conditions. IEEE Trans. Power Deliv. **29**(4), 1639–1647 (2014)
5. K. Seethalekshmi, S. Singh, S. Srivastava, A classification approach using support vector machines to prevent distance relay maloperation under power swing and voltage instability. IEEE Trans. Power Deliv. **27**(3), 1124–1133 (2012)
6. A. Swetapadma, A. Yadav, Data mining based fault during power swing identification in power transmission system. IET Sci. Meas. Technol. **10**(2), 130–139 (2016)
7. D. Pal, B. Mallikarjunna, R. Reddy, Synchrophasor assisted adaptive relaying methodology to prevent zone-3 mal-operation during load encroachment. IEEE Sens. J. **17**(23), 7713–7722 (2017)
8. P. Kundu, A.K. Pradhan, Enhanced protection security using the system integrity protection scheme (SIPS). IEEE Trans. Power Deliv. **31**(1), 228–235 (2016)
9. P. Gawande, S. Dambhare, New predictive analytic aised response based system integrity protection scheme. IET Gener. Transm. Distrib. **13**(8), 1204–1211 (2019)
10. S.S. Samantaray, A. Sharma, Supervising zone-3 operation of the distance relay using synchronised phasor measurements. IET Gener. Transm. Distrib. **13**(8), 1238–1246 (2018)
11. D. Kumar, J. Savier, Synchrophasor based system integrity protection scheme for an ultra-mega-power project in India. IET Gener. Transm. Distrib. **13**(8), 1220–1228 (2019)
12. S. Das, R. Dubey, B. Panigrahi, S. Samantaray, Secured zone-3 protection during power swing and voltage instability: an online approach. IET Gener. Transm. Distrib. **11**(2), 437–446 (2016)
13. Network Protection & Automation Guide, Alstom Grid (2011)
14. J. Khodaparast, M. Khederzadeh, *Adaptive Concentric Power Swing Blocker. Protection and Control of Modern Power Systems*, vol. 1, no. 16 (Springer, 2016)
15. Line Protection Relay User Manual Version 3.3, ERL Phase Power Technologies (2003)
16. P. Kundur, *Power System Stability and Control* (Mc-Graw-Hill, New York, 1994)
17. R. Sodhi, S. Shrivastav, S. Singh, A simple scheme for wide area detection of impeding voltage instability. IEEE Trans. Smart Grid **3**(2), 818–827 (2012)
18. Power swing and out-of-step considerations on transmission lines, in *D6, IEEE Power System Relay Committee Working Group* (IEEE, 2005)
19. Increase line loadability by enabling load encroachment functions of digital relays, in *NERC Planning Committee* (New Jersey, 2005)
20. Methods to increase line relay loadability, in *NERC Planning Commission* (New Jersey, 2006)
21. H. Song, B. Lee V. Ajarappu, Control strategies against voltage collapse considering undesired relay operations, in *Proceeding of the Institute Engineering Technology Generation Transmission and Distribution,* vol. 3, no. 2, pp. 164–172 (2009)
22. N. Rajalwal, P. Mishra, Impact of DFIG on voltage stability of a network in smart grid: an analysis, in *Technologies for Smart City Energy Security and Power*, India, 2018
23. PSCAD, IEEE-30 Bus System, (Manitoba Hydro International Ltd., Manitoba, Canada, 2018)

Real-Time Voltage Stability Monitoring Using Machine Learning-Based PMU Measurements

Mohammed Amroune, Arif Bourzami, Mohamed Zellagui, and Ismail Musirin

Abstract Recently, due to the increasing demand with scarcity in installed production capacities, power systems are being operated closer to voltage stability limits resulting in a higher eventuality of voltage collapse. Thus, fast and accurate monitoring of voltage stability has become an important factor in the efficient operation of modern power systems. In this chapter, two approaches based on the combination of multi-layer perceptron (MLP) neural network and adaptive neuro-fuzzy inference system (ANFIS) with moth swarm algorithm (MSA) have been proposed to monitor voltage stability of power systems using phasor measurement units (PMUs) data. In the proposed hybrid MLP–MSA and ANFIS–MSA models, the MSA algorithm is adopted to optimize the connection weights and biases of the MLP network and to determine the tuning parameter in ANFIS model. To evaluate the prediction capability and efficiency of the proposed models, several statistical indicators such as root mean square error (RMSE), correlation coefficient (R) and root mean square percentage error (RMSPE) are used. Numerical studies are carried out on two standard power systems. The obtained results indicate that the proposed ANFIS–MSA model has the most reliable and accurate prediction ability and deemed to be the effective method to estimate the voltage stability margin of the power system based on measurements from PMU devices.

M. Amroune (✉) · A. Bourzami
Department of Electrical Engineering, University of Ferhat Abbas, Setif 1, 19000 Setif, Algeria
e-mail: mohammed.amroune@yahoo.fr

M. Zellagui
Département de Génie Électrique, École de Technologie Supérieure, Université de Québec, Montréal, Canada

Department of Electrical Engineering, Faculty of Technology, University of Batna 2, Fesdis 05078, Batna, Algeria

I. Musirin
Faculty of Electrical Engineering, Universiti Teknologi MARA, 40450 Shah Alam, Selangor, Malaysia

H. Haes Alhelou et al. (eds.), *Wide Area Power Systems Stability, Protection, and Security*, Power Systems, https://doi.org/10.1007/978-3-030-54275-7_16

1 Introduction

The rapid growth of electricity demand, the environmental restrictions and the deregulation policies coerce the power systems to operate close to their stability limits. In such conditions, any contingency such as the loss of a generator or transmission line in the system may cause voltage collapse. Therefore, it is necessary to continuously monitor the voltage stability of the power system to avoid the risk of large blackouts. To monitor the voltage stability of the power system in real-time, the process of measurements collection and voltage stability margin (VSM) computation must be accomplished within the required time frame. Traditionally, supervisory control and data acquisition (SCADA) system have been used to collect measurements regularly every few minutes [1]. Therefore, real-time voltage stability monitoring is impractical with using traditional SCADA system. In recent years, the wide-area measurement system (WAMS) is increasingly being deployed in modern power systems (smart grids). With the prevalence of WAMS based on phasor measurement units (PMUs), the power system stability issues can be treated more efficiently. PMUs overcome the disadvantages of SCADA by providing the synchronized measurements of voltage and current phasors and frequency at a very high speed. The synchrophasor measurements gathered from PMUs can support the tracking of fast event and provide sufficient information for voltage stability monitoring. Voltage stability margin (VSM) estimation is one of the commonly used techniques for real-time voltage stability monitoring based on the provided PMU measurements. Many studies have been developed based on machine learning techniques to estimate the VSM in real-time. In [2], multi-layered perceptron (MLP) network based-back-propagation algorithm is introduced to estimate the VSM using the energy method. Joya et al. [3] utilized a sequential learning strategy to design a single MLP network to estimate the line voltage stability index for different load conditions. Venkatesan and Jolad [4] proposed the application of an MLP based model for fast voltage contingency ranking. The load flow equations are adopted, in this work, to determine the minimum singular values and the findings of the load flow analysis are used to train the MLP network. Authors in [5] proposed a novel MLP-based algorithm that involves a reduced number of inputs to estimate the voltage magnitude of weakest buses in the system. In [6], an effective technique based on Gram–Schmidt orthogonalization is proposed to find the optimal number of MLP inputs required to assure a good assessment of voltage stability. Adaptively trained MLP network is used in [7] as a mapping tool to approximate the available loading margin of the system. In this work, Z-score technique is applied to find and process any bad variable in the training dataset for the MLP network. Generally, ANN is considered as a powerful method for performing nonlinear regression. However, ANNs suffer from some drawbacks such as the amount of training time, the functional relationship which gets changed from one topology to another and the requirement of the appropriate values of weights and bias parameters [8, 9].

In the last years, many studies have been reported in the literature, exploiting the ability of support vector machine (SVM) technique for voltage stability monitoring.

Reference [10] discuss the evaluation of voltage stability using the regression version of SVM or the so-called support vector regression (SVR). Suganyadevi and Babulal [11] proposed the use of ν and ε types of SVR model with various kernel functions to estimate the VSM. In [12], least squares SVM (LS–SVM) with a reduced set of inputs is adopted to estimate the power system load-ability margin. Sajan et al. [9] developed a hybrid model integrating SVR with genetic algorithm (GA) for voltage stability evaluation. In the same way, we proposed in our previous works [13, 14] two hybrid models combining SVR with ant lion optimization (ALO) and dragonfly optimization (DFO) algorithms for voltage stability assessment. In these works, the ALO and DFO algorithms are adopted to find the appropriate SVR parameters. It was stated that the developed GA–SVR, ALO–SVR and DFO–SVR models have better performance compared to the MLP network. Although that SVM is a powerful and promising classification and regression tool, it suffers from overrun time and necessitates more memory for a big training dataset. On the other hand, the efficiency of the SVM model is highly depending upon the selected internal parameters [15].

Adaptive neuro-fuzzy inference system (ANFIS) is another powerful and flexible method proposed by some researchers for voltage stability monitoring. Modi et al. [16, 17] proposed the application of the ANFIS model to monitor the voltage stability of power systems incorporating FACTS devices. In [18], a fuzzy inference model is established and optimized by ANN and GA algorithm to assess the power system security margins. Authors of [19] adopted the subtractive clustering (SC) technique and ANFIS model to evaluate the VSM of the power system. Amroune et al. [14] introduce a method of utilizing ANFIS model-based synchrophasor measurements for on-line prediction of VSM. Even though with the good performance of the ANFIS model, its application in voltage stability analysis is still limited. The high computational costs and the complex set of its parameters are the major drawbacks of this method [11]. Therefore, the application of efficient methods to adjust ANFIS parameters will be of great importance since the unsuitable selection of these parameters can lead to inaccurate classification/regression.

The main purpose of this chapter is to present two real-time voltage stability monitoring approaches for secure and reliable power system operation. In the first approach, an improved multi-layer perceptron (MLP) neural network based on PMUs measurements is proposed to estimate the VSM of the power system in a real-time manner. In the proposed model, the moth swarm algorithm (MSA) [20] is integrated with the MLP network to optimize the connection weights and biases of the network to improve its performance. In the second approach, a novel hybrid model combining the adaptive neuro-fuzzy inference system (ANFIS) and MSA is proposed to monitor the voltage stability of power system. In the proposed hybrid model, the MSA algorithm is adopted to obtain proper parameter settings for the ANFIS-based subtractive clustering (SC) technique.

The rest of the chapter is organized as follows: Sect. 2 explains the standard structure of the MLP neural network, ANFIS and MSA algorithm. Section 3 describes the proposed hybrid MLP–MSA and ANFIS–MSA models. Section 4 describes the implementation of the proposed hybrid models for voltage stability monitoring. In

Sect. 5, the proposed hybrid models are validated and compared in IEEE 30-bus and IEEE 118-bus standard test systems. Finely a conclusion is drawn in Sect. 6.

2 Methods

This section presents the basic information of artificial neural network (ANN), adaptive neuro-fuzzy inference system (ANFIS) and moth swarm algorithm (MSA).

2.1 *Artificial Neural Network (ANN)*

2.1.1 Model of Neuron

Artificial neural network (ANN) is a type of machine learning techniques that simulates the mechanism of information management of the human brain system. A diagram of a neuron model with a single *n*-element input vector, which forms the modern basis for ANNs, is illustrated in Fig. 1. The main elements of the neural model are listed below:

- An input vector connected to a summation node via connecting links. Each of these links has an associated weight (w_i);
- A summation node in which the weighted input $w_i x_i$ is added to the scalar bias b to form the network input;
- An activation function (threshold) for limiting the amplitude of the output of the neuron.

The main objective of an activation function is to confirm that the neuron's response is bounded or limited. The activation functions are generally divided into

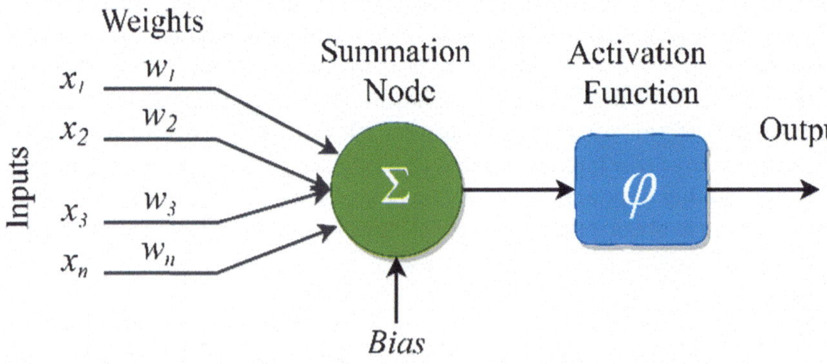

Fig. 1 Model of neuron

two main types linear and non-linear activation functions. Nonlinear functions such as logarithmic sigmoid, hyperbolic tangent sigmoid functions and pure linear function are the frequently utilized functions. The output (y) of the neuron governed by the activation function (φ) can be expressed as follows:

$$y = \varphi\left(\sum_{i=1}^{n} w_i x_i + b\right) \tag{1}$$

2.1.2 Multilayer Perceptron (MLP) Neural Network

One of the most used neural networks in engineering applications is the multilayer perceptron (MLP) neural network [21]. The structure of MLP includes one input layer, one output layer, and one or more hidden layers. The connections between the neurons are performed through some pre-specified weights. Figure 2 shows the typical planning of neurons in an MLP neural network. In this Figure, every node represents an artificial neuron. The neurons are organized in layers, there are one input layer, one output layer and multiple hidden layers. The relationship between the layers can be given by the following equations [22]:

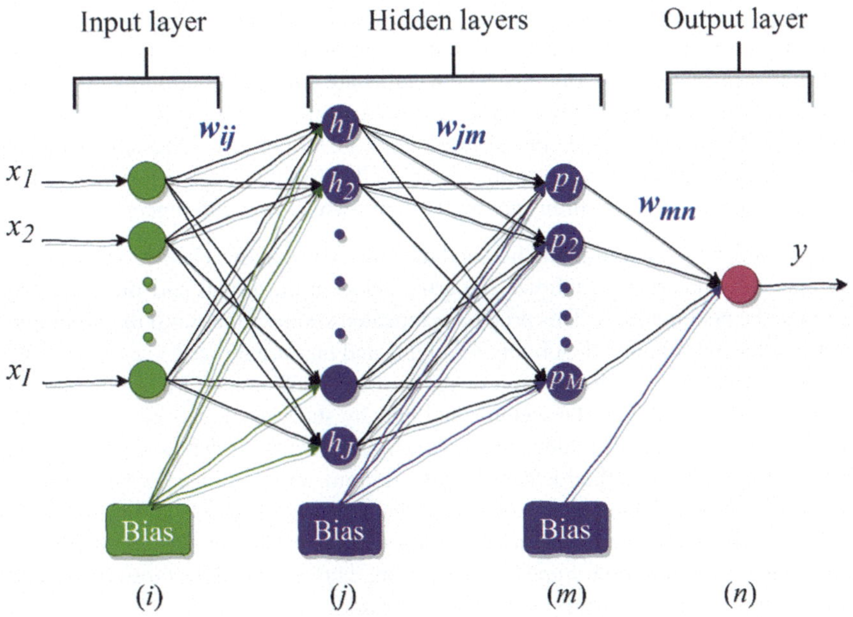

Fig. 2 Multilayer perceptron neural network

$$h_j = \varphi_{h1}\left(\sum_{i=1}^{I} w_{ij}x_i + b_{1j}\right),\ j = 1, \ldots K \tag{2}$$

$$p_m = \varphi_{h2}\left(\sum_{j=1}^{K} w_{jm}h_j + b_{2m}\right),\ m = 1, \ldots, M \tag{3}$$

$$y_n = \varphi_{out}\left(\sum_{m=1}^{M} w_{mn}p_m + b_{3n}\right),\ n = 1, \ldots, N \tag{4}$$

where w_{ij}, w_{jm} and w_{mn} are the associated weights, b_{1j}, b_{2m} and b_{3n} are the biases, $\varphi(x)$ denotes the activation function.

2.1.3 Training of MLP

Artificial neural networks are trained based on the relevant data by learning algorithms. During the training process, the weight and bias parameters are optimized. Then, these parameters are employed to process test dataset to obtain the final output. The MLP network learning can be divided into two main groups: supervised and unsupervised learning.

- Supervised learning—In this group, the system is presented with a set of inputs and the correct outputs, an external trainer controls the learning to learn a general rule that maps inputs to outputs. The weights are adjusted to minimize the error between the network outputs and the desired outputs;
- Unsupervised learning—In this group, there is no trainer involved and no labelled responses are given to the learning algorithm. Here the network is just exposed to a set of inputs and algorithms are left to their own to draw inferences.

In the supervised learning method and the one we use in this chapter, the weights and bias are adjusted to minimize the error between the actual and the predicted values in the next iteration. This process is repeated several times until the minimum error is achieved. Finally, the obtained weights and bias are utilized to carry out any tasks of the ANN i.e. classification or regression. There are several techniques to find the optimal values of the weights and bias by supervised learning. One of the vastly applied learning algorithms to train the MLP networks is the back-propagation (BP) algorithm. This method is based on the minimization of the error between the predicted and actual outputs by adjusting the weights. Notwithstanding its wide utilization, BP algorithm has some drawbacks such as the slow error convergence rate and the local minimum trap [23]. Therefore, there is a need for more robust and efficient optimization algorithms for MLP network training.

2.2 *Adaptive Neuro-Fuzzy Inference System (ANFIS)*

2.2.1 Overview of ANFIS

ANFIS was introduced by Jang in 1993 [24]. It is a machine learning technique incorporates the advantages of ANNs and fuzzy system. The fuzzy part generates a relationship between inputs and outputs, and the parameters associated with the membership part are specified by the neural network. Thence, the main features of both fuzzy and ANN methods are combined in this system.

The sample design of the ANFIS model with two inputs and two rules is shown in Fig. 3. It consists of five main layers; each layer contains several nodes designated by the node function. The functionality of these five layers is given as follows [24]:

Layer 1 (Fuzzification): In this layer, the inputs x and y are subjected to a membership function (e.g., triangle, trapezoidal, Gaussian). The generated output $O_{1,i}$, using generalized Gaussian membership function, can be expressed as follows:

$$O_{1,i} = \mu_{A_i}(x), i = 1, 2, O_{1,i} = \mu_{B_{i-2}}(y), i = 3, 4 \quad (5)$$

where μ_{Ai} and μ_{Bi} are Gaussian membership function given by:

$$\mu(x; c, \sigma) = e^{-\frac{1}{2}\left(\frac{x-c}{\sigma}\right)} \quad (6)$$

where A_i and B_i are the membership values of the μ; c and σ are the centre and width, respectively.

Layer 2 (Product): The output of each node in this layer is the product of all the received signals that are coming to this layer. This product can be computed using the following equation:

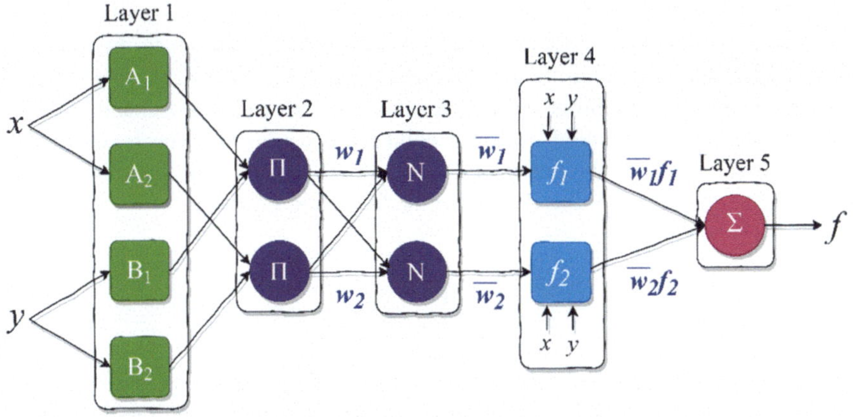

Fig. 3 The structure of ANFIS model for two inputs and two rules

$$O_{2,i} = \mu_{A_i}(x) \times \mu_{B_{i-2}}(y),\ i = 1, 2 \quad (7)$$

Layer 3 (Normalization): In this layer, the output of the layer 2 is normalized using the following equation:

$$O_{3,i} = \bar{\omega}_i = \frac{\omega_i}{\sum_{i=1}^{2} \omega_i},\ i = 1, 2 \quad (8)$$

Layer 4 (Defuzzification): The output of layer 3 is passed through the adaptive nodes of layer 4 as follows:

$$O_{4,i} = \bar{\omega}_i f_i = \bar{\omega}_i (p_i x + q_i y + r_i),\ i = 1, 2 \quad (9)$$

where p, q and r are the consequent parameters of the ith node. These parameters are determined throughout the training phase.

Layer 5 (Overall output): Consists of a single node, which produce the overall output of the model.

$$O_5 = \sum_{i=1}^{2} \bar{\omega}_i f_i = \frac{\sum_{i=1}^{2} \bar{\omega}_i f_i}{\omega_1 + \omega_2} \quad (10)$$

2.2.2 Subtractive Clustering (SC)

The most crucial step in the developing of the ANFIS model is the generation of fuzzy inference system (FIS) with an optimum number and form of fuzzy rules to reduce the computational complexities. Thus, several methods such as grid partitioning, fuzzy c-means and subtractive clustering have been proposed to automate this process. Compared to the other algorithms, subtractive clustering (SC) [25] gives a better distribution of cluster centres and reduces the amount of data associated with the given problem. In this method, each data point is taken as a cluster centre candidate, afterwards, it computes the potential *Pi* of each data point *xi* by determining the density of neighbouring points data using the following Equation.

$$P_i = \sum_{j=1}^{m} \exp\left(-\frac{\|x_i - x_j\|^2}{(r_a/2)^2}\right) \quad (11)$$

where m is the total number of data points in the N-Dimensional space. x_i and x_j are the data points, r_a is a positive constant defining a neighbourhood radius, and || || represents the Euclidean distance. The data point with the highest potential value is chosen as the first cluster centre x_{c1} and its density is P_{c1}. For the next cluster centre, the influence of the first cluster centre is subtracted to define the novel density values, as given by the Eq. 12:

$$P_i = P_i - P_{c1} \exp\left(-\frac{\|x_i - x_{c1}\|^2}{\left(r_b/2\right)^2}\right) \tag{12}$$

$$r_b = \eta \times r_a \tag{13}$$

where η is a positive number greater than 1.

According to Eq. 12 all the points close to the measured cluster centre x_{c1} will have low potential values and therefore they will not be taken as the next cluster centres. The next cluster centre x_{c2} is chosen after the recalculation of the potential of each data point. This process is repeated until sufficient cluster centres are produced.

2.3 *Moth Swarm Algorithm (MSA)*

Moth swarm algorithm (MSA) is a novel meta-heuristic optimization method proposed in 2017 by Ali Mohamed et al., [26] as a developed version of moth flame optimizer [20]. This algorithm is inspired by the navigational behaviour of moths in nature. The position of the light is expressed as the optimal solution, and the brilliance of this light is considered as the objective function. MSA algorithm comprises three collections of moths, which are defined as follows:

Pathfinders: A small group of moths that has the aptitude to find out the new areas over the optimization space and to discover the best position as the light source and to lead other individuals in the population to this position.

Prospectors: This second group is taking charge of wandering into arbitrary spiral paths set by the first group.

Onlookers: This group of the moths drift directly toward the best global solution which has been determined by prospectors.

Through the iterations, each moth is integrated into the optimization problem to search for the luminescence intensity of its corresponding light source. Pathfinders' positions are taken as the best fitness values, while the second and third best fitness take the names of prospectors and onlookers, respectively. The MSA is represented in four main phases [20]:

2.3.1 Initialization

The initial position of moths is selected randomly as:

$$X_{ij} = rand[1, 0] \times \left(X_j^{\max} - X_j^{\min}\right) + X_j^{\min} \forall i \in \{1, 2, \ldots, n\},\ j \in \{1, 2, \ldots, d\} \tag{14}$$

where n is the number of populations and d is the dimension of the problem.

After initialization, the type of moth in the swarm is chosen based on the calculation of objective function. The best value of the objective function is selected to be pathfinders, and others are selected to be prospectors and onlookers.

2.3.2 Reconnaissance

In this phase, the pathfinders are updating their positions through the following five steps. In the first step, a proposed diversity index is employed to select the crossover points. The normalized dispersal degree at t iteration can be expressed as follows:

$$\sigma_j^t = \sqrt{\frac{\frac{1}{N_p}\sum_{i=1}^{N_p}\left(X_{ij}^t - \overline{X_j^t}\right)^2}{\overline{X_j^t}}} \tag{15}$$

The variation coefficient can be computed as follows:

$$\overline{X_j^t} = \frac{1}{N_p}\sum_{i=1}^{N_p} X_{ij}^{t2} \tag{16}$$

where N_p is the number of pathfinders:

$$\mu^t = \frac{1}{d}\sum_{j=1}^{d}\sigma_j^t \tag{17}$$

where μ^t is the variation degree of the relative dispersion.

In the second step, the random processes based on α-stable distribution are explained as Lévy flights [27].

The third step is called difference vectors Lévy mutation in which the sub-trial vectors are generated based on host vectors and donor vectors.

$$v_{pj}^t = \begin{cases} v_{pj}^t \text{ if } j \in c_p \\ x_{pj}^t \text{ if } j \notin c_p \end{cases} \tag{18}$$

In the fourth step, the position of each pathfinder is updated based on an adaptive crossover.

In the final step, a selection strategy is applied to define the best solutions to survive the next generation as follows:

$$\overrightarrow{x_p^{t+1}} = \begin{cases} \overrightarrow{x_p^t} \text{ if } f\left(\overrightarrow{v_p^t}\right) \geq f\left(\overrightarrow{x_p^t}\right) \\ \overrightarrow{v_p^t} \text{ if } f\left(\overrightarrow{v_p^t}\right) < f\left(\overrightarrow{x_p^t}\right) \end{cases} \tag{19}$$

The probability value P_p is estimated as follows:

$$P_p = \frac{fit_p}{\sum_{p=1}^{np} fit_p} \tag{20}$$

The luminescence intensity is computed from the fitness function of the problem f_p as follows:

$$fit_p = \begin{cases} \frac{1}{1 + f_p} \text{ for } f_p \geq 0 \\ 1 + |f_p| \text{ for } f_p < 0 \end{cases} \tag{21}$$

2.3.3 Transverse Orientation

In this phase, the moths with the minimal luminosity of light are taken as prospectors, and their numbers n_f reduced throughout iterations as follows:

$$n_f = round\left((n - n_p) \times \left(1 - \frac{t}{T}\right)\right) \tag{22}$$

where n_p is the number of pathfinders, t is the current iteration, T is the number of iterations.

The position of each prospector is updated according to the spiral flight path as follows:

$$x_i^{t+1} = \left|x_i^t - x_p^t\right| \cdot e^{\theta} \cdot \cos 2\pi\theta + x_p^t \tag{23}$$

where $\theta \in [r, 1]$ is a random number to define the spiral shape and $r = -1 - t/T$.

2.3.4 Celestial Navigation

In this phase, the number of prospectors is decreased and the number of onlookers is increased. Moth with low fitness value is considered the onlooker and it can be computed using Eq. (24).

$$N_o = n - N_s - N_p \tag{24}$$

The onlooker contains the two following groups:
(1) The first group fly according to Gaussian distribution with $N_G = N_o/2$

$$f(q) = \frac{1}{\sqrt{2\pi uG}} \exp\left(\frac{(q_u)^2}{2\sigma_G^2}\right) - \infty < q < \infty$$
$$\left(q \sim N\left(\mu, \sigma_G^2\right)\right) \tag{25}$$

$$x_i^{t+1} = x_i^t + \varepsilon_1 + \left[\varepsilon_2 \times best - \varepsilon_3 \times x_i^t\right]$$
$$\forall i \in \{1, 2, \ldots, N_G\} \tag{26}$$

$$\varepsilon_1 \sim random(size(d)) \oplus \left(best_g^t, \frac{\log t}{t} \times \left(x_i^t - best_g^t\right)\right) \tag{27}$$

where ε_1 is the random sample from Gaussian distribution, *best* is the global best solution (moonlight) which is obtained by transverse orientation and ε_2 and ε_3 are random numbers that range from [0, 1].
(2) The second group with size $N_A = N_o - N_G$
The updating equation for this group can be given as:

$$x_i^{t+1} = x_i^t + 0.001 \cdot G\left[x_i^t - x_i^{\min}, x_i^{\max} - x_i^t\right] + \left(1 - {}^{g}/{}_{G}\right) \cdot r_1$$
$$\cdot\left(best_p^t - x_i^t\right) + {}^{2g}/{}_{G.r_2} \cdot \left(best_g^t - x_i^t\right) \tag{28}$$

where $i \in \{1, 2, \ldots, N_A\}$.

2 g/G is the social factor, 1–g/G is the cognitive factor, r_1 and r_2 are randomly chosen numbers in the space [0, 1], $best_p$ is the arbitrarily chosen light source from the novel pathfinders group based on the probability value of its corresponding solution. At the end of every iteration, the type of each moth is redefined for the upcoming iteration.

3 Proposed Hybrid Models

3.1 MSA for Training MLP Network

As the MLP training is one of the main challenges in the use of this method, the appropriate values of weights and bias parameters must be defined to improve the efficiency of the MLP network [28]. One of the widely applied learning algorithms to find the optimum values of the weights and bias parameters is the back-propagation (BP) algorithm. However, the BP algorithm has some drawbacks such as the slow error convergence rate and the local minimum trap [28]. Several optimization methods have been proposed in the literature to enhance the performance of the neural networks, such as simulated annealing (SA), tabu search (TS), genetic algorithm (GA) and others. Therefore, the improvement of the performance of the MLP network can be achieved by replacing the conventional algorithms used in the training of MLP by more efficient optimization algorithms. In this section, a detailed description of the training process of the MLP network using MSA algorithm is presented. Two main phases are considered when the MSA algorithm is adopted to train MLP network, the first one is the representation of the search agents in the MSA and the last one is the choice of the fitness function. In MSA algorithm each search agent (moth) is encoded to represent the MLP candidate (weights and bias). Therefore, control vectors include a set of weights and a set of biases. The length of each vector is equal to the total number of weights and biases which depends on the number of input variables and the number of hidden layer neurons. The root mean square error (*RMSE*) is used as a fitness function. This assessment metric computes the difference between actual and predicted values by MLP–MSA model. *RMSE* is given by the following equation:

$$RMSE = \sqrt{\frac{1}{n}\sum_{i=1}^{n}(a_i - p_i)^2} \tag{29}$$

where n designates the total number of data, a and p represent the actual and the predicted outputs, respectively.

The flowchart of the proposed MLP–MSA prediction model is shown in Fig. 4.

3.2 Hybrid ANFIS–MSA Model

Although ANFIS is a powerful mathematical tool for data regression and function estimation. Compared to other algorithms such as k-means clustering and fuzzy c-means, ANFIS-based SC gives a better distribution of the cluster centre and reduces the amount of data associated with the given problem. However, there is no standard rule to select its parameters, which is considered as the main blowbacks of this

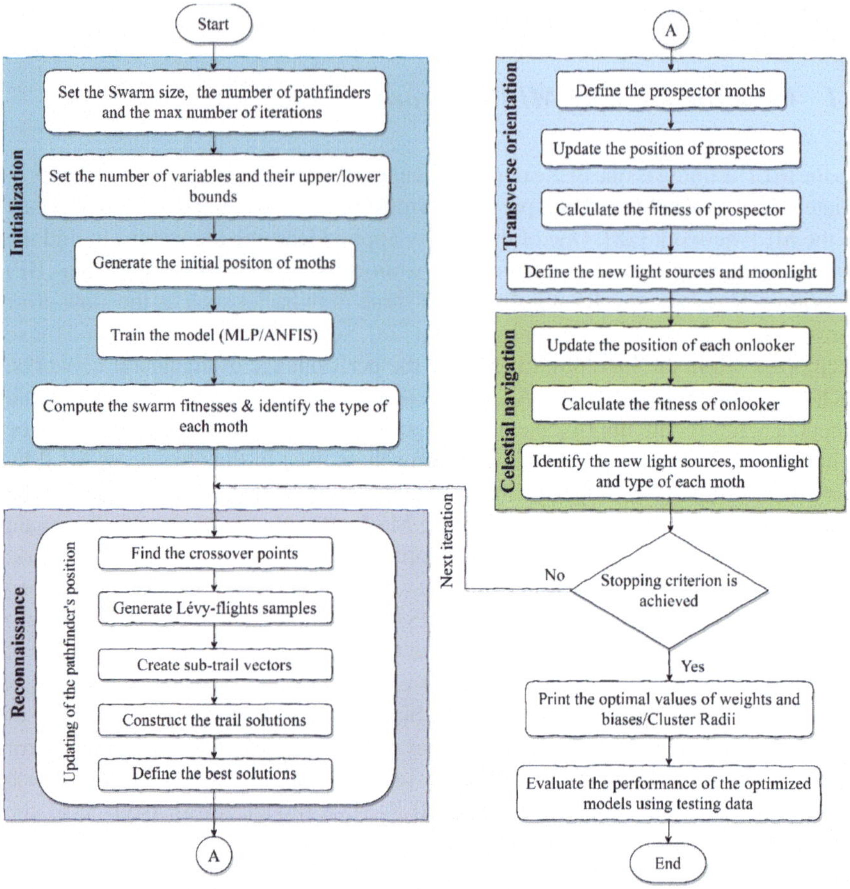

Fig. 4 Flowchart of the proposed hybrid models

method. Cluster radius parameter is one of these parameters that highly influenced on the complexity and generalization abilities of the ANFIS model. A small cluster radius results in small clusters in the data and, hence, many fuzzy rules. Large cluster radius yields few large clusters in the data which means fewer fuzzy rules [29]. Therefore, the application of an efficient method to adjust cluster radii will be of great importance. In this study, the MSA algorithm will be used to tune the cluster radii parameter of the ANFIS model. The proposed hybrid ANFIS–MSA model starts by generating the initial position of moths, includes the initial values of cluster radii, using Eq. (14). The next step involves the training of the ANFIS model, the calculation of the swarm fitness and the identification of the type of each moth. Before these steps, the data is divided into two sets of training and testing. The *RMSE*, expressed by Eq. (29), is used as a fitness function. The detailed flow diagram of the proposed model is given in Fig. 4.

3.3 Accuracy Assessment Criteria

To assess the accuracy of the proposed models, four performance criteria are used. These performance criteria are summarized as follows:

3.3.1 Root Mean Square Error (*RMSE*)

RMSE is the most commonly utilized measure of the differences between the values predicted by a model and the actual values. The model with the small value of *RMSE* is considered the best. The *RMSE* index is given by the Eq. (29).

3.3.2 Correlation Coefficient (*R*)

The correlation coefficient (R), with a value in the range [0, 1], delivers good information about the accuracy of the machine learning models. A value closer to 1 designates a good accuracy of the model. The R index is expressed as follows:

$$R = \frac{\sum_{i=1}^{n}(a_i - \bar{a})(p_i - \bar{p})}{\sqrt{\sum_{i=1}^{n}(a_i - \bar{a})^2 \sum_{i=1}^{n}(p_i - \bar{p})^2}} \tag{31}$$

where $\bar{a}$ and $\bar{p}$ are the rate of the actual and the estimated values, respectively.

3.3.3 Percent Root Mean Square Error (*PRMSE*)

PRMSE measures the accuracy of a machine learning method as a percentage, and it can be given by:

$$PRMSE = \frac{RMSE}{\sqrt{\frac{1}{n}\sum_{i=1}^{n} p_i^2}} \times 100 \tag{32}$$

The model accuracy is excellent for PRMSE < 10%, good for 10% < PRMSE < 20%, reasonable for 20% < PRMSE < 30% and low for PRMSE > 30%.

4 Models Implementation for Voltage Stability Monitoring

4.1 Voltage Stability Indicator

In recent years, several indices have been developed to evaluate voltage stability status, to predict voltage stability margin and to identify the weak buses/area in the system. According to [30], the on-line voltage stability index (VSI) proposed by Yanfeng et al. [31], can be considered as one of the best line voltage stability indices. This index indicates the variation of voltage stability margin in the power system for real, reactive and apparent powers transmitted in the line. VSI is expressed as follows [31]:

$$VSI = \min\left(\frac{P\text{max} - P_r}{P\text{max}}, \frac{Q\text{max} - Q_r}{Q\text{max}}, \frac{S\text{max} - S_r}{S\text{max}}\right) \tag{33}$$

where

$$P\text{max} = \sqrt{\frac{V_s^4}{4X} - Q_r\frac{V_s^2}{X}} \tag{34}$$

$$Q\text{max} = \frac{V_s^2}{4X} - \frac{P_r^2X}{V_s^2} \tag{35}$$

$$S\text{max} = \frac{(1 - \sin(\theta))V_s^2}{2\cos(\theta)^2X} \tag{36}$$

where P_{max}, Q_{max}, and S_{max} are, respectively, the maximum transferred real, reactive and apparent powers, V_s and V_r are, respectively, the sending and the receiving end voltages, X is the line reactance, θ is the line impedance angle. The value of VSI index must be greater than 0 for stable systems and the branch with the lower value is considered to be weak compared to the branch with a higher value.

4.2 Generation of Training and Testing Data

As PMUs have been widely implemented in many power systems, application of wide-area PMU measurements in real-time voltage stability monitoring has been of great interests. Many studies confirmed that PMU data are good indicators for voltage stability monitoring and they can be taken as inputs to the prediction models [1, 13, 14]. Since PMU devices can provide synchronized measurements, which include the magnitude and phase angle of voltages, both of them are chosen as inputs of the developed MLP–MSA and ANFIS–MSA models i.e., inputs = $\{(|V_i|, \delta_i), i \in$ PMU buses$\}$. On the other hand, the minimum VSI values, computed using load flow

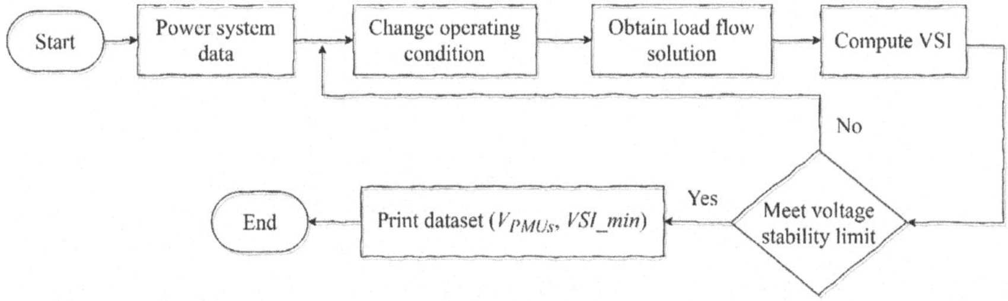

Fig. 5 Flowchart of the generation of training and testing data

equations, at each operating point are used as the output variables. The generation of training data is carried out through off-line simulation processes by varying both the active and reactive power simultaneously on each load bus in the system. The load is increased with a constant load factor from the base case until the system reaches the voltage stability limit. The voltage magnitudes and angles of PMU buses are obtained by solving conventional load flow at each load generating sample. The flowchart of the generation of training and testing data is depicted in Fig. 5. The collected dataset will be then applied to train and to evaluate the proposed prediction models. Once the training process is accomplished and the stopping condition is reached, model testing is required to verify the performance of the models over the actual and predicted data.

4.3 Real-Time Prediction of VSI

The last phase deals with the implementation of the developed MLP–MSA and ANFIS–MSA models to predict VSI in a real-time manner. In this phase, VSI is predicted using real-time measurements provided by PMUs. These provided data may support the tracking of dynamic phenomena and provide the necessary information for power system voltage stability monitoring. The time-synchronized data taken from throughout the distributed PMU units will be sent to the control system in which the well trained MLP and ANFIS models are employed to predict voltage stability margin for each operating point. The precise and synchronized real-time measurements obtained by PMUs with the fast evaluation of voltage stability offered by the proposed models can help system operators to take the required control action, such as load shedding or emergency demand response [32], to prevent voltage collapse.

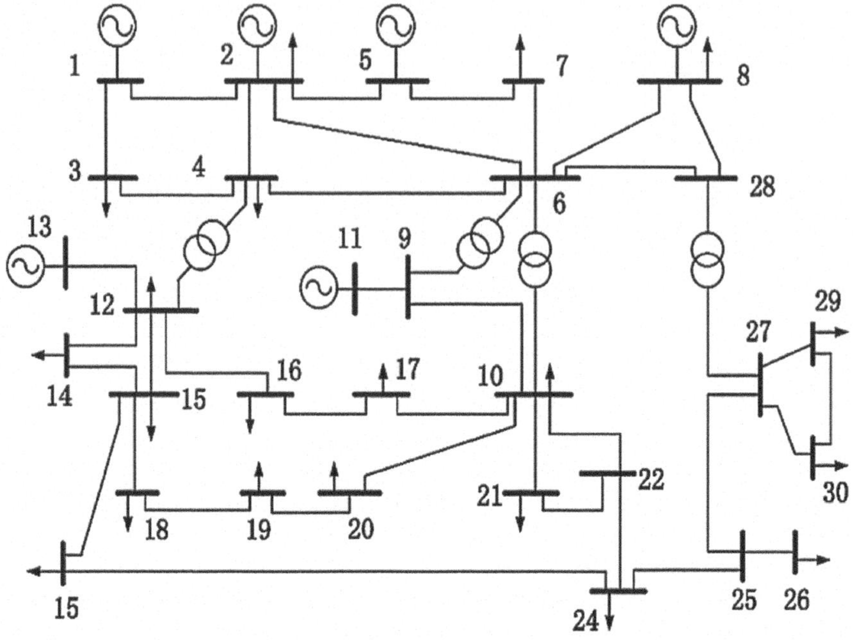

Fig. 6 Single line diagram of the IEEE 30-bus test system

5 Case Study and Simulation Results

5.1 Test Systems

5.1.1 IEEE 30-Bus Test System

The first test system used to validate the performance of the proposed models is the standard IEEE 30-bus test system [33]. This system is shown in Fig. 6, and it consists of 30 buses, 6 thermal units, 41 branches and 21 loads.

5.1.2 IEEE 118-Bus Test System

The performance of the proposed models for voltage stability monitoring has been validated also on the IEEE 118-bus test system shown in Fig. 7 [33]. This system contains 118 buses, 51 thermal units, 196 branches and 91 loads.

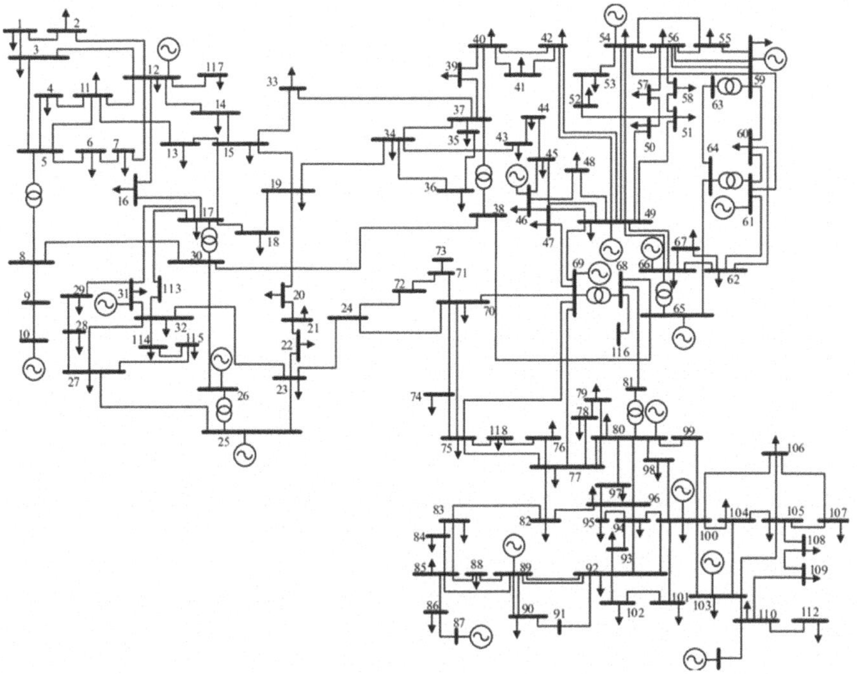

Fig. 7 Single line diagram of the IEEE 118-bus test system

5.2 Data Preparation

One of the crucial factors for successful implementation of any machine learning technique is the generation of proper training data. To generate the training data for the proposed models, all loads of both test systems are uniformly increased, with constant load power factors, from their base case loadings to the voltage collapse. As aforementioned, the gathered voltage magnitudes and angles by the distributed PMUs (voltages where PMUs are installed) will be used as the inputs for the proposed models, while the min values of VSI as the outputs. The optimal number and locations of PMUs for both test systems are obtained using simulated annealing (SA) method in PSAT (Power System Analysis Toolbox) software [34]. The optimal number and placement of PMUs are represented in Table 1. Afterwards, the collected data is divided into 80% for training and 20% for testing the models. The first portion is

Table 1 Number and locations of PMUs

Test system	Number of PMUs	Location of PMUs
IEEE 30-bus system	7	3, 5, 10, 12, 19, 23, 27
IEEE 118-bus system	28	2, 8, 11, 12, 17, 21, 25, 28, 33, 34, 40, 45, 49, 52, 56, 62, 72, 75, 77, 80, 85, 86, 90, 94, 101, 105, 110, 114

employed for initial training and models parameter optimization, and the second portion to further adjust and evolve the predictors to simulate real conditions.

5.3 Performance Comparison

The performance of MLP–MSA and ANFIS–MSA is outlined in this section. The developed programs are written using MATLAB software and the simulations are carried out on a computer with Intel Core i5 CPU @2.7 GHz, 4 GB RAM and Windows7 as the operating system. The load flow is obtained using MATPOWER [33]. In all simulations, the MSA parameters are tuned as follows: the number of search agents (candidate solutions) is fixed to 30 and the number of pathfinders to 18. For the MLP network, the number of hidden neurons is set to 20 (determined by a trial-and-error process). The MSA is used to optimize the weights connecting the input layer with the hidden layer, the weights connecting the hidden layer with the output layer and the biases. For the ANFIS model development, the SC (genfis2) technique based on the Gaussian type of membership function is used to generate fuzzy rules. According to [35] Gaussian membership function can be considered as the best fit to use with ANFIS model. The squash factor, the accept ratio and the reject ratio were set, by default in MATLAB toolbox, to 1.25, 0.5 and 0.15, respectively. The MSA algorithm is adopted to find the best value of cluster radii in the range of [0.2 0.5] [36].

5.3.1 Application to the IEEE 30-Bus Test System

This section demonstrates the effectiveness of the proposed methods on the IEEE 30-bus system. The results of the proposed methods have been compared with those found in the literature using the same data as in [14]. Figure 8a shows the convergence curve of MSA algorithm seeking for the optimal values for MLP's weights and biases. It can be seen that the MSA has a slow convergence rate due to the problem

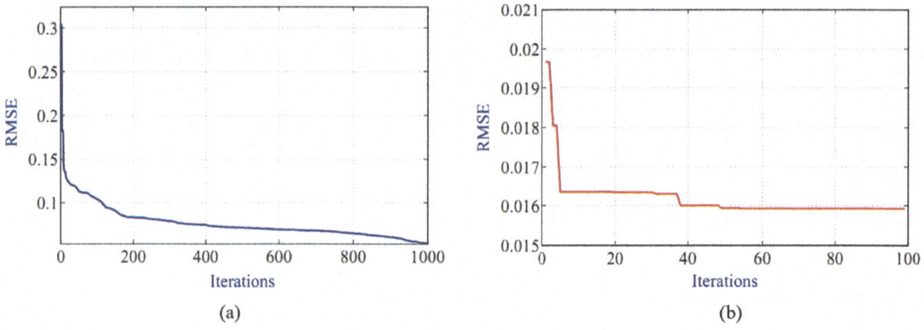

Fig. 8 Convergence curves of MSA for **a** MLP training, **b** ANFIS training

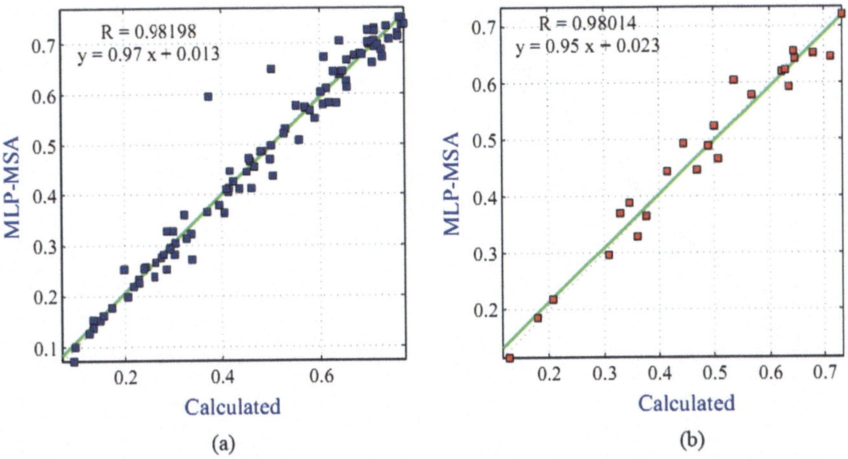

Fig. 9 Predicted values by the MLP–MSA versus actual values **a** Training phase **b** Testing phase

complexity. Figure 9a, b show the plots of the actual and predicted values of VSI via MLP–MSA method in the training and testing phases. It is seen that the MLP–MSA predictions are in agreement with the actual values. The prediction accuracy of the models was measured using *RMSE*, *PRMSE* and *R* indices. In the training phase, the computed values of *RMSE*, *PRMSE* and *R* were 0.0372, 6.1137 and 0.98198, respectively. In the testing phase, these indices are found to be 0.0380, 7.1458 and 0.98014 respectively. It is seen that the MLP–MSA model can estimate VSI with a good accuracy which means that the MSA algorithm has performed efficiently in tuning the MLP's weights and biases.

MSA algorithm is used also to find the optimal cluster radius of ANFIS-based SC technique, then, the performance of the trained ANFIS model was evaluated in VSI prediction. The convergence curve of the MSA algorithm is illustrated in Fig. 8b and the smallest error was obtained with the cluster radius of 0.2518. Figure 10a, b depicts the scatter plot of the predicted VSI values using ANFIS–MSA against actual ones in the training and testing phases. For the training step, the correlation coefficient (*R*) was found to be 0.99685. In the testing phase, the predictions result in a correlation coefficient of 0.99378. The correlation between the actual VSI values and ANFIS–MSA predictions is much better than MLP–MSA predictions. The comparison of different statistical indices for the proposed models and other models in the literature is shown in Table 2. As shown in this table, ANFIS–MSA outperforms all other prediction models. For this model, the values of *RMSE* and *PRMSE*, in the training phase, are calculated to be 0.0160 and 2.7873, respectively. In the testing phase, these values are found to be 0.0206 and 3.9442, respectively. It is seen that all used criteria confirmed that the proposed MLP–MSA and ANFIS–MSA methods give a good prediction accuracy compared to ANFIS and DFO–SVR [14]. On the other hand, the ANFIS–MSA performance culminates in giving the best prediction results than the MLP–MSA model.

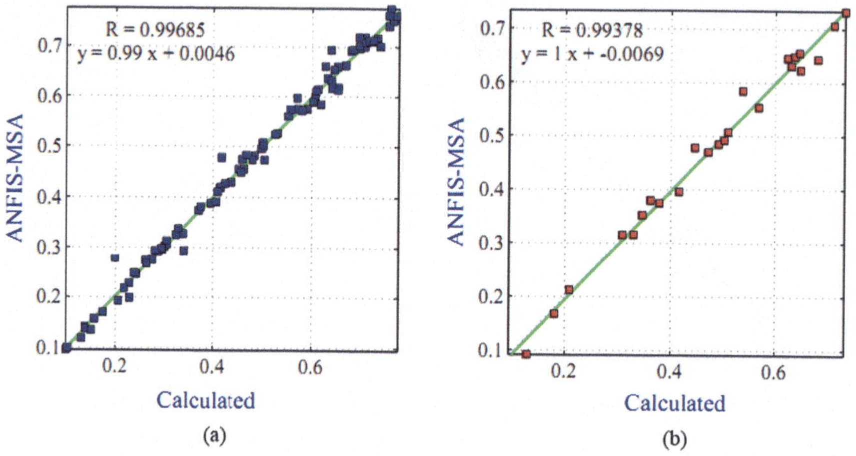

Fig. 10 Predicted values by the ANFIS–MSA versus actual values **a** Training phase **b** Testing phase

Table 2 Statistical performance of MLP–MSA and ANFIS–MSA

Method	Training			Testing		
	RMSE	*PRESE*	*R*	*RMSE*	*PRMSE*	*R*
MLP–MSA	0.0372	6.1137	0.98198	0.0380	7.1458	0.98014
ANFIS–MSA	**0.0160**	**2.7873**	**0.99685**	**0.0206**	**3.9442**	**0.99378**
ANFIS [14]	0.0228	–	0.99325	0.0385	–	0.97492
DFO–SVR [14]	0.0166	–	0.99641	0.0273	–	0.98776

5.3.2 Application to the IEEE 118-Bus Test System

The proposed new models are evaluated in this section on the IEEE 118-bus power system to compare their efficiency. Figure 11a illustrates the convergence curve of MSA algorithm searching for the optimal values for MLP's weights and biases. Correlations between actual and predicted values of VSI for training and testing phases are shown in Fig. 12a, b. As seen from this Figure, the MLP–MSA model gives a correlation coefficient of 0.99814 for the training data while this coefficient for the testing data is equal to 0.94607. The proposed hybrid model managed to produce an *RMSE* and *PRMSE* of 0.0087411 and 1.5056, respectively, in the training phase. In the testing phase, this model yield to the values of *RMSE* and *PRMSE* of 0.1381 and 9.6441, respectively.

The effectiveness of the proposed ANFIS–MSA model in the prediction of VSI was also checked in the case of the IEEE 118-bus system. Figure 11b displays the optimization process of the *RMSE* versus iterations number for MSA to find the optimum value of ANFIS cluster radius and to predict VSI. The optimal found value of cluster radius is 0.4599. The scatter plots for this model are illustrated in Fig. 13a, b.

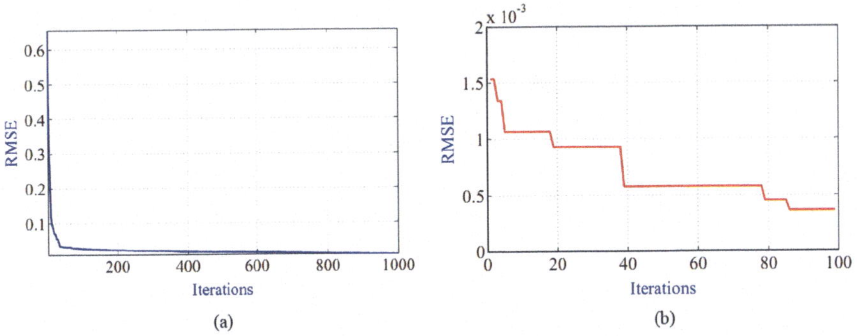

Fig. 11 Convergence curves of MSA for **a** MLP training, **b** ANFIS training

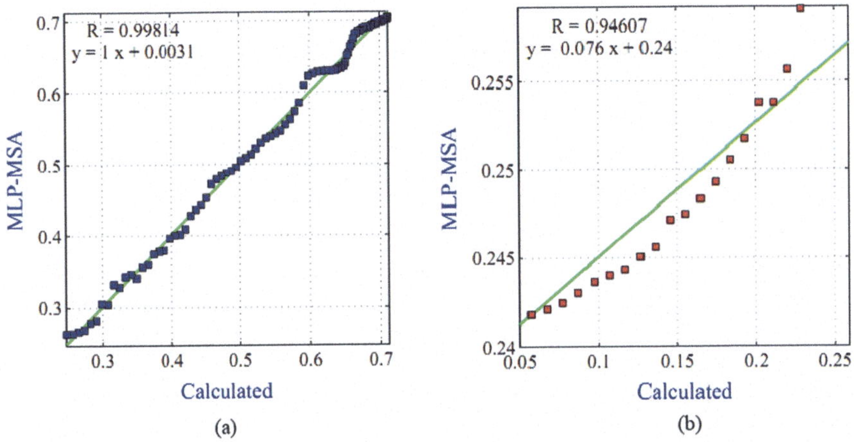

Fig. 12 Predicted values by the MLP–MSA versus actual values **a** Training phase **b** Testing phase

From this Figure, it is apparent that the ANFIS–MSA model gave the most scattered estimates. The results of different statistical indices for both proposed prediction models are listed in Table 3. It is shown that the *RMSE* and *PRMSE* of the ANFIS–MSA prediction method are computed to be 3.6708e^{-4} and 0.0632, in the training phase and 0.0042 and 2.8645 in the testing phase. By using ANFIS–MSA method, we observed that the *RMSE* in the testing phase was reduced by 0.1339. As for the *R*, it increased by 0.05392. From these results, we can note clearly that the prediction of VSI using ANFIS–MSA model gives better results compared to the MLP–MSA model. It is seen from the comparison of the statistical indicators values, in both case studies of IEEE 30-bus and IEEE 118-bus test systems, that the proposed models give a good prediction accuracy of VSI. However, ANFIS–MSA model acquired relatively lower values of *RMSE* and *PRMSE*, and higher values of *R*. Accordingly, it can drown the conclusion that the proposed MLP–MSA and ANFIS–MSA models

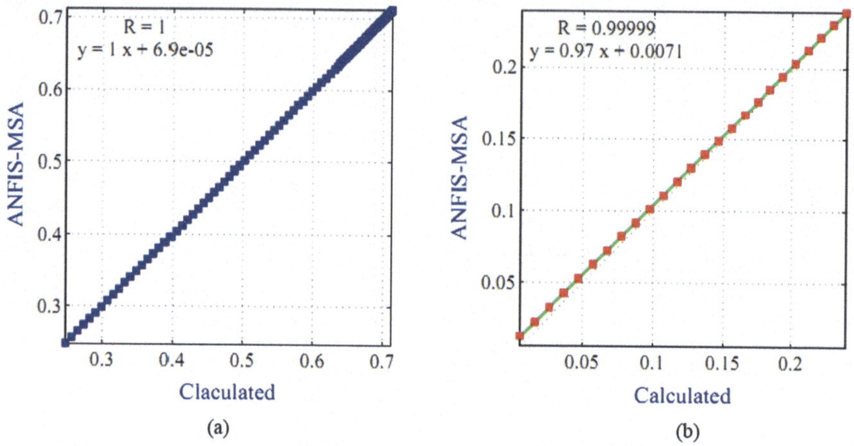

Fig. 13 Predicted values by the ANFIS–MSA versus actual values **a** Training phase **a** Testing phase

Table 3 Statistical results of the developed models

Method	Training			Testing		
	RMSE	*PRMSE*	*R*	*RMSE*	*PRMSE*	*R*
MLP–MSA	0.0087411	1.5056	0.99814	0.1381	9.6441	0.94607
ANFIS–MSA	$3.6708e^{-4}$	0.0632	1	0.0042	2.8645	0.99999

would be an appealing option for voltage stability monitoring since the obtained results are superior to those from the other models for the considered case studies.

6 Conclusions

This chapter proposes novel measurement-based methods for the real-time monitoring of voltage stability using PMU data. The proposed methods are based on the training of multi-layer perceptron (MLP) neural networks and adaptive neuro-fuzzy inference systems (ANFIS) deploying a moth swarm algorithm (MSA). The problem of training the MLP and ANFIS was first formulated as a minimization problem. The objective was to minimize the root mean square error (RMSE), and the parameters were linking weights and biases for the MLP network and the cluster radius for ANFIS-based subtractive clustering. The proposed MLP–MSA and ANFIS–MSA methods require only the voltage phasors' data provided by the PMU units, to predict the voltage stability of the power system. The applicability and the performance of the proposed hybrid models have been investigated using standard IEEE 30-bus and IEEE 118-bus test systems and compared with existing methods in the literature.

The results obtained from the two test systems clearly showed the efficiency and accuracy of the proposed MLP–MSA and ANFIS–MSA methods as compared to other existing methods. On the other hand, the simulation results reveal that the ANFIS–MSA performance culminates in the best prediction results compared to the MLP–MSA model.

References

1. H. Su, T.Y. Liu, Enhanced-Online-Random-Forest Model for Static Voltage Stability Assessment Using Wide Area Measurements. IEEE Transactions in Power Systems **33**(6), 6696–6704 (2018)
2. B. Jeyasurya, Artificial neural networks for power system steady-state voltage instability evaluation. Electr. Power Syst. Res. **29**, 85–90 (1994)
3. G. Joya, F. García-Lagos, F. Sandoval, Contingency evaluation and monitorization using artificial neural networks. Neural Comput. Appl. **19**(1), 139–150 (2007)
4. M. Venkatesan, B. Jolad, Artificial neural network-based contingency ranking. Comput Netw Inf Technol **1**(142), 33–38 (2011)
5. P. Aravindhababu, G. Balamurugan, ANN-based voltage estimation. Appl. Soft. Comput. **12**, 313–319 (2012)
6. A. Bahmanyar, A. Karami, Power system voltage stability monitoring using artificial neural networks with a reduced set of inputs. Electr Power Energy Syst **58**, 246–256 (2014)
7. S. Ashraf, A. Gupta, D. Choudhary, S. Chakrabarti, Voltage stability monitoring of power systems using reduced network and artificial neural network. Electrical Power and Energy Systems **87**, 43–51 (2017)
8. P. Duraipandy, D. Devaraj, Extreme learning machine approach for on-line voltage stability assessment. Swarm Evol Emotic Comput **8298**, 397–405 (2013)
9. K.S. Sajan, V. Kumar, B. Tyagi, Genetic algorithm-based support vector machine for on-line voltage stability monitoring. Electrical Power and Energy Systems **73**, 200–208 (2015)
10. M. Suganyadevi, C. Babulal, Fast assessment of voltage stability margin of a power system. J. Electr Syst **10**(3), 305–316 (2014)
11. M. Suganyadevi, C. Babulal, Support vector regression model for the prediction of load-ability margin of a power system. Appl. Soft Comput. **24**, 304–315 (2014)
12. P. Duraipandy, D. Devaraj, On-Line Voltage Stability Assessment using Least Squares Support Vector Machine with Reduced Input Features, International Conference on Control, Instrumentation, Communication and Computational Technologies, Kanyakumari, India (2014)
13. M. Amroune, I. Musirin, T. Bouktir, M.M. Othman, The amalgamation of SVR and ANFIS models with synchronized phasor measurements for on-line voltage stability assessment. Energies **10**(11), 1693 (2017)
14. M. Amroune, I. Musirin, T. Bouktir, Power system voltage stability assessment using a hybrid approach combining dragonfly Optimization algorithm and support vector regression. Arab J Sci Eng **43**, 3023 (2018)
15. M. Amroune, Machine Learning Techniques Applied to On-Line Voltage Stability Assessment: A Review. Arch Computat Methods Eng (2019). https://doi.org/10.1007/s11831-019-09368-2
16. P. Modi, S. Singh, J. Sharma, Voltage stability evaluation of power system with FACTS devices using fuzzy neural network. Eng. Appl. Artif. Intell. **20**, 481–491 (2007)
17. P. Modi, S. Singh, J. Sharma, Fuzzy neural network-based voltage stability evaluation of power systems with SVC. Appl. Soft Comput. **8**, 657–665 (2008)
18. A. Berizzi, C. Bovo, D. Cirio, M. Delfanti, M. Merlo, M. Pozzi, Online fuzzy voltage collapse risk quantification. Electr. Power Syst. Res. **79**, 740–749 (2009)

19. P. Torres, H. Peralta, Castro a power system loading margin estimation using a neuro-fuzzy approach. IEEE Trans. Power Syst. **22**(4), 1955–1964 (2007)
20. S. Mohamed, M. El-Gaafary, A.M. Hemeida, Optimal power flow using moth swarm algorithm. Electr. Power Syst. Res. **142**, 190–206 (2017)
21. A. Parcheco-Vega, M. Sen, K.T. Yang, R.L. McClain, Neural Network Analysis of in-Tube Refrigerating Heat Exchanger with Limited Experimental Data. Int. J. Heat Mass Transf. **44**(4), 763–770 (2001)
22. D. Leila, K. Babak, K. Farshad, Optimal design approach for heating irregular-shaped objects in three-dimensional radiant furnaces using a hybrid genetic algorithm–artificial neural network method. Engineering Optimization (2017). https://doi.org/10.1080/0305215X.2017.1323889
23. P.P. Sarangi, A. Sahu, M. Panda, Training a Feed-Forward Neural Network Using Artificial Bee Colony with Back-Propagation Algorithm. (2014) In: Mohapatra D., Patnaik S. (eds) Intelligent Computing, Networking, and Informatics. Advances in Intelligent Systems and Computing, vol 243. Springer, New Delhi
24. Jang, Adaptive-network-based fuzzy inference system, IEEE Trans-Syst Man Cybern, 23(3), 665–685 (1993)
25. S. L. Chiu, Fuzzy model identification based on cluster estimation. Journal of Intelligent and Fuzzy Systems, 267–278 (1994)
26. M. Seyedali, Moth-flame optimization algorithm: A novel nature-inspired heuristic Paradigm. Knowl.-Based Syst. **89**, 228–249 (2015)
27. R.N. Mantegna, Fast accurate algorithm for numerical simulation of Levy stable stochastic processes. Phys. Rev. **49**(5), 4677–4683 (1994)
28. S. Mirjalili, How effective is the Grey Wolf optimizer in training multi-layer perceptrons. Appl Intell **43**(1), 150–161 (2015)
29. M. Shahram, Optimal design of adaptive Neuro-fuzzy inference system using genetic algorithm for electricity demand forecasting in Iranian industry. Springer-Verlag, Berlin Heidelberg **20**, 4897–4906 (2016)
30. V. Salehi, M. Ossama, Real-Time Voltage Stability Monitoring and Evaluation Using Synchrophasor. IEEE North American Power Symposium (2011)
31. G. Yanfeng, N. Schulz, A. Guzman, Synchrophasor-Based Real-Time Voltage Stability Index. Power Systems Conf, PSCE. IEEE PES, 1029–1036 (2006)
32. M. Amroune, T. Bouktir, I. Musirin, Power system voltage instability risk mitigation via emergency demand response-based whale optimization algorithm. Protection and Control of Modern Power Systems, 4–25 (2019)
33. R. D. Zimmerman, C. E. Murillo-S'anchez, and R. J. Thomas, Matpower: SteadyState Operations. Planning and Analysis Tools for Power Systems Research and Education, Power Systems, IEEE Transactions on, 26(1), 12–19 (2011)
34. F. Milano, Power system analysis toolbox (PSAT), (Available at:) http://faraday1.ucd.ie/psat.html
35. M.N.M. Salleh, N. Talpur, K.H. Talpur, A Modified Neuro-Fuzzy System Using Metaheuristic Approaches for Data Classification, Artificial Intelligence - Emerging Trends and Applications, Marco Antonio Aceves-Fernandez. IntechOpen (2018). https://doi.org/10.5772/intechopen.75575
36. K. Gopalakrishnan, H. Ceylan, O. Attoh-Okine, *Intelligent and Soft Computing in Infrastructure Systems Engineering* (Springer, Berlin, 2009)

Wide-area Transmission System Fault Analysis Based on Three-Phase State Estimation with Considering Measurement Errors

Alireza Ghaedi and Mohammad Esmail Hamedani Golshan

Abstract In this chapter, a wide-area integrated method including a set of algorithms for transmission lines fault analysis is introduced. The proposed method is based on extension and modification of state estimation formulation. Thus, the method is applicable to both symmetrical and asymmetrical networks as well as all fault types including symmetrical and asymmetrical ones. The method exploits the capacities of state estimation formulation and the solution algorithm of weighted least squares (WLS) to reduce the effect of inherent errors on the fault location accuracy and detection and elimination of bad data in the measurement vectors. For this purpose, an error model of the measurement chain including instrument transformers and PMUs is proposed. This model is used to design measuring errors covariance matrix in the state estimation formulation. The performance of the proposed method has been investigated through numerous fault events simulated on different locations of all transmission lines of the IEEE 118-bus test system.

Keywords Asymmetrical faults · Bad data detection · Fault location · Measurement chain error · State estimation · Untransposed transmission lines

1 Introduction

Expansion of transmission lines in vast geographical areas has caused these lines to be more exposed to natural phenomena such as lightning, windstorm, severe winds, floods, and human factors. These phenomena can lead to permanent faults

---

Springer-Verlag Tiergartenstr. 17, 69121 Heidelberg, Germany http://www.springer.com/lncs.

A. Ghaedi (✉) · M. E. Hamedani Golshan
Department of Electrical and Computer Engineering, Isfahan University of Technology, Isfahan 84156-83111, Iran
e-mail: a.ghaedi@ec.iut.ir

M. E. Hamedani Golshan
e-mail: hgolshan@cc.iut.ac.ir

H. Haes Alhelou et al. (eds.), *Wide Area Power Systems Stability, Protection, and Security*, Power Systems, https://doi.org/10.1007/978-3-030-54275-7_17

in transmission systems and long duration electricity interruption. That is why fault location studies have received much attention from the first days of transmission systems development. The higher accuracy of fault location speed up transmission line repair process. Consequently, network is restored to the normal state and power supply is reconnected for subscribers in the shortest possible time leading to higher system reliability.

A faulty line can usually be identified using protection systems or supervisory systems such as SCADA.[1] After that, finding the exact location of the fault is the main challenge lying ahead. The motivation behind this chapter is to present an integrated method which is able to estimate fault location, faulty line discrimination and fault type identification by using a number of wide-are phasor measurements. Another motivation is using a known and systematic framework for implementation of the integrated method and decreasing the effect of the inherent errors related to the measuring components on the method's results.

1.1 Literature Review and Background

So that the preliminary research dates back to the 1930s. However, the importance of these methods became much more evident during World War II, and countries such as the United States, Canada, Belgium, France, and Japan obtained great achievements in this field [1]. An intensive plan was organized to develop fault location methods in Japan so that a team of professors from the University of Tokyo and representatives of nine power generating companies started collaborating as a "committee for fault location on transmission lines" [1]. Alongside these coherent activities in Japan, industrial companies each conducted separate research in the United States; however, according to a pre-arranged agreement between companies, there was a completely free exchange of information obtained from research. A comprehensive report by the AIEE[2] was published for the first time in order to review the existing methods and present future horizons of fault location studies [1]. Primary fault location methods can be classified into two main categories of visual inspection and fault location using fixed measurements [1]. Figure 1 summarizes this classification.

Primary methods of automatic fault location were unable to locate the fault. Some methods were only applicable to lines without power, so they could only detect permanent faults. However, to detect transient faults, the algorithms need either to be applied automatically to power lines or use the recorded data before the breakers can be opened. As such, fault location methods can be divided into two main groups:

- Traditional fault location methods
- Automatic fault location methods

[1]Supervisory Control and Data Acquisition.

[2]American Institute of Electrical Engineers.

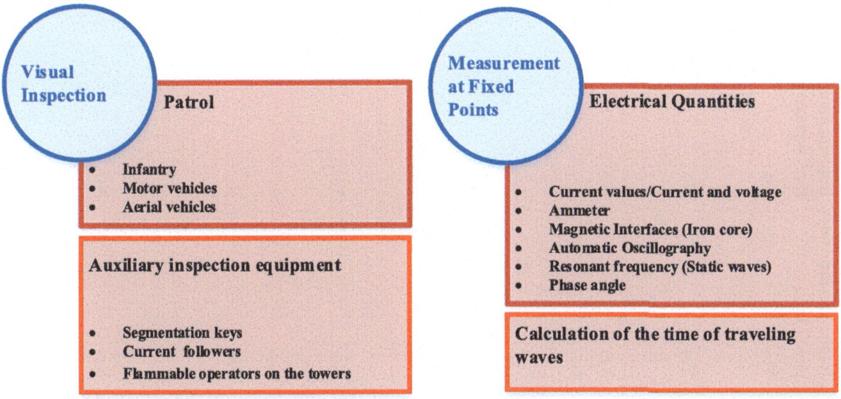

Fig. 1 Primary fault location methods

1.1.1 Traditional Fault Location Methods

As mentioned earlier, traditional methods include a set of methods that cannot automatically detect the location of the fault in the aftermath of the fault, but patrols and maintenance teams should locate the fault by the use of equipment. Table 1 summarizes the traditional fault location methods used for overhead lines. Although many methods have been introduced so far, none of them were reliable methods for locating the fault. With the passage of time and the installation of equipment called "fault indicator" on the posts or transmission line towers, valuable information was provided to the users about the location of the fault. Then, the addition of radio features to fault indicators has made it possible to transmit fault-related information in impassable locations and bad weather conditions.

1.1.2 Automatic Fault Location Methods

Despite all the efforts made in different and unusual fault location methods, automatic fault location methods are still the most effective. These methods determine the physical location of the fault by processing the voltage and current waveforms. Most of the methods known today are among the automatic fault location methods. Automatic fault location methods can be categorized in terms of different aspects such as selection of appropriate frequency of under-study quantities, location of electrical quantities extraction, and choice of time/frequency domains. Meanwhile, in general, they can be divided into three main groups [2, 3], including traveling waves based methods, artificial intelligence-based methods, and main-frequency component-based methods. Each of the above-mentioned groups can be divided into two sub-sets of single-ended and two-ended in terms of used electrical quantities in the algorithm.

Table 1 Summary of traditional fault location methods used for overhead lines [1]

Row	Fault location method	Maximum algorithm error (%)	Implementation on power lines	Implementation on lines without power
1	Ammeter The ammeters are faulty at the terminals leading to the line, so they roughly calculate the faulty area	20%	Yes	No
2	Automatic oscillography This is similar to the above method, with the exception that both voltage and current signals are used	10%	Yes	No
3	Waves generated by the fault The wave generated by a sudden voltage drop due to an occurred fault moves to the line terminals and is reflected continuously between the line post and the fault impedance. At the time of displacement, the wave is electrically measured and converted to circuit miles	1%	Yes	No
4	Magnetic interfaces (Iron core) It is similar to the automatic oscillographs method, except that the fault voltage and current are calculated by measuring the magnitude of the magnetic residues inside an iron core	20%	Yes	No
5	Pulse radar A pulse is applied to the line and the length of time the pulse is sent to and returns from the fault location is measured electrically. Then, the measured time is converted to circuit miles	1%	Yes/No	Yes/Yes
6	Resonant frequency A variable frequency source is connected to the line. At regular frequency intervals, the line itself exhibits a series resonance characteristic. The distance between these intervals is directly correlated with the distance of the source from the fault location	2%	No	Yes
7	Flammable operators on the towers The lightning current on the tower's legs creates visible light for patrol groups. This method uses currents caused by lightning or electrical faults to ignite some of the gunpowder embedded in the towers. Sparks created by the ignition of gunpowder remain in the air for some time, causing the tower connected to the fault location to be seen	0%	Yes	No
8	Followers A tracking signal is applied to the transmission line, which can be easily detected by portable devices. At the time of patrolling the line, a decrease in signal strength indicates a proximity to the fault point	0%	Yes	No

Decision making in traveling waves approaches is based on the analysis of the position-time graphs and the motion of the voltage and current waves [4, 5]. In this group of methods, several factors such as hard detection of initial traveling wave front and reflection wave front are considered as major challenges [4]. Numerous methods have been introduced to solve these challenges with defining new criteria [4, 6, 7]. In recent years, with the development of artificial intelligence-based methods in power system studies, these methods have also been used in fault location algorithms. These methods estimate fault location using tools such as support-vector machine (SVM) [8, 9] and extracting different characteristics of network signals [10]. The computational difficulties as well as the costs incurred to the power system in traveling wave-based methods and artificial intelligence-based ones have made impedance-based methods still very popular among researchers. Impedance method is considered as the most famous main-frequency component-based method. Impedance methods calculate the fault location directly through voltage and current phasors [11]. They calculate the fault location through the obtained data and without requiring any special hardware/software and as a result are highly economical [12]. However, they have disadvantages due to their dependence on faulty transmission line parameters and fault impedance [13]. Due to this dependence on various parameters, many studies have been conducted on the vulnerability of this group of algorithms [14]. Various types of impedance methods are investigated in [15] and different strategies are presented to improve their performance and reduce their errors. In contrast to two-ended methods, impedance-based methods that just use faulty single-ended data, do not require a communication link between the sending and receiving ends. On the other hand, less dependence on parameters in the two-ended methods results in an increase in the accuracy [15–18]. In addition to the introduced methods, a number of impedance-based methods have focused on determining fault location in the presence of high fault impedances. The main focus of these methods is on the phase to ground faults. As an instance, the method used in [19] uses the electrical quantities of both terminals and the faulty line parameters to estimate the fault location. Some methods such as [20, 21] estimate fault location by determining the Thevenin equivalent circuit of the network [20, 21]. Although these methods are the one-ended methods, they are able to greatly reduce the parameter dependency by taking advantage of equivalent circuit and statistical theories. Recently, impedance methods have been extended for multi-terminal transmission lines [22, 23] and nonhomogeneous transmission lines [24]. In the 80's, after introducing phasor measurement units (PMUs) [25], access to concurrent synchronous data was provided throughout the network. In addition to measuring the magnitude of voltages and currents, the equipment is also able to extract their phases. Additionally, the accuracy of PMUs measurements is far higher than that of other conventional measurement devices. The emergence of PMUs has led to the creation of another group of fault location algorithms called wide-are fault location methods (WAFL). This group of methods is able to estimate the fault location by measuring some of the network voltages and currents from different points, not necessarily the bus connected to the faulty line [26–30]. The wide area methods gained their popularity mainly due to their capability to decide on the location of the occurred fault based on scattered information from the voltage and current phasors in the network [31, 32]. Some WAFL approaches have been introduced such as hybrid (synchronized and unsynchronized) measurements and non-iterative methods [33–35].

1.2 Contributions

This study was carried out to overcome some common challenges of the fault location methods proposed over the last few decades. The error in estimating the fault location introduced by bad data and inherent errors in the measurement chains are some of these challenges. For this purpose, a novel algorithm based on the well-known formulation of system state estimation is proposed for fault location. The other challenge in fault location algorithms is how to encounter network asymmetries and asymmetric faults. In order to determine the fault location in the cases where the network is asymmetric, e.g. due to lack of fully transmission system transposition, and occurrence of asymmetric faults (single-line-to-ground, line-to-line, and double line-to-ground), the proposed algorithm is based on three-phase state estimation formulation. The proposed method is also able to determine the fault occurrence, fault type, and phases that contribute to fault, along with fault location. Therefore, the proposed method includes a set of algorithms based on the three-phase state estimation.

To tackle the mentioned challenges and goals, the basic three-phase state estimation algorithm is modified by considering the fault location as a hypothetical bus on which is not feasible to install a PMU. The voltage phasor of this hypothetical bus, the injected current (fault current) and the fault location are also added to the problem variables. Thus, the achievements presented in this innovative study are summarized as follows:

- Proposing a set of algorithms based on the modified three-phase state estimation for fault detection, faulty line discrimination, fault location estimation and fault type and faulty phase(s) identification.
- Exploiting the capacities of three-phase state estimation for locating all kinds of symmetrical and asymmetrical faults on transposed and untransposed lines.
- Presenting an error model of the measurement chain including instrument transformers and PMUs.
- Reducing the effect of inherent error measurement chain on the fault location accuracy and detecting and eliminating bad data in the measurement vectors based on the design of measuring errors covariance matrix in the state estimation formulation.

2 Three-Phase State Estimation

The most common faults in power systems are asymmetrical faults [36] and also transmission systems are mainly asymmetric due to lack of fully lines transposition. Thus, the formulation of three-phase state estimation has been expanded in this study for the possibility of considering asymmetric faults, incomplete lines transposition and network asymmetries.

The purpose of power system state estimation is to calculate the magnitude and phase angle of the voltage of all network buses (state variables) using the quantities measured at different points in the network. The general form of three-phase state estimation equations are as follows:

$$\overline{z} = \begin{bmatrix} \overline{z}_1 \\ \overline{z}_2 \\ \cdot \\ \cdot \\ \cdot \\ \overline{z}_m \end{bmatrix} = \begin{bmatrix} \overline{h}_1(x_1, x_2, ..., x_n) \\ \overline{h}_2(x_1, x_2, ..., x_n) \\ \cdot \\ \cdot \\ \cdot \\ \overline{h}_m(x_1, x_2, ..., x_n) \end{bmatrix} + \begin{bmatrix} \overline{e}_1 \\ \overline{e}_2 \\ \cdot \\ \cdot \\ \cdot \\ \overline{e}_m \end{bmatrix} = \overline{h}(\overline{x}) + \overline{e} \tag{1}$$

In (1), m shows the number of three-phase PMU measurement blocks. Thus, the vector of system measurements $\overline{z}$ includes magnitude and phase angle of buses voltages and lines currents and has the order of $3m \times 1$. The vector of measurement functions, $\overline{h}(\overline{x})$, includes relationships between measurements and state variables. The order of $\overline{h}(\overline{x})$ is $3m$. Assuming there are N_{bus} buses in the network, the number of the state variables n is equal to $6N_{bus}$ including three magnitudes and three-phase angles for voltage phasor of each bus. The vector of state variables is denoted by $\overline{x} = \left[x_1\ x_2\ ...\ x_n\right]^T$. The measurement errors vector is of the order of $3m \times 1$ and is denoted by $\overline{e}$.

One of the most commonly used methods for solving state estimation problem is weighted least squares (WLS) method [37]. The purpose of WLS estimator is to minimize the value of objective function of (2):

$$J(\overline{x}) = \left[\overline{z} - \overline{h}(\overline{x})\right]^T.R^{-1}.\left[\overline{z} - \overline{h}(\overline{x})\right] \tag{2}$$

where, matrix R is the measurement errors covariance matrix.

$$R = Cov(e) = E\left[\overline{e}.\overline{e}^T\right] = diag(\overline{\sigma_1^2}, \overline{\sigma_2^2}, \ ..., \ \overline{\sigma_m^2}) \tag{3}$$

Diagonal matrices $\overline{\sigma^2}_i$s indicate the variance of the measurements related to the i-th three-phase measurement block.

$$\overline{\sigma^2}_i = \begin{bmatrix} \sigma_{i_a}^2 & 0 & 0 \\ 0 & \sigma_{i_b}^2 & 0 \\ 0 & 0 & \sigma_{i_c}^2 \end{bmatrix}. \tag{4}$$

The necessary condition for optimal solution of $J(x)$ is as (5):

$$g(\overline{x}) = \frac{\partial J(\overline{x})}{\partial \overline{x}} = 0 \tag{5}$$

or

$$-H^T.R^{-1}.\left[\overline{z} - \overline{h}(\overline{x})\right] = 0 \tag{6}$$

where $H(\overline{x}) = \frac{\partial \overline{h}(\overline{x})}{\partial \overline{x}}$ is the Jacobian matrix of measurement functions. Using Taylor expansion and considering the linear approximation of $\overline{h}(\overline{x}^k + \Delta\overline{x}^k) \cong \overline{h}(\overline{x}^k) + H(\overline{x}^k).\Delta\overline{x}^k$, (6) can be written as:

$$H^T(\overline{x}^k).R^{-1}.H(\overline{x}^k).\Delta\overline{x}^k = H^T(\overline{x}^k).R^{-1}.\left[\overline{z} - \overline{h}(\overline{x}^k)\right] \tag{7}$$

where $\Delta\overline{x}^k = \overline{x}^{k+1} - \overline{x}^k$ and $\overline{x}^{k+1}$ and $\overline{x}^k$ denote the solutions of $(k+1)$-th and k-th steps of (7). By defining the gain or information matrix as

$$G(\overline{x}) = H^T.R^{-1}.H \tag{8}$$

and rearranging (7), the unknown $\overline{x}$ is obtained by the iterative solution of (9):

$$\overline{x}^{k+1} = \overline{x}^k - G(\overline{x}^k)^{-1}.g(\overline{x}^k) \tag{9}$$

The iteration process is continued until $\left|\Delta\overline{x}^k\right| < \varepsilon$ where ε and k are the convergence threshold of the problem solution and the iteration number in problem solution process, respectively.

3 The Proposed Method to Determine Matrix R

Determining the power system status requires measuring a number of network electrical quantities. Figure 3 shows how to extract the electrical quantities. The process shown in Fig, 3 is called "measurement process" or "measurement chain". Each measurement chain consists of two main components, namely an instrument transformer and a measuring device. Therefore, the measurement accuracy is directly affected by the accuracy of both instrument transformers and measuring devices. Equipment errors include the errors related to their inherent limitations and systematic or biased errors due to their failure or other reasons.

The inherent errors of the equipment in the measurement chain are declared by the equipment manufacturer. These errors should be in accordance with the national or international standards. The measurement error exists on both magnitude and phase angle of voltage and current signals (Fig. 2).

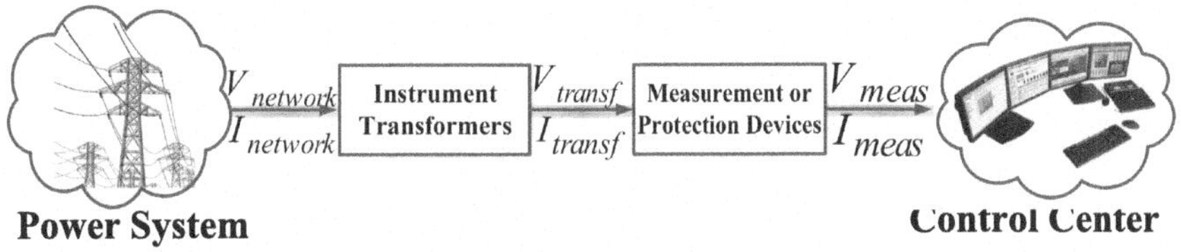

Fig. 2 A typical measurement chain in power system

Table 2 Maximum error of measurement devices [40]

Conventional measurements		PMU measurements		
Active power	Reactive power	Voltage	Current	Phase angle
±3%	±3%	±0.02%	±0.03%	±0.54°

IEEE C57.13 standard has provided accuracy classes for instrument transformers [38]. Accordingly, the maximum error of voltage and current transformers depends on the power system operating condition.

For fault location, it is necessary to examine the network quantities during the fault. The protective instrument transformers are used to obtain the abnormal voltage and current during the fault. The maximum permissible error and the accuracy class for this equipment are given in [39]. Table 2 also shows the maximum possible error of measurement equipment in accordance with IEEE C37.118.1 standard [40]. To identify the accuracy of measurement chain, it is necessary to consider the maximum possible error occurring in both the intermediary transformers and measuring devices to determine the error of measurement chains. In the following, we examine the error of each measurement chain in terms of errors of its components.

For this purpose, popular statistical distributions for equipment error are used. Due to the features such as continuity, symmetric feature, and ultimately zero mean in standard mode, standard normal distribution has been used for modeling the errors. In other references, the standard normal distribution is also called error distribution [36]. Many engineering applications also use standard normal distribution to model the errors.

In the measurement chain shown in Fig. 3, the relationship between converted quantities (F_{transf}) and actual quantities of the network ($F_{network}$) is given by (10). The final value of quantities in the output of the measurement process used in the algorithms is also expressed as (11):

$$F_{transf} = F_{network} + N(0, u^F_{IT}) \tag{10}$$

$$F_{meas} = F_{transf} + N(0, u^F_{PMU}) \tag{11}$$

where $N(0, \sigma^2)$ expresses a normal distribution with the mean 0 and the variance σ^2. In (10) and (11), u^F_{IT}, and u^F_{PMU} represent the standard uncertainty in the output

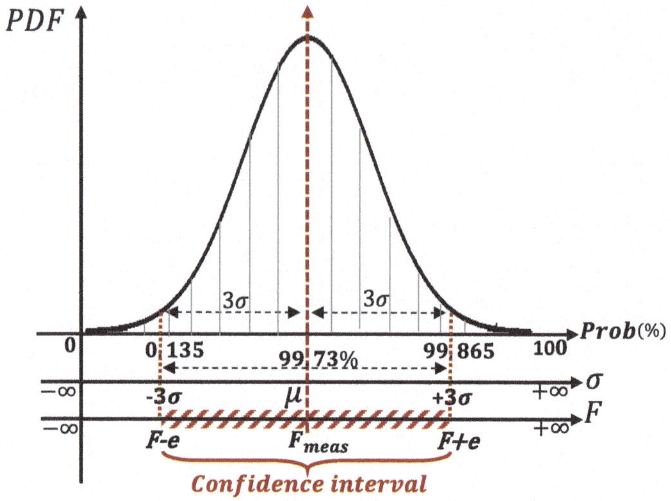

Fig. 3 Standard normal distribution and parameters definition

quantities of voltage/current transformers and PMUs, respectively. Furthermore, F is used to represent the measured quantities including V (voltage magnitude), I (current magnitude), θ^V (voltage phase angle), and θ^I (current phase angle). IT is also used to represent voltage transformer (VT) or current transformer (CT). Therefore, it is expected that the converted quantities and measured values lie within the confidence intervals:

$$(1 - e_{IT})F_{network} < F_{transf} < (1 + e_{IT})F_{network} \tag{12}$$

$$F_{transf}(1 - e_{PMU}) < F_{meas} < F_{transf}(1 + e_{PMU}) \tag{13}$$

where e^F_{IT} and e^F_{PMU} are the maximum possible error of magnitude or phase angle for the instrument transformers and PMUs, respectively. It should be noted that due to the stochastic nature of the measuring process, the obtained quantities may not be contained in the confidence interval. The probability of the obtained quantities being within the confidence interval is defined as level of confidence (p). For example, consider a voltage magnitude measurement, it is expected that the obtained value reflects the confidence interval $0.98\left|V_{transf}\right| < |V_{meas}| < 1.02\left|V_{transf}\right|$ that is 99.73%. In the other words:

$$Prob(0.98\left|V_{transf}\right| < |V_{meas}| < 1.02\left|V_{transf}\right|) = 99.73\% \tag{14}$$

It means that from each 1,000 measured samples, 3 samples are out of the confidence interval. Since in a power system, samples are continuously extracted, their behavior can be described by the standard normal distribution. The confidence interval, confidence level and its relation are shown in Fig. 3.

Table 3 Value of the coverage factor k_p that produces an interval having level of confidence p assuming a standard normal distribution

Level of confidence p (%)	Coverage factor k_p
90	1.645
95	1.96
95.45	2
99	2.576
99.73	3

Uncertainties of voltage and current magnitudes in the standard normal distribution are obtained by the following equations [36]:

$$u_{IT}^F.k_p = e_{IT}^F\,|F_{meas}| \tag{15}$$

$$u_{PMU}^F.k_p = e_{PMU}^F\,|F_{meas}| \tag{16}$$

where k_p represents coverage factor and $|F_{meas}|$ is the magnitude of voltage or current sent to the control center. The different k_ps corresponding to various values of p are presented in Table 3. If F means the voltage or current phase angle, uncertainties in the standard normal distribution are presented as follows:

$$u_{IT}^F.k_p = e_{IT}^F \tag{17}$$

$$u_{PMU}^F.k_p = e_{PMU}^F \tag{18}$$

By substituting (10) in (11), equation (19) is obtained.

$$F_{meas} = F_{network} + N(0, u_{IT}^F) + N(0, u_{PMU}^F) \tag{19}$$

As instrument transformers and PMUs are separate equipment with different physical structures and tasks, it is very close to reality to assume that they are independent. Thus, according to (19), the standard uncertainty of a measurement chain can be obtained from the two related normal distributions using the moment-generating function of those normal distributions. The moment-generating function for a typical normal distribution with the mean μ and the variance σ^2 is defined as $e^{\mu.t+\frac{\sigma^2.t^2}{2}}$. According to the statistical theorems, if two independent random variables are combined, the moment-generating function for the new random variable is obtained from the product of moment-generating functions of the independent variables [36].

$$\varphi_{chain}(t) = \varphi_{IT}(t).\varphi_{PMU}(t) \tag{20}$$

In (20), $\varphi_{IT}(t)$ and $\varphi_{PMU}(t)$ are moment-generating functions for instrument transformers and PMUs, respectively. For all equipment, the probability distribution

Table 4 Integrated error of measurement chain, including the accuracy classes of protection transducers and PMUs

Voltage Measurement Chain			Current Measurement Chain		
Accuracy class	Voltage error (%)	Phase displacement (min)	Accuracy class	Current error at rated primary current (%)	Phase displacement at rated primary current (min)
3P	±3.67	±120.0013	5P	±1.00045	±60
6P	±6.33	±240.00061	10P	±3.00015	–

function is a standard normal distribution, meaning that the mean value for all of them is zero ($\mu = 0$). Therefore, the moment-generating function of a measurement chain can be expressed as:

$$\varphi_{chain}(t) = e^{0.5\sigma_{IT}^2.t^2}.e^{0.5\sigma_{PMU}^2.t^2} \tag{21}$$

or

$$\varphi_{chain}(t) = e^{0.5(\sigma_{IT}^2+\sigma_{PMU}^2)t^2} \tag{22}$$

It concludes that the distribution of combination of two normally distributed independent variables is another normal distribution which has mean μ_{chain} and standard deviation σ_{chain} as:

$$\begin{aligned} \mu_{chain} &= \mu_{IT} + \mu_{PMU} = 0 \\ \sigma_{chain} &= \sqrt{\sigma_{IT}^2 + \sigma_{PMU}^2} \end{aligned} \tag{23}$$

Therefore, using (10) to (23), the error of a measurement chain can be calculated directly via (24) and the related weight coefficients can be obtained through (15) to (18).

$$e_{Total}^F = \sqrt{(e_{IT}^F)^2 + (e_{PMU}^F)^2} \tag{24}$$

The maximum errors of the measurement chains given in Table 4 have been calculated by (24) based on the maximum allowable errors of voltage and current given in Table 2. In most of the studies in the literature, just the PMU errors have been considered to evaluate the performance of fault location algorithms. However, according to Table 4, the main part of measurement error is related to the IT error, i.e. the integer part of the combined error is related to the IT errors, while its decimal part is due to the PMU errors. As shown in Table 4, IT errors are much higher than PMU errors and it is unreasonable to ignore them.

The new approach proposed in this section can be used for calculating the maximum combinational error of each measurement chain.

4 Modified Three-Phase State Estimation Formulation for Fault Location

In this section a new method for recognition of fault occurrence, discrimination of the faulty line, estimation of fault location and identification of fault type is proposed based on three-phase state estimation for all fault types including three-phase faults, line-to-line faults, double line-to-ground faults, and single-line-to-ground faults.

Installed PMUs provide the magnitude and phase angle values of the measured quantities to the algorithm. To explain the proposed algorithms, Fig, 4 is presented. Figure 4a shows transmission line ij when a fault f occurs at distance d_f from bus i. Figure 4b shows three-phase details of Fig. 4a for a single-line-to-ground fault including phase c.

4.1 Fault Location as a Hypothetical Bus

In the proposed algorithm, the fault point (point f) on the faulty line (line ij) is considered as a hypothetical bus during the fault. As a result, the original N_{bus}-bus system can be considered as a $N_{bus}+1$-bus system and the fault current is recognized as the injected current to the hypothetical bus.

Since there is no measurement equipment on the hypothetical bus, in the modified state estimation problem seven new state variables including the magnitude and phase angle of three-phase voltages of the hypothetical bus along with the fault position are added to the original state variables vector. Therefore, in the modeling of the

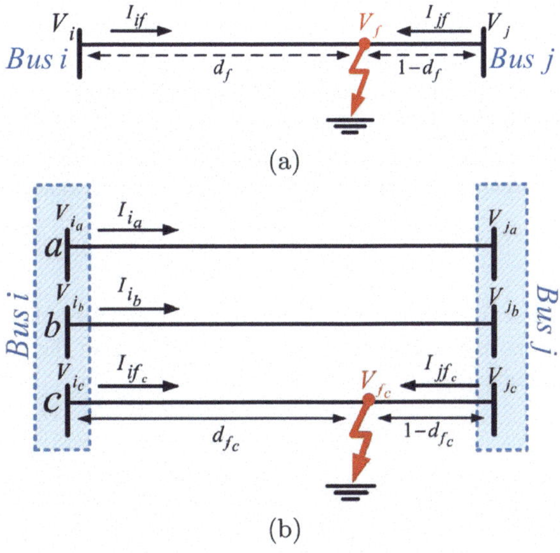

Fig. 4 Transmission line ij during a fault occurred at distance d_f from the i-th bus **a** Single line diagram **b** Single-line-to-ground fault on phase c

fault location problem based on the state estimation procedure, the vector of the state variables, $\overline{x}_{modified}$ is expressed as:

$$\overline{x}_{modified} = \begin{bmatrix} \overline{x}^{T}{}_{original} \mid \underbrace{\overline{\theta}^{V}_{f_{ph}}{}^{T} \ \overline{V}^{T}_{f_{ph}} \ d_f}_{\overline{x}_{added}} \end{bmatrix}_{6N_{bus}+7} \tag{25}$$

where $\overline{\theta}^{V}{}_{f_{ph}}$, $\overline{V}_{f_{ph}}$, and d_f are the phase angle and magnitude vectors of three-phase voltages at the hypothetical bus f and the fault location, respectively. It is observed that the number of state variables added to the modified problem is constant for all fault types. In this case, the Jacobian matrix is modified as (26).

$$H_{modified} = \left[H_{original} \ \middle| \ \frac{\partial \overline{h}(\overline{x}_{modified})}{\partial \overline{x}_{added}} \right] \tag{26}$$

where H is the Jacobian matrix of original N_{bus}-bus system and $H_{modified}$ is the modified Jacobian matrix corresponding to $\overline{x}_{modified}$ for the fault location purpose. As the number of network measurements before and during the fault are constant, the number of rows of matrix $H_{modified}$ does not change with respect to H, and the number of columns which are added to matrix H are equal to the number of variables added to the state variables.

4.2 Bad Data Detection

One of the advantages of state estimation formulation is the systematic capability for dealing with measurement errors. The inherent errors of the measurement chain are modeled by the covariance matrix of errors. On the other hand, due to reasons such as components failure, being under vibration or inappropriate installation environment, some measurement equipment might encounter errors much higher than those listed in Table 4. This kind of resulted data is called bad data. In this study, the largest residual vector is used to identify bad data. For this purpose, after solving the modified state estimation problem, the measurement residual vector is calculated using (27):

$$r(\widehat{\overline{x}}) = \overline{z} - \overline{h}(\widehat{\overline{x}}) \tag{27}$$

where $\widehat{\overline{x}}$ indicates the final value of state variable vector. In order to determine the normalized residual vector, the residual of the covariance matrix must first be calculated as:

$$\Omega(\widehat{\overline{x}}) = R - H(\widehat{\overline{x}}).G^{-1}(\widehat{\overline{x}}).H^{t}(\widehat{\overline{x}}) \tag{28}$$

Then, the elements of the normalized measurement residual vector are calculated as:

$$r^N_{i_{ph}}(\widehat{\overline{x}}) = \frac{r_{i_{ph}}(\widehat{\overline{x}})}{\sqrt{\Omega_{ii_{ph}}(\widehat{\overline{x}})}} \quad (29)$$

where, $\Omega_{ii_{ph}}$ indicates the elements of residual covariance diagonal matrix (Ω) corresponding to the i-th measurement and $r_{i_{ph}}$ is the i-th element of the measurement residual vector. After calculating the normalized measurement residual vector, its largest element is determined and selected.

If the value of the largest element of the normalized residual vector ($r^N_{\max}$) is greater than the pre-determined threshold ($\left|r^N_i\right| > \beta$), this means that the measurement corresponding to this element contains bad data. For bad data detection by the largest residual vector, the distinctive threshold β is usually selected equal to 3 in the literature such as [36, 37]. The choice of $\beta = 3$ as the threshold in references is based on the experiences related to huge tests and stochastic theories. In addition we have tested the threshold β equal to 3 by too much fault scenarios on different test system for the measurement chains errors more than those given in Table 4. For all studied scenarios, the elements corresponding to the failed measurement in the measurement residual normalized vector have been greater than 3. When the bad data is detected, the two following actions can be performed to improve the state estimation results; correcting bad data using estimated values of the state variables or solving the modified state estimation problem again by removing the faulty measurement from the measurement values vector, in case of data redundancy.

4.3 *Observability and Data Redundancy*

The magnitude and phase angle of three-phase voltages at N_{bus} buses, magnitude and phase angle of the hypothetical bus at fault point and the distance of fault point from one end of the faulty line for three phases should be determined by WLS algorithm. Thus, the number of the variables in the proposed fault location formulation is $6N_{bus} + 7$. In the fault location problem based on the state estimation technique, similar to a conventional state estimation problem, the best solution is obtained when the problem is in the over-determined condition. The over-determined condition refers to the condition in which the number of the measurements ($3m$) in the network is greater than the number of the variables ($n = 6N_{bus} + 7$). However, in order to take advantage of detecting and removing bad data, the number of measurements must be more than the variables ($3m \gg 6N_{bus} + 7$). Thus the redundancy of measurements determined the number of bad data which can be detected. The redundancy technique can be realized by more measurements or using two or several data sets related to different time instants during the fault period.

4.4 Proposed Algorithm

The proposed fault location algorithm is based on three-phase state estimation which requires three-phase model of the network and transmission lines and three-phase PMU measurements including voltage and current phasors. Thus, the proposed algorithm uses the network data and PMU measurements for detection of faulty line, accurate estimation of fault location and identification of fault type according to the following steps as shown in Fig. 5.

4.4.1 Detection of Faulty Line

Step 1: If some transmission lines are disconnected from the network, the probable fault incident investigation and possible faulty line discrimination processes are initiated.
Step 2: Under normal operation (prior to the line trip), the system situation including the network configuration (breakers status) and system operation quantities are specified by the measurement and monitoring systems. When a fault is occurred on a transmission line, the network configuration becomes different with the initial network configuration. The step 2 of the algorithm considers the initial network configuration along with the last set of data before fault clearing by opening the faulty line circuit breakers for achieving three-phase state estimation.

For more explanation, assume the circuit breakers are opened after 5 cycles from the fault occurrence instant. Thus, if the reporting rate of PMUs be one sample per cycle, then there are 5 set of phasor samples provided by PMUs during the fault. Since there is not compatibility between the system configuration (related to the normal situation) and the measurements used in the state estimation study, the bad data are detected and consequently the current measurements corresponding to the maximum normalized residual vector indicates the faulty line.

Briefly, using the initial network configuration (prior to lines disconnection) and the last set of data before circuit breakers tripping, three-phase state estimation is executed according to (1)–(9). According to (27)–(29) and using the bad data detection process for state estimation results, the normalized residual vector of current measurements would be calculated. The existence of bad data can be due to fault occurrence and the lack of conformity between the network configuration and data used in the executed state estimation. Therefore, there are two possible cases:

1. If bad data is not detected in current measurements, it can be concluded that the fault occurrence is not the cause of circuit breakers tripping.

2. If bad data is detected in current measurements, it means a fault has occurred and the current measurements corresponding to the maximum normalized residual vector indicate the faulty line.

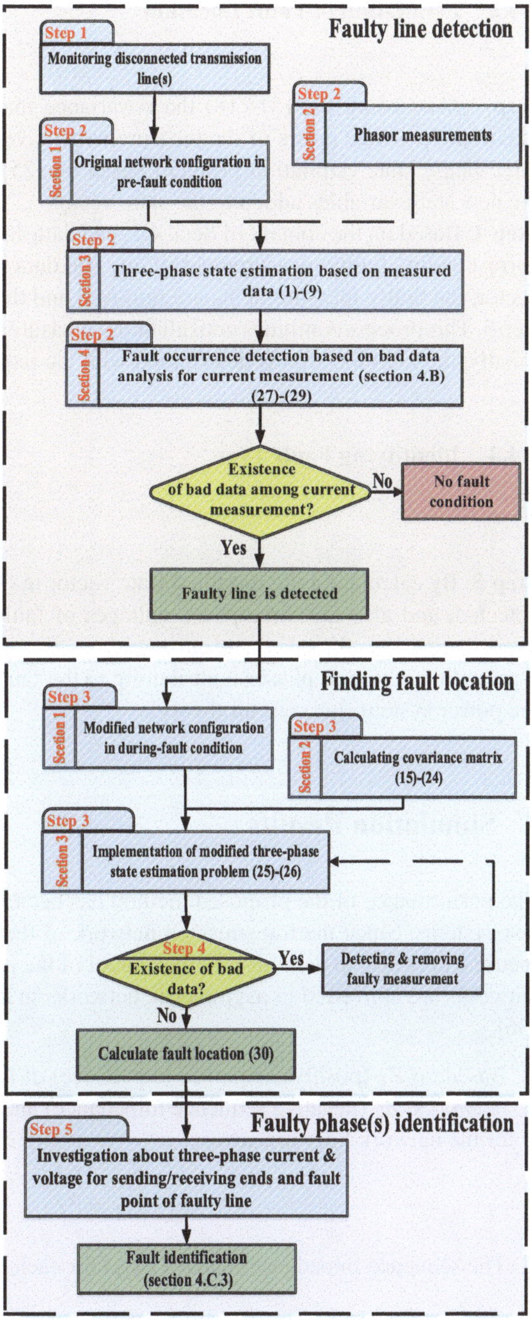

Fig. 5 Flowchart of proposed method

4.4.2 Estimation of Fault Location

Step 3: According to (15)–(18) the covariance matrix is calculated based on the maximum possible errors of the measurements given in Table 4. Then the modified three-phase state estimation problem based on (25)–(26) is solved by considering the new state variables added to the state vector.
Step 4: Based on the content of Sect. 4.2, bad data analysis is performed to detect the corresponding faulty measurement. If any bad data is detected in the measurements vector, the faulty measurements are removed and the calculation begins again from step 3. This process continues until all faulty measurements are detected and removed. Finally, the fault location is determined with the required accuracy.

4.4.3 Identifying Fault Type

Step 5: By calculating the modified state vector in (25), the fault location is exactly detected, and also the three-phase voltages of fault point $V_{f_{ph}}$ are calculated. By investigating $V_{f_{ph}}$, $V_{i_{ph}}$, $V_{j_{ph}}$ and currents $I_{if_{ph}}$, $I_{jf_{ph}}$ (Fig. 4) during the fault, the fault type and the phase or phases contributing to the fault can be identified according to the power system short circuit theory.

5 Simulation Results

The performance of the proposed method has been investigated using several IEEE test systems. Since the transmission network of these systems are inherently symmetrical network, to demonstrate the benefit of the proposed method, the symmetric networks are converted to asymmetric networks in the following steps according to [36].

1. Based on Z_1 (positive sequence impedance) of the lines given by the data set of networks, Z_2 (negative sequence impedance) and Z_0 (zero sequence impedance) of the network lines are set as:

$$\begin{aligned} Z_2 &= Z_1 \\ Z_0 &= 3Z_1 \end{aligned} \tag{30}$$

2. The sequence impedance matrix (Z_{012}) for each transmission line is formed as:

$$Z_{012} = \begin{bmatrix} Z_0 & 0 & 0 \\ 0 & Z_1 & 0 \\ 0 & 0 & Z_2 \end{bmatrix} \tag{31}$$

3. The phase impedance matrix (Z_{abc}) for each transmission line is calculated as:

$$Z_{abc} = \begin{bmatrix} Z_{aa} & Z_{ab} & Z_{ac} \\ Z_{ba} & Z_{bb} & Z_{bc} \\ Z_{ca} & Z_{cb} & Z_{cc} \end{bmatrix} = T^{-1}.Z_{012}.T \tag{32}$$

where

$$T = \begin{bmatrix} 1 & 1 & 1 \\ 1 & e^{j120} & e^{j240} \\ 1 & e^{j240} & e^{j120} \end{bmatrix} \tag{33}$$

In the Z_{abc} calculated by (32), the mutual impedance between phases, Z_{ab}, Z_{bc}, and Z_{ac} are equal. To model non-perfect transposition of the lines phases according to [36]:

$$\begin{aligned} Z_{ab} &= Z_{bc} \\ Z_{ac} &= 0.6Z_{ab} \end{aligned} \tag{34}$$

The Z_{abc} of lines calculated as above presents asymmetric lines and are used in the three-phase state estimation. The fault location algorithm need to knowing the network topology and parameters. About 1 million fault studies on different test systems (IEEE 9, 14, 39, 57 and 118-bus test networks) have been demonstrated excellent performance of the proposed method. In this section, some results related to the modified IEEE 118-bus test system [42, 43] are only presented. The fault simulation was performed using the PowerFactory environment [41] and the related data were processed using appropriate softwares. The simulations were performed on a PC with an Intel Core i7 CPU including 32GB of RAM. Usually, the PMUs are located to attain observability of network under normal operation. For example, to realize complete observability conditions, studies of [44] have been led to locate 28 PMUs on buses 3, 8, 11, 12, 17, 21, 25, 28, 34, 35, 40, 45, 49, 53, 56, 62, 72, 75, 77, 80, 85, 86, 90, 94, 102, 105, 110 and 114 of 118-bus system as shown in Fig. 6. However, these PMUs cannot guarantee the observability during faults, thus, it is required to complete the initial set with other PMUs for attaining fault observability and encountering bad data, inherent error of measurement chains and parameter errors. These purpose are achieved by adding three PMUs at buses 47, 61 and 63 in the 118-bus system. However, this choice is not unique and the accurate results in fault location and detection of some measurement and parameter errors can be alternatively obtained by adding three PMUs to buses 51, 64 and 70. It is evident that the incorrect choices of PMUs cannot guarantee the results for the proposed algorithm. For example, adding PMUs at buses 57, 74 and 84 in addition to 28 initial PMUs leads to unacceptable errors in fault location for some cases such as faults occurred on line 64-65. In addition, the studies have been demonstrated that adding only one or two PMUs cannot provide full fault observability in presence of measurement errors. We have specified the required PMUs with heuristic methods; however, the optimal locations can be specified by the systematic methods such as ones in [45]. Consequently, the optimal number and locations of PMUs are determined based on

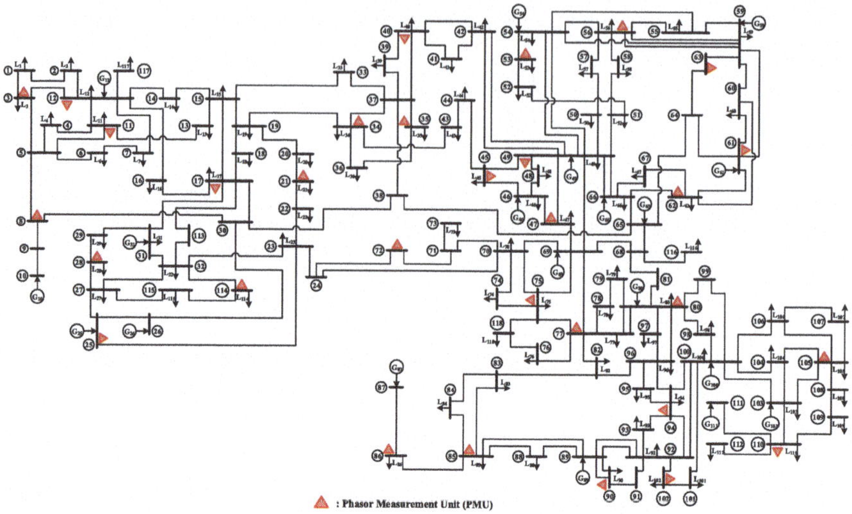

Fig. 6 IEEE 118-bus system

defining an appropriate optimization problem with required rules related to the state observability and fault observability.

In simulations, measurement chain errors have randomly been generated using the normal distribution based on the maximum allowable value given in Table 4. The simulated errors very closely reflect real measuring conditions. Numerous fault events have been simulated on different locations of all network transmission lines of the test systems. For each fault, 100 cases of added random errors to the measurement vector were considered. Then, for each location scenario, the relative error of fault location algorithm is calculated as the average of 100 simulated cases according to (35).

$$FLE(\%) = \frac{1}{100}\sum_{s=1}^{100}\frac{\left|d_{f,s}^{est} - d_{f}^{true}\right|}{d_{Line}} \times 100\% \tag{35}$$

where FLE is the relative error of the algorithm's result for each location scenario, $d_{f,s}^{est}$ is the estimated location of fault for the s-th case, d_{f}^{true} is the actual location of the fault, and d_{Line} is the length of the line.

5.1 Faulty Line Discrimination

The performance of the proposed method for discrimination of the faulty line was investigated by simulation of different fault scenarios on all 177 lines of the IEEE 118-bus system. It is observed that the method correctly performs for all the tested

Table 5 Results for a fault on line (3-5)

Faulty line discrimination	Bad data results for current measurements							
	Bad data solution	$r_{\max}^{N}$	Corresponding measurement	measured in bus				
	1st	18.742	I_{3-5}	3				
	2nd	16.449	I_{5-3}	5				
	Faulty line	Line (3-5)						
Fault identification	Estimated faulty line quantities (p.u.)							
	Quantities	Voltage magnitude				Current magnitude		
	Phase	a	b		c	a	b	c
Bus								
	Sending	0.951	0.198	0.981	0.381	9.523	0.275	
	Hypothetical	0.941	0.000		0.993	–	–	
	Receiving	0.848	0.191		0.971	0.396	6.317	0.302
	Fault data	True				Estimated		
	d_f (p.u.)	0.2				0.20041		
	Type	SLG				SLG		
	Faulty phase	b				b		
	FLE (%)	0.041%						

scenarios. For example, consider a single-phase-to-ground short-circuit including phase b occurred on line connected between buses 3 and 5 at 20% distance from bus 3. At first, the measured phasors related to during fault (before opening of circuit breakers of two sides of the line) and the basic network configuration in the three-phase state estimation problem are used to solve the problem. In this case, the maximum normalized residual is obtained as 18.742 which corresponds to the measured current I_{3-5}. The second maximum normalized residual vector is 16.449 which corresponds to the measured current I_{5-3}. These results indicate that line (3–5) can be a faulty line candidate. Then, the bad data are eliminated by solving the modified three-phase estimation problem with the hypothetical bus on line (3–5). The results of solving the problem for line 5 are shown in Table 5. These results demonstrate very high accuracy of the fault location and fault type identification.

5.2 Importance of Matrix R and Measurement Chain Error Modeling in the Fault Location Algorithm

In Sect. 3, we discussed about the calculation method of the maximum error of measurement chains using the maximum allowable errors of PMUs and transducer transformers. The maximum error of measurement chains is used to correctly construct the measurement errors covariance matrix, R, which is required in the state estimation formulation. In this section, the importance of correct construction of matrix R and also modeling the measurement chains error is studied. For this purpose, random measurement errors are generated using the measurement chain error model proposed in Sect. 3. Then matrix R is constructed considering two different assumptions as follow.

$Case1$: The error of instrument transformers are not considered in the measurement chains error and matrix R is only constructed based on PMUs errors.

$Case2$: Matrix R is accurately constructed based on the measurement chains errors.

Table 6 gives the average of FLEs obtained from the proposed fault location algorithm for two cases 1 and 2 applied on numerous different fault scenarios on the transmission lines of IEEE 118-bus test system. It is observed that the errors of fault location algorithm are not acceptable in case 1. The large FLEs in this case are due to the mismatch between the real measurement error and incorrect assumption in the construction of matrix R. The obtained outcomes confirm significance of accurate measurement chain modeling in obtaining truthful results which is properly performed in this chapter.

5.3 Study of Bad Data Detection in the Measurement Vector

One of the important aspects of the proposed method is detecting bad data. A comprehensive study of bad data detection has been performed by adding various biased errors to different PMU measurements. The proposed method correctly identifies the failed measurements for all cases. Here the results of a few studied cases are presented. These cases are related to the occurrence of all fault types on line (16-17) with added biased error to measurement chain of bus 11 as:

$Case1$: Adding simultaneous errors to magnitude and phase angle of voltage phasor of bus 11. The errors of magnitude and phase angle are set as 20% and 10 degrees, respectively.

$Case2$: Adding simultaneous errors to either magnitude or phase angle of current phasors of lines (11-12) at bus 11 and (11–13) at bus 11. The errors of magnitude and phase angle are considered as 20% and 10 degrees, respectively.

Table 7 summarizes the results of these case studies. For each case, the two left columns provide the maximum of the normalized residual vector corresponding to the failed measurements. After removing the failed measurements from the data set

Table 6 Impact of matrix R and measurement chains error modeling in the fault location algorithm

Line No./ FLE (%)	Case 1	Case 2	Line No./FLE (%)	Case 1	Case 2	Line No./ FLE (%)	Case 1	Case 2	Line No. /FLE (%)	Case 1	Case 2	Line No./ FLE (%)	Case 1	Case 2	Line No./ FLE (%)	Case 1	Case 2
Line 1–2	3.614	0.257	Line 19–34	3.312	0.169	Line 40–41	2.918	0.089	Line 55–59	2.289	0.11	Line 76–77	3.646	0.173	Line 92–94	3.703	0.118
Line 1–3	4.679	0.161	Line 20–21	2.418	0.334	Line 40–42	2.177	0.114	Line 56–57	3.094	0.149	Line 77–78	3.587	0.073	Line 93–94	3.067	0.17
Line 2–12	2.705	0.31	Line 21–22	4.234	0.223	Line 41–42	4.76	0.308	Line 56-58	4.506	0.271	Line 77–80 C1	2.855	0.126	Line 94–100	3.602	0.153
Line 3–12	3.432	0.263	Line 22–23	3.14	0.055	Line 42–49 C1	2.865	0.19	Line 56–59 C1	2.565	0.043	Line 77–80 C2	2.06	0.251	Line 94–95	3.335	0.125
Line 3-5	4.567	0.272	Line 23–24	4.564	0.197	Line 42–49 C2	2.455	0.065	Line 56–59 C2	4.404	0.225	Line 77–82	2.036	0.307	Line 94–96	3.841	0.201
Line 4–11	3.845	0.316	Line 23–25	2.945	0.094	Line 43–44	3.894	0.271	Line 59–60	2.712	0.2	Line 78–79	4.419	0.014	Line 95–96	4.001	0.163
Line 4–5	4.196	0.086	Line 23–32	2.097	0.241	Line 44-45	4.449	0.249	Line 59–61	2.568	0.222	Line 79-80	3.84	0.215	Line 96–97	2.601	0.135
Line 5-11	2.828	0.14	Line 24–70	2.433	0.078	Line 45–46	2.573	0.097	Line 60–61	4.183	0.15	Line 80–96	4.414	0.102	Line 98–100	3.255	0.276
Line 5–6	3.036	0.318	Line 24–72	4.703	0.147	Line 45-49	3.401	0.104	Line 60–62	4.473	0.218	Line 80-97	3.179	0.114	Line 99–100	2.757	0.286
Line 6–7	3.304	0.312	Line 25–27	2.747	0.062	Line 46-47	3.256	0.074	Line 61–62	3.845	0.263	Line 80-98	4.567	0.27	Line 100–101	4.503	0.25
Line 7–12	3.717	0.105	Line 26–30	4.245	0.16	Line 46–48	4.073	0.055	Line 62–66	3.269	0.017	Line 80–99	2.18	0.098	Line 100–103	4.841	0.102

(continued)

Table 6 (continued)

Line No./FLE (%)	Case 1	Case 2	Line No./FLE (%)	Case 1	Case 2	Line No./FLE (%)	Case 1	Case 2	Line No. /FLE (%)	Case 1	Case 2	Line No./FLE (%)	Case 1	Case 2	Line No./FLE (%)	Case 1	Case 2
Line 8–9	4.022	0.273	Line 27–115	4.465	0.315	Line 47–49	3.733	0.334	Line 62–67	2.982	0.269	Line 82–83	2.514	0.251	Line 100–104	3.138	0.127
Line 8–30	4.055	0.013	Line 27–28	2.768	0.266	Line 47–69	4.238	0.282	Line 63–64	2.581	0.147	Line 82–96	2.215	0.189	Line 100–106	4.022	0.133
Line 9–10	2.617	0.159	Line 27–32	4.307	0.081	Line 48–49	2.081	0.199	Line 64–65	3.908	0.28	Line 83–84	2.146	0.153	Line 101–102	3.517	0.076
Line 11–12	2.904	0.155	Line 28–29	4.161	0.324	Line 49–50	3.048	0.274	Line 65–68	2.314	0.288	Line 83–85	2.282	0.234	Line 103–104	2.724	0.158
Line 11–13	2.467	0.146	Line 29–31	4.191	0.23	Line 49–51	3.259	0.313	Line 66–67	2.853	0.196	Line 84–85	2.102	0.272	Line 103–105	3.527	0.259
Line 12–117	4.411	0.115	Line 30–38	3.701	0.207	Line 49–54 C1	2.22	0.203	Line 68–116	2.987	0.218	Line 85–86	2.384	0.202	Line 103–110	4.368	0.209
Line 12–14	4.347	0.055	Line 31–32	2.319	0.082	Line 49–54 C2	3.689	0.023	Line 68–81	4.778	0.1	Line 85–88	3.464	0.281	Line 104–105	2.967	0.304
Line 12–16	3.337	0.126	Line 32–113	3.159	0.3	Line 49-66 C1	3.961	0.2	Line 69–70	2.979	0.176	Line 85-89	2.662	0.104	Line 105–106	4.181	0.058
Line 13–15	3.892	0.334	Line 32-114	3.024	0.202	Line 49–66 C2	3.094	0.167	Line 69–75	2.633	0.324	Line 86–87	3.271	0.306	Line 105–107	4.324	0.037
Line 14–15	3.092	0.229	Line 33–37	4.133	0.287	Line 49–69	3.701	0.077	Line 69–77	2.344	0.25	Line 88–89	2.63	0.161	Line 105–108	3.077	0.269

(continued)

Table 6 (continued)

Line No./ FLE (%)	Case 1	Case 2	Line No./FLE (%)	Case 1	Case 2	Line No./ FLE (%)	Case 1	Case 2	Line No. /FLE (%)	Case 1	Case 2	Line No./ FLE (%)	Case 1	Case 2	Line No./ FLE (%)	Case 1	Case 2
Line 15–17	3.622	0.1	Line 34–36	4.326	0.114	Line 50–57	3.642	0.242	Line 70–71	2.847	0.246	Line 89–90 C1	3.995	0.257	Line 106–107	4.144	0.193
Line 15–19	2.365	0.032	Line 34–37	3.861	0.328	Line 51–52	3.011	0.133	Line 70–74	4.347	0.312	Line 89–90 C2	4.633	0.173	Line 108–109	2.036	0.338
Line 15–33	3.062	0.023	Line 34–43	4.304	0.217	Line 51–58	3.182	0.056	Line 70–75	2.646	0.122	Line 89-92 C1	2.578	0.1	Line 109-110	4.146	0.268
Line 16–17	2.776	0.14	Line 35–36	3.637	0.047	Line 52–53	3.702	0.06	Line 71–72	3.85	0.136	Line 89–92 C2	2.866	0.135	Line 110–111	3.571	0.193
Line 17–113	4.204	0.095	Line 35–37	2.612	0.14	Line 53–54	2.517	0.236	Line 71–73	3.362	0.057	Line 90–91	2.666	0.058	Line 110–112	3.7	0.111
Line 17–18	4.682	0.218	Line 37–39	4.354	0.317	Line 54-55	2.888	0.244	Line 74–75	4.254	0.296	Line 91-92	4.747	0.279	Line 114–115	4.39	0.193
Line 17–31	2.963	0.208	Line 37–40	3.38	0.013	Line 54–56	4.735	0.07	Line 75–118	3.714	0.301	Line 92–100	3.695	0.185			
Line 18–19	4.737	0.104	Line 38–65	2.655	0.166	Line 54–59	2.695	0.293	Line 75–77	4.569	0.319	Line 92–102	3.916	0.289			
Line 19-20	2.257	0.074	Line 39–40	2.075	0.244	Line 55–56	2.949	0.316	Line 76–118	3.991	0.157	Line 92–93	4.821	0.123			

Table 7 Results of bad data detection in the measurement vector

Fault Type	Case 1				Case 2			
	1st iteration FLE(%)	Failed magnitude measurement	Failed phase angle measurement	Last iteration FLE(%)	1st iteration FLE(%)	Failed magnitude measurement	Failed phase angle measurement	Last iteration FLE(%)
		$r^N_{\max}$	$r^N_{\max}$			$r^N_{\max}$	$r^N_{\max}$	
A-G	9.854	18.971	17.93	0.286	12.688	23.272	21.342	0.388
B-G	10.692	21.459	18.173	0.198	11.782	23.585	19.697	0.253
C-G	8.715	19.391	17.242	0.268	12.145	23.848	19.491	0.306
AB	8.392	19.23	16.839	0.267	12.172	21.878	17.223	0.217
AC	9.717	20.023	18.231	0.093	12.468	21.005	18.961	0.174
BC	9.633	19.806	18.108	0.044	13.436	24.092	18.59	0.224
AB-G	9.148	20.212	18.143	0.125	11.195	20.998	18.712	0.382
AC-G	9.355	18.36	17.008	0.165	14.542	24.087	20.387	0.222
BC-G	9.053	19.924	18.894	0.223	12.131	20.182	17.543	0.272
ABC	9.692	20.934	16.139	0.232	14.139	21.855	18.506	0.215
ABC-G	8.857	20.434	18.171	0.13	13.17	20.374	18.889	0.365

and solving the modified state estimation using the sound data, the average FLE of fault location algorithm is given in the right column for each case.

For example, considering an A-G fault in the first case, when the solving process is performed, the $r^N_{\max}$, which corresponds to the magnitude of voltage measurement of bus 10, is obtained as 24.536, which is significantly higher than the distinctive threshold ($\beta = 3$). Then, after removing this faulty measurement, the solution process is repeated. The value of $r^N_{\max}$ in the second repetition is also greater than $\beta = 3$ and equals to 21.044 which corresponds to the phase error of voltage measurement of bus 10. In the third repetition, the value of $r^N_{\max}$ reaches to less than 3 and FLE becomes equal to 0.024%. The calculated FLE is much less than 1%, which is usually considered as an acceptable error for the fault location algorithms.

As shown, in all simulated cases, the failed measurements were detected correctly and after removing the faulty measurement from the solving process, the fault location was estimated with an acceptable accuracy. In all cases, the FLEs are less than 0.37%.

5.4 *Impact of Different Data Redundancy Methods on the Proposed Algorithm*

As a necessary condition, there should be always redundancy in the measurements vector to obtain the best estimation for the fault location in the presence of measure-

Table 8 Comparison of $FLE(\%)$ for different methods of providing data redundancy

	Fault type	A-G	B-G	C-G	AB	AC	BC	AB-G	AC-G	BC-G	ABC	ABC-G
Methods												
Method A		0.011	0.017	0.084	0.086	0.097	0.084	0.081	0.042	0.017	0.087	0.041
Method B		0.025	0.264	0.099	0.289	0.308	0.136	0.127	0.308	0.297	0.143	0.197

ment chain inherent errors and bad data. One of the ways to create data redundancy is access to more voltage and current phasors. An alternative way is using multiple samples of measured quantities during the fault interval with the same number of measurements.

In the first method (method A), we assume that the measurements of voltage phasors and current phasors in the network of IEEE 118-bus test system are available according to the aforementioned 31 PMUs as shown in Fig. 6, while in the second method (method B), there are only 28 PMUs located at buses 3, 8, 11, 12, 17, 21, 25, 28, 34, 35, 40, 45, 49, 53, 56, 62, 72, 75, 77, 80, 85, 86, 90, 94, 102, 105, 110 and 114 according to [44]. In both methods, the number of measurements is more than the minimum requirement for full observability of the system. In method A, one phasor sample of each measurement is used but in the second method, two phasors related to the time interval before circuit breaker opening are used as inputs to the algorithm.

According to Table 8, the obtained results for various types of faults in the 118-bus system show that in both methods, data redundancy has enabled the algorithm to detect fault location with a high degree of accuracy. Furthermore, it can be realized that in the proposed algorithm, only the existence of measurement data redundancy is sufficient for detecting and eliminating faulty measurements, but how to generate data redundancy has no major effect on the algorithm process and the accuracy of the fault location algorithm.

5.5 A Discussion on the Speed of the Proposed Algorithm

The fault location algorithms are developed to accurate estimation of the fault point on transmission lines to identify the weak network points in the case of the temporary faults or reduce the repair time in the case of the permanent faults. Thus, in contrast to protective algorithms, the speed of the fault location algorithm and related algorithms such as the algorithm of identifying the faulted line is not a main and important aspect. In fact, these algorithms are offline ones and executed after the protection system operation. However, Table 9 compares the speed and accuracy of the proposed algorithm with a number of the existing PMU-based fault location algorithms. The average execution time of the proposed algorithm and six existing PMU-based fault location algorithms for different fault scenarios on IEEE 118-bus system [33, 46–

Table 9 Average execution time of proposed algorithm in comparison with six existing PMU-based fault location algorithms for different fault scenarios on IEEE 118-bus system

Methods	Proposed method	[52]	[46]	[47]	[48]	[49]	[50]
Execution time	0.088 *	0.49	0.079	7.7	0.61	1.09	0.328
Average FLE (%)	0.23	0.47	0.65	0.61	0.31	0.51	0.84

* The execution time is calculated for average execution time of scenarios with three bad measurements detection and fault location calculation

Table 10 Minimum, maximum and average execution time of proposed algorithm on different test networks

Test network	Proposed method		
	Min. execution time*	Max. execution time**	Average execution time
IEEE 14 bus	0.028 s	0.056 s	0.044 s
IEEE 39 bus	0.051 s	0.081 s	0.073 s
IEEE 57 bus	0.061 s	0.088 s	0.079 s
IEEE 118 bus	0.078 s	0.97 s	0.089 s

* The execution time is calculated for average execution time of scenarios with two bad measurements detection and fault location calculation
** The execution time is calculated for average execution time of scenarios with five bad measurements detection and fault location calculation

50] gives in Table 9 demonstrate the execution time of the most methods such as the proposed method is fraction of 1 Sec. The minimum, average and maximum times of the proposed algorithm execution for four test systems with different sizes are given in Table 10, also demonstrates the execution time increases with the system size although the times are still small.

Table 10 demonstrates the execution time increases with the system size although the times are still small. The presented methods in this chapter can be modified for different applications in power systems operation, control, protection and security [33, 51, 53–56].

6 Conclusions

In this chapter, a fault location algorithm for transmission lines based on the three-phase state estimation formulation is proposed. The proposed modified formulation integrates all tasks including fault occurrence recognition, faulty line discrimination, fault location estimation and fault type and faulty phase(s) identification. It exploits the excellent capacities of three-phase state estimation for analyzing all symmetrical and asymmetrical faults on transposed and untransposed lines, reducing the effect of inherent errors of measurements chain on the fault location accuracy and detecting and eliminating the bad data in the measurement vector. The algorithm's performance has been investigated by simulating numerous fault cases on different locations of the transmission lines of the modified 39-bus test system with an asymmetric network. For each of the fault cases, the inherent errors of the measurements chain are considered in the measurement vector based on the proposed error model of the measurements chain. It was demonstrated that for all cases, the error of fault location algorithm is much less than 0.5%. In addition, the importance of correct construction of measurement error covariance matrix as one of the effective parameters on the state estimation calculation has been investigated. The proposed error model of the measurement chain uses the maximum allowable errors of PMUs and transducer transformers to calculate the maximum errors of the measurement chains utilized to correctly construct matrix R.

References

1. AIEE Committee Report,:Bibliography and summary of fault location methods [includes discussion]., Trans. American Institute of Electrical Engineers. Part III: Power Apparatus and Systems, vol. 74, no. 3, pp. 1423– 1428. (1955)
2. R.J. Hamidi, H. Livani, Traveling-wave-based fault-location algorithm for hybrid multiterminal circuits. IEEE Trans. Power Del. **32**(1), 135–144 (2017)
3. IEEE Std. C37.114-2014 (Rev. IEEE Std. C37.114-2004),: IEEE guide for determining fault location on ac transmission and distribution lines, pp. 1–76, (2015)
4. F. Deng, X. Zeng, X. Tang, Z. Li, Y. Zu, L. Mei, Travelling-wave-based fault location algorithm for hybrid transmission lines using three-dimensional absolute grey incidence degree. International Journal of Electrical Power & Energy Systems **114**, 105306 (2020)
5. Akmaz, D., Salih Mami, M., Arkan, M., Emin Taluk, M.,: Transmission line fault location using traveling wave frequencies and extreme learning machine, Electric Power Systems Research, vol. 155, pp. 1–7, (2018)
6. O. Naidu, A.K. Pradhan, A Traveling Wave-Based Fault Location Method Using Unsynchronized Current Measurements. IEEE Transactions on Power Delivery **34**, 505–513 (2019)
7. E.J.S. Leite, F.V. Lopes, F.B. Costa, W.L.A. Neves, Closed-Form Solution for Traveling Wave-Based Fault Location on Non-Homogeneous Lines. IEEE Transactions on Power Delivery **34**, 1138–1150 (2019)
8. O.A. Gashteroodkhani, M. Majidi, M. Etezadi-Amoli, A.F. Nematollahi, B. Vahidi, A hybrid SVM-TT transform-based method for fault location in hybrid transmission lines with underground cables. Electric Power Systems Research **170**, 205–2014 (2019)
9. C. Fei, G. Qi, C. Li, Fault location on high voltage transmission line by applying support vector regression with fault signal amplitudes. Electric Power Systems Research **160**, 173–179 (2018)
10. Y.Q. Chen, O. Fink, G. Sansavini, Combined Fault Location and Classification for Power Transmission Lines Fault Diagnosis With Integrated Feature Extraction. IEEE Transactions on Industrial Electronics **65**, 561–569 (2018)
11. J. Izykowski, R. Molag, E. Rosolowski, M.M. Saha, Accurate location of faults on power transmission lines with use of two-end unsynchronized measurements. IEEE Trans. Power Del. **21**(2), 627–633 (2006)
12. Ahmed, A. S., Attia, M. A., Hamed, N. M., Abdelaziz, A. Y.,: Modern optimization algorithms for fault location estimation in power systems, international journal of Engineering science and technology, vol. 20, no. 5, pp. 1475–1485, (2017)

13. D. Akmaz, M.S. Mamiş, M. Arkan, M.E. Tğluk, Transmission line fault location using traveling wave frequencies and extreme learning machine. Elec. Power Sys. Research **155**, 1–7 (2018)
14. Dalcastagne, A. L., Zimath, S. L.,: A study about the sources of error of impedance-based fault location methods. In: Transmission and distribution conference and exposition, pp. 1–6, IEEE/PES, Latin America (2008)
15. R.A. de Aguiar, A. Dalcastagnê, H.H. Zürn, R. Seara, Impedance-based fault location methods: Sensitivity analysis and performance improvement. Elec. Power Sys. Research **155**, 236–245 (2018)
16. Dalcastagnê, A., Noceti Filho, S., Zurn, H., Seara, R.,: An iterative two-terminal fault-location method based on unsynchronized phasors, IEEE Trans. Power Del., vol. 23, no. 4, pp. 2318–2329, (2008)
17. Y. Liao, N. Kang, Fault-location algorithms without utilizing line parameters based on the distributed parameter line model. IEEE Trans. Power Del. **24**(2), 579–584 (2009)
18. J. Izykowski, E. Rosolowski, P. Balcerek, M. Fulczyk, M.M. Saha, Accurate noniterative fault-location algorithm utilizing two-end unsynchronized measurements. IEEE Trans. Power Del. **26**(2), 547–555 (2011)
19. J. Doria-Garcia, C. Orozco-Henao, L.U. Iurinic, J.D. Pulgarn-Rivera, High impedance fault location: Generalized extension for ground faults. International Journal of Electrical Power & Energy Systems **114**, 105387 (2020)
20. Didehvar, S., Mohammadi Chabanloo, R.,: Accurate estimating remote end equivalent impedance for adaptive one-ended fault location, Electric Power Systems Research, vol. 170, pp. 194–204, (2019)
21. T. Spielbck, A. Bel, Design of a one-sided, impedance-based transmission line fault locator using line topology and source impedances. Electric Power Systems Research **161**, 123–128 (2018)
22. M.A. Elsadd, A.Y. Abdelaziz, Unsynchronized fault-location technique for two- and three-terminal transmission lines. Electric Power Systems Research **158**, 228–239 (2018)
23. M. Ghazizadeh-Ahsaee, Accurate arcing fault location method for M-terminal transmission lines. International Journal of Electrical Power & Energy Systems **98**, 147–155 (2018)
24. Y. Lee, T. Lin, C. Liu, Multi-Terminal Non-homogeneous Transmission Line Fault Location Utilizing Synchronized Data. IEEE Transactions on Power Delivery **34**, 1030–1038 (2019)
25. A.G. Phadke, J.S. Thorp, K.J. Karimi, State estimation with phasor measurements. IEEE Trans. Power Sys. **1**(1), 233–238 (1986)
26. S. Azizi, M. Sanaye-Pasand, A straightforward method for wide-area fault location on transmission networks. IEEE Trans. Power Del. **30**(1), 264–272 (2015)
27. Saha, M. M., Izykowski, J., Rosolowski, E.,: Fault location on power networks, Springer Science & Business Media, (2009)
28. Q. Jiang, B. Wang, X. Li, An efficient PMU-based fault-location technique for multi terminal transmission lines. IEEE Trans. Power Del. **29**(4), 1675–1682 (2014)
29. D. Sayari, P.S. Shiv, K.P. Bijaya, Transmission line fault detection and location using Wide Area Measurements. Elec. Power Sys. Research **151**, 96–105 (2017)
30. A.H. Al-Mohammed, M.A. Abido, An adaptive fault location algorithm for power system networks based on synchrophasor measurements. Elec. Power Sys. Research **108**, 153–163 (2014)
31. S. Azizi, M. Sanaye-Pasand, M. Paolone, Locating faults on untransposed, meshed transmission networks using a limited number of synchrophasor measurements. IEEE Trans. Power Sys. **31**(6), 4462–4472 (2016)
32. Mahamedi, B., Sanaye-Pasand, M., Azizi, S., Zhu, J. G.,: Unsynchronised fault-location technique for three-terminal lines, IET Gen., Trans. & Distrib., vol. 9, no. 15, pp. 2099–2107, (2015)
33. M. Majidi, M. Etezadi-Amoli, M.S. Fadali, A sparse-data-driven approach for fault location in transmission networks. IEEE Trans. Smart Grid **8**(2), 548–556 (2017)
34. M.A. Elsadd, A.Y. Abdelaziz, Unsynchronized fault-location technique for two- and three-terminal transmission lines. Elec. Power Sys. Research **158**, 228–239 (2018)

35. S. Hussain, A.H. Osman, Fault location on series and shunt compensated lines using unsynchronized measurements. Elec. Power Sys. Research **116**, 166–173 (2014)
36. A. Ghaedi, M.E.H. Golshan, M. Sanaye-Pasand, Transmission line fault location based on three-phase state estimation framework considering measurement chain error model. Electric Power Systems Research **178**, 106048 (2020)
37. A. Gomez-Exposito, A. Abur, *Power system state estimation: theory and implementation* (CRC Press, 2004)
38. IEEE Std. C57.13-2016 (Revision of IEEE Std C57.13-2008): IEEE standard requirements for instrument transformers, pp. 1–96, (2016)
39. Alstom,: Network protection & automation guide. Alstom grid, (2011)
40. IEEE Std. C37.118.1-2011 (Revision of IEEE Std. C37.118-2005),: IEEE standard for synchrophasor measurements for power systems, pp. 1–61, (2011)
41. DIgSILENT GmbH, PowerFactory v15.2.6 software, https://www.digsilent.de/en/powerfactory.html
42. Information Trust Institute Grainger College of Engineering, IEEE 118-Bus System, https://icseg.iti.illinois.edu/ieee-118-bus-system
43. University of Washington, Power Systems Test Case Archive, http://labs.ece.uw.edu/pstca/pf118/pg_tca118bus.htm
44. F. Aminifar, A. Khodaei, M. Fotuhi-Firuzabad, M. Shahidehpour, Contingency-constrained PMU placement in power networks. IEEE Trans. Power Sys. **25**(1), 516–523 (2010)
45. M.E.H. Golshan, S.H.H. Dolatabadi, S.M. Tabatabaei, Determining minimum number and optimal placement of PMUs for fault observability in one-terminal algorithms. IET Generation, Transmission & Distribution **12**(121), 5789–5797 (2018)
46. G. Feng, A. Abur, Identification of faults using sparse optimization, *52nd Annual Allerton Conference on Communication*, Control, and Computing (Allerton) (IEEE, Allerton, 2014), pp. 1040–1045
47. Q. Jiang, X. Li, B. Wang, H. Wang, PMU-based fault location using voltage measurements in large transmission networks. IEEE Trans. on power Del. **27**(3), 1644–1652 (2012)
48. A.S. Dobakhshari, Fast accurate fault location on transmission system utilizing wide-area unsynchronized measurements. International Journal of Electrical Power & Energy Systems **101**, 234–242 (2018)
49. S. Das, S.P. Singh, B.K. Panigrahi, Transmission line fault detection and location using wide area measurements. Electric Power Systems Research **151**, 96–105 (2017)
50. C. Wang, C.X. Dou, X.B. Li, Q.Q. Jia, A WAMS/PMU-based fault location technique. Electric Power Systems Research **77**(8), 936–945 (2007)
51. H.H. Alhelou, M.H. Golshan, J. Askari-Marnani, Robust sensor fault detection and isolation scheme for interconnected smart power systems in presence of RER and EVs using unknown input observer. International Journal of Electrical Power & Energy Systems **99**, 682–694 (2018)
52. H.H. Alhelou, M.E. Hamedani-Golshan, R. Zamani, E. Heydarian-Forushani, P. Siano, Challenges and opportunities of load frequency control in conventional, modern and future smart power systems: A comprehensive review. Energies **11**(10), 2497 (2018)
53. Haes Alhelou, H., Hamedani Golshan, M. E., Hajiakbari Fini, M.,: Wind driven optimization algorithm application to load frequency control in interconnected power systems considering GRC and GDB nonlinearities, Electric Power Components and Systems, 46(11-12), 1223-1238, (2018)
54. Haes Alhelou, H., Hamedani-Golshan, M. E., Njenda, T. C., Siano, P.,: A survey on power system blackout and cascading events: Research motivations and challenges, Energies, 12(4), 682, (2019)
55. H.H. Alhelou, M.E.H. Golshan, N.D. Hatziargyriou, A decentralized functional observer based optimal LFC considering unknown inputs, uncertainties, and cyber-attacks. IEEE Transactions on Power Systems **34**(6), 4408–4417 (2019)
56. H.H. Alhelou, M.E.H. Golshan, N.D. Hatziargyriou, Deterministic dynamic state estimation-based optimal lfc for interconnected power systems using unknown input observer. IEEE Transactions on Smart Grid (2019)

Data-Driven Wide-Area Situation Analyzer for Power System Event Detection and Severity Assessment

Divya Rishi Shrivastava, Shahbaz Ahmed Siddiqui, and Kusum Verma

Abstract Real-time power system monitoring and assessment leads to two major concern, prediction and evaluation of security and stability of power system. This assists in determination of in-time probable anomaly of the system. However, at the same time it requires real—time technological applications to measure network data at all strategic geographical locations. Synchrophasor technology based wide-area situational awareness ensures power system real—time monitoring and assessment. The chapter proposes real—time data driven Wide-area Situation Analyzer (WASA). WASA first detects an event in the system using synchrophasor measurements and then assesses its vulnerability posed to power network. The vulnerability is measured as severity in terms of first swing transient instability. Level of severity index is developed in terms of generator going out of step. The bus voltage trajectories going away with rest of the system due to generator(s) transient instability are considered. The proposed new approach is based on Center of Frequency (COF) formulated from limited Phasor Measurement Unit measurements. To check for an event existence in the system, a new decision based COF concept is defined. In order to determine the severity of the identified event, a new Predictor Indices (PI) is proposed using COF and PMU measurements. These predictor indices are used in assessment methodology, based on Adaptive Boosting (AdaBoost) of decision estimators. Furthermore, comparative results of proposed wide-area situational analyzer with other machine learning algorithms are also shown. The proposed WASA is instigated on IEEE New England 39 Bus system, successfully validating analyzer performance. The different type of events considered are generation outage, bus outage, load outage and line

D. R. Shrivastava (✉)
Department of Electrical Engineering, Manipal University Jaipur, Jaipur 303007, India
e-mail: divyarishi.shrivastava@jaipur.manipal.edu

S. A. Siddiqui
Department of Mechatronics Engineering, Manipal University Jaipur, Jaipur 303007, India
e-mail: shahbazahmed.siddiqui@jaipur.manipal.edu

K. Verma
Department of Electrical Engineering, Malaviya National Institute of Technology Jaipur, Jaipur 302017, India
e-mail: kverma.ee@mnit.ac.in

H. Haes Alhelou et al. (eds.), *Wide Area Power Systems Stability, Protection, and Security*, Power Systems, https://doi.org/10.1007/978-3-030-54275-7_18

events. Additionally, if any bus outage occurs due to line faults then it is considered as single severity. The results reflects the efficacy of the proposed analyzer in event detection and its assessment efficiently and effectively with very less computational burden. The ability of the proposed analyzer to identify events quickly and correctly makes it appropriate for real—time applications.

Keywords Event detection · Severity assessment · Data-driven method · Wide-area situation analyzer · Predictor indices · WAMS · Power system security

1 Introduction

Modern power system is highly dynamic in nature, is continuously subjected to varied operating conditions. These varying operating conditions may cause failure in the power system. Hence, control center are required to make use of state –of –art tools to timely detect probable failures and increase reliability of electric grids. A secure power system operation thus safeguards continuous monitoring and assessment. With implementation of synchrophasor technology, the real time disturbance monitoring and assessment is adapted [1]. This assessment requires mathematical modelling of the network that can incorporate network changes. This model are to be capable enough for timely say of 'events occurrence' and assess the reach of event in terms of 'severity'. This capability of timely event detection and its assessment is suitable to develop a mechanism for real-time applications. These mechanisms using synchronized measurements are classified in data driven based and physics based approach [2]. Measurements like bus frequency, rate of change of frequency, bus voltage magnitude, rate of change of voltage magnitude, and angles are available from synchrophasor measurements. These measurements either directly or derived quantities of these fundamental signals can be utilized to assess the real-time situation of the power network. Results from [3–13] presents power system event detection diagnostics and their assessment in terms of either classification or localization. These results are formulated by making use of machine learning techniques, statistical tests, signal processing techniques, data transformation techniques, energy methods and combinations of these methods. In majority of these techniques timely detection is achieved, however assessment of such events in terms of their effect on, 'systems health' is not show cased. Transient instability is one of the major aspect that leads the system to collapse. Therefore, transient stability assessment is crucial and has to be continuously evaluated real-time to ensure system remains in safe limits. In [14–31], various methods are proposed to identify and assess transient stability status of power system. In these methods different tools like machine learning, statistical, pattern recognition, energy methods and/or their combinations with synchronized measurements are also applied.

Largely, event detection methods proposed in literature are evaluated in terms of their response time and accuracy. On detection of events, control centers are interested in such events that can cause system to go in 'un-safe limits'. Hence, it is paramount that severity should be determined once the system is detected to have event. The severity assessment help the system operator to develop robust remedial measure if the system is moving towards insecure operation. In general, system inertia is used to assess transient instability status of power system. Assessment of real-time inertia is not easy and a small variation in inertia value may alter transient stability results [32, 33]. To overcome problem of real-time inertia assessment, Centre of Power (COP) is proposed in [27]. Power is not directly available through phasor estimates and has to be additionally calculated. This additional calculation is not easy as due to any event, power network changes and a small variation in calculation due to change in data may lead to different power calculation. This may result in inaccurate transient stability results. Hence, calculation of such indices with minimum additional computation efforts is more feasible.

Therefore, it can be imperative to develop a composite methodology for detecting event timely and assess its severity on power system. The develop method should have a good success rate and minimum response time such that, if required control or remedial actions may be initiated timely. This chapter proposes Wide Area Situation Analyzer (WASA), it utilizes PMU measurements to determine event detection and its severity assessment within first few cycles of event occurrence. A new notion of Center of Frequency (COF) to detect event and using this new notion novel indices, used as predictor indices for severity assessment are proposed. These predictors are statistical measures and hence are very effective and easily computed. The proposed WASA uses Adaptive Boosting of Decision Trees (AdaBoost) regression to amount of severity an event hold in the power network. To calculate the level of severity for an event, a methodology based on probable bus voltage trajectories that move away with rest of the system due to generator(s) transient instability, is proposed. The analyzer takes synchronized bus frequency and voltage magnitude as input and first detects event followed by severity assessment for occurred event, as its assessment is paramount to maintain system intact following event. In this work, events like short circuit, double line outage leading to bus outage, generator outage and load outage are used. These events poses major challenge to system in maintaining stability. Additionally, if any bus outage occurs due to double line faults then it is considered as single severity. The proposed analyzer is data driven and requires less computational effort, can be easily implemented on any test power system. This chapter investigates proposed WASA on IEEE 39 bus power system to predict severity within ~300 ms and ~0.08 root mean square error.

2 Proposed Wide Area Situation Analyzer (WASA)

Power System equipped with PMU infrastructure can capture real-time networks pre, during and post event dynamics [32]. The time tagged network data are stored in data concentrators and retrieved as need arises. In general, bus voltage magnitude and frequency measurements available from PMU are used to monitor real-time power system dynamics [31].

A. *Concept of Center of Frequency-Voltage*

This sub-section describes a new concept of Center of Frequency (COF). The concept is to highlight the combined impact of frequency and voltage variation on the system operation. In general, power system severity assessment utilizes bus voltage angles measurements. Algorithms based on bus voltage magnitudes for severity assessment are simpler and faster [25]. Therefore, in the presented work synchronized bus voltage magnitudes and bus frequency are considered to be input quantities for event detection and its severity assessment. In order to define the concept of method, let assume bus frequency f and voltage magnitude $|v|$ are available from PMUs and represented as given by (1). First and last elements for each column in two matrices is length of synchronized data sample from bus 1 to nth bus. Bus measurements recorded are with respect to reference bus of the power system.

$$f = \begin{pmatrix} f_{11} & \cdots & f_{1n} \\ \vdots & \ddots & \vdots \\ f_{m1} & \cdots & f_{mn} \end{pmatrix}; |\text{v}| = \begin{pmatrix} v_{11} & \cdots & v_{1n} \\ \vdots & \ddots & \vdots \\ v_{m1} & \cdots & v_{mn} \end{pmatrix} \tag{1}$$

Unlike [27] where concept of center of power used for system vulnerability assessment. The concept of center of frequency do not require additional calculation, instead it uses bus frequency measurements available from PMUs. The objective of the chapter is to develop a situation analyzer that analyzes power system event and its severity in minimum time and with least error. This concept presents real-time system's overall status in terms of frequency and voltage as in (2),

$$COFV = \frac{\sum_{i=1}^{n} f_i |v|_i}{\sum_{i=1}^{n} f_i} \text{ p.u.} \tag{2}$$

where, n is total bus in power network. Equation (2) gives time-stamped Center of Frequency-Voltage (*COFV*) for any system. When we calculate *COFV*, as event occurs pre event values prior to an event is changed, and may take *COFV* to either lie around pre fault value or it may lie far from pre fault values during post fault conditions.

B. *Algorithm for Event Detection*

Corresponding to any load event there will be variation in *COFV*. This variation is shown in (3) as event detection that is used to detect event in power system. This change in (3) is shown as Event Detection (*ED*) index and is used for detect event detection in power system.

$$ED = \frac{\Delta COFV|_{(i-(i-1))}}{COFV_i} \tag{3}$$

where, *i* is *ith* time-stamp, *(i-1)* is previous time-stamp and *ED* is event detection index. In order to distinguish between random load perturbation and events, a threshold (ε) is defined in (4). If index $ED > |\varepsilon|$, there is an event and if $ED \leq |\varepsilon|$ there is either no event or it's a case of random load perturbation.

$$\varepsilon = \frac{l}{n}(|\Delta V_l|) \text{ p.u.} \tag{4}$$

In (4) l is number of load bus, n is total bus in power network and $|\Delta V_l|$ is per unit change in voltage at load buses at any instant. For any power system, the event detection criteria is calculated by using network bus topology in terms of load bus and total buses in power system.

C. *Severity Calculation using COF and Voltage magnitudes*

After event detection, the next crucial step is to assess its effect on system health. The system may behave abnormally due to one of the following instabilities viz: rotor angle, voltage or frequency instability. Rotor angle instability may be first swing or multi swing transient instability problem. This chapter focuses severity assessment based on first swing transient instability problem. The system health depends on severity level, i.e. if event is more severe it may lead to system collapse rapidly. In this context, first swing transient instability is significant and its timely and fast assessment can prevent the system from catastrophic failure. The bus voltage trajectories going away with rest of the system due to generator(s) transient instability are considered. In this chapter, only generator bus and load bus are considered (zero power bus are not considered) as important nodes. That is due to generator(s) transient instability, generator bus and load bus voltage trajectory are taken as important buses. Relative bus voltage magnitude w.r.t center of frequency-voltage for these important nodes is measured and noted as given by (5)

$$SA_b = |v|_b - COFV \tag{5}$$

where *b* is (load bus + generator buses) and *SA* is severity assessment for important nodes. The value of SA for each *b* is calculated and compared with pre-defined threshold (μ). If SA for any bus is greater than μ, then bus is 'critical'. For such cases Severity Matrix (*SM*) is calculated and is mathematically shown as:

$$\begin{aligned} &If\ SA_b > |\mu| \rightarrow SM_b = 1 \\ &else \qquad\quad\ \rightarrow SM_b = 0 \end{aligned} \tag{6}$$

For SM = '1' signifies 'critical bus', is counted and weighted for all total important nodes giving overall Severity Index (*SI*) by (7).

$$SI = \frac{\sum SA_b}{N_b} \tag{7}$$

A pseudo code for severity index calculation below elaborates systematic computation. Severity computed is in terms of first swing transient stability. The value of *SI* lies in range [0,1]. In the range '0' indicates no severity whereas '1' indicates maximum severity as if all important nodes are transient unstable.

Pseudocode: Calculation of Severity Index

```
Initialize: Call |v| and COFV
 for each  monitored bus
    | Calculate |v| - COFV
    | If  (Calculate |v| - COFV) > absolute(μ)
    |   | Severity Count = 1
    | else
    |   | Severity Count = 0
    | end
 end
```

$$\text{Severity Index} = \frac{\text{count(severity Count=1)}}{\text{Total monitored bus}}$$

Record: Severity Index

Pseudocode discussed above is for those buses for which voltage trajectory move away from rest of system due to generator(s) transient instability. Additionally, due to double line faults, bus outage occurs. This single bus outage is also considered as single critical bus and can be considered as minimum severity.

D. *Data mining through Adaptive Boosting of decision trees (AdaBoost)*

AdaBoost [34] is a boosting ensemble of decision trees regressor. Model learns from increasing weights of previous mistakes. It is adaptive as following weak learners stay in favor of the instances that are misclassified by previous classifiers. Root Mean Square error (RMSE) given by (8) is used to evaluate performance of AdaBoost regressor. RMSE is standard deviation of the prediction errors.

$$RMSE = \sqrt{\frac{\sum_{i=1}^{p}\left(\hat{y}_i - y_i\right)}{p}} \tag{8}$$

where, y_i is actual value, $\hat{y}_i$ is predicted value and p is total unseen testing samples. As event is detected and relative bus voltage magnitude is computed using (5), standard deviation of post event data is used as input features for regressor as in (9). These calculated features are taken as Predictor Indices (PI). Standard deviation measures disturbance present in the signal. The features calculated from archived synchronized database is used to train a regressor for associated severity in offline mode. The trained regressor then predicts severity level in real-time for different unseen operating conditions. The feature are given by following relation:

$$feature = S.D(SA_b)\, \forall b \in \text{ important bus} \tag{9}$$

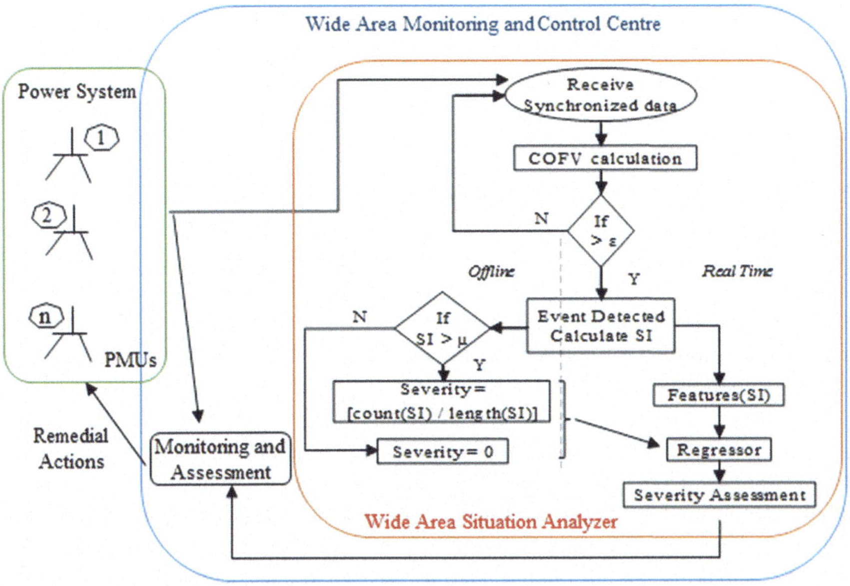

Fig. 1 Proposed Wide Area Situation Analyzer (WASA)

The overall structure of proposed Wide Area Situation Analyzer (WASA) is shown in Fig. 1.

The frequency and voltage magnitude inputs from data center are used to give severity assessment and share with control center for further processing and initiating appropriate measures if required.

3 Performance of WASA: Result and Discussion

The proposed WASA is validated through numerical simulation and analysis on IEEE 39 bus system implemented on DIgSILENT powerfactory. For effective test of proposed analyzer, number of events at different power system loading is used.

A. *Test System Data*

The study is carried for four types of events as short circuit cleared with line outage, double line outage leading to bus outage, generator outage and load outage. Bus voltage magnitude and bus frequency measurements from PMU with 60 samples in 1 s are recorded. As two consecutive value of *COFV* is sufficient to detect any event in the system. Two data samples of bus frequency and bus voltage magnitude are used to detect event. The data window for severity analysis is of post disturbance 18 cycles (300 ms). For events like short circuits, clearance time used in this study is of 6 to 12 cycle. This fault clearance time is sufficient for protection scheme to come into action. Since the test system have 19 load buses and 10 generator buses, while 2 loads are on generator buses, thus in all 27 vital buses are considered that take care of either generator bus, load bus or both.

B. *Event Detection*

The proposed analyzer is implemented on IEEE 39 bus system, having 19 buses with dispatchable loads. At any given instant, random load variation at all the buses is in such a way that voltage variation is $\pm 1\%$. The threshold ε is calculated by substituting these values in (4), with $l = 19$, $n = 39$ and $\Delta V = \pm 1\%$ (here the threshold value of ε is taken as ~0.005 p.u). This threshold value is system dependent. Visualization of event detection is presented graphically in Figs. 2 and 3. The voltage angle-time series is shown for 27 important buses. Figure 2a depicts short circuit case initiated at 0 s at bus 03 and cleared in 6 cycles by opening line 02-03. Time response of bus voltage angle is shown for 3 s. During short circuit events, frequency and voltage value decreases, *COFV* value goes down, and again increases after fault clearance. Event detection index crosses threshold, as at the time of fault inception and at time of fault clearance the ED values is more than threshold. Figure 2b shows double line fault at 0 s making bus 28 outage. Due to this bus outage, network flow changes and accordingly voltage and frequency values changes. The resulting *COFV* and *ED* changes from pre fault values and event is detected at index crosses the threshold. Figure 3a visualizes generator 01 outage connected to bus 30. Due to generator outage, power imbalance takes place and after some time this imbalances minimizes and network functions with decreased generation. As visualized, event detection index identifies the event as soon as generator outage takes place. Figure 3b shows load outage at bus 20. It is visualized that effect of this load outage on bus voltage is less than other events as post event *COFV* value decreases but this decreases is less than other events. Other load outage may have large effects on system voltage and depends on network operating conditions. It is found that Event Detection (*ED*) crosses threshold (ε) for all the four types of events. Similarly, other events are detected and visualized. On detection of event, severity assessment methodology is implemented and is discussed in following sub-section.

C. *Severity Assessment*

Once event is detected, post event data is used to first calculate relative bus voltage magnitude. Relative bus voltage magnitude is used to construct severity index using pseudo code as discussed above. Severity index is then computed that gives level of severity due to event occurred. The level of severity gives number of buses, who's voltage trajectories move away with rest of the system due to generator(s) transient instability. The deviation of bus voltage magnitude from *COFV* shows behavior of bus voltage with rest of the system. With event inception, each bus voltage magnitude will deviate from system *COFV*. The more it deviates, chances of system going towards insecure limits increases. In general, there are difference among stable and unstable faults, post-fault voltage trajectories. For a post-fault stable system, voltage trajectories tends towards flat following a recovery. In case of unstable system, voltage

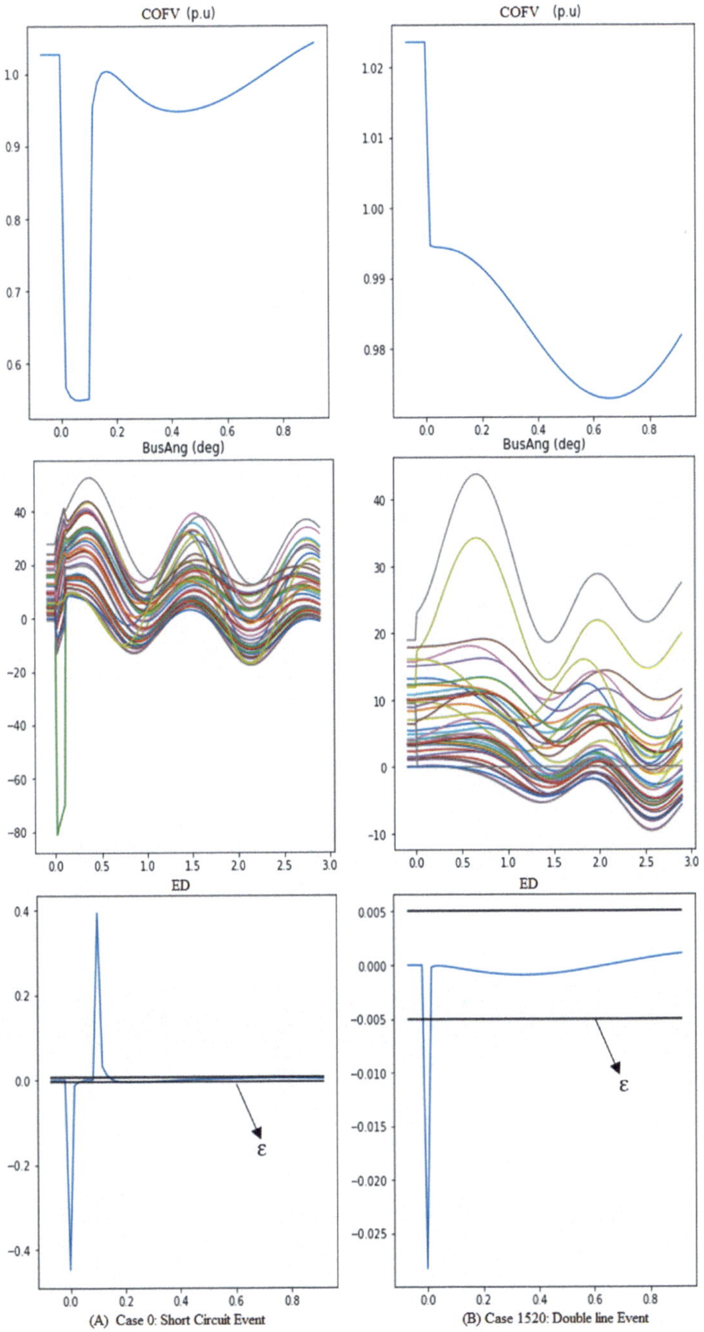

Fig. 2 Visualization of event detection for short circuit and double line faults

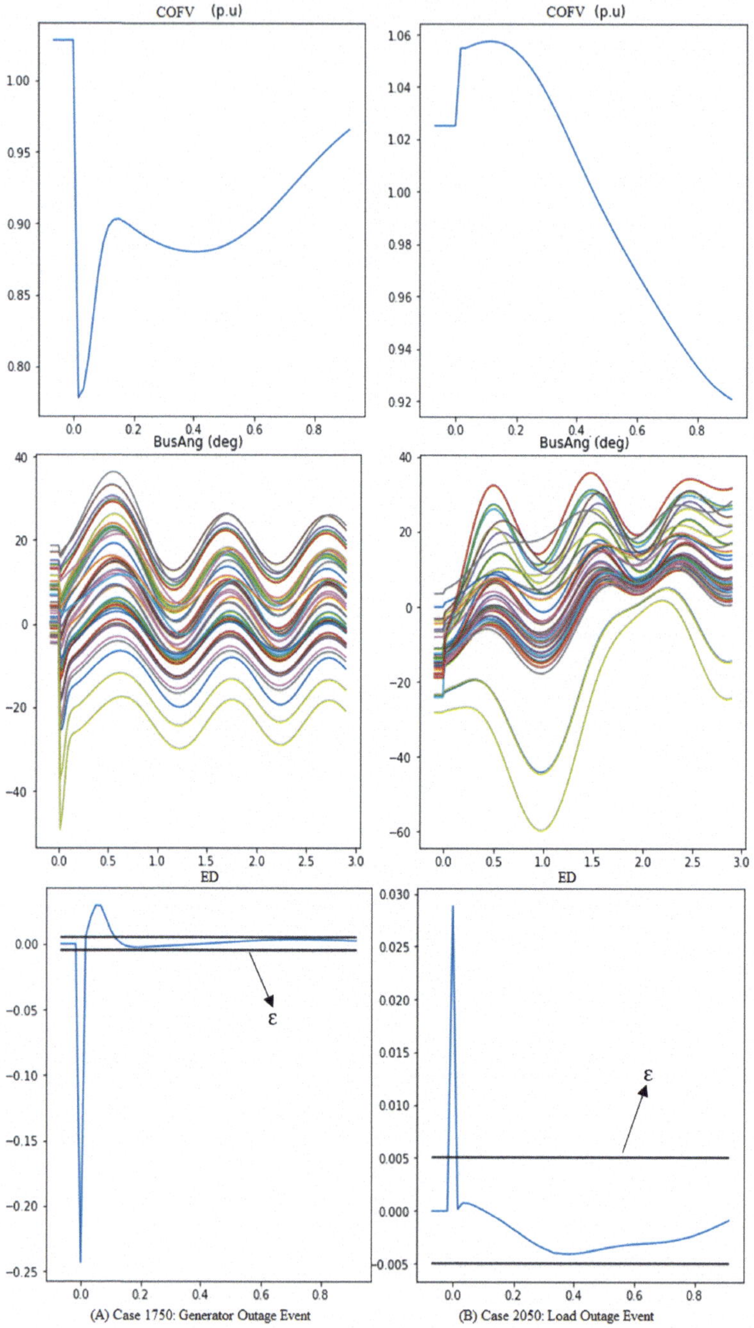

Fig. 3 Visualization of event detection for generator outage and load outage events

trajectories gradually gives rising or falling trend following initial recovery. It is observed during multiple time domain simulation of test power system that at least one bus crosses the threshold (μ greater than 0.2), and simultaneously at least one monitored bus goes transient unstable. Hence, deviation of 0.2 p.u of bus voltage with respect to center of frequency voltage is used as threshold for system qualifying in severity status. In order to understand how event can pose severity to system, Fig. 4 shows an example cases, where event does not affects system operation and can be considered to be less severe event. In Fig. 4, following a short circuit event system remains stable as load and generator bus voltage magnitude do not crosses the threshold. The first row shows generator bus and load bus relative voltage magnitude. While second row shows time series generator bus and load bus voltage angles.

In Fig. 5 double line fault event, which leads to bus outage is shown. In this figure for generator and load bus case, it can be clearly visualized that since generators are transient stable no important bus crosses threshold. Although one load bus do crosses threshold, is due to fact that double line outage event has lead this load bus outage from system. As discussed in section II, sub-section C can be coined minimum severity.

Figure 6 shows case of generator transient unstable case where, occurred event poses severity to the system. In such cases, voltage of critical bus will be most diverged from *COFV*. If for any event, the voltage trajectory of buses diverge from *COFV* and crosses the threshold. The respective bus then enters in 'critical bus' category and is noted in severity index and counted with '1 critical bus' out of total important bus. Figure also shows early detection of critical buses that will move out of step. It can also be shown that the critical load bus move out of threshold before critical generator bus. These critical bus shows instability around 0.8 s whereas bus

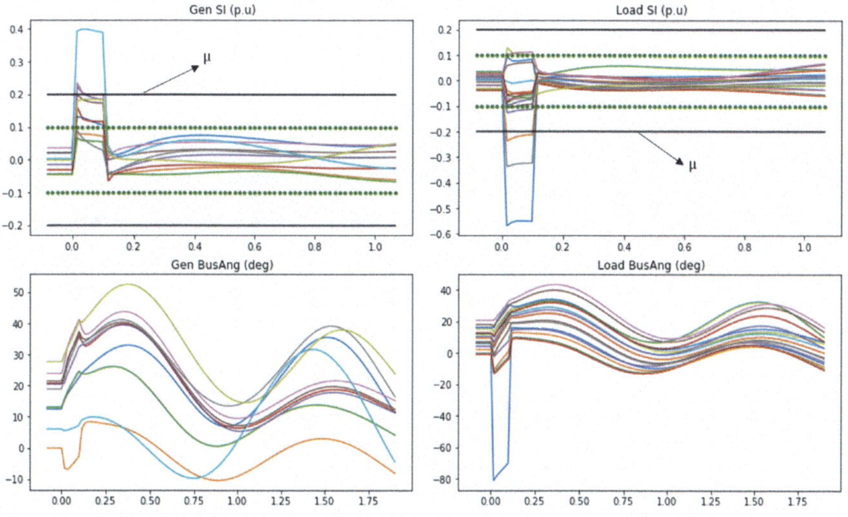

Fig. 4 Visualization of Stable case for short circuit event

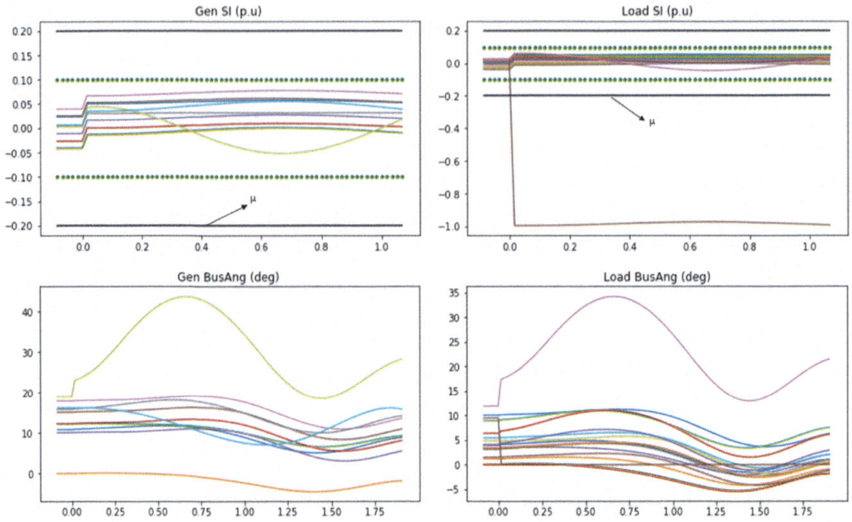

Fig. 5 Visualization of double line event leading to bus outage

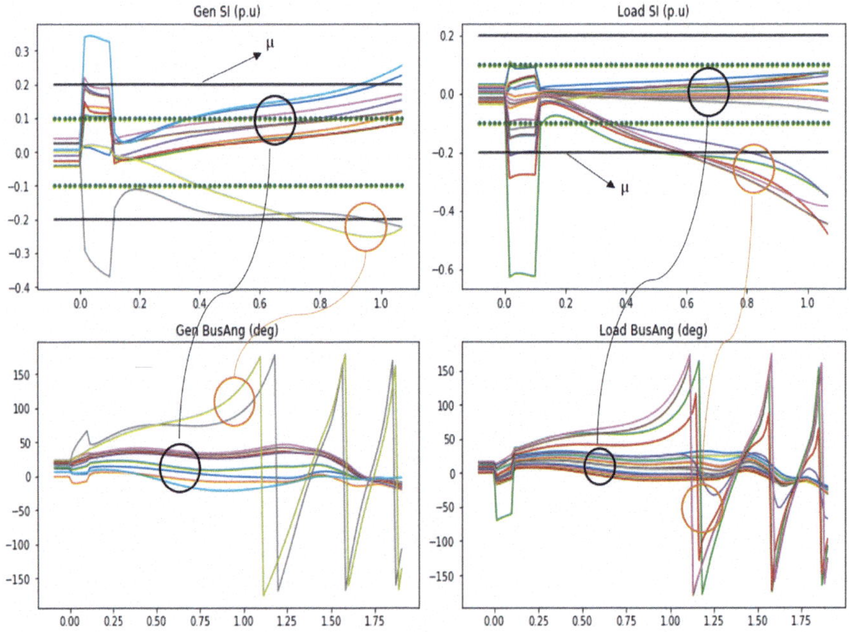

Fig. 6 Visualization of transient unstable case for short circuit event

actually move out of step after 1 s. Thereby giving additional margin for initiating emergency actions.

D. *Results of AdaBoost Regressor*

To calculate the severity of an event in real-time, AdaBoost regressor is implemented. A large dataset of 2221 cases is generated with different operating conditions for modelling and testing unseen cases. Apart from four types of events, no event cases are also incorporated in the study to make the WASA familiar with no event data. These cases overall contribute to zero severity case of 1121 and non-zero severity cases to 1100 as shown in Table 1. For each case, the analyzer takes total 1560 samples out of which 156 for event detection and 1404 for severity assessment. The standard deviation of relative bus voltage magnitude computed using (5) are used as features to the regressor to build and test it for severity ranging between [0, 1]. The overall details are tabulated in Table 2. This data is reduced to 27 Indices as input features. These indices are equal to number of nodes of interest in the system. The AdaBoost Regressor is modeled for various number of trees as shown in Fig. 7 and ensemble of 78 trees is made to build the Regressor to predict severity due to an event. As at this ensemble, RMSE is minimum. Response time taken by proposed WASA for severity assessment is outlined in Table 3. The delay in receiving synchronized data is considered as 0.2 s [13].

Results infers that proposed analyzer computes severity in less than 0.3 s for all the considered cases. This timely detection and assessment of event ensures timely

Table 1 Data Info for operating conditions

S.No	Disturbance type	Total operating conditions	Severity cases	
1	Short circuit cleared with line outage	1499	Zero Severity = 1121	Non-zero Severity = 1100
2	Double line outage leading to bus outage	250		
3	Generator Outage	172		
4	Load outage	280		
5	Normal (no event)	20		
	Total	2221		

Table 2 Details of data from PMU, features and target for Regressor

Synchronized data from PMU	Features	Target
Event detection: Bus voltage magnitude (39×2) + bus frequency $(39 \times 2) = 156$ Severity assessment: Bus voltage magnitude (39×18) + bus frequency $(39 \times 18) = 1404$ Total = 1560	Monitored buses = 27	Severity = [0,1]

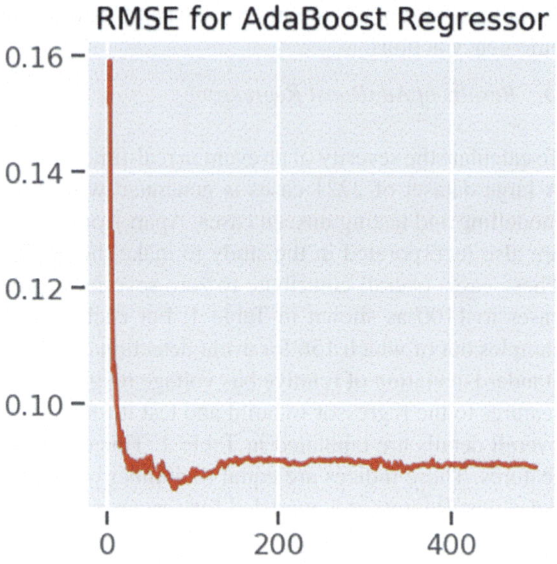

Fig. 7 Root mean square error with different ensemble of trees

Table 3 Response time for severity assessment

Operation	Time (s)
Receiving synchronized data	0.2
COFV calculation	0.003
Event detection	0.0006
SI calculation	0.005
Feature extraction	0.03
Prediction Time (per case)	0.06
Total Time	0.2986

initiation of emergency or remedial actions, if required. The performance of WASA with other Regressor available in literature is also computed and shown in Fig. 8.

The time of action of proposed algorithm with classifiers like Random Forest (RF), Gradient Boosting (GBR), Support Vector Regressor (SVR), Decision Tree Regressor, k-Nearest Neighbor and Multi-Layer Perceptron Regressor is within ~0.35 s. However, the error of these classifiers is more than AdaBoost Regressor. A visualized result for random twenty sample unseen test cases is presented in Fig. 9.

These cases contains both zero and non-zero severity and shows actual and predicted severity for different cases are shown. The analyzer can hence provide probable solution for severity assessment an event contains upon its occurrence and can help control operator to initiate any remedial or control action.

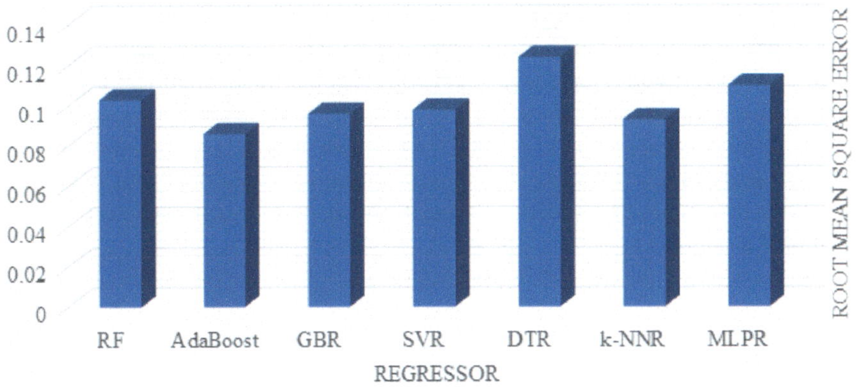

Fig. 8 Proposed WASA performance with Regressors

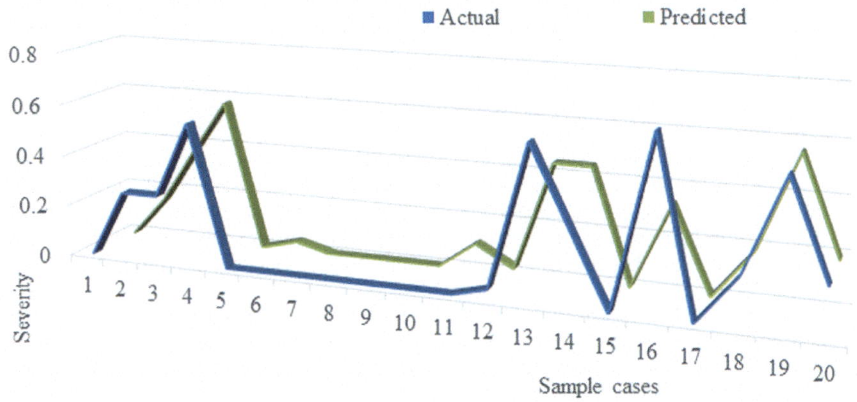

Fig. 9 WASA performance with AdaBoost

4 Conclusions

The real-time situational analyzer WASA is developed to detect and assess the vulnerability of events in terms of severity using synchrophasor data. New concept of Center of Frequency-Voltage (*COFV*) based on real-time bus frequency and voltage measurements is proposed. Successive change in *COFV* detects event. Once event is identified, severity of the event is calculated by formulating new predictor indices. These predictor indices are standard deviation of severity assessment calculated from voltage trajectories and center of frequency based bus voltage magnitude. Level of severity index is also developed in terms of generator going out of step. The bus voltage trajectories going away with rest of the system due to generator(s) transient instability are considered. Due to double line faults, bus outage occurs and is taken as minimum severity. This single bus outage can also be considered as single critical bus These Predictor Indices early detects probable number of generator going out of step. In order to predict probable severity in real-time AdaBoost Regressor is trained using statistical features extracted from these indices. The proposed Wide Area Severity Analyzer is applied on IEEE 39-bus test system. The response time for proposed analyzer is ~ 300 ms and root mean square error of ~ 0.08, which ensures early event detection and severity assessment, and increases real-time power system situational awareness.

References

1. J. Ree, D. La, V. Centeno, J.S. Thorp, et al., Synchronized phasor measurement applications in power systems. IEEE Trans. Smart Grid **1**(1), 20–27 (2017)
2. S. Brahma, R. Kavasseri, H. Cao et al., Real-time identification of dynamic events in power systems using PMU data, and potential applications- models, promises, and challenges. IEEE Trans. Power Deliv. **32**(1), 294–301 (2017)
3. M.K. Jena, B.K. Panigrahi, S.R. Samantaray, A new approach to power system disturbance assessment using wide-area postdisturbance records. IEEE Trans. Ind. Inform. **14**(3), 1253–1261 (2018)
4. H.H. Alhelou, M.E. Hamedani Golshan, J. Askari-Marnani, Robust sensor fault detection and isolation scheme for interconnected smart power systems in presence of RER and EVs using unknown input observer. Int. J. Electr. Power Energy Syst. **99**, 682–694 (2018)
5. H.H. Alhelou, Fault detection and isolation in power systems using unknown input observer. in *Advanced Condition Monitoring and Fault Diagnosis of Electric Machines* (IGI global, 2019), pp. 38–58
6. M. Rafferty, X. Liu, D. Laverty et al., Real – time event detection and classification using moving window PCA. IEEE Trans. Smart Grid **7**(5), 2537–2548 (2016)
7. W. Gao, J. Ning, Wavelet-based disturbance analysis for power system wide-area monitoring. IEEE Trans. Smart Grid **2**(1), 121–130 (2011)
8. H. Haes Alhelou, M. Esmail Hamedani Golshan, T. Cuthbert Njenda, P. Siano. Wams-based online disturbance estimation in interconnected power systems using disturbance observer. Appl. Sci. **9**(5), 990 (2019)
9. R. Meier, B. McCamish, E. Cotilla-Sanchez, et al., Event detection using correlation within arrays of streaming PMU data, in *IEEE Power & Energy Society General Meeting*, 2018, pp. 1–5
10. H. Haes Alhelou, M. Esmail Hamedani-Golshan, T. Cuthbert Njenda, P. Siano, Wide-area measurement system-based optimal multi-stage under-frequency load-shedding in interconnected smart power systems using evolutionary computing techniques. Appl. Sci. **9**(3), 508 (2019)
11. T.C. Njenda, M.E.H. Golshan, H.H. Alhelou, WAMS based under frequency load shedding considering minimum frequency predicted and extrapolated disturbance magnitude. in *2018 Smart Grid Conference (SGC)* (IEEE, 2018), pp. 1–5
12. P. Gopakumar, M.J.B. Reddy, D.K. Mohanta, Transmission line fault detection and localization methodology using PMU measurements. IET Gener. Transm. Distrib. **9**(11), 1033–1042 (2015)
13. H.H. Alhelou, M.-E. Hamedani-Golshan, R. Zamani, E. Heydarian-Forushani, P. Siano, Challenges and opportunities of load frequency control in conventional, modern and future smart power systems: a comprehensive review. Energies **11**(10), 2497 (2018)
14. S.A. Siddiqui, K. Verma et al., Real – time monitoring of post – fault scenario for determining generator coherency and transient stability through ANN. IEEE Trans. Ind. App. **54**(1), 685–692 (2018)

15. H. Haes Alhelou, M. Esmail Hamedani Golshan, M. Hajiakbari Fini, Wind driven optimization algorithm application to load frequency control in interconnected power systems considering GRC and GDB nonlinearities. Electr. Power Compon. Syst. **46**(11–12), 1223–1238 (2018)
16. H. Haes Alhelou, M. Esmail Hamedani-Golshan, T. Cuthbert Njenda, P. Siano, A survey on power system blackout and cascading events: Research motivations and challenges. Energies **12**(4), 682 (2019)
17. H.H. Alhelou, M. Esmail Hamedani Golshan, N.D. Hatziargyriou, A decentralized functional observer based optimal LFC considering unknown inputs, uncertainties, and cyber-attacks. IEEE Trans. Power Syst. **34**(6), 4408–4417 (2019)
18. S. Wang, J. Yu, W. Zhang, Transient stability assessment using individual machine equal area criterion PART I: unity principle. IEEE Access **6**, 77065–77076 (2018)
19. J.S. Ramirez, A.B. Salas, M. Esparza et al., Frequency domain methods for accuracy assessment of wideband models in electromagnetic transient stability studies. IEEE Trans. Power Deliv. **35**(1), 71–83 (2020)
20. H.H. Alhelou, M. Esmail Hamedani Golshan, N.D. Hatziargyriou, Deterministic dynamic state estimation-based optimal lfc for interconnected power systems using unknown input observer. IEEE Trans. Smart Grid (2019)
21. T.L. Vu, K. Turitsyn, Lyapunov functions family approach to transient stability assessment. IEEE Trans. Power Syst. **31**(2), 1269–1277 (2016)
22. H.H. Alhelou, S. Jamal Mirjalili, R. Zamani, P. Siano, Assessing the optimal generation technology mix determination considering demand response and EVs. Int. J. Electr. Power Energy Syst. **119**, 105871 (2020)
23. H.H. Alhelou, M. Esmail Hamedani Golshan, T. Cuthbert Njenda, N.D. Hatziargyriou, An overview of UFLS in conventional, modern, and future smart power systems: challenges and opportunities. Electr. Power Syst. Res. **179**, 106054 (2020)
24. M. Oluic, M. Ghandhari, B. Berggren, Methodology for rotor angle transient stability assessment in parameter space. IEEE Trans. Power Syst. **32**(2), 1202–1211 (2017)
25. A.D. Rajapakse, F. Gomez, K. Nanayakkara et al., Rotor angle instability prediction using post-disturbance voltage trajectories. IEEE Trans. Power Syst. **25**(2), 947–956 (2010)
26. M. Anghel, J. Anderson, A. Papachristodoulou, Stability analysis of power systems using network decomposition and local gain analysis, in *Proceedings of IREP Symposium-Bulk Power System Dynamics and Control* (2013), pp. 978–984
27. H. Suprême, L.A. Dessaint, I. Kamwa et al., Development of new predictors based on the concept of center of power for transient and dynamic instability detection. IEEE Trans. Smart Grid. **9**(4), 3605–3615 (2018)
28. D.R. Gurusinghe, A.D. Rajapakse, Post-disturbance transient stability status prediction using synchrophasor measurements. IEEE Trans. Power Syst. **31**(5), 3656–3664 (2016)
29. J. Ying, X. Yuan, J. Hu, Inertia characteristic of DFIG-based WT under transient control and its impact on the first-swing stability of SGs. IEEE Trans. Energy Conver. **32**(4), 1502–1511 (2017)
30. J.-Y. Ruan, Z.-X. Lu, Y. Qiao et al., Transient stability of wind turbine adopting a generic model of DFIG and singularity-induced instability of generators/units with power–electronic interface. IEEE Trans. Energy Conver. **30**(3), 1069–1080 (2015)
31. S.A. Siddiqui, K. Verma et al., A unified control scheme for power system transient stability enhancement through preventive and emergency control. Int. Trans. Elect. Energy Syst. **26**, 365–383 (2015)
32. D.R. Shrivastava, S.A. Siddiquiand, K. Verma, Optimal PMU placement for coordinated observability of power system under contingencies, *in Proceedings of International Conference Circuits and Systems*, December 2017, pp. 334–339

33. P.K. Naik, N.-K.C. Nair, A.K. Swain, Impact of reduced inertia on transient stability of networks with asynchronous generation. Int. Trans. Elect. Energy Syst. **26**(1), 175–191 (2016)
34. J.L. Cremer, I. Konstantelos, S.H. Tindemans et al., Data-driven power system operation: exploring the balance between cost and risk. IEEE Trans. Power Syst. **34**(1), 791–801 (2019)

Techno-Economic Analysis of WAMS Based Islanding Detection Algorithm for Microgrids with Minimal PMU in Smart Grid Environment

R. Rohikaa Micky, R. Sunitha, and S. Ashok

Abstract With large scale deployment of non dispatchable renewable energy sources, distributed generators (DGs) have paved way for multiple microgrids. Existing standards stipulate disconnection of DGs in the event of any fault in either utility or microgrid side. However, to ensure reliable power supply from these microgrids, the proposed scheme operates an islanding detection algorithm (IDA) employed at the microgrid control center (MGCC) that acts on the switch at the utility point of common coupling (PCC). The whole microgrid is switched over to an islanded mode on detection of an islanding event. Phasor measurement unit (PMU) data from the utility PCC enables accurate islanding detection using islanding detection monitoring factor (IDMF) and rate of change of inverse hyperbolic cosecant function of voltage (ROCIHCF) along with voltage at the PCC. Mathematical morphological filters are employed to detect any persistent short circuit fault in the microgrid side which may island the DG. For such faults, decision for disconnecting the DG or islanded operation is based on probability of power balance (PoB) and probability of islanding duration in the sub-microgrids, computed at the MGCC. Further, performance of the IDA in microgrids is assessed using a proposed, microgrid performance index (MGPI) considering the uncertainties in NDRES. Suitability of the proposed indices to predict events leading to islanded operation in real time is also validated using decision tree (DT) method. The discrimination capability between islanding and other transient events of DTs, yielded an accuracy of approximately 99.9% for minimum detection time, which proves the prowess of the method in real time scenario. Compared to existing methods, the proposed method promises reduced islanding detection time. Another distinct feature is that time for islanding detection remains same irrespective of power mismatch ratios. The proposed method

R. Rohikaa Micky (✉) · R. Sunitha · S. Ashok
Electrical Engineering Department, National Institute of Technology Calicut, Calicut, Kerala, India
e-mail: rohikaaeee@gmail.com

R. Sunitha
e-mail: rsunitha@nitc.ac.in

S. Ashok
e-mail: ashoks@nitc.ac.in

H. Haes Alhelou et al. (eds.), *Wide Area Power Systems Stability, Protection, and Security*, Power Systems, https://doi.org/10.1007/978-3-030-54275-7_19

prevents false alarms for critical non-islanding events with zero non detection zone. Utilizing the proposed method, any redundant DG outage can be avoided, minimizing their down time. An economic analysis of the proposed islanding detection method using wide area monitoring system (WAMS) with minimal PMU deployment has also been studied to reduce the cost of PMU installation without sacrificing the reliability benefit for the customers.

Keywords Islanding detection · Smart grids · Microgrids · PMU · Islanding detection monitoring factor · Microgrid performance index · Smart grid

1 Introduction and Chapter Overview

Owing to rapid development of microgrids promoting non-dispatchable renewable energy sources (NDRES) worldwide due to the envisioned economic and environmental benefits, problems like voltage variations, reverse power flow, power fluctuations and power quality issues are encountered in the power system. Another such problem which arises with huge deployment of NDRES is occurrence of islanding events which if unnoticed is hazardous to service personnel, cause voltage and frequency changes and synchronization issues while reconnecting to utility grid (UG). In islanding condition, though the microgrid (MG) is electrically disconnected from the UG, the DG units remain still energized to meet the local load [1].

As per [2] DG should be discontinued from service in 2 s at the initiation of an islanding event (IE). However, this current practice is not a viable option as it disrupts reliable power supply to customers [3]. Instead, it would be pragmatic to permit islanded operation which ensure continuous supply to consumers. However, for this, safety of MG has to be studied at UG and microgrid levels. For fault in the UG, MG may be switched over to an isolated operation. An islanding detection algorithm (IDA) should have the ability to distinguish between UG and local microgrid faults [4]. During microgrid faults, islanding detection technique (IDT) needs to island only the smallest part of the faulted area as far as possible [5].

IDTs are broadly categorized into local and remote techniques. The former includes passive and active techniques. Passive IDTs do not function when there is zero power exchange between the utility and microgrid. Recent researches on passive IDT aimed to reduce non detection zone (NDZ) by utilizing rate of change of sequence components of currents [6] and mathematical morphology filters [7]. However, main shortcomings of these works are non- zero NDZ and spurious tripping during three phase faults involving ground. Also, many have not been tested for all the events occurring in the power system. Active IDTs have negligible NDZ, but it degrades power quality. Remote IDTs have null NDZ and depends on communication of data between the DGs and UG through power line carrier communication (PLCC), transfer trip and supervisory control and data acquisition (SCADA) schemes. Hybrid IDTs combine superior qualities of passive and active IDTs to reduce NDZ and power quality issues.

In the IDTs discussed previously, each distributed generator (DG) is lodged with an IDA. For faults in the MG, if healthy DGs are permitted to cater to the isolated network, blackouts may be prevented. However, control schemes in the isolated MG should be equipped with proper power sharing, frequency and voltage control algorithms. Majority of the existing techniques are based on faults in the UG side. In [8] only grid side fault is examined. Islanding was executed by opening UG circuit breaker in [9]. But, load was shed unnecessarily for minimising NDZ for low power mismatches and also the DG is disconnected for feeder loss within the MG.

Reference [10] suggests utilising a control centre while considering the whole MG as a "single DG block". Thus, efficient communication and intelligence protocols are required as an interface between UG, the islanded network and the microgrid control centre (MGCC). In this regard, [11] defined PCC as the point of coupling between utility and MG and the point of connection of utility with distributed energy resource (DER) as PoC. Therefore, it is essential to recognise every single IE so that MG may execute the isolated operation by coordinated action of inverter control and storage devices. Some of the few works which investigated possible islanded operation due to an IDT functioning from utility PCC can be found in [12, 13]. But these IDTs lacked proper definition of chosen threshold values and did not consider DG islanding situations arising due to feeder loss or faults remaining uncleared in the MG. Thus it is imperative to explore reliability of a robust IDA acting at the utility grid PCC and its potential to remove false alarms for events occurring in the UG or MG. Therefore, the work presented in this chapter proposes an IDA using passive technique and minimal number of phasor measurement units (PMUs) with reliable communication infrastructure which is presumed to be existing in the smart grid.

The remaining sections of the chapter are organised as follows. Section 2 presents an overview of the concept behind the proposed islanding detection algorithm. Implementation of the algorithm with the developed indices are discussed in Sect. 3. Result analysis for various islanding scenarios are demonstrated in Sect. 4 and in Sect. 5, an economic analysis has been performed for the proposed islanding detection with minimal PMU deployment. Real time islanding detection is carried out using decision tree technique in Sect. 6 and finally, conclusions are drawn in Sect. 7.

2 Concept of Proposed Islanding Detection Algorithm

Block diagram representation of the proposed IDA is depicted in Fig. 1. Power system parameters like voltage and current magnitudes, real, reactive and apparent powers are collected at the MGCC from PMUs. Further in MGCC, two proposed indices are calculated and the islanding detection algorithm is executed. Probability of power balance which indicates towards possible islanded operation is also estimated in the control centre. Based on the probability of power balance, microgrid performance index is obtained and this is further used for analysis of benefits accrued with regard to reduction in customer outage costs. Decision tree models are developed using

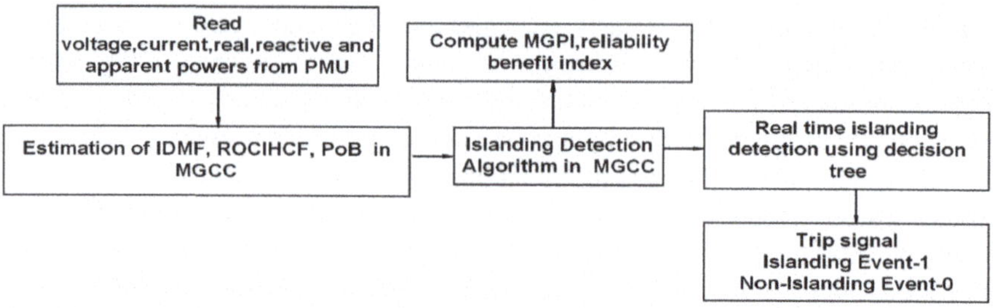

Fig. 1 Basic block diagram of the proposed IDA

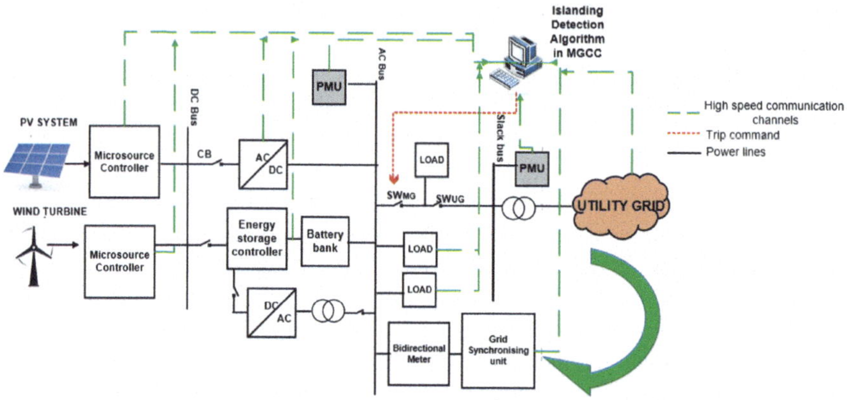

Fig. 2 Schematic representation of proposed IDA

training and test data obtained from the proposed IDA from several scenarios in the system for real time islanding detection.

Figure 2 illustrates a typical MG consisting of one wind turbine, one PV panel, local loads, converters, and controllers for the sources and the load. Here, two phasor measurement units are deployed in the system one at the reference bus of the UG and other at the PCC of the whole microgrid. Time synchronized measurements are transferred to the MGCC where decisions are taken based on the IDA. In this work, IEs are analyzed for a whole MG which has multiple sub-microgrids within it. IDA in the MGCC sends signal to MG switch (SW_{MG}), so that the whole microgrid can be managed as an autonomous unit for disconnection of UG switch (SW_{UG}) and during any persistent fault in the utility side. In islanded mode, control techniques suggested in [14] can be utilized to maintain frequency and voltage within the constraints. Switches, SW_{MG} and SW_{UG} also prevent reconnection of the unsynchronized MG with the UG. Upon detection of a loss of feeder connected to DG either the control mode has to be switched to isolated mode or the DG needs to be decoupled. It is decided depending on probability of power balance (PoB) in the sub-microgrids. Voltage and current relays at the PoC send status of DGs to the MGCC. PoB of the sub-micrograms is determined at the MGCC to command them for islanded operation. Authentic communication for two-way data exchange between utility, MGCC and PMUs is supposed to be existing in the smart grid as depicted in Fig. 2. Synchrophasor units transfer time synchronized current and voltage measurements from PCC to control centre through wide area communication technologies such as WiMAX, cellular, digital subscriber line and fibre optic networks.

Thus main features of the IDA are (1) islanding detection time is reduced as measurements are taken at the PCC of SW_{UG} instead of PoC of DG and the UG. (2) down time of sub islands can be minimised based on PoB. (3) On detection of loss of connection with the UG, the IDA switches open SW_{MG} at the PCC for islanded operation of MG. Another distinct feature of the proposed IDA is that only the fault affected DGs need to be disconnected when islanded operation is not feasible in the identified sub-microgrids. If PoB in the sub-microgrids is within the thresholds, the concerned DG along with its loads and energy units can be managed with proper reactive power/voltage control or real power/frequency control techniques in the sub-microgrids.

3 Implementation of the IDA

Here, effectiveness of the proposed IDA is demonstrated using 3 cases that might lead to islanding condition in a grid connected MG.

Case 1: IDA detects any inception of opening of utility side CB, isolators or switches. For fault in the MG, each DG is observed separately to prevent spurious tripping. PoB is monitored at each DG terminal. Cases II and III are considered which if undetected might threaten the microgrid security and reliability.

Case 2: Loss of feeders connecting DGs which might lead to DGs being disconnected from the utility.

Case 3: Permanent feeder fault in sub-microgrids which might lead to CB operation resulting in DG islanding. Control algorithm in DG switches over to autonomous mode if probability of power balance satisfies the threshold range in Cases II and III.

3.1 *Case 1*

The proposed IDA utilises two proposed indices along with voltage at the utility PCC. The indices along with their threshold and the algorithm is explained here. On islanding detection, control schemes may be adopted for islanded operation as in [14, 15].

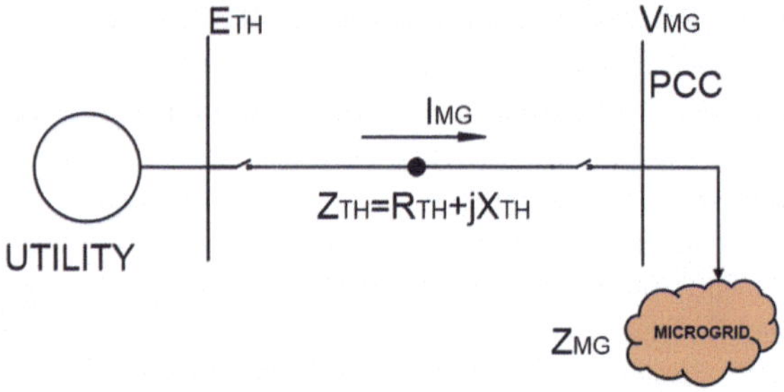

Fig. 3 Thevenin equivalent model of grid connected microgrid

A. *Islanding Detection Monitoring Factor*

The whole grid with the DGs is modelled as an equivalent Thevenin network as in Fig. 3. If utility grid voltage and microgrid impedance are E_{TH} and Z_{MG} respectively. Then,

$$E_{TH} = V_{MG} + Z_{TH} I_{MG} \tag{1}$$

Here, V_{MG} and I_{MG} denote voltage and current at utility PCC of the MG. These are measurable variables from phasor measurement units. E_{TH} and Z_{TH} denote utility grid Thevenin voltage and impedance. (1) can be written in terms of Z_{TH} as:

$$E_{THr} + jE_{TH\text{Im}} = (V_{MGr} + jV_{MG\text{Im}}) + (R_{TH} + jX_{TH})(I_{MG} + jI_{MG\text{Im}}) \tag{2}$$

E_{THr} and E_{THIm} denote real and imaginary parts of E_{TH}, V_{MGr} and V_{MGIm} represent real and imaginary parts of V_{MG}. R_{TH}, X_{TH} are resistive and reactive parts of Z_{TH} and I_{MGr}, I_{MGIm} are real and imaginary parts of current drawn by microgrid.

For a time *t*, four unknown variables namely, E_{THr}, E_{TIm}, R_{TH} and X_{TH} are solved using V_{MG} and I_{MG} and by representing (2) in matrix form as (3) [16],

$$\begin{bmatrix} 1 & 0 & -I_{MGr,1} & I_{MGIm,1} \\ 0 & 1 & -I_{MGIm,1} & -I_{MGr,1} \\ 1 & 0 & -I_{MGr,2} & I_{MGIm,2} \\ 0 & 1 & -I_{MGIm,2} & -I_{MGr,2} \end{bmatrix} \begin{bmatrix} E_{THr} \\ E_{THIm} \\ R_{TH} \\ X_{TH} \end{bmatrix} = \begin{bmatrix} V_{MGr,1} \\ V_{MGIm,1} \\ V_{MGr,2} \\ V_{MGIm,2} \end{bmatrix} \tag{3}$$

Apparent power, S_{MG} drawn by MG isiven by (4) and (5).

$$S_{MG} = 3 Z_{MG} I_{MG}^2 \tag{4}$$

If $\bar{Z}_{MG} = Z_{MG} \angle \alpha$ and $\bar{Z}_{TH} = Z_{TH} \angle \beta$

$$S_{MG} = \frac{E_{TH}^2 Z_{MG}}{Z_{TH}^2 + Z_{MG}^2 + 2 Z_{TH} Z_{MG} cos(\beta - \alpha)} \tag{5}$$

where α and β are the angles of Z_{MG} and Z_{TH} respectively.

Maximum apparent power of MG, represented as S_{MMG}, can be obtained as,

$$S_{MMG} = \frac{E_{TH}^2}{2V_{MG}I_{MG}[1 + \cos(\beta - \alpha)]} \quad (6)$$

S_{MMG} gives maximum loadability limit which depends on Thevenin parameters which vary with operating scenarios.

Islanding detection monitoring factor (IDMF) is formulated as difference between the microgrid's apparent power and its maximum apparent power and is given as,

$$IDMF = S_{MG} - S_{MMG} \quad (7)$$

$$\begin{aligned} IDMF = E_{TH}^2\big((V_{MG}I_{MG}(1 + 2V_{MG}I_{MG}) - Z_{TH}^2\big) \\ +2V_{MG}I_{MG}cos(\beta - \alpha)(Z_{TH} + (V_{MG}I_{MG}))) \end{aligned} \quad (8)$$

Thus with two bus Thevenin equivalent circuit any system parameter variations in UG and microgrid can be observed at PCC. Normally, IDMF has a high magnitude and it is network dependent. IDMF with negative sign indicates an islanding event. With increase in load, Thevenin impedance increases and under islanding conditions, IDMF shows sudden change in the value. IDMF acts a gauge for maximum loading capability for any event in the UG or MG. However, simulation studies showed that IDMF is incapable to distinguish between islanding and other transient events in case of power export especially when the power mismatches are low. Analyzing this behavior of IDMF, to make the IDA free of nuisance trippings, a new index is proposed and considered along with it.

B. *Rate of Change of Inverse Hyperbolic Cosecant Function*

Inverse hyperbolic functions involve natural logarithmic function. Even minute variations in voltage can be maximized using inverse hyperbolic function which would make IDA prompt. Compared to inverse hyperbolic secant function [17] inverse hyperbolic cosecant function (IHCF) [18] gives a more magnified variation in voltage.

IHCF of voltage is expressed as,

$$\text{IHCF} = \ln\left(\frac{1 + \sqrt{1 + V^2}}{V}\right) \quad (9)$$

where, V is rms voltage $\forall\ V \neq 0$.

Superiority of IHCF is its proportional additive changes in the variable as in (9), compared to inverse hyperbolic secant function expressed in (10).

$$\text{IHSF} = \ln\left(\frac{1 + \sqrt{1 - V^2}}{V}\right) \quad (10)$$

By taking Δt as the sampling interval, rate of change of IHCF (ROCIHCF) is given by,

$$\mathrm{ROCIHCF} = \mathrm{IHCF}/\Delta \mathrm{t} \tag{11}$$

C. *Islanding detection indicator*

Though combined use of IDMF and ROCIHCF, results in reduced NDZ and can discriminate IEs from non-islanding events, it resulted in false alarms during short circuit faults involving ground in nearby feeders. Thus, voltage is also considered along with the indices. Proposed IDA may be logically summed up as an indicator named, islanding detection indicator and is defined in (12) as,

Islanding detection indicator (IDI) =

$$\begin{aligned} &\big((IDMF < \mu) OR \big(\kappa_{Lo} \le ROCIHCF \le \kappa_{Up}\big)\big) \\ &AND \\ &\big(\big(v_{minUp} \ge V \ge v_{minLo}\big) OR (V \ge v_{Ul})\big) \end{aligned} \tag{12}$$

where μ is IDMF threshold, κ_{Lo}, κ_{Up} are lower and upper limits of ROCIHCF. v_{minUp}, v_{minLo} are the ranges for minimum values of voltage and v_{Ul} represents the maximum value of voltage for the precise distinction between IEs and non-islanding event (NIEs). IDI is either 1or 0: where one indicates an IE whereas zero indicates NIE.

Threshold values are decided through exhaustive simulations for several events and operating scenarios to prevent spurious alarms. To detect an IE, threshold values of the indices are obtained as IDMF < (−400) and 0.1 ≤ ROCIHCF ≤ 1.1. Voltage reduced drastically for three phase and line faults involving ground which were cleared after 0.1 s on initiation of the fault. However, it is a transient condition and should not be misinterpreted as an IE. Therefore, minimum value of voltage should lie in the range of 0.86 p.u to 0.8 p.u. This constraint eliminates false alarms and meets the standards of voltage constraint of 0.85 p. u and 1.1p.u. In the system studied, during power export, it was noticed that voltage at the PCC bus maintained a voltage near to 1.0 p.u during islanding conditions. Therefore, here, in order to recognize IEs during zero, and other different power mismatch ratios, upper threshold of voltage is fixed at 1.0 p.u. All the indices are computed in every half cycle.

3.2 Case 2: Loss of Feeder Connected to DG in MG

Any feeder loss connected to a DG causes its separation from the utility and therefore needs to be monitored. If DG's output power matches with the load powers in the sub-island, DG need not be disconnected and can have an islanded operation within the system voltage and frequency constraints. However, in this situation resulting in minute changes in voltage and frequency cannot be detected by voltage and frequency relays. In this scenario, current relays at the DG terminals can be utilised to communicate to the control center. Then, at the MGCC, PoB in that sub-microgrid is compared to the threshold as would be explained in Sect. 6. For islanded management of the sub-microgrid the transition time to isolated management mode should be in the time span for which the power balance is sustained in the islanded network. Else, the DG should be tripped.

3.3 Case 3: Persistent Fault in MG

A persistent fault in any feeder in MG may result in CB operation leading to DG tripping. Therefore, such faults should be detected faster. For early detection of persistent faults within the MG, concept of mathematical morphological filters (MMF) is adopted for time series analysis of fault current. Dilation and erosion operators for fault current with structural element (SE) of length and height of 3 and 0.01 respectively are given by (13) and (14).

Dilation operator for fault current, $I_f(k)$ [19] is given as,

$$\left(I_f \oplus SE_n\right)(k) = max\left\{I_f(k-s) + SE(s)\right\} \tag{13}$$

Erosion operator for $I_f(k)$ is given by (14) as,

$$\left(I_f \ominus SE_n\right)(k) = min\left\{I_f(k+s) - SE(s)\right\} \tag{14}$$

where $I_f(k)$ and SE(k) represent the function of the fault current and structural element with SE_n = [0.01, 0.01, 0.01]; k and s are integers with k = (1, 2 … N), N being number of samples in the input signal, s = (1, 2 … m) denoted the elements in the SE and m < N, m being length of SE.

Average of the dilation and erosion operators gives the dilation and erosion median filter (DEMF) and is given as,

$$DEMF_n(k) = \frac{1}{2} * \left(I_f \oplus SE_n + I_f \ominus SE_n\right) \tag{15}$$

where n = (1,2, …, m).

Then ΔI_f is the difference of $DEMF_n$ from $I_f(k)$ and is expressed as,

$$\Delta I_f(k) = I_f(k) - \left(\frac{DEMF_1(k) + DEMF_2(k) + \cdots + DEMF_m(k)}{m}\right) \tag{16}$$

Thus fault detector is defined as the absolute value of the difference between ΔI_f at every consecutive intervals given by (17),

$$\Delta DEMF_n(k) = \left|\Delta I_f(k+1) - \Delta I_f(k)\right| \tag{17}$$

From the simulation studies it has been concluded that a persistent short circuit fault is identified when ΔDEMF exceeds 111% of its nominal value for more than 6 cycles. Then the DG can be initiated for an islanded operation satisfying the PoB constraints. Depending on PoB, DG may be disconnected or isolated. This also avoids spurious disconnection for transient short circuit faults. Flow diagram of the proposed IDA is shown in Fig. 4. System parameters read from PMUs installed in two buses are used to calculate the indices. If IDI is 1, an IE is detected to manage the whole microgrid for an islanded operation. At the same time, IDA checks for any loss of feeder in the MG and feeder current is monitored. A trend of current magnitude prone to zero is communicated to MGCC by current relay. Any persistent short circuit fault is recognized rapidly by the fault detector operator given by (12) and communicated to the MGCC. For case 2 and case 3, islanded operation is decided based on PoB in the sub-microgrids.

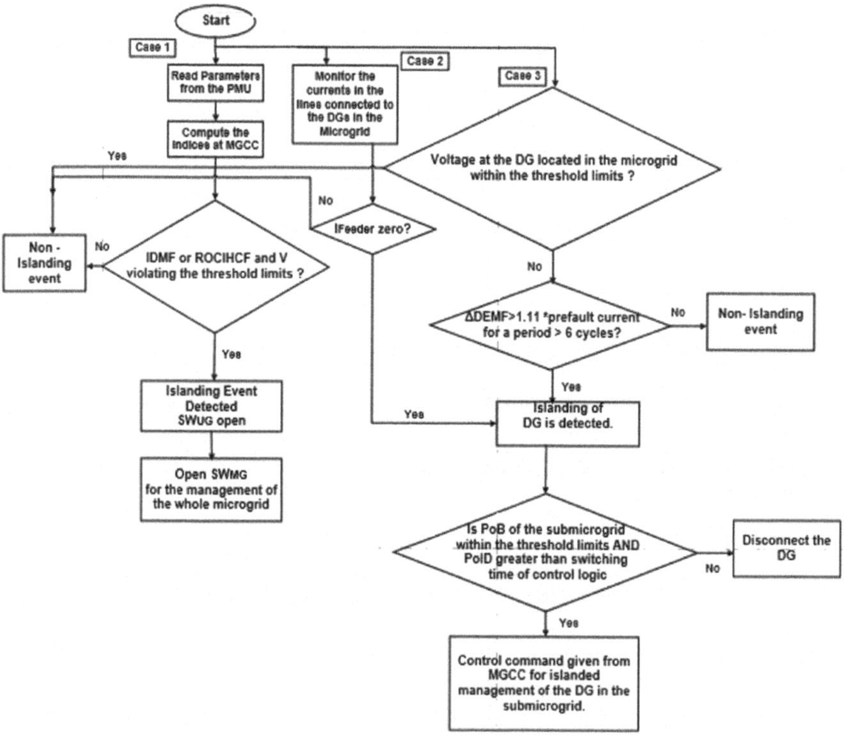

Fig. 4 Functional flowchart depiction of the proposed IDA

4 Result Analysis

System [7] shown in Fig. 5 is a balanced network of 120 kV with 2 step down transformers of rating 120 kV/25 kV. Three distribution feeders spread from 25 kV bus, which acts as the utility PCC. The DG system consists of two 9 MW wind farms at 575 V, 60 Hz with six wind turbines of 1.5 MW and 1 synchronous diesel engine with Woodward governor model of 2.5 MW,3.125 MVA,575 V,60 Hz. Loads are modelled as constant impedance models. Rated short circuit MVA is 2500 MVA. The whole microgrid can be considered as a superset of networked sub-microgrids. Three sub-microgrids are microgrid 1, microgrid 2 and microgrid 3 as represented in Fig. 5. The sampling rate is 3.6 kHz at 60 Hz nominal frequency. The simulation period is 1 s. Modelling of test system and simulations are conducted in PowerFactory and IDA in MATLAB 2012b in an Intel Core i5 3470, 8 GB RAM, 3.2 GHz processor.

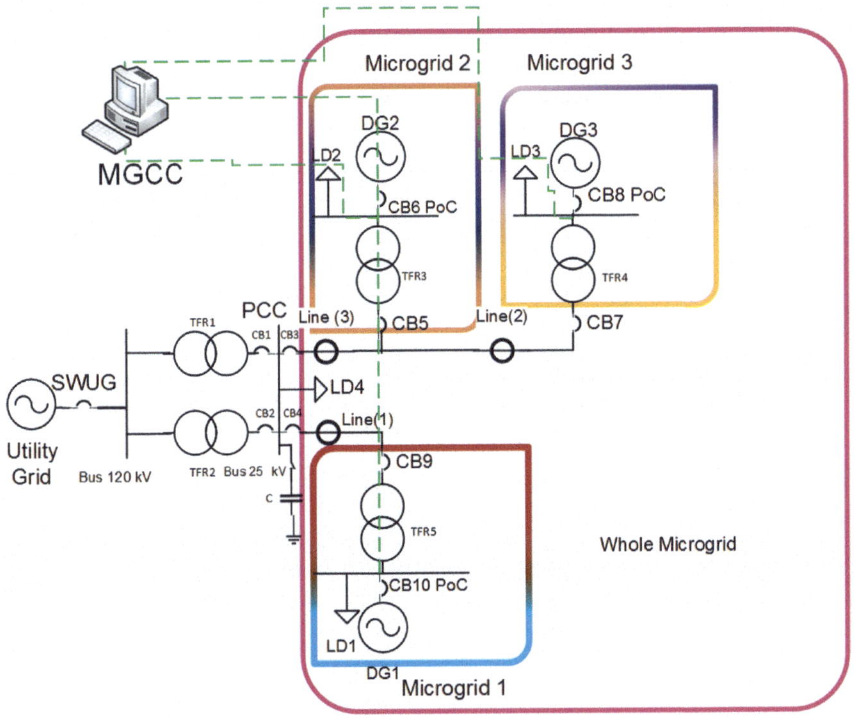

Fig. 5 Test system under study

4.1 Case 1

For IE such as opening of CB1 and CB2 or SW_{UG}, a trip command is given to island the whole MG at t = 0.5 s. Simulation studies have been carried out while importing active power by keeping DG generation levels fixed and varying the feeder load. Zero active power mismatch is observed at 40% feeder load. 5, 10 and 20% of power mismatches are observed when load is 50%, 60% and 70% of its nominal value. The IDA is able to identify the islanding conditions for the selected threshold values of ROCIHCF. Magnitude of ROCIHCH increases on islanding. IDMF also reduces below its threshold for all the power mismatches. Figure 6 shows change in voltage at the PCC for various for the above IEs. It can be observed that for these events voltage infringes the threshold. It may be noted that, voltage too falls between 0.86 and 0.8 along with IDMF and ROCIHCF indicating an IE. All the above IEs performed are tabulated in Table 1.

If IDMF or ROCIHF violates its threshold constraints and if voltage too breaches its constraints, trip command is signalled to the whole MG as IDI is 1 for all the power scenarios. From Table 1, it is concluded that islanding detection time is 5 ms irrespective of power mismatch ratios. The negative power magnitudes in the table

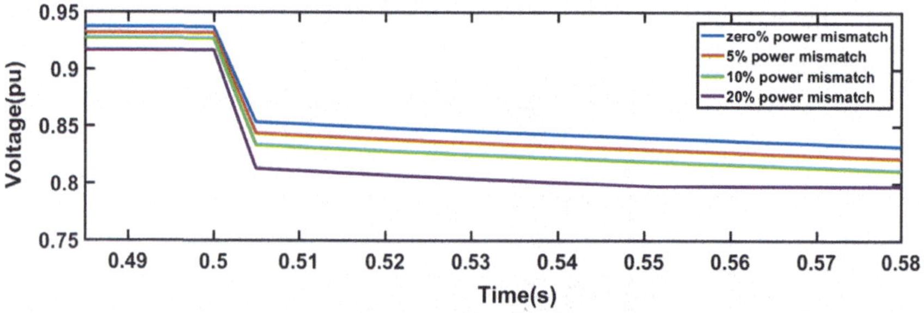

Fig. 6 Voltages at PCC for various active power mismatches

Table 1 Parameters for islanding condition at different power mismatches

Active power mismatch (%)	Detection Time (ms)	IDMF	ROCIHCF	V(p.u)	IDI
−20	5	−464157	0.15	1	1
−10	5	−53793	0.07	1	1
−5	5	−6975	0.03	1	1
0	5	−2428	0.30	0.85	1
5	5	−2086	0.32	0.84	1
10	5	−1652	0.33	0.83	1
20	5	−985	0.36	0.81	1

Table 2 IDI for cases 2 and 3 in the microgrid

Line connected	IDI for case 2	IDI for case 3	IDI for case 3
Line1	1	0	0
Line2	0	1	1
Line3	0	0	1

indicate export of power. The chosen threshold ensures minimum islanding detection time. Likewise, simulation studies have been done for reactive power export scenarios.

4.2 *Case 2*

Loss of Line 1, Line 2 or Line 3 leading to open conditions of CB6, CB8 or CB10 respectively are considered here. Column 2 of Table 2 indicates islanding detection index at PoC for loss of Line 1. For these faults, IDI at the PCC would be zero. On detecting Line 1 current to be zero, protective relay at the DG terminal signals to MGCC for islanded management. For loss of Line1 connected to microgrid 1, IDI is high only for Line1. Current and voltage are within their threshold limits for other lines as per the algorithm. Thus sub-microgrids 2 and 3 remain intact and PoB of sub-microgrid 1 decides its further operation.

4.3 *Case 3*

This case is evaluated for two scenarios and tabulated as columns 3 and 4 in Table 2. It is observed in column 3 that IDI is 1 for Line 2 outage. For the other two lines IDI is zero. Last column indicates IDI for simultaneous outage of Line 2 and Line 3.

Figure 7 shows current in Line1 normalized with respect to its peak value. A 3 phase fault in Line1 at 0.5 s is shown in Fig. 8a. MMF is applied to the normalized current as explained before. Then, DEMF and ΔDEMF are computed. It may be seen that ΔDEMF is exceeding 0.1 pu which is 111% greater than its nominal value for more than 0.1 s. However, for a fault which is cleared in 0.1 s depicted in Fig. 8b, shows that ΔDEMF exceeds the threshold at the fault initiation but thereafter lowers to 0.099 p. u. when fault is cleared. Thus IE can be detected even before circuit breaker opens. Further examination of ΔDEMF output showed that it remained under the threshold for other oscillatory events in the MG. Thus the MMF needs to be functioning for each DGs existing in the grid.

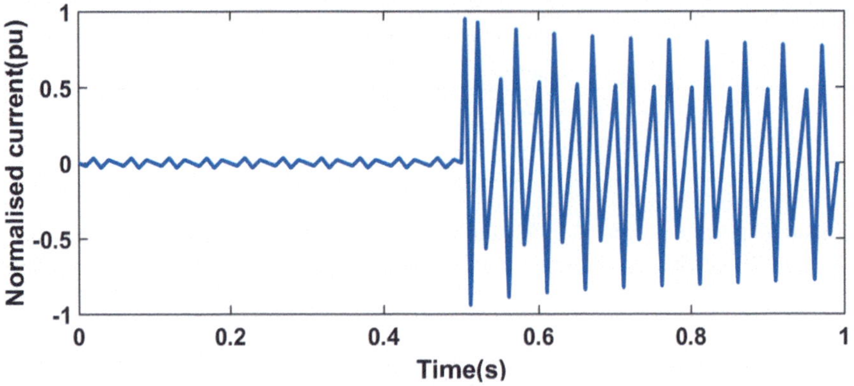

Fig. 7 Normalized current in Line1

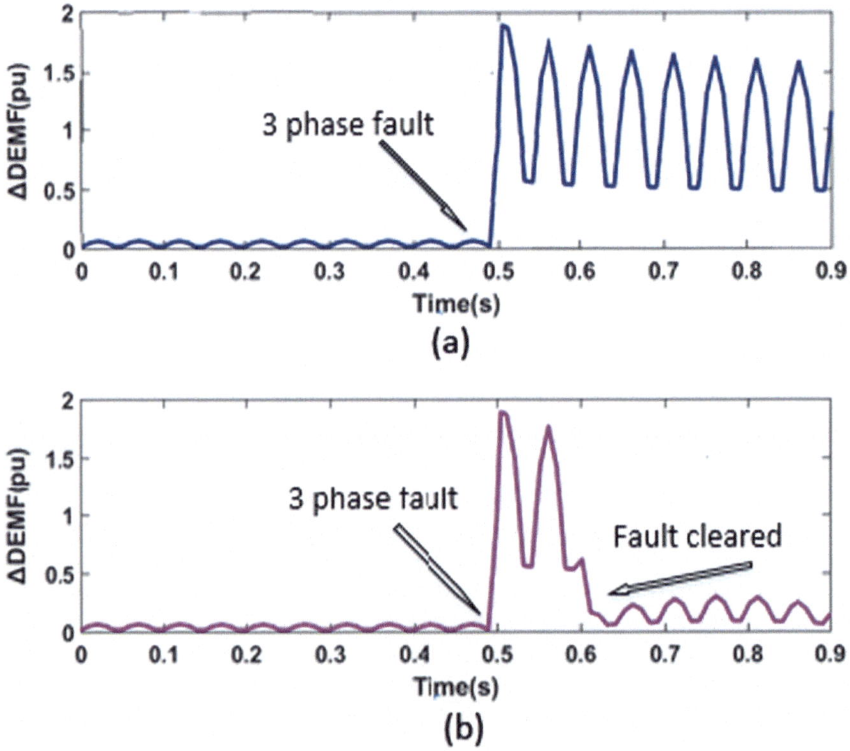

Fig. 8 Current in Line1 **a** 3 phase persistent fault; **b** 3 phase temporary fault

4.4 Performance Evaluation of the IDA for Transient Events

Transient events such as transformer and capacitor switching, network topology change, parallel feeder loss, and connection/disconnection of single and multiple DGs have been simulated to evaluate robustness of the IDA for non-islanding events. Figure 9a depicts variation in voltage for different DG disconnections. The figure confirms that voltage satisfies the threshold range for each case. Capacitor bank switching to adjust the power factor can give rise to voltage swing. This phenomenon may be misinterpreted as an IE. To check stability of the IDA, a capacitor of 8.5 μF connected at the PCC is stepped up to 850 μF in steps of three. Figure 9b proved IDA's prowess to distinguish IE from transient events.

Islanding detection techniques usually misconstrue three-phase fault involving ground in nearby feeders as IEs [7]. Robustness of the IDA in eliminating false alarms for faults which are cleared at different locations along the feeder length, each with different values of fault resistance have been evaluated. The algorithm did not create false alarms during these scenarios. Figure 10 shows that, for single line to ground (LG faults in line (2) and line (3)) and three phase faults, voltage reduces extremely. However, an event shall be classified as islanding only if magnitude of voltage is between 0.86 and 0.8 p.u.

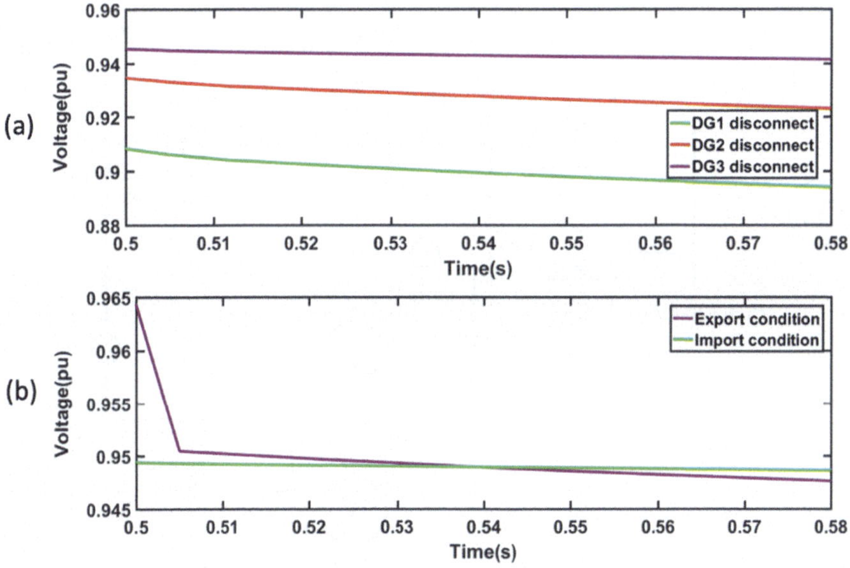

Fig. 9 PCC voltage in MG **a** DG disconnection; **b** Capacitor bank switching during export/import conditions

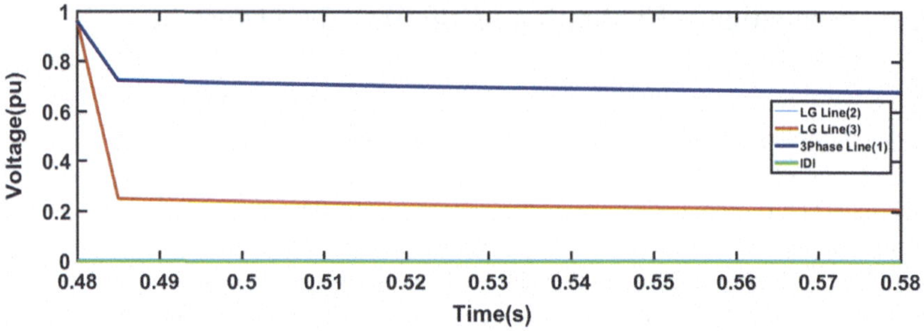

Fig. 10 PCC voltage for various faults

For analysing the performance during change in network configuration, tie-switch SW1 which is normally open, is closed as shown by red ellipse in Fig. 11. IDI remains zero and hence the algorithm precisely identified it as a non-islanding event.

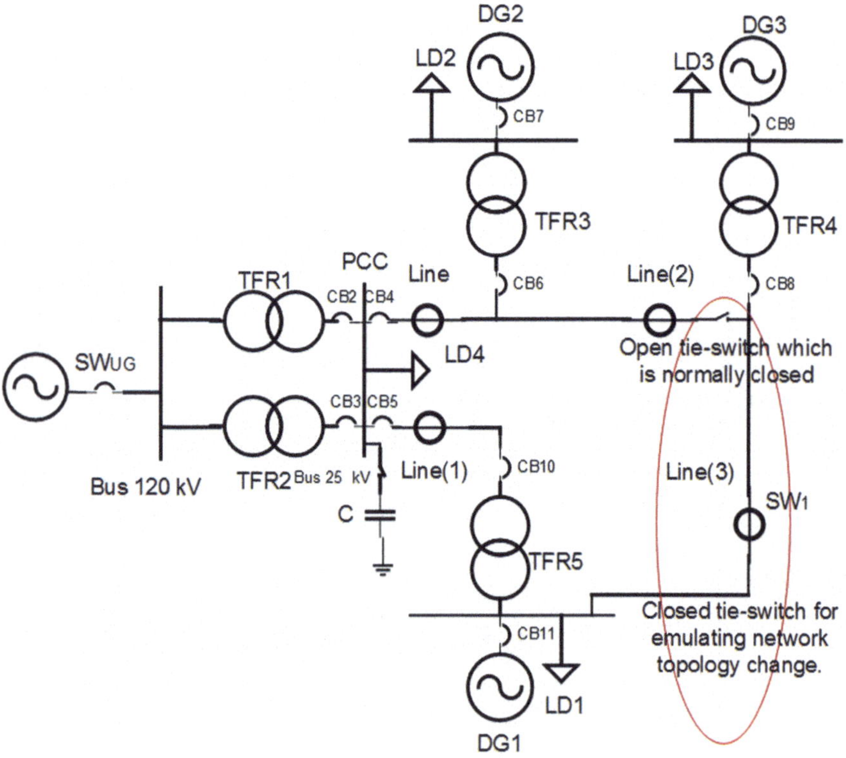

Fig. 11 Microgrid depicting network topology change in DIgSILENT/PowerFactory

5 Economic Analysis of the Proposed IDA with Minimal Placement of PMUs

As per [20] reducing number of buses where PMUs are deployed does not lead to significant decrease in its deployment cost. However, the authors have performed an economic analysis with deployment of two PMUs with respect to number of sub-microgrids going to islanded operation instead of being disconnected and analysed the reduction in cost of consumer outage with their islanded operation. Thus here minimal PMU placement refers to the two PMUs deployed at the PCC of the whole MG with the UG and the reference bus in the UG. When load and generation in the microgrid matches, at the instant of disconnection from utility, an islanded operation of the DG is possible [21]. Hence it becomes necessary for a probabilistic analysis considering the intermittent and uncertain nature of NDRES in microgrid. To avoid redundant DG disconnection in the sub-microgrids, probability of power balance (PoB) is introduced [22].

5.1 Probability of Power Balance and Microgrid Performance Index

PoB is defined as the probability density function of the ratio of total power demand to the total power generation in the islanded sub-microgrid. PoB in sub-microgrids 1, 2 and 3 can be expressed by (18)–(20). Sub-microgrid 1 consists of 9 MW wind plant and load of 8 MW, sub-microgrid 2 represents 9 MW wind plant with 2.5 MW load. A 2.5 MW diesel generator feeding 8 MW load constitute sub-microgrid 3. The sources and loads are assumed to maintain a power factor of 1.

$$PoB_1 = PDF1\left[\frac{\sum_{i=1}^{N} PL_{SMG1} + jQL_{SMG1}}{P_{SMG1} + jQ_{SMG1}}\right] \tag{18}$$

$$PoB_2 = PDF2\left[\frac{\sum_{i=1}^{N} PL_{SMG2} + jQL_{SMG2}}{P_{SMG2} + jQ_{SMG2}}\right] \tag{19}$$

$$PoB_3 = PDF3\left[\frac{\sum_{i=1}^{N} PL_{SMG3} + jQL_{SMG3}}{P_{SMG3} + jQ_{SMG3}}\right] \tag{20}$$

where PDF1, PDF2 and PDF3 are probability density functions with a normal distribution as given in (21) for sub-microgrids 1, 2 and 3 respectively.

$$\text{PDF} = \frac{1}{\sigma\sqrt{2\pi}} * e^{\frac{-(A-\mu)^2}{2\sigma^2}} \tag{21}$$

where A defines ratio of output power of DG to the load, μ and σ denote mean and standard deviation respectively.

PL_{SMG}, QL_{SMG} denote real and reactive power demands in the microgrid and P_{SMG}, Q_{SMG} are the real and reactive power generation from the sources in the microgrid.

Islanded operation is possible when PoB lies between 0.95 and 1.5 to accommodate any NDZ [22]. Lower threshold of 0.95 is adopted to include the situation where generation exceeds load. The upper value accommodates system losses in the sub-microgrids. PoB threshold is then chosen as,

$$0.95 < \text{PoB} < 1.5 \tag{22}$$

Time span during which the island satisfies the PoB limits can be defined as duration of probability of operation (PoID) in islanded mode,

$$PoID = n * t \tag{23}$$

where n represents instants of time when PoB satisfies the constraints, t being sampling rate. Time required to change from grid tied to islanded mode for converters has been approximated to be as under 640 ms as per [14].

Sampling rate is 10 ms with delay in communication assumed as 20 ms [14]. PoB obtained for sub-microgrids 1 and 2 are illustrated in Fig. 12a. Relay senses and communicates to the MGCC when CB5 connecting to sub-microgrid 2 is opened. As PoB is not satisfying the threshold, instantaneous measure would be to detach the wind plant. To evaluate the proposed IDA for IEs in the microgrid an index is formulated. Microgrid Performance Index (MGPI) is the ratio of number of autonomous sub-microgrids on islanding detection to number of IEs identified at the PoC.

$$MGPI = \frac{\sum_{i=1}^{n} N\_SMG_i}{\sum_{j=1}^{m} W_j} \tag{24}$$

where *N_SMG* is number of autonomous sub-microgrids determined based on PoB, n is total number of sub-microgrids, Wj represents total number of breaker openings and m denotes number of PoCs.

Thus MGPI yields the following conditions,

$$MGPI = \begin{cases} > 1; forsuccessfulsubmicrogridmanagement \\ 0; otherwise \end{cases} \tag{25}$$

PoB for sub-microgrid 1 is 0.96 when CB9 is opened at 0.52 s. It falls inside PoB limits. PoID has been found to be 60 ms as shown in Fig. 12b. Likewise, for circuit breaker openings of CB5, CB7 and CB9, two sub-microgrids could be managed in islanded mode out of the three sub-microgrids in the test system considering the PoB.

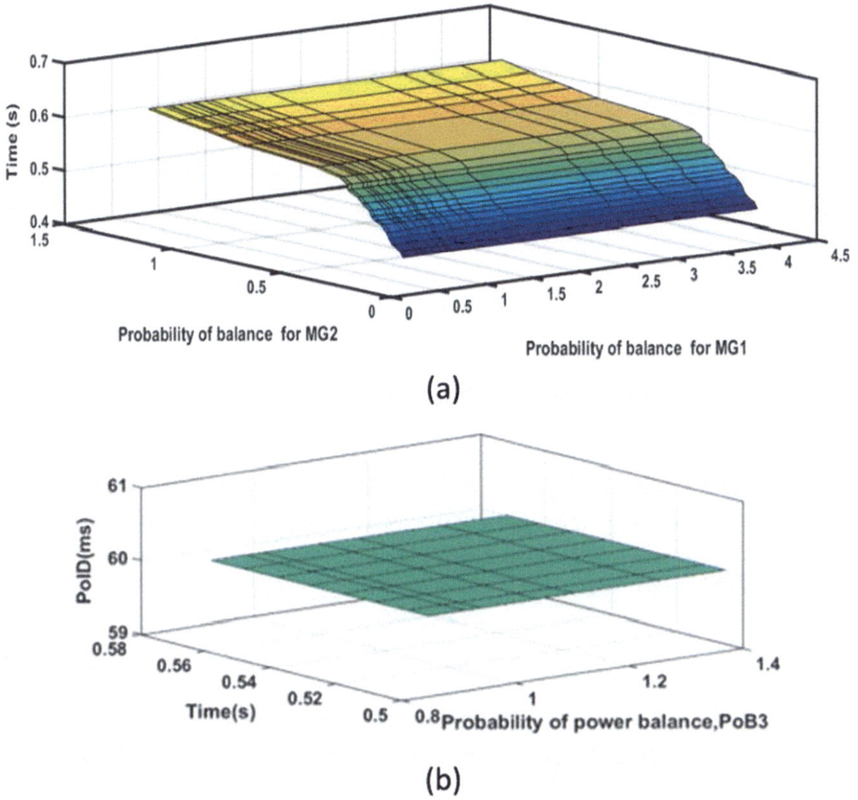

Fig. 12 **a** PoB of sub-microgrid 1, sub-microgrid 2; **b** PoID with respect to islanding time and PoB of MG1

DG in sub-microgrid 2 only had to be disconnected. This results in an MGPI value of two, stipulating proper microgrid utilization instead of disconnecting all the DGs.

5.2 Reliability Benefit for Microgrid Under Islanded Operation

A lower MGPI indicates increased average energy not supplied (AENS) by the sub-microgrid. Thus, to assess economic benefit for microgrid with the proposed IDA, reliability benefit is defined as the reduction in cost of consumer power outage. It also involves reduction in installation cost of PMU at each bus. Thus reliability benefit with islanded sub-microgrids (RB_{ISMG}) is expressed as,

$$RB_{ISMG} = COC_r + CPMU_r \tag{26}$$

where COC_r is reduced customer outage cost and $CPMU_r$ is reduced PMU installation cost expressed as in (27) and (29)

$$COC_r = (COC - COC_{ISMG} * t_d \quad (27)$$

COC and COC_{ISMG} refer to customer outage cost when all the sub-islands are disconnected and customer outage cost for sub-microgrid disconnected based on MGPI.

Where

$$COC = \sum_{j=1}^{l} RIE_j * ENS_j \, \$/t_d \quad (28)$$

$$L = l - N\_SMG$$

RIE is rate of interrupted energy \$/KWh, ENS is energy not supplied KWh/t_d, t_d is duration of islanded operation assumed as 1 h for the study.

Reduced PMU installation cost is defined as,

$$CPMU_r = CPMU - CPMU_{ISMG} \quad (29)$$

$CPMU$ and $CPMU_{ISMG}$ refer to PMU installation cost at all the buses and PMU installation cost in two buses as envisaged in the proposed IDA respectively.

where

$$CPMU = \sum_{i}^{b} C_{iPMU} * x_i \quad (30)$$

C_{iPMU} is PMU installation cost at bus i, $x_i = 1$; if PMU deployed in bus i.

RIE for load is assumed as 1000 \$/KWh and PMU installation cost is 40,000 \$ [23].

For the case considered, an economic analysis has been carried out using the proposed reliability benefit index. In Table 3, COC_r gives the economic benefit when all three sub-microgrids are disconnected versus two and one possible sub-microgrids that can be autonomously managed during the islanded duration. $CPMU_r$ indicates the difference in installation cost when PMUs are installed at all the buses and PMUs installed at two buses as proposed in this work. Thus it may be concluded that, as

Table 3 Reliability benefit index for microgrid

MGPI	COC_r (10^4\$)	$CPMU_r$ (10^4\$)	$\boldsymbol{RB_{ISMG}}$(10^4\$)
2	10,240	16	1040
1	512	16	528

the number of islanded sub-microgrids increases or as MGPI increases, reduction in customer outage costs increase profoundly. Moreover, minimal placement of PMUs minimizes total cost associated with minimize a total cost associated with PMU failures and cost of its restoration and cost of communication links between PMUs.

6 Decision Tree Technique for Islanding Detection in Real Time

For quick islanding detection in real time, decision tree (DT) technique has been adopted. Initially, this study has used 50% of the data as training and remaining 50% as test set using percent split validation method. The two indices and voltage at the utility PCC are used as attributes with islanding detection index (IDI) as target obtained from offline testing and given as input to SPSS®IBM statistics for building DT model. Predictive scoring for real time analysis is performed using the DT model built. For IE at zero power balance situation, out of 5043 instants, 4038 non islanding cases are identified as NIE and one non islanding case has been misclassified as IE. However, all the 1004 islanding cases are classified as IEs as shown in Table 4. To compare the model predictions with the actual data, a correlation table is built as in Table 5. From the table, it may be observed that, a correlation of 0.984 between the model IDI and the predicted IDI for the real data using Pearson coefficient indicates a very high positive correlation between the actual and predicted islanding index. This also substantiates the selection of the threshold value. Thus the model is successful with a reliability of 100% with an accuracy of 99.97%.

Table 4 Classification table

Predicted				
Data	Observed	Non islanding	Islanding	%correct
	Non islanding	4038	1	99.97
	Islanding	0	1004	100
	Overall Percentage	80.1%	19.9%	100

Table 5 Correlation table of a real event with simulated DT model

Correlation		IDI	Predicted IDI
IDI	Pearson	1	0.984
Predicted IDI	Pearson	0.984	1

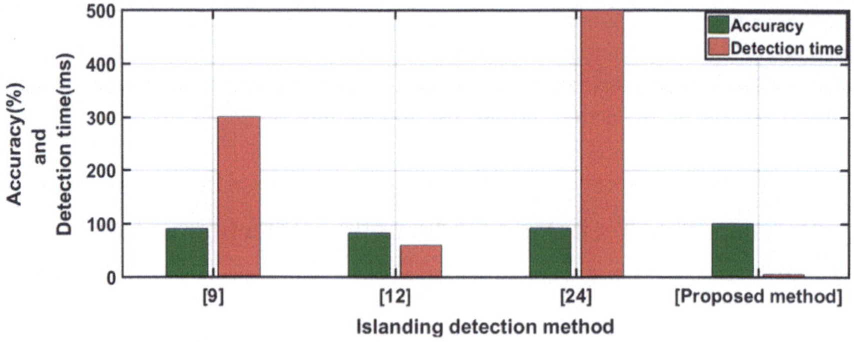

Fig. 13 Comparison of accuracy and detection times of the proposed method with contemporary works

7 Comparison of the Proposed IDA with Recent Works

The proposed IDA_{UPCC} has been compared with recently published schemes and presented in Fig. 13. Among these [9] have IDTs deployed at the PCC as well as the PoC of each DG terminal while Ref. [12] has the IDT deployed in the PCC. Islanding detection algorithm in [24] works at each DG terminal using an ANN classifier based on transient response data from synchrophasor units. It is observed from the figure that the proposed IDA has better features with faster response and higher accuracy compared to others. Reference [24] has the lowest response time with 0.5 s while IDA_{UPCC} detects islanding event in 5 ms.

As a future work and research gaps, the applications of PMU and wide-area monitoring systems in modern power systems operation, control, stability, protection and security should be well-investigated [25–30]. PMUs are the backbone of wide are monitoring system that can improve the overall security and stability of power systems. Therefore, it is suggested to make in-depth evaluation of the power system security under the new environment control based one PMUs data. Likewise, new WAMS-based methods are in demand for improving the islanding detection in future smart grids [31–33].

8 Conclusion

A reliable and accurate islanding detection algorithm is proposed here with minimal PMU deployment and maximum reduction in customer outage cost. The IDA works from the utility PCC with sturdy communication infrastructure, instead of being implemented at the DG terminals. Islanding detection monitoring factor, rate of change of inverse hyperbolic cosecant function of voltage are utilized along with the magnitude of voltage at utility PCC to make the whole microgrid manage an islanded operation. Distinctive features of the IDA are its zero non-detection zone, islanding detection time of 5 ms irrespective of percentage of power mismatches and ability to reduce false alarms. Thus the IDA with two PMUs deployed in the system prevents redundant DG disconnection for fault events in the sub-microgrids. For persistent fault in the sub-microgrid, probability of power balance in the sub-microgrids determines whether the DG has to be disconnected or considered for islanded operation. Economic analysis for reduction in customer outage costs has been performed which indicates greater reliability benefits with higher microgrid performance index. Real time DT testing proved robustness of the proposed IDT with hundred percent accuracy.

References

1. B. Matic-Cuka, M. Kezunovic, Islanding detection for inverter- based distributed generation using support vector machine method. IEEE Trans. Smart Grid. **5**(6), 2676–2686 (2014)
2. IEEE standard for interconnecting distributed resources with electric power systems – amendment 1, IEEE Std 1547a-2014 (Amendment to IEEE Std 1547–2003) (2014)
3. C. Seung-Tae, Real-time analysis of an active distribution network - coordinated frequency control for islanding operation, Dissertation for the Doctoral Degree, Technical University of Denmark (2012)
4. G. Buigues, A. Dysko, V. Valverde, I. Zamora, E. Fernandez, Microgrid protection: technical challenges and existing techniques, in *Proceedings of International Conference Renewable Energies Power Quality*, Bilbao, Spain, Mar 2013, pp. 222–227
5. S. Chowdhury, S.P. Chowdhury, P. Crossley, *Microgrids and Active Distribution Networks* (IET Renewable Energy Series 6) (Institute of Engineering Technology, London, U.K., 2009)
6. K. Sareen, B.R. Bhalja, R.P. Maheshwari, Universal islanding detection technique based on rate of change of sequence components of currents for distributed generations. IET Renew. Power Gener. **10** (2), 228–237 (2016)
7. M.A.Farhan, K.S Swarup, Mathematical morphology-based islanding detection for distributed generation. IET Gener. Transm. Distrib. **10**(2), 518–525 (2016)
8. R. Nale, M. Biswal, N. Kishor, A Transient component based approach for islanding detection in distributed generation. IEEE Trans. Sustain. Energy (2018). https://ieeexplore.ieee.org/document/8424028
9. A.G. Abd-Elkader, S.M. Saleh, M.B. Magdi Eiteba, A passive islanding detection strategy for multi-distributed generations. Electr. Power Energy Syst. **99**, 146–155 (2018)
10. O. Palizban, K. Kauhaniemi, J.M. Guerrero, Microgrids in active network management- Part II. Renew. Sustain. Energy Rev. **36**, 440–451 (2014)
11. IEEE standard for Interconnection and Interoperability of Distributed Energy resources with Associated Electric Power Systems Interfaces, IEEE Std. 1547 (2018)
12. C.N. Papadimitriou,V.A. Kleftakis, N.D. Hatziargyriou, A novel islanding detection method for microgrids based on variable impedance insertion. Electr. Power Syst. Res. **121**, 58–66 (2014)
13. M. Vatani, M.J. Sanjari, G.B. Gharehpetian, Islanding detection in multiple-DG microgrid by utility side current measurement. Int. Trans. Electr. Energy Syst. **25**(9), 1905–1922 (2015)
14. D. Shi, R. Sharma, Y. Ye, Adaptive control of distributed generation for microgrid islanding, in *4th IEEE PES Innovative Smart Grid Technologies Europe (ISGT Europe)* (Copenhagen, 2013), pp. 1–5
15. M.A. Hossain, H. Pota, M.J. Hossain, A.M.O. Haruni, Active power management in a low-voltage islanded microgrid. Int. J. Electr. Power Energy Syst. **98**, 36–48 (2018)
16. K. Vu, M. Begovic, D. Novosel, Use of local measurements to estimate voltage stability margin. IEEE Trans. Power Syst. **14**(3), 1029–1035 (1999)

17. K. Sareen, B.R. Bhalja, R.P. Maheshwari, Islanding detection technique based on inverse hyperbolic secant function. IET Renew. Power Gener. **10**(7), 1002–1009 (2016)
18. M. Trott, Inverse trigonometric and hyperbolic functions. In Mathematica Guide book for programming, 2004 ed. (New York, Springer-Verlag, 2004)
19. S. Gautam, S.M. Brahma, Detection of high impedance fault in power distribution systems using mathematical morphology. IEEE Trans. Power Syst. **28**(2), 1226–1233 (2013)
20. A. Pal, A.K.S. Vullikanti, S.S. Ravi, A PMU placement scheme considering realistic costs and modern trends in relaying. IEEE Trans. Power Syst. **32**(1), 552–561 (2017)
21. R. Bründlinger, B. Bletterie, Unintentional islanding in distribution grids with a high penetration of inverter-based DG: Probability for islanding and protection methods, in *Proceedings of IEEE Russia Power Technology Conference*, St. Petersburg, Russia, 2005, pp. 1–7
22. R. Rohikaa Micky, A. Sankar, R. Sunitha, A combined islanding detection algorithm for grid connected multiple microgrids for enhanced microgrid utilization. Int. Trans. Electr. Energy Syst. e12232 (2019). https://doi.org/10.1002/2050-7038.12232
23. U.S. Department of Energy, Office of Electricity Delivery and Energy Reliability, Factors affecting PMU installation costs (Oct 2014). https://www.smartgrid.gov/files/PMU-cost-study-final10162014_1.pdf
24. D. Kumar, P. Sarathee Bhowmik, Artificial neural network and phasor data based islanding detection in smart grid. IET Gener. Transm. Distrib. **12**(21), 5843–5850 (2018)
25. H.H. Alhelou, M.E. Hamedani-Golshan, R. Zamani, E. Heydarian-Forushani, P. Siano, Challenges and opportunities of load frequency control in conventional, modern and future smart power systems: a comprehensive review. Energies **11**(10), 2497 (2018)
26. H.H. Alhelou, M.E. Golshan, J. Askari-Marnani, Robust sensor fault detection and isolation scheme for interconnected smart power systems in presence of RER and EVs using unknown input observer. Int. J. Electr. Power Energy Syst. **99**, 682–694 (1 Jul 2018)
27. H. Haes Alhelou, M.E. Hamedani Golshan, M. Hajiakbari Fini, Wind driven optimization algorithm application to load frequency control in interconnected power systems considering GRC and GDB nonlinearities. Electr. Power Compon. Syst. **46**(11–12), 1223–38 (21 Jul 2018)
28. H.H. Alhelou, M.E. Golshan, N.D. Hatziargyriou, Deterministic dynamic state estimation-based optimal lfc for interconnected power systems using unknown input observer. IEEE Trans. Smart Grid (9 Sep 2019)
29. H.H. Alhelou, M.E. Golshan, N.D. Hatziargyriou, A decentralized functional observer based optimal LFC considering unknown inputs, uncertainties, and cyber-attacks. IEEE Trans. Power Syst. **34**(6), 4408–4417 (2019). (May 13)
30. H.H. Alhelou, M.E. Golshan, T.C. Njenda, N.D. Hatziargyriou, An overview of UFLS in conventional, modern, and future smart power systems: challenges and opportunities. Electr. Power Syst. Res. **1**(179), 106054 (2020). (Feb)
31. R. Zamani, M.-E. Hamedani-Golshan, H. Haes Alhelou, P. Siano, H.R. Pota, Islanding detection of synchronous distributed generator based on the active and reactive power control loops. Energies **11**(10), 2819 (2018)
32. R. Zamani, M.E.H. Golshan, H.H. Alhelou, N. Hatziargyriou, A novel hybrid islanding detection method using dynamic characteristics of synchronous generator and signal processing technique. Electr. Power Syst. Res. **175**, 105911 (2019)
33. A. Abyaz, H. Panahi, R. Zamani, H. Haes Alhelou, P. Siano, M. Shafie-khahand, M. Parente. An effective passive islanding detection algorithm for distributed generations. Energies **12**(16), 3160 (2019)

Independent Estimation of Generator Clustering and Islanding Conditions in Power System with Microgrid and Inverter-Based Generation

Edgar Gómez, Ernesto Vázquez, Nohemí Acosta, and Manuel A. Andrade

Abstract The use of phasor measurement units (PMU) allows us to obtain synchronized measurements of various points in the network and whit them analyze the stability of power systems. This chapter presents an algorithm based on participation factors to estimate generator clustering and to evaluate its application on controlled islanding on a power system, with distributed generation, using the data from PMUs after a severe disturbance. The proposed islanding detection method uses the data obtained from PMUs to represent the dynamics of the entire power system and form a measurement matrix, updated using a sliding window, containing the angles of the voltage phasors. Then, a covariance matrix is computed, and the eigenvalues and eigenvectors of this matrix are obtained. Subsequently, the most energetic eigenvalue is identified, and its participation factors are calculated. The participation factors are used as a contribution measurement of each generator into the most energetic eigenvalue, i.e., they will show the contribution made by each one after a disturbance. The clusters will be formed by generators sharing the same participation level. Controlled islanding condition of the system will be evaluated by using the clustering schemes proposed in the literature.

Keywords Generator clustering · Inverter-based generation · Smart grid · Islanding detection · Distributed generation · Modern power systems

E. Gómez · E. Vázquez · N. Acosta · M. A. Andrade (✉)
School of Mechanical and Electrical Engineering, Universidad Autónoma de Nuevo León, Monterrey, Nuevo Le ón, Mexico
e-mail: manuel.andradest@uanl.edu.mx

E. Gómez
e-mail: egomezd@uanl.edu.mx

E. Vázquez
e-mail: evazquezmtz@gmail.com

N. Acosta
e-mail: martha.acostamnt@uanl.edu.mx

H. Haes Alhelou et al. (eds.), *Wide Area Power Systems Stability, Protection, and Security*, Power Systems, https://doi.org/10.1007/978-3-030-54275-7_20

1 Introduction

In traditional electrical power systems, the generation of electrical energy is in charge of few large-capacity generation plants, basing their operation on synchronous machines, which are dispatched in a coordinated manner to supply all the demand of the system. With the technological advance and to improve the operation of the network, microgrids with distributed generation (DG) resources have been introduced to the system, this mostly at the distribution level to take advantage of the small local generation. The DG's are mostly renewable generation sources and are of small capacity compared to the traditional plant. DG's are strategically placed to distribute additional energy near the local load. The integration of the DG's has been increased due to the advantages it presents, such as the environmental benefit due to the use of mainly renewable energy, allowing the primary network transmission capacity to be released. Besides, losses in transmission and distribution lines are reduced due to its location near the load, thereby improving network efficiency. DGs improve system reliability by reducing dependency on large plants and allowing continuity of service in the event of a failure in the main network.

However, DG presents some disadvantages that should be considered: the availability of the source determines its location; the intermittency of its primary source; power quality problems. Also, DG does not provide inertia to the system and the time constants of DG units are smaller than synchronous generators constants, making the power system weak and less tolerant of disturbances.

Although power systems are designed to be robust and tolerant of contingencies, they can become vulnerable to them, and even more so with the addition of DG. Simple events (i. e. the loss or operation output of an element) can occur up to severe events (as the loss of a group of elements), causing the activation of the system protections and leading to the disconnection and isolation of one or more generators and loads from the rest of the network. The afore is known as unintentional island formation; this, in turn, can lead to a partial or total collapse of the system. In many occasions, to safeguard the integrity of the system against severe disturbance, the disconnection of individual elements is performed to form an independent microgrid which can subsist. This action is known as intentional island formation. Under these conditions, the presence of DG's enables island operation. The electric islanding phenomenon is one of DG's main problems.

Unintentional island formation may trigger several problems in the network in terms of stability, security and energy quality. Since, in this scenario, the island may have a power deficit or excess accompanied by the dynamic response of synchronous generators, variations in both voltage and frequency are often found. For these reasons, it is necessary to estimate how generators will be grouped and to evaluate island conditions once the frequency cannot be recovered to operating values. Likewise, an islanding operation can cause risks both for the electricity companies and for consumers since the event that originated the island, may not be detected by the DG and it continues to function and in turn, feeds the event (as in the case of a failure). On the other hand, if generators are out of sync with the main network, the DG's and the

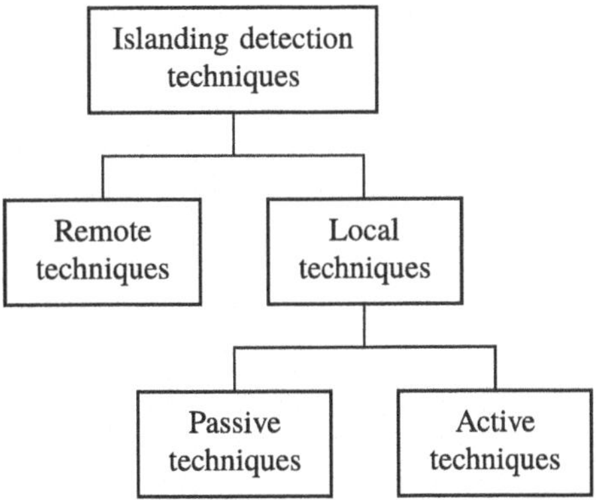

Fig. 1 Classification of islanding detection techniques

generators can be damaged if a reconnection with the primary system is attempted [1].

In an island condition, it is essential to take control actions to avoid disconnecting generators and subsequently, analyze controlled islanding to improve the performance of the power system after it suffers a severe disturbance. Several methods have been proposed to estimate generators clustering after a disturbance, and they meet its objective. However, some of them, to some extent, have disadvantages that cause estimating the generators clustering and hence controlled islanding become complex when power system topology or some parameters are unknown. The electric island phenomenon has been a subject of study for many years and has been mainly intended for the increase in distributed generation. That is why they have developed techniques for the detection and protection of electric islands. Among these, they are more distinguished in 2 main groups: remote and local methods (see Fig. 1).

The first group are remote techniques that consist of communication systems between elements of the electrical system, as well as connection points with the DG. For this, these techniques require advanced hardware and with it a higher price, which makes it not profitable to implement. In [2] they use distribution feeders as routes where they transmit a coded signal from a substation to the DG to monitor it when the signal is interrupted, they detect the island condition. Other of these techniques are based on the use of monitoring, control and data acquisition systems (SCADA)[3]. Voltage sensors are installed on the DG side and the measurements obtained are transmitted to the network and by monitoring them determine if the DG is connected or in an island condition. Under this same principle, techniques based on the use of PMU have been developed [4] to monitor the angular difference of the voltages between two points where an island can be formed. This difference is derived to obtain the frequency and determine a threshold in normal operation. If this threshold is exceeded, island formation is detected [5].

The second group consists of local techniques divided into two subgroups (see Fig. 1): passive and active methods. Passive methods are based on the monitoring of some network parameters such as frequency, voltage, phase changes, or harmonic distortion at the common connection points (PCC); precisely the point where the electric island can form. In turn, these techniques can be time- or frequency-domain based. Those that stand out in the time domain are those based on over-low voltage and over-low frequency detection [3]. Voltage or frequency are monitored where the island may occur or at the DG connection point. When an island condition occurs, due to power imbalances, the voltage and frequency values vary. Therefore, when these variables exceed a predetermined value, island formation is detected. They are low-cost methods, and generally, the DG's converters have this protection system.

Likewise, another widely used technique that detects the formation of islands faster than the previous ones is the one based on the measurement of the rate of change of frequency (ROCOF) [6]. This technique uses the derivative of the frequency calculated from the voltage measurements at the susceptible point of island formation; it uses the principle that when a disturbance occurs the value of this derivative changes and when it is an island condition, it increases. Based on this, for each system, an adjustment value of this derivative is assigned in a stable state, and when it exceeds this value, island formation is detected.

Ref. [7] presents a time-domain technique based on the analysis of the ripple content of the voltage at the connection point between the DG, and the network. When an island condition occurs, this content increases considerably due to the commutation of the high inverter frequency. The island formation is determined by utilizing the level of undulation established in a stable state.

Among the most reported techniques in the frequency domain are those that monitor the content of harmonic distortion (THD) [8]. This method calculates the THD from the monitoring of the generator current. It is monitored by calculating an average for each cycle and evaluating the difference between each cycle, defining a threshold limit of difference. When this threshold is exceeded, the generator disconnection is detected, and island formation is recognized. The method is based on the commutation of the DG's converters: when the DG has disconnected the THD level decreases. In [8], this methodology is used to test the correct operation of wind turbines. However, there are load variations in the system —many of them are non-linear— and affect the THD levels. Therefore, for complex systems, the selection of the threshold setting can become complicated.

Techniques have also been reported that use the wavelet transform to estimate the frequency components of the monitored signals. In [9], authors use local current and voltage measurements and, with the help of the wavelet transform, calculate the rate of change of the power defined in the frequency domain. When a value of change is exceeded, the island is identified.

The other group of local techniques are active methods. These techniques inject signals to change the amplitude, phase or frequency of the current or voltage waveforms in the CCP. When the DG is connected to the system, the injected distortions are absorbed by the power system. When the DG is disconnected, these distortions cause the network protections to activate and detect the island.

The most-reported methods are those based on the measurement of impedance [10]. These methodologies are based on the injection of frequency components into the output current of the DG inverter. The selection of these frequency components is aligned with the system so that it does not affect its THD level. With this modified current and measuring the voltage at the same point, it is possible to calculate the impedance. When an impedance variation is detected, an island operation is determined.

Ref. [11] presents another active technique based on the injection of currents in the dq frame. This technique uses the Park's transform to refer the triphase signals to the dq reference frame in which i_d and i_q currents are injected at a specific frequency. When the power system operates under normal conditions, these currents do not cause effects on the electrical variables. However, when the island separation occurs, these current injections cause a deviation in the system frequency. By monitoring this frequency and determining a threshold, island formation is detected.

An additional methodology used is the active frequency drift technique (AFD) [12]. The injection of currents causes the frequency of the system in CCP to deviate from the established limits for the correct operation of installed protections in the DG. Under the same operating principle, other techniques were developed: the reactive power variation (RPV) is an active technique used for island detection that introduces reactive power variances to generate reactive power mismatches between the load and the inverter output, which in turn generates a frequency deviation from the voltage measured in CCP. Like the previous methodologies, the imbalance of this frequency is monitored, and if it exceeds the established range, the island is detected.

The methodologies mentioned above fulfil the objective of detecting the island formation; however, most of them do it so only at a specific point where the separation of the system is foreseen, or at the connection point of the DG. Besides, many of the techniques only consider the separation of the DG. I. e., they do not take into account the rest of the system, nor the synchronous machines that support the network. Some of the methodologies developed modify some parameters of the network that, if not correctly adjusted, can harm the power system operation and many more require much investment for its implementation. An important point to note about the presented techniques is that they only consider the formation of the island, but not the grouping of it.

Islanding detection in systems with DG penetration is a challenging task. This chapter aims to present a methodology that allows detecting island formation. However, it covers both the estimation of the generator's clustering and the evaluation of its application on controlled islanding on a power system with distributed generation, using data obtained from phasor measurement units (PMUs) after a severe disturbance.

The chapter is organized as follows. Section 2 recalls some definitions of eigenvalue theory and participation factors. Section 3 presents and explains the method to detect islands. Further, on Sect. 4 the methodology performance is assessed. Finally, the conclusion is presented in Sect. 5.

2 Theoretical Foundation

In this section, some critical issues necessary to understand the proposed methodology will be introduced. In the following subsections, a brief exposition of the topic of eigenvalues and eigenvectors will be made, as well as their sensitivity, in order to reach the necessary foundation of the presented algorithm, which are the participation factors.

2.1 *Eigenvalues and Eigenvectors*

For a matrix $\boldsymbol{A} \in \mathcal{M}_n(\mathbb{C})$, the scalar λ is called the characteristic value or eigenvalue of $\boldsymbol{A}$ if there exists a nonzero vector $\boldsymbol{x}$ that satisfies the following:

$$\boldsymbol{A}\boldsymbol{x}_i = \lambda_i \boldsymbol{x}_i. \tag{1}$$

The vector $\boldsymbol{x}_i \neq 0 \in \mathbb{C}^n$ is known as the rigth eigenvector of $\boldsymbol{A}$ associated with the eigenvalue λ_i, for $i = 1, 2, \ldots, n$ [13, 14]. This eigenvector has the form:

$$\boldsymbol{x}_i = \begin{bmatrix} x_{1,i} \\ x_{2,i} \\ \vdots \\ x_{n,i} \end{bmatrix}. \tag{2}$$

Similarly, the column vector $\boldsymbol{y}_j \in \mathbb{C}^n$ which satisfies

$$\boldsymbol{y}_j \boldsymbol{A} = \lambda_j \boldsymbol{y}_j \quad \text{for } j = 1, 2, ..., n \tag{3}$$

is called the left eigenvector of $\boldsymbol{A}$ associated with the eigenvalue λ_j. $\boldsymbol{y}_j$ has the following form:

$$\boldsymbol{y}_j = \begin{bmatrix} y_{i,1} \; y_{i,2} \cdots y_{i,n} \end{bmatrix}. \tag{4}$$

The set of eigenvalues of the matrix $\boldsymbol{A}$ can be found rearranging (1) as

$$(\boldsymbol{A} - \lambda \boldsymbol{I})\boldsymbol{x} = 0. \tag{5}$$

and solving for λ the polynomial $\det(\boldsymbol{A} - \lambda \boldsymbol{I}) = 0$.

The left and right eigenvectors corresponding to the different eigenvalues are orthogonal. That is, for $i \neq j$, we have:

$$\boldsymbol{y}_j \boldsymbol{x}_i = 0. \tag{6}$$

However, when the eigenvectors are associated to the same eigenvalue, we have:

$$y_i x_i = c, \tag{7}$$

where c is a nonzero constant. If both eigenvectors are normalized, we have:

$$y_i x_i = 1. \tag{8}$$

In terms of matrices, the set of right eigevectors of $\boldsymbol{A}$ can be expressed as

$$\boldsymbol{X} = \left[\boldsymbol{x}_1 \mid \cdots \mid \boldsymbol{x}_n\right], \tag{9}$$

and the set of left eigenvectors as

$$\boldsymbol{Y} = \left[\boldsymbol{y}_1^T \mid \cdots \mid \boldsymbol{y}_n^T\right]^T, \tag{10}$$

Likewise, if $\boldsymbol{A}$ has distinct eigenvalues $(\lambda_1, \ldots, \lambda_p)$, $1 \leq p \leq n$, there exist a matrix

$$\boldsymbol{\Lambda} = \begin{bmatrix} \lambda_1 & & 0 \\ & \ddots & \\ 0 & & \lambda_n \end{bmatrix} \tag{11}$$

such that (1) and (8) can be expressed as

$$\boldsymbol{AX} = \boldsymbol{\Lambda X}, \tag{12}$$

$$\boldsymbol{YX} = \boldsymbol{I}. \tag{13}$$

From (12), the matrix $\boldsymbol{A}$ can be reduced to diagonal form [14] by

$$\boldsymbol{X}^{-1}\boldsymbol{AX} = \boldsymbol{\Lambda}. \tag{14}$$

Eigenvalues and their respective eigenvectors provide relevant information about the dynamics of the matrix $\boldsymbol{A}$. I. e., the eigenvector measures the rate of change of the magnitude of the eigenvalues.

2.2 *Eigenvalue Sensitivity*

Once the eigenvalues and eigenvectors have been defined, the sensitivity of the eigenvalue will be explained briefly, since this is where the analysis of the participation factors is derived, which are the foundation of the work carried out.

The mathematical modelling of the dynamic behaviour of a physical system is referred to as a set of nonlinear differential equations. These equations require to be linearized at an operating point of the system to perform some dynamic studies in particular; obtaining a set of ordinary differential equations. The solution of such equations is governed by the eigenvalues of the system, algebraically related to the system parameters. Any variation in the system parameters will produce changes in the behaviour of the eigenvalues. These changes are dependent on the sensitivity of the eigenvalues to said parameters.

In particular, for an electric power system, the eigenvalues are a function of all the design and control parameters of the power system. A change in any of these parameters affects the performance of the system. Therefore, this will cause a change in the behaviour of the eigenvalue. The change magnitude depends on the sensitivity of the eigenvalues to the parameter. Moreover, it depends on the change of said parameter [15].

The eigenvalue problem associated with a matrix $\boldsymbol{A}$ is defined by the algebraic equation (1). A differential change in the elements of $\boldsymbol{A}$ directly generates a change in the eigenvalues.

To explain the sensitivity of the eigenvalues to the system parameters, assume a change in the elements of $\boldsymbol{A}$. For this, we take the partial derivative of (1) with respect to element $a_{k,j}$ from $\boldsymbol{A}$, obtaining:

$$\frac{\partial \boldsymbol{A}}{\partial a_{k,j}}\boldsymbol{x}_i + \boldsymbol{A}\frac{\partial \boldsymbol{x}_i}{\partial a_{k,j}} = \frac{\partial \lambda_i}{\partial a_{k,j}}\boldsymbol{x}_i + \lambda_i \frac{\partial \boldsymbol{x}_i}{\partial a_{k,j}}. \tag{15}$$

Multiplying on the left by $\boldsymbol{y}_i$ and noting that $\boldsymbol{y}_i\boldsymbol{x}_i = 1$, (15) is simplified to:

$$\boldsymbol{y}_i \frac{\partial \boldsymbol{A}}{\partial a_{k,j}}\boldsymbol{x}_i = \frac{\partial_i \lambda}{\partial a_{k,j}}. \tag{16}$$

Because

$$\frac{\partial \boldsymbol{A}}{\partial a_{k,j}} = \begin{cases} 1 & \text{for the elements in the } k\text{-th row and } j\text{-th column,} \\ 0 & \text{otherwise;} \end{cases} \tag{17}$$

Eq. (16) can be written as

$$\frac{\partial \lambda_i}{\partial a_{k,j}} = y_{i,k} x_{j,i}, \tag{18}$$

i. e., the sensitivity of the eigenvalue λ_i with respect to the element $a_{k,j}$ of $\boldsymbol{A}$ is equal to the product of the element $y_{i,k}$ of the left eigenvector and the element $x_{j,i}$ from the right eigenvector [16].

2.3 *Participation Factors*

From the sensitivity of an eigenvalue concept, the participation matrix ($\boldsymbol{P} \in \mathcal{M}_n(\mathbb{C})$) combines right and left eigenvectors as a measure of association between the variables of a matrix and its eigenvalues. Defined as

$$\boldsymbol{P} = \left[\boldsymbol{p}_1 | \cdots | \boldsymbol{p}_n\right], \tag{19}$$

where

$$\boldsymbol{p}_i = \begin{bmatrix} p_{1,i} \\ p_{2,i} \\ \vdots \\ p_{n,i} \end{bmatrix} = \begin{bmatrix} x_{1,i}\, y_{i,1} \\ x_{2,i}\, y_{i,2} \\ \vdots \\ x_{n,i}\, y_{i,n} \end{bmatrix}. \tag{20}$$

where $x_{k,i}$ is the k, i-th element of the the matrix $\boldsymbol{X}$ (the k-th element of the right eigenvector $\boldsymbol{x}_i$), and $y_{i,k}$ is the i, k-th element of the matrix $\boldsymbol{Y}$ (the k-th element of the left eigenvector $\boldsymbol{y}_i$).

The element $p_{k,i} = x_{k,i} y_{i,k}$ is called a participation factor. If the eigenvectors are normalized, the sum of the participation factors associated with any eigenvalue is 1.

Taking up and analyzing (18), it is observed that the participation factor $p_{k,i}$ is equal to the sensitivity of the eigenvalue λ_i with respect to the element $a_{k,k}$ of $\boldsymbol{A}$.

$$p_{k,i} = \frac{\partial \lambda_i}{\partial a_{k,k}}. \tag{21}$$

Therefore, the participation factors of λ_d will be all the elements of the diagonal of the sensitivity matrix of λ_d; this is generated by calculating $\partial \lambda_i$ with respect to all the elements of the matrix.

2.4 Covariance Matrix

A covariance matrix concentrates the variances and covariances of a set of variables with respect to the samples. The covariance matrix is widely used for multivariate statistical issues. Let us see in more detail what it is about.

Suppose a data set V of p variables with n samples. The variables are denoted by the set $(v_1, v_2, \ldots, v_p)$. Therefore this data set can be viewed as a rectangular matrix $\boldsymbol{V} \in \mathcal{M}_{n,p}(\mathbb{R})$:

$$\boldsymbol{V} = \begin{bmatrix} v_{1,1} & \ldots & v_{1,p} \\ \vdots & \ddots & \vdots \\ v_{n,1} & \ldots & v_{n,p} \end{bmatrix}. \tag{22}$$

From this, the variance σ^2 of the variable v is defined as the average of the squared differences with respect to its mean, that is:

$$\sigma_v^2 = \frac{1}{n} \sum_{i=1}^{n} (v_i - \overline{v})^2. \tag{23}$$

Similarly, given two variables v and w the covariance can be defined as:

$$\sigma_{vw} = \frac{1}{n} \sum_{i=1}^{n} (v_i - \overline{v})(w_i - \overline{w}). \tag{24}$$

where $\bar{\cdot}$ denotes here the arithmetic mean of $\mathbb{R}^n$.

In matrix notation, the covariance matrix $\boldsymbol{S} \in \mathcal{M}_p(\mathbb{R})$ is then expressed as:

$$\boldsymbol{S} = \begin{bmatrix} \sigma_1^2 & \dots & \sigma_{1p} \\ \vdots & \ddots & \vdots \\ \sigma_{p1} & \dots & \sigma_p^2 \end{bmatrix}, \tag{25}$$

where σ_i^2 is the variance of the variable v_i, and σ_{ij} is the covariance between the variables v_i and v_j. This matrix is square and symmetric and summarizes the variability of the data and the information related to the linear relationships between the variables. If the covariances are not equal to zero, this indicates that there is a linear relationship between these two variables [17].

3 Estimation Method of Generation Clustering and Islanding Condition in Power System

In this section, each of the steps of the proposed algorithm for the identification of generator clusters and the detection of islanding condition in an electrical power system will be explained in detail through the analysis of participation factors.

This algorithm is implemented in three stages: the first stage consists of data acquisition; the second stage, of data processing; while the last one consists of the calculation of the participation factors, the detection of the islanding condition, and the determination of the grouping of the machines. These stages are presented in the flow diagram shown in Fig. 2.

In the following sections, a detailed description of each of the algorithm stages is presented below.

3.1 First Stage: Data Acquisition

The first stage of the algorithm corresponds to the reading of the input signals as well as their processing. As seen in Fig. 2, this stage, in turn, has two steps: phasor measurement and data window sampling. Let us analyze each one.

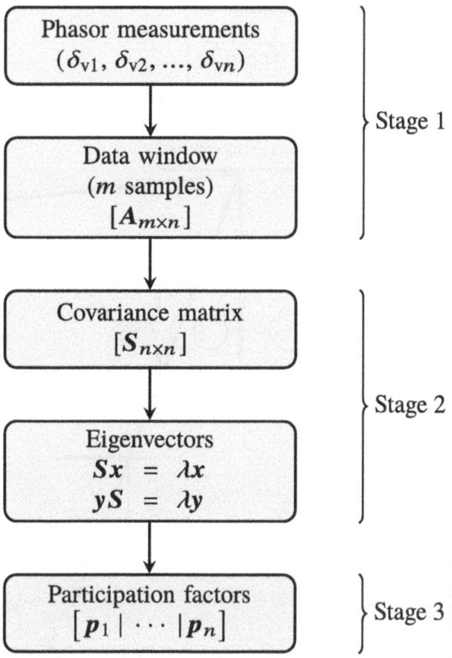

Fig. 2 Stages of the proposed algorithm

3.1.1 Phasor Measurements

To perform transient stability analysis and obtain reliable results, an accurate model of the power system containing the information network connection and parameters of the elements that comprise it is necessary. Currently, having a realistic power system model represents a significant challenge due to constant changes in the network connection, the dynamics of the loads and the complex models of the transmission lines, and even more so today with the connection of DG [18].

The emergence and application of the wide-area measurement systems (WAMS) in power systems has facilitated the monitoring of the stability of the system since it provides information on its dynamics. They also allow monitoring the entire system at the same time, reducing computational time.

WAMS uses devices to collect data; these are known as PMUs. PMUs allow obtaining synchronized measurements at various points of the network, with high precision and speed. Therefore, it can be used for real-time applications. These devices calculate the current and voltage phasors of the location where they are installed and are synchronized with a global time reference, which makes it possible to compare the phasors measured at that point with the other phasors obtained from the PMUs installed in the different locations of the network.

For the present algorithm, the input data is obtained from the PMU devices of the WAMS. These signals come from the PMUs installed in the generation buses or buses above (Fig. 3), this because PMUs can generally be embedded into protection

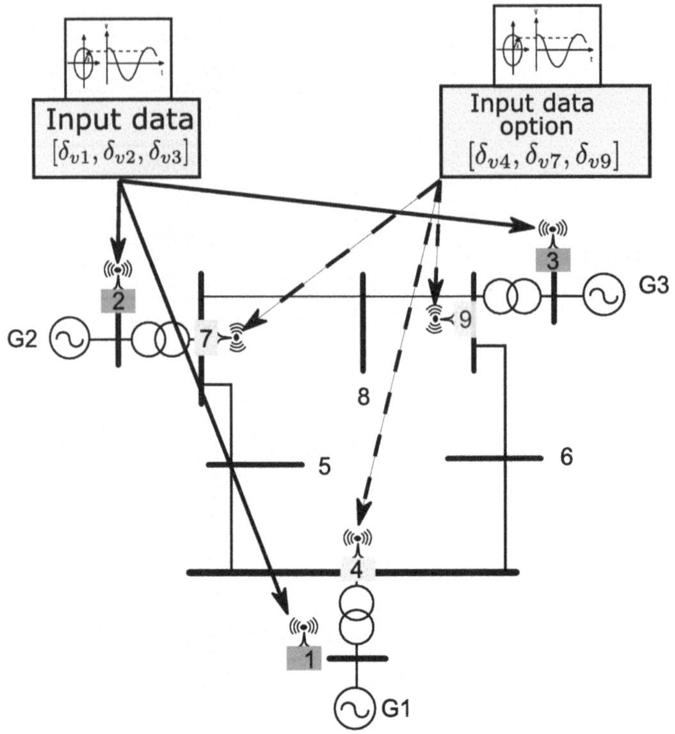

Fig. 3 Algorithm's input data

devices. However, as will be seen later, the measurement can be performed on any of the network buses.

Usually, the network state estimation and the detection of events of the power system are based on measurements of voltages and currents in the buses of interest. This mainly to the difficulty of measuring the internal angle of the machines, which would be the most direct measurement of the dynamics of the system. The dynamics of the voltage phasor's angle at the buses, obtained from the PMUs, represents the dynamics of the machine internal angle in terms of the behaviour that the system will have subjected to a disturbance. In addition to this measurement, there are less noise-contaminated signals [19].

Figure 4 shows the simulation of the voltage phasor angle at the machine terminals and the internal angle of the machine when the power system is subjected to a three-phase failure; Fig. 4a a stable case and Fig. 4b an unstable case are presented. As noted, the angle of the voltage phasor at machine terminals reflects the dynamic behaviour of the machine.

The input signals are arranged in a vector:

$$s_{\text{in}} = \left[\delta_{\text{v1}}\ \delta_{\text{v2}}\ \cdots\ \delta_{\text{v}n}\right], \tag{26}$$

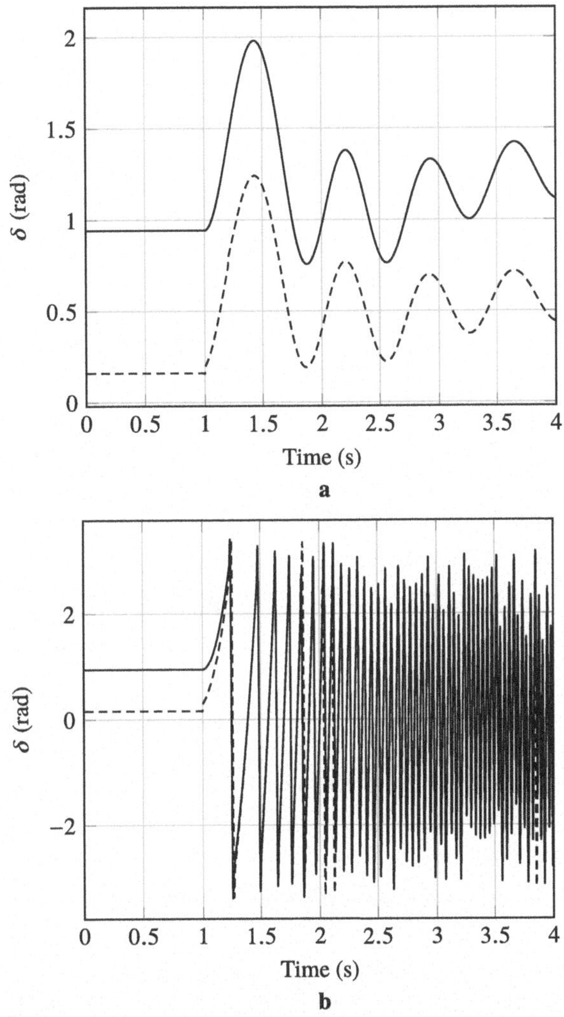

Fig. 4 The behaviour of the machine angle (*solid lines*) and behaviour of the angle of the voltage phasor at machine terminals (*dashed lines*) in the event of a three-phase fault in the power system. **a** stable case. **b** unstable case

where $\delta_{\mathrm{v}i}$ represents the signals measured on the buses and n the number of buses measured. For an initial approach n is the number of generation buses.

3.1.2 Data Window

Once the phasor measurements are read in real-time, using a mobile data window, they are stored in a data matrix $\boldsymbol{A}$, made up of 32 samples per cycle, with a sampling frequency of 256 Hz, as seen in Fig. 5.

For each data window, a matrix $\boldsymbol{A} \in \mathcal{M}_{m,n}(\mathbb{R})$ is formed:

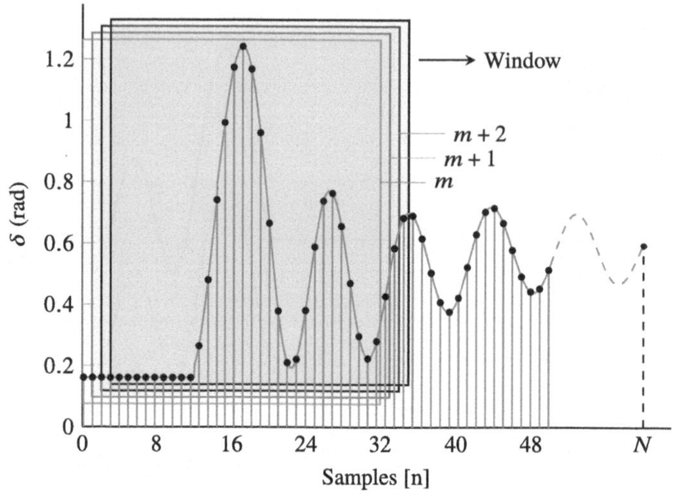

Fig. 5 Sliding data window

$$A = \begin{bmatrix} \delta_{1,1} & \delta_{1,2} & \cdots & \delta_{1,n} \\ \delta_{2,1} & \delta_{2,2} & \cdots & \delta_{1,n} \\ \vdots & \vdots & \ddots & \vdots \\ \delta_{m,1} & \delta_{m,2} & \cdots & \delta_{m,n} \end{bmatrix} \tag{27}$$

consisting of m samples in each data window and n measured buses.

3.2 Second Stage: Data Processing

As mentioned in Sect. 3, the second stage consists of the processing of the data window generated in the previous step. Each of the procedures that comprise it is explained in detail below.

3.2.1 Covariance Matrix

The covariance matrix of the data matrix $\boldsymbol{A}$ obtained in the previous step is calculated to obtain the most synthesized information on the variability of the data.

The formation of this covariance matrix is crucial since it concentrates the mutual and proper relationship of the input variables, which in this case are the voltage angles of the buses measured. Likewise, a square symmetric matrix is obtained.

Once the measurement matrix is formed from the most recent sliding window, the covariance matrix of $\boldsymbol{A}$ is obtained, applying (24), from which the matrix $\boldsymbol{S} \in \mathcal{M}_n(\mathbb{R})$, where n represents the number of buses measured, is obtained:

$$\boldsymbol{S} = \begin{bmatrix} \sigma_1^2 & \sigma_{12} & \dots & \sigma_{1n} \\ \sigma_{21} & \sigma_2^2 & \dots & \sigma_{2n} \\ \vdots & \vdots & \ddots & \vdots \\ \sigma_{n1} & \sigma_{n2} & \dots & \sigma_n^2 \end{bmatrix}. \tag{28}$$

This matrix is symmetric since $\sigma_{ij} = \sigma_{ji}$.

3.2.2 Eigenvector Calculation

With the previous step, a square matrix was obtained, and its eigenvectors can already be calculated. The eigenvalues and eigenvectors of a matrix represent its dynamics since they contain information on the behaviour of the matrix.

To gain insight into (1) and (3) lets note that a matrix $\boldsymbol{A}$ can increase or decrease the magnitude of both left and right eigenvectors without changing their direction. The eigenvalue represents this rate of change of the magnitude of the eigenvector [20]. Thus, the covariance matrix eigenvalues indicate the variance of the variables in the direction of the eigenvectors.

In this proposed methodology, it is interesting to observe the behaviour of the eigenvalue that generates the maximum change in the eigenvectors. Therefore, the dominant eigenvalue λ_d (defined as the eigenvalue with the maximum magnitude) will be used. λ_d indicates the highest variance in the direction of the dominant eigenvector [21]. This eigenvalue represents the dynamics of the angles of the power system machines, as well as the voltage angles dynamics provided by the PMUs.

Therefore, it is proposed to calculate the eigenvectors according to (1) and (3) associated with the dominant eigenvalue. An important aspect here is that, as mentioned, the covariance matrix is a symmetric positive definite matrix; therefore, the left and right eigenvectors concerning the dominant eigenvalue are the same [14]:

3.3 Third Stage: Calculation of Participation Factors

The next and last step of the proposed algorithm is the calculation of the participation factors to measure the relative participation of the variables of the covariance matrix $\boldsymbol{S}$ associated with the dominant eigenvalue. Moreover, it is necessary to calculate the eigenvectors associated with the dominant eigenvalue in the previous step.

As we saw in Sect. 2.2 and Sect. 2.3, participation factors can be calculated in two ways: calculating the sensitivity matrix concerning the dominant eigenvalue (21) and taking the values of the diagonal or directly obtaining the eigenvectors of the covariance matrix $\boldsymbol{S}$ corresponding to λ_d and applying (20).

Taking the eigenvalue problem defined by (1):

$$\boldsymbol{S}\boldsymbol{x} = \lambda \boldsymbol{x}. \tag{29}$$

Thus,

$$\begin{aligned}(\boldsymbol{S}\boldsymbol{x})^T &= \lambda \boldsymbol{x}^T, \\ \boldsymbol{x}^T \boldsymbol{S}^T &= \lambda \boldsymbol{x}^T.\end{aligned}$$

Since the covariance matrix $\boldsymbol{S} \in \mathcal{M}_n(\mathbb{R})$ is a symmetric matrix ($\boldsymbol{S} = \boldsymbol{S}^T$), then:

$$\boldsymbol{x}^T \boldsymbol{S} = \lambda \boldsymbol{x}^T, \tag{30}$$

i. e., the right eigenvectors are also left eigenvectors of the covariance matrix $\boldsymbol{S}$.

Applying (20), and taking into account (30), the participation factors associated to $\boldsymbol{x}_d$

$$\boldsymbol{p}_d = \begin{bmatrix} p_{d1} \\ p_{d2} \\ \vdots \\ p_{dn} \end{bmatrix} \tag{31}$$

can be calculated as

$$p_{di} = x_{di}^2, \tag{32}$$

where $\boldsymbol{x}_d$ is the eigenvector associated to the dominant eigenvalue $\lambda_d(\boldsymbol{S})$ and n is the number of acquired measurements. The element p_{di} is related to the participation of the δ_i measurement into the system dynamics.

Where d refers to the column of the dominant eigenvector, n is the number of buses measured. As can be seen, each element of this vector corresponds to the participation factor of each of the measured angles (δ_i). Therefore, the behaviour of the participation factors could be used to identify the island formation and the grouping mainly of generation buses, when this condition occurs.

The analysis of these participation factors also allows us to identify which machines or buses are the most sensitive concerning the disturbances present in a power system.

By analyzing how the participation factors of each generation bus change, it is possible to observe the grouping of buses that present a similar sensitivity factor, along with the buses with the highest sensitivity tend to separate themselves from the others.

4 Testing and Validation of the Proposed Methodology

This section will evaluate the capacity of the presented method to detect and recognize the grouping of system generators in an electrical island condition. This section presents a brief description of the two test systems used, the tests carried out, and the results obtained at the simulation level.

4.1 Test Systems

This section presents the characteristics of the systems used to carry out the corresponding tests to validate the proposed algorithm. For this, two test systems are modelled and simulated using DIgSILENT PowerFactory simulation software.

4.1.1 First Test System

The first test system corresponds to the IEEE 39-bus system, also known as the New England system, consisting of 39 buses, ten synchronous generators, 19 loads with constant impedance, 34 lines and 12 transformers. The generators have AVR controls and governors. Also, each machine has a PSS stabilizer. Ref. [22] presents comprehensive information about the system.

The nominal frequency of the system is 60 Hz, and the voltage level of the network is 345 kV. Figure 6 shows the one-line diagram of this system.

4.1.2 Second Test System

The second test system corresponds to Anderson's 9-bus test system, shown in Fig. 7. The original system consists of 3 synchronous machines with IEEE Type 1 exciters [23]. Generation buses have different voltage values, so they are connected to the system by means of 3 transformers, with a ring of the nominal voltage of 230 kV interconnected by six lines. Detailed data can be found in [24].

This classical 9-bus system was modified replacing the generator connected to bus four by a rural medium-voltage (MV) distribution network: a CIGRE benchmark introduced for DG integration studies (see Fig. 8). This benchmark is a modification from a German MV distribution network [25]. In Fig. 8, it is observed that the MV network is made up of several types of DG sources, so it provides a more realistic picture of DG penetration. The characteristics and parameters of the CIGRE network modelling are detailed in [26–28].

The systems presented are those that will use to verify the functionality of the proposed methodology. The following sections show the test cases and the results obtained.

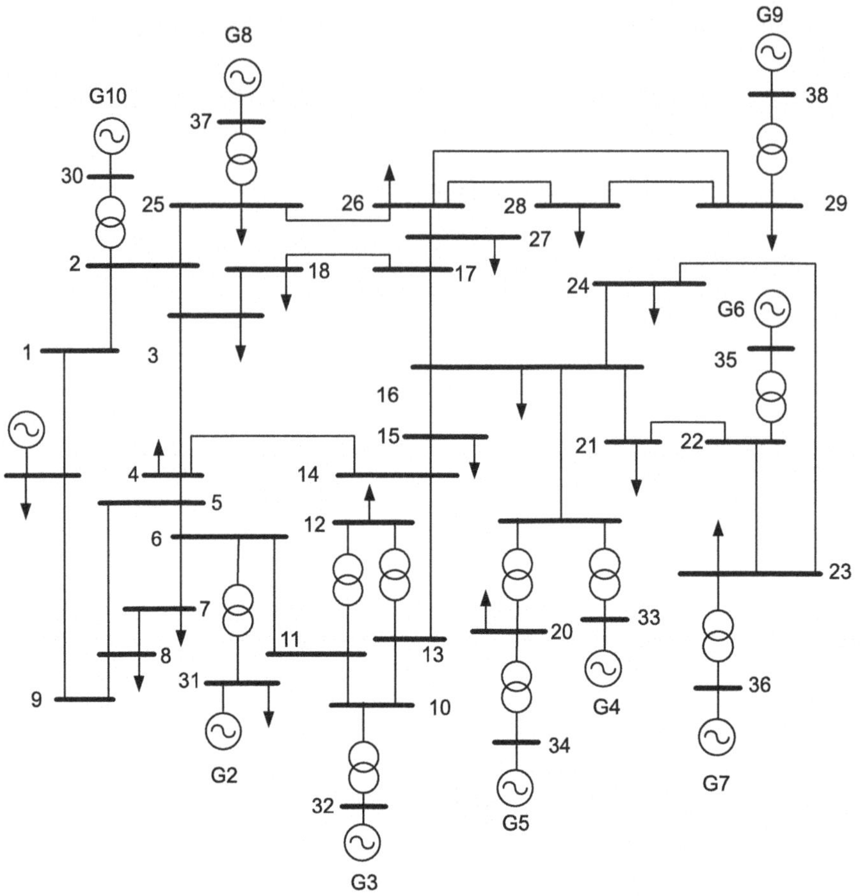

Fig. 6 Single-line diagram of the IEEE 39-bus power system

Fig. 7 Anderson's 9-bus modified test system

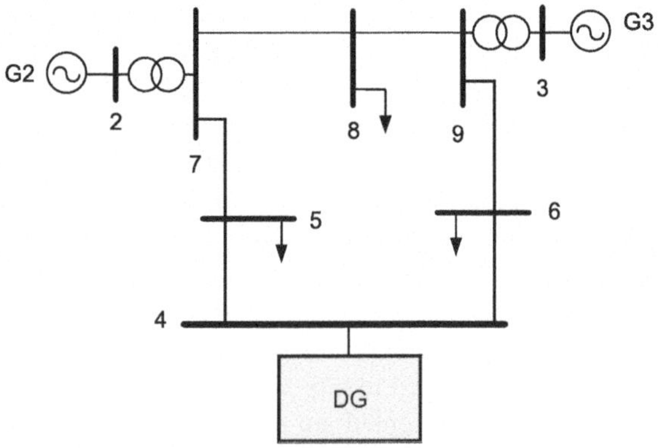

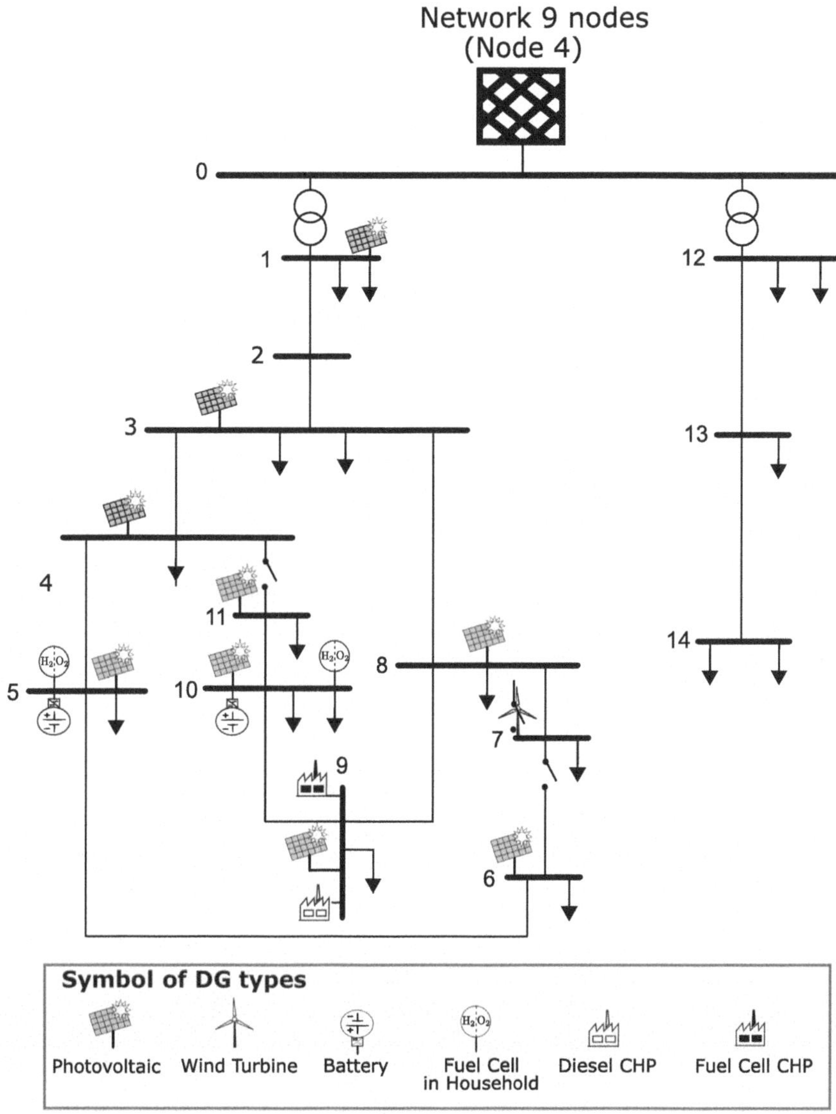

Fig. 8 Single-line diagram of MV distribution benchmark network

Table 1 Test cases in 39-bus system

Case	Number of groups	Measurement point
1	2	Generation bus
2	3	Generation bus
3	3	All buses in the system

Table 2 Generator groups for Case 1

Group	Generators
1	G1, G2, G3, G6, G7, G8, G9, G10
2	G4, G5

4.2 Tests and Results for the 39-Bus System

In order to evaluate the performance of the algorithm, three simulation scenarios for the first test system are analyzed (Table 1). In each situation, the islands were created by opening switches at a simulation time $t = 2\,\text{s}$. It is worthy to note for simulation cases that because generator one is taken as a reference; its behaviour will not be so noticeable.

4.2.1 Case 1

In Case 1, as observed in the Table 1, the formation of two groups was confirmed. That is, an island was formed by disconnecting the line that goes from bus 16 to bus 19, causing two groups of generators to run coherently, as shown in Table 2 and Fig. 9.

The methodology presented in Sect. 3 was applied to obtain and analyze the behaviour of the participation factors. For this case, since the measurements were made only on the generation bus, the method returns a vector with ten elements for each sampling window. Each element of this vector represents the behaviour of each synchronous machine for which the groupings are determined. Graphically, Fig. 10 shows the dynamics of the participation factors. It can be seen that the system is divided into two groups.

As observed in Fig. 10, the dynamic behaviour of the participation factors with respect to time highlights the formation of the two groups or islands. At time $t = 2\,\text{s}$, when the line is opened, the factors' dynamics change, indicating a disturbance occurring in the power system. At a time $t = 500\,\text{ms}$ after the disturbance occurred, it is observed how the factors of each group converge to a single value.

Table 3 presents the numerical values of the partition factors at three different times: at time $t_1 = 1\,\text{s}$ when the line is still closed; at time $t_2 = 2.5\,\text{s}$ when the line

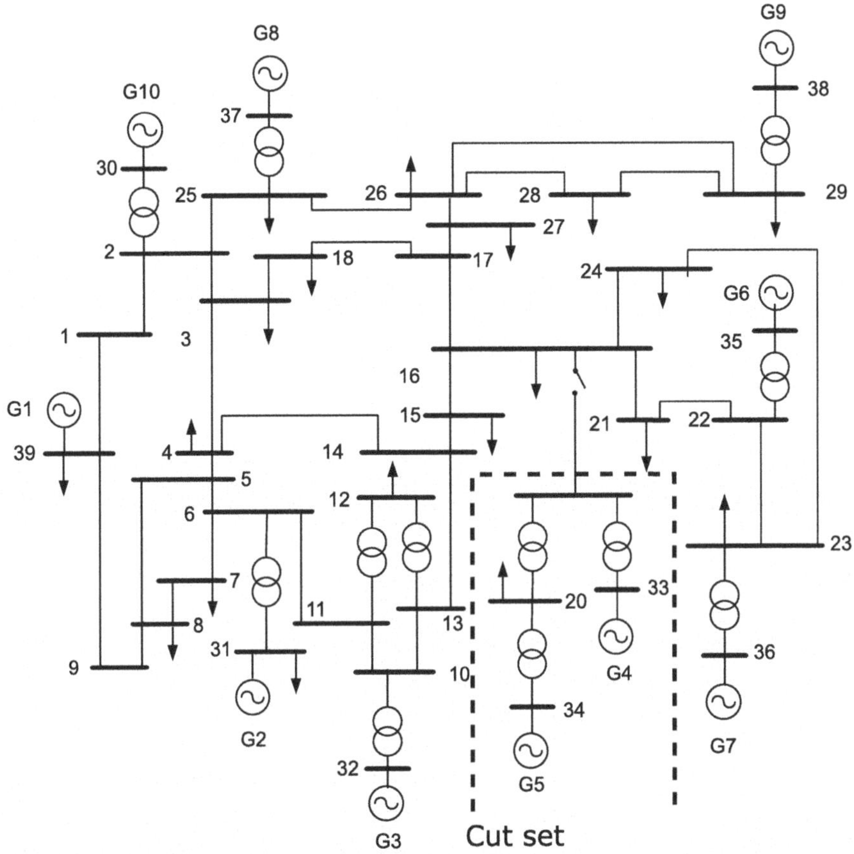

Fig. 9 Islanding formation: Case 1 (the dotted lines represent the cut necessary to produce the two groups)

had already been disconnected; and at time $t_3 = 3$ s when the participation factors had already stabilized.

It is observed how numerically G5 and G4 present similar values at all times. Further, the fact that, before the line is opened, they have similar values, does not determine that they are going to be in the same group as G2 has a similar initial value. The above shows that the algorithm does determine a correct grouping, even though at the beginning, some generators have the same participation factor magnitude. Likewise, as mentioned in Sect. 2.3, the sum of the participation factors at every time instant is equal to 1, as can be corroborated in Table 3.

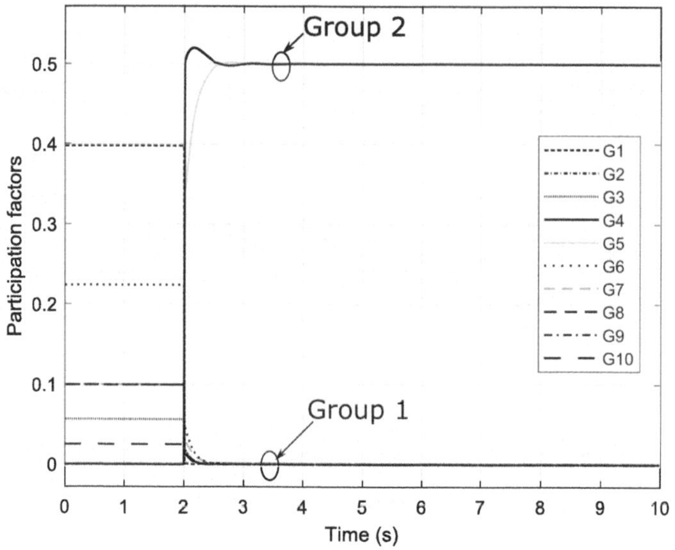

Fig. 10 Dynamics of the participation factors for Case 1

Table 3 Numerical values of the participation factors over time for Case 1

Generator	Participation factor		
	t_1	t_2	t_3
G1	0.3975	0.001	6.1548×10^{-6}
G2	0	6.1911×10^{-6}	5.8689×10^{-7}
G3	0.0559	9.2554×10^{-7}	2.006×10^{-6}
G4	0	0.5021	0.4996
G5	0	0.4955	0.5002
G6	0.2236	0.0005	3.4603×10^{-5}
G7	0.0993	0.0006	4.039×10^{-5}
G8	0.0993	3.7215×10^{-7}	1.8468×10^{-7}
G9	0.0993	7.4794×10^{-7}	1.359×10^{-5}
G10	0.0248	8.4285×10^{-8}	2.1399×10^{-6}

Table 4 Generator groups for Case 2

Group	Generators
1	G1, G2, G3, G6, G7
2	G8, G9, G10
3	G4, G5

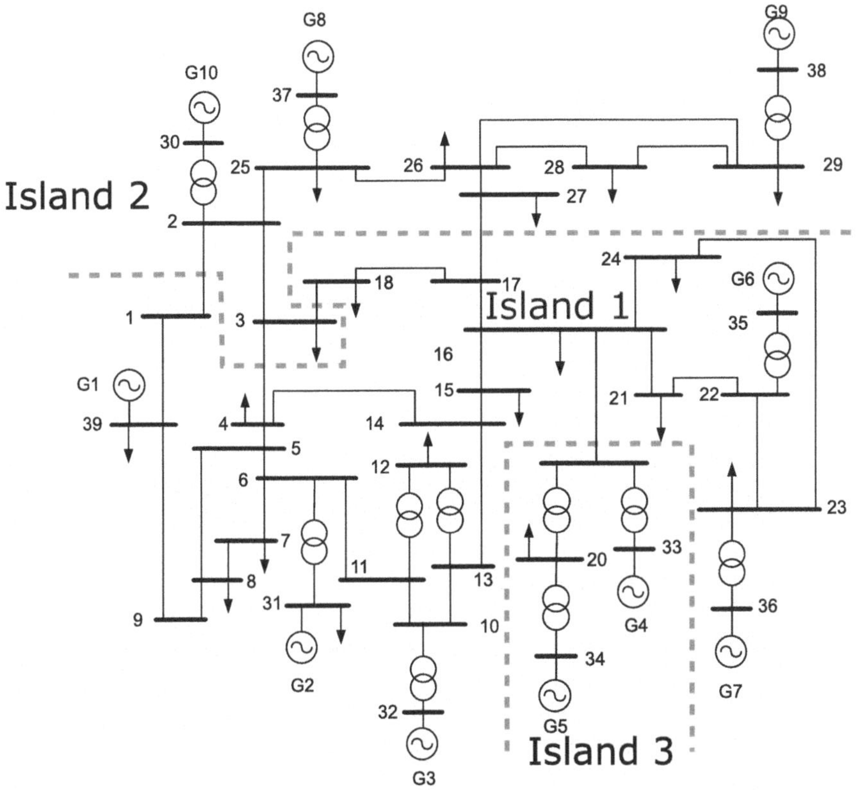

Fig. 11 Islanding formation: Case 2 (the dotted lines represent cuts necessary to produce the three groups)

4.2.2 Case 2

For the second test case, the system was divided into three islands accordingly to a previous analysis [29]. However, a modification was made to the predefined grouping for the island of the previous case can be incorporated.

Table 4 shows the generators in each formed group. Additionally, Fig. 11 graphically shows the line openings made for the formation of the islands.

Accordingly, using the proposed methodology the dynamic behaviour of the generator was obtained, as shown in Fig. 12. It is observed that the algorithm detects the formation of the islands and the correct grouping of generators.

Further, Fig. 12 shows the grouping of the generators was readily determined before the 500 ms mark, as the island condition presented. Table 5 shows numerically the results obtained at specific points in time ($t_1 = 1$ s, $t_2 = 2.5$ s, and $t_3 = 3$ s). It is verified numerically how the representative participation factors for each group of generators tend to converge to the same value.

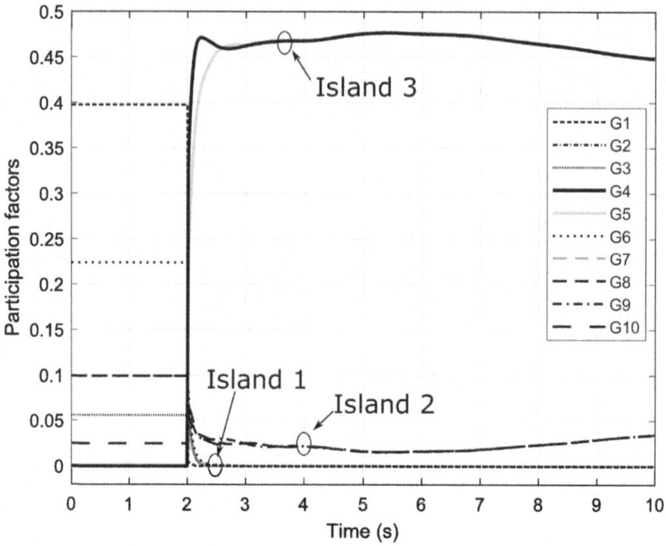

Fig. 12 Dynamics of the participation factors for Case 2

Table 5 Numerical values of the participation factors over time for Case 2

Generator	Participation factor		
	t_1	t_2	t_3
1	0.3975	0.0009	0.0001
2	0	1.0354×10^{-5}	1.5725×10^{-5}
3	0.0559	1.6517×10^{-9}	1.2002×10^{-5}
4	0	0.462	0.4622
5	0	0.4558	0.4625
6	0.2236	0.0007	5.3902×10^{-6}
7	0.0993	0.0009	3.6004×10^{-6}
8	0.0993	0.0262	0.0251
9	0.0993	0.0291	0.0247
10	0.0248	0.0239	0.0251

4.2.3 Case 3

The purpose of the third case is to evaluate the performance of the method measuring not just the generation buses of the system. For this case, the same partition configuration was used as in the previous case (Fig. 11). The difference with Case 2 is that the measurements were acquired from the 39 buses of the system. Thirty-nine participation factors were computed (one for each bus). Figure 13 shows the obtained results. It can be seen after the island separation the participation factors display

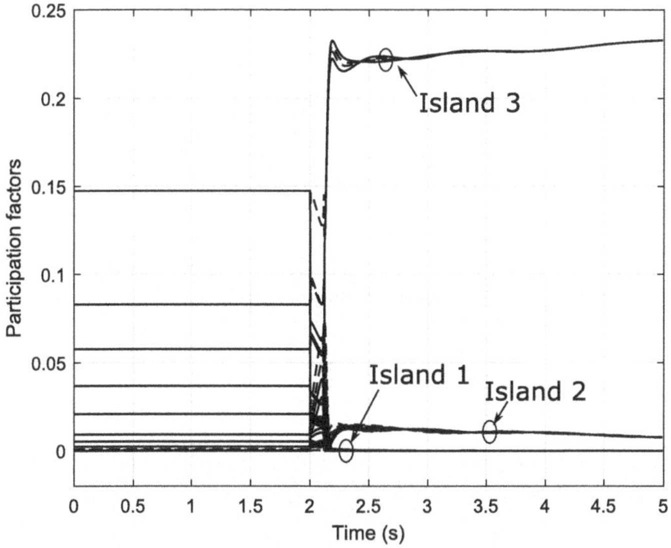

Fig. 13 Dynamics of the participation factors for Case 3

different behaviour than the previous cases. This because, for this case, the covariance matrix has 39 rows and 39 columns, i. e., there are more variables involved, and the covariance between them highlights this behaviour. However, the detection of the grouping was performed approximately 250 ms after the event occurred. After this time, it is seen how each island's participation factors converge to the same value.

4.3 Tests and Results for the 9-Bus Extended System

In a follow-up with the tests to evaluate the presented method, this section presents the tests performed on the second system. The objective is to analyze the operation of the algorithm when the system has a mix of DG sources, where the majority are unconventional sources. The extensive use of converters for their connection makes the tests more interesting. There are two simulation cases, which are detailed below.

4.3.1 Case 1

For the first simulation scenario, the original 9-bus system is evaluated (Fig. 14). The literature shows that this system can be optimally divided into two islands [29]. In Fig. 15 is observed that the method determines the grouping correctly. Note that the magnitude of the participation factors associated with each generator before the event is in general very different to the magnitude shown after the event. Moreover,

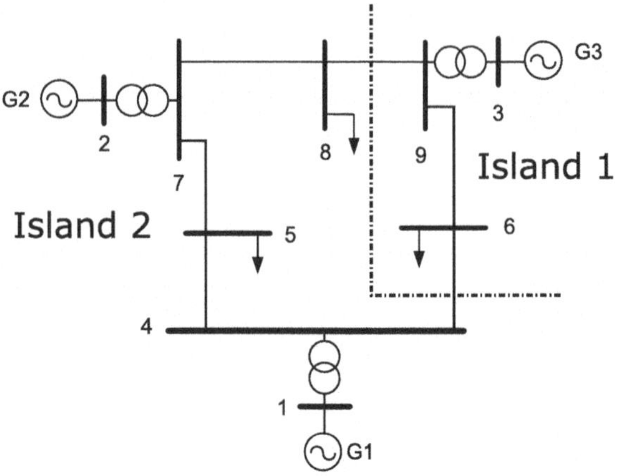

Fig. 14 Island formation for the unmodified Anderson's 9-bus system (the dotted lines represent the cut necessary to produce the two islands)

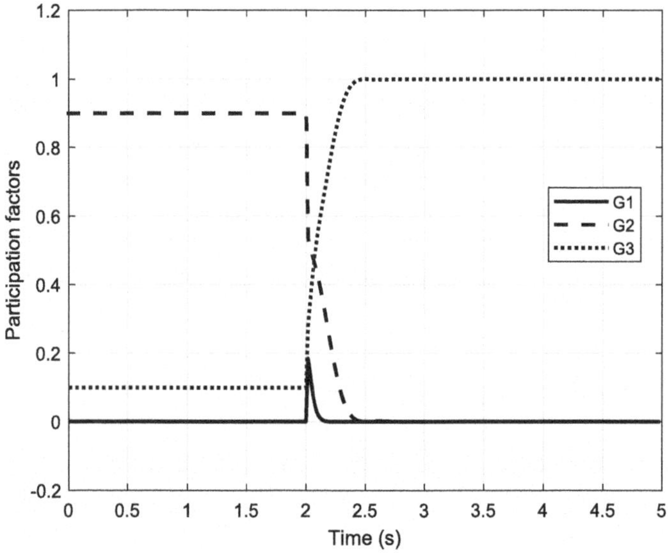

Fig. 15 Dynamics of the participation factors of Case 1 for the 9-bus system

the factors associated to generator one and two tends to similar magnitudes. It is also confirmed that the sum of all the factors is equal to one every time instant.

Using measurements at the generators terminals (buses 1–3), the method correctly determines one island consisting of buses 3, 6, and 9; and a second island consisting of buses 1,2,4,5,7, and 8. For this case, a 3-by-3 covariance matrix was obtained at each sampling time.

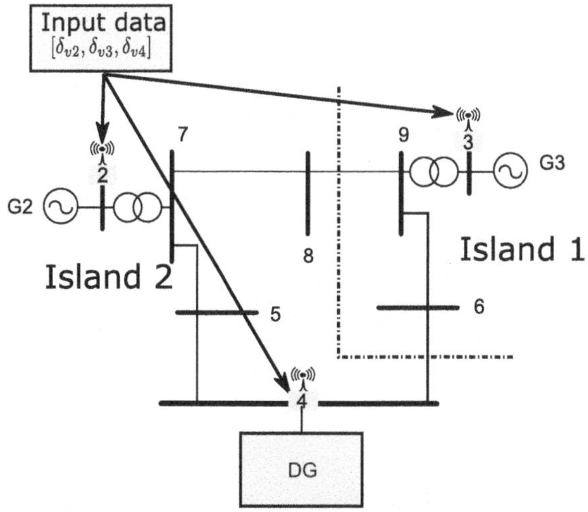

Fig. 16 Island formation: Case 2 (the dotted lines represent the cut necessary to produce the two islands)

4.3.2 Case 2

For this case, the generation grouping of the extended system (including distributed generation sources) shown in Fig. 7 is evaluated.

The objective of this test is to asses how the method's performance is affected by the presence of DG. For this scenario, bus 4, where the medium voltage network is connected (Fig. 8), is considered for the sake of analysis as a controlled-voltage bus. The system is divided as in the previous case (Fig. 16); however, now the measurements are acquired at the terminals of generators 2 and 3 (buses 2 and 3) and bus 4 (Fig. 16).

Figure 17 shows the behaviour of the participation factors. It can be observed that, as in the previous cases, the method correctly identifies the formation of the island and accurately groups the generation buses. Further, the method performance is not affected by the DGs connected to the system.

As mentioned above, measurements can be acquired either on the generation buses or buses above. For this scenario, a second test was performed where measurements were taken after the transformer, as seen in Fig. 18. Figure 19 shows the dynamic behaviour of the participation factorsfor this measurement arrangement. The method performance is not altered. The grouping of the generation buses was correctly identified.

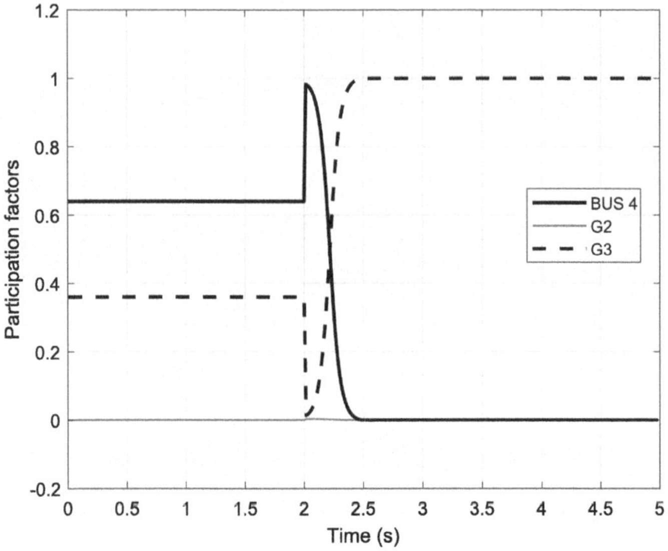

Fig. 17 Dynamics of the participation factors of Case 2 for the 9-bus extended system

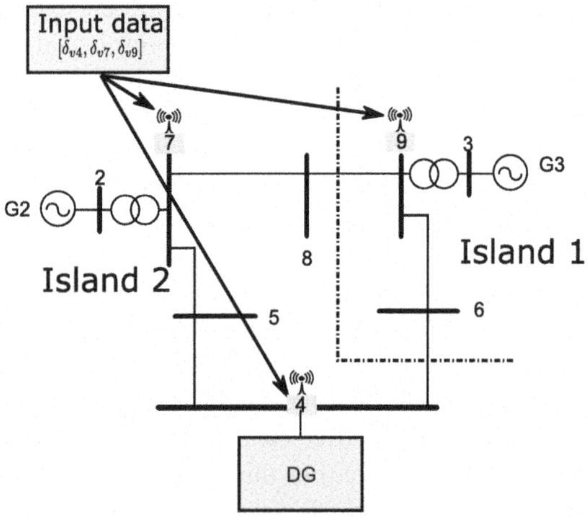

Fig. 18 Island formation: Case 2 measuring data after transformers

5 Conclusion

Although many of the algorithms proposed in literature detect the formation of an electrical island, they are not capable of distinguishing the grouping dynamics of the generators. Further, many of them require *a priori* system configuration and analysis. The sweeping change in electrical power systems' operational paradigm requires algorithms capable of adapting to new generation technologies such as photovoltaic, wind, and battery storage.

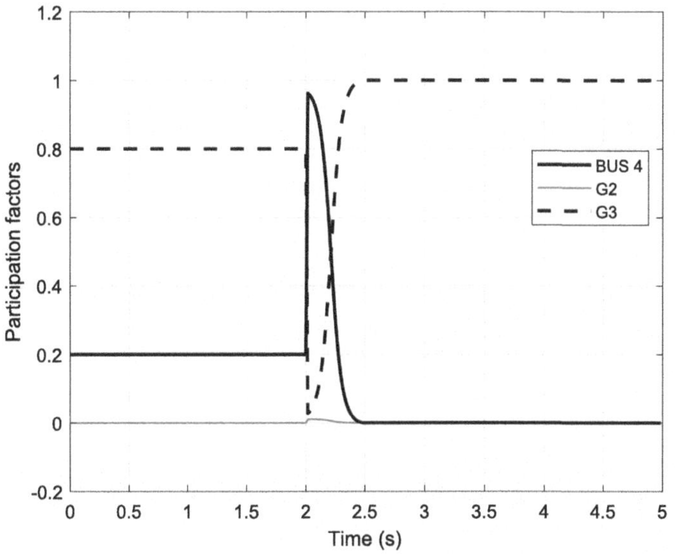

Fig. 19 Dynamics of the participation factors of Case 2 for the 9-bus extended system measuring data after transformers

A significant challenge of islanding detections methodologies is the ability to detect and estimate the grouping of generators during and after an islanding condition. The analysis of the participation factors proved to be useful to identify the generator's grouping since it does not need the modelling of the network and adapts to the dynamics of the operation of the power system. Both aggregated and individual DGs are grouped correctly in ~250–500 ms. However, work is necessary to real-time track the formed generator's groups dynamically.

It is worthy to note the advantage of using measurements at generator terminals as well as measurements non-generation buses on the network. This feature allowed the aggregation of DG presented in Fig. 19.

References

1. IEEE standard for interconnecting distributed resources with electric power systems. IEEE Std 1547-2003 pp. 1–28 (2003)
2. W. Xu, G. Zhang, C. Li, W. Wang, G. Wang, J. Kliber, A power line signaling based technique for anti-islanding protection of distributed generators-Part I: Scheme and analysis. IEEE Transactions on Power Delivery **22**(3), 1758–1766 (2007)
3. Bower, W.I., Ropp, M.: Evaluation of islanding detection methods for utility-interactive inverters in photovoltaic systems. Tech. Rep. SAND2002-3591, Sandia National Labs., Albuquerque, NM; Livermore, CA (2002)
4. Sykes, J., Koellner, K., Premerlani, W., Kasztenny, B., Adamiak, M.: Synchrophasors: A primer and practical applications. In: 2007 Power Systems Conference: Advanced Metering, Protec-

tion, Control, Communication, and Distributed Resources, pp. 213–240 (2007)
5. Pena, P., Etxegarai, A., Valverde, L., Zamora, I., Cimadevilla, R.: Synchrophasor-based anti-islanding detection. In: 2013 IEEE Grenoble Conference, pp. 1–6 (2013)
6. Wright, P.S., Davis, P.N., Johnstone, K., Rietveld, G., Roscoe, A.J.: Field testing of ROCOF algorithms in multiple locations on Bornholm Island. In: 2018 Conference on Precision Electromagnetic Measurements (CPEM 2018), pp. 1–2 (2018)
7. Guha, B., Haddad, R.J., Kalaani, Y.: A novel passive islanding detection technique for converter-based distributed generation systems. In: 2015 IEEE Power Energy Society Innovative Smart Grid Technologies Conference (ISGT), pp. 1–5 (2015)
8. Kazemi Karegar, H., Shataee, A.: Islanding detection of wind farms by THD. In: 2008 Third International Conference on Electric Utility Deregulation and Restructuring and Power Technologies, pp. 2793–2797 (2008)
9. Morsi, W.G., Diduch, C.P., Chang, L.: A new islanding detection approach using wavelet packet transform for wind-based distributed generation. In: The 2nd International Symposium on Power Electronics for Distributed Generation Systems, pp. 495–500 (2010)
10. M. Tedde, K. Smedley, Anti-islanding for three-phase one-cycle control grid tied inverter. IEEE Transactions on Power Electronics **29**(7), 3330–3345 (2014)
11. P. Gupta, R.S. Bhatia, D.K. Jain, Average absolute frequency deviation value based active islanding detection technique. IEEE Transactions on Smart Grid **6**(1), 26–35 (2015)
12. Li, X., Balog, R.S.: Analysis and comparison of two active anti-islanding detection methods. In: 2014 IEEE 57th International Midwest Symposium on Circuits and Systems (MWSCAS), pp. 443–446 (2014)
13. S.I. Grossman, *Elementary Linear Algebra*, 5th edn. (Saunders College Pub, Fort Worth, 1994)
14. G. Allaire, S.M. Kaber, *Numerical Linear Algebra* (Springer, New York, 2008)
15. H. Nicholson, Eigenvalue and state-transition sensitivity of linear systems. Proceedings of the Institution of Electrical Engineers **114**(12), 1991–1995 (1967)
16. P. Kundur, *Power System Stability and Control* (McGraw-Hill, New York, 1994)
17. J.E. Jackson, *A User's Guide to Principal Components* (Wiley-Interscience, Hoboken, N. J., 2003)
18. Zhou, H., Tang, F., Jia, J., Ye, X.: The transient stability analysis based on WAMS and online admittance parameter identification. In: 2015 IEEE Eindhoven PowerTech, pp. 1–6 (2015)
19. Yuan, Z., Xia, T., Zhang, Y., Chen, L., Markham, P.N., Gardner, R.M., Liu, Y.: Inter-area oscillation analysis using wide area voltage angle measurements from FNET. In: IEEE PES General Meeting, pp. 1–7 (2010)
20. C.T. Chen, *Linear System Theory and Design*, 4th edn. (Oxford University Press, New York, 2013)
21. Georgiou, G.M., Voigt, K., Qiao, H.: Stochastic computation of dominant eigenvalue and the law of total variance. In: 2015 International Joint Conference on Neural Networks (IJCNN), pp. 1–4 (2015)
22. Athay, T., Podmore, R., Virmani, S.: A practical method for the direct analysis of transient stability. IEEE Transactions on Power Apparatus and Systems **PAS-98**(2), 573–584 (1979)
23. IEEE recommended practice for excitation system models for power system stability studies. IEEE Std 421.5-2016 (Revision of IEEE Std 421.5-2005) pp. 1–207 (2016)
24. P.M. Anderson, A. Fouad, *Power system control and stability*, 2nd edn. (IEEE Press; Wiley-Interscience, Piscataway, N. J., 2003)
25. B. Buchholz, C. Schwaegerl, T. Stephanblome, H. Frey, N. Lewald, N. Lewaldi, Advanced planning and operation of dispersed generation ensuring power quality, security and efficiency in distribution systems. CIGRÉ Session **2004**, 1–8 (2004)
26. Rudion, K., Orths, A., Styczynski, Z.A., Strunz, K.: Design of benchmark of medium voltage distribution network for investigation of dg integration. In: 2006 IEEE Power Engineering Society General Meeting, pp. 6 pp.– (2006)
27. Strunz, K.: Developing benchmark models for studying the integration of distributed energy resources. In: 2006 IEEE Power Engineering Society General Meeting, pp. 2 pp.– (2006)

28. A.H. Kasem Alaboudy, H.H. Zeineldin, J. Kirtley, Microgrid stability characterization subsequent to fault-triggered islanding incidents. IEEE Transactions on Power Delivery **27**(2), 658–669 (2012)
29. L. Ding, F.M. Gonzalez-Longatt, P. Wall, V. Terzija, Two-step spectral clustering controlled islanding algorithm. IEEE Transactions on Power Systems **28**(1), 75–84 (2013)

Resilience in Wide Area Monitoring Systems for Smart Grids

Mohammad Shahraeini and Panayiotis Kotzanikolaou

Abstract WAMS infrastructures consist of various elements such as digital metering devices, communication and processing systems, in order to facilitate the operation, monitoring and control of power grids. For smart grids, resilience is a high-priority design requirement, since they must be able to resist in failures at any layer, caused by intentional attacks or unintentional events. In this chapter we review existing approaches in the literature for WAMS resilience. Based on our recent work on dependency analysis for WAMS resilience [23], we describe methodologies that take into consideration both optimization and resilience metrics during WAMS design. We explain how WAMS resilience can be increased by reducing the dependencies of WAMS components and by selectively adding controlled redundancy of measurement units and communication links. Finally, we describe how this resilience model can be extended to also take into account the dynamic structure of the smart grid caused by the integration of renewable energy sources.

Keywords Wide Area Measurement Systems · Resilience · Renewable resources · Centrality metrics

1 Introduction

Smart grids facilitate the bidirectional flow of both energy and information among different entities, by using advanced information and communication services, in order to increase network efficiency and stability [8, 22]. In addition, they simplify the integration of renewable energy sources. By continuously measuring the state from the system buses, they support the real-time distribution of the electricity

M. Shahraeini (✉)
Department of Electrical Engineering, Golestan University, Gorgan, Iran
e-mail: m.shahr@gu.ac.ir

P. Kotzanikolaou
Department of Informatics, University of Piraeus, Piraeus, Greece
e-mail: pkotzani@unipi.gr

H. Haes Alhelou et al. (eds.), *Wide Area Power Systems Stability, Protection, and Security*, Power Systems, https://doi.org/10.1007/978-3-030-54275-7_21

supply, for example in cases where electricity demand experiences dynamic changes. In order to move from traditional power grids towards smart grids, different kinds of measurement and applications have been used to improve the reliability, security, and efficiency of electricity supply systems. Wide Area Measurement Systems (WAMS) is a key technology for achieving this goal [21]. WAMS infrastructures consist of advanced digital metering devices, real-time communication systems and processing facilities, in order to continuously monitor, operate and control power systems in wide geographic area [21]. In particular, WAMS support the operation of three main interconnected processes [19]: data acquisition, data transmission and data processing. The first process of WAMS, i.e. data acquisition, is implemented using measuring devices, such as traditional RTU/SCADA systems, or more sophisticated *Phasor Measurement Units* (PMU). The use of PMUs in smart grids is continuously increasing, since they provide high resolution and accuracy on measuring and also support synchronized phasor data [23]. In order to implement the second process of WAMS (i.e. data transmission), communication systems are installed in the grid to support the delivery of data. In general, new communication systems have been designed based on Open System Interconnection (OSI) layer model. In such models, the lowest layer is generally known as "*transmission media*" which creates a shared medium between different nodes in the network. Historically, many kinds of media have been used in power grids as transmission media and *OPtical Ground Wire* (OPGW) is one of the more desirable media, due to its high capacity and also the flexibility of installation in the smart grids [20]. The last process of WAMS (i.e. data processing) performs by WAMS applications, which has been also known as the Energy Management Systems (EMS) software package. WAMS applications are mainly designed to perform the operation, protection, control, and also optimization of the power systems. State estimation (SE), load power flow (LPF), optimal power flow (OPF), load forecast (LF), and online low-frequency oscillation (LFO) analyses are some examples of WAMS applications [18].

State estimation has been known as the most important WAMS application that obtains creditable data and system states from measured raw and noisy data. Then, these creditable data can be used by other WAMS applications. Therefore, SE is the most important WAMS application, since it is the basis for other WAMS applications [19]. In this chapter, PMU-based state estimation is assumed as the basis of the processing unit same as our previous works [19, 23]. Based on three processes described above, WAMS design may aim either to independently obtain one of the processes, or to simultaneously achieve two or more of those at the same time. Comprehensive design of WAMS has been known as achieving two first processes at the same time, i.e. simultaneous placing of PMUs and their related OPGWs in the smart grids [19].

Although cost optimization is considered as the most important concern when designing WAMS, increasing the resilience of the WAMS layers is also an important design challenge. The raise of advanced threats against smart grids, SCADA and other critical infrastructures [24] indicates the importance of designing resilient WAMS architectures; it must be possible to achieve a minimum acceptable level of WAMS operation, even when unwanted events undermine the availability of several WAMS components. Ideally, no component should be 'important enough' to significantly affect the operation of WAMS in case of a failure [23].

Comprehensive designing of WAMS creates the opportunity of planning both optimal and resilient WAMS. Although resilience and cost optimization are contradicting requirements. However, an interesting problem is to concurrently achieve both properties: balance between infrastructure resilience and cost optimization. The result will be a suboptimal solution, which will offer adequate resilience, with the minimum cost overhead in comparison to the cost optimal solution. We have proposed this approach in our previous work [23].

In this chapter we extend our previous method for designing resilient WAMS for smart grids by also considering the application of renewable resources. To do this, we first use our proposed method in [23] and map WAMS infrastructures to dependency graphs; all the fundamental components of the electrical (buses), the measurement (PMUs), and the communication (OPGWs) layers are modeled as nodes of the graph, while their informational dependencies are modeled as directed weighted edges. In order to quantify these dependencies, we classify the importance of system buses, based on the placement of renewable resources in the power networks. The integration of renewable source will allow for small modifications in the generation topology of the power network structure and will assist in achieving our ultimate goal of distributing the importance of all WAMS components; ideally no element should be (significantly) more important than any other component of the same type.

This chapter is organized as follows: In Sect. 2 we provide a brief overview of the research efforts for WAMS resilience. Also, we will briefly describe our previous approach for WAMS resilience in this section. In Sect. 3 we extend this approach to capture renewable energy sources. Section 4 concludes this chapter.

2 Related Work on WAMS Resilience

We provide a brief overview of the research efforts for WAMS resilience. Then, we briefly describe our previous approach for WAMS resilience [23], which is based on dependency analysis. This approach will be extended in Sect. 3 to capture renewable resources.

2.1 Resilience in Critical Infrastructures

For critical infrastructures (CIs), resilience is a design requirement of high proprity. Because CIs provide the underlying infrastructure for services which are vital in nature, critical infrastructures must be able to withstand unwanted events, such as faults, natural hazards or even deliberate attacks. Informally speaking, resilience is an overloaded term, covering many different aspects such as resistance in unwanted events, the ability to continue to operate during or after an unwanted event, the ability to adopt and absorb disruptions, or the ability to recover in a small time frame.

Several organizations have provided variations of definitions for critical infrastructure resilience. According to the US National Infrastructure Advisory Council [2] infrastructure resilience is defined as: *"the ability of critical infrastructure systems, networks, and functions to withstand and rapidly recover from damage and disruption and adapt to changing conditions."* In the literature, various features have been proposed as 'metrics' or characteristics for infrastructure resilience. A widely used taxonomy by [2] proposes the use of the following features to measure resilience: robustness, resourcefulness, recovery and adaptability. The combination of all these features, explained below, can be used to quantify the resilience level of an infrastructure.

- *Robustness*: It refers to the ability to continue system operation in case of interruptions, at least to some minimum level. Robustness design should take into consideration not only interactive events that are expected to happen, but also events with low probability of occurrence but with high impact.
- *Resourcefulness*: This property reflects the ability of systems to effectively manage disruptive events, by implementing and prioritizing the appropriate mitigation controls, to be applied in case of a disruptive event. Example of such controls may include redundant systems that may be activated in case of failure of the relevant primary systems (e.g. backup energy sources).
- *Rapid recovery*: It refers to the capability of a system to quickly restore after an interruption or failure and return to its normal condition. This can be implemented for example, by applying procedures and/or technical controls, such as an automated data restoration procedure.
- *Adaptability*: It refers to the ability of a system to absorb the consequences (or impact) of a disaster. This is mainly implemented with mitigation controls aiming at restoring the state of a system under an attack, during the event. An example of an adaptable mitigation control, would be to distribute the network traffic on-the-fly through alternative links, if a network link is currently 'under pressure'.

The UK Cabinet Office provides a similar definition in [9]. Here resilience is defined based on properties that are similar to the above definition, including: *Resistance*, i.e. strengthening/applying those mitigation controls that can minimize the potential consequence for a system, after a disruptive event has been realized; *Reliability*, i.e. designing an infrastructure or system in such a way that it can inherently withstand abnormal events and maintain its operation; *Redundancy*, i.e. considering redundant parts (e.g. over-scaling critical parts or foreseeing backup parts) during the design; and finally *Recovery and Response*, i.e. the adoption of proper recovery controls that will ensure the quick recovery of the system, in case of disruptions. Obviously, the various definitions are conceptually close and are essentially describing similar requirements [11]. For example, in [2] robustness is defined in a way that is very similar with the requirement of resistance as defined in [9]. In this paper we will use the terminology as defined in [9] and we will use resistance and robustness as synonymous terms.

It is obvious that since resilience implies the use of redundancy and robustness techniques when designing an infrastructure, a resilient infrastructure can not, by definition, be the cost optimal solution. However, since both cost optimization and resilience may be design requirements for critical infrastructure planning, several works attempt to provide solutions that balance these properties (e.g. [23]). Such sub-optimal solutions aim to provide a solution having a significantly increased resilience level but with the minimum possible increase in the cost, when compared to the cost-optimal solution.

2.2 Approaches for WAMS Resilience

Since WAMS support the monitoring, operation and control of smart grids, they are a vital element of smart grids. For this reason the resilience of WAMS systems has been extensively studied in the literature. One way to classify existing solutions in the literature for WAMS resilience is based on the underlying layers of WAMS, i.e. the processing, the communication and the measuring infrastructures [23].

WAMS Processing. The main goal of several works is to increase WAMS resilience by targeting at its processing units. For example, this is achieved in [16] by applying an algorithm that estimates electro-mechanical oscillation modes, which adds a fault tolerance mechanism to the alternating direction method of multipliers (ADMM). Another approach is proposed in [6], which combines dynamic state estimation with energy functions. It proposes a resilient Wide Area Protection System (WAPS), which utilizes an analytical method that uses energy functions for state estimation in order to supervise the relay. In [25] an attack-resilient Wide Area Monitoring, Protection, and Control (WAMPAC) framework is proposed. The goal is to protect the bulk power system from typical cyber attacks.

WAMS Communications. The communication layer of WAMS is responsible for the exchange of all the relevant information, such as the transmission of buses' states to the control center or the transmission of control signals from the control center towards the other WAMS elements. This, the resilience of the communication infrastructure of WAMS must be assured. In [26], a solution based on a redundant damping controller is proposed. The controller makes use of both local signals and wide area signals, in order to mitigate communication failures and to increase the WAMS resilience in such failures. A software based, distributed storage mechanism is proposed in [15], to enforce Quality of Service (QoS) in the network communication of WAMS systems. It involves, among others, controls for self-recoverability and for tolerance of network failures in WAMS infrastructures. In [3], a solution that prevent typical network-layer attack agains WAMS communication systems, such a Distributed Denial of Service (DDoS) attacks is presented. This solution is based on the MultiPath-TCP (MPTCP) protocol, which is extended by applying a steam hopping mechanism, in order to identify potential attack signals at the upper (transport and application) layers of the WAMS communication elements.

WAMS Measuring. In this line of works, the topological placement of the measurement units is modeled and formulated as an optimization problem. an [1] a model for the contingency-constrained PMU placement is presented. It considers various contingency conditions, such as the loss of measurements and the loss of communications. A different approach is proposed in [12], where redundant PMUs are placed is selected areas, in order to increase the resilience of the measuring layer of WAMS. In the same line, the work of [5], proposes a method for the strategic placement of measuring units (PMUs) in the power grid. The goal is to achieve robustness against gross measurement errors, using a so called, Least Absolute Value state estimator. The work of [13], deals with the problem of robustness in Distribution System State Estimation (DSSE). Their approach applies different sources of uncertainty using a weighted least square (WLS) approach. In the work of [14], the goal is to select the most appropriate weights to achieve robust state estimation, by using Fuzzy Inference Systems (FIS).

Although, as described above, WAMS resilience may be categorized based on the particular layer of WAMS they are targeting (i.e. processing, communication or measuring), the relevant works may also be categorized according to the main resilience features that each proposal is applying (i.e., robustness, redundancy, recovery or adaptability). The majority of the works target on improving either the robustness (aka resistance) or the redundancy (aka resourcefulness—see the discussion regarding the resilience terminology in Sect. 2.1 above). For example, the works of [5, 13] mainly aim to increase the robustness of WAMS, by carefully selecting the placement of measurement units on the design phase. On the other hand, other solutions mainly aim to increase the redundancy (or resourcefulness) of WAMS, e.g., [1, 12], and to support continuous operation in case where some PMUs are out of service.

2.3 A Dependency Analysis Model for Assessing WAMS Resilience

The approach presented in [23] aims to increase WAMS resilience based on a dependency analysis model. We briefly describe this approach and we refer to the original paper for a detailed analysis. Dependency analysis has been applied in critical infrastructures in order to measure the level of dependencies between assets of such infrastructures. If an asset (or complete infrastructure) A depends on another asset B, then it is possible to affect the proper operation of asset A by affecting the operation of B. Clearly, the more an asset affects other assets, the more important it is in terms of resilience, since its unavailability will cause a high impact to the infrastructure operation.

The observability of power networks depends on various elements of WAMS. State measurement depends on the PMUs, while the transmission of the measured data (state information) to the control center depends on the underlying communication infrastructure (OPGWs). In [23] dependency graphs are used to model all the elements of WAMS, such as the buses, the measuring (PMU), the communication (OPGW) and the control center, as the nodes of the dependency graph. In [23], *informational dependency* is used to model WAMS dependencies. This is based on the fact that the system state observation depends on continuous information flows (i.e. the bus states) which originate from the system buses, are collected by the PMUs and are then transferred through the OPGWs to the control center for processing. The weight of theses dependencies is based on an extension of the risk-based approach defined in [7].

Importance Metrics for WAMS. Based on the characteristics of WAMS, structural importance was used in [23] as an appropriate type for WAMS elements, in contrast to dynamic importance metrics. Structural importance can be measured based on graph centrality metrics. In [4] different types of centrality metrics (such as *degree*, *eigenvector*, *betweenness* and *closeness* centrality) have been formulated, in order to rank all the network vertices and edges. In particular, degree centrality was chosen as the most appropriate centrality metric, while its computation was based on electrical degree centrality. Although degree centrality suffices for typical WAMS components, as we explain in Sect. 3, this is not sufficient for measuring renewable resources.

Quantifying the Resilience Level. Quantifying the resilience gain of the proposed method, is based on the computation of the deviation of the overall resilience level that a test solution (i.e. a WAMS architecture and its resulting dependency graph) under examination has, in comparison to the "ideal" resilience level of an "ideal" power graph in terms of dependencies, as described bellow.

Let $G(V, E)$ and $G_{com}(V_{com}, E_{com})$ denote the power infrastructure and its required WAMS communication infrastructure, and let $\overrightarrow{PMU}$ be a vector denoting the location of all the PMUs. Hence, the dependency graph of the WAMS ($\mathcal{G}$) can be constructed by $G(V, E)$, $G_{com}(V_{com}, E_{com})$ and $\overrightarrow{PMU}$.

Let $\mathcal{I}_{PMU}$ (resp. $\mathcal{I}_{OPGW}$) denote the ideal importance values, i.e. the lowest possible importance value that can be achieved *by all* PMUs (resp. *all* OPGWs) for a given graph. The resilience of an examined network $\mathcal{G}$ is defined as:

$$
\begin{aligned}
Res(\mathcal{G}) = {} & \frac{1}{2}\Big[100 - \sum_{p \in PMU} |Imp(p) - \mathcal{I}_{PMU}|\Big] \\
& + \frac{1}{2}\Big[100 - \sum_{o \in E_{CC}} |Imp(v) - \mathcal{I}_{OPGW}|\Big]
\end{aligned}
\tag{1}
$$

Informally, Eq. (1) will output the cumulative deviation of all the PMUs and all the OPGWs from the overall minimum possible importance value. For the PMUs, this depends on the total number of PMUs, while for the OPGWs it depends on the number of OPGW lines that end up to the control center. Note that an ideal network $\mathcal{G}$, from a resilience perspective, having PMUs and OPGWs with zero deviation from the ideal values $\mathcal{I}_{PMU}$ and $\mathcal{I}_{OPGW}$ respectively, has, by definition, a resilience level equal to 100. Obviously achieving the optimal resilience level is very hard in real world cases. However, the ideal resilience can be used as a metric to theoretically examine the deviation of a test solution from the ideal case. For a detailed analysis, again we refer to [23].

3 Extending Our Dependency Analysis WAMS Resilience Model for Renewable Resources

In smart grids, placing renewable resources improves reliability and resiliency of electrical infrastructure, but at the same time, control of such a grid faces some challenging issues e.g. stability issues, bidirectional power flows, low inertia and uncertainty [10]. In order to overcome these control challenges, real-time monitoring of such resources is essential and therefore, resilience should also consider their monitoring. In other words, the resilience gain in the power generation offered by placement of renewable resources, depends on the resilience in the monitoring process of these resources. This can be considered as the main difference between the algorithm proposed in [23] and the current study. In [23], the main goal is to design a WAMS that has more resilience for states of the nodes with more structural importance, while in this study we have aimed to design a WAMS with added resilience, for those locations that renewable resource are more likely to be installed.

As discussed before, our proposed model constructs cyber dependency graph from information flow of the WAMS. In such a model, impacts (importance) of the nodes have been assumed as input for algorithm. It is also mentioned that based on the resilience goal, different kinds of "*bus impact*" may be defined in order to achieve resilience WAMS. The aim of current study is to design a WAMS that has enough resiliency to monitor renewable resources which are installed in the power grid (resilient WAMS hereafter).

3.1 Quantifying System Buses Based on Installation of Renewable Resources

The first step of proposed method is to rank system buses based on the defined goal of resiliency. It has been previously discussed that two classes of node importance (or impact) can be defined for system buses: "*Structural*" and "*Dynamic*" impacts. Since renewable resources are located in the fixed places and do not changed dynamically, structural importance is preferable and is considered in this study.

In order to rank system buses for choosing the optimal location for renewable resources, the method proposed in [17] is used. There, three different sets of system buses are chosen based on three different complex network criterion, i.e. "*Center of Mass*", "*Local Cluster Coefficient*", and nodes with higher "*Betweenness* and *Closeness*". They have shown that the set of nodes with higher betweenness and closeness metrics are the most appropriate for placing renewable resources and have the highest grant/request in all the areas (i.e., more loads are being supplied during a power outage).

It is worth nothing that the main difference between our approach and the method presented in [17] is that we aim to rank all the nodes, while they only choose a set with specific number of nodes. Consequently, we perform the following approach to rank system buses: For each node, we define a complex number named "*Complex Network Index*" (CNI) as follows:

$$CNI_i = X_i + \mathrm{i}Y_i = \frac{B_i + \mathrm{i}C_i}{\sqrt{B_i{}^2 + C_i{}^2}} \tag{2}$$

where, CNI_i is complex network index of ith node, and B_i and C_i are respectively normalized values (in %) of the betweenness and closeness centralities.

As an example, the CNI values of the nodes in the IEEE 30 bus system are depicted in Fig. 1. As it can be seen in the figure, all the nodes are located within the circle with the radius of 100 in the first quarter of the complex plane. It can also be observed that most of the nodes have low or zero betweenness value.[1]

Based on our examination, in order to rank the nodes for placement of renewable resources, priority should be given to the betweenness, but at the same time high values of closeness must also be considered. To achieve this, we have used a novel approach to define four different areas by using three different oval curves described below:

$$\left(\frac{4X}{k}\right)^2 + Y^2 = 100^2 \qquad k = 1, 2, 3 \tag{3}$$

Using the above curves, four impact (importance) areas have been created in the complex plane in the scale [Very Low, Low, High, and Very High]. It can be seen that in such leveling approach, an increase in the closeness value of a node may level up such a node to higher impact areas. Using this area partitioning and depending on where the CNI of a node is located, the impact of such a node will be determined. The numerical impact values of 1 to 4 have been assigned to the four different impact levels in the qualitative scale defined above (1: Very Low to 4: Very High). The impact values of IEEE 30 bus system are shown in Table 1.

3.2 Designing Resilience WAMS for IEEE 30 Bus System

In order to validate our proposed methodology for designing resilent WAMS, we design and assess two independent scenarios for IEEE 30 bus test case: In the first scenario, we examine the design of a cost optimal WAMS infrastructure, where the

[1] The zero betweenness value is for "*Pendant*" nodes. A pendant node is the node that only connects to rest of graph with a single link.

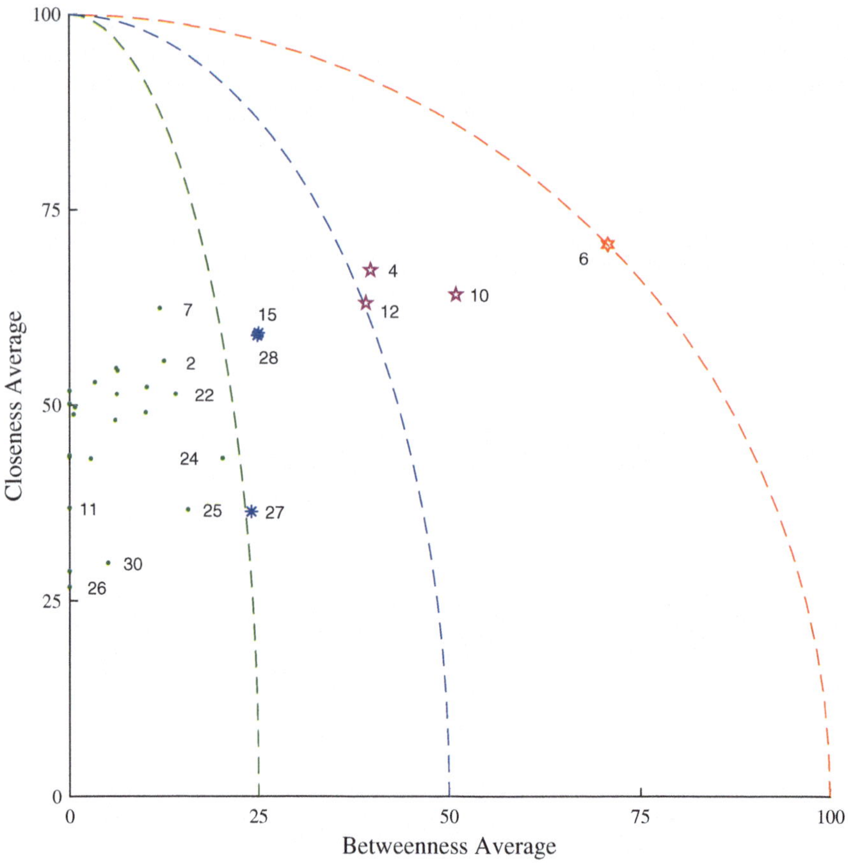

Fig. 1 Quantifying IEEE 30 bus nodes based on importance of placing renewable resources

Table 1 Impact values of buses for IEEE 30 bus system

Impact Level Impact Value	Very Low 2.38%	Low 4.76%	High 7.14%	Very High 9.52%
Bus No.	1–2–3–5–7–8–9–11–16–17–18–19 20–21–22–23–24–25–26–29–30	15–27–28	4–10–12	6

total cost of WAMS implementation is considered as initial design parameters. To achieve cost optimal WAMS, we use the optimization problem which have been previously proposed in [19] as follows:

$$\begin{cases} \min(cost_{PMU} + cost_{OPGW}) \\ s.t. \begin{cases} \overset{+}{A} \cdot \overrightarrow{PMU} > \hat{1} \\ G_{CO}(V_{CO}, E_{CO}) \text{ is connected} \end{cases} \end{cases} \tag{4}$$

where, $cost_{PMU}$ and $cost_{OPGW}$ respectively represent the normalized value of the total cost of PMUs and OPGW links (including their installation costs). $\overset{+}{A}$ is generalized adjacency matrix of power system represented by graph $G(V, E)$ ($\overset{+}{A} = A + I_{n \times n}$). $\overrightarrow{PMU}$ is a vector that shows the location of the PMUs in $G(V, E)$. $G_{CO}(V_{CO}, E_{CO})$ is connected subgraph of $G(V, E)$ in which V_{CO} and E_{CO} represents the location of routers and OPGW links, respectively. Finally $\hat{1}$ is the n-dimension vector whose all arrays equal to 1.

In the second scenario, we design the resilient WAMS structure in the presence of renewable resources. In this scenario the design approach is similar to the previous optimization (i.e. Eq. (4)), except that the resilient threshold is also added to the problem as a hard constraint [23]. The optimization problem is as follows:

$$\begin{cases} \min(cost_{PMU} + cost_{OPGW}) \\ s.t. \begin{cases} \overset{+}{A} \cdot \overrightarrow{PMU} > \hat{1} \\ G_{RC}(V_{RC}, E_{RC}) \text{is connected} \\ Res(\mathcal{G}) T_{Res} \end{cases} \end{cases} \tag{5}$$

where, $G_{RC}(V_{RC}, E_{RC})$ stands for communication infrastructure; $Res(\mathcal{G})$ is dependency graph constructed by $G(V, E)$, $G_{RC}(V_{RC}, E_{RC})$ and $\overrightarrow{PMU}$; and T_{Res} stands for threshold value for resiliency of WAMS.

For IEEE 30 bus test case, first we apply proposed algorithm using the obtained impact scale described in 3.1. We apply optimization methods presented in Eq. (4) and Eq. (5) to obtain the cost optimal and resilient solutions for this test case ($T_{Res} = 45\%$). Then, in order to show the robustness of the resilient solution, we have used our proposed method represented in Eq. (1) to calculate the resilience level and to compare both solutions.

Figure 2 shows the cost optimal WAMS for IEEE 30 bus test system. This WAMS is designed by placing 10 PMUs and installing 22.21% of transmission lines with OPGW fiber. The resilience level of this solution is 40.63%. Figure 3 shows the resilient WAMS for this system that is implemented by 13 PMUs and 41.99% OPGW fiber links. By adding 3 extra PMUs and 19.78% extra OPGWs, the resilience level of this solution is equal to 47.19% and is above the defined threshold. Note that in both Figs. 2 and 3, PMU enabled buses are in blue, the red values denote PMU impact values, while the blue values correspond to OPGW impact.

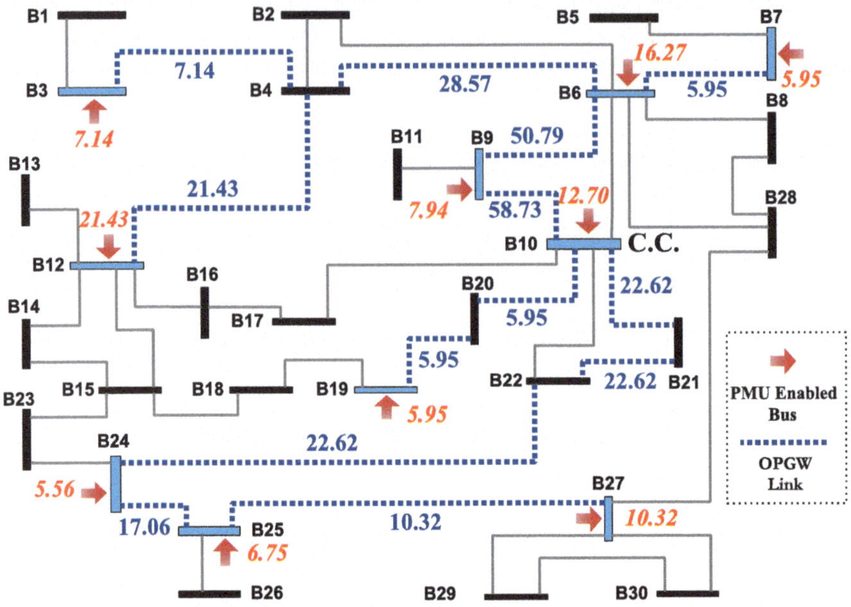

Fig. 2 Cost-optimal WAMS for for IEEE 30 bus test system—Res: 40.63%.

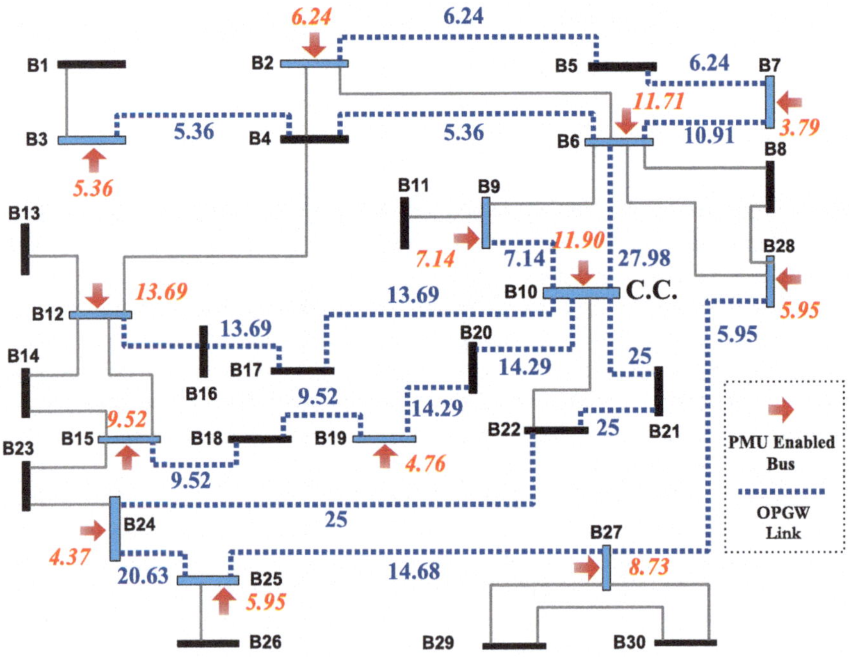

Fig. 3 Resilience WAMS for for IEEE 30 bus test system—Res: 47.19%

Table 2 Importance of WAMS Components for the Two Solutions—Left: Cost Optimal WAMS Shown in Fig. 2—Right: Resilience Solution Shown in Fig. 3

Cost Optimal Res = 40.63%				Final Case Res = 47.19%					
PMU	Imp	OPGW	Impact	PMU	Imp	OPGW	Impact	OPGW	Impact
3	7.14	(3,4)	7.14	2	6.94	(3,4)	5.36	(19,20)	14.29
6	16.27	(4,6)	28.57	3	5.36	(2,5)	6.94	(10,21)	25
7	5.95	(6,7)	5.95	6	11.71	(4,6)	5.36	(21,22)	25
9	7.94	(6,9)	50.79	7	3.97	(5,7)	6.94	(22,24)	25
10	12.7	(9,10)	58.73	9	7.14	(6,7)	10.91	(24,25)	20.63
12	21.43	(4,12)	21.43	10	11.9	(6,10)	27.98	(25,27)	14.68
19	5.95	(10,20)	5.95	12	13.69	(9,10)	7.14	(27,28)	5.95
24	5.56	(19,20)	5.95	15	9.52	(12,16)	13.69		
25	6.75	(10,21)	22.62	19	4.76	(10,17)	13.69		
27	10.32	(21,22)	22.62	24	4.37	(16,17)	13.69		
		(22,24)	22.62	25	5.95	(15,18)	9.52		
		(24,25)	17.06	27	8.73	(18,19)	9.52		
		(25,27)	10.32	28	5.95	(10,20)	14.29		

Table 2 represents all impact values for both solutions. As in can be seen in resilient WAMS solution, robustness is achieved by 3 extra PMUs and also 19.78% extra OPGW links in the system. It is also can be observed that the extra PMUs are placed on buses with high impact values (e.g. buses 15 and 28) or on the neighboring of important nodes (e.g. bus 12). This is due to the fact that the most important (high impact) nodes should be over-determined by more than one PMU. As a consequence of above facts, the results confirm the effectiveness of the proposed algorithm.

4 Conclusion

Resilience is a main design goal for Wide Area Monitoring Systems in smart grids. Research efforts found in the literature for WAMS resilience stem from resilience approaches used in critical infrastructures, and may target to increase one or more of the resilience characteristics, such as the robustness or the redundancy of critical elements. In addition, WAMS resilience approaches may target to one or more layers of WAMS, such as the processing, communications or measuring layer. Recent studies attempt to combine both robustness with controlled redundancy and increase the resilience in all the layers of WAMS.

In this chapter we have reviewed the current literature in WAMS resilience. We have also extended our previous work [23], in order to consider the resilience of WAMS structure in the presence of renewable resources. The main goal is to design a WAMS infrastructure that inherently has added resilience for those locations that renewable resource are more likely to be installed. Our simulation results demonstrate that such an approach may provide added resilience in the complete infrastructure. Possible extensions that may enhance the proposed model, involve the consideration of the electricity market conditions, which may be used to enhance the cost optimization model. Another aspect that is interesting to examine is to incorporate into our model the special characteristics of renewable resources (e.g. reheat or superheat of wind turbines) which may also affect the expected level of resilience. Also, in this study PMU-based state estimator is only consiered as WAMS. The proposed methodology can be extended for resiliency of the hybrid state estimators, by concurrently utilizing PMU and SCADA information.

References

1. F. Aminifar, A. Khodaei, M. Fotuhi-Firuzabad, M. Shahidehpour, Contingency-constrained pmu placement in power networks. IEEE Transactions on Power Systems **25**(1), 516–523 (2009)
2. BERKELEY, A., WALLACE, M., AND COO, C. A framework for establishing critical infrastructure resilience goals. *Final Report and Recommendations by the Council, National Infrastructure Advisory Council* (2010)
3. DEMIR, K., AND SURI, N. Towards ddos attack resilient wide area monitoring systems. In *Proceedings of the 12th International Conference on Availability, Reliability and Security* (2017), ACM, p. 99
4. L.C. Freeman, Centrality in social networks conceptual clarification. Social networks **1**(3), 215–239 (1978)
5. GÖL, M., AND ABUR, A. Optimal pmu placement for state estimation robustness. In *IEEE PES Innovative Smart Grid Technologies, Europe* (2014), IEEE, pp. 1–6
6. KAVASSERI, R., AND BANERJEE, A. Warp: Wide area resilient protection, national science foundation (nsf), 2017
7. KOTZANIKOLAOU, P., THEOHARIDOU, M., AND GRITZALIS, D. Assessing n-order dependencies between critical infrastructures. *International Journal of Critical Infrastructures 6 9*, 1-2 (2013), 93–110
8. D. Mah, P. Hills, V.O. Li, R. Balme, *Smart grid applications and developments* (Springer, 2014)
9. OFFICE, C. Keeping the country running: natural hazards and infrastructure, 2011
10. D.E. Olivares, A. Mehrizi-Sani, A.H. Etemadi, C.A. Cañizares, R. Iravani, M. Kazerani, A.H. Hajimiragha, O. Gomis-Bellmunt, M. Saeedifard, R. Palma-Behnke et al., Trends in microgrid control. IEEE Transactions on smart grid **5**(4), 1905–1919 (2014)
11. M. Panteli, P. Mancarella, Modeling and evaluating the resilience of critical electrical power infrastructure to extreme weather events. IEEE Systems Journal **11**(3), 1733–1742 (2015)
12. J. Paudel, X. Xu, K. Balasubramaniam, E.B. Makram, A strategy for pmu placement considering the resiliency of measurement system. Journal of Power and Energy Engineering **3**(11), 29 (2015)
13. P.A. Pegoraro, S. Sulis, Robustness-oriented meter placement for distribution system state estimation in presence of network parameter uncertainty. IEEE Transactions on Instrumentation and Measurement **62**(5), 954–962 (2013)
14. PEREIRA, J. C., MIRANDA, V., AND SARAIVA, J. T. Fuzzy control of state estimation robustness. In *Proceedings of the 14th Power Systems Computation Conference* (2002)
15. QIAN, T., XU, H., ZHANG, J., CHAKRABORTTY, A., MUELLER, F., AND XIN, Y. A resilient software infrastructure for wide-area measurement systems. In *2016 IEEE Power and Energy Society General Meeting (PESGM)* (2016), IEEE, pp. 1–5

16. Rajabi, A., and Bobba, R. B. A resilient algorithm for power system mode estimation using synchrophasors. In *Proceedings of the 2nd Annual Industrial Control System Security Workshop* (2016), ACM, pp. 23–29
17. M. Saleh, Y. Esa, A. Mohamed, Applications of complex network analysis in electric power systems. Energies **11**(6), 1381 (2018)
18. M. Shahidehpour, Y. Wang, *Communication and control in electric power systems: applications of parallel and distributed processing* (John Wiley & Sons, 2004)
19. M. Shahraeini, M.S. Ghazizadeh, M.H. Javidi, Co-optimal placement of measurement devices and their related communication infrastructure in wide area measurement systems. IEEE Transactions on Smart Grid **3**(2), 684–691 (2012)
20. Shahraeini, M., Javidi, M., and Ghazizadeh, M. A new approach for classification of data transmission media in power systems. In *2010 International Conference on Power System Technology* (2010), IEEE, pp. 1–7
21. M. Shahraeini, M.H. Javidi, Wide area measurement systems. Advanced topics in measurements 303–322 (2012)
22. M. Shahraeini, M.H. Javidi, M.S. Ghazizadeh, Communication infrastructure planning for wide area measurement systems in power systems. International Journal of Communication Networks and Distributed Systems **10**(4), 319–334 (2013)
23. M. Shahraeini, P. Kotzanikolaou, A dependency analysis model for resilient wide area measurement systems in smart grid. IEEE Journal on Selected Areas in Communications **38**(1), 156–168 (2019)
24. I. Stellios, P. Kotzanikolaou, M. Psarakis, C. Alcaraz, J. Lopez, A survey of IoT-enabled cyberattacks: Assessing attack paths to critical infrastructures and services. IEEE Communications Surveys & Tutorials **20**(4), 3453–3495 (2018)
25. Wang, J. Design of wide-area power system damping controllers resilient to communication failures. *Cybersecurity for Energy Delivery Systems Peer Review* (2016)
26. S. Zhang, V. Vittal, Design of wide-area power system damping controllers resilient to communication failures. IEEE Transactions on Power Systems **28**(4), 4292–4300 (2013)

Cyber Kill Chain-Based Hybrid Intrusion Detection System for Smart Grid

Vivek Kumar Singh and Manimaran Govindarasu

Abstract Today's electric power grid is a complex, automated, and interconnected cyber-physical system (CPS) that relies on supervisory control and data acquisition (SCADA)-based communication infrastructure for operating wide-area monitoring, protection, and control (WAMPAC) applications. With a push towards making the grid smarter, the critical SCADA infrastructure like power system is getting exposed to countless cyberattacks that necessitate the development of state-of-the-art intrusion detection systems (IDS) to provide comprehensive security solutions at different layers in the smart grid network. While considering the continuously evolving attack surfaces at physical, communication, and application layers, existing conventional IDS solutions are insufficient and incapable to resolve multi-dimensional cybersecurity threats because of their specific nature of the operation, either a data-centric or protocol-centric, to detect specific types of attacks. This chapter presents a hybrid intrusion detection system framework by integrating a network-based IDS, model-based IDS, and state-of-the-art machine learning-based IDS to detect unknown and stealthy cyberattacks targeting the SCADA networks. We have applied the cyber-kill model to develop and demonstrate attack vectors and their associated mechanisms. The hybrid IDS utilizes attack signatures in grid measurements and network packets as well as leverages secure phasor measurements to detect different stages of cyber-attacks while following the kill-chain process. As a proof of concept, we present the experimental case study in the context of centralized wide-area protection (CWAP) cybersecurity by utilizing resources of the PowerCyber testbed at Iowa State University (ISU). We also describe different classes of implemented cyber-attacks and generated heterogeneous datasets using the IEEE 39 bus system. Finally, the performance of the hybrid IDS is evaluated based in terms of detection rate in real-time cyber-physical environment.

V. K. Singh (✉) · M. Govindarasu
Department of Electrical and Computer Engineering, Iowa State University, Ames, IA, USA
e-mail: vsingh@gmail.com

M. Govindarasu
e-mail: gmani@iastate.edu

H. Haes Alhelou et al. (eds.), *Wide Area Power Systems Stability, Protection, and Security*, Power Systems, https://doi.org/10.1007/978-3-030-54275-7_22

Keywords Cyber-physical system · Intrusion detection systems · Centralized wide-area Protection · Smart grid · Cyber kill chain · PMUs · WAMs · Cyber-attack

1 Introduction

Today's energy infrastructure is undergoing a massive transformation across all generation, transmission, and distribution systems to provide reliability, efficiency, and sustainability to the power system network. With a high dependence on advanced communications, as well as the increasing integration of smart meters and sophisticated controls, electric power systems have evolved into densely interconnected cyber-physical systems and the existing information technology (IT)-based cybersecurity measures are often ineffective at preventing them. In recent years, several WAMPAC applications, such as state estimation, wide-area protection scheme (WAPS), wide-area voltage controller (WAVC), etc., are developed to provide real-time monitoring and control as necessary to maintain the stability and reliability of the power system [1]. Since these WAMPAC applications are not conventionally designed to handle unexpected cybersecurity threats, any unusual malfunction or signicant operational delays, triggered through cyber-attacks, can affect the system observability, reliability, and stability of power system. This motivates the need to go beyond the traditional paradigm of "security by obscurity" and "bolt-on" security measures of retrofitting the existing system with conventional security solutions and develop a suite of layered innovative security solutions to enhance the grid resiliency against possible cybersecurity threats. Several efforts, such as the National Institute of Standards and Technology Interagency Report (NISTIR) 7628 [2], DOE Cyber Security Roadmap for Energy Delivery Systems [3], DOE Electricity Subsector Cybersecurity Capability Maturity Model (ES-C2M2) [4], North American Electric Reliability Corporation Critical Infrastructure Protection (NERC CIP) Standards [5], and National Electric Sector Cybersecurity Organization Resource (NESCOR) reports [6], have provided an in-depth understanding to identify cybersecurity vulnerabilities and develop mitigation and preventive strategies. Further, the IEEE guide published by the Power System Relaying Committee [7] recommends strong cybersecurity practices and measures, including access controls, firewalls, cryptography; however, poor security key management, weak cryptography, and misconfigured firewall rules can degrade the secure operation of the power system. Therefore, it is highly imperative to thoroughly investigate the cybersecurity vulnerabilities and develop state-of-the-art detection and defense techniques to neutralize possible threats and make the grid *attack-resilient*.

1.1 Challenges for IDS in Smart Grid

The current IDS solutions in the power system face several challenges in detecting anomalies accurately and timely by analyzing multi-dimensional data at physical, network, and application layers, which are elaborated in greater detail below.

(1) ***Detection Latency***: Several WAMPAC applications have a stringent timing requirement to perform their optimal control operations. Therefore, in order to support the seamless integration of IDS with the grid infrastructure, detection latency has to be minimized so that it does not affect the normal operation of the power system. Also, the operational timing requirement of different applications varies widely. For example, the centralized wide-area protection scheme (CWAPS) has a strict timing requirement, typically in the order of 50–150 ms, AGC operates in approximately 4–8 s, and economic dispatch operates every 5 min. Further, the development of attack-resilient infrastructure requires an immediate incident response that can quickly restore the grid condition to the normal state.

(2) ***Robustness and Consistency***: Since cyber-attacks vary from a nave level to a sophisticated level, it is not justified to completely rely on the specific types of IDS to detect all classes of attacks. Although the SIDS shows a high accuracy without signaling false alarms as compared to the ABIDS, it is only able to detect known cyber-attacks, and thus it has to be updated regularly with newly discovered attack signatures. The current existing ABIDS and SIDS show good accuracy; however, it is difficult to obtain detailed information from them about different classes of attacks. Further, the advanced persistent threat (APT) actors can leverage their skill sets, expertise knowledge, intellectual capabilities, and operational tools resources to perform sophisticated and coordinated cyber-attacks by manipulating cyber and physical information in spatial and temporal domains. These stealthy cyber-attacks are difficult to be detected by a signature or anomaly-specific IDS at the network or host level.

(3) ***Data Management***: The current operational technology (OT) environment is inundated with an overload of network information and power measurements that frequently lead to the challenge of Big data, which is difficult to handle from the conventional IDS perspective. The current big data challenge is driven by volume, velocity, and veracity of data. Volume in the smart grid environment includes line flows, relay status, phasor measurement, network packets, etc. that significantly contribute to the large volume of data. Velocity refers to the sampling rate at which data is processing at substation and control center networks; and variety refers to the complex data problem, including multi-dimensional structure data, high dimensional data, and data from multiple independent sources. Further, since the current grid infrastructure depends on multiple hardware and software resources for handling grid measurements and network traffic, there exists no real-time sensing platform that can allow the processing of heterogeneous datasets to facilitate the operation of different types of IDSs.

1.2 Related Work

In recent years, there has been a strong urge in the development of IDS pertinent to the cyber-physical security in the smart grid environment. Several researchers with different backgrounds have proposed different types of IDSs, such as model-based, rule-based, protocol-specific, machine learning or data-mining-based IDS, etc. for the smart grid cyber-physical security. In [8] and [9], the authors have proposed a model-based IDS using real-time load forecast information to detect faulty SCADA measurements in the context of automatic generation control (AGC) in the power system. In [10], a decision tree-based supervised machine learning algorithm is applied to detect malicious tripping of relays. In [11], the authors have shown how the model-based IDS can be developed in a centralized manner using load forecasts and secure phasor measurements for the state estimation. Although the centralized IDS is developed pertinent to the energy management system (EMS) applications like state estimation, automatic generation controller (AGC), etc., the multi-agents-based distributed IDS is also proposed in [12] to avoid a single point of failure while detecting anomalies in the decentralized protection scheme. Apart from the anomaly-based IDSs, several signature-based IDSs [13]–[16] are also proposed that perform deep packet inspection on the SCADA and synchrophasor communication protocols to detect cyber intrusions in real-time. Further, in [14], the authors show how two open-source IDS tools-Snort and BRO, can be utilized in detecting a data integrity attack using the timing information of two consecutive network packets and compared their performances in terms of accuracy and latency rates. Although the rules-based IDS works well in detecting malicious network traffic using cyber logs, it requires an intensive knowledge and rigorous analysis for developing rules and is apposite for the big-data problem. Meanwhile, several research efforts have shown the application of machine learning algorithms and data mining techniques in detecting malicious and non-malicious events like line faults. Pan et al. presents a learning-based IDS using the common path mining technique to classify cyber-attacks, normal operations, and physical disturbances [17]. The common path mining technique learns temporal patterns for different scenarios using synchrophasor measurements and audit logs in an automated way.

2 Intrusion Detection System in Smart Grid

2.1 Smart Grid Communication Architecture

The smart grid network consists of multiple communication architectures that are interconnected through physical and application layers to facilitate real-time monitoring of the grid network and protect the grid's health on spatial and temporal levels. These communication architectures can be classified into three major categories: SCADA network, synchrophasor network, and advanced metering infrastructure (AMI) network.

(1) **SCADA Network**: Consists of a remote terminal unit (RTU), as a substation gateway that receives measurements from sensors, transducers, and instruments, which are located at remote grid stations as field devices and transmit data to the control center for real-time monitoring and control of generation, transmission, and distribution systems every few seconds. Several communication protocols like IEC 61850, IEC 60870, Modbus, Distributed Network Protocol 3 (DNP3), etc. are utilized to support SCADA-based wide-area applications.

(2) **Synchrophasor Network**: Includes phasor measurement units (PMUs), phasor data concentrators (PDCs), global positioning system (GPS) clocks, Transmission Control Protocol/Internet Protocol (TCP/IP) network infrastructure, and data storage and collection system. The traditional SCADA system fails to provide faster and high-resolution measurements that are necessary for wide-area dynamic monitoring and control of the power system. These limitations can be overcome by deploying PMUs in the field-area network (FAN) that provides phasor measurements at a sampling rate of 30 to 120 samples/second to support mission-critical applications. Further, there has been a rapid shift in extending the SCADA EMS to incorporate synchrophasor-based wide-area control applications like a wide-area voltage controller (WAVC), oscillation damping, etc. [18]. The authors of [19] present the design and architecture of synchrophasor-based WAPS for different types of applications, including voltage instability, oscillation monitoring, thermal overloading, etc.

(3) **AMI Network**: Includes smart meters, data aggregator, data manager, and communication network that facilitates the bi-directional communication between smart meters and grid utilities for exchanging information related to power consumption, outage reporting and awareness, and price updates. During normal operation, data aggregators (DAs) receives real-time consumption information from smart meters and sends necessary commands through the neighborhood-area network (NAN). Further, smart meters also communicate with devices, located on the customer's premises, through the home-area network (HAN). ANSI C12 series is the most commonly used communication protocol in the US, while IEC62056 protocols dominate the AMI market in the EU to support the information exchange in a wide-area network (WAN), NAN, and HAN of AMI [20].

Fig. 1 clearly illustrates attack surfaces, as shown by lightning bolt symbols, in the grid network that can be exploited by attackers based on the existing vulnerabilities at device, network, and application levels. Since the communication protocols in the grid network are not encrypted, there is numerous possible scope of cyber-attacks, despite the existing defense mechanisms, such as firewall, a virtual private network (VPN), etc. For example, an attacker can sniff the clear text communication packets going between the control center and substation networks to perform data integrity attacks over WAN. An attacker can also sniff wide-area communication packets to develop a network footprint, and later perform Man-in-the-Middle (MITM) or denial of service (DoS) attacks to undermine the system observability and controllability. In AMI, the deployed smart meters operate as an interface between the utility's network and HAN or NAN, which makes them an ideal target for cyber-attacks.

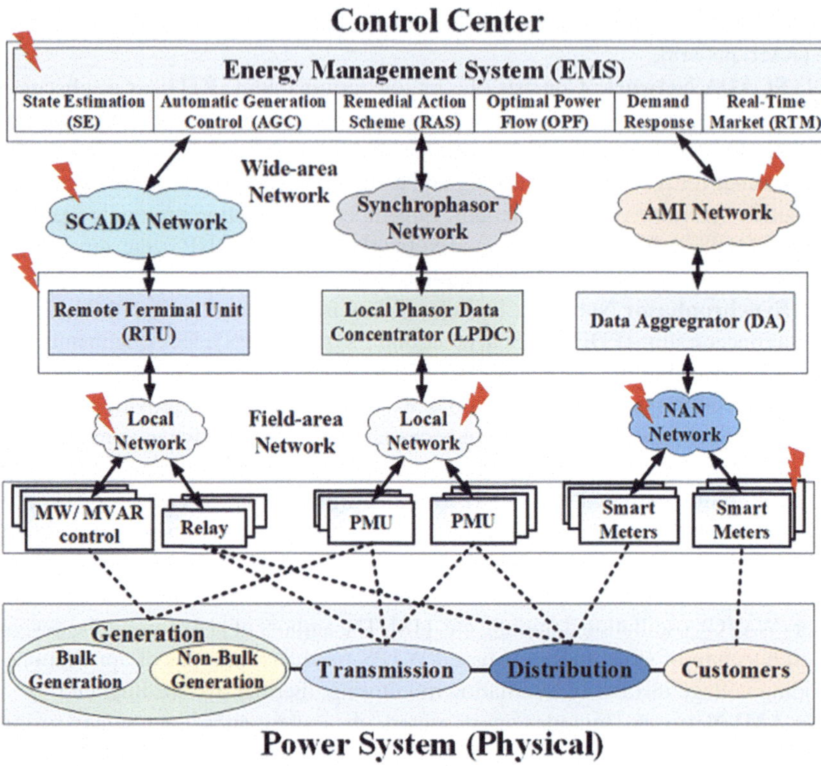

Fig. 1 High-level schematic architecture of the smart grid

Also, several vulnerabilities have been reported in smart meters installation. For example, a control unit system in smart meters is subjected to reverse engineering, side-channel, and data-integrity attacks. Further, at the national level, the advanced and nation-sponsored attackers can perform stealthy and coordinated cyber-attacks, such as compromising control centers that are difficult to detect using conventional security methods.

2.2 Intrusion Detection System Taxonomy

IDS is based on the notion that the system behavior during cyber-attacks would be different from legitimate behavior. Several types of IDSs are developed to detect anomalies accurately and timely at physical, cyber, and application layers in the power system. Fig. 2 shows the taxonomy of IDS based on their locations and the nature of operations in the grid network. Based on the location-based taxonomy, it can be classified into two different types: Network and Host-based IDSs.

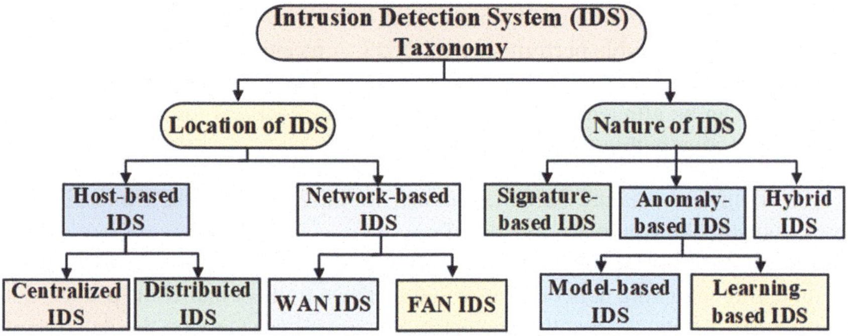

Fig. 2 Taxonomy of intrusion detection system (IDS) in the power system

(1) **Network-based IDS (NIDS)**: This IDS inspects the communication traffic to detect security breaches in the grid network. Different techniques like port-mirroring, network taps, switched port analyzer, etc. are developed by the network infrastructure vendors to facilitate packet sniffing in the context of NIDS. It can be classified into two broad categories: WAN IDS and FAN IDS.

- WAN IDS (WIDS): This IDS is deployed over WAN that sniffs the network traffic and detects anomalies using the statistical or baseline models and attack signatures. Potential challenges related to the WIDS are network data overload, quality of service (QoS), and encryption, and signature lag time.
- FAN IDS (FIDS): This IDS passively monitors network traffic of smart meters and actuators in a local area network (LAN), FAN, or NAN and detects anomalies in real-time. It is generally deployed at the field level, made tamper-resistant, and supported extra computational power and memory to forward the detected alerts to gateways and central IDS deployed at the utility control center.

(2) **Host-based IDS (HBIDS)**: It is deployed at operating and application systems and operated through an internal computing system to monitor and analyze traffic patterns at the system level. Based on its location in the grid network, it can be classified into broad categories: Centralized IDS and distributed IDS.

- Centralized IDS (CIDS): It requires global measurements to analyze cyber and physical events; and hence, it is deployed at the control center level and provides much accurate and consistent detection performance because of its global view of the grid network. It can also be utilized to develop application-specific IDS for the EMS like bad data detection in the synchrophasor network [21] by monitoring incoming measurements and outgoing control signals. However, it is also an ideal target for cyber-attacks and can lead to a single point of failure, if compromised that can render the whole EMS control center vulnerable to cyber threats.
- Distributed IDS (DIDS): It overcomes the limitation of a *"single point of failure"* by introducing multiple autonomous IDS agents that are deployed at the substation levels. In DIDS, it is crucial to consider an efficient communication topology like a

mesh network and communication standard with an anomaly detection algorithm for optimal and reliable performance of DIDS. A recent survey in [22] has shown the efficiency of a mesh network topology and communication standard like Zigbee [23] can be efficiently utilized to provide low-cost and low power-standard communication within the wireless network.

According to the nature-based taxonomy, we have classified IDS into different types: signature and anomaly-based IDSs.

(1) **Signature-based IDS (SIDS)**: This IDS detects anomalies based on the notion of comparing incoming network traffic to the known trails of malicious packets that are stored in the attack signature database. Since it relies solely on the database of known attack signatures, it cannot detect unknown or new attacks that do not match with the existing attack signatures. Several IDS tools, such as Snort, BRO (Zeek), Firestorm, Spade, etc., can be utilized in developing signature-based IDS in real-time based on the defined rules.

(2) **Anomaly-based IDS (ABIDS)**: It identifies malicious events based on deviations in the normal system behavior instead of looking into the library of known attack patterns. Based on the statistical profiling, it develops a baseline of normal cyber and physical activities and sends alert messages in real-time to the control center operators if the system deviates from the defined baseline. Different types of IDSs, such as model-based IDS, machine learning-based IDS, multi-agents-based IDS, etc. can be an integral part of anomaly-based IDS for detecting attacks in the power system.

- Model-based IDS (MIDS): This IDS leverages protocol information and historical and redundant measurements for developing a prediction model, and malicious and unknown attacks are detected based on behavior-based rules that are defined during the statistical and temporal correlation analysis of incoming data streams. Behavior-based rules include timing-based rules, range-based rules, transmission line status-based rules as well as rules defined based on the coordination and correlation of different events.
- Learning-based IDS (LIDS): This IDS applies several state-of-the-art machine learning algorithms like supervised and unsupervised algorithms, and data mining techniques to detect unknown, stealthy, and coordinated cyber-attacks. It relies on an immense volume of power system data to develop a non-linear complex relationships as necessary to distinguish between natural disturbances, malicious, and non-malicious events. This IDS involves data pre-processing, input feature selection, training, and real-time testing of different participating classifiers, and based on their performances, the best classifier is selected for optimal decision making.

(3) **Hybrid IDS (HIDS)**: This IDS integrates the conventional signature-based IDS with anomaly-based IDS on a common platform to provide better accuracy in detecting multi-level intrusions by utilizing both network and power system information while exhibiting a minimum detection latency. This IDS also overcomes the weakness of other previously discussed IDSs by leveraging the best qualities of other

IDSs while simultaneously monitoring physical, network, and application layers to minimize the attack surface in the grid network.

3 Hybrid Intrusion Detection System: A Potential Solution

Since it is imperative to thoroughly investigate possible cyber-attacks in the grid network to minimize attack surfaces, it is imperative to develop a comprehensive and robust IDS that can provide an optimal detection performance with a minimum detection latency while addressing the discussed IDS challenges. The HIDS provides one of the promising solutions that incorporates both grid measurements and network traffic information to minimize blind spots from the traditional IDS and capture possible intricacies at physical, cyber, and application layers [24]. In particular, the mechanism of HIDS combines network logs-based and data-driven approaches to detect different types of malicious events in the system operations.

Figure 3 shows the high-level schematic architecture of HIDS that integrates the existing conventional SIDS with the state-of-the-art LIDS and MIDS to accurately detect multi-level intrusions at physical, network, and application layers in real-time, and also minimize detection latency by assessing the network integrity in real-time. It consists of four layers: Layer 1 presents a SIDS that detects anomalies based on

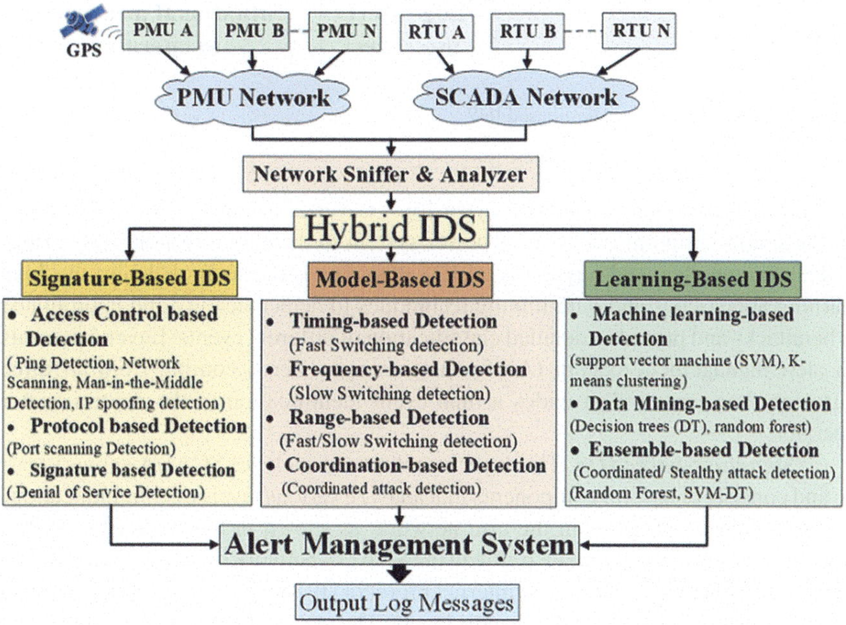

Fig. 3 High-level schematic architecture of HIDS in the smart grid

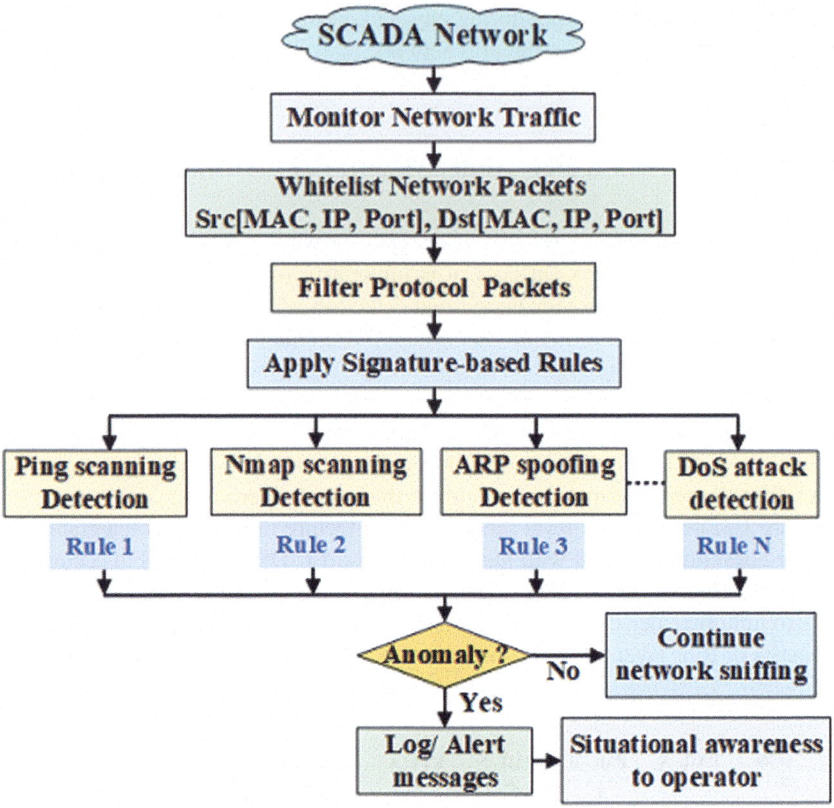

Fig. 4 Methodology of signature-based IDS

the known attack signatures using cyber logs; Layer 2 presents a MIDS that relies on the spatio-temporal behavior of power system to develop threshold-based rules to detect anomalies, and Layer 3 is a LIDS that applies the state-of-the-art machine learning approaches and data-mining techniques to detect stealthy and coordinated cyberattacks and provide a detailed classification of different events; Layer 4 presents an alert management system (AMS) that manages alerts, coming from all three intrusion detectors, and provides a final event identification to the control center operator.

(a) ***Signature-based IDS***: This layer monitors and analyzes SCADA network traffic and consists of several components that analyze various levels of communications traffic to detect anomalies in the grid network, as shown in Fig. 4. Access-control white-listing ensures legitimate traffic in the SCADA network by white-listing media access control (MAC) addresses, internet protocol (IP) addresses, and port numbers in the hardware, network, and transport layers. The protocol white-listing filters specific SCADA protocols and related function codes. Further, signature-based rules are applied to detect anomalies based on the known attack signatures like ping scanning detection, Nmap scanning detection, address resolution protocol (ARP) spoofing detection, denial of service (DoS) attack detection, etc., which are publicly available in open-source IDS databases. Finally, the generated alert messages during network intrusions are directly forwarded to the AMS to provide situational awareness about network intrusions to the control center operator.

(b) ***Model-based IDS***: This layer performs an in-depth analysis of SCADA communication protocols and analyzes spatio-temporal behaviors of power systems to define normal and legitimate behavior models. In particular, it filters the SCADA communication packets, computes incoming packet rate, and extracts digital and analog values to develop behavior-based rules. Several behavior-based rules can be developed based on the timing of packets, range of power system variables, relays status, and a combination of behavior-based rules in a coordinated fashion, as shown in Fig. 5, to detect malicious and unknown threats. In this layer, once an anomaly is detected, the generated alert messages are fed to the machine learning classifiers as input features to accurately detect and classify cyber-physical events, including cyber-attacks and line faults.

(c) ***Learning-based IDS***: This layer applies machine-learning algorithms and data mining techniques to detect stealthy and coordinated cyber-attacks using multi-source heterogeneous system data and also differentiates cyber-attacks from natural disturbances like line faults, to provide intelligent decision support to the control center operator. For building the classification model, the SCADA and synchrophasor measurements, along with the cyber logs, are collected from the grid network and forwarded to the data aggregator at the control center. Fig. 6 shows the high-level methodology for developing machine learning-based IDS that includes two phases: offline process and online process. During the offline process, a library of heterogeneous datasets from a multi-source system has been generated for different events that are labeled later in the integer format to facilitate the supervised learning process. Afterward, data pre-processing steps are carried out to improve the data quality by formatting and sampling it to develop approximate models with data cleaning to filter inconsistent values and eliminate rows with a missing data. Further, the data transformation module is applied to normalize datasets for enhancing smoothness and homogeneity among samples followed by features selection and extraction, which filter irrelevant information and unreliable data that may affect the learning process and events prediction. Several feature selection techniques, such as filter (Pearson Correlation, Chi-Square, etc.), wrapper (best-first search method, backward elimination, etc.), and embedded methods (decision tree, L1 (LASSO) regularization) can be applied for selecting relevant features. The selected input features are utilized for developing, training, and updating machine learning models with new scenarios or cases when models are not online. Finally, during the online process, the trained model is deployed for testing multi-events classification, detecting malicious and benign events, and sending output logs to the AMS for final events identification and visualization.

(d) ***Alert management system (AMS)***: It receives alert messages from all three IDSs: NIDS, MIDS, and LIDS, and manages them through log parsers by performing real-time logical processing based on defined logic rules to prioritize alert types.

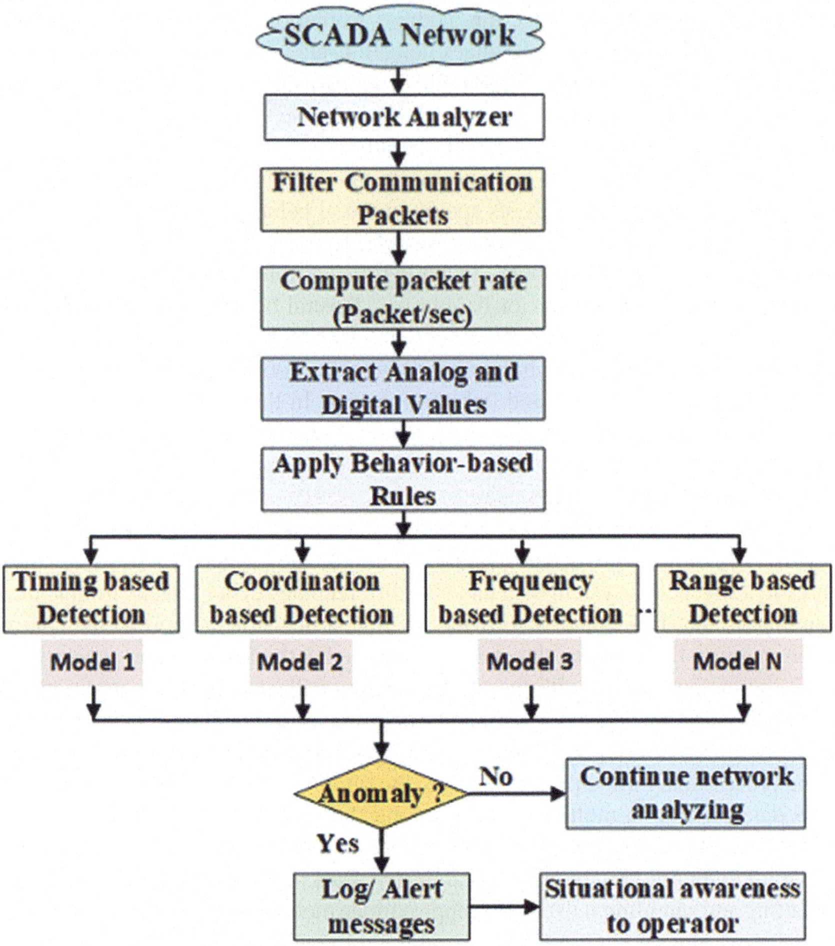

Fig. 5 Methodology of model-based IDS

Fig. 7 shows the log publishers X, Y, and Z, which are generated from network-based, model-based, and learning-based IDSs, and fed to the decision logic. The decision logic consists of two rules, i.e. rule 1 and rule 2.

Rule 1: It receives alerts from the Log Publisher X for SIDS and forwards them to the aggregator (B, C) as an output B. For example, if the output of rule 1 is $x1$, then B is also set to $x1$.

Rule 2: It receives alerts from the Log Publishers Y and Z for MIDS and LIDS and compares their alert logs to provide a final identification of events. If the alert outputs are conflicting, then LIDS is given a higher preference because of its advanced capability and sophistication in learning different types of events. For example, if Y is $y1$ and Z is $z1$, if $y1 = z1$, then C is set to $y1$ or $z1$. If $y1$ is not equal to $z1$, then C is set to $z1$ and $z1$ alert log is forwarded to the aggregator (B, C).

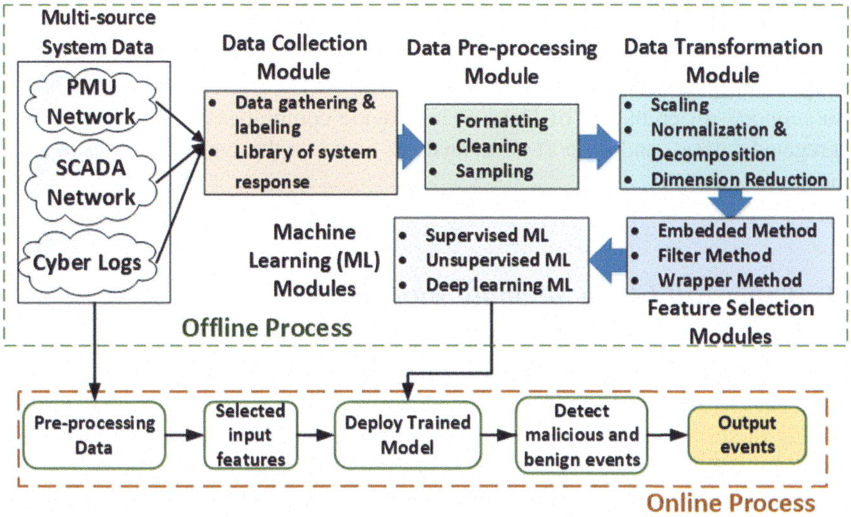

Fig. 6 Methodology of learning-based IDS

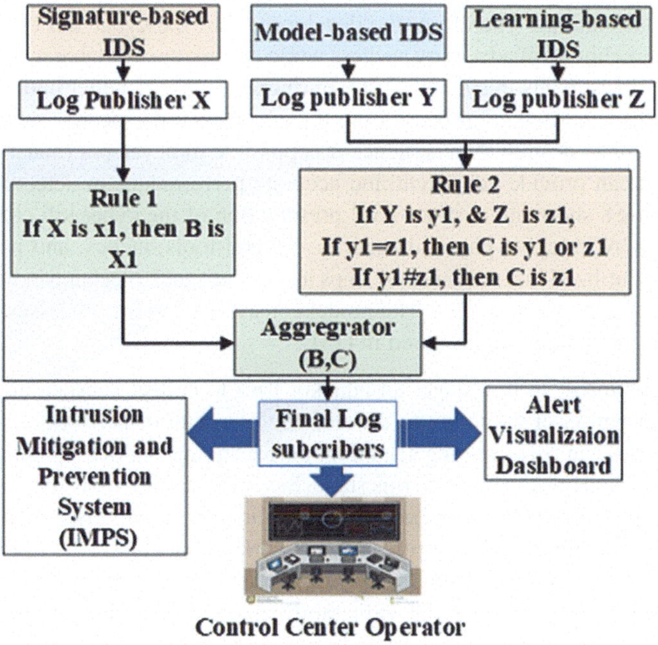

Fig. 7 High-level methodology of alert management system

The final alert logs that are coming from the aggregator (B, C) are displayed through real-time visualization dashboard and it can also be utilized for developing intrusion mitigation and prevention system (IMPS). The visualization dashboard also supports different sets of features to provide comprehensive visibility of the aggregated datasets and support event analysis based on the aggregated logs in real-time.

3.1 Cyber Kill Chain Mapping with HIDS

Since the attacker's skills, intellectual capabilities, and operational resources vary from a nave to a sophisticated level, it is imperative to have a comprehensive understanding of several attack processes and mechanisms. The cyber kill-chain model presents possible footsteps that can be utilized by attackers to successfully execute severe and stealthy cyber attacks. For example, the real cyber attack on Ukraine's power grid in 2015 [25] followed a similar cyber kill chain model, where the attackers performed a sequence of steps to understand the operational technology (OT) network and SCADA distribution system followed by shutting down multiple online distributed substations. Therefore, the development of the cyber kill chain model reduces the likelihood of adversary success while optimizing available resources and minimizing investments in cybersecurity. Further, it assists to better understand the end-to-end decision-making process from the adversary's perspective while engaging them to create desired effects; hence it is possible to develop a robust intrusion detector that can provide consistent and accurate performance in detecting cyber-attacks. Figure 8 shows an abstract-level presentation of the cyber kill chain in the context of SCADA cyber-physical security. Several tools, tactics, and procedures (TTP) can be utilized in a sequence of steps as per the attack mechanism to perform successful stealthy cyber-attacks. The model consists of various processes or stages that are elaborated here, as discussed in [13].

1. ***Reconnaissance***: In this stage, an attacker tries to collect substantial and relevant information of the target to develop the blueprint of network architecture. The attacker can perform ping scanning, port scanning, service scanning, etc. as attack mechanisms to complete this stage. Several scanning tools like Ping Scanner, Nmap, Zenmap, etc. can be leveraged to identify alive hosts, map network addresses, and figure out the up-to-date network architecture.
2. ***Access and Exploitation***: In this stage, an attacker tries to communicate or connect to a target to discover potential vulnerabilities. Later, the obtained information about the existing vulnerabilities can be exploited to gain a foothold or the privilege escalation to launch a successful attack. The vulnerability assessment or penetration testing can be used as an attack mechanism; and tools like OpenVAS, Metasploit, Nessus, etc. can be utilized to complete this stage.
3. ***Attack Launch/ Execution***: Before reaching this stage, an attacker must ensure that he has obtained the necessary privileges to execute or launch different types

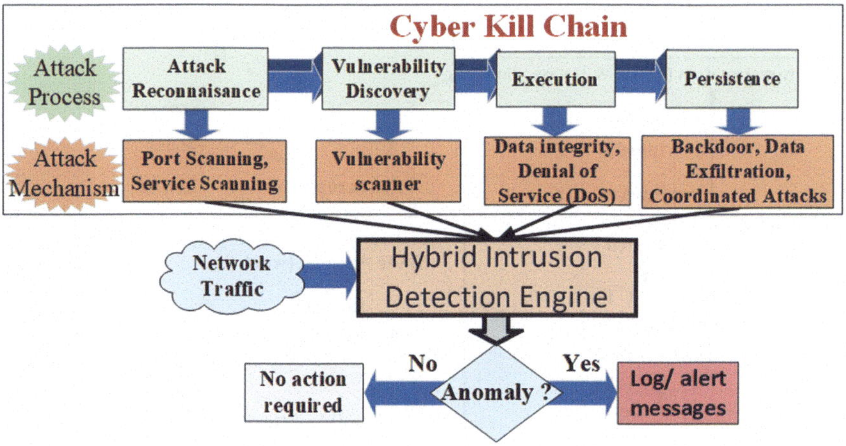

Fig. 8 Cyber kill-chain mapping with the hybrid IDS

of attacks on system measurements, control signals, wide-area communication network or operating field devices to disrupt the grid stability.

4. ***Persistence***: This is the final stage where an attacker creates an additional back-door or access channel to maintain his persistence access to the compromised system that can be exploited later for attack repetition or launching multiple attacks in a coordinated fashion.

Since our main objective is to detect all kinds of attacks, irrespective of the attacker's intelligence, different components of HIDS are developed around the kill chain and mapped with its different stages to detect attackers at an initial stage and predict their next move. Note that any disruption in the process/stage can break the chain process, and thus, it may interrupt the attacker's objective of destabilizing the grid network. Also note that the sequence of chain model can be modified, changed, and expanded depending upon the scenario, security investigation, and OT organization.

4 Case Study: Hybrid IDS for Wide-Area Protection Scheme

4.1 Problem Formulation

A centralized wide-area protection scheme (CWAPS), also known as a remedial action scheme (RAS), is an automatic protection system that performs corrective actions to prevent widespread outages and maintain the system's stability and reliability during disturbances. The corrective actions, as defined by the NERC guideline

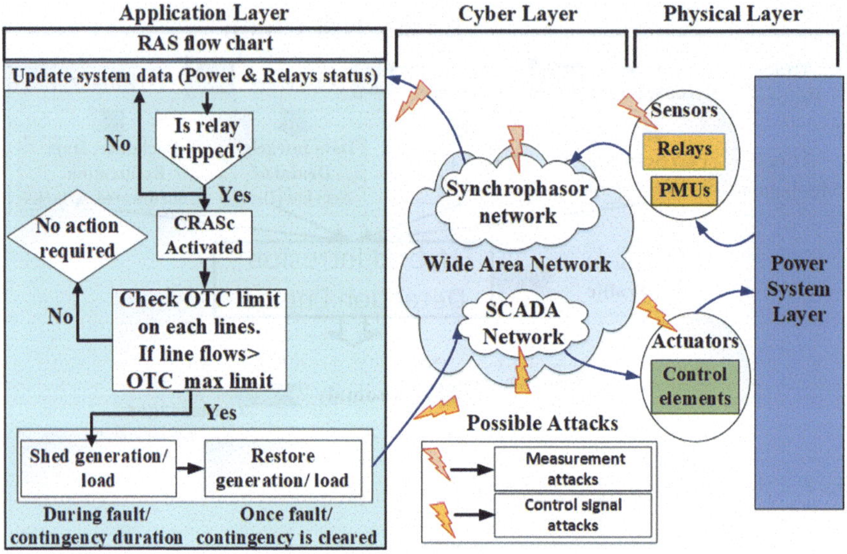

Fig. 9 Attack surface representation in the CWAPS

[26], include a change in generation and load (MW or MVAR) and system configuration. Seethalekshmi et al. presents the requirements of WAPS and explains how the WAPS overcomes the major drawbacks of the SCADA-based localized protection scheme [27]. According to the 2008 design guide of the Western Electricity Coordinating Council (WECC), RAS is divided into four different types: event-based, parameter-based, response-based, and a combination of the above [28]. The event and parameter-based schemes are open-loop faster schemes, which take inputs as relays status, line currents, voltages, etc. and perform corrective actions by shedding the load, generation, and other pre-defined actions. The response-based scheme is a close-loop slower scheme that examines the dynamic behavior while performing corrective actions. In particular, the RAS is mainly utilized during transient instability, voltage instability, and thermal overloads.

Fig. 9 shows possible attack surfaces in CWAPS architecture at physical, cyber, and application layers, as well as measurement and control sides. Since the CWAPS relies on the SCADA and synchrophasor communication networks that are interacting with data sharing devices for normal operation, the existing cybersecurity vulnerabilities can be exploited by attackers to launch simple or elaborated classes of cyber-attacks like denial of service (DoS), data integrity, etc. [29]. Moreover, it cannot prevent themselves from legitimate users who misuse their privileges to perform malicious activities. Therefore, there is a compelling urge to develop a HIDS for CWAPS that can detect attacks at an earlier stage to break the life-cycle of cyber-attacks.

In this work, we have implemented an event and parameter-based CWAPS that consists of a centralized RAS controller (CRASc). The CRASc, initially at an armed stage, collects phasor data at a regular interval in terms of relays status, power line flows, and generator output. During a single line outage, the CRASc is triggered, it checks the operational transfer capability (OTC) of the remaining adjacent lines that are directly connected to the generator. If the current line flows exceed its maximum operational transfer capability (OTC_max) limit, it performs corrective action by shedding generation to prevent the thermal overloading in other adjacent lines. Note that we have considered a thermal overload limit while computing the OTC_max, other factors like voltage and angular stabilities are ignored in this work. The OTC_max limit of each transmission line is provided through the predefined action table that also provides information about how much generation has to be reduced for the specific line contingency. Apart from the generation shedding, it is also allowed to restore the generation once the fault/contingency is cleared, as mentioned in [26].

4.2 Scenarios and Data Generation

This subsection presents several types of events, including malicious (cyber intrusions) and non-malicious events (line faults), which are considered to develop a robust hybrid IDS that can provide accurate and consistent results. Moreover, a library of the system database is generated for different scenarios for testing, validation, and evaluation.

4.2.1 Physical Line Faults

It involves different types of faults, including symmetrical and asymmetrical faults that can happen on transmission lines. In this case, we have considered 5 different types of faults: line to ground (L-G), double line to ground (LL-G), three phases to ground (LLL-G), line to line (L-L), and 3 phase faults (L-L-L). The unsymmetrical faults, (L-G), (LL-G), (L-L) are more frequent and cause uneven flows of current and phase shifts in a 3-phase power system. The symmetrical faults, (LLL-G) and (L-L-L) cause the short-circuiting of three phases and often to the ground. These faults are very rare but have a severe impact on the system's stability.

4.2.2 Cyber Attacks

We have considered several types of cyber intrusions around the kill-chain model, irrespective of the attacker's intelligence that can have a potential impact on system stability. We have classified attack vectors into two different types: IT-based attacks and SCADA-based attacks. Both are discussed in details in the remainder of this section

1. ***IT-Based Attacks***: IT-based attacks include traditional host and network-based attacks-including scanning attacks (e.g., ping and Network Mapper (NMAP) scanning), DoS attacks, and spoofing attacks (e.g., IP spoofing, ARP poisoning)–that can be deployed in the SCADA environment to develop a blueprint of the network architecture and compromise power system devices.
2. ***SCADA-Based Attacks***: SCADA-based attacks include those attacks that are defined in the OT environment pertinent to the SCADA power system. These attack vectors target insecure SCADA communications protocols, field devices, computers, and several other digital access points to inflict severe damage on the grid infrastructure. We consider three different attack vectors:
(a) **Malicious tripping attack**: This attack vector involves the malicious tripping of a physical relay that can be performed in several ways, such as unauthorized access to the control center, altering the setting of physical relays, etc. During a MITM attack between substation and control center networks, the false tripping command packets are injected to trip a circuit breaker and disconnect power system components.
(b) **Pulse attack**: This attack vector involves periodically changing an input control signal by adding the pulse attack parameter, λ_{pulse}, for a small-time interval, $(t1)$. It retains the original input for a remaining interval, $(T - t1)$, for the given time period, (T), as shown in Eq. 1.
(c) **Ramp attack**: This attack vector involves adding a time-varying ramp signal to the input control signal based on a ramp signal parameter, λ_{ramp}, as shown in Eq. 2.

$$P_{pulse} = \begin{cases} P_i(1 + \lambda_{pulse})(t = t_1) \\ P_i(t = T - t_1) \end{cases} \tag{1}$$

$$P_{ramp} = P_i + \lambda_{ramp} * t \tag{2}$$

4.3 Hybrid IDS Components

This subsection presents the three components of HIDS where SIDS is utilized to detect IT-based attacks, and MIDS and LIDS are deployed to detect SCADA-based attacks as well as physical disturbances.

4.3.1 Signature-Based IDS Component

For SIDS, we have defined several rules for detecting IT attacks, as shown in Table 1. Table 1 shows detailed information about IDS rules corresponding to different stages of attacks based on the kill chain model. In this table, rule 1 and rule 2 belong to the reconnaissance (stage 1), rule 3 belongs to the access (stage 2), and rule 4 belongs to

Table 1 Snort rules for signature-based IDS

Rules	Attack	Snort IDS Rules
Rule1	Ping Scanning (Reconnaissance)	*Alert icmp* $ *EXTERNAL_NET any -> (IP of your substation RTU) any (msg:ICMP to Substation; content:*| 10 11 12 13 14|; *sid: 9000547; rev:1;)*
Rule 2	Nmap Scanning (Reconnaissance)	*alert tcp any any -> (IP of your substation RTU) 22 (msg:NMAP TCP Scan;sid:10000005; rev:2;)*
Rule 3	Telnet Access (Access)	*Alert tcp* $ *EXTERNAL_NET any-> (IP of your substation RTU) 23 (msg:Incoming Telnet ; content; root; nocase; sid: 9000546; rev:1;)*
Rule 4	DOS Attack (launch)	*Alert tcp* $ *EXTERNAL_NET any -> (IP of your substation RTU) 20000 (msg:Warning DoS attack incoming; threshold:type threshold, track by src, count 100, seconds 5; sid: 9000547; rev:1;)*

the launch stage (stage3). Note that we have utilized the Snort IDS tool to analyze the SCADA traffic through network interfaces and later develop these rules as discussed in [13].

Rule 1: It detects the ping scanning attack on the substation network by capturing incoming network traffic on the specified network internet protocol (IP) address for the Internet Control Message Protocol (ICMP) protocol.

Rule 2: This rule detects the Nmap scanning attack whenever an attacker performs TCP-based Nmap scanning on the substation RTU on port 22.

Rule 3: This rule detects an unauthorized Telnet session through a root login to the substation RTU at port 23.

Rule 4: This rule detects a DoS attack on the substation network targeting the distributed network protocol 3 (DNP3) communication on port 20000. In this rule, an alert is generated after the first 100 SYN packets (SYN flood) within a sampling period of 5 s.

4.3.2 Model-Based IDS Component

For MIDS, we have considered three behavior-based rules: range-based detection, status-based detection, and timing-based detection, by defining thresholds in the Zeek (BRO) analyzer function and Snort IDS for the DNP3 communication to detect pulse attack, tripping attack, and ramp attack, as shown in Table 2 and 3, also discussed in [13, 14, 24].

Table 2 Zeek scripts for model-based IDS in DNP3

<table>
<tr><th>Alert ID</th><th>Model-based Rules (Attack)</th><th>Zeek IDS Scripts</th></tr>
<tr><td>Alert ID 1</td><td>Range-based Detection (Pulse Attack)</td><td>event dnp3_analog_input_SPwFlag(c: connection, is_orig: bool, flag: count, value: count){ if(value != 0 && value < 3000000000 && value != 1065353216) { if (value < 1123679256 |||| value > 1200000000){ c $dnp3$alert = 1; }}}</td></tr>
</table>

Table 3 Snort rules for model-based IDS in DNP3

<table>
<tr><th>Alert ID</th><th>Model-based Rules (Attack)</th><th>Snort IDS Rules</th></tr>
<tr><td>Alert ID 2</td><td>Status-based Detection (Tripping Attack)</td><td>Alert tcp !(IP from your control center) any -> (IP of your substation RTU) 20000 (msg:Unauthorized Relay Trip; content :|00 81| ;rev:1;)</td></tr>
<tr><td>Alert ID 3</td><td>Timing-based Detection (Ramp Attack)</td><td>Alert tcp (IP from your control center) any -> (IP of your substation RTU) 20000 (msg:Ramp attack ; content : |00 81|; threshold:type threshold, track by src, count 2, seconds 0.3; sid: 9000547; rev:1;)</td></tr>
</table>

Range-based detection: In this case, a minimum threshold value is defined based on the analog value that is extracted from the DNP3 SCADA communication protocol. Table 2 shows a range-based detection in the Zeek script that generates an alert with an alert ID 1 if the generated output power goes below the defined initial generation. For example, if the output power of generator 1 is 135.4 MW, then an alert is triggered if the generation reduces to the minimum threshold of 108.32 MW (0.8 *135.4) that is equivalent to 1123679256 as an unsigned 32-bit integer, one of the few available data types in Zeek (Bro) IDS [23].

Status-based detection: It triggers an alert in Snort IDS with an alert ID 2 if the line status changes from 1 to 0 to notify the control center operator that the specific relay is tripped.

Timing-based detection: This rule is defined based on the statistical analysis of two consecutive control signal packets. We assign the minimum threshold value to 0.3 seconds for two consecutive normal DNP3 packets based on the high-speed auto-reclosing time, as discussed in [14], and an alert ID 3 is triggered in the Snort IDS.

4.3.3 Learning-Based IDS Component

This component of hybrid IDS applies machine-learning algorithms and data mining techniques to detect the last stage (Persistence) of kill chain model that includes stealthy and coordinated cyber-attacks. This approach utilizes secure phasor measurements that are communicated over a separate WAN like the NASPI network (NASPInet) with inherent cybersecurity features, and alert logs, obtained from the model-based IDS, to accurately detect attacks and provide the detailed classification of them while distinguishing them from natural disturbances like line faults. The detailed classification and clarification of cyber-physical events provide a comprehensive understanding of incidents and also assist the control center operator to take intelligent decisions.

Fig. 10 shows the methodology for developing LIDS to perform multi-events classification. For building the classification model, the phasor measurements, including generator bus voltage magnitude (Vm_g) and line bus voltage magnitudes (Vm_i, Vm_j), where subscripts i and j represent the sending and receiving ends of the transmission lines, are collected from the deployed PMUs. Apart from PMU measurements, the learning-based IDS also receives the generated alert logs (Y) from model-based IDS as input features. The offline process is applied, as outlined in Fig. 10 from connector A to B, for training and updating model and developing the final machine learning model. During the offline process, a library of the dataset has been generated for different events, including cyberattacks, line faults, and nor-

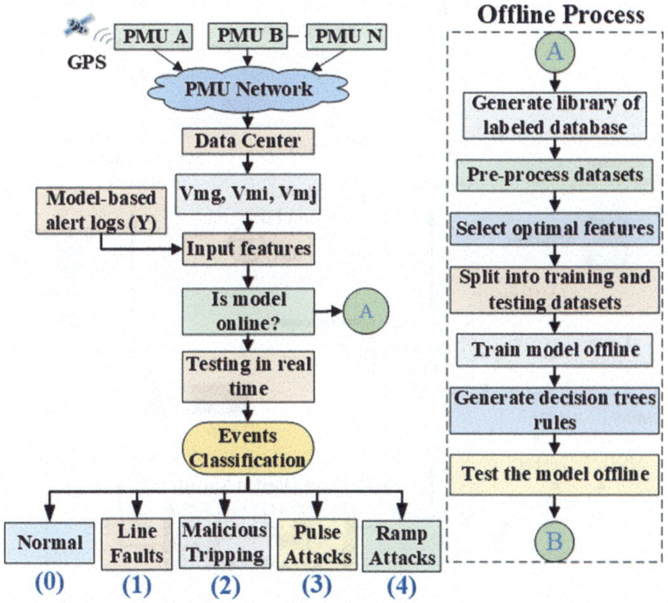

Fig. 10 Machine learning-based IDS using PMU measurements

mal events, which are labeled later in the integer format, as part of the supervised learning. Afterward, data preprocessing steps are carried out that involve normalization, transformation, and features selection and extraction, which filter out the irrelevant information and unreliable data that may affect the learning process and events prediction. Further, the Pearson Correlation-based feature selection technique is applied to select the relevant features. The obtained dataset is split into training and testing datasets. It is appropriate to note that due to the space limitation, we are not discussing the details of different scenarios required for generating the labeled datasets. Overall, we have generated datasets for the five events: normal (0), line faults (1), malicious tripping attack (2), pulse attack (3), and ramp attack (4). Finally, the trained model is deployed for performing multi-events classification and sending output logs to the operator.

4.4 Experimental Setup

Fig. 11 shows the hardware in the loop (HIL)-based cyber-physical system (CPS) testbed for attack-detection experiments in CWAPS. We have modeled the IEEE 39

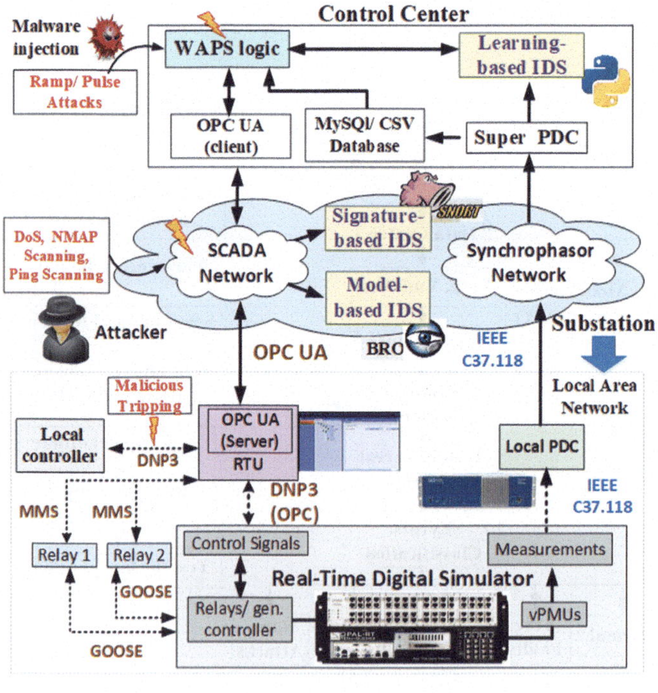

Fig. 11 Experimental setup for attack-detection experiments

bus system in ARTEMiS/SSN (eMEGASIM), and simulated in the real-time digital simulator (OPAL-RT). We have deployed virtual PMU models (vPMUs) to generate synthetic phasors, and computed the relevant features inside the simulator. The computed relevant features are sent to the hardware local phasor data concentrator (LPDC), deployed at the substation network, which forwards the data to the software-based super PDC (Open PDC) at the control center over the WAN. The super PDC saves the data in a local comma-separated values (CSV) historian as well as in the MySQL database. The stored data is used for generating the labeled database and further training and testing machine learning model, as a part of the LIDS. In this experiment, the CRAS controller is running in the python script, which is communicating with the substation RTU through the Kepserver's OPC Unified Architecture (UA) client-server interfaces. The substation RTU, as shown in a pink box, is communicating with a simulator using the DNP3 (OPC server) SCADA communication. The software OpenPDC collects the phasor measurements and forwards it to the LIDS. Also, the CWAPS controller receives phasor measurements through the MySQL in real-time, and sends the control signal back to the substation network through the SCADA communication to provide an appropriate response, if necessary, to close the loop. Also, SIDS and MIDS are deployed over the WAN that sniffs the SCADA traffic and detects possible anomalies.

Fig. 12 shows the CRAS-enabled modified IEEE 39 bus system that is divided into two major areas, where the area 2, working as a primarily generation area, is supplying generation to the area 1 through the tie-lines L15-16 and L16-17. During the tripping of line L16-17, the line L15-16 gets overloaded. Therefore, the CRAS controller sheds the generation at bus 35, as shown by black colored arrow, and the equal amount of load is shed at bus 18 to maintain the system frequency. To perform the HIL experiment, relay 1 and relay 2 is mapped to lines L15-16, and L16-17 [30].

For implementing IT-based attacks, the installed Kali Linux machine is listening to the network traffic between the control center and substation network. We have utilized the pre-installed tools, *Nmap*, and *ping* command, in the Kali machine to perform the attack reconnaissance. The DoS attack is performed by sending a huge number of random packets to the RTU through the TCP SYN flooding attack using *hping* tool.

For implementing SCADA-based attacks, we have performed the malicious tripping attack on the relay 2 to trip the line L16-17 by replaying the tripping packet using the python script through the MITM between the substation and local control center. For executing ramp and pulse attacks on generator 35 (G35), the malware, Trojan Horse, is installed in the OPC server-based substation RTU, which provides backdoor access to the attacker. The attacker closes the legitimate RTU program and initiates python script-based malicious logic, which periodically sends the control signal to the simulator targeting the generator (G35) to initiate ramp and pulse attacks. We have also simulated 3 phase to ground faults followed by the normal tripping of the line L16-17, and multiple simulations are performed for different cases as discussed in [6]. Note that in this work, we assume that the attacker is only looking to compromise insecure SCADA network and the synchrophasor network is

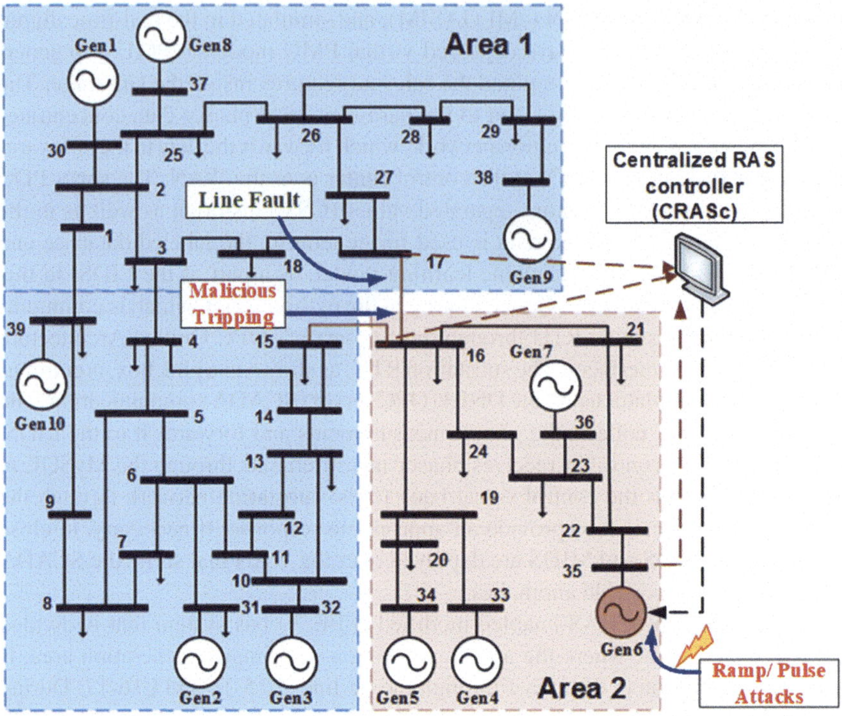

Fig. 12 CRAS enabled modified IEEE 39 bus system

secure with inherent cybersecurity features and the detection results are provided in the next section.

4.5 Results and Discussions

Table 4 shows the performance of HIDS as well as its comparison with an individual SIDS in terms of accuracy rate. It can be observed that the SIDS is able to detect IT attacks including ping scanning, Nmap scanning, DoS attack, and Telnet access attack with an accuracy rate of 100%; however, it fails to identify stealthy SCADA-related attacks and physical disturbances like line-to-ground faults. The HIDS merges LIDS with SIDS and MIDS to detect IT and OT attacks with an accuracy rate of 100% and 98.71%. Further, it is also able to detect physical disturbances with an accuracy rate of 97.94% using machine learning-based random forest classifier during 70% training and 30% testing datasets.

Note that while developing the LIDS, different machine learning classifiers were applied—such as decision tree (DT), random forest (RF), and support vector machine

Table 4 Accuracy Rate of several IDSs

IDSs	IT Attacks	OT Attacks	Line Faults
Network-based IDS (%)	100%	×	×
Hybrid IDS (%) (Network+Model+Learning)	100%	98.71%	97.94%

Table 5 Average accuracy rate for different classifiers during Case 1 and Case 2 experiments

Parameters	Case 1	Case 2
70% Training and 30% Testing		
Decision Tree (DT)	88.5	79.8
Support Vector Machine (SVM)	91.3	89.2
Random Forest (RF)	96.33	95.4

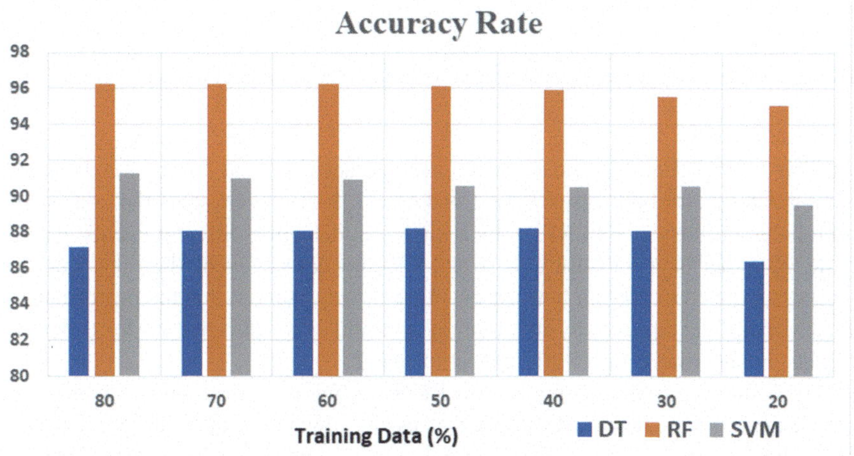

Fig. 13 Average accuracy rate for different classifiers during case 2

(SVM)—to select the best classifier. Table 5 shows the average accuracy rate of different classifiers during case 1 and case 2 experiments for 70% training and 30% testing datasets. Note that case 1 represents the scenario when PMU measurements and cyber alerts, generated from MIDS, are utilized as input features, whereas input features for case 2 include only PMU measurements. Table 5 also shows that the RF exhibits a higher accuracy rate as compared to other classifiers with an average accuracy of 96.33% in case 1 and 95.4% in case 2.

Fig. 13 shows the average accuracy rate of different classifiers during testing in case 2 where the training and testing datasets were varied from 80% training and 20% testing to 20% training and 80% testing datasets to analyze the robustness of different classifiers and it clearly illustrates the consistent performance of RF as compared

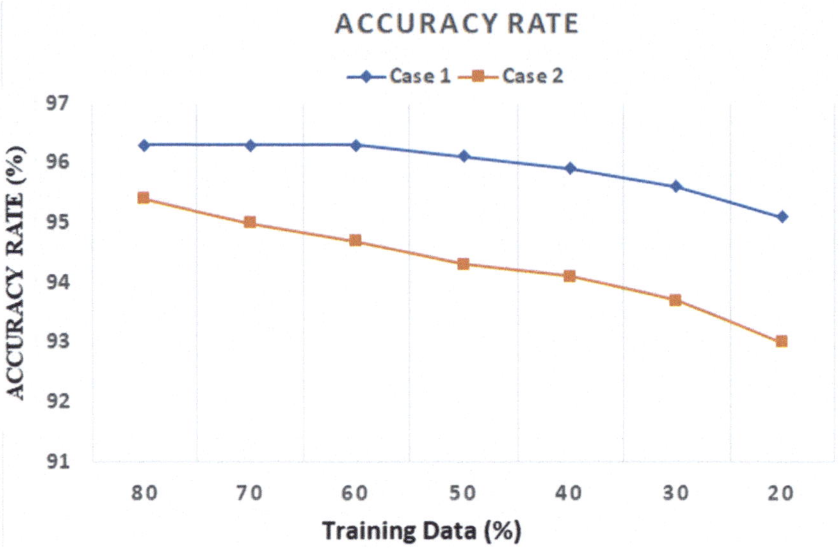

Fig. 14 Average accuracy rate of random forest (RF) for different training datasets

to DT and SVM. Fig. 14 represents the average accuracy rate of RF for different training datasets for both cases: case 1 and case 2. It clearly presents the consistent and reliable performance of RF, as its accuracy in detecting different events during testing is higher than 95% for case 1 and 93% for case 2, even when the training dataset is reduced to 20% in both cases. Further, we observed that the performance of each classifier improves by including power and cyber information for different datasets, as shown in case 1 with respect to case 2.

Fig. 15 shows the processing time of RF for different % of training datasets for both cases. It can be observed that the processing time (sec) for training the model was higher in case 1 as compared to case 2, which is amplified during the higher % of training datasets. Note that the RF exhibits a larger processing time in classifying events as compared to the DT because of a large number of associated decision trees in RF and the final prediction is made based on the majority vote, as discussed in [24].

5 Conclusion

Developing an intrusion detection system for the smart grid cybersecurity is a challenging task as it requires an in-depth understanding of power system related applications, grid network architectures, and comprehensive knowledge of cutting-edge technologies. In this chapter, we presented a systematic approach for developing a hybrid IDS by integrating conventional network security solutions with state-of-

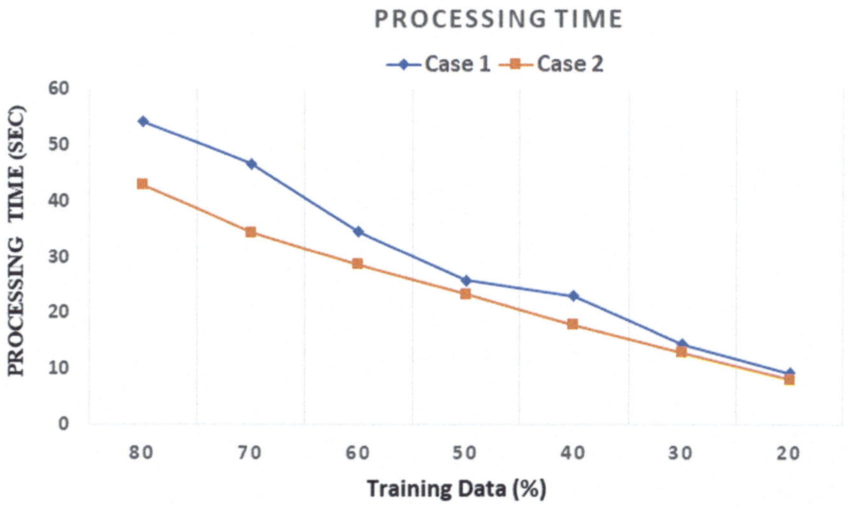

Fig. 15 Processing time of random forest (RF) for different training datasets

the-art machine learning and model-based intrusion detection approaches to detect advanced and persistent intruders at different stages while following the kill-chain model. Initially, this chapter discussed different types of IDSs and highlighted the existing challenges of developing IDS in the smart grid network. As a proof of concept, one case study was presented where hybrid IDS is applied in a centralized wide-area protection scheme to detect different types of cyberattacks, including IT- and SCADA-based attacks. In particular, Snort and Zeek IDS tools were applied in developing signature and model-based IDSs, and machine learning-based classification algorithms, including decision tree, random forest, and support vector machine, were applied for developing the learning-based IDS. Further, several steps were described to implement these cyber-attacks in the CPS testbed environment by utilizing the resources available at Iowa State University PowerCyber laboratory. Experimental results showed the superior performance of hybrid IDS to accurately detect different classes of anomalies and physical disturbances. Our case study also showed that the random forest-based classifier exhibited a higher accuracy rate as compared to the other machine learning classifiers, and a combination of grid measurements and alert logs, generated from model-based IDS, also assisted in providing a detailed classification of different events. The potential avenue for future work is to develop a library of novel IDSs for other SCADA protocols, such as GOOSE, Modbus, and smart energy profile (SEP) 2.0 protocols.

References

1. V. Terzija et al., Wide-Area Monitoring, Protection, and Control of Future Electric Power Networks. Proceedings of the IEEE **99**(1), 80–93 (2011)
2. National Institute of Standards and Technology (NIST), "NISTIR 7628 Revision 1: Guidelines for Smart Grid Cyber Security", September 2014
3. U.S. Department of Energy (DOE) Energy Sector Control Systems Working Group, 'Roadmap to Achieve Energy Delivery Systems Cybersecurity', Technical Report, 2011
4. U.S. Department of Energy (DOE) 'Cybersecurity Capability Maturity Model (C2M2)', February 2014
5. NERC, 'Critical Infrastructure Protection (CIP) Standards', 2015
6. National Electric Sector Cybersecurity Organization Resource (NESCOR), 'Wide Area Monitoring, Protection, and Control Systems (WAMPAC)-Standards for Cyber Security Requirements', 2012
7. U.S. Department of Energy (DOE) 'Cybersecurity Capability Maturity Model (C2M2)', February 2014
8. S. Sridhar et al., Model-Based Attack Detection and Mitigation for Automatic Generation Control. IEEE Transactions on Smart Grid **5**(2), 580–591 (2014)
9. S. Sarangan, V.K. Singh, M. Govindarasu, "Cyber Attack-Defense Analysis for Automatic Generation Control with Renewable Energy Sources,", North American Power Symposium (NAPS). Fargo, ND **2018**, 1–6 (2018)
10. V.K. Singh, M. Govindarasu, "Decision Tree Based Anomaly Detection for Remedial Action Scheme in Smart Grid using PMU Data,", IEEE Power & Energy Society General Meeting (PESGM). Portland, OR **2018**, 1–5 (2018)
11. A. Ashok et al., "Online Detection of Stealthy False Data Injection Attacks in Power System State Estimation," in IEEE Transactions on Smart Grid, vol. 9, no. 3, pp. 1636-1646
12. V.K. Singh, A. Ozen, M. Govindarasu, "A Hierarchical Multi-Agent Based Anomaly Detection for Wide-Area Protection in Smart Grid,", Resilience Week (RWS). Denver, CO **2018**, 63–69 (2018)
13. V.K. Singh, S.P. Callupe, M. Govindarasu, "Testbed-based Evaluation of SIEM Tool for Cyber Kill Chain Model in Power Grid SCADA System,", North American Power Symposium (NAPS). Wichita, KS, USA **2019**, 1–6 (2019)
14. V.K. Singh, H. Ebrahem, M. Govindarasu, "Security Evaluation of Two Intrusion Detection Systems in Smart Grid SCADA Environment,", North American Power Symposium (NAPS). Fargo, ND **2018**, 1–6 (2018)
15. V.K. Singh, E. Vaughan, J. Rivera, "SHARP-Net: Platform for Self-Healing and Attack Resilient PMU Networks,", IEEE Power & Energy Society Innovative Smart Grid Technologies Conference (ISGT). Washington, DC, USA **2020**, 1–5 (2020)
16. Y. Yang et al., "Intrusion Detection System for network security in synchrophasor systems," IET International Conference on Information and Communications Technologies (IETICT 2013), Beijing, China, 2013, pp. 246-252
17. S. Pan et al., Developing a Hybrid Intrusion Detection System Using Data Mining for Power Systems. IEEE Transactions on Smart Grid **6**(6), 3104–3113 (2015)
18. NERC, Reliability Guideline: PMU placement and Installation, December 2016
19. M. Begovic et al., Wide-Area Protection and Emergency Control. Proceedings of the IEEE **93**(5), 876–891 (2005)
20. Berthier, R., Sanders, W. H. (2011). Specification-based intrusion detection for advanced metering infrastructures. In Proceedings - 2011 17th IEEE PRDC 2011 (pp. 184-193)
21. M. Wu, S. Member, L. Xie, S. Member, Online detection of low-quality synchrophasor measurements: A data-driven approach. IEEE Trans. Power Syst. **32**(4), 2817–2827 (2016)
22. C.F. Garcia-Hernandez et al., "Wireless sensor networks and applications: A survey," IJCSNS Int. J. Comput. Sci. Netw. Security **7**(3), 264–273 (2007)
23. S. Safaric and K. Malaric, "ZigBee wireless standard," in Proc. 48th Int. Symp. ELMAR-2006, Zadar, Croatia, Jun. 07-09, 2006, pp. 259-262

24. V.K. Singh, E. Vaughan, J. Rivera, A. Hasandka, "HIDES: Hybrid Intrusion Detector for Energy Systems,", IEEE Texas Power and Energy Conference (TPEC). College Station, TX, USA **2020**, 1–6 (2020)
25. ICS-CERT, Cyber-Attack Against Ukranian Critical Insfrastructure
26. NERC,Remedial Action Development Definition Development project 2010-05.2 -Special Protection System
27. K. Seethalekshmi et al., "Wide-area protection and control: Present status and key challenges," in Proc. 15th Nat. Power Syst. Conf., Mumbai, India, Dec. 2008, pp. 169-175
28. WECC remedial action scheme catalog summary [Internet]; 2008
29. V. Kumar Singh, A. Ozen, M. Govindarasu, "Stealthy cyber attacks and impact analysis on wide-area protection of smart grid,", North American Power Symposium (NAPS). Denver, CO **2016**, 1–6 (2016)
30. V.K. Singh, "Evaluation of Anomaly Detection for Wide-Area Protection Using Cyber Federation Testbed,", et al., IEEE Power & Energy Society General Meeting (PESGM). Atlanta, GA, USA **2019**, 1–5 (2019)

The manufacturer's authorised representative in the EU is Springer Nature Customer Service Centre GmbH, Europaplatz 3, 69115 Heidelberg, Germany. If you have any concerns regarding our products, please contact ProductSafety@springernature.com

Printed and bound by CPI Group (UK) Ltd, Croydon, CR0 4YY
30/09/2026
02225697-0017